# Lecture Notes in Physics

## Monographs



Springer
*Berlin*
*Heidelberg*
*New York*
*Barcelona*
*Hong Kong*
*London*
*Milan*
*Paris*
*Singapore*
*Tokyo*

## The Editorial Policy for Monographs

The series Lecture Notes in Physics reports new developments in physical research and teaching - quickly, informally, and at a high level. The type of material considered for publication in the monograph Series includes monographs presenting original research or new angles in a classical field. The timeliness of a manuscript is more important than its form, which may be preliminary or tentative. Manuscripts should be reasonably self-contained. They will often present not only results of the author(s) but also related work by other people and will provide sufficient motivation, examples, and applications.
The manuscripts or a detailed description thereof should be submitted either to one of the series editors or to the managing editor. The proposal is then carefully refereed. A final decision concerning publication can often only be made on the basis of the complete manuscript, but otherwise the editors will try to make a preliminary decision as definite as they can on the basis of the available information.
Manuscripts should be no less than 100 and preferably no more than 400 pages in length. Final manuscripts should be in English. They should include a table of contents and an informative introduction accessible also to readers not particularly familiar with the topic treated. Authors are free to use the material in other publications. However, if extensive use is made elsewhere, the publisher should be informed. Authors receive jointly 30 complimentary copies of their book. They are entitled to purchase further copies of their book at a reduced rate. No reprints of individual contributions can be supplied. No royalty is paid on Lecture Notes in Physics volumes. Commitment to publish is made by letter of interest rather than by signing a formal contract. Springer-Verlag secures the copyright for each volume.

## The Production Process

The books are hardbound, and quality paper appropriate to the needs of the author(s) is used. Publication time is about ten weeks. More than twenty years of experience guarantee authors the best possible service. To reach the goal of rapid publication at a low price the technique of photographic reproduction from a camera-ready manuscript was chosen. This process shifts the main responsibility for the technical quality considerably from the publisher to the author. We therefore urge all authors to observe very carefully our guidelines for the preparation of camera-ready manuscripts, which we will supply on request. This applies especially to the quality of figures and halftones submitted for publication. Figures should be submitted as originals or glossy prints, as very often Xerox copies are not suitable for reproduction. For the same reason, any writing within figures should not be smaller than 2.5 mm. It might be useful to look at some of the volumes already published or, especially if some atypical text is planned, to write to the Physics Editorial Department of Springer-Verlag direct. This avoids mistakes and time-consuming correspondence during the production period.
As a special service, we offer free of charge LaTeX and TeX macro packages to format the text according to Springer-Verlag's quality requirements. We strongly recommend authors to make use of this offer, as the result will be a book of considerably improved technical quality.
Manuscripts not meeting the technical standard of the series will have to be returned for improvement.
For further information please contact Springer-Verlag, Physics Editorial Department II, Tiergartenstrasse 17, D-69121 Heidelberg, Germany.

Marcus Kriele

# Spacetime

## Foundations of General Relativity and Differential Geometry



Springer

**Author**

Marcus Kriele
Technische Universität Berlin
Fachbereich Mathematik, Sekr. MA 8-3
Strasse des 17. Juni 136
D-10623 Berlin, Germany



Library of Congress Cataloging-in-Publication Data.

Die Deutsche Bibliothek - CIP-Einheitsaufnahme

**Kriele, Marcus:**
Spacetime : foundations of general relativity and differential geometry / Marcus Kriele. - Berlin ; Heidelberg ; New York ; Barcelona ; Hong Kong ; London ; Milan ; Paris ; Singapore ; Tokyo : Springer, 1999
(Lecture notes in physics : N.s. M, Monographs ; 59)
ISBN 3-540-66377-0

ISSN 0940-7677 (Lecture Notes in Physics. Monographs)
ISBN 3-540-66377-0 Springer-Verlag Berlin Heidelberg New York



Typesetting: Camera-ready by the author
Cover design: *design & production*, Heidelberg

SPIN: 10644466 55/3144/du - 5 4 3 2 1 0 – Printed on acid-free paper

# Preface

Space and time are the two most fundamental concepts in our world because all else is unimaginable without assuming that space (or time) exists. It is therefore not surprising that the sophisticated Euclidean model of space already existed more than 2000 years. For centuries it was a common belief by scientists and philosphers alike that the Euclidean structure of space was one of the very few eternal truths. It was only at the beginning of the 20th century that this belief was shattered with the introduction of Albert Einstein's theories of special and general relativity. Today, Einstein's theory of general relativity is completely established, and there are many textbooks which explain it at all levels of mathematical sophistication. What is missing, however, is a modern textbook on general relativity for mathematicians and mathematical physicists with emphasis on the physical justification of the mathematical framework. This book aims to fill this gap.

Knowledge of physics is *not* assumed. While physical and heuristic arguments are given, they are not used as substitutes for any proofs. The book is also suitable as an introduction to pseudo-Riemannian geometry with emphasis on the intuition for geometrical concepts.

## The physical theme of the book

Modern textbooks on general relativity typically start with a more or less formal introduction to pseudo-Riemannian geometry. In such textbooks some knowledge of special relativity is usually assumed, and the reader is expected to accept the geometrical framework presented on trust. This approach is very economical but obscures the extent to which classical general relativity succeeds in describing our universe, and also where it may fail. This is a point that is of particular relevance to those attempting to quantize gravity. From a physical point of view it is important to realize which parts of the theory reflect genuine physical insights, and which are dispensible. One way this can be achieved is through a critical introduction that stresses foundational matters. There are no modern textbooks taking this approach, and I hope to fill this gap with my book.

One of the most exciting aspects of general relativity is the prediction of black holes and the Big Bang. Such predictions gained weight through the singularity theorems pioneered by Penrose. In various textbooks on general relativity singularity theorems are presented and then used to argue that black holes exist and that the universe started with a big bang. To date what has been lacking is a critical analysis of what these theorems really predict. We give a proof of a typical singularity theorem and use this theorem to illustrate problems arising through the possibilities of "causality violations" and very weak "shell crossing singularities". These problems add weight to the point of view that the singularity theorems alone are not sufficient to predict the existence of physical singularities.

## The mathematical theme of the book

In order to gain both a solid understanding of and good intuition for any mathematical theory, one should try to realize it as a model of a familiar non-mathematical concept. Physical theories have had an especially important impact on the development of mathematics, and conversely various modern physical theories require rather sophisticated mathematics for their formulation. Today, both physics and mathematics are so complex that it is often very difficult to master the theories in both subjects. However, in the case of pseudo-Riemannian differential geometry or general relativity the relationship between physics and mathematics is especially close, and it is therefore possible to profit from an interdisciplinary approach.

Euclidean geometry had its origins as the description of shapes in *physical* space. It is generally considered a mathematical discipline rather than a physical theory, because it is possible to derive it from a small set of physical *postulates*, which can alternatively be viewed as mathematical *axioms*. Since the concept of space is basic to our everyday experience, Euclidean geometry combines mathematical rigor with intuitiveness — a combination which has proved to be extremely fruitful for both mathematics and physics. Riemannian geometry is abstracted from the study of surfaces in Euclidean space and inherits much of the intuitiveness of Euclidean geometry. Hence Riemannian geometry is very well developed, and a growing number of geometers have branched out to Lorentzian or even pseudo-Riemannian geometry. In my experience, these fields (and even Riemannian geometry) appear quite abstract to the majority of students.

A careful analysis of space, time, and free fall — the most fundamental (classical) physical concepts — leads almost automatically to Lorentzian geometry. With respect to Lorentzian geometry, we are therefore in a similar situation as ancient geometers were with respect to

Euclidean geometry. What's more, virtually no physical background is required for this approach. Since Riemannian geometry comes to play in the study of submanifolds representing an instant in time, it is completely straightforward to extrapolate pseudo-Riemannian geometry from the special and physically motivated cases of Lorentzian and Riemannian geometry.

While some modern textbooks present pseudo-Riemannian geometry (and general relativity) to mathematicians (an example of this is that by O'Neill (1983)), they have not motivated the geometry from basic properties of space and time. Instead they have developed it as an abstract mathematical theory. To ensure that the mathematical description mirrors the physical concepts, all definitions have a justification in this book. *This approach also leads to a careful treatment of the structural aspects of the mathematics.*

## How to read this book

This book is not designed so that it is necessary for the reader to start at page 1 and then to read on until she or he arrives at page 420. People who take this approach will very likely give up before they reach page 10! The material is orderd in such a way as to allow the text to be used as a reference source. It is an unfortunate fact that many parts of the theory that logically belong to the preliminaries are not of immediate interest to a reader who is interested in space and time, and so the reader is urged to follow the guides in the margins, which provide a shortcut. As an example, the text in the margin denotes the beginning of a passage belonging to the shortcut: **p. 111** ↓ denotes the page number where the last shortcut passage ended and [↓ **p. 222**] the page number where the present passage will end. Additional explanations in the footnotes are indicated by →1, where 1 refers to the number of the corresponding footnote. The end of shortcut passages is marked similarly. Having understood the material leading to Einstein's equation it is then not difficult to return to the parts that have been skipped on an earlier reading. In addition, hints are given at the beginning of most sections as to what is important and should be read .

**p. 111** ↓ →1 [↓ **p. 222**]

This book, with its 420 pages is meant to cover both general relativity and pseudo-Riemannian differential geometry. It is therefore clear that some important topics had to be omitted.

For mathematicians, the most important omissions are certainly some topics peculiar to Riemannian geometry, such as the Hopf-Rinow theorem (O'Neill 1983, Theorem 5.21) and the Myers theorem (O'Neill 1983, Theorem 10.24). Because these results are contrary to intuition

[1] Explanations referring to the guide in the margin.

one should obtain for Lorentzian (or general pseudo-Riemannian) geometry and since they are not needed for the description of space and time, they have been omitted from this book.

Physicists may find that the presentation of this book is only loosely linked to other physical theories. This loose linkage is possible since the theory of space and time is fundamental to any other physical theory. The book is therefore accessible to mathematicians and physicists alike. Physicists who are interested in applications to astrophysics may wish to consult the bookd by Weinberg (1972). Weinberg's approach is opposite to the one used in this book, and personally I believe that it should ideally be read after the reader has a solid knowledge of the conceptional aspects of relativity as presented in this book. Most other books on general relativity also present the "Kerr solution", which is supposed to model the exterior of a rotating black hole. It has been omitted since it is not essential to understanding general relativity. Moreover, it is well described in other books. People interested in this solution should probably first read Chap. 12 of the book by Wald (1984). The purely mathematical aspects of this solutions are clearly presented in O'Neill's book (1995).

## Acknowledgements

The reader will undoubtedly notice that this book owes much to excellent text books and survey articles. For the philosphical aspects of this book I wish to mention especially the classic book by Weyl (1923) and the survey article by Ehlers (1973).

I have also freely used material which appears elsewhere (O'Neill 1983; Wald 1984; Beem and Ehrlich 1981; Hawking and Ellis 1973; Sachs and Wu 1979; Karcher 1994; De Felice and Clarke 1990; Abraham and Marsden 1978; Garabedian 1986) without always acknowledging this fact.

I warmly thank Bernd Wegner, who encouraged me and recommended the book project to Springer-Verlag. I wish especially to thank Volker Perlick not only for introducing me to relativity but also for reading through the whole manuscript and for his many important improvements.

*This book is dedicated to two Australian relativity students who on their way to gaining their doctorates courageously stood up against the immoral behaviour of their supervisor and the highhandedness of their university.*

Göttingen, 19th July 1999 M. Kriele

# Table of Contents

# List of Figures

# List of postulates

# 1. Local theory of space and time

> This book is **not** meant to be read in the order the material is presented. Please follow the guide in the margins or skip material as proposed in the italic text at the beginning of most sections.

p. 1 ↓ →1 [↓ p. 3]

In this chapter we will develop those aspects of space and time which can be locally observed, say in a laboratory. We will start with Euclid's description of space and then incorporate time in to the picture. The path we take is rather historical. It starts with intuitive but surprisingly complicated concepts (Newton's theory of absolute space and absolute time) and ends with the not so intuitive but mathematically simpler theory of special relativity. The guiding principle of this book will *not* be mathematical simplification, but the solution of problems occurring in earlier theories.

The mathematical description in this chapter seems to be global and leads to extrapolations which are not validated by any experiments and which are not generally true. In the following chapters we will take up this point again, and show that the description given in this chapter should be considered infinitesimally rather than globally. This is the theme of the book.

## 1.1 Space

In this section we consider space and introduce Euclidean geometry. This material is assumed to be familiar to the reader and is therefore presented in a rather concise way.

[1] Readers who wish to learn the essentials of the theory of space and time quickly and do not mind skipping some mathematical proofs can use the guide in the margins.

### 1.1.1 Affine space

*In this section we introduce affine space as our most elementary description of space. Affine space is just $\mathbb{R}^n$ where the special properties of $0 \in \mathbb{R}^n$ are ignored.*[2]

It is a basic experience that we can uniquely describe any point in space by three real numbers. This seems to be the idea of Descartes (1637) who developed analytic geometry as an example of his *Discours de la Méthode*. While it is therefore plausible to identify $\mathbb{R}^3$ with (physical) space, $\mathbb{R}^3$ contains a distinguished point 0 whereas space apparently does not. Hence by using $\mathbb{R}^3$ as a description of space we introduce a mathematical structure which has no physical counterpart. This would lead to constructions which cannot be realized in space. For instance, there is the unique negative of a vector $v \in \mathbb{R}^3$ but there is no way to assign the negative to a point in space. As another example, addition of vectors has no direct interpretation in terms of points in space. If we want to have a reliable description of space with the property that all phenomena exhibited in this description are mirrored by physically verifiable phenomena, we have to abstract from these additional structures.

We will now isolate those structures of $\mathbb{R}^3$ which have an intuitive meaning in terms of space. Given two points $x, y$ we can construct an arrow $v$ which points from $x$ to $y$. This arrow induces a map from space to space. We just move the arrow (without rotating) such that its untipped end coincides with a given point $z$. The point $z$ is then mapped to the tip of the arrow. It appears that — as long as we don't rotate $v$ — this definition is independent of the path which we use to move $v$ from $x$ to $z$. In $\mathbb{R}^3$, this *parallel transport* is just given by the map $R_{y-x}\colon \mathbb{R}^3 \to \mathbb{R}^3$, $z \mapsto z + y - x$. Observe that $y - x$ stands here for the arrow $v$ which is *not* a point. Thus the geometric interpretation is different from a simple addition of vectors. in order to separate the concepts involved we define the concept of real affine space.

We do not yet know what "rotating" should mean in mathematical terms — so far we have simply a physical picture in mind. However, the following definition (for the $n$-dimensional case) reflects some properties of our naive notion of parallel transport and does not introduce any additional properties. The set of points is denoted by $\mathbb{A}^n$.

**Definition 1.1.1.** *An* $n$-dimensional real affine space *is a set* $\mathbb{A}^n$ *and a collection* $\{R_v\colon \mathbb{A}^n \to \mathbb{A}^n \mid v \in \mathbb{R}^n\}$ *of bijective maps such that the following holds.*

*(i)* $R_{v+w} = R_v \circ R_w \quad \forall v, w \in \mathbb{R}^n$,

[2] Readers who already have some knowledge of manifold theory (cf. Chap. 2) and connections (cf. Sect. 2.6) can take affine space to be the usual $\mathbb{R}^n$ considered as a manifold together with the flat connection.

*(ii) for all $x, y \in \mathbb{A}^n$ there exists a unique $v_{x,y} \in \mathbb{R}^n$ such that $R_{v_{x,y}}(x) = y$.*

*We denote $R_v(x)$ by $x + v$ or $v + x$ and $v_{x,y}$ by $y - x$.*

We will see below that this definition is so restrictive that it does not allow for more affine spaces than there are vector spaces.

*Remark 1.1.1.* Of course, space is 3-dimensional. We work with general dimension $n$ for several reasons. Firstly, most of the the theory we will be developing does not depend on the dimension — with the exception of a key-result whose proof, however, is too complicated to be presented in this book (cf. Theorem 5.3.1). Secondly, it is often convenient to compare the theory with lower dimensional analogues which are easier to visualize. It is therefore advisable to formulate it in a way which encompasses these analogues and also shows the restriction of the analogy. Thirdly, there exist modifications of Einstein's general theory of relativity to higher dimensions ("Kaluza-Klein theories").

Note that for $x \in \mathbb{A}^n$, $u, v \in \mathbb{R}^n$ the associative law

$$x + (u + v) = (x + u) + v$$

holds. It is easy to see that all $n$-dimensional real affine spaces are isomorphic, and can be realized by $\mathbb{R}^n$ in the following way. Choose any $o \in \mathbb{A}^n$ and define $\phi_o\colon \mathbb{A}^n \to \mathbb{R}^n$, $x \mapsto \phi_o(x) = v_{o,x}$. Now identify $\mathbb{A}^n$ with $\phi_o(\mathbb{A}^n)$ and define for $x \in \phi_o(\mathbb{A}^n)$, $v \in \mathbb{R}^n$ the bijection $\tilde{R}_v(x) = \phi_o(R_v(\phi_o{}^{-1}(x)))$. Clearly, $\tilde{v}_{x,y} = v_{\phi_o{}^{-1}(x),\phi_o{}^{-1}(y)}$. Observe that these definitions are independent of the arbitrarily chosen point $o$. Conversely, choosing an $o \in \mathbb{A}^n$ we can recover the structure of $\mathbb{R}^n$ by identifying $o$ with the zero vector 0.

[p. 1 ↓] →3 ↓ p. 10

### 1.1.2 The fundamental theorem in affine geometry and doubly ruled surfaces

*In this section we present some results of affine geometry which will be needed in the proof of Theorem 1.4.1. This section is very technical and should be omitted on first reading.*

Let $o, x_1, \ldots, x_k \in \mathbb{A}^n$ and $\alpha^1, \ldots, \alpha^k \in \mathbb{R}$ such that $\sum_{i=1}^k \alpha^i = 1$. Then the *barycentre*

$$\alpha^1 x_1 + \alpha^2 x_2 + \ldots + \alpha^k x_k := o + \sum_{i=1}^{k} \alpha^i (x_i - o)$$

[3] Section 1.1.2 is needed for the proof of Theorem 1.4.1 which is central to our interpretation of the Michelson-Morley experiment. However, the reader is strongly advised against reading this part now.

is independent of $o$ and therefore an affine invariant. The symbol $+$ is defined via the right hand side and can only be applied to "linear combinations" where the real factors add to 1. An *affine subspace* $B$ of $\mathbb{A}^n$ is a set of points $\{x = \alpha^1 x_1 + \alpha^2 x_2 + \ldots + \alpha^k x_k \mid \sum_{i=1}^k \alpha^i = 1\}$, where $x_1, \ldots, x_k$ are pairwise different, fixed points. The *affine dimension* of $B$ is $k-1$. It follows that an affine subspace is an affine space. An affine subspace of dimension 1 is called an *affine line*. We call points lying on a single line *collinear*. Observe that lines are the smallest sets which are invariant under parallel transport.

**Definition 1.1.2.** *An* affine map *is a map* $f\colon \mathbb{A}^n \to \mathbb{A}^n$, $f(x) = A(x-o) + b$, *where* $A$ *is a linear map,* $o \in \mathbb{A}^n$, *and* $b \in \mathbb{R}^n$. *If* $A$ *is bijective then* $f$ *is called an* affine transformation.

*A* collineation *is a bijection* $\phi\colon \mathbb{A}^n \to \mathbb{A}^n$ *such that* $\phi$ *and* $\phi^{-1}$ *map affine lines into affine lines.*

It follows that all bijective, affine maps are collineations. Conversely, the fundamental theorem in affine geometry asserts that any collineation must be affine.

**Theorem 1.1.1.** *Let* $\mathbb{A}^n$ *be an affine space over* $\mathbb{R}$ *and fix* $o \in \mathbb{A}$. *Let* $f\colon \mathbb{A}^n \to \mathbb{A}^n$ *be a bijection which takes each three collinear points into collinear points. Then there exists a point* $b \in \mathbb{A}^n$ *and an invertible linear map* $A$ *such that* $f(x) = A(x-o) + b$ *for all* $x \in \mathbb{A}^n$.

The proof is elementary but lengthy and requires some preparatory lemmas. Here we give an algebraic proof and refer to (Berger 1987, p. 52–55) for a particularly geometric proof.

**Lemma 1.1.1.** *Let* $x, y, z \in \mathbb{A}^n$. *Then* $x, y, z$ *lie on an affine line if and only if there exists a* $\lambda \in \mathbb{R}$ *such that* $x = y + \lambda(z-y)$.

*Proof.* $x$ lies on the line generated by $x, y$ if and only if there exists an $\beta \in \mathbb{R}$ with $x = \beta y + (1-\beta) z = y + \beta(y-y) + (1-\beta)(z-y) = y + (1-\beta)(z-y)$. ∎

**Lemma 1.1.2.** *Let* $o, x_1, \ldots, x_k \in \mathbb{A}^n$, *be* $\phi$ *a collineation,* $\lambda^1, \ldots, \lambda^k \in \mathbb{R}$, *and*

$$x = o + \sum_{i=1}^{k} \lambda^i \, (x_i - o) \in \mathbb{A}^n .$$

*Then there exist* $\mu^1, \ldots, \mu^k \in \mathbb{R}$ *such that*

$$\phi(x) = \phi(o) + \sum_{i=1}^{k} \mu^i \, (\phi(x_i) - \phi(o)) .$$

*Proof.* For $k = 1$ the claim is clear by the definition of a collineation. Assume now, the assertion is true for all $m \in \{1, \dots, k\}$. For

$$x = o + \sum_{i=1}^{k+1} \lambda^i (x_i - o) \qquad \text{let} \qquad x' = o + \sum_{i=1}^{k} \lambda^i (x_i - o)\,.$$

Then we have

$$x = x' + \lambda^{m+1}(x_{m+1} - o). \tag{1.1.1}$$

Define also

$$y = o + \lambda^{m+1}(x_{m+1} - o), \tag{1.1.2}$$

$$z = \frac{1}{2}y + \frac{1}{2}x'. \tag{1.1.3}$$

By Lemma 1.1.1 the triples $\{x, x', o\}$, $\{y, o, x_{m+1}\}$, and $\{z, o, x\}$ consist each of collinear points. Thus there exist $\alpha, \beta, \gamma$ such that

$$\begin{aligned}
\phi(z) &= \alpha\phi(x') + (1 - \alpha)\phi(y),\\
\phi(x) &= \beta\phi(o) + (1 - \beta)\phi(z),\\
\phi(y) &= \phi(o) + \gamma(\phi(x_{m+1}) - \phi(o)).
\end{aligned}$$

This implies

$$\begin{aligned}
\phi(x) &= \beta\phi(o) + (1 - \beta)\phi(z)\\
&= \beta(\phi(o) - \phi(o)) + (1 - \beta)(\phi(z) - \phi(o)) + \phi(o)\\
&= (1 - \beta)((\alpha\phi(x') + (1 - \alpha)\phi(y)) - \phi(o)) + \phi(o)\\
&= (1 - \beta)(\alpha(\phi(x') - \phi(o)) + (1 - \alpha)(\phi(y))) + \phi(o)\\
&= (1 - \beta)(\alpha \sum_{i=1}^{k} \mu'^i\, (\phi(x_i) - \phi(o))\\
&\quad = (1 - \alpha)\gamma(\phi(x_{m+1}) - \phi(o))) + \phi(o)\\
&= \sum_{i=1}^{k+1} \mu^i\, (\phi(x_i) - \phi(o)) + \phi(o).
\end{aligned}$$

∎

**Lemma 1.1.3.** *Let $o, x_1, \dots, x_n \in \mathbb{A}^n$ such that $\{x_1 - o, \dots, x_n - o\}$ is a basis of $\mathbb{R}^n$. Then $\{\phi(x_1) - \phi(o), \dots, \phi(x_n) - \phi(o)\}$ is also a basis of $\mathbb{R}^n$.*

*Proof.* Let $\tilde{x} \in \mathbb{A}^n$ be any point and let $x = \phi^{-1}(\tilde{x})$. Since $\{x_1 - o, \ldots, x_n - o\}$ is a basis of $\mathbb{R}^n$ there exist $\xi^i \in \mathbb{R}$ such that $x - o = \sum_{i=1}^n \xi^i (x_i - o)$. But then Lemma 1.1.2 implies that there exist $\mu^1, \ldots, \mu^n \in \mathbb{R}$ such that

$$\tilde{x} - \phi(o) = \phi(x) - \phi(o) = \sum_{i=1}^{k} \mu^i \left(\phi(x_i) - \phi(o)\right).$$

Since $\tilde{x}$ was arbitrary the assertion follows. ∎

**Lemma 1.1.4.** *Let $k: \mathbb{R} \to \mathbb{R}$ an automorphism, i.e., $k(\alpha\beta) = k(\alpha)k(\beta)$ and $k(\alpha+\beta) = k(\alpha)+k(\beta)$ for all real numbers $\alpha, \beta$. If $k \neq 0$ then $k = \mathrm{id}$*

*Proof.* $k(0) = k(0+0) = k(0) + k(0)$ implies $k(0) = 0$. Assume, there is an $\alpha \neq 0$ with $k(\alpha) = 0$. Then $k(\beta) = k(\alpha)k(\beta/\alpha) = 0$ for all $\beta$ and $k$ must vanish. Hence $k(\alpha) \neq 0 \forall \alpha \neq 0$. $k(1) = k(1 \cdot 1) = k(1)k(1)$ implies $k(1) = 1$. By induction we obtain $k(n) = n$ for all natural numbers. $k(-n) = k(0 - n) = k(0) - k(n) = -k(n)$. Similarly, we have $k(1/n) = 1/k(n) = 1/n$. For $n, m \in \mathbb{Z}$ we have now $k(n/m) = n/m$ and the lemma is proved for all rational numbers. $\alpha \leq \beta$ implies $k(\alpha) \leq k(\beta)$ since for any positive number $\gamma^2$ we have $k(\gamma^2) = k(\gamma)k(\gamma) >\geq 0$. Let now $\gamma$ be any number. Then there exists a monotonously increasing sequences $\alpha_i \to \gamma$ of rational numbers and likewise a monotonously decreasing sequence $\beta_i \to \gamma$. Hence $\alpha_i = f(\alpha_i) \leq f(\gamma) \leq f(\beta_i) = \beta_i$ which implies $f(\gamma) = \gamma$. ∎

**Lemma 1.1.5.** *Let $o, x_1, \ldots, x_n \in \mathbb{A}^n$ such that $\{x_1 - o, \ldots, x_n - o\}$ is a basis of $\mathbb{R}^n$. Let $\psi: \mathbb{A}^n \to \mathbb{A}^n$ be a map which satisfies the conditions of Theorem 1.1.1 and, in addition, $\psi(o) = o, \psi(x_1) = x_1, \ldots, \psi(x_n) = x_n$. Then $\psi = \mathrm{id}$.*

*Proof.* We prove the lemma by induction. For $n = 1$ the claim is trivial since by Lemma 1.1.4 the only automorphism of the field $\mathbb{R}$ is the identity. Assume now the assertion holds for all subspaces of dimension $< n$. Let $x = \sum_{i=1}^n \xi^i (x_i - o) + o$ and $x'_k = \sum_{i=1, i\neq k}^n \xi^i (x_i - o) + o$. Lemma 1.1.2 Implies that there exist $\eta_k^i \in \mathbb{R}^n$ such that

$$\psi(x'_k) - \psi(o) = \sum_{\substack{i=1 \\ i \neq k}}^{n} \eta_k^i (\psi(x_i) - \psi(o)).$$

It follows that

$$\psi_1 = \psi_{|o + \mathrm{span}\{x_2 - o, \ldots, x_n - o\}}$$

maps $o + \mathrm{span}\{x_2 - o, x_n - o\}$ into itself. By the same argument $\psi^{-1}$ maps $o + \mathrm{span}\{x_2 - o, \ldots x_n - o\}$ into itself also. Thus, by induction hypothesis, we have $\psi_1 = \mathrm{id}$ and therefore $\psi(x) = x$ if $\xi^1 = 0$. Hence we can assume $\xi^1 \neq 0$. Let $\eta^\pm$ be numbers such that $\eta^+, \eta^-, \xi^1$ are pairwise different and let

$$\begin{aligned} x_\pm &= o + \eta^\pm (x_1 - o), \\ y_\pm &= o + \frac{\xi^2 \eta^\pm}{\eta^\pm - \xi^1}(x_2 - o) + \cdots + \frac{\xi^n \eta^\pm}{\eta^\pm - \xi^1}(x_n - o). \end{aligned}$$

Then

$$x = x_\pm + \frac{\eta^\pm - \xi^1}{\eta^\pm}(y_\pm - x_\pm),$$

i.e., $\{x_+, x, y_+\}$ and $\{x_-, x, y_-\}$ are collinear triples of points which generate lines that are not parallel. They must therefore intersect in $x$ and we obtain $\psi(x) = x$. ∎

*Proof of Theorem 1.1.1.* By Lemma 1.1.3 there exists a regular,linear map $A\colon \mathbb{R}^n \to \mathbb{R}^n$ such that $\phi(x_i) - \phi(o) = A(x_i - o)$. Let

$$\tilde{\phi}(x) := A(x - o) + \phi(o).$$

Then $\psi := \tilde{\phi}^{-1} \circ \phi$ is a collineation which satisfies $\psi(o) = \tilde{\phi}^{-1}(\phi(o)) = o$, $\psi(x_i) = \tilde{\phi}^{-1}(\phi(x_i))\tilde{\phi}^{-1}(A(x_i - o) + \phi(o)) = x_i$. hence we can apply Lemma 1.1.5 and the theorem is proved. ∎

We will now turn our attention to special subsets of affine space which become important in the proof of Theorem 1.4.1.

**Definition 1.1.3.** *Let $\mathcal{U} \subset \mathbb{R}^2$ be an open set and $x\colon \mathcal{U} \to \mathbb{A}^n$ be a $C^\infty$ map such that at each point $(s,t) \in \mathcal{U}$ the differential* $\mathrm{D}x(s,t)$ *is injective. Then $x$ is called an* immersed surface. *If $x$ is also injective then it is simply called a* surface.

A surface should be envisaged by its image, a two-dimensional, smooth subset. An immersed surface may have self-intersections.

Since lines have such a fundamental meaning in affine geometry, surfaces which are generated by lines are of special interest.

**Definition 1.1.4.** *A* ruled surfaces *surface is a surface which can be parametrized by a function of the form $x(s,t) = c(s) + tw(s)$. Such a parametrization is called a* ruling *of the surface.*

*Example 1.1.1.* The simplest ruled surfaces are those which admit two different rulings.

(i) A trivial example would be any plane.

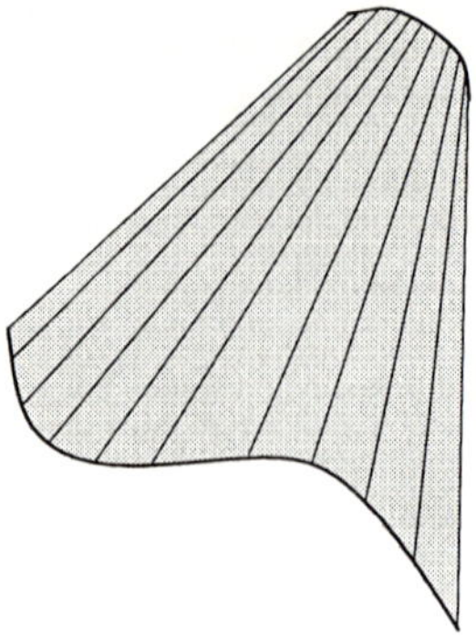

**Fig. 1.1.1.** A ruled surface

(ii) A slightly more sophisticated example is given by the rotational hyperboloid. Let $c(s) = (\cos(s), \sin(s), 0)^\top$ be the unit circle in $\mathbb{R}^3$ and consider $x_{\text{hyp}}(s,t) = c(s) + t(\dot{c}(s) + (0,0,1)^\top)$. Clearly, $x$ is a ruled surface and explicitly given by $x_{\text{hyp}}(s,t) = (\cos(s) - t\sin(s), \sin(s) + t\cos(s), t)^\top$. Since it satisfies the equation $(x^1_{\text{hyp}})^2 + (x^2_{\text{hyp}})^2 - (x^3_{\text{hyp}})^2 = 1$ it must be a rotational hyperboloid. The same surface is described by $\tilde{x}_{\text{hyp}}(s,t) = c(s) + t(-\dot{c}(s) + (1,0,0)^\top)$ which is a different ruling.

(iii) A third example is given by the hyperbolic paraboloid. Let $c(s) = (s,0,0)^\top$ and $w(s) = \frac{1}{\sqrt{1+k^2s^2}}(0,1,ks)^\top$. Then $x_{\text{par}}(s,t) = c(s) + tw(s)$ satisfies $kx^1_{\text{par}}x^2_{\text{par}} = x^3_{\text{par}}$ and $x_{\text{par}}$ parametrizes a hyperbolic paraboloid. We can interchange $x^1_{\text{par}}$ and $x^2_{\text{par}}$ to obtain a different ruling of the same surface, $\tilde{x}_{\text{par}}(s,t) = (0,s,0)^\top + \frac{1}{\sqrt{1+k^2s^2}}(1,0,ks)^\top$.

**Theorem 1.1.2.** *Let $M \subset \mathbb{A}^n$ be a surface which admits two different rulings. Then — up to an affine transformation — $M$ is a subset of either a plane, a rotational hyperboloid, or a hyperbolic paraboloid.*

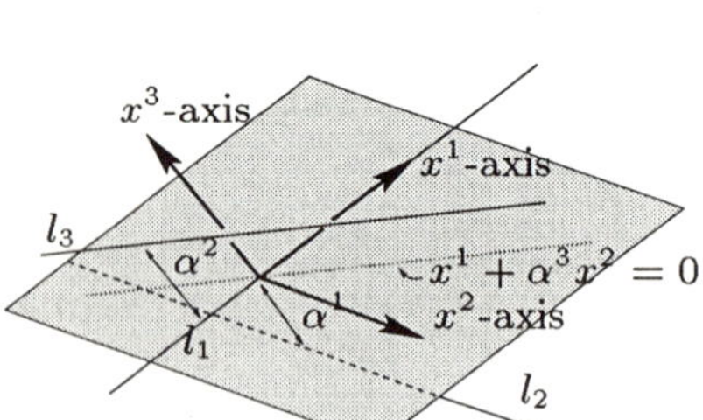


**Fig. 1.1.2.** Proof of Theorem 1.1.2 — first case

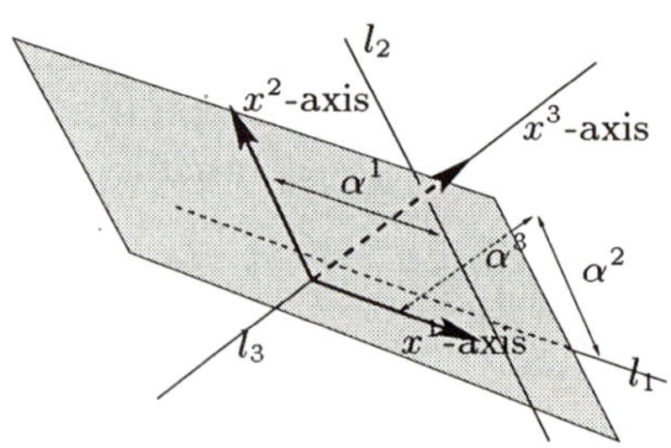


**Fig. 1.1.3.** Proof of Theorem 1.1.2 — second case

*Proof.* It is easy to see that any surfaces $M \subset \mathbb{A}^n$ with two rulings can locally be embedded into $\mathbb{A}^3$. One just has to consider a line $l_1$ of the first ruling which intersects a line $l_2'$ of the second ruling. Choose another line $l_3'$ of the second ruling which also intersects $l_1$. Then all three lines span a 3-dimensional affine subspace. At least locally, any further line of the first ruling must intersect both $l_2'$ and $l_3'$ whence it is contained in the same affine subspace. Since $M$ is generated by the lines of the first ruling we have proved the assertion.

If there are any two generators of the first ruling which lie in a plane then the ruled surface must be this plane. Hence we can assume that any two generators are linearly independent. There are now two possibilities. Either there exist three generators which are all parallel to a single plane or any three generators are linearly independent.

In the first case let $l_1, l_2, l_3$ be different generators of the first ruling which are all parallel to a single plane. We can now find linear coordinates $\{x^1, x^2, x^3\}$ such that the $x^1$-Axis coincides with $l_1$ and the $x^2$-axis is parallel to $l_2$. By choosing the origin appropriately, $l_1$ is given by $x^2 = x^3 = 0$, $l_2$ by $x^1 = 0$, $x^3 = \alpha^1$, and $l_3$ is given by $x^3 = \alpha^2$, $x^1 + \alpha^3 x^2 = 0$, where $\alpha^1, \alpha^2, \alpha^3 \in \mathbb{R}$. Let $P$ be a two-plane which contains $l_1$. Then there exists an $s \in \mathbb{R}$ such that $P$ is given by $x^2 + sx^3 = 0$. Any generator $l'$ of the second ruling which is contained in $P$ must intersect both $l_2$ and $l_3$. We obtain for the intersection points:

$$\begin{aligned} \{P \cap l_2\}&: x^1 = 0, \quad x^2 = -\alpha^1 s, \quad x^3 = \alpha^1; \\ \{P \cap l_3\}&: x^1 = \alpha^2\alpha^3 s, \quad x^2 = -\alpha^2 s, \quad x^3 = \alpha^2. \end{aligned}$$

It follows that $l'$ has the parameter form

$$l' = \left\{ \begin{pmatrix} 0 \\ -\alpha^1 s \\ \alpha^1 \end{pmatrix} + t \begin{pmatrix} \alpha^2\alpha^3 s \\ -s\left(\alpha^2 - \alpha^1\right) \\ \alpha^2 - \alpha^1 \end{pmatrix} \,\middle|\, t \in \mathbb{R} \right\}.$$

Since the ruled surface is generated by such these lines $l'$, we have obtained a parametrization $(s,t) \mapsto x(s,t)$ of it. Eliminating the parameters $s, t$ we obtain $x^1 x^3 = -\alpha^3\alpha^2 x^2$, whence the surface must be a hyperbolic paraboloid.

For the second case we choose linear coordinates $\{x^1, x^2, x^3\}$ such that the $x^3$-Axis coincides with $l_3$, the $x_2$-axis is parallel to $l_2$, and the $x^1$-axis is parallel to $l_1$. We can chose the origin 0 of the coordinate system such that it lies in $l_3$ and such that $l_2$ lies in the plane $x^3 = 0$. Then there exist numbers $\alpha^1, \alpha^2, \alpha^3 \in \mathbb{R}$ such that $l_3 = \{x \,|\, x^1 = x^2 = 0\}$, $l_2 = \{x \,|\, x^1 = \alpha^1, x^3 = 0\}$, and $l_1 = \{x \,|\, x^2 = \alpha^2, x^3 = \alpha^3\}$. Let $P$ be a plane which contains $l_3$ and is not parallel to $x^2 = 0$. Then there exists an $s \in \mathbb{R}$ such that $P$ is given by $x^1 - sx^2 = 0$. Any line $l'$ of the second family which lies in $P$ must intersect $l_1$ and $l_2$. We calculate

$$\{P \cap l_1\}: x^1 = s\alpha^2, \quad x^2 = \alpha^2, \quad x^3 = \alpha^3;$$
$$\{P \cap l_2\}: x^1 = \alpha^1, \quad x^2 = \frac{\alpha^1}{s}, \quad x^3 = 0.$$

This gives the line

$$l' = \left\{ \begin{pmatrix} \alpha^1 \\ \alpha^1/s \\ 0 \end{pmatrix} + t \begin{pmatrix} \alpha^1 - s\alpha^2 \\ \alpha^1/s - \alpha^2 \\ -\alpha^3 \end{pmatrix} \middle| t \in \mathbb{R} \right\}.$$

It follows that $(s, t)$ are parameters of the ruled surface. If we eliminate $(s, t)$ we obtain the equation

$$-\alpha^3 x^1 x^2 + \alpha^2 x^1 x^3 - \alpha^1 x^2 x^3 + \alpha^1 \alpha^3 x^2 = 0.$$

This is a quadric. We could use the Gram-Schmidt-procedure to show that this quadric is affinely equivalent to a hyperboloid. But since any quadric in $\mathbb{R}^3$ is affinely equivalent either to the sphere $(x^1)^2 + (x^2)^2 + (x^3)^2 = 1$, the two-dimensional pseudo-hyperbolic space, $-(x^1)^2 + (x^2)^2 + (x^3)^2 = 1$, the rotational hyperboloid, $-(x^1)^2 + (x^2)^2 + (x^3)^2 = -1$, the cone $(x^1)^2 + (x^2)^2 - x^3 = 0$, the hyperbolic paraboloid $-(x^1)^2 + (x^2)^2 - x^3 = 0$, or a plane, we can infer without any further calculation that our surface must be affinely equivalent to a rotational hyperboloid. ∎

### 1.1.3 Euclidean geometry

p. 3 ↓
[↓ p. 11]

*Euclidean geometry gives the local model of space. In the following sections we will obtain models of space & time which incorporate Euclidean geometry as description of space. Unless otherwise stated, here and in the following space has dimension $n-1$. We assume that Euclidean geometry is known to the reader and therefore only summarize a few facts.*

In affine space, we have no definition for "length" or "angle". Since these are fundamental concepts for our perception of space, we must endow affine space with an additional structure. The first scientific and experimentally well tested description and axiomatization of space involving these notions culminated in the "Elements of *Euclid*" (ca. 340b.C.–270b.C.). In modern terminology, Euclid's theory of space can be identified with Euclidean geometry.

The central object of Euclidean geometry is the scalar product.

**Definition 1.1.5.** *A* scalar product *on a real vector space* $\mathbb{V}$ *is a map*

$$\mathbb{V} \times \mathbb{V} \to \mathbb{R} \tag{1.1.4}$$

$$(u, v) \mapsto \langle u, v \rangle \tag{1.1.5}$$

*such that for any* $u, v, w \in \mathbb{V}$, $\lambda, \mu \in \mathbb{R}$ *the properties*

*(i)* $\langle u, \lambda v + \mu w \rangle = \lambda \langle u, v \rangle + \mu \langle u, w \rangle$,
*(ii)* $\langle u, v \rangle = \langle v, u \rangle$,
*(iii)* $\langle u, u \rangle \geq 0$,
*(iv)* $\langle u, u \rangle = 0 \Rightarrow u = 0$

*hold.*

We can now define an Euclidean space as an affine space equipped with a scalar product.

**Definition 1.1.6.** *An* Euclidean space *is a pair* $(\mathbb{A}^{n-1}, \langle \cdot, \cdot \rangle_{\mathbb{R}^{n-1}}$, *where*

*(i)* $\mathbb{A}^{n-1}$ *is the* $(n-1)$*-dimensional, real affine space with associated vector space* $\mathbb{R}^{n-1}$,
*(ii)* $\langle \cdot, \cdot \rangle_{\mathbb{R}^{n-1}}$ *is a scalar product on* $\mathbb{R}^{n-1}$.

*A map* $\phi \colon \mathbb{A}^{n-1} \to \mathbb{A}^{n-1}$ *is an* isometry *if and only if*

$$\langle \phi(y_1) - \phi(x_1), \phi(y_2) - \phi(x_2) \rangle_{\mathbb{R}^{n-1}} = \langle y_1 - x_1, y_2 - x_2 \rangle_{\mathbb{R}^{n-1}}$$

*for all* $y_1, x_1, y_2, x_2 \in \mathbb{A}^{n-1}$.

The physical notions we whish to capture with our mathematical definitions are "distance" and "angele". The *distance* between two points $x, y \in \mathbb{A}^{n-1}$ should only depend on the connecting vector $u = y - x$. It is plausible to demand that the distance of $x$ and $x + \lambda u$ is $\lambda$ times the distance between $x$ and $x + u$. Hence, given a scalar product, the definition $\operatorname{dist}(x, y) := ||x - y||_{\mathbb{R}^{n-1}} := \sqrt{\langle x - y, x - y \rangle_{\mathbb{R}^{n-1}}}$ seems to be a reasonable choice.

It is clear, however, that in order to measure the angle between two vectors one needs a map $\mathbb{R}^n \times \mathbb{R}^n \to \mathbb{R}$ which is symmetric in both entries and remains unchanged if one of the vector is multiplied by a real number. The *angle* between two directions $u$, $v$ may therefore be defined by $\angle(u, v) = \arccos\left(\frac{\langle u, v \rangle_{\mathbb{R}^{n-1}}}{||u||_{\mathbb{R}^{n-1}} ||v||_{\mathbb{R}^{n-1}}}\right)$.

It is not a priori clear that a scalar product is indeed the appropriate additional structure for defining lengths and angles. See (Weyl 1923, §19) for a theoretical justification of the usage of scalar products.

**Proposition 1.1.1.** *A map* $\psi \colon \mathbb{A}^{n-1} \to \mathbb{A}^{n-1}$ *leaves the Euclidean structure* $(\mathbb{A}^{n-1}, \langle \cdot, \cdot \rangle_{\mathbb{R}^{n-1}})$ *invariant if and only if there exist a linear map* $A \colon \mathbb{R}^{n-1} \to \mathbb{R}^{n-1}$ *and points* $o, b \in \mathbb{A}^{n-1}$ *such that* $\psi(x) = A(x - o) + b$ *and* $\langle Au, Au \rangle_{\mathbb{R}^{n-1}} = \langle u, u \rangle_{\mathbb{R}^{n-1}}$ *for all* $u \in \mathbb{R}^{n-1}$.

[p. 10 ↓]
→4
↓ p. 12

[4] In the proof of Proposition 1.1.1 we appeal to Theorem 1.1.1.

*Proof.* Observe first that an application of $\langle Au, Au\rangle_{\mathbb{R}^{n-1}} = \langle u, u\rangle_{\mathbb{R}^{n-1}}$ for all $u \in \mathbb{R}^{n-1}$ to the vector $u = v + w$ implies $\langle Av, Aw\rangle_{\mathbb{R}^{n-1}} = \langle v, w\rangle_{\mathbb{R}^{n-1}}$ for all $v, w \in \mathbb{R}^{n-1}$. Hence the map $x \mapsto \psi(x) = A(x-o) + b$ satisfies

$$\begin{aligned}
\langle \psi(y_1) - \psi(x_1), &\psi(y_2) - \psi(x_2)\rangle_{\mathbb{R}^{n-1}} \\
&= \langle A(y_1-o) - A(x_1-o), A(y_2-o) - A(x_2-o)\rangle_{\mathbb{R}^{n-1}} \\
&= \langle A(y_1 - x_1), A(y_2 - x_2)\rangle_{\mathbb{R}^{n-1}} \\
&= \langle y_1 - x_1, y_2 - x_2\rangle_{\mathbb{R}^{n-1}} .
\end{aligned}$$

Conversely, any map $\psi$ which preserves the Euclidean structure preserves in particular the affine structure. Hence Theorem 1.1.1 implies that there is a linear map $A$ and a point $b$ such that $\psi(x) = A(x-o) + b$ for all $x \in \mathbb{A}$. Since $\psi(y) - \psi(x) = A(y - x)$ it is clear that $A$ must satisfy $\langle Au, Au\rangle_{\mathbb{R}^{n-1}} = \langle u, u\rangle_{\mathbb{R}^{n-1}}$ for all $u \in \mathbb{R}^{n-1}$. ∎

p. 11 ↓
[↓ p. 29]
*Remark 1.1.2.* At first sight our definition of a Euclidean space may seem to be too general. The reader may feel that in space there is a subset of physically distinguished scalar products:

Let $e$ be a vector which we use as measuring stick defining unit length and $E$ a plane which contains $e_1$. Using a pair of compasses we can construct a line $l_e \subset E$ which is orthogonal to $e_1$ and therefore also a vector $e_2$ of the same length as $e_1$ but perpendicular to $e_1$. We may now construct a second plane $E_{e_1}$ by rotating $e_2$ around $e_1$ and a third plane $E_{e_2}$ by rotating $e_1$ around $e_2$. The intersection $E_{e_1} \cap E_{e_2}$ is a line orthogonal to $e_1$ and $e_2$. Using again or pair of compasses we can construct a third vector $e_3$ which is of unit lenght and orthogonal to $e_1$ and $e_2$. Our distinguished scalar product is now given by $\langle e_i, e_j\rangle_{\mathbb{R}^3} = \delta_{ij}$.

It follows from the Theorem of Pythagoras that the length of a vector $u$ is given by $||u||_{\mathbb{R}^3}$. We can use a pair of compasses to approximately (but arbitrarily well) divide the circle into a fixed number of arcs thereby introducing an approximate measure of angle. From the definition of the cosine it is clear that (up to a constant factor depending on the number of arcs) the size of an angle is given by the definition above.

However, this introduction of the standard scalar product is based on procedures which are intuitive but which cannot be defined in mathematical terms without having a scalar product in the first place. In fact, if we had started with any given scalar product $\langle \cdot, \cdot\rangle$ and had defined

(i) a rotation as a linear map which leaves the scalar product invariant and
(ii) a pair of compasses as a device which for each given plane $E$ containing a given vector $e$ produces all vectors $e' \in E$ with the same length as $e$,

then using our construction we would just have recovered $\langle \cdot, \cdot\rangle$. The following proposition gives a mathematical explanation of this fact.

**Proposition 1.1.2.** *Let* $(\mathbb{A}^{n-1}, \langle\cdot,\cdot\rangle_{\mathbb{R}^{n-1}})$ *and* $(\mathbb{A}^{n-1}, \tilde{\langle}\cdot,\cdot\tilde{\rangle}_{\mathbb{R}^{n-1}})$ *be two Euclidean spaces. Then there exists an affine map* $\psi\colon \mathbb{A}^{n-1} \to \mathbb{A}^{n-1}$ *which satisfies*

$$\langle\psi(y_1) - \psi(x_1), \psi(y_2) - \psi(x_2)\rangle_{\mathbb{R}^{n-1}} = \tilde{\langle} y_1 - x_1, y_2 - x_2 \tilde{\rangle}_{\mathbb{R}^{n-1}}$$

*for all* $x_1, y_1, x_2, y_2 \in \mathbb{A}^{n-1}$.

*Proof.* Choose any points $o, b \in \mathbb{A}^{n-1}$ and let $\{e_1, \ldots, e_{n-1}\}$ (respectively $\{\tilde{e}_1, \ldots, \tilde{e}_{n-1}\}$) be an orthonormal basis with respect to $\langle\cdot\cdot\rangle_{\mathbb{R}^{n-1}}$) (respectively $\tilde{\langle}\cdot\cdot\tilde{\rangle}_{\mathbb{R}^{n-1}}$). We define the linear map $A$ by $A\tilde{e}_i = e_i$. Then the affine map $\psi(x) = A(x-o) + b$ is the desired isomorphism. ∎

THe map $\psi$ is often referred to as an *Euclidean transformation.*

**Corollary 1.1.1.** *Let* $\langle\cdot,\cdot\rangle_{\mathbb{R}^{n-1}}$ *be a scalar product of* $\mathbb{R}^{n-1}$. *Then there is a basis* $\{e_1, \ldots, e_{n-1}\}$ *of* $\mathbb{R}^{n-1}$ *such that* $\langle e_i, e_j\rangle_{\mathbb{R}^{n-1}} = \delta_{ij}$, *where*

$$\delta_{ij} = \begin{cases} 1 & \textit{for } i = j, \\ 0 & \textit{otherwise} \end{cases}$$

*is the* Kronecker symbol.

Today, Euclidean geometry is often taught as a prime example for a closed and consistent mathematical theory. This obscures the fact that angles and distances are physically measurable and that therefore Euclidean geometry can be falsified as a physical theory. (For instance, one of the most influential philosophers since the time of enlightenment, Kant (1781), wrongly considered space as given "a priori").

In modern times, *Carl Friedrich Gauß* (1777–1855) seems to have been the first to realize the possibility that Euclidean geometry may not be the correct description of our world — though the lengend that he tried to verify Euclidean geometry by measuring the angles between three mountain summits is not true (Osserman 1995, page 66). He has developed a non-Euclidean geometry in which the parallel axiom does not hold but did not publish it. This geometry was also independently discovered by the Hungarian mathematician *János Bolyai* (1802–1860). Later in this book we will conclude that space should be described by geometries which are far more general than those considered by Gauß and Bolyai.

## 1.2 Absolute space and absolute time

*In this section we present the "naive" model of space and time. We will take care to show how complicated it really is. We will also give a short account of Newton's theory of particles which is the main physical justification of this spacetime concept.*

Time seems to have striking similarities with space but nevertheless to be something which is very different. Like space time is a continuum. However, space is a 3-dimensional continuum while time is 1-dimensional. Moreover, we can freely move in space but merely drift in time. It is often practical to treat space and time in a unified manner. For instance, spacetime diagrams are used to describe movements (cf. Fig. 1.2.1).

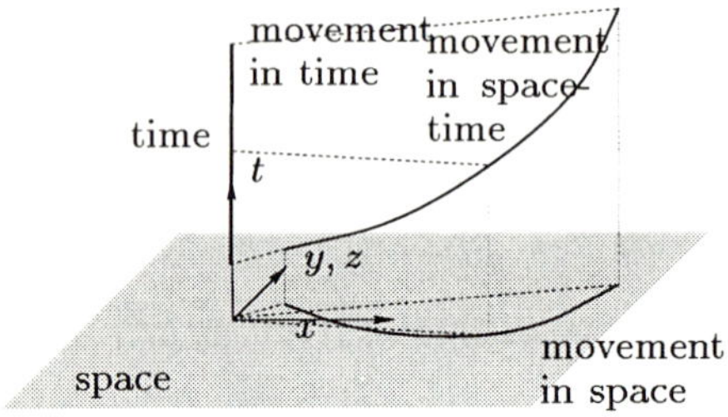


**Fig. 1.2.1.** A curve in a spacetime diagram

*Hermann Minkowski* (1864–1909) (Minkowski 1909) has pointed out that nobody has ever experienced space without time or time without space. This observation is borne out by characterizing space & time in the following way.

**Definition 1.2.1.** *A* primitive spacetime *is set. The points of a spacetime are called* events.

Of course, this definition does not tell anything about the relation of space and time or even allows to distinguish between these concepts. In order to do so we must supplement the primitive spacetime with a geometrical structure.

In the preceding section we have recalled that space can well be described by $(n-1)$-dimensional Euclidean space. The fact that time is 1-dimensional indicates that spacetime can be considered as an $n$-dimensional affine space which is foliated by $(n-1)$-dimensional subspaces each of them carrying a Euclidean structure. Any foliation with affine hyperspaces corresponds to a linear map $\tau\colon \mathbb{R}^n \to \mathbb{R}$, where $x, y \in \mathbb{A}^n$ are in the same hyperspace if and only if $\tau(x-y) = 0$. Denote by $E_x = \{y \in \mathbb{A}^n \,|\, \tau(y-x) = 0\}$ the affine hyperplane through $x$ and let $o, o' \in \mathbb{A}^n$. Then the vector spaces associated with all these affine hyperplanes $E_x$ $(x \in \mathbb{A}^n)$ are identical. In fact, they are given by $\tau^{-1}(0) = \{v \in \mathbb{R}^n \,|\, \tau(v) = 0\}$. Hence we only have to specify one single Euclidean scalar product $\langle\cdot,\cdot\rangle_{\tau^{-1}(0)}$ on the vector space $\tau^{-1}(0)$ in order to get a foliation of $(n-1)$-dimensional Euclidean spaces. This is in accordance with our experience that the geometry of space does not change from one instant of time to another.

The map $\tau$ can be interpreted as a *world clock*: The time difference between to events $x$ and $z$ is just $\tau(z-x)$. Observe that $\tau$ is uniquely defined up to a factor. This factor corresponds to the physical unit in which time is measured.

We still need to link events in different hypersurfaces which correspond to the same point in space. The simplest way to do so is to introduce as a second structure a vector $\mathfrak{t}$ and interpret all points lying on the line $x + \mathbb{R}\mathfrak{t}$ as the same point in space at different times. If we normalize $\mathfrak{t}$ by $\tau(\mathfrak{t}) = 1$ then the time difference between $x$ and $y = x + t\mathfrak{t}$ is just $t$.

**Definition 1.2.2.** *A* Newton spacetime *is a quadrupel*

$$(\mathbb{A}^n, \mathfrak{t}, \tau, \langle\cdot,\cdot\rangle_{\tau^{-1}(0)}), \tag{1.2.6}$$

*where* $\mathfrak{t} \in \mathbb{R}^n$ *is a distinguished vector,* $\tau\colon \mathbb{R}^n \to \mathbb{R}$ *a linear map such that* $\tau(\mathfrak{t}) = 1$*, and* $\langle\cdot,\cdot\rangle_{\tau^{-1}(0)}$ *is a scalar product on the vector space* $\tau^{-1}(0)$.

This definition is just the content of *Isaac Newton's* (1642–1727) theory of absolute time and absolute space.[5]

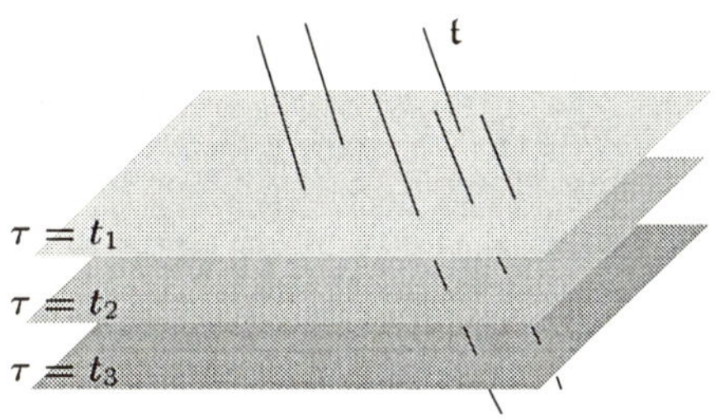


**Fig. 1.2.2.** Absolute space, absolute time

We see that spacetime is fibred twice, By lines parallel to $\mathfrak{t}$ and by hyperspaces of the form $E_{o+t\mathfrak{t}}$ where $o$ is some fixed event. This structure may appear quite cumbersome but it captures our naive point of view in a geometrical way.

One can think of $\mathfrak{t}$ as defining a time axis and therefore an absolute notion of *rest*, and think of $\tau$ as defining an absolute notion of *instant of time*. The pair $(\tau, \mathfrak{t})$ induces a projection $\vec{\cdot}\colon \mathbb{R}^n \to, \mathbb{R}^n$, $v \mapsto \vec{v}$, where $v = v^0\mathfrak{t} + \vec{v}$ and $\tau(\vec{v}) = 0$.

A map which leaves the structure of a Newton spacetime invariant consists of a spatial Euclidean transformation as given in Proposition 1.1.2 and a spacetime translation.

**Proposition 1.2.1.** *A map* $\psi\colon \mathbb{A}^n \to \mathbb{A}^n$ *leaves the Newton spacetime* $(\mathbb{A}^n, \mathfrak{t}, \tau, \langle\cdot,\cdot\rangle_{\tau^{-1}(0)})$ *invariant if and only if there exist a linear map* $A\colon \tau^{-1}(0) \to \tau^{-1}(0)$ *and points* $o, b \in \mathbb{A}^n$ *such that*

[5] In the next section we will discuss an improved spacetime model which is named after Galileo Galilei who lived before Newton. The reason for this is that Galilei emphasized different points than Newton, points which are more important to us nowadays. However, what will be referred to as a Galilei spacetime also incorporates ideas due to Newton.

*(i)* $\psi(x) = A\overrightarrow{(x-o)} + \tau(x-o)\mathrm{t} + b$ *and*
*(ii)* $\langle Au, Au\rangle_{\tau^{-1}(0)} = \langle u, u\rangle_{\tau^{-1}(0)}$ *for all* $u \in \tau^{-1}(0)$.

*Proof.* It is easy to check that maps of this form are isomorphisms of Newton structures. Conversely, observe that any affine map $\psi$ which maps any affine hyperplane $E_x$ into some affine hyperplane$E_{x'}$ is necessarily of the form

$$\psi(x) = A\overrightarrow{(x-o)} + \tau(x-o)v + b$$

where $A$ is a linear map of $\tau^{-1}(0)$ into itself, $v \in \mathbb{R}^n$ and $o, b \in \mathbb{A}^n$.

That $A$ satisfies $\langle Au, Au\rangle_{\tau^{-1}(0)} = \langle u, u\rangle_{\tau^{-1}(0)}$ for all $u \in \tau^{-1}(0)$ follows from Proposition 1.1.1 and the fact that $\psi$ restricted to $E_x$ is an isometry of Euclidean spaces.

Since the vector $\mathrm{t}$ is an invariant of the Newton spacetime the equation $o+\mathrm{t}-o = \mathrm{t}$ implies $\psi(o+\mathrm{t})-\psi(o) = \mathrm{t}$ and therefore $(\tau(\mathrm{t})v + b) - b = \mathrm{t}$. But this equation is equivalent to $v = \mathrm{t}$. ∎

Observe that the choice of $o$ is irrelevant, it can always be absorbed by $b$.

We call the set of all isomorphisms $\psi$ of the Newton spacetime the *Newton group* $\mathcal{N}$.

Given a Newton spacetime we can find a basis $\{e_1, \ldots, e_n\}$ of $\mathbb{R}^n$ with respect to which $\tau = (1, 0, \ldots, 0)$, $\mathrm{t} = (1, 0, \ldots, 0)^{\mathsf{T}}$, and $\langle u, v\rangle_{\tau^{-1}(0)} = \sum_{i,j=1}^{n-1} \delta_{ij} u^i v^j$.

### 1.2.1 Non-relativistic particles

*Here we very briefly indicate elementary aspects of Newton's theory of particle mechanics. We will only touch on those features which are necessary for later sections. This section is included for the benefit of mathematicians.*

A particle is thought to be a small material object without interior or exterior structure. This is of course a gross idealization of many macroscopic objects, but for some purposes surprisingly good. Billiard balls are typical examples. On the other hand, one cannot neglect the internal structure of a football. It will be noticibly deformed when hit. This contributes to its springiness and at the same time shows that the particle model is not adequate. An American football has a shape which contributes to its movement when it rolls on a flat surface. Again, a particle description would be a bad approximation.

Newton observed that even if all its structure can be neclected, a material object does carry a parameter which characterizes its movement in spacetime. This parameter is its mass.

**Definition 1.2.3.** *A* non-relativistic particle with mass $m$ *is a pair* $(m, \gamma)$ *where* $m \in \mathbb{R}^+$ *and* $\gamma: t \mapsto \gamma(t) \in \mathbb{A}^n$ *satisfies* $\tau(\dot{\gamma}(t)) = 1$. *The curve* $\gamma$ *is called its* world line *in spacetime.* [6]

It has been first expressed by Galilei (cf. Sect. 1.3) that under ideal conditions a particle which is not subjected to any external force moves along a straight line.[7]

**Definition 1.2.4.** *A* non-relativistic inertial particle *is a non-relativistic particle* $(m, \gamma)$ *which satisfies* $\ddot{\gamma} = 0$.

It is clear that a non-relativistic particle is inertial if and only if $\gamma(t) = x + t(\mathfrak{t} + \vec{v})$ for some point $x \in \mathbb{A}^n$ and some constant vector $\vec{v}$ with $\tau(\vec{v}) = 0$.

Of special interest to us are *collisions* of inertial particles. We understand under a collision of particles any interaction of them which is confined to a compact subset of spacetime. It is best to think of collisions in the sense of colliding billard balls. But we explicitly allow that particles break up or stick together. Since the collision is confined in space and time it is possible to speak in connection with a collision of *incoming* and *outgoing* inertial particles. Let $(m_i, \gamma_i)_{i=1,\ldots,k}$ denote the incoming inertial particles and $(m'_j, \gamma'_j)_{j=1,\ldots,l}$ the outgoing inertial particles. Then the following laws are experimentally well justified.[8]

(i) *Conservation of mass.* $\sum_{i=1}^k m_i = \sum_{j=1}^l m'_j$.

(ii) *Conversation of spatial momentum.* $\sum_{i=1}^k m_i \vec{\dot{\gamma}}_i = \sum_{i=1}^l m'_j \vec{\dot{\gamma}}'_j$.

(iii) *Conservation of kinetic energy.*

$$\sum_{i=1}^{k} \frac{1}{2} m_i \left\langle \vec{\dot{\gamma}}_i, \vec{\dot{\gamma}}_i \right\rangle_{\tau^{-1}(0)} = \sum_{j=1}^{l} \frac{1}{2} m'_j \left\langle \vec{\dot{\gamma}}'_j, \vec{\dot{\gamma}}'_j \right\rangle_{\tau^{-1}(0)}.$$

It is easy to see that these laws are invariant with respect to isomorphisms in the Newton group $\mathcal{N}$.

It is clear that most particles do not move along straight lines. In this case an external *force* must act on the particle in order to force it to take a different path.

**Definition 1.2.5.** *A (time dependent)* force field $\vec{F}$ *is a map* $F: \mathbb{A}^n \to \tau^{-1}(0)$.

---

[6] We only need $\tau(\dot{\gamma}(t)) > 0$ in order to guarantee that the particle moves into its future. The normalization $\tau(\dot{\gamma}(t)) = 1$ synchronizes each particle with the world clock $\mathfrak{t}$.

[7] It is not absolutely clear whether Galilei really meant straigt lines or more complicated curves which take into account the shape of the earth.

[8] These laws are intimately linked to the homogeneity of space and time. This is the content of the *Noether Theorem*. For further details cf. any textbook on (theoretical) mechanics.

In a given force field $\vec{F}$ a particle $\gamma$ moves according to the differential equation

$$m\ddot{\vec{\gamma}} = \vec{F}. \tag{1.2.7}$$

In particular, vanishing force implies that $\gamma$ is an inertial particle.

According to the physical interactions under consideration a particle may also carry a variety of other parameters besides $m$. As an example consider an electrical field $\vec{E}\colon \mathbb{A}^n \to \tau^{-1}(0)$. Every particle $(m, \gamma)$ carries another parameter $q$ which determines the force with which the electical field acts on the particle, $\vec{F} = q\vec{E}$.

## 1.3 Galilei's theory of relativity

*In this section we drop some of the structure of Newton spacetime in order to arrive at Galilei's theory of relativity. We also argue that his theory was revolutionary given the paradigms of the time.*

Galilei's theory of relativity has been motivated by cosmology. We do not feel that the earth moves into any preferred direction. It is therefore plausible to believe that the earth is at rest and that all objects at the sky are moving around it: The sun rises in the East and during the course of a day moves to the West, and there are analogous descriptions of the movements of the moon and the stars. It was already well known that planets are not moving along strictly circular orbits. In the traditional cosmology of the Greek astronomer *Claudius Ptolemeaus* (ca. 100–160) this was accounted for by an elaborate construction using epicycles.

It was a revolutionary act of *Nicolaus Copernicus* (1473–1543) to assert that the sun is the centre of the universe and that the earth is moving around the sun just like any other planet or star (Copernicus 1543). He did so in order to arrive at a model in which movements would be theoretically more uniform and which would therefore be in better accordance with the teaching of the ancient greek philosophers *Pythagoras* (ca. 570 b.C.–500 b.C.) and *Platon* (ca. 428 b.C.–347 b.C.) (cf. (Kanitscheider 1984)). However, his model was not only technically more complicated (using more epicycles than Ptolemeaus) but also encountered a number of serious problems.

(i) If Copernicus was right one should be able to discover a parallaxis effect at the sphere of fixed stars. If the fix star sphere and the earth rotate both around the sun with different velocities, then one should observe different angles $\alpha, \beta$ between two neighbouring stars according to the time of the year. (Cf. Fig. 1.3.1).

(ii) Some passages in the bible seem to contradict the theory of Copernicus. In particular, it states that Josua stopped the sun for a few hours. This statement would not make sense if the sun would not have moved before.

(iii) The model of Copernicus is inhomogeneous. While all planets circle around the sun the moon definitely moves around the earth. No other exception was known. According to Ptolemeaus, every object in the sky moves around the earth. Hence the traditional system seems to be more homogeneous on a large scale and therfore to be advantageous.

(iv) The laws of mechanics seem to contradict Copernicus' hypothesis. Imagine a stone falling from the top of a tower. Since the tower (being fixed to the ground) would move together with the earth, one would not expect the freely falling stone to hit the ground at the foot of the tower. However, exactly this is everyday experience. (Cf. Fig. 1.3.2).

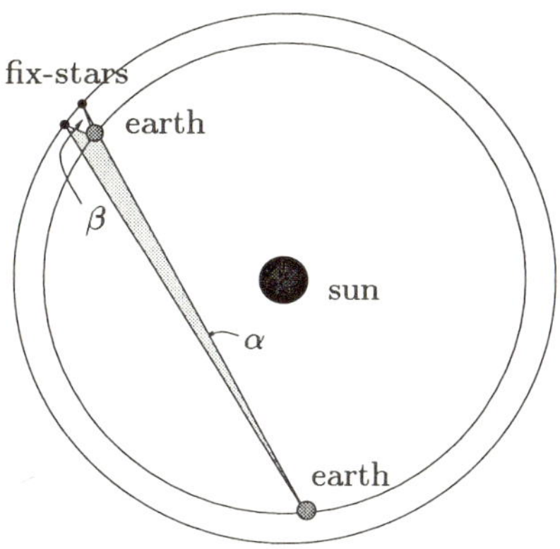


**Fig. 1.3.1.** Parallaxis effect

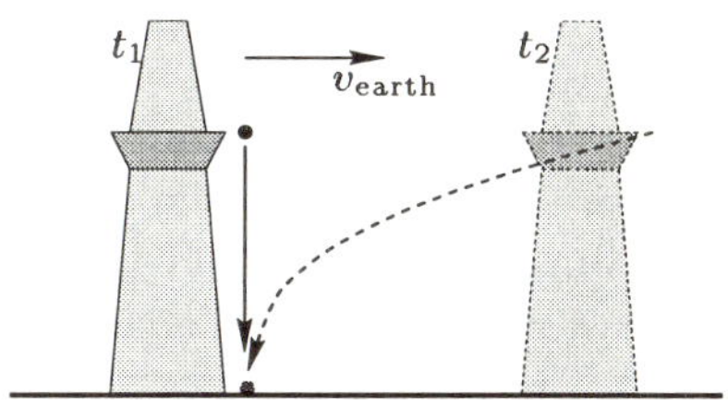


**Fig. 1.3.2.** Tower example

Problem (i) has been addressed by Copernicus himself. He just assumed that the sphere of fixed stars is so large that the parallaxis effect cannot be measured. Ironically, the true radius is orders of magnitudes larger than the radius he proposed. (He was just concerned with making the effect unobservable). The other three objections have been answered by *Galileo Galilei* (1564–1642) some 60 years later.

Galilei was least successful with Problem (ii). While he could quote church authorities (for instance, *Aurelius Augustinus* (354–430)) to the effect that one should not interpret the bible literarily when it comes to questions of physics, the establishment remained unconvinced. One of the reasons has been the fear to set a precedence. If people started to doubt any part of the writing they could as well start to be sceptical about other parts which are closer to the main doctrine. Hence there was a major threat to the whole building of Christian belief. The theory of Copernicus was put on the index and Galilei — after having written a brillant but rather defiant semi-popular book (Galilei 1632)[9] on the matter — was sentenced to house arrest. He obtained this comparatively mild punishment after a public but insincere abdication of his scientific assertions.

Galilei solved Problem (iii) by careful observation (Galilei 1610). The telescope had just been invented and Galilei was one of the first to use it as a scientific tool. He observed that the planet Jupiter also has moons and used this oservation to show that the cosmological system of Ptolemeaus system of the universe was not more homogeneous than the system of Copernicus. On the other hand, since it was believed that beyond the moon the world was filled with a medium very different from air, many philosphers doubted the accuracy of the telescope. They claimed therefore that it was doubtful that the telescope which was acknowledged to work well on earth could be trusted when applied to the position of planets. Galilei argued that the telescope was accurate with respect to all the known phenomena in the sky and that it was therefore justified to use it as a scientific tool.

Galilei solved Problem (iv) by asserting a *law of inertia* which asserts that a constant movement had no influence on physical processes. Galilei supplemented this law with the important physical assertion that complicated velocities can be decomposed into simpler ones. According to this law the stone would keep its initial tangential velocity while falling down and therefore come to rest at the foot of the tower — regardless of the velocity of the earth. It can be argued that this solution of the problem was the most revolutionary act in natural sciences and started physics as a scientific discipline in the modern sense. Recall that everyday experience seems to point against Galilei's law of inertia: If we set a wagon into motion it will certainly come to a stop after some while. Moreover, there was a generally accepted physical theory by *Aristotle* (384 b.C.–322 b.C.) which explained this experimental fact. (The wagon has an initial *impetus* which is responsible for the movement and which is used up during the motion.) Galilei gave many examples to make his law of inertia plausible and to show that it is a law for a limiting case without friction. For instance, he claimed that a stone falling from the mast top of a smoothly sailing ship would also reach the ground at the foot of the mast.[10]

---

[9] This book is a literary and physical master piece. Even today it is well worth reading!

[10] As compelling this example may appear to us, at the time there were some good reasons to doubt it. Since the velocities involved are rather small it would be difficult to verify Galilei's claim experimentally. Also, while the ship moves wind is blowing into the same direction. It is conceivable that the stone is just blown to the right position. (To value the merit of such counter arguments one has to be aware that at this time, good, quantitative physics

Galilei realized that his law of inertia is not compatible with the notion of absolute rest. Instead he postulated a fundamental princple of relativity.

**Postulate 1.3.1 (Galileian relativity).** *For any two observers which move relative to each other with constant velocity all physical processes*[11] *are the same.*

It follows that the vector t in the definition of Newton spacetime which defines absolute rest does not have a physical meaning. (It is another sign of the originality of Galilei that Newton thought he had to re-introduce a concept which had already been shown to be superfluous).

While we have lost the notion of absolute space we can still retain absolute time. Spacetime is then fibred by hyperplanes $t = \text{const}$ and we obtain the following simpler strucure of spacetime.

**Definition 1.3.1.** *A* Galilei spacetime *is a triple*

$$\left(\mathbb{A}^n, \tau, \langle\cdot,\cdot\rangle_{\tau^{-1}(0)}\right), \tag{1.3.8}$$

*where $\tau\colon \mathbb{R}^n \to \mathbb{R}$ is a non-zero linear map and $\langle\cdot,\cdot\rangle_{\tau^{-1}(0)}$ is a scalar product on the vector space $\tau^{-1}(0)$.*

**Fig. 1.3.3.** Relative space, absolute time

The linear map $\tau$ defines a world clock by defining $\tau(y - x)$ to be the time difference between to events $x$ and $y$ — exactly as in the Newtonian model presented above. In contrast to Newton's spacetime we do not have the vector field t at our disposal and therefore there is no absolute rest space. We have replaced "absolute space" by a distinguished *family* of "intertial systems" or "inertial observers". The notion of "Rest" can only be defined relative to an "inertial observer":

---

has not yet been available). Some of these arguments have already been answered by Galilei, who, for instance, circumvented the wind argument by claiming that the physics in a *cabin* of a smoothly sailing ship would be exactly the same as on earth.

[11] Strictly speaking, he only considered mechanical processes.

**Definition 1.3.2.** *Let* $\left(\mathbb{A}^n, \tau, \langle\cdot,\cdot\rangle_{\tau^{-1}(0)}\right)$ *be a Galilei spacetime.*

(i) *An* non-relativistic observer *is a curve* $\gamma\colon t \mapsto \gamma(t) \in \mathbb{A}^n$ *such that* $\tau(\dot\gamma(t)) = 1$.
(ii) *A* non-relativistic inertial observer $\gamma$ *is a curve of the form* $\gamma(t) = x + t\mathrm{t}$, *where* $\mathrm{t} \in \mathbb{R}^n$, $\tau(\mathrm{t}) = 1$.
(iii) *A non-relativistic observer* $\mu$ *is at* rest with respect to a non-relativistic inertial observer $\gamma(t) = x + t\mathrm{t}$ *if* $\dot\mu(t) = \mathrm{t}$.

Hence given a non-relativistic inertial observer $\gamma(t) = x + t\mathrm{t}$ we obtain a splitting of spacetime into space and time *relative* to $\gamma$. Physically, this amounts to regarding the observer $\gamma$ as being at rest. We can also interpret $\mathrm{t}$ as a relative time axis. Relative to the non-relativistic inertial observer $\gamma$ we have thus recovered the structure of a Newton spacetime. Notice, however, that this is only possible by arbitrarily distinguishing one non-relativistic inertial observer. This motivates the following definition.

**Definition 1.3.3.** *Let* $\mathrm{t} \in \mathbb{R}^n$ *be a vector with* $\tau(\mathrm{t}) = 1$. *The pair* $(\mathrm{t}, \tau)$ *is called a* non-relativistic reference frame.

For any given reference frame $(\mathrm{t}, \tau)$ we obtain a map $\vec{(\cdot)}\colon \mathbb{R}^n \mapsto \mathbb{R}^{n-1}$ via the unique decomposition $v = \vec{v} + v^0\mathrm{t}$ where $\tau(\vec{v}) = 0$.

**Proposition 1.3.1.** *A map* $\psi\colon \mathbb{A}^n \to \mathbb{A}^n$ *leaves the Galilei spacetime* $(\mathbb{A}^n, \tau, \langle\cdot,\cdot\rangle_{\tau^{-1}(0)})$ *invariant if and only if there are a linear map* $A\colon \tau^{-1}(0) \to \tau^{-1}(0)$, *a vector* $v \in \mathbb{R}^n$, *and points* $o, b \in \mathbb{A}^n$ *such that*

(i) $\psi(x) = A\overrightarrow{(x-o)} + \tau(x-o)v + b$,
(ii) $\tau(v) = 1$, *and*
(iii) $\langle Au, Au\rangle_{\tau^{-1}(0)} = \langle u, u\rangle_{\tau^{-1}(0)}$ *for all* $u \in \tau^{-1}(0)$.

*Proof.* It is straigtforward to check that maps of this form are isomorphisms of Galilei spacetimes. Conversely, observe that any affine map $\psi$ which maps each affine hyperplane $E_x$ into some other affine hyperplanes $E_{x'}$ is necessarily of the form

$$\psi(x) = A\overrightarrow{(x-o)} + \tau(x-o)v + b$$

where $A$ is a linear map of $\tau^{-1}(0)$ into itself, $v \in \mathbb{R}^n$ and $o, b \in \mathbb{A}^n$.

Since $\psi$ preserves $\tau$ we have $\tau(\psi(x)-\psi(o)) = \tau(x-o)$ for all $x, o \in \mathbb{A}^n$. Hence we obtain

$$\tau\left(A\overrightarrow{(x-o)} + \tau(x-o)v + b - b\right) = \tau(x-o)\tau(v) = \tau(x-o)$$

which in turn implies $\tau(v) = 1$.

The third property follows since $A$ must preserve the Euclidean scalar product $\langle\cdot,\cdot\rangle_{\tau^{-1}(0)}$. ∎

The *Galilei group* $\mathcal{G}$ is the group of maps which leaves the Galilei spacetime invariant.

It should be noted that the Galilei spacetime is compatible with Newton's theory of particles as described at the end of Sect. 1.2. The Galilei spacetime was well accepted as the correct model of space and time for more than 200 years. However, in the 19th century a theorey of electro-magnetism emerged which, together with this spacetime model, was incompatible with Postulate 1.3.1. Still, scientists continued to think that the postulate would hold for mechanical processes.

## 1.4 Einstein's special theory of relativity

*We start with a discussion of the fundamental Michelson-Morley Experiment which indicates that the velocity of light has an absolute value c. These findings indicate that the set of all possible light rays form a further invariant of nature. We will see that this leads to the structure of a Minkowski spacetime (Theorem 1.4.1), or, equivalently, to Einstein's special theory of relativity. We use the results from the two preceding sections to show that there is no need for additional structures in spacetime. In Sect. 1.4.2 we give a short discussion of some consequences of special relativity such as the "Twin paradox" (which, of course, is not paradoxical at all).*

*The proof of the fundamental Theorem 1.4.1 requires section 1.1.2*

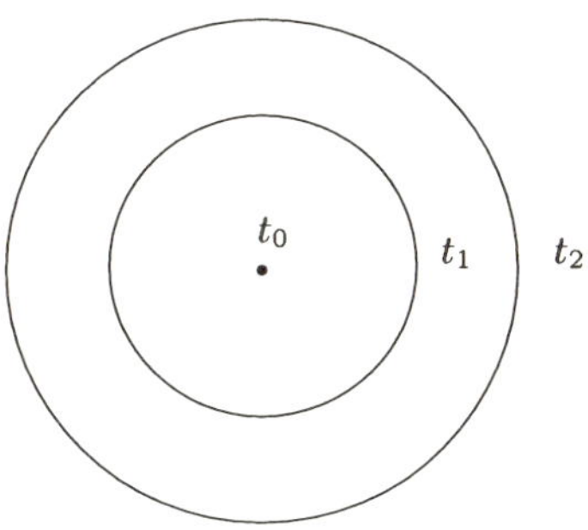


**Fig. 1.4.1.** A flash of light at times $t_0 = 0$, $t_1$, $t_2$

In the 17th century two important properties of light emerged.

In 1676 *Olaf Römer* discovered that the velocity of light is finite. He did this by noticing that there was a yearly aberration in the periods of the moons of Jupiter.

The Dutch physicist *Christian Huygens* (1629–1695) developed a wave theory of light (Huygens 1690). In a very superficial way, we may view light as an analogon to water waves.[12]

[12] The following two paragraphs should not be taken too seriously by the reader. We give an overly simplified version of the wave theory of light

In water waves each "drop of water" individually moves in a circle thereby inducing a similar movement (with a small time delay) of neighbouring drops. All these moving drops together form a wave (cf. Fig. 1.4.2) Since each drop is influenced by the neighbouring drops these time delays accumulate and the whole wave seems to move. If two different waves meet then (in a very rough approximation) they simply linearly superpose each other.[13] This will result in a characteristic (and often complicated) pattern, the "interference pattern". In particular, this superposition will result in a much larger wave if both waves are synchronized and in the other extreme they may cancel

Diffraction experiments indicate that this crude picture qualitatively also applies to light for which, however, matters are mathematically simpler. Again using a very rough model, one may think of the electrical field $E$ at each point as oscillating up and down with respect of a fixed direction. The influence of neighbouring points gives rise of to a wave as described above. The wave length $\lambda$ is the distance between two consecutive maxima and very small. It specifies the colour of light. If two waves are superimposed then the result may be brighter if they are synchroniced. In the other extreme, the waves may even cancel altogether if the setup is arranged such that maxima and minima (of the same size) are superimposed. In this case the result is darkness. (cf. Fig. 1.4.3).

In order to explain the wave nature of light one used to believe that space is filled with a substance called "ether" which plays the same rôle as the water for the water waves. An important problem would then be to determine the movement of the earth with respect to the ether.

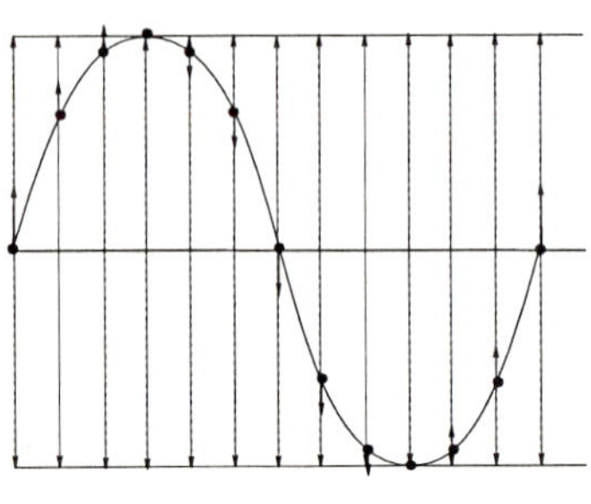

**Fig. 1.4.2.** A wave consisting of linked oscillations

Since the velocity of light should not be directionally dependent, in the non-relativistic reference frame connected with the ether a flash of light should propagate in concentric spheres (cf. Fig. 1.4.1). The corresponding

— just enough in order to understand the Michelson-Morley experiment presented below. Moreover, today the theory of quantum electro dynamics provides a much deeper understanding.

[13] For water waves, this linear superposition is in fact a rather bad approximation.

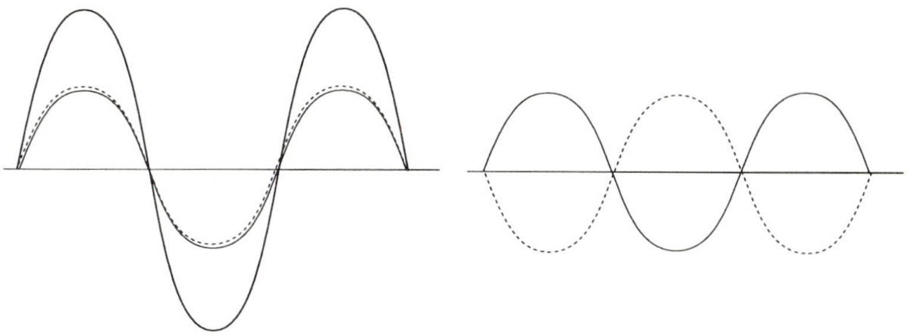

**Fig. 1.4.3.** Superposition of waves

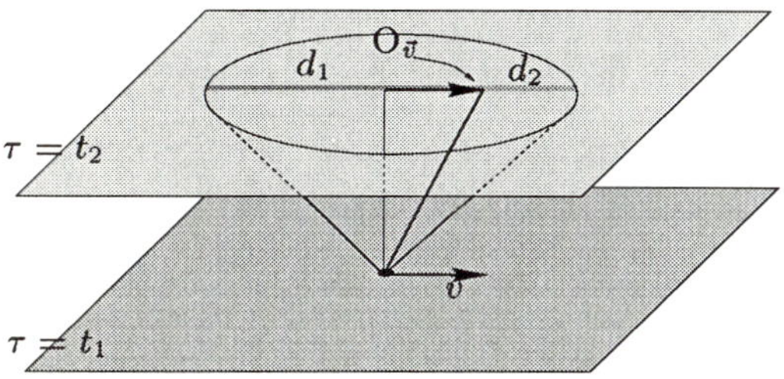


**Fig. 1.4.4.** Future light cone and Galileian relativity. The observer moving with spatial velocity $\vec{v}$ measures a different centre $O_{\vec{v}}$ of the flash of light and therefore different radii $d_1$, $d_2$ for its wave front

picture in spacetime would be a cone. To be more precise, consider the non-relativistic reference frame of the ether, given by the pair $(\mathsf{t}, \tau)$. Let $o$ be the event at which the flash of light is emitted and $E_o = \{x \in \mathbb{A}^n \mid \tau(x - o) = 0\}$ be the instant of time defined by $o$. We have a foliation $\{E_{o+t\mathsf{t}}\}_{t\in\mathbb{R}}$ of $\mathbb{A}^n$ with spatial hyperspaces. Each vector $v$ can be uniquely decomposed into spatial and temporal components, $v = \tau(v)\mathsf{t} + \vec{v}$ where $\tau(\vec{v}) = 0$. A light ray which is sent out at $x \in E_o$ with the spatial velocity $\vec{c}$ describes the curve $o + \mathbb{R}(\mathsf{t} + \vec{c})$ in $\mathbb{A}^n$. Hence with respect to a reference frame fixed to the ether a flash of light corresponds to the *future light cone*

$$C_o^+ = \left\{ y \in \mathbb{A}^n \mid \|c\|^2_{\tau^{-1}(0)} (\tau(y-o))^2 = \langle \overrightarrow{y-o}, \overrightarrow{y-o} \rangle_{\tau^{-1}(0)}, \tau(y-x) \geq 0 \right\}$$

in spacetime. The fact that the field $x \mapsto C_x^+ = C_o^+ + x - o$ of future light cones are not invariant with respect to Galilei transformations (cf. Fig. 1.4.4) would enable one to measure the reference frame of the ether, i.e, the movement of the earth with respect to the ether. This was the aim of *Albert Abraham Michelson* (1852–1931) and *Edward Williams Morley* (1838–1923) (Michelson 1881), (Michelson and Morley 1887) in their famous interference experiment (cf. Figs. 1.4.5, 1.4.6). A light ray is partially reflected at a half silvered mirror $H$. The part of the light ray which is not reflected at $H$ is reflected at a mirror $M$ and then partially reflected at $H$ before reaching the observer $O$. The part of the ray which is immediately reflected at $H$ is reflected by a mirror $M'$ and then (partially) passes through $H$ to arrive at the observer $O$. The distance

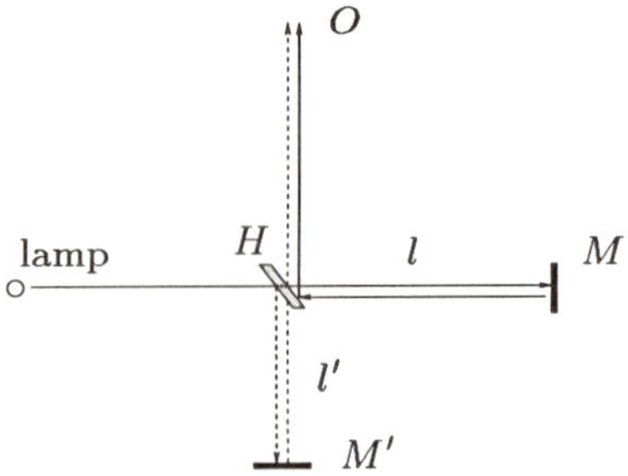


**Fig. 1.4.5.** Michelson-Morley experiment, at rest relative to the ether

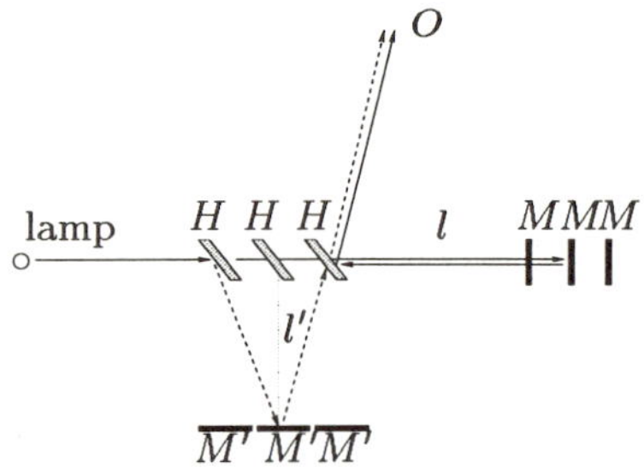


**Fig. 1.4.6.** Michelson-Morley experiment, moving relative to the ether

between $M$ and $H$ is $l$ whereas the distance between $H$ and $M'$ is $l'$. Both light rays have the same intensity when they arrive at $O$. Here they produce an *interference pattern* which allows to measure the difference of the distances which each light ray has travelled (Here one makes use of the fact that the wave lenghts of visible light are extremely small and that the superposition effect allows to measure the distance which a light ray has travelled at an accuracy of half a wave length). Since the laboratory is at rest with respect to the earth, according to Galileian relativity the interference pattern should depend on the angle between $\overrightarrow{HM}$ and the velocity of the earth. Let $c \in \mathbb{R}^+$ be the modulus of the velocity of the light in the ether, $\vec{v}$ be the velocity of the earth relative to the ether and assume first that $\overrightarrow{HM} \parallel \vec{v}$ (cf. Fig. 1.4.6). The first part of the light ray will travel from $H$ to $M_1$ in time $t_1$ and cover the distance $ct_1 = l + ||\vec{v}||_{\tau^{-1}(0)} t_1$. If it travels from $M$ to $H$ in time $t_2$, it will cover the distance $ct_2 = l - ||\vec{v}||_{\tau^{-1}(0)} t_2$. These equations imply

$$t_1 + t_2 = \frac{2l/c}{1 - ||\vec{v}||^2_{\tau^{-1}(0)}/c^2}. \tag{1.4.9}$$

The other part of the light ray travels in time $t'$ from $H$ to $M'$ and in the same time back to $H$, thereby covering the distance

$$2ct' = 2\sqrt{l'^2 + ||\vec{v}||^2_{\tau^{-1}(0)} t'^2}.$$

This gives

$$2t' = \frac{2l'/c}{\sqrt{1 - ||\vec{v}||^2_{\tau^{-1}(0)}/c^2}}. \tag{1.4.10}$$

We are interested in the time difference $\Delta t = 2t' - (t_1 + t_2)$ for both paths. Since the the number $||\vec{v}||_{\tau^{-1}(0)}/c$ is very small we only need to calculate the time difference to second order in $||\vec{v}||_{\tau^{-1}(0)}/c$.

$$
\begin{aligned}
\Delta t = t_1 + t_2 - 2t' &= \frac{2}{c}\frac{l - l'\sqrt{1 - \|\vec{v}\|^2_{T^{-1}(0)}/c^2}}{1 - \|\vec{v}\|^2_{T^{-1}(0)}/c^2} \\
&\approx \frac{2}{c}\frac{l + \frac{1}{2}(\|\vec{v}\|^2_{T^{-1}(0)}/c^2)l' - l'}{1 - \|\vec{v}\|^2_{T^{-1}(0)}/c^2} \\
&= \frac{2}{c}\frac{\delta l\left(-1 + \frac{1}{2}(\|\vec{v}\|^2_{T^{-1}(0)}/c^2\right) + \frac{1}{2}\|\vec{v}\|^2_{T^{-1}(0)}/c^2 l}{1 - \|\vec{v}\|^2_{T^{-1}(0)}/c^2} \\
&\approx \frac{2}{c}\left(\delta l\left(-1 + \frac{1}{2}(\|\vec{v}\|^2_{T^{-1}(0)}/c^2\right) + \frac{1}{2}\|\vec{v}\|^2_{T^{-1}(0)}/c^2 l\right) \\
&\quad \times \left(1 + \|\vec{v}\|^2_{T^{-1}(0)}/c^2\right) \\
&= \frac{l}{c}\frac{\|\vec{v}\|^2_{T^{-1}(0)}}{c^2} - \frac{2\delta l}{c}\left(1 + \frac{1}{2}\frac{\|\vec{v}\|^2_{T^{-1}(0)}}{c^2}\right),
\end{aligned}
$$

where $l' = l + \delta l$. This gives a displacement per wave length $\lambda$ of

$$
\Delta Z_{||} = \frac{c\Delta t}{\lambda} \approx \frac{l}{\lambda}\frac{\|\vec{v}\|^2_{T^{-1}(0)}}{c^2} - \frac{2\delta l}{c}\left(1 + \frac{1}{2}\frac{\|\vec{v}\|^2_{T^{-1}(0)}}{c^2}\right).
$$

It follows that the interference depends crucially on the length difference $\delta l$ which cannot be measured accurately enough. In order to overcome this difficulty Michelson and Morley turned the whole setup by $\pi/2$ and measured the interference difference. For the rotated setup we must set $\Delta t = 2t' - t_1 - t_2$ and interchange $l, l'$. An analogous (but in the details slightly differnt) calculation gives

$$
\Delta Z_{\perp} = -\frac{c\Delta t}{\lambda} \approx \frac{l}{\lambda}\frac{\|\vec{v}\|^2_{T^{-1}(0)}}{c^2} + \frac{2(-\delta l)}{c}\left(1 + \frac{\|\vec{v}\|^2_{T^{-1}(0)}}{c^2}\right).
$$

The relative displacement depends on $\delta l$ only up to second order and is given by

$$
\Delta Z = \Delta Z_{||} - \Delta Z_{\perp} \approx \frac{2l}{\lambda}\frac{\|\vec{v}\|^2_{T^{-1}(0)}}{c^2}\left(1 - \frac{\delta l}{l}\right).
$$

If one assumes that the sun rests relative to the ether, uses the Hg spectral line with $\lambda \approx 5.461 \cdot 10^{-10}m$ and has $l \approx 21m$ then one would obtain $\Delta Z \approx 0.4$ which is well in the range which can be observed. However, all such experiments had a negative outcome.

A possible explanation of this negative outcome is that the earth rests with respect to the ether. But the earth circles around the sun which itself rotates in our galaxy. Since in the course of the year the earth changes its velocity direction relative to these other velocities, it is inconceivable that all year round the velocity of the earth relative to the ether can be neglected. Another explanation would be that light moves like particles and that therefore Galileian relativity would apply to light as well. Since the light was from earth bound sources and the observations have been made on earth, this assumption would explain the negative outcome of the experiment. The Michelson-Morley experiment has therefore been repeated using star light — again with negative results. Now one could argue that as soon as the starlight is reflected at mirrors fixed to the earth, the light reflected light should be viewed as being produced on the earth. This would explain even negative results for star light in the framework of Galileian relativity (Hasse 1995). While this explanation is conceivable, it would demand a new theory of reflection. It is much simpler to assume that the velocity of light is independent of the movement of its source. This is the traditional interpretation which we will adopt in this book. It has been given further support by many consequences of the resulting theory (for instance, the possibility of obtaining huge amounts of energy from nuclear fission and nuclear fusion).

Since our interpretation of the experiment of Michelson and Morley is in contradiction to Galilei's theory of relativity we have to reconsider the foundations of spacetime. In order to do so we start with our new insight about the nature of light propagation, i.e. that the set of possible future light cones is an invariant structure of spacetime. Here and in the following we will chose units such that $c = 1$. (In the SI-system, one has $c = 2.99792458 \cdot 10^8 \mathrm{m/s}$.)

**Postulate 1.4.1 (Invariance of the future light cones).**
*Spacetime can be identified with $\mathbb{A}^n$ together with an invariant field of future light cones $C_x^+ = C_o^+ + x - o$, $x \in \mathbb{A}^n$.*

We start the investigation of this postulate, by first determining all maps which leave this futre light cone structure invariant. To simplify the discussion we choose again a non-relativistic reference frame $(\tau, \mathrm{t})$ and denote by $\vec{(\cdot)}$ the induced projection of $\mathbb{R}^n$ to $\tau^{-1}(0)$. Defining the bilinear form

$$\eta_0 \colon \mathbb{R}^n \times \mathbb{R}^n \to \mathbb{R}, \quad (u,v) \mapsto \eta(u,v) = -\tau(u)\tau(v) + \langle \vec{u}, \vec{v} \rangle_{\tau^{-1}(0)}$$

we can write

$$C_x = \{y \in \mathbb{A}^n \,|\, \eta(y-x, y-x) = 0 \text{ and } C_x^{\pm} = \{y \in C_x \,|\, \tau(y-x) \gtrless 0\}.$$

We call $C_x$ the *light cone* and $C_x^-$ the *past light cone* at $x$. It is clear that a transformation which leaves the field the future light cones invariant must also leave the field of light cones $x \mapsto C_x$ invariant. The bilinear form $\eta_0$ is a Minkowski metric as defined below.

**Definition 1.4.1.** *A* Minkowski metric $\eta$ *is a constant bilinear form on $\mathbb{R}^n$ with signature* $(-,+,\ldots,+)$.[14]

[14] This means that there is an "orthonormal basis" as defined directly below.

*A basis* $\{e_0, \ldots, e_{n-1}\}$ *of* $\mathbb{R}^n$ *is called an* orthonormal basis *with respect to* $\eta$ *if* $\eta(e_0, e_a) = -\delta_{0a}$ *and* $\eta(e_i, e_j) = \delta_{ij}$ *for all* $i, j \in \{1, \ldots, n-1\}$, $a \in \{0, \ldots, n-1\}$.

*A* Lorentz transformation *is a linear map* $L \colon \mathbb{R}^n \to \mathbb{R}^n$ *such that* $\eta(u, v) = \eta(Lu, Lv)$ *for all* $u, v \in \mathbb{R}^n$. *The set of all Lorentz transformation is the* the Lorentz group *and denoted by* $O(n, 1)$.

It is now easy to find a class of maps which leave the light cone structure invariant. Let $\alpha \in \mathbb{R} \setminus \{0\}$, $b \in \mathbb{R}^n$, $o \in \mathbb{A}^n$, and $\phi \colon \mathbb{R}^n \to \mathbb{R}^n$ be given by $\phi(x) = \alpha L(x - o) + b$. It is immediate that $\phi(C_y) = C_{\phi(y)}$ for all $y \in \mathbb{R}^n$ and that therefore the transformation $\phi$ satisfies our invariance requirement. The following theorem due to Alexandrov (1950) implies that all isomorphisms are of this special form.

**Theorem 1.4.1.** *Let* $n \geq 3$ *and* $\phi \colon \mathbb{R}^n \to \mathbb{R}^n$ *be a bijective map such that* $\phi$ *and* $\phi^{-1}$ *map lightlike vectors into lightlike vectors,* $\eta(x - y, x - y) = 0 \Rightarrow \eta(\phi(x) - \phi(y), \phi(x) - \phi(y)) = 0$ *and analogously for* $\phi^{-1}$. *Then there exist an* $L \in O(n, 1)$, *an* $\alpha \in \mathbb{R} \setminus \{0\}$, *an* $o \in \mathbb{A}^n$, *and a* $b \in \mathbb{R}^n$ *such that* $\phi(x) = \alpha L(x - o) + b$ *for all* $x \in \mathbb{A}^n$.

[p. 12 ↓] →15 ↓ p. 30

There are several proofs of this result. For instance, Benz (1992) shows that the theorem follows from the fundamental theorem of Laguerre geometry. We will follow Alexandrov (1975) who gave a particularly elegant proof. His proof rests on results in affine geometry which have been given in Sect. 1.1.1.

*Proof of Theorem 1.4.1.* Let $y \in C_x$. Then $\eta(y - x, y - x) = 0$ and by assumption $\eta(\phi(y) - \phi(x), \phi(y) - \phi(x)) = 0$. The last equality implies $\phi(y) \in C_{\phi(x)}$ and therefore that $\phi$ maps generators of the light cone $C_x$ into the light cone $C_{\phi(x)}$.

Now assume that $P$ is a two-plane which intersects $C_x$ in two generators $l_x$ $l'_x$. We will show that $\phi(P)$ is also a two-plane. Since for any $y \in l_x$ the cone $C_y$ intersects $P$ in generators $l_y = l_x$ and $l'_y$ which is paralle to $l'_x$ (and similarly for $y' \in l'_x$), $P$ is ruled by two different families of parallel generators. Since $\phi$ maps generators into generators $\phi(P)$ must also be a surface with two different rulings and, by Theorem 1.1.2, be affinely equivalent to either a plane, a rotational hyperboloid, or a hyperbolic paraboloid. Each generator in $P$ of one family intersects all generators of the other family. Since in a rotational hyperboloid the generators at opposite points of the circle $c(s)$ lie in parallel planes, $\phi(P)$ cannot be affinely equivalent to a rotational hyperboloid. To see that $\phi(P)$ cannot be affinely equivalent to a hyperbolic paraboloid note that in a hyperbolic paraboloid all generators of a given family are parallel to a single 2-plane and that this property is affinely invariant. Consider $\phi(y) \in l_{\phi(x)}$

[15] The proof of this theorem requires the material presented in Sect. 1.1.2. The theorem but not its proof is essential for the following.

and fix any 2-plane $Q$. Then there are exactly two generators $l_1$, $l_2$ of $C_{\phi(y)}$ which are parallel to $Q$. If we now (parallelly) translate the cone along $l_{\phi(x)}$ we see that the generators of the translated cone which are parallel to $Q$ must also be parallel to $l_1$ and $l_2$. This implies that $\phi(P)$ is generated by a family of parallel lines. Hence the hyperbolic paraboloid degenerates to a plane.

We can now show that $\phi$ is an affine map. Let $l$ be any line and $x \in l$ and consider two different planes $P_1$, $P_2$ which contain $l$ and intersect $C_x \setminus \{x\}$. The intersections of these planes with $C_x$ consist of two generators each. Then $\phi(P_1)$ and $\phi(P_2)$ are also planes and their intersection $\phi(l)$ a line. It follows from Theorem 1.1.1 that $\phi$ is affine.

Since the property $\eta(y-x, y-x) = 0 \Rightarrow \eta(\phi(y)-\phi(x), \phi(y)-\phi(x)) = 0$ is translation invariant we can without loss of generality consider assume that $\phi$ is linear. Let $\mathfrak{t}$ be a vector with $\eta(\mathfrak{t}, \mathfrak{t}) = -1$ and let $\mathfrak{e}$ be a vector with

$$\eta(\mathfrak{e}, \mathfrak{e}) = 1, \qquad \eta(\mathfrak{e}, \mathfrak{t}) = 0. \tag{1.4.11}$$

Then $0 = \eta(\mathfrak{e}\pm\mathfrak{t}, \mathfrak{e}\pm\mathfrak{t})$ implies $0 = \eta(\phi(\mathfrak{e})\pm\phi(\mathfrak{t}), \phi(\mathfrak{e})\pm\phi(\mathfrak{t}))$ and therefore $\eta(\phi(\mathfrak{e}), \phi(\mathfrak{e})) = -\eta(\phi(\mathfrak{t}), \phi(\mathfrak{t})) =: \alpha$, $\eta(\phi(\mathfrak{e}), \phi(\mathfrak{t})) = 0$. Any vector $v$ can be decomposed into $v = v_{\mathfrak{e}}\mathfrak{e} + v_{\mathfrak{t}}\mathfrak{t}$, where $\mathfrak{e}$ satisfies Equations (1.4.11) and $v_{\mathfrak{e}}, v_{\mathfrak{t}} \in \mathbb{R}$. We obtain $\eta(\phi(v), \phi(v)) = \alpha\eta(v, v)$ which implies that $\frac{1}{\alpha}\phi$ leaves the quadratic form associated with $\eta$ invariant. But then it must also leave $\eta$ invariant. ∎

p. 29 ↓
[↓ p. 36]

**Definition 1.4.2.** *Let $\eta$ be a Minkowski metric. The pair $(\mathbb{A}^n, \eta)$ is called* Minkowski spacetime.

Using orthonormal bases it is easy to see that all Minkowski spacetimes are isomorphic, i.e., we can speak of "the" Minkowski spacetime.

In a Minkowski spacetime there is no designation of future and past. (Observe that we needed the 1-form $\tau$ in order to define the future direction.) Observe that the set $C_o \setminus \{o\}$ consists of two connected components, say $C_o^+ \setminus \{o\}$ and $C_o^- \setminus \{o\}$. We may now choose $C_o^+$ (respectively, $C_o^-$) as the set of events in $C_x$ to the future (respectively, past) of $o$ (including $o$). Hence $C_0^+$ is the set of all events which can be reached by a light ray with source in $o$. By continuity, this also determines the future direction at any other event $x \in \mathbb{A}^n$ where $C_x^+ = (x-o) + C_o^+$. Observe that this definition coincides with our previous definition in the case $\eta = \eta_0$. Let $v \in \mathbb{R}^n$ be a vector with $\eta(v, v) =< 0$. Then we have

(i) either $\eta(v, w) < 0$ for all $w \in C_o^+$ and $\eta(v, w) > 0$ for all $w \in C_o^-$
(ii) or $\eta(v, w) > 0$ for all $w \in C_o^+$ and $\eta(v, w) < 0$ for all $w \in C_o^-$.

Hence we can alternatively define the future direction by singling out a vector $v \in \mathbb{R}^n$ with $\eta(v, v) < 0$. Since it is more practical to work with vectors than with connected components of light cones one usually chooses this alternative definition.

**Definition 1.4.3.** *Let* $(\mathbb{A}^n, \eta)$ *a Minkowski spacetime. A* time orientation *is an equivalence class* $[v]$ *of vectors in* $\mathbb{R}^n$ *such that* $\eta(v, w) < 0$ *for all* $v, w \in [v]$.

*Given a time orientation* $[v]$*, we say that the future light cone at* $o \in \mathbb{A}$ *is the set* $C_o^+ = \{o + w \in C_o \,|\, \eta(w, v) \leq 0\}$ *and the past light cone is the set* $C_o^- = \{o + w \in C_o \,|\, \eta(w, v) \geq 0\}$.

Let $\tilde{\mathcal{P}}$ be the invariance group of the light cone structure and $\mathcal{P}$ be the group of Poinaré transformations, $\mathcal{P} = \{x \mapsto L(x - o) + b \,|\, b \in \mathbb{R}^n, o \in \mathbb{A}^n, L \in \mathrm{O}(m, 1)\}$. Given an orientation $[v]$, we call $\mathcal{P}^+ = \tilde{\mathcal{P}}^+ \cap \mathcal{P}$ the group of *time orientation preserving Poincare transformations.* The discussion above suggests to reduce the group $\tilde{\mathcal{P}}$ to a subgroup $\tilde{\mathcal{P}}^+$ by asserting that the elements of $\tilde{\mathcal{P}}^+$ map future light cones into future lightcones.

**Lemma 1.4.1.** *Let* $(\mathbb{A}^n, \eta)$ *a Minkowski spactime and* $[v]$ *a time orientation. Then* $\tilde{\mathcal{P}}^+ = \{\psi \in \tilde{\mathcal{P}} \,|\, \psi(C_o^+) = C_{\psi(o)}^+\}$ *is a subgroup of* $\tilde{\mathcal{P}}$.

*Proof.* Clear. ∎

**Lemma 1.4.2.** *The transformation* $\phi$ *is an element of* $\tilde{\mathcal{P}}$ *if and only if there is an* $\alpha \in \mathbb{R} \setminus \{0\}$ *and a* $\psi \in \mathcal{P}$ *such that* $\phi = \alpha\psi$; $\phi \in \tilde{\mathcal{P}}^+$ *if and only if there is an* $\alpha \in \mathbb{R}^+ \setminus \{0\}$ *and a* $\psi \in \mathcal{P}^+$ *such that* $\phi = \alpha\psi$.

*Proof.* This follows directly form Theorem 1.4.1. ∎

A priori it is conceivable that there exist other fundamental invariants of space and time which would restrict the group even further. We will now show that because of the validity of Euclidean geometry and the principle of Galileian relativity 1.3.1 this is not the case.

**Proposition 1.4.1.** *Fix a non-relativistic, inertial observer* $t \mapsto o + t\mathfrak{t}$ *and consider its associated Newton spacetime* $(\mathbb{A}^n, \mathfrak{t}, \tau, \langle\cdot,\cdot\rangle_{\tau^{-1}(0)})$. *For each* $x \in \mathbb{A}$ *let* $E_x = \{y \in \mathbb{A}^n \,|\, \tau(y - x) = 0\}$.

- *(i) There exists a Minkowski metric* $\eta$ *which generates the light cones as measured by the observer* $t \mapsto o + t\mathfrak{t}$. *Further, this Minkowsi metric is unique up to a multiple.*
- *(ii) Let* $x \in \mathbb{A}^n$. *The map* $\phi \in \tilde{\mathcal{P}}$ *restricted to the Euclidean space* $E_x, \langle\cdot,\cdot\rangle_{\tau^{-1}(0)}$ *is an isometry if and only if* $\phi \in \mathcal{P}$.
- *(iii) Let* $\mathcal{P}'$ *be a subgroup of the Poincaré group* $\mathcal{P}$ *such that*
  - *(a) for each Euclidean isometry* $\psi\colon E_o \to E_{o'}$ *there exists a* $\phi \in \mathcal{P}'$ *with* $\phi_{|E_o} = \psi$,

*(b) for each non-relativistic observer $t \mapsto o' + t\mathrm{t}'$ with $\eta(\mathrm{t}', \mathrm{t}') < 0$ there is a $\phi \in \mathcal{P}'$ with $\phi(o + \mathbb{R}^+\mathrm{t}) = o' + \mathbb{R}^+\mathrm{t}'$,*

*(c) All $\phi \in \mathcal{P}'$ preserve the time orientation.*

*Then $\mathcal{P}' = \mathcal{P}^+$.*

*Proof.* (i): The adapted Minkowski metric $\eta$ is just given by $\eta(u, v) = -\tau(u)\tau(v) + \langle \vec{u}, \vec{v} \rangle_{\tau^{-1}(0)}$. Observe that $\tau(w) = -\eta(\tau, w)$ for all $w \in \mathbb{R}^n$.

(ii): This follows directly form Lemma 1.4.2.

(iii): We will first show that for any two different vectors $e_0$, $e_0'$ with $\eta(e_0, e_0) = \eta(e_0', e_0') = -1$ and $\eta(e_0, e_0') < 0$ there is a Lorentz transformation $L$ which leaves $\mathrm{span}\{e_0, e'0\}$ and its $\eta$-orthogonal complement $\mathrm{span}\{e_0, e'0\}^\perp = \{w \in \mathbb{A} | \ \eta(w, e_0) = \eta(w, e_0') = 0\}$ invariant. There is a vector $v \in \mathbb{R}^n$ with $e_0' \parallel e_0 + v$ and $\eta(e_0, v) = 0$. Since $\eta(e_0, v)$ we have $\eta(v, v) > 0$ and the vector $e_1 = v/\sqrt{\eta(v, v)}$ is well defined. By definition it satisfies $\eta(e_0, e_1) = 0$ and $\eta(e_1, e_1) = 1$. We complete the set of linearly independent vectors $\{e_0, e_1\}$ to an $\eta$-orthonormal basis $\{e_0, e_1, \dots, e_{n-1}\}$ of $\mathbb{R}^n$. Let $L$ be the Lorentz transformation defined by

$$Le_0 = (e_0 + v)/\sqrt{1 - \|v\|_{\mathbb{R}^{n-1}}^2} = e_0',$$

$$Le_1 = (e_1 + \|v\|_{\mathbb{R}^{n-1}} e_0)/\sqrt{1 - \|v\|_{\mathbb{R}^{n-1}}^2}, \text{ and}$$

$$Le_i = e_i \quad \forall i \in \{2, \dots, n\}.$$

It maps $e_0$ into $e_0'$ and leaves the subspaces $\mathrm{span}\{\mathrm{t}, \vec{e}_1\}$, $\mathrm{span}\{e_i\}$ ($i \in 2, \dots, n-1\}$) invariant. This transformation is called a *Lorentz boost.*

We will now show that the group $G$ generated by all Newton transformations with respect to the inertial frame $(e_0, \eta(e_0, \cdot))$ and all Poincaré transformations $\phi$ of the form $\phi(x) = o + L(x - o)$ where $L$ is a Lorentz boost coincides with the group $\mathcal{P}^+$ of time orientation preserving Poincaré maps. It is clear that $G$ is a subgroup of $\mathcal{P}^+$. To show the converse Let $o, o' \in \mathbb{A}^n$ and $\{e_0, \dots, e_{n-1}\}$, $\{e_0', \dots e_{n-1}'\}$ two orthonormal bases with respect to $\eta$. We have to show that the Poincare transformation $\psi$ which maps $o$ into $o'$ and $o + e_i$ into $o' + e_i'$ for all $i$ is a composition of maps in $\mathcal{N}$ and $G$. In each basis there is exactly one vector $e_k$ (respectively, $e_l'$) with $\eta(e_k, e_k) = -1$ (respectively, $\eta(e_l', e_l') = -1$) and $\eta(e_k, \mathrm{t}) < 0$ (respectively, $\eta(e_l', \mathrm{t}) < 0$). We can renumber the basis vectors such that $e_k = e_0$, $e_l' = e_0'$. If $e_0 = e_0'$ let $\psi_1$ be the map $x \mapsto x + (o' - o)$. If $e_0 \neq e_0'$ then there is a vector $v \in \mathbb{R}^n$ with $\eta(v, e_0) = 0$, $0 < \eta(v, v) < 1$, and $e_0' \parallel e_0 + v$. In this case let $\psi_1$ be an element of the Newton group with respect to non-relativistic inertial frame $(e_0, \eta(e_0, \cdot))$ which maps $o$ to $o'$ and $e_1$ to $v/\|v\|_{\tau^{-1}(0)}$. Consider the Poincaré transformation $\psi_2(x) = o + L(x - o)$ where $L$ is Lorentzian boostwhich maps $e_0$ into $e_0'$ and leaves the plane $\mathrm{span}\{e_0, e'0\}$ invariant. Observe that $\{e_0, \psi_1(e_1), \dots, \psi_1(e_{n-1})\}$ and $\{(\psi_2)^{-1}(e_0'), (\psi_2)^{-1}(e_1'), \dots, (\psi_2)^{-1}(e_{n-1}')\}$ are both orthonormal with respect to $\eta$ and that

$$\begin{aligned}\operatorname{span}\{\psi_1(e_1),\dots,\psi_1(e_{n-1})\} &= \operatorname{span}\{(\psi_2)^{-1}(e'_1),\dots,(\psi_2)^{-1}(e'_{n-1})\}\\ &= \{x \,|\, \eta(x-o,e_0)=0\}.\end{aligned}$$

Hence there is element $\psi_3$ of the Newton group with respect the to non-relativistic inertial frame $(e_0, \eta(e_0,\cdot))$ which leaves $o'$ invariant and satisfies $\psi_3 \circ \psi_1(e_i) = (\psi_2)^{-1}(e'_i)$ $(i = 1,\dots,n-1)$. This implies $\psi = \psi_2 \circ \psi_3 \circ \psi_1$.

Finally we can show that $\mathcal{P}^+$ is the only subgroup of $\mathcal{P}$ which satisfies (a), (b), (c).

Let $\psi\colon E_o \to E_{o'}$ be an Euclidean isometry. Then $\psi$ extends to a unique transformation $\phi_\psi \in \mathcal{P}$ which leaves $\mathfrak{t}$ invariant, $\phi(x) = \psi(\overrightarrow{x-o}) + \tau(x-o)\mathfrak{t}$. Hence the Newton group $\mathcal{N}$ is a sub-group of $\mathcal{P}'$.

Let $o' \in \mathbb{A}^n$ and let $\mathfrak{t}'$ be a vector with Since $\tau(\mathfrak{t}') = 1$ and $\eta(\mathfrak{t}',\mathfrak{t}') < 0$. By assumption there is a $\psi \in \mathcal{P}$ which satisfies $\phi(o + \mathbb{R}^+\mathfrak{t}) = o' + \mathbb{R}^+\mathfrak{t}'$. In particular, the associated Lorentz transformation maps $\mathfrak{t}$ into $\mathbb{R}^+\mathfrak{t}'$. We have seen above that there are Newton transformations $\psi_1$, $\psi_3$ with respect to $(\mathfrak{t},\tau)$ such that $\psi = \psi_2 \circ \psi_3 \circ \psi_1$, where a $\psi_2$ is the Poincaré transformation that corresponds to a Lorentz boost which maps $\mathfrak{t}$ into $\mathbb{R}^+\mathfrak{t}'$ and leaves the spaces $\operatorname{span}\{\mathfrak{t},\mathfrak{t}'\}$, and its $\eta$-orthogonal complement $\operatorname{span}\{\mathfrak{t},\mathfrak{t}'\}^\perp$ invariant. Hence $\psi_2$ is the composition of maps provided by assumption (a) and (b). Since $\mathfrak{t}'$ was arbitrary we obtain that $\mathcal{P}'$ contains all Lorentzian boosts and all elements of the Newton group associated with $(\mathfrak{t},\tau)$. Consequently, $\mathcal{P}^+ = \mathcal{P}'$. ∎

Let $(\mathfrak{t},\tau)$ be a non-relativistic reference frame and $E_o$ be a hyperspace in spacetime which represents an instant of time. If we assume the axioms of Euclidean geometry for $E_o$ then, to be consistent, we have to assume that this structure is invariant with respect to any transformation which is an isomorphism of our physical structure. Hence (ii) of Proposition 1.4.1 implies that we must restrict to the Poincaré group $\mathcal{P}$. Preservation of time orientation reduces this group to $\mathcal{P}^+$. By Proposition 1.4.1 (iii), the axioms of Euclidean geometry, and Postulate 1.3.1, the Poincaré group cannot be further reduced. Hence we conclude that space and time are well described by a Minkowski spacetime together with a time orientation.

The (arbitrary) non-relativistic inertial frame $(\mathfrak{t},\tau)$ we have started with satisfies $\eta(\mathfrak{t},\mathfrak{t}) = -1$. Recall that $\mathfrak{t}$ was the velocity vector of the inertial observer $t \mapsto o + t\mathfrak{t}$ who was supposed to be at rest. Since the Poincaré transformations are the isomorphisms of Minkowski spacetime, any other inertial observer $t \mapsto o' + t\mathfrak{t}'$ who can be supposed to be at rest must be linked to $t \mapsto o + t\mathfrak{t}$ by a time orientation preserving Poincaré transformation $\phi$. If we denote the associated linear transformation by

$L_\phi$ then $\mathfrak{t}' \in \mathbb{R}^+ L_\phi(\mathfrak{t})$ holds.[16] In particular, all admissable observers $\mathfrak{t}'$ must satisfy $\eta(\mathfrak{t}', \mathfrak{t}') < 0$.

**Definition 1.4.4.** *Let $(\mathbb{A}^n, \eta)$ be a Minkowski spacetime with time orientation $[v]$.*

- *(i) A* (special-relativistic) infinitesimal observer *is a vector* $\mathfrak{t} \in \mathbb{R}^n$ *with* $\eta(\mathfrak{t}, \mathfrak{t}) = -1$, $\eta(v, \mathfrak{t}) < 0$.
- *(ii) A* (special-relativistic) observer *is a curve* $\gamma\colon t \mapsto \gamma(t)$ *such that the velocity vector* $\dot\gamma(t)$ *is a special-relativistic infinitesimal observer for all* $t$.
- *(iii) A* (special-relativistic) inertial observer *is a curve* $\gamma\colon t \mapsto x + t\mathfrak{t} \in \mathbb{A}^n$ *where* $x \in \mathbb{A}^n$ *and* $\mathfrak{t}$ *is a special-relativistic infinitesimal observer.*

For any special-relativistic infinitesimal observer $\mathfrak{t}'' \in \mathbb{R}^n$ and any any $o'' \in \mathbb{A}^n$ there is a time orientation preserving Poincare transformation $\psi$ which maps $t \mapsto o + t\mathfrak{t}$ to $o'' + t\mathfrak{t}''$ for all $t$.

The rest space with respect to $(\mathfrak{t}, \tau)$ at the event $o \in \mathbb{A}^n$ coincides with $E_o = \{x \in \mathbb{A}^n \mid \eta(x - o, \mathfrak{t}) = 0\}$. Let $\phi \in \mathcal{P}^+$ such that its associated Lorentz transformation maps $\mathfrak{t}$ to $\mathfrak{t}'$. Since $\phi$ maps $E$ into the set $E' = \{x' \in \mathbb{A}^n \mid \eta(x' - o', \mathfrak{t}') = 0\}$ this set must be interpreted as the rest space with respect to the special-relativistic inertial observer $t \mapsto o' + t\mathfrak{t}'$. In general, this space does *not* coincide with the non-relativistic rest space $E_{o'}$. Hence we arrive at the following definition.

**Definition 1.4.5.** *Let $\mathfrak{t}$ be a special-relativistic infinitesimal observer.*

- *(i) The* infinitesimal rest space *with respect to* $\mathfrak{t}$ *is the set* $\mathfrak{t}'^{\perp} := \{w \in \mathbb{R}^n \mid \eta(\mathfrak{t}', w) = 0\} \subset \mathbb{R}^n$.
- *(ii) The* affine rest space containing $o' \in \mathbb{A}^n$ *with respect ot* $\mathfrak{t}$ *is given by* $o' + \mathfrak{t}'^{\perp} \subset \mathbb{A}^n$.

The affine rest space inherits the Euclidean scalar product $\langle u, v\rangle_{(\mathfrak{t}')^\perp} = \eta(u, v)$ $(v, w \in (\mathfrak{t}')^\perp)$.

Similarly, the time difference $\Delta t_{\mathfrak{t}'}(x, y)$ of two events $x, y$ with respect to the special-relativistic infinitesimial observer $\mathfrak{t}'$ is given by $t$ where $t \in \mathbb{R}$ is the (unique) number such that $x + t\mathfrak{t}'$ lies in the affine rest space of $\mathfrak{t}'$ which contains $y$.

The original non-relativistic observer $t \mapsto o + t\mathfrak{t}$ is also a special-relativistic observer and for this observer the non-relativistic and special-relativistic definitions for rest space with associated Euclidean scalar product and time difference coincide. The relativity Postulate 1.3.1 implies that for every other special-relativistic inertial observer $o' + t\mathfrak{t}'$

[16] Observe that $\mathfrak{t}' \neq L_\phi(\mathfrak{t})$ in general since $\tau$ which is used to normalize $\mathfrak{t}'$ is not preserved by Lorentz transformations.

lengths and angle should be measured by $\langle \cdot, \cdot \rangle_{(\mathfrak{t}')^\perp}$ and time differences by $\Delta t_{\mathfrak{t}'}(\cdot, \cdot)$.

In conclusion, a Minkowski spacetime together with a time orientation contain all the geometric information of spacetime. This geometric struture is mathematically simpler and more elegant than the structure of a Galilei spacetime.

*Remark 1.4.1.* The geometric structure of spacetime was discovered by Minkowski (1909). But before him *Albert Einstein* (1879–1955) (Einstein 1905) had realized that absolute time does not exist and came to an equivalent but less elegant description of spacetime. His work contains the main physical discovery which justifies to speak of *Einstein's special theory of relativity.* An important precursor of Einstein was *Hendrik Antoon Lorentz* (1853–1928) whose explanation of the Michelson-Morley experiment anticipated the length contraction[17].

### 1.4.1 Causality in special relativity

We start with some terminology which will be justified below.

**Definition 1.4.6.** *Let* $(\mathbb{A}^n, \eta)$ *be the Minkowski spacetime.*

*(i) A vector* $w$ *is called* spacelike, *if* $\eta(w,w) > 0$, timelike *if* $\eta(w,w) < 0$, *and* lightlike *(or* null*) if* $\eta(w,w) = 0$. *A vector* $w$ *is called causal if it is timelike or lightlike.*

*(ii) Let* $[\mathfrak{t}]$ *be a time orientation. A causal vector* $u$ *is called* future directed *(*past directed*) if* $\eta(u, \mathfrak{t}) < 0$.

*(iii) A curve* $\gamma$ *is called* spacelike *(resp.,* timelike, lightlike *(or* null*),* causal, future directed, past directed*) if all its velocity vectors* $\dot{\gamma}$ *are* spacelike *(resp.,* timelike, lightlike *(or* null*),* causal, future directed, past directed*).*

Let $w$ be a vector and $x, y \in \mathbb{A}^n$ with $y = x + w$. If $w$ is spacelike then there is an infinitesimal observer $\mathfrak{t}$ such that $\eta(w, \mathfrak{t}) = 0$. This implies that the events $x$ and $y = x + w$ lie in the same affine rest space with respect to $\mathfrak{t}$, in particular, these events are taking place at the same time with $\mathfrak{t}$. Hence there cannot be any causal process which links $x$ to $y$ or vice versa.

On the other hand, if $w = y - x$ is timelike and future directed then either $w/\sqrt{-\eta(w,w)}$ or $-w/\sqrt{-\eta(w,w)}$ is an infinitesimal observer. For definiteness assume that $w/\sqrt{-\eta(w,w)}$ is in the time orientation. The inertial observer $t \mapsto x + t\frac{\pm w}{\sqrt{-\eta(w,w)}}$ connects $x$ with $y$. Hence the event $x$ definitely must take place before the event $y$. This motivates our causality definitions above (see also Postulate 8.0.1).

[17] His interpretation was different from Einstein's, however.

This discussion also implies that the field of light cones $x \mapsto C_x$ serves as a *causal boundary*.

**Corollary 1.4.1.** *Let* $(\mathbb{A}^n, \eta)$ *be a Minkowski spacetime.*

*(i) The set of all events* $y \in x$ *which can be causally influenced by processes taking place at* $x$ *are given by* $J^+(x) := \{y \in \mathbb{A}^n \,|\, y$ *can be reached from* $x$ *by a causal, future directed curve*$\}$.

*(ii)* $\partial J^+(x) = C_x^+$.

Above we have seen that the most important ingredient of our discussion is the fact that the isomorphisms of spacetime are just the elements of the time orienation preserving Poincaré group $\mathcal{P}^+$. The main step to arrive at $\mathcal{P}^+$ is contained in Theorem 1.4.1 in connection with Postulate 1.4.1. One may object that the constancy of the velocity of light is a rather awkward postulate. However, it is closely linked to the fundamental notion of causality. The following theorem which has also been obtained by Alexandrov (1975) allows to replace the light cone structure in our fundamental postulate by the assumption of causality.

**Theorem 1.4.2.** *Let* $n \geq 3$ *and consider the Minkowski space* $(\mathbb{A}^n, \eta)$. *Let* $\phi\colon \mathbb{A}^n \to \mathbb{A}^n$ *be a bijective map such that* $\phi$ *and* $\phi^{-1}$ *both respect causality:* $y \in J^+(x) \Leftrightarrow \phi(y) \in J^+(\phi(x))$ *for all* $x, y \in \mathbb{A}^n$.

*Then there exist an* $L \in \mathrm{O}(n,1)$, *an* $\alpha \in \mathbb{R} \setminus \{0\}$, *an* $o \in \mathbb{A}^n$, *and a* $b \in \mathbb{R}^n$ *such that* $\phi(x) = \alpha L(x - o) + b$ *for all* $x \in \mathbb{A}^n$.

[p. 30 ↓]
↓ p. 36

*Proof.* This theorem is a corollary to Theorem 1.4.1. We only have to show that lightlike vectors are mapped into lightlike vectors by $\phi$ and $\phi^{-1}$.

We will first prove that for $y \in J^+(x)$, $x \neq y$ the condition $\eta(y-x, y-x) = 0$ is equivalent to the assertion $A(x,y) :\Leftrightarrow$ "for all $z_1, z_2 \in J^+(x) \cap J^-(y)$ we have either $z_1 \in J^+(z_2)$ or $z_2 \in J^+(z_1)$".

If $\eta(y-x, y-x) = 0$ then $J^+(x) \cap J^-(y)$ is a part of a single light ray. Clearly, any two points on a light ray are causally related. If $\eta(x-y, x-y) < 0$ then $J^+(x) \cap J^-(y)$ contains an open set $U \subset \mathbb{A}^n$. It follows that $U$ must contain points $z_1, z_2$ which are connected via a spacelike vector $w$. It is clear that $z_1 \notin J^+(z_2)$ and $z_2 \notin J^+(z_1)$.

Now we will prove the theorem. If $x-y$ is lightlike then $A(x,y)$ holds. Since $A(x,y)$ is formulated entirely in terms of causal relationships, $A(\phi(x), \phi(y))$ must also hold. But this is equivalent to

$$\eta(\phi(x) - \phi(y), \phi(x) - \phi(y)) = 0.$$

Analogously for $\phi^{-1}$. ∎

p. 36 ↓
[↓ p. 41]

It may be philosophically more appealing to demand causality than invariance of the light cones but it should be noted that our original version is closer to the actual experiments motivating special relativity.

### 1.4.2 Length contraction and time dilatation

Since there are no absolute time or absolute space it should not come at a surprise that lengths in space and time-differences are observer-dependent. To simplify notation let $||w|| = \sqrt{\eta(w,w)}$ for any spacelike vector $w$. Fix an event $x \in \mathbb{A}^n$ and an infinitesimal observer $\mathfrak{t}$ and consider a rod which rests in the affine rest space $x + \mathfrak{t}^{\perp}$ of $\mathfrak{t}$. If with respect to the observer its endpoints are given by $x$ and $x + \ell$ at a time $t_0$, then it will sweep out the subset

$$S = \{z \in \mathbb{A}^n \,\big|\, z = x + \lambda\ell + \mu\mathfrak{t}, \lambda \in [0,1], \mu \in \mathbb{R}\}$$

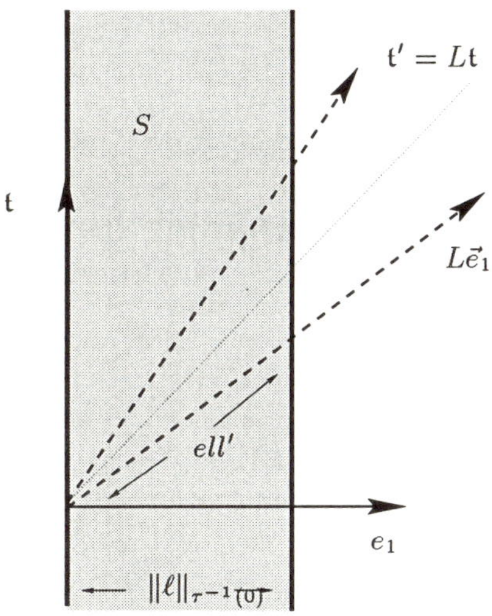

**Fig. 1.4.7.** Length contraction

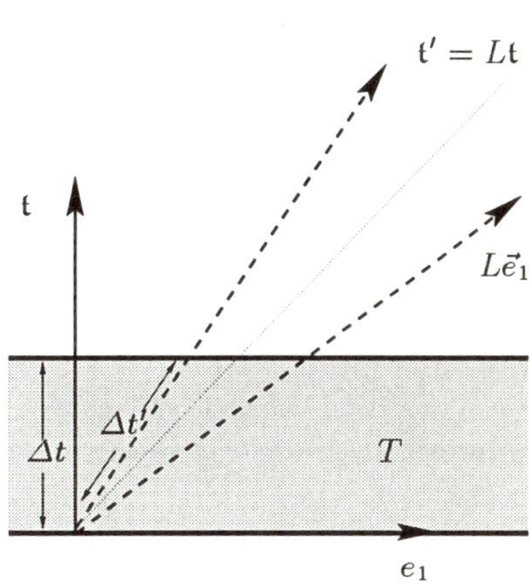

**Fig. 1.4.8.** Time dilatation

in spacetime (cf. Fig. 1.4.7). We complete $\{\mathfrak{t}, e_1 = \ell/||\ell||$ to an orthonormal basis $\{\mathfrak{t}, e_1, \ldots, e_{n-1}\}$ of $\mathbb{A}^n$. Now consider a second infinitesimal observer $\mathfrak{t}'$ who moves (relative to $\mathfrak{t}$) with the velocity $v = ||\vec{v}|| \cdot e_1$. Let $L$ be a Lorentz transformation which maps $\mathfrak{t}$ into $\mathfrak{t}'$ and leaves $\mathrm{span}\{\mathfrak{t}, e_1\}$ invariant. We have $\mathfrak{t}' = L\mathfrak{t} = (\mathfrak{t}+v)/\sqrt{1-||\vec{v}||^2}$, $Le_1 = (e_1 + ||\vec{v}||\mathfrak{t})/\sqrt{1-||\vec{v}||^2}$, and $Le_i = e_i \quad \forall i \in \{2, \ldots, n-1\}$. The rest spaces relative to $\mathfrak{t}'$ are all parallel to $L\mathfrak{t}^{\perp}$. In order to determine the length of the rod with respect to $\mathfrak{t}'$ we must calculate the length of

$$S \cap (x + L\mathfrak{t}^{\perp}) = \left\{ x + \alpha \frac{e_1 + ||\vec{v}||\mathfrak{t}}{\sqrt{1-||\vec{v}||^2}} \Big| \, \frac{\alpha}{\sqrt{1-||\vec{v}||^2}} \in [0, ||\ell||] \right\}.$$

It follows that $||\ell'|| = \sqrt{1-||\vec{v}||^2} \cdot ||\ell|| < ||\ell||$. Hence the infinitesimal obserer $\mathfrak{t}'$ measures a shorter length than $\mathfrak{t}$. This *Lorentz contraction* is one of the reasons why initially many physicists found special general realtivity hard to understand. It should be remarked that an investigation of 3-dimensional objects shows that the Lorentz contraction is more like a rotation. In particular, a moving sphere looks like a sphere at rest.

There is a similar effect with respect to time, the *time dilatation*. In spacetime, a time intervall $\Delta t$ with respect to $\mathfrak{t}$ is given by the set

$$T = \left\{x + t\mathfrak{t} + a \,\middle|\, t \in [0, \Delta t], a \in \mathfrak{t}^{\perp}\right\}$$

(cf. Fig. 1.4.8). In order to calculate $\Delta t'$ we must consider the subset

$$T \cap (x + \mathbb{R}L\mathfrak{t}) = \left\{x + t\frac{\mathfrak{t} + v}{\sqrt{1 - \|\vec{v}\|^2}} \,\middle|\, \frac{t}{\sqrt{1 - \|\vec{v}\|^2}} \in [0, \Delta t]\right\}$$

of spacetime. It follows that $\Delta t' = \sqrt{1 - \|\vec{v}\|^2}\Delta t < \Delta t$. This is the reason for the twin "paradox": Consider two twins, one of them staying at home, the other one travelling with high velocity away, and then, after some years coming home. Afterwards the twin who had travelled will be younger. Let $\mathfrak{t}$ be the infinitesimal observer associated with the first twin, $\mathfrak{t}_o$ be the infinitesimal observer of the second twin at her outward journey and $\mathfrak{t}_r$ be the infinitesimal observer of the second twin during her return journey. As a consequence of time dilatation, the time lapse between outset and return with respect to the travelling twin will be shorter then the time lapse with respect to the twin at rest. Hence after her return the twin who has travelled will be younger than the other one. If the time interval and the involved velocities are large enough the effect can be spectacular.

This effect is purely geometrical and has nothing to do with acceleration. The fact that the situation is not completely symmetric, i.e. that the the twin who had travelled changes her direction and therefore is not an inertial observer is only needed to get her back but has nothing to do with the effect. To make this point clearer imagine that spacetime would be cylindrical. To be concrete, let $y$ be an event such that $v = y - x$ is spacelike and consider the hyperspaces $F_x = \{z \in \mathbb{A} |\, \eta(z-x, v) = 0\}$ and $F_y = \{z \in \mathbb{A} |\, \eta(z-y, v) = 0\}$. These two hyperspaces enclose a region $M$ which is bounded in the direction of $\pm v$. Now assume that spacetime is just the set $M$ where $F_x$ and $F_y$ are identified. It can be shown that locally it is impossible to differentiate between $(M, \eta)$ and $(\mathbb{A}^n, \eta)$. (We will not prove this fact. It will become clear in the next chapter). If the twins lived in $(M, \eta)$ instead of $(\mathbb{A}^n, \eta)$, the sister who travels would not need to turn back. As soon as she arrives at $F_y$ (or $F_x$) she would (by our identification) be at the other hyperspace and therefore on the other side of her sister. Just travelling on she would eventually meet her sister again. If $n = 2$ then $M$ is just an ordinary cylinder and it is easy to visualize the the whole setup (cf. Fig. 1.4.10).

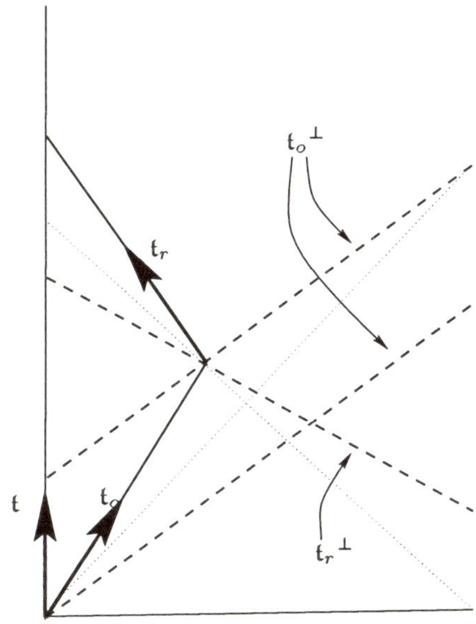


**Fig. 1.4.9.** Twin paradox

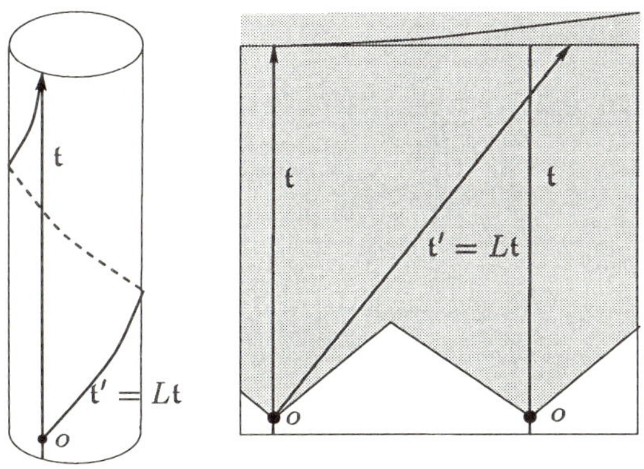


**Fig. 1.4.10.** The twin paradox in a cylindrical universe

### 1.4.3 Relativistic particles and photons

*We give a brief outline of the elementaryy concepts of photons, particles and their collisions. Their non-relativitic analogues have been introduced in the section on the non-relativistic theory of particles (cf. p 16). The content of this section will be used to motivate definitions in Chap. 5. Part of the physical discussion in Chap. 6 also requires this section.*

The following is the direct analogue of Definition 1.2.3 for a non-relativistic particle.

**Definition 1.4.7.** *A* special-relativistic particle with mass $m$ *is a pair* $(m, \gamma)$ *where* $m \in \mathbb{R}^+$ *and* $\gamma$ *is a curve satisfying* $\eta(\dot{\gamma}, \dot{\gamma}) = -1$. *The curve* $\gamma$ *is called its* world line *in spacetime.*

*A* special-relativistic inertial particle *is a particle* $\gamma$ *which satisfies* $\ddot{\gamma} = 0$.

Unlike in the non-relativistic case, we have now only one law which covers collisions. Let $(m_i, \gamma_i)_{i=1,\ldots,k}$ denote the incoming and $(m'_j, \gamma'_j)_{j=1,\ldots,l}$ outgoing particles of a collision. Then *conservation of momentum* is expressed by the single equation

$$\sum_{i=1}^{k} m_i \dot{\gamma}_i = \sum_{j=1}^{l} m'_j \dot{\gamma'}_j. \tag{1.4.12}$$

Choose any infinitesimal observer $\mathsf{t}$ and denote the projection to the orthogonal complement of $\mathsf{t}$ by $\vec{\cdot}\colon v \mapsto \vec{v}$. Then the momentum $m\dot{\gamma}$ splits into spatial and temporal parts as follows:

$$m\dot{\gamma} = \frac{m}{\sqrt{1 - ||\vec{\dot{\gamma}}||}} \left(\mathsf{t} + \vec{\dot{\gamma}}\right).$$

Conservation of the spatial momentum takes the form

$$\sum_{i=1}^{k} m_i / \sqrt{1 - ||\vec{\dot{\gamma}}_i||^2}\; \vec{\dot{\gamma}}_i = \sum_{j=1}^{l} m'_j / \sqrt{1 - ||\vec{\dot{\gamma}'}_j||^2}\; \vec{\dot{\gamma}'}_j. \tag{1.4.13}$$

which is a modified form of the non-relativistic conservation law (ii) on page 17. Energy conservation takes the form

$$\sum_{i=1}^{k} m_i / \sqrt{1 - ||\vec{\dot{\gamma}}_i||^2} = \sum_{j=1}^{l} m'_j / \sqrt{1 - ||\vec{\dot{\gamma}'}_j||^2}. \tag{1.4.14}$$

Since

$$\frac{m}{\sqrt{1 - ||\vec{v}||^2}} = m + \frac{1}{2} m ||\vec{v}||^2 + o(||\vec{v}||^2)$$

we recover as an approximation the non-relativistic law of mass conservation. If the equality $\sum_{i=1}^{k} m_i = \sum_{j=1}^{l} m'_j$ holds exactly we also recover an approximation of the conservation law for kinetic energy. The relativistic version of the conservation laws is not only more transparent and economical, it also leads to important applications:

*Example 1.4.1.* If one bombards lithium $^7$Li with hydrogen $^1$H one obtains helium $^2$He according to the nuclear reaction

$$^1\mathrm{H} + {}^7\mathrm{Li} \to 2\,{}^2\mathrm{He} + \text{energy}.$$

The weights of 1 mol (i.e, $6.02213 \cdot 10^{23}$) atoms of hydrogen (respectively, lithium, helium) are 1, 00783g (respectively, 7.01601g, 4, 00260g). It follows that the final product is lighter by about 0, 01864g/mol hydrogen atoms. According to the energy conservation theorem above, this mass difference corresponds to the energy $E \approx 1,864 \cdot 10^{-5}\text{kg} \cdot c^2 \approx 1,864 \cdot (2.9979) \cdot 10^{11}\ \text{m kg/s} \approx 1.67525 \cdot 10^{12}\text{J}$. This energy is huge in comparison to the amount of material involved. The energy amount of an adult is about 8000 kJ per day or 2920000kJ per year. It follows that $1.67525 \cdot 10^{12}\text{J}$ would last a human being for more than 500 years. Observe that the amount of energy which can be obtained from nuclear fusion is huge because the velocity of light $c$ is extremely large. In this book we chose *natural units* where a length unit is defined as $c$·time unit. These units are appropriate to discuss relativistic effects on a theoretical level but obscure the fact that velocities in everyday live are neglegible with respect to $c$.

A *photon* is (classically) characterized by its velocity and its frequency. From elementary quantum mechanics one has the relationship $E = h\nu$ where $h = 6.62608 \cdot 10^{-34}\text{J s}$ is the Planck constant and and $\nu$ the frequency of the photon. These quantities uniquely determine the momentum $p$ of the photon. Let $\mathfrak{t}$ be an infinitesimal observer and $E = h\nu$ be the energy of the photon as measured by $\mathfrak{t}$. If $\vec{c}$ is the spatial velocity of the photon relative to $\mathfrak{t}$ then its momentum is completely determined and given by $p = E(\mathfrak{t} + \vec{c}/||\vec{c}||)$. Any other infinitesimal observer $\mathfrak{t}'$ measures the light frequency $\nu' = -\eta(\mathfrak{t}', p)/h$. The frequency of visible light light ranges from about $4 \cdot 10^{14}$ oscillations per second (infra red light) to $8 \cdot 10^{14}$m oscillations per second (ultra violet light. [p. 36 ↓] ↓ p. 43

# 2. Analysis on manifolds

p. 41 ↓
→1
[↓ p. 50]

*Special relativity can only be expected to be a good description locally. We will assume that special relativity is an exact "infinitesimal" description, i.e. that it holds as a first order approximation near any given point. A rigorous formulation of this idea requires manifolds and tensor fields which will be introduced in this chapter.*

*This chapter contains much more material than is necessary for understanding the following Chap. 3. While in an ideal world, all this material would be standard knowledge of mathematicians and theoretical physicists,we give a self-contained treatment for those readers who still have to learn about analysis of manifold. We will be a little more general than necessary. Instead of using the field* $\mathbb{R}$ *we will develop the theory for both fields* $\mathbb{R}$ *and* $\mathbb{C}$ *and write* $\mathbb{K}$ *if a statement is valid in both cases. This generality is not needed for the main purpose of the book, i.e., for presenting the theory of spacetime. However, both physicists who go on to study gauge theory and mathemticians who are interested in differential geometry as a mathematical discipline will benefit from this generality. It is also instructive to see which concepts depend on the real structure. Writing* $\mathbb{K}$ *instead of* $\mathbb{R}$ *if possible will not introduce any additional difficulty.*

*While everything presented here will be used somewhere in the book, readers primarily interested in space and time may want to skip material where possible and come back later to it when needed. The minimal amount the reader should know in order to pass on to the next chapter is*

1. *The definition of manifolds: Sect. 2.1 up to Sect. 2.1.1;*
2. *The tangent bundle: Sect. 2.2;*
3. *Tensors and tensor fields: Sect. 2.3 up to including Definition 2.3.7, Sect. 2.3.2;*
4. *Connections: Sect. 2.6 up to including Definition 2.6.2;*
5. *Examples of connections: Sect. 2.7.*

There is a conceptional problem with the theory developed so far. From our local experiments we implicitly extrapolated a structure of spacetime which has only been tested in a small part of the small spacetime region which is inhabitated by human beings. The Michelson Morley experiment only indicates that Minkowski spacetime is nowadays a good description

[1] The guide in the margins assumes that the reader has no knowledge of analysis on manifolds.

of the spacetime structure of a (comparatively small) earthbound laboratory. A weak form of the philosophical *Copernican principle* states that our position in spacetime is in no way special. In particular, at any other event in spacetime one would observe the same physical laws. Natural sciences which go beyond the mere cataloguing of phenomena would be impossible without adopting this principle. Hence we feel justified to *extrapolate* that

> every event in the universe has a neighbourhood whose spacetime structure is well described by Minkowski spacetime.

We will see that this extrapolation is very different from the naive postulate that the universe has globally the structure of a Minkowski spacetime.

As a first step we will have to find out how to connect our local Minkowski spacetimes. To give a simple example which exhibits part of the problem consider the *torus* $\mathbb{T}^n$ which can be obtained from $\mathbb{A}^n$ via the identification $x \sim x + ae_i$ for all $a \in \mathbb{Z}, x \in \mathbb{A}^n$, where $\{e_i\}_{i\in\{1,\ldots,n\}}$ is a fixed standard basis of $\mathbb{R}^n$. We can equip $\mathbb{T}^n$ in a natural way with a Minkowski metric induced by $\mathbb{A}^n$. While locally there is no possibility to differentiate between $(\mathbb{T}^n, \eta)$ and $(\mathbb{A}^n, \eta)$, both spaces are globally very different (cf. Fig. 2.0.1).

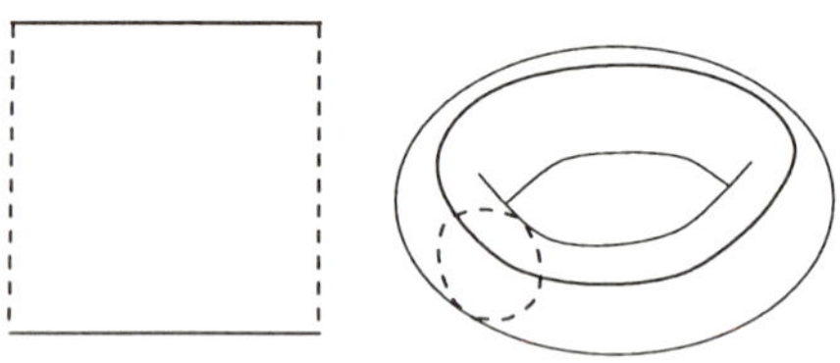

**Fig. 2.0.1.** The torus $\mathbb{T}^2$

We will now develop the mathematical techniques needed to globalize the structure given by our collection of Minkowski spacetimes. First note that we cannot even expect that locally Minkowski spacetime is an *exact* description. But it is reasonable to expect that Minkowski spacetime is the better an approximation the smaller the subset of events we are considering is. This means that we will have to formulate the theory *infinitesimally*.

In Sect. 2.1 we will generalize (part of) the structure of $\mathbb{A}^n$ to *manifolds* which may be thought of as a collection of local $\mathbb{A}^n$s. We will then construct *tangent spaces* to a manifold which are the infinitesimal approximations of it (Sect. 2.2).

## 2.1 Manifolds

*In this section we will localize part of the structure of $\mathbb{A}^n$ and lay the foundation for calculus.*

One of the most important structures of $\mathbb{R}^n$ is given by the collection of all open subsets, because this collection is needed when one defines limits, the most basic notion of analysis. Since $\mathbb{A}^n$ can be thought of as $\mathbb{R}^n$ with the special properties of the vector 0 been forgotten about there is a straightforward way to define open sets in $\mathbb{A}^n$. The subset $\mathcal{U} \subset \mathbb{A}^n$ is open if and only if there is a point $o \in \mathbb{A}$ and an open set $\tilde{\mathcal{U}} \subset \mathbb{R}^n$ such that $\mathcal{U} = \{o + v \mid v \in \tilde{\mathcal{U}}\}$.

We will now localize the topological structure of $\mathbb{A}^n$, i.e., the part of the structure of $\mathbb{A}^n$ which tells us which subsets of $\mathbb{A}^n$ are open.

**Definition 2.1.1.** *A* topological space $(M, \tau)$ *is a set $M$ together with a collection $\tau$ of subsets of $M$ which satisfies the following properties.*

*(i)* $\emptyset \in \tau$, $M \in \tau$;
*(ii)* $\mathcal{U}, \mathcal{V} \in \tau \Rightarrow \mathcal{U} \cap \mathcal{V} \in \tau$;
*(iii) if $A$ is an index set and $\mathcal{U}_a \in \tau$ for each $a \in A$ then $\bigcup_{a\in A} \mathcal{U}_a \in \tau$.*

*The collection $\tau$ is called the* topology *of $M$. A set $\mathcal{U} \subset M$ is* open *if $\mathcal{U} \in \tau$ and is* closed *if $M \setminus \mathcal{U} \in \tau$. A set $\mathcal{V} \subset M$ is a* neighbourhood *of a point $x \in M$ if there is an $\mathcal{U} \in \tau$ with $x \in \mathcal{U} \subset \mathcal{V}$.*

It is clear that the collection of open sets of $\mathbb{K}^n$ (and therefore also of $\mathbb{A}^n$) satisfies properties (i)–(iii) of Definition 2.1.1. This justifies the definition of a topological space. On the other hand, a general topological space may have properties which are quite pathological. This can be seem from the following two extremal examples. Let $M$ be any set and define $\tau_{\text{fine}}$ to be the set of all subsets of $M$ and $\tau_{\text{coarse}} = \{\emptyset, M\}$. Then $(M, \tau_{\text{fine}})$ and $(M, \tau_{\text{coarse}})$ are topological spaces.

Definitions which can be stated purely in terms of open sets carry over to topological spaces.

**Definition 2.1.2.** *Let $(M, \tau)$ and $(\tilde{M}, \tilde{\tau})$ be topological spaces.*

*(i) A map $f \colon M \to \tilde{M}$ is called* continuous *if $f^{-1}(\tilde{\mathcal{U}}) \in \tau$ for all $\tilde{\mathcal{U}} \in \tilde{\tau}$. A bijective, continuous map whose inverse is also continuous is called a* homeomorphism.

*(ii) A subset $\mathcal{U} \subset M$ is* compact *if for every collection of sets $\{\mathcal{U}_a\}_{a\in A}$ with $\mathcal{U}_a \in \tau$ and $\mathcal{U} \subset \bigcup_{a\in A} \mathcal{U}_a$ there are finitely many $\mathcal{U}_{a(1)}, \ldots \mathcal{U}_{a(k)}$ with $\mathcal{U} \subset \bigcup_{i=1}^{k} \mathcal{U}_{a(i)}$.*

*(iii) A topological space $(M, \tau)$ is* connected *if $\mathcal{U}, \mathcal{V} \in \tau$ with $\mathcal{U} \cap \mathcal{V} = \emptyset$ implies $\mathcal{U} = M$, $\mathcal{V} = \emptyset$ or $\mathcal{V} = M$, $\mathcal{U} = \emptyset$. For the topological spaces we are interested in (cf. Definition 2.1.4) below) this is equivalent to the requirement that any two points can be connected by a continuous curve $[0, 1] \to M$.*

*(iv) A collection of open subsets* $\{\mathcal{U}_a\}_{a\in A}$ *is a* basis *of the topology* $\tau$ *if for every open set* $\mathcal{V}$ *there is a subset* $B \subset A$ *with* $\bigcup_{b\in B} \mathcal{U}_b = \mathcal{V}$.

*(v) A subset* $\mathcal{U} \subset \mathcal{V}$ *is called dense if* $\mathcal{V}\overline{\mathcal{U}}$ *where* $\overline{\mathcal{U}}$ *is the* closure *of* $\mathcal{U}$, *i.e., the smallest closed set containing* $\mathcal{U}$.

*(vi) A collection of open subsets* $\{\mathcal{U}_a\}_{a\in A}$ *is a* sub-basis *if all finite intersections of sets* $\mathcal{U}_a$ *form a basis of the topology* $\tau$.

**Lemma 2.1.1.** *Let* $M$ *be a set and* $\{\mathcal{U}_a\}_{a\in A}$ *be any collection of subsets of* $M$ *which satisfies* $\bigcup_{a\in A} \mathcal{U}_a = M$. *Then there is uniquely defined topology* $\tau$ *of* $M$ *such that* $\{\mathcal{U}_a\}_{a\in A}$ *together with the empty set* $\emptyset$ *are a sub-basis of* $\tau$.

*Proof.* This follows immediately from the definition of a topology. ∎

We can now describe those topological spaces which cannot be locally distinguished from $\mathbb{A}^n$. Let $(M, \tau)$ be a topological space which is *Hausdorff*, i.e has the property that for any two different points $x, y \in M$ there are neighbourhoods $\mathcal{U}$ of $x$, $\mathcal{V}$ of $y$ which satisfy $\mathcal{U} \cap \mathcal{V} = \emptyset$. The topological space $(M, \tau)$ is locally indistinguishable from $\mathbb{A}^n$ (considered as a topological space) if each $x \in M$ has an open neigbourhood $\mathcal{U}$ such that there is a homeomorphism $\varphi\colon \mathcal{U} \to V \subset \mathbb{R}^n$.[2] The pair $(\mathcal{U}, \varphi)$ is called a *topological chart of* $M$. Since for each open subset $\hat{U} \subset U$ the restriction of $\varphi$ to $\hat{\mathcal{U}}$ is also a homeomorphism onto its image, we have the following *compatibility property.*

> Let $(\mathcal{U}, \varphi)$ and $(\tilde{\mathcal{U}}, \tilde{\varphi})$ be topological charts of $(M, \tau)$ with $\mathcal{U} \cap \tilde{\mathcal{U}} \neq \emptyset$. Then the map $\tilde{\varphi} \circ \varphi^{-1}\colon \varphi(\mathcal{U}) \cap \tilde{\varphi}(\tilde{\mathcal{U}}) \to \varphi(\mathcal{U}) \cap \tilde{\varphi}(\tilde{\mathcal{U}})$ is a homeomorphism.

We would like to have not only a topological structure on $M$ but also a structure which allows us to use the tools of analysis. Unfortunately, there is not an independent definition of a "differentiable space" which is analogous to the definition of a topological space. To get an idea how this difficulty can be overcome we can view the charts $(\mathcal{U}, \varphi)$ as a way to induce the local topological structure of $\mathbb{R}^n$ on $M$. To be more precise, let $M$ be any set, $\{\mathcal{U}_i\}_{i\in I}$ be a set of subsets of $M$ with $\bigcup_{i\in I} \mathcal{U}_i = M$, and $\varphi_i : \mathcal{U}_i \to \mathcal{V}_i$ bijective maps onto open subsets of $\mathbb{R}^n$. We can now attempt to define a topology on $M$ by using the sets $\{\varphi_i^{-1}(W_i) \mid W_i \subset U_i \text{ is open}\}$ as a sub-basis for the topology of $M$. In order to get a topology consistent with a local description we have now also to *demand* the compatibility property above. Still, the resulting topological space could fail to be Hausdorff (cf. Fig. 2.1.1). Since this is a local property, we will demand it in addition. We have now defined a topological structure on $M$ which is locally indistinguishable from the topological structure of $\mathbb{R}^n$. This definition can be carried over to the differentiable structure.

[2] Recall from the definition of the topology of $\mathbb{A}^n$ that the map $\psi_o\colon \mathbb{A}^n \to \mathbb{R}^n$, $x \to x - o$ is a homeomorphism.

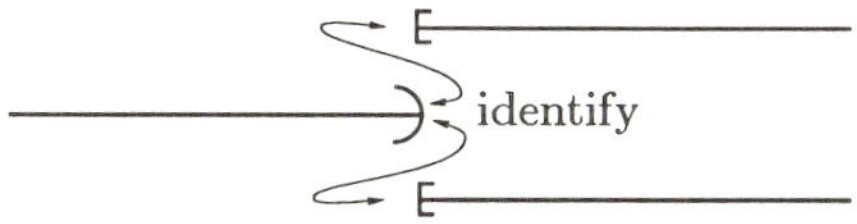


**Fig. 2.1.1.** A topological space which is locally homeomorphic to $\mathbb{R}$ but fails to be Hausdorff

**Definition 2.1.3.** *Let $I$ be an index set and $(M,\tau)$ be a topological space which is Hausdorff. A* $C^k$-atlas of $M$ *is a collection* $\varphi_i : \mathcal{U}_i \subset M \to \mathbb{K}^n$ $(i \in I)$ *of local homeomorphisms such that*

*(i) each $\mathcal{U}_i$ is open and connected,*
*(ii) each $x \in M$ is is contained in some $\mathcal{U}_i$,*
*(iii) for each $i,j$ with $\mathcal{U}_i \cap U_j \neq \emptyset$ the map $\varphi_i \circ \varphi_j^{-1} : \varphi_j(\mathcal{U}_i \cap \mathcal{U}_j) \to \varphi_i(\mathcal{U}_i \cap \mathcal{U}_j)$ is a $C^k$-diffeomorphism.*

*The pairs $(\mathcal{U}_i, \varphi_i)$ are called* charts *of $M$. A chart $(\mathcal{U}_i, \varphi_i)$ is* centered *at $x \in M$ if $x \in \mathcal{U}_i$ and $\varphi_i(x) = 0$. Two charts $(\mathcal{U}_a, \varphi_a)$ $(a = 1, 2)$ are called* compatible *if they satisfy the compatibility condition (iii). Two atlasses are* compatible *if each pair of charts in their union is compatible. A $C^k$-atlas $A$ is called* maximal *if any $C^k$-atlas containing $A$ coincides with $A$.*

*Remark 2.1.1.* In the case $\mathbb{K} = \mathbb{C}$ every $C^k$-atlas $(k \geq 1)$ has the property that $\varphi_i \circ (\varphi_j)^{-1}$ is analytic, i.e., is locally given by its Taylor series. This follows immediately from the fact that $\mathbb{C}$-differentiable functions are analytic.

For technical reasons (cf. Sect. 2.1.2) we will also demand that the topology of $M$ has a countable basis. This means that there are countably many sets $\{\mathcal{V}_i\}_{i \in \mathbb{N}}$ such that any open set $\mathcal{W}$ is the union of sets $\mathcal{V}_i$.

**Definition 2.1.4.** *Let $(M,\tau)$ be a connected topological space which is Hausdorff and which has a countable basis. $(M,\tau)$ together with a maximal $C^k$-atlas is called a* $C^k$-manifold. *A $C^\infty$-manifold is also called a* smooth manifold. *We will often refer to smooth manifolds simply as* manifolds.

*A subset $N \subset M$ is an* $m$-dimensional submanifold *of $M$ if for each $x \in N$ there is a chart $(\mathcal{U}, \varphi)$ of $M$ centered at $x$ such that $\varphi(N \cap \mathcal{U}) = \varphi(\mathcal{U}) \cap \{y \in \mathbb{K}^n \mid y^{m+1} = \cdots = y^n = 0\}$. An $(n-1)$-dimensional submanifold is often called a* hypersurface.

Observe that a subset $N \subset M$ can be a manifold without being a submanifold of $M$ (cf. Fig. 2.1.2).

The following lemma guarantees that a manifold is determined by any (not necessarily maximal) $C^k$-atlas which is compatible with the given maximal $C^k$-atlas. In practice, it is therefore sufficient to work with any given atlas. It can be shown (Hirsch 1976) that any maximal

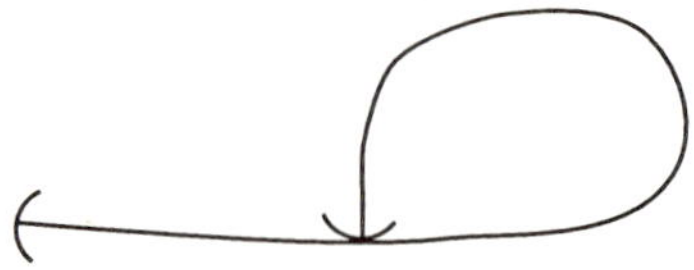

**Fig. 2.1.2.** A manifold $M \subset \mathbb{R}^2$ which is not a submanifold of $\mathbb{R}^2$

$C^k$-atlas contains a subatlas which is $C^\infty$. Hence for most questions it is no loss of generality to consider only smooth (i.e., $C^\infty$) manifolds.[3]

**Lemma 2.1.2.** *Let $A$ be a $C^k$-atlas. Then there is a unique, maximal $C^k$-atlas containing $A$.*

*Proof.* Let $B$ be the set of $C^k$-charts which are compatible which each chart in $A$. Clearly, $A \subset B$. Any two charts $(\mathcal{V}_1, \psi_1)$ and $(\mathcal{V}_2, \psi_2)$ in $B$ are compatible. To see this let $x \in \mathcal{V}_1 \cap \mathcal{V}_2$ and $(\mathcal{U}, \varphi)$ be a chart in $A$ with $x \in \mathcal{U}$. Then the maps $\varphi \circ (\psi_1)^{-1}$ and $\psi_2 \circ \varphi^{-1}$ are $C^k$ in the open set $\psi_1(\mathcal{U} \cap \mathcal{V}_1 \cap \mathcal{V}_2)$. It follows by composition that $\psi_2 \circ (\psi_1)^{-1}$ is also $C^k$. That $\psi_1 \circ (\psi_2)^{-1}$ is $C^k$ can be shown in the same way. It remains to prove that $B$ is maximal and unique. The first assertion follows from the definition of $B$. Assume that $B'$ is another atlas containing $A$. Since every chart of $B'$ is compatible with each chart of $A$, it must belong to $B$ by the definition of $B$. Hence $B' \subset B$ and the second assertion follows as well. ∎

*Example 2.1.1.* Consider the cylinder which can be obtained by identifying opposite sides of the rectangle $[a_1, b_1] \times (a_2, b_2)$,

$$\{(a_1, y) \,|\, y \in [a_2, b_2]\} \sim \{(b_1, y) \,|\, y \in [a_2, b_2]\}.$$

As charts we can take the maps

$$\begin{aligned}
\varphi_1 \colon \ & ([a_1, b_1] \times (a_2, b_2)) \setminus \left(\left\{\frac{b_1 - a_1}{2}\right\}\right) \times (a_2, b_2) \to \mathbb{R}^2 \\
& (x, y) \mapsto \begin{cases} (x, y) & \text{for } x < \frac{b_1 - a_1}{2}, \\ (x - (b_2 - a_2), y) & \text{for } x > \frac{b_1 - a_1}{2} \end{cases} \\
\varphi_2 \colon \ & (a_1, b_1) \times (a_2, b_2) \to \mathbb{R}^2 \\
& (x, y) \mapsto (x, y).
\end{aligned}$$

Consider a (long) rectangular strip of paper. The two shorter opposite edges can be glued together in two ways. Either one obtaines a cylinder (cf. Example 2.1.1) or a figure which looks like a cylinder with a twist. This *Möbius band* can mathematically be constructed as follows (cf. Fig. 2.1.3).

[3] It can in fact be shown that there is always an analytic subatlas. However, it is not a good idea to restrict to analytic atlasses because then some important technical tools do not work (cf. Sect. 2.1.2.)

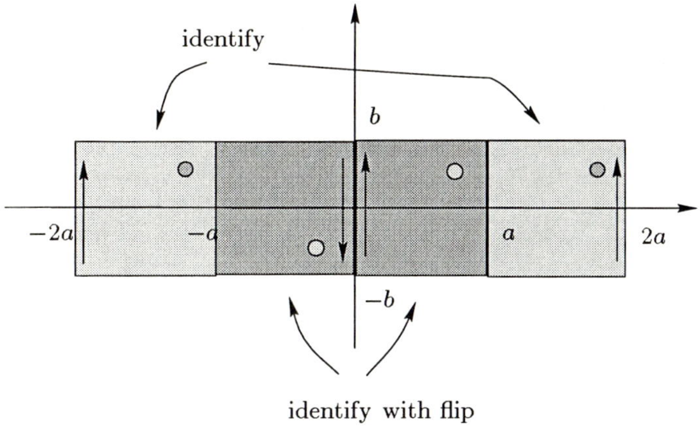


**Fig. 2.1.3.** The construction of a Möbius band

*Example 2.1.2 (Möbius band).* Let $\mathcal{V}_1 = (0, 2a) \times (-b, b) \subset \mathbb{R}^2$ and $\mathcal{V}_2 = (-2a, 0) \times (-b, b) \subset \mathbb{R}^2$. We define on $\mathcal{V}_1 \cup \mathcal{V}_2$ an equivalence relation $\sim$ by

$$(x^1, x^2) \sim (x^1 - 3a, x^2) \text{ if } (x^1, x^2) \in (a, 2a) \times (-b, b),$$
$$(x^1, x^2) \sim (x^1 - a, -x^2) \text{ if } (x^1, x^2) \in (0, a) \times (-b, b)$$

and the manifold $M$ by $M = (\mathcal{V}_1 \cup \mathcal{V}_2)/\sim$. Denote the canonical projection $\mathcal{V}_1 \cup \mathcal{V}_2 \to M$, $(x^1, x^2) \mapsto [(x^1, x^2)]$ by $\pi$ and let $\mathcal{U}_i = \pi^{-1}(\mathcal{V}_i)$, $\varphi_i = (\pi^{-1})_{|\mathcal{U}_i}$ $(i \in \{1, 2\})$. Then $\{(\mathcal{U}_1, \varphi_1), (\mathcal{U}_2, \varphi_2)\}$ is an atlas of $M$.

Since manifolds have a differentiable structure we can define differentiable maps on manifolds.

**Definition 2.1.5.** *Let $M, N$ be $C^l$-manifolds and $k \leq l$. A map $f\colon M \to N$ is* $C^k$-differentiable *if for every two charts $(\mathcal{U}_a, \varphi_a)$, $(\mathcal{V}_b, \psi_b)$ of $M, N$ the composed map $\psi_b \circ f \circ (\varphi)^{-1}$ is a $C^k$-map. A $C^\infty$-differentiable map is called* smooth. *The set of all $C^k$-differentiable maps from $M$ to $N$ is denoted by $C^k(M, N)$.*

*The maps $g\colon M \to \mathbb{K}^m$ and $h\colon \mathbb{K}^m \to M$ are* $C^k$-differentiable *if for every chart $(\mathcal{U}_a, \varphi_a)$ the composed maps $g \circ (\varphi_a)^{-1}\colon \mathbb{K}^n \to \mathbb{K}^m$ and $\varphi_a \circ g\colon \mathbb{K}^m \to \mathbb{K}^n$ are $C^k$-differentiable. The set of all $C^k$-differentiable maps from $M$ to $\mathbb{K}^m$ and from $\mathbb{K}^m$ to $M$ are denoted by $C^k(M, \mathbb{K}^m)$ and $C^k(\mathbb{K}^m, M)$.*

It is easy to see that this definition is satisfied if $f$ is $C^k$ with respect to any given atlas. For $N = \mathbb{K}^n$, $M = \mathbb{K}^m$, the definition coincides with the usual definition of differentiability in elementary analysis.

Recall from analysis that a continously differentiable map $F\colon \mathbb{K}^n \to \mathbb{K}^m$ has rank $r$ at $x \in \mathbb{K}^n$ if the the subspace $\mathrm{D}F(x)\mathbb{K}^n \subset \mathbb{K}^m$ has the dimension $r$.

Let $f\colon M \to N$ be a $C^k$-differentiable map and $x \in M$. If $(\mathcal{U}_1\varphi_1)$, $(\mathcal{U}_2, \varphi_2)$ are charts centered at $x$ and $(\mathcal{V}_1\psi_1)$, $(\mathcal{V}_2, \psi_2)$ charts centered at $f(x)$, then the rank of the maps $\psi_1 \circ f \circ {\varphi_1}^{-1}$ at $\varphi_1(x)$ and $\psi_2 \circ f \circ {\varphi_2}^{-1}$ at $\varphi_2(x)$ coincide. We can therefore speak of the rank of $f$ at $x$ and the following definition is independent of the chosen charts.

**Definition 2.1.6.** *Let $f\colon M \to N$ be a $C^k$-differentiable map and $x \in M$. Let $(\mathcal{U}, \varphi)$ be (any) chart of $M$ centered at $x$ and $(\mathcal{V}, \psi)$ be (any) chart of $N$ centered at $f(x)$. The map $f$*

*(i) has* rank $r$ *at $x$ if $\psi \circ f \circ \varphi^{-1}$ has rank $r$ at $\varphi(x)$,*
*(ii) is an* immersion *at $x$ if $\mathrm{D}(\psi \circ f \circ \varphi^{-1})$ is an injective linear map at $\varphi(x)$,*
*(iii) is an* submersion *at $x$ if $\mathrm{D}(\psi \circ f \circ \varphi^{-1})$ is a surjective linear map at at $\varphi(x)$,*
*(iv) is an* local diffeomorphism *at $x$ if $\mathrm{D}(\psi \circ f \circ \varphi^{-1})$ is a bijective linear map at $\varphi(x)$.*

**Lemma 2.1.3.** *Let $f\colon M \to N$ be a $C^k$-differentiable map of rank $r$ at $x \in M$. Then there is a neighbourhood $\mathcal{W}$ of $x$ such that for each $y \in \mathcal{W}$ the map $f$ has rank $r_y \geq r$ at $y$.*

*In particular, if $f$ is an immersion (respectively, a submersion, a local diffeomorphism at $x$) then it is also an immersion (respectively, a submersion, a local diffeomorphism) at any $y \in \mathcal{W}$.*

*Proof.* Since $\mathrm{D}(\psi \circ f \circ \varphi^{-1})$ is continous the existence of $r$ linearly independent vectors in $\mathrm{D}(\psi \circ f \circ \varphi^{-1})(\varphi(x))\mathbb{K}^n$ implies that for $y$ close enough to $x$ there are also $r$ linearly independent vectors in $(\mathrm{D}\psi \circ f \circ \varphi^{-1})(\varphi(y))\mathbb{K}^n$. Hence the rank of f cannot fall in a sufficiently small neighbourhood of a given point. For the second statement observe that immersions, submersions, and local diffeomorphisms all have maximal rank. ∎

[p. 43 ↓]
↓ p. 57

### 2.1.1 Construction of manifolds

*In analysis, the inverse function theorem plays a fundamental rôle because it allows to draw local conclusions from infinitesimal assumptions. In this section we show that the inverse function theorem and also similar theorems carry over to manifolds. A special case (Proposition 2.1.1) allows a construction of submanifolds without specifying an Atlas.*

*We will occasionally use the results of this section. But the reader who is not primarily interested in analysis on manifolds is advised to skip this section and to return to it when needed.*

The following lemma is a consequence of the inverse function theorem.

**Lemma 2.1.4.** *Let $\mathcal{U} \subset \mathbb{K}^n$, $\mathcal{V} \subset \mathbb{K}^m$ be open, $x \in \mathcal{U}$, and $f\colon \mathcal{U} \to \mathcal{V}$ be a continuous, differentiable map with constant rank $r$ in $\mathcal{U}$.*

*Then there exist*

*(i) an open neigbourhood $\tilde{\mathcal{U}} \subset \mathcal{U}$ of $x$,*
*(ii) a homeomorphism $\phi\colon \tilde{\mathcal{U}} \to \{y \in \mathbb{K}^n \mid |y| < 1\}$,*
*(iii) an open neighbourhood $\tilde{\mathcal{V}} \supset f(\mathcal{U})$ of $f(x)$,*
*(iv) and a homeomorphism $\psi\colon \{z \in \mathbb{K}^m \mid |z| < 1\} \to \tilde{\mathcal{V}}$*

*such that $f = \psi \circ p_r \circ \phi$ where $p_r(y^1, \ldots, y^n) = (y^1, \ldots, y^r, 0, \ldots, 0)$.*

*Proof.* Let $E$ be the $(n-r)$-dimensonal vector space $E = \{v \in \mathbb{K}^n \mid \mathrm{D}f(x)v = 0\}$ and $F \subset \mathbb{K}^n$ be an $r$-dimensional vector space with $\mathbb{K}^n = E \oplus F$. Let $\{e_{r+1}, \ldots, e_n\}$ be a basis of $E$ and $\{e_1, \ldots, e_r\}$ be a basis of $F$.[4] For each $y \in \mathbb{K}^n$ let $\lambda_e^i(y)$ be the $i$th component of $y$ with respect to the basis $\{e_1, \ldots, e_n\}$. The vectors $f_i = \mathrm{D}f(x)e_i$ $(i \in \{1, \ldots, r\})$ are a basis of the $r$-dimensinal vector space $\mathrm{D}f(x)\mathbb{K}^n \subset \mathbb{K}^m$. We choose linearly independent vectors $f_{r+1}, \ldots f_m$ such that $\{f_1, \ldots, f_m\}$ is a basis of $\mathbb{K}^m$. For $z \in \mathbb{K}^m$ let $\mu_f^i(z)$ be the $i$th component of $z$ with respect to this basis. We define

$$\lambda(y) = \begin{pmatrix} 0 \\ \vdots \\ 0 \\ \lambda_e^{r+1}(y) \\ \vdots \\ \lambda_e^n(y) \end{pmatrix} \in \mathbb{K}^n, \quad \mu(z) = \begin{pmatrix} \mu_f^1(z) \\ \vdots \\ \mu_f^r(z) \\ 0 \\ \vdots \\ 0 \end{pmatrix} \in \mathbb{K}^n.$$

The map

$$g\colon \mathcal{U} \mapsto \mathbb{K}^n, \quad y \mapsto \mu \circ f(y) + \lambda(y)$$

is continuously differentiable and $\mathrm{D}g(x)\colon \mathbb{K}^n \mapsto \mathbb{K}^n$ is invertible. Hence by the inverse function theorem there is a neighbourhood $\hat{\mathcal{U}} \subset \mathcal{U}$ of $x$ such that $g$ is a diffeomorphism from $\hat{\mathcal{U}}$ onto the open set $g(\hat{\mathcal{U}})$. There is an $\epsilon > 0$ such that $B_\epsilon(x) := \{z \in \mathbb{K}^n \mid |z - g(x)| < \epsilon\} \subset g(\hat{\mathcal{U}})$. We define $\tilde{\mathcal{U}} := g^{-1}(B_\epsilon(x))$ and the bijective map

$$\phi\colon \tilde{\mathcal{U}} \to B_1(0) \subset \mathbb{K}^n, \quad y \mapsto \frac{1}{\epsilon}(g(y) - g(x)).$$

The rank of $f$ is constant on $\mathcal{U}$ which implies that $\dim(\mathrm{D}f(y)\mathbb{K}^n) = r$ for all $y \in \mathcal{U}$. Since $\mathrm{D}g(y)$ is bijective for all $y \in \tilde{\mathcal{U}}$ and $\mathrm{D}g(y)(v) = \mu(\mathrm{D}f(y)v)$ for all $v \in F$ the maps $\mathrm{D}f(y)\colon F \to \mathrm{D}f(y)F$ and

$$\mu\colon \mathrm{D}f(y)F \to \mathbb{K}^r \oplus \{0\} \subset \mathbb{K}^r \oplus \mathbb{K}^{n-r} = \mathbb{K}^n$$

[4] This numeration of basis vectors of $E \oplus F$ may seem slightly odd but will prove to be more practical later on in the proof.

are both bijective. Let $\nu_y \colon \mathbb{K}^r \oplus \{0\} \to \mathrm{D}f(y)F$ be the inverse to the latter map.

We will now show that the map $h = f \circ \phi^{-1}$ does not depend on the variables $y^{r+1}, \ldots, y^n$. We write $\mathbb{K}^n = \mathbb{K}^r \oplus \mathbb{K}^{n-r}$ and $v = v_1 \oplus v_2$ for any $v \in \mathbb{K}^n$. Since $f(y) = h(\frac{1}{\epsilon}\mu \circ f(y) \oplus \frac{1}{\epsilon}\lambda(y) - \frac{1}{\epsilon}\mu \circ f(x) \oplus \frac{1}{\epsilon}\lambda(x))$ we have

$$\begin{aligned}\epsilon \mathrm{D}f(y)v = {} & \mathrm{D}_1 h_{|\frac{1}{\epsilon}\mu\circ f(y)\oplus\frac{1}{\epsilon}\lambda(y)-\frac{1}{\epsilon}\mu\circ f(x)\oplus\frac{1}{\epsilon}\lambda(x)} \circ \mu \circ \mathrm{D}f(y)v \\ & + \mathrm{D}_2 h_{|\frac{1}{\epsilon}\mu\circ f(y)\oplus\frac{1}{\epsilon}\lambda(y)-\frac{1}{\epsilon}\mu\circ f(x)\oplus\frac{1}{\epsilon}\lambda(x)} \circ \lambda v.\end{aligned}$$

Inserting $\mathrm{D}f(y) = \nu_y \circ \mu \circ Df(y)$ into this equation implies

$$\mathrm{D}_2 h_{|\frac{1}{\epsilon}\mu\circ f(y)\oplus\frac{1}{\epsilon}\lambda(y)-\frac{1}{\epsilon}\mu\circ f(x)\oplus\frac{1}{\epsilon}\lambda(x)} \circ \lambda v = \sigma_y \circ \mu \circ \mathrm{D}f(y)v,$$

where $\sigma_y$ is the linear map given by

$$\sigma_y \colon \mathbb{K}^r \oplus \{0\} \to \mathbb{K}^m, \quad y \mapsto \epsilon\nu_y - \mathrm{D}_1 h_{|\frac{1}{\epsilon}\mu\circ f(y)\oplus\frac{1}{\epsilon}\lambda(y)-\frac{1}{\epsilon}\mu\circ f(x)\oplus\frac{1}{\epsilon}\lambda(x)}.$$

Since $y \mapsto \frac{1}{\epsilon}\mu \circ f(y) \oplus \frac{1}{\epsilon}\lambda(y)$ is invertible and $\lambda$ maps $\mathbb{K}^n$ onto $\{0\} \oplus \mathbb{K}^{n-r}$ we have only to show that $\sigma_y = 0$ in order to prove $\mathrm{D}_2 h = 0$. For $v \in F$ we have $\lambda(v) = 0$ and and therfore $\sigma_y \circ \mu \circ \mathrm{D}f(y)v = 0$. The map $\mu \circ \mathrm{D}f$ coincided on $F$ with $\mathrm{D}g(y)$ and is therefore bijective. In particular, $\mu \circ \mathrm{D}fF = \mathbb{K}^r \oplus \{0\}$ which is the domain of $\sigma_y$. Hence $\sigma_y$ vanishes.

Since we have proven that $h(y)$ does not depend on $y^{r+1}, \ldots y^n$ we can write $\hat{h}(y_1)$ instead of $h(y_1 \oplus y_2)$.

We identify $\mathbb{K}^m$ with $\mathbb{K}^r \oplus \mathbb{K}^{m-r}$ and write $w = w_1 \oplus w_3$. Let $\{b_1, \ldots, b_{m-r}\}$ be the canonical basis of $\mathbb{K}^{m-r}$ and let $\tau \colon \mathbb{K}^{m-r} \to \operatorname{span}\{f_{r+1}, \ldots, f_m\}$ be the linear map which is defined by $\tau(b_i) = f_i$. We define the map $\psi$ by

$$\psi \colon B_1(0) \subset \mathbb{K}^m \to \mathbb{K}^m \quad z_1 \oplus z_3 \mapsto \hat{h}(z_1 + \mu(f(x))) + \tau(z_3).$$

This implies

$$\begin{aligned}\psi \circ p_r \circ \phi(y) &= \psi \circ p_r(\frac{1}{\epsilon}g(y) - \frac{1}{\epsilon}g(x)) \\ &= \psi \circ p_r(\frac{1}{\epsilon}\mu \circ f(y) - \frac{1}{\epsilon}\mu \circ f(x) + \frac{1}{\epsilon}\lambda(y) - \frac{1}{\epsilon}\lambda(x)) \\ &= \frac{1}{\epsilon}\psi \circ p_r(\mu \circ f(y) - \mu \circ f(x)) \\ &= \frac{1}{\epsilon}h(\mu \circ f(y) - \mu \circ f(x) + \mu(f(x))) = \frac{1}{\epsilon}h(\mu \circ f(y)).\end{aligned}$$

Denote the projection of $\mu$ to the span of the first $r$ canonical basis vectors of $\mathbb{K}^n$ by $\hat{\mu}$. Then we have $\hat{h} \circ \hat{\mu} = h \circ \mu$. Equation $\mathrm{D}_2 h = 0$ implies $f(y) = \hat{h}(\frac{1}{\epsilon}\hat{\mu}(f(y)))$ which in turn gives $\psi \circ p_r \circ \phi(y) = f(y)$.

We still have to show that $\psi$ is a homeomorhism. But this follows since both $\hat{h}$ and $\tau$ are homeomeorphisms. ∎

**Corollary 2.1.1.** *Let* $M$ *be an* $n$*-dimensional,* $N$ *be an* $m$*-dimensional* $C^k$*-manifold, and* $f\colon M \to N$ *be a* $C^k$*-differentiable map which has constant rank in a neighbourhood of* $x \in M$*. Then there exist charts* $(\mathcal{U}, \varphi)$*,* $(\mathcal{V}, \psi)$ *centered at* $x$*,* $f(x)$ *such that* $\varphi(x) = 0$*,* $\psi(f(x)) = 0$*, and*

$$\psi \circ f \circ \varphi^{-1}\colon \mathbb{K}^n \to \mathbb{K}^m, \quad (x^1, \ldots, x^n) \mapsto (x^1, \ldots, x^r, 0, \ldots, 0).$$

**Proposition 2.1.1.** *Let* $M$ *be an* $n$*-dimensional,* $N$ *be an* $m$*-dimensional* $C^k$*-manifold, and* $f\colon M \to N$ *be a* $C^k$*-differentiable map which has constant rank in a neighbourhood of* $x \in M$*. Let* $y \in f(M)$*. Then the set* $f^{-1}(y)$ *is a closed* $(n-r)$*-dimensional submanifold of* $M$*.*

*Proof.* Let $x \in f^{-1}(y)$ and take charts as given by Corollary 2.1.1. Then we have

$$\begin{aligned}\varphi(\mathcal{U} \cap f^{-1}(y)) &= \varphi(\mathcal{U}) \cap (\psi \circ f \circ \varphi^{-1})^{-1}(0) \\ &= \varphi(\mathcal{U}) \cap \{z \in \mathbb{K}^n \,|\, z^1 = \cdots = z^r = 0\}.\end{aligned}$$

The assertion follows directly from the Definition 2.1.4 of a submanifold. ∎

Proposition 2.1.1 is a powerfull tool which is used to construct manifolds.

*Example 2.1.3.* Consider the submersion $f\colon \mathbb{R}^n \setminus \{0\} \to \mathbb{R}$, $x \mapsto |x|^2$. Proposition 2.1.1 implies that the sphere of radius $\rho > 0$ in $M$ which coincides with the set $f^{-1}(\rho^2)$ is an $(n-1)$-dimensional submanifold of $\mathbb{R}^n$. The reader try for her- or himself to write down directly an atlas of the sphere of radius $\rho$. It is much more work.

### 2.1.2 Partition of unity

*In this section we provide a method of localization using functions which is practical if one wants to define a global object using charts. The prime example is the definition of integration in Sect. 2.5.4. This method works only for real manifolds:* $\mathbb{K} = \mathbb{R}$*.*

*This subsection is somewhat technical. The reader may therefore want to skim (or skip) this section on first reading and to return to its proofs when the results of this section are needed.*

The aim of this section is to construct an atlas and for each chart a function with support in this chart such that

(i) at any point there are only finitely many charts which intersect
(ii) the sum of all functions (which is well defined by (i)) equals 1.

We start with two topological lemmas.

**Lemma 2.1.5.** *Let* $M$ *be a manifold. If* $M$ *is not compact then there exists a sequence of open sets* $\{\mathcal{W}_i\}_{i \in \mathbb{N}}$ *with compact closure sucht that* $\overline{\mathcal{W}_i} \subset \mathcal{W}_{i+1}$ *and* $\bigcup_{i \in \mathbb{N}} \mathcal{W}_i = M$*.*

*Proof.* Let $\{\mathcal{U}_i\}_{i\in\mathbb{N}}$ be a countable basis of the topology of $M$ such that all $\overline{\mathcal{U}_i}$ are compact. Let $\mathcal{W}_1 := \mathcal{U}_1$. Since the closure of this set is compact there is a $k_1 \in \mathbb{N}$ with $\overline{\mathcal{W}_1} \subset \bigcup_{i=1}^{k_1} U_i =: \mathcal{W}_2$. We proceed now by induction. Assume, we have already constructed $\mathcal{W}_1, \ldots, \mathcal{W}_p$, where $\mathcal{W}_p = \bigcup_{i=1}^{k_p} U_i$. Then there is a $k_{p+1} > k_p$ such that $\overline{\mathcal{W}_p} \subset \bigcup_{i=1}^{k_{p+1}} U_i =: \mathcal{W}_{p+1}$. ∎

**Lemma 2.1.6.** *Let $M$ be a manifold. And $\{\mathcal{U}_a\}_{a\in A}$ be open sets which cover all of $M$. Then there exists a countable collection $\{\mathcal{V}_j\}_{j\in\mathbb{N}}$ of open sets such that*

*(i) each $\mathcal{V}_j$ lies in some $\mathcal{U}_a$ and has compact closure,*
*(ii) each $\mathcal{V}_j$ intersects only finitely many $\mathcal{V}_i$.*
*(iii) $M = \bigcup_{j\in\mathbb{N}} \mathcal{V}_j$.*

*Proof.* We will first show that we can restrict to the case that $A$ is countable. Let $\mathcal{O} = \{\mathcal{O}_i\}_{i\in\mathbb{N}}$ be a countable basis and let $\{(\mathcal{V}_c, \varphi_c)\}_{c\in C}$ be an atlas of $M$. For each $\mathcal{V}_c$ there are countably many sets $\mathcal{O}_{i,c} \in \mathcal{O}$ such that $\mathcal{V}_c = \bigcup_{i=1}^{\infty} \mathcal{O}_{i,c}$. Since $\mathcal{O}$ is countable so is the set $\{\mathcal{O}_{i,c}\}_{i\in\mathbb{N},c\in C} \subset \mathcal{O}$. Let $\mathcal{O}^j$ be a re-numbering of these sets and choose for each $j \in \mathbb{N}$ an index $c(j) \in C$ with $\mathcal{O}^j \subset \mathcal{V}_{c(j)}$. Then the collection $\{(\mathcal{O}^j, \varphi_{c(j)})\}_{j\in\mathbb{N}}$ is a countable atlas of $M$. Since $\mathcal{O}^j$ is homeomorphic to $\mathbb{K}^n$ there is for each $j$ a dense sequence $\{x_{i,j}\}_{i\in\mathbb{N}}$ of points in $\mathcal{O}^j$. For each $(i,j) \in \mathbb{N}\times\mathbb{N}$ let $a(i, j \in A)$ be an index with $x_{i,j} \in \mathcal{U}_{a(i,j)}$. Then the countable subset $\{\mathcal{U}_{a(i,j)}\}_{i,j\in\mathbb{N}}$ of $\{\mathcal{U}_a\}_{a\in A}$ covers all of $M$. It follows that we can assume without loss of generality that $A$ is countable.

Let $\mathcal{W}_0 = \emptyset$ and let $\{\mathcal{W}_j\}_{j\in\mathbb{N}}$ be the sequence of sets constructed in Lemma 2.1.5. The set $\overline{\mathcal{W}_{k+1}} \setminus \mathcal{W}_k$ is compact and can be covered by finitely many $\mathcal{U}_{a_1(k)}, \ldots, \mathcal{U}_{a_{m(k)}(k)}$ for every $k \in \mathbb{N}$. We set $\mathcal{V}_{k,l} = \mathcal{U}_{a_l(k)} \cap (\mathcal{W}_{k+2} \setminus \overline{\mathcal{W}_{k-1}})$ which defines a countable family of sets since $\mathbb{N}\times\mathbb{N}$ is countable. Property (i) is clear from the definition of $\mathcal{V}_{k,l}$ and property (iii) follows from the fact that the sets $\overline{\mathcal{W}_{k+1}} \setminus \mathcal{W}_k$ cover all of $M$. Finally, property (ii) follows because each set $\overline{\mathcal{W}_k}$ is only intersected by finitely many $\mathcal{V}_{k,l}$. ∎

**Lemma 2.1.7.** *Let $M$ be a real manifold and $\mathcal{U}, \mathcal{V}$ open sets with $\overline{\mathcal{U}} \subset \mathcal{V}$. Then there is a $C^\infty$-function $h\colon M \to [0,1]$ with*

*(i) $h(x) = 1$ for all $x \in \overline{\mathcal{U}}$,*
*(ii) $h(x) = 0$ for all $x \in M \setminus \mathcal{V}$.*

*Proof.* We consider first the special case of two balls with origin $0 \in \mathbb{R}^n$ but different radii: Let $0 < r_1 < r_2$ and $B_{r_1}(0) = \{x \in \mathbb{R}^n | \, |x| < r_1\}$, $B_{r_2}(0) = \{x \in \mathbb{R}^n | \, |x| < r_2\}$. The map

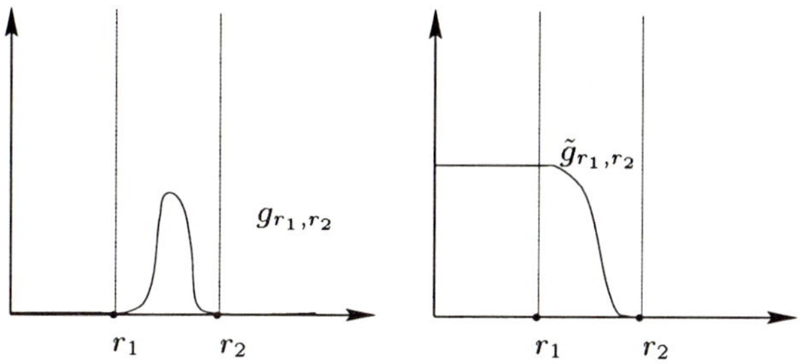


**Fig. 2.1.4.** The proof of Lemma 2.1.7

$$g_{r_1,r_2} : \mathbb{R}^+ \to \mathbb{R}^+, \quad t \mapsto \begin{cases} \exp\left(\frac{1}{(t-r_1)(t-r_2)}\right) & \text{for } r_1 < t < r_2, \\ 0 & \text{otherwise} \end{cases}$$

is $C^\infty$ and has the support $[r_1, r_2]$. It follows that

$$\tilde{g}_{r_1,r_2}(t) = 1 - \frac{\int_{r_1}^{t} g_{r_1,r_2}(s)\mathrm{d}s}{\int_{r_1}^{r_2} g_{r_1,r_2}(s)\mathrm{d}s}$$

is also $C^\infty$. The properties

$$\begin{aligned} \tilde{g}_{r_1,r_2}(t) &= 1 \text{ for all } t \in [0, r_1], \\ \tilde{g}_{r_1,r_2}(t) &\in (0,1) \text{ for all } t \in (r_1, r_2), \\ \tilde{g}_{r_1,r_2}(t) &= 0 \text{ for all } t \in [r_2, \infty) \end{aligned}$$

imply that the $C^\infty$-function

$$\tilde{h}_{r_1,r_2} : [0,1] \to \mathbb{R}^+, \quad x \mapsto g_{r_1,r_2}(\|x\|)$$

is well defined and satisfies $\mathrm{supp}(\tilde{h}_{r_1,r_2}) = \overline{B_{r_2}(0)}$, $h_{r_1,r_2}(x) = 1$ for all $x \in B_{r_1}(0)$.

Consider now open sets $\mathcal{U}$, $\mathcal{V}$ with $\overline{\mathcal{U}} \subset \mathcal{V}$ and let $\{\mathcal{V}_j\}_{j \in \mathbb{N}}$ be a collection of open sets as provided by Lemma 2.1.6. For each $j$ we will now construct a smooth function $h_j$ which satisfies

(i) $h_j(x) \in [0,1]$ for all $x \in M$,
(ii) $h_j(x) = 1$ for all $x \in \mathcal{U} \cap \mathcal{V}_j$,
(iii) $\mathrm{supp}(h_j) \subset \mathcal{V} \cap \mathcal{V}_j$.

Let $x \in \mathcal{U} \cap \mathcal{V}_j$ and $(\mathcal{W}_x, \varphi_x)$ be a chart centered at $x$ with $\varphi_x(x) = 0$ and $\mathcal{W}_x \subset \mathcal{V} \cap \mathcal{V}_j$. There are positive numbers $r_1(x) < r_2(x)$ such that $\overline{B_{r_2(x)}(0)} \subset \varphi_x(\mathcal{W}_x)$. The map

$$\tilde{h}_x^j : \mathcal{W}_x \to \mathbb{R}^+, \quad y \mapsto \begin{cases} h_{r_1(x),r_2(x)} \circ \varphi_x(y) & \text{for } y \in \mathcal{W}_x \\ 0 & \text{otherwise} \end{cases}$$

is well defined and smooth. Since $\overline{\mathcal{U} \cap \mathcal{V}_j}$ is compact there exist finitely many points $x_1, \ldots, x_K$ such that the open sets

$$\{\varphi_{x_k}{}^{-1}(B_{r_1(x_k)}(0))\}_{k=1,\ldots,K}$$

cover $\overline{\mathcal{U} \cap \mathcal{V}_j}$. Hence the map $h_j(y) := 1 - \prod_{k=1}^{K}(1 - h_x^j(y))$ is also well defined and smooth. Clearly, we have $h_j(y) \in [0,1]$ for all $y \in M$. Since $h_{x_k}(y) = 0$ for all $y \in M \setminus \overline{\mathcal{V} \cap \mathcal{V}_j}$ we have also $h_j(y) = 0$ for all $y \in M \setminus \overline{\mathcal{V} \cap \mathcal{V}_j}$. If $y \in \mathcal{U} \cap \mathcal{V}_j$ then there exists an $x_l$ with $h_{x_l}^j(y) = 1$ which implies $h_j(y) = 1$.

The function $\tilde{h}(x) := \sum_{j=1}^{\infty} h_j(x)$ is well defined since for each $x$ all but finitely many $h_j(x)$ vanish. $\tilde{h}$ satisfies $\operatorname{supp}(\tilde{h}) \subset \mathcal{V}$, $\tilde{j}(x) \geq 0$ for all $x \in M$, and $\tilde{h}(x) \geq 1$ for all $x \in \overline{\mathcal{U}}$. Hence in order to prove the lemma we only need to normalize $\tilde{h}$ appropriately. Let $\hat{\mathcal{U}} = \{x \in M \,|\, \tilde{h}(x) < 1/2\}$. This set is open and its closure is contained in $M \setminus \overline{\mathcal{U}}$. Hence by the same construction there exists another smooth function $\hat{h}$ with $\hat{h}(x) \geq 0$ for all $x \in M$, $\hat{h}(x) \geq 1$ for all $x \in \hat{\mathcal{U}}$, and $\operatorname{supp}(\hat{h}) \subset M \setminus \overline{\mathcal{U}}$. Observe that $\hat{h}$ and $\tilde{h}$ do not vanish both at any given point. Hence

$$h(x) = \frac{\tilde{h}(x)}{\tilde{h}(x) + \hat{h}(x)}$$

is well defined and smooth. Further, $h(x) \in [0,1]$ for all $x \in M$. If $x \in \overline{\mathcal{U}}$ then $\tilde{h}(x) \geq 1$ and therefore $\hat{h}(x) = 0$ which in turn implies $h(x) = 1$. If $x \in M \setminus \mathcal{V}$ then $\tilde{h}(x) = 0$ and therefore $h(x) = 0$. ∎

**Definition 2.1.7.** *A smooth* partition of unity *is a set of functions* $\{f_a \colon M \to [0,1]\}_{a \in A}$ *such that*

*(i) each point $x \in M$ has a neighbourhood which is only intersected by the support of finitely many $f_a$,*
*(ii) $\sum_{a \in A} f_a(x) = 1$ for all $x$.*

*A partition of unitity is subordinate to an open covering $\{\mathcal{U}_b\}_{b \in B}$ if for every $a \in A$ there is a $b \in B$ with* $\operatorname{supp}(f_a) \subset U_b$.

**Theorem 2.1.1 (Existence of a partition of unity).** *If $\{\mathcal{U}_b\}_{b \in B}$ is an open covering of a real manifold $M$ then there exists a countable partition of unity $\{f_j\}_{j \in \mathbb{N}}$ which is subordinate to $\{\mathcal{U}_b\}_{b \in B}$.*

*Proof.* For each $x \in M$ let $b(x)$ be an index with $x \in \mathcal{U}_{b(x)}$ and $\widetilde{\mathcal{U}}_{b(x)}$ be a neighbourhood of $x$ with $\overline{\widetilde{\mathcal{U}}_{b(x)}} \subset \mathcal{U}_{b(x)}$. Lemma 2.1.6 implies that there exists a sequence $\{x_j\}_{j \in \mathbb{N}}$ and a countable collection of open sets $\{\mathcal{V}_j\}$ such that

(i) each $\mathcal{V}_j$ lies in some $\widetilde{\mathcal{U}}_{b(x_j)}$ and therefore $\overline{\mathcal{V}_j} \subset \mathcal{U}_{b(x_j)}$;

(ii) each $\mathcal{V}_j$ intersects only finitely many $\mathcal{V}_i$;
(iii) $M = \bigcup_{j\in\mathbb{N}} \mathcal{V}_j$.

We can now apply the same argument to the collection of open sets $\mathcal{V}_j$ to find a countable collection of open sets $\{\mathcal{W}_k\}_{k\in\mathbb{N}}$ such that

(i) each $\overline{\mathcal{W}_k}$ lies in some $\mathcal{V}_{j(k)}$;
(ii) each $\mathcal{W}_k$ intersects only finitely many $\mathcal{W}_l$;
(iii) $M = \bigcup_{k\in\mathbb{N}} \mathcal{W}_k$.

By Lemma 2.1.7 there is for each $k$ a function $h_k\colon M \to [0,1]$ with $h_k(x) = 1$ for all $x \in \mathcal{W}_k$, $h_k(x) = 0$ for all $x \in M \setminus \overline{\mathcal{V}_{j(k)}}$. Since each $\mathcal{V}_j$ intersects at most finitely many $\mathcal{V}_i$ for each $x \in M$ we have $h_k(x) = 0$ for all but finitely many $k$. This implies that the sum $x \mapsto \sum_{k=1}^{\infty} h_k(x)$ is well defined and smooth. Since each $x$ lies in some $\mathcal{W}_k$ we have $\sum_{k=1}^{\infty} h_k(x) \geq 1$ for all $x \in M$. Thus

$$f_k\colon M \to [0,1], \quad x \mapsto f_k(x) = \frac{h_k(x)}{\sum_{l=1}^{\infty} h_l(x)}$$

is well defined and smooth. Property (i) of Definition 2.1.7 is satisfied because each $\mathcal{V}_j$ intersects only finitely many $\mathcal{V}_i$ and $\operatorname{supp}(f_k) \subset \mathcal{V}_j(k)$. Property (ii) follows directly from the definition of $f_k$. Hence $\{f_k\}_{k\in\mathbb{N}}$ is a partition of unity. That it is subordinate to the open covering $\{\mathcal{U}_b\}_{b\in B}$ follows from $\operatorname{supp}(f_k) \subset \mathcal{V}_j(k) \subset \mathcal{U}_{b(x_{j(k)})}$. ∎

*Remark 2.1.2.* In Lemma 2.1.7 and Theorem 2.1.1 it was necessary to restrict to the case $\mathbb{K} = \mathbb{R}$. If $\mathbb{K} = \mathbb{C}$, both results are wrong in general. This is so because complex-differentiable maps are automatically analytic, i.e., can locally be written as a power series. This would be impossible for the function $h$ in Lemma 2.1.7.

## 2.2 Vector bundles and the tangent bundle

p. 50 ↓
[↓ p. 58]

In ordinary calculus, the derivative $\mathrm{D}f$of a map $f\colon \mathbb{K}^n \to \mathbb{K}^m$ is the linear approximation of $f$, i.e., it is defined by $f(x) = f(a) + \mathrm{D}f_{|a}(x-a) + o(|x-a|)$ where $o(|x-a|)/|x-a| \to 0$ $(x \to a)$. To study the linear approximation of a map rather than the map itself is certainly one of the most powerful approaches in mathematics and physics. Because of the limit $x - a \to 0$ this approach is often referred to as working *infinitesimally.* Until the middle of this century people spoke of infinitesimal (or infinitely small) displacements (meaning the vector $x-a$ if it was 'small'). This terminology can lead to misunderstandings but stresses the main idea of analysis. While we will give a modern presentation, it is a good idea to keep the 'infinitesimal way of thinking' in mind.

The definition of a linear approximation of a function $f$ rests on the linear structure of $\mathbb{K}^n$. In the general case of a manifold, such a structure is not at hand. But it is possible to define a linear approximation of a map in two steps. First, we linearize the manifold itself. This gives rise to the the *tangent space* $T_aM$ at a point $a$ of a manifold $M$. Then we linearize the map thereby obtaining a linear map $T_af\colon T_aM \to T_{f(a)}N$ between (linear) tangent spaces.

[p. 57 ↓] →5 ↓ p. 59

We will linearize the manifold $M$ by attaching an $n$-dimensional vector space to $M$ at each point $x \in M$. At first one may think that it is sufficient to consider the product $M \times \mathbb{K}^n$ and to define $T_xM = \{x\} \times \mathbb{K}^n$. However, this would introduce a global structure via the (global) product $\times$. In order to keep within the spirit of localization, we can only demand that such a product exists locally. This motivates the following definition.

**Definition 2.2.1.** *A* $k$-vector bundle $(E, \pi_E, M)$ *over an $n$-dimensional manifold $M$ is a triple consisting of a $(k+n)$-dimensional manifold $E$, and submersion $\pi_E\colon E \to M$ such that*

*(i) For each $x \in M$ is $(\pi_E)^{-1}(x)$ a $k$-dimensional vector space over* $\mathbb{K}$,

*(ii) for each $x \in M$ there is a neighbourhood $\mathcal{U}$ and a diffeomorphism*

$$\psi\colon U \times \mathbb{K}^k \to (\pi_E)^{-1}(\mathcal{U}),$$

*where for each $y \in \mathcal{U}$ the restricted map*

$$\psi_y : \mathbb{K}^k \to (\pi_E)^{-1}(y), \vec{v} \mapsto \psi(y, \vec{v})$$

*is a linear isomorphism.*

*$M$ is called the* base manifold, *$\pi_E$ the* projection, *$E_y := (\pi_E)^{-1}(y)$ the* fibre *over $y$, $E$ the* total space, *and $\psi$ the* bundle chart *or* local trivialisiation.

We will often just speak of the vector bundle $E$ instead of $(E, \pi_E, M)$.

Notice that this is just a localisation of $M \times \mathbb{K}^k$. We call $(E, \pi_E, M)$ *trivial* if there exists a local trivialisation of the form $\psi\colon M \times \mathbb{K}^k \to E$. In this case, $E$ can be identified with $M \times \mathbb{K}^k$. An example of a vector bundle which is not trivial is given by the *Möbius band.*

*Example 2.2.1 (Möbius band, continued from page 49)).* The Möbius band $M$ is also a vector bundle. The bundle projection is given by $\pi_M(x) = (\varphi_i)^{-1} \circ p_1 \circ \varphi_i(x)$ where $p_1 : \mathbb{R}^2 \to \mathbb{R}^2$ is the projection

[5] The set of tangent spaces of a Manifold forms again a manifold of a special type, a vector bundle. While we will construct many special vector bundles and general vector bundles are of importance in gauge theory, their general definition is not central to our discussion.

$(x^1, x^2) \mapsto (x^1, 0)$ and $i \in \{1, 2\}$ is an index with $x \in \mathcal{U}_i$. It is clear that this vector bundle is not trivial.

**Definition 2.2.2.** *Let $E$ be a vector bundle. A map $\sigma : \mathcal{U} \subset M \to E$ wich satisfies $\pi_E(\sigma(x)) = x$ for all $x \in \mathcal{U}$ is called a* section of the vector bundle $E$.

*A collection $\{\sigma_1, \dots, \sigma_k\}$ of sections such that*

$$\operatorname{span}\{\sigma_1(x), \dots, \sigma_k(x)\} = E_x$$

*for all $x \in \mathcal{U}$ is called a* frame of $E$.

The following definition will play a role later on (cf. Theorem of Frobenius 2.5.3).

**Definition 2.2.3.** *Let $(E, \pi_E, M)$ be a vector bundle. A vector subbundle $(F, \pi_F, N)$ of the vector bundle $E$ consists of a submanifold $F$ of $E$ and submanifold $N$ of $M$ such that*

*(i) $\pi_F := (\pi_E)_{|F} : F \to N$ defines a vector bundle structure with base manifold $N$,*
*(ii) $F_y$ is a vector subspace of $E_y$ for all $y \in N$.*

### 2.2.1 Construction of the tangent bundle

p. 58 ↓
[↓ p. 60]

In affine geometry we had distinguished between points $x \in \mathbb{A}^n$ and translations (or vectors) in the associated vector space $\mathbb{K}^n$. A translation $v : x \mapsto x + v$ is a global concept. Notice that a translation is originally thought of as moving the point $x$ along the curve $\gamma_v : [0, 1] \mapsto \mathbb{A}^n$, $t \mapsto x + tv$ to its endpoint. The velocity vector of the curve is $v$ which may be regarded as the infinitesimal (but in this case exact) approximation of $\gamma_v$. Given an arbitrary smooth curve $\gamma : (a, b] \mapsto \mathbb{A}^n$, $t \mapsto \gamma(t)$, we take its derviative as its infinitesimal approximation at a given point $x = \gamma(t_0)$. Taking all these velocity vectors at $x$ we obtain a vector space $T_x\mathbb{A}^n$ which is attached to $\mathbb{A}^n$ at $x$. This vector space is in a natural way isomorphic to the associated vector space given by the translations. In the following we will transfer these ideas to manifolds. The main difficulty we have to solve is the lack of an associated vector space in which we can take the derivative of a curve.

Let $M$ be a smooth manifold, $x \in M$ and $\gamma : (-\epsilon, \epsilon) \to M$ be a smooth curve in $M$ which passes through $x$ at parameter value $0$. Another smooth curve $\hat{\gamma} : (-\hat{\epsilon}, \hat{\epsilon})$ with $x = \hat{\gamma}(0)$ is called *$x$-equivalent* to $\gamma$ if $\frac{\mathrm{d}}{\mathrm{d}t}(\varphi \circ \gamma)_{|0} = \frac{\mathrm{d}}{\mathrm{d}t}(\varphi \circ \hat{\gamma})_{|0}$. This definition is independent of the chosen chart and puts all curves through $x$ with the same velocity into one equivalence class $[\gamma_x]$.

**Definition 2.2.4.** *Let $M$ be a smooth manifold and $x \in M$. The space of all equivalence classes $[\gamma_x]$ is called the* tangent space of $M$ at $x$ *and denoted by $T_xM$. Its elements are called* tangent vectors.

We must show that $T_xM$ has indeed the structure of a vector space. Choose any chart $(\mathcal{U}, \varphi)$ centered at $x$. This chart induces a bijective map

$$\Theta_x^\varphi : T_xM \to \mathbb{K}^n, \quad [\gamma_x] \mapsto \frac{\mathrm{d}}{\mathrm{d}t}(\varphi \circ \gamma)_{|0}$$

which we can use to induce on $T_xM$ the vector space structure of $\mathbb{K}^n$. Let $\alpha \in \mathbb{K}$ and $[\gamma_x], [\mu_x] \in T_xM$. Then we define

$$\alpha[\gamma_x] := (\Theta_x^\varphi)^{-1}\left(\alpha\Theta_x^\varphi([\gamma_x])\right)$$

and

$$[\gamma_x] + [\mu_x] := (\Theta_x^\varphi)^{-1}\left(\Theta_x^\varphi([\gamma_x]) + \Theta_x^\varphi([\gamma_x])\right).$$

This vector space structure is independent of the chosen chart. In fact, let $(\mathcal{V}, \psi)$ be another chart centered at $x$ and denote $\Theta_x^\psi \circ (\Theta_x^\varphi)^{-1}$ by $\Theta_x^{\psi,\varphi^{-1}}$. Then we have

$$\begin{aligned}
(\Theta_x^\varphi)^{-1}\left(\alpha\Theta_x^\varphi([\gamma_x])\right) &= (\Theta_x^\psi)^{-1} \circ \Theta_x^\psi \circ (\Theta_x^\varphi)^{-1}\left(\alpha\Theta_x^\varphi \circ (\Theta_x^\psi)^{-1} \circ \Theta_x^\psi([\gamma_x])\right) \\
&= (\Theta_x^\psi)^{-1}\Theta_x^{\psi,\varphi^{-1}} \circ \left(\alpha(\Theta_x^{\psi,\varphi^{-1}})^{-1} \circ \Theta_x^\psi([\gamma_x])\right) \\
&= (\Theta_x^\psi)^{-1}\left(\alpha\Theta_x^\psi([\gamma_x])\right)
\end{aligned}$$

and analogously

$$\begin{aligned}
&(\Theta_x^\varphi)^{-1}\Big(\Theta_x^\varphi([\gamma_x]) + \Theta_x^\varphi([\mu_x])\Big) \\
&\quad = (\Theta_x^\psi)^{-1} \circ \Theta_x^\psi \circ (\Theta_x^\varphi)^{-1}\Big(\Theta_x^\varphi \circ (\Theta_x^\psi)^{-1} \circ \Theta_x^\psi([\gamma_x]) \\
&\qquad\qquad + \Theta_x^\varphi \circ (\Theta_x^\psi)^{-1} \circ \Theta_x^\psi([\mu_x])\Big) \\
&\quad = (\Theta_x^\psi)^{-1}\Theta_x^{\psi,\varphi^{-1}} \circ \Big((\Theta_x^{\psi,\varphi^{-1}})^{-1} \circ \Theta_x^\psi([\gamma_x]) \\
&\qquad\qquad + (\Theta_x^{\psi,\varphi^{-1}})^{-1} \circ \Theta_x^\psi([\mu_x])\Big) \\
&\quad = (\Theta_x^\psi)^{-1}\Big(\Theta_x^\psi([\gamma_x]) + \Theta_x^\psi([\mu_x])\Big).
\end{aligned}$$

We call the set $TM := \cup_{x \in M} T_xM$ the *tangent bundle* of the manifold $M$. The vector spaces $T_xM$ can be understood as infinitesimal models of $M$ at $x$.

[p. 59 ↓]
↓ p. 61

**Proposition 2.2.1.** *Let $M$ be a smooth manifold. There is a natural vector bundle $(TM, \pi_{TM}, M)$ associated with $M$ such that $(\pi_{TM})^{-1}(x) = T_xM$ for all $x \in M$.*

*Proof.* Let $TM = \bigcup_{x \in M} T_x M$ and define $\pi_{TM}([\gamma_x]) = x$. We choose an atlas $\{(\mathcal{U}_a, \varphi_a)\}_{a \in A}$ of $M$ and define the structure of a smooth manifold on $M$ through the atlas $\{((\pi_{TM})^{-1}(\mathcal{U}_a), \Psi_a)\}$ where

$$\Psi_a \colon (\pi_{TM})^{-1}(\mathcal{U}_a) \to \mathbb{K}^n \oplus \mathbb{K}^n, \quad [\gamma_x] \mapsto \varphi_a(x) \oplus \Theta_x^{\varphi_a}(\gamma[x]).$$

The bijections bijections

$$\psi_a \colon \mathcal{U}_a \times \mathbb{K}^n \to (\pi_{TM})^{-1}(\mathcal{U}_a), \quad (x, v) \mapsto (\Theta_x^{\varphi_a})^{-1}(v).$$

are then diffeomorphisms such that $(\psi_a)_x \colon \mathbb{K}^n \to T_x M$, $v \mapsto (\Theta_x^{\varphi_a})^{-1}(v)$ are linear isomorphisms. ∎

p. 60 ↓
[↓ p. 71]

It is not always practical to work with equivalence classes. We will therefore also give an equivalent definition which is better suited for calculations at the cost of being less intuitive. The key observation is that each tangent vector $[\gamma_x] \in T_x M$ induces a map

$$D_{[\gamma]_x} \colon \{f \in C^\infty(\mathcal{U}, \mathbb{K}) \,|\, \mathcal{U} \text{ is an open neighbourhood of } x\} \to \mathbb{K},$$
$$f \mapsto \frac{\mathrm{d}}{\mathrm{d}t} f \circ \gamma(0).$$

This map has the following properties.

(i) $D_{[\gamma]_x}$ is $\mathbb{K}$-linear,
(ii) for any smooth functions $f, g \colon M \to \mathbb{K}$ the derivation property

$$D_{[\gamma]_x}(fg) = D_{[\gamma]_x}(f)g(x) + f(x)D_{[\gamma]_x}(g)$$

holds,
(iii) for any open neighbourhood $\mathcal{U}$ of $x$ and any two smooth functions $f, g$ which coincide in $\mathcal{U}$ we have $D_{[\gamma]_x}(f) = D_{[\gamma]_x}(g)$.

This motivates the following

**Definition 2.2.5.** *A map*

$$v_x \colon \{f \in C^\infty(\mathcal{U}, \mathbb{K}) \,|\, \mathcal{U} \textit{ is an open neighbourhood of } x\} \to \mathbb{K}$$

*which satisfies properties (i)–(iii) above is called a* derivation. *The vector space of derivations at $x$ with addition and multiplication being defined pointwise, $(av_x + bw_x)(f) = a(v_x(f)) + b(w_x(f))$, is denoted by $\mathcal{D}_x$.*

*Remark 2.2.1.* The reader may wonder why in our definitions derivations act on $\{f \in C^\infty(\mathcal{U}, \mathbb{K}) \,|\, \mathcal{U}$ is an open neighbourhood of $x\}$ instead of the simpler set $C^\infty(M, \mathbb{K})$. For $\mathbb{K} = \mathbb{R}$ we could indeed have chosen $C^\infty(M, \mathbb{K})$, but in the case $\mathbb{K} = \mathbb{C}$ there exist only very few globally defined differentiable maps in general. In fact, in the extreme case of the complex torus only the constant functions are smooth. However, locally there is always an abundance of smooth functions.

**Lemma 2.2.1.** *If $f\colon M \to \mathbb{K}$ is constant in a neighbourhood of $x$, then $v_x(f) = 0$ for all derivations $v_x \in \mathcal{D}_x$.*

*Proof.* Let $f(x) = a$. $v_x(f) = av_x(f/a) = av_x(1) = av_x(1 \cdot 1) = av_x(1) \cdot 1 + a \cdot 1v_x(1) = 2av_x(1) = 2v_x(f)$. ∎

**Theorem 2.2.1.** *$\mathcal{D}_x$ is an $n$-dimensional vector space.*

*Proof.* We know already that $\mathcal{D}_x$ is a vector space and have therefore only to show that $\dim(\mathcal{D}_x) = n$. Let $(\mathcal{U}, \varphi)$ be a chart centered at $x$ and let $x^i : \mathcal{U} \to \mathbb{K}$ the $i$th coordinate component of $\varphi^{-1}$. Observe that we have for any $h\colon \varphi(U) \to \mathbb{K}$ the identity $h(y) = h(0) + \sum_{i=1}^n h_i(y)y^i$, where $h_i$ is defined by $h_i(y) := \int_0^1 \frac{\partial h(ty)}{\partial y^i} dt$ and $y^i$ is the $i$th coordinate in $\mathbb{K}^n$. Applying this identity to the function $f \circ \varphi^{-1}\colon \varphi(\mathcal{U}) \to \mathbb{K}$ we obtain

$$f(y) = f(x) + \sum_{i=1}^n x^i(y) \left(f \circ \varphi^{-1}\right)_{;i} \circ \varphi^{-1}(y).$$

Hence for any derivation $v_x \in \mathcal{D}_x$ to $f$ we get

$$v_x(f) = v_x(f(x)) + \sum_{i=1}^n \Big( v_x(x^i) \left(f \circ \varphi^{-1}\right)_{;i} \circ \varphi(x) + x^i(x) v_x(\left(f \circ \varphi^{-1}\right)_{;i} \circ \varphi)\Big).$$

The first summand vanishes by Lemma 2.2.1 and the last term vanishes since $x^i(x) = 0$. From $v_x(f) = \sum_{i=1}^n \left(f \circ \varphi^{-1}\right)_{;i} \circ \varphi(x) v_x(x^i)$ we see that $v_x$ is uniquely determined by the $n$ numbers $v_x(x^1), \ldots, v_x(x^n)$. ∎

**Definition 2.2.6.** *Let $M$ be a manifold, $(\mathcal{U}, \varphi)$ be a chart, and*

$$p^i : \mathbb{K}^n \to \mathbb{K}, \quad \begin{pmatrix} v^1 \\ \vdots \\ v^n \end{pmatrix} \mapsto v^i.$$

*Then $x^i : \mathcal{U} \to \mathbb{K}$, $y \mapsto p^i \circ \varphi(y)$ is the* $i$th coordinate function *with respect to $(\mathcal{U}, \varphi)$ and the collection $(x^1, \ldots, x^n)$ a* coordinate system. *The pointwise basis $\{\partial_{x^1}, \ldots, \partial_{x^n}\}$ defined by $\partial_{x_i}(x^j) = \delta_i^j$ is called the* Gaußian basis associated with *$(\mathcal{U}, \varphi)$ and the vector fields $\partial_{x^i}$ are called* Gaußian vector fields. *We will often simply write $\partial_i$ instead of $\partial_{x^i}$.*

**Corollary 2.2.1.** *Let $M$ be a smooth manifold and $(\mathcal{U}, \varphi)$ be a chart centered at $x \in M$. For any derivation $v_x \in \mathcal{D}_x$ and any function $f \in C^\infty(M)$ we have*

$$v_x(f) = \sum_{i=1}^{n} v^i \partial_{x^i} f = \sum_{i=1}^{n} v^i \frac{\partial f \circ \varphi^{-1}}{\partial x^i},$$

*where* $\frac{\partial}{\partial x^i}$ *is the usual partial derivative in* $\mathbb{K}^n$.

**Proposition 2.2.2.** *The tangent space* $T_x M$ *is canonically isomorphic to* $\mathcal{D}_x$. *The isomorphism is given by the map* $i\colon T_x M \to \mathcal{D}_x$, $i([\gamma]_x)(f) = \frac{\mathrm{d}}{\mathrm{d}t} f \circ \gamma(t_0)$ *and well defined.*

*Proof.* Clear by construction. ∎

Hence we can dispend with the symbol $\mathcal{D}_x$ and always use $T_x M$ instead. Our first definition using curves has the advantage to work also in infinite dimensional settings. However, we are only concerned with finite dimensional manifolds and derivations are more practical to work with than equivalence classes of curves.

**Definition 2.2.7.** *Let* $M$ *be a manifold,* $x \in M$ *and* $\mathcal{U}$ *be a neighbourhood of* $x$. *For any* $v_x \in T_x M$ *and* $f \in C^\infty(\mathcal{U}, \mathbb{K})$ *the number* $v_x \bullet f := \mathrm{d}f(v_x) := v_x(f)$ *is called the* derivative of $f$ in direction $v_x$. *The map* $\mathrm{d}f\colon TM \to \mathbb{K}$, $v_x \mapsto \mathrm{d}f(v_x)$ *is the* differential *of* $f$.

### 2.2.2 The derivative of maps between manifolds

In the preceding section we have linearized the manifold. We can now linearize differentiable maps $f\colon M \to N$ between manifolds thereby obtaining linear maps $T_x f\colon T_x M \to T_{f(x)} N$.

**Definition 2.2.8.** *Let* $M, N$ *be manifolds,* $x \in M$, *and* $\psi\colon M \to N$ *be a differentiable map. Then* $T_x\psi\colon T_x M \to T_x N$, $T_x\psi(v)(f) = v(f \circ \psi)$ *is called the* tangent map *(or simply the* derivative*) of* $\psi$. *We will often denote* $T_x\psi$ *by* $\psi_*$.

Observe that in terms of equivalence classes of curves, the derivative of $\psi$ is given by $[\gamma_x] \mapsto [\psi \circ \gamma_{\psi(x)}]$ which is clearly the linear approximation of the curve $\psi \circ \gamma$ at $\psi(x)$.

It is instructive to calculate the tangent map with respect to a coordinate system. Let $(\mathcal{U}, \varphi)$ and $(\tilde{\mathcal{U}}, \tilde{\varphi})$ be charts of $M$ and $N$ respectively, and denote by $(x^1, \ldots, x^n)$ and $(\tilde{x}^1, \ldots, \tilde{x}^k)$ the associated coordinate systems. In these charts $\psi$ has the representation $\Psi = \tilde{\varphi} \circ \psi \circ \varphi^{-1}$ and we calculate

$$\begin{aligned} \psi_* v(f) &= \sum_{i=1}^{n} v^i \partial_{x^i}(f \circ \psi) = v^i \frac{\partial f \circ \psi \circ \varphi^{-1}}{\partial x^i} \\ &= v^i \frac{\partial f \circ \tilde{\varphi}^{-1}}{\partial \tilde{x}^j} \frac{\partial(\tilde{\varphi} \circ \psi \circ \varphi^{-1})^j}{\partial x^i} = v^i \frac{\partial \Psi^j}{\partial x^i} \partial_{\tilde{x}^j} f \end{aligned}$$

Hence with respect to a coordinate system, the tangent map $\psi_*$ is just the derivative of the map $\Psi$. Again we see the tangent map $T_x\psi$ is the linearization of $\psi$ at $x$.

The following is an immediate corollary of Definition 2.1.6 and Lemma 2.1.3.

**Corollary 2.2.2.** *Let $M$ be a manifold and $x \in M$. A continuously differentiable map $\psi\colon M \to N$ is a submersion (respectively, immersion, local diffeomorphism) near $x$ if and only if $T_x\psi\colon T_xM \to T_{\psi(x)}N$ is surjective (respectively, injective, bijective).*

## 2.3 Tensors and tensor fields

*Tensor fields play a central rôle in geometry and physics.*

*Differential forms are not absolutely necessary for the theory of space and time. However, their usage has many advantages and they also provide a very natural way to define integration. In particular, the integral theorems of Stokes and Gauß have a very simple, common form when stated in terms of differential forms (cf. Theorem 2.5.5). Unfortunately, the introduction of differential forms requires some technical preparation.*

*Some readers may therefore wish to skip the sections dealing with differential forms on first reading.*

The tangent bundle of a manifold is the collection of its linear approximations $\{T_xM\}$. In order to make use of the simplifications arising from linearization we need to express physical and geometrical objects in terms of maps which are adapted to the linear structure of $T_xM$ for all $x \in M$. We will see later in the book that the notion of a *tensor field* provides a good framework for this (here still rather vague) idea. In the present section we simply introduce tensor fields as mathematical concepts.

### 2.3.1 Algebraic preliminaries: tensors

In linear algebra, the concept of a tensor unifies vectors, linear forms, bilinear forms, linear maps, determinants etc. Let $V$ be an $n$-dimensional vector space over $\mathbb{K}$. Then its *dual space* $V^*$ is the vector space of all linear maps $V \to \mathbb{K}$. It is easy to see that $V^*$ is isomorphic to $V$. In fact, if $\{e_1, \ldots, e_n\}$ is a basis of $V$ then the set $\{\theta^1, \ldots, \theta^n\} \subset V^*$, defined by $\theta^i(e_j) = \delta^i_j$, is a basis of $V^*$. It is uniquely defined by $\{e_1, \ldots, e_n\}$ and called the *dual basis*.

While the isomorphism of $V$ and $V^*$ defined by $e_i \mapsto \theta_i$ depends on the choice of basis $\{e_1, \ldots, e_n\}$, there is a *canonical* isomorphism $\iota_V$ of $V$ and $V^{**}$.

$$\begin{aligned}\iota_V : V &\to V^*,\\ v &\mapsto \iota_V(v) : f \mapsto f(v) \in \mathbb{K}\end{aligned}$$

In the following we will freely make use of this canonical identification $V \approx V^{**}$ given by $\iota_V$ and write $v$ instead of $\iota_V(v)$.

Using this identification we not only can view a vector $v$ as a linear map $V^* \to \mathbb{K}$ but also a linear map $A: V \to V$ as a bilinear map $\tilde{A}: V^* \times V \to \mathbb{K}$, $(f, v) \mapsto \tilde{A}(f, v) = f(Av)$. This reasoning can be generalized and we are let to the following unifying concept.

**Definition 2.3.1.** *An* $\binom{r}{s}$-tensor *is a map*

$$\phi: \underbrace{V \times \cdots \times V}_{s \text{ copies}} \times \underbrace{V^* \times \cdots \times V^*}_{r \text{ copies}} \to \mathbb{K}$$

*which is linear in each of its entries. We say that* $\phi$ *is an* $r$ times covariant *and* $s$ times contravariant *tensor or a tensor of* order $\binom{r}{s}$. *The space of all* $\binom{r}{s}$*-tensors is denoted by* $T^r_s(V)$.

The most important special cases are $T^0_0(V) = \mathbb{K}$, $T^1_0(V) = V$, and $T^0_1(V) = V^*$. A bilinear form such as a scalar product is an element of $T^0_2(V)$ and a linear map $V \to V$ is an element of $T^1_1(V)$.

For an explanation of the terminology "covariant/contravariant" see Remark 2.3.1 below. First we will need to define the *components* of a tensor. This in turn requires the introduction of the (natural) "tensor product" $\otimes$ of tensors which generalizes the usual product of numbers.

**Definition 2.3.2.** *Let* $\phi \in T^r_s(V)$, $\psi \in T^q_p(V)$. *Then we define the* tensor product $\phi \otimes \psi \in T^{r+q}_{s+p}(V)$ *by*

$$\begin{aligned}\phi \otimes \psi\,(v_1, \ldots, v_s, w_1, \ldots, w_p, \omega^1, \ldots, \omega^r, \eta^1, \ldots \eta^q)& \\ := \phi(v_1, \ldots, v_s, \omega^1, \ldots, \omega^r)\, \psi(w_1, \ldots, w_p, \eta^1, \ldots \eta^q)&.\end{aligned}$$

Observe that the tensor product is *not* commutative.

**Lemma 2.3.1.** *The tensor product is associative.*

*Proof.* Let $\phi, \phi_2, \phi_3$ be tensors of order $\binom{r_1}{s_1}$, $\binom{r_2}{s_2}$, $\binom{r_3}{s_3}$ respectively. Then we have

$$\begin{aligned}&(\psi_1 \otimes \psi_2) \otimes \psi_3(v_1, \ldots, v_{s_1}, v_{s_1+1}, \ldots, v_{s_1+s_2}, v_{s_1+s_2+1}, \ldots, v_{s_1+s_2+s_3}, \\ &\qquad \omega^1, \ldots, \omega^{r_1}, \omega^{r_1+1}, \ldots, \omega^{r_1+r_2+1}, \omega^{r_1+r_2+1}, \ldots, \omega^{r_1+r_2+r_3}) \\ &\quad = (\psi_1 \otimes \psi_2)(v_1, \ldots, v_{s_1}, v_{s_1+1}, \ldots, v_{s_1+s_2}, \omega^1, \ldots, \omega^{r_1}, \\ &\qquad \omega^{r_1+1}, \ldots, \omega^{r_1+r_2}) \\ &\qquad \times \psi_3(v_{s_1+s_2+1}, \ldots, v_{s_1+s_2+s_3}, \omega^{r_1+r_2+1}, \ldots, \omega^{r_1+r_2+r_3}) \\ &\quad = \psi_1(v_1, \ldots, v_{s_1}, \omega^1, \ldots, \omega^{r_1})\psi_2(v_{s_1+1}, \ldots, v_{s_1+s_2}, \omega^{r_1+1}, \ldots, \omega^{r_1+r_2}) \\ &\qquad \times \psi_3(v_{s_1+s_2+1}, \ldots, v_{s_1+s_2+s_3}, \omega^{r_1+r_2+1}, \ldots, \omega^{r_1+r_2+r_3})\end{aligned}$$

and analogously

$$\begin{aligned}
&\psi_1\otimes(\psi_2 \otimes \psi_3)(v_1,\ldots,v_{s_1},v_{s_1+1},\ldots,v_{s_1+s_2},v_{s_1+s_2+1},\ldots,v_{s_1+s_2+s_3},\\
&\qquad \omega^1,\ldots,\omega^{r_1},\omega^{r_1+1},\ldots,\omega^{r_1+r_2+1},\omega^{r_1+r_2+1},\ldots,\omega^{r_1+r_2+r_3})\\
&\quad = \psi_1(v_1,\ldots,v_{s_1},\omega^1,\ldots,\omega^{r_1})\\
&\qquad (\psi_2 \otimes \psi_3)(v_{s_1+1},\ldots,v_{s_1+s_2},v_{s_1+s_2+1},\ldots,v_{s_1+s_2+s_3},\\
&\qquad \omega^{r_1+1},\ldots,\omega^{r_1+r_2},\omega^{r_1+r_2+1},\ldots,\omega^{r_1+r_2+r_3})\\
&\quad = \psi_1(v_1,\ldots,v_{s_1},\omega^1,\ldots,\omega^{r_1})\psi_2(v_{s_1+1},\ldots,v_{s_1+s_2},\\
&\qquad \omega^{r_1+1},\ldots,\omega^{r_1+r_2})\\
&\qquad \times \psi_3(v_{s_1+s_2+1},\ldots,v_{s_1+s_2+s_3},\omega^{r_1+r_2+1},\ldots,\omega^{r_1+r_2+r_3}).
\end{aligned}$$

∎

**Lemma 2.3.2.** *If $\{e_1,\ldots,e_n\}$, $\{\theta^1,\ldots,\theta^n\}$ is a pair of dual bases then the set*

$$\{\theta^{i_1}\otimes\cdots\otimes\theta^{i_s}\otimes e_{j_1}\otimes\cdots\otimes e_{j_r}\}_{i_1,\ldots,i_s,j_1,\ldots,j_r\in\{1,\ldots,n\}}$$

*forms a basis of the space $T^r_s(V)$ of all $\binom{r}{s}$-tensors. In particular, we have $\dim(T^r_s(V)) = n^r n^s$.*

*Proof.* The set of tensors $\{\theta^{i_1}\otimes\cdots\otimes\theta^{i_s}\otimes e_{j_1}\otimes\cdots\otimes e_{j_r}\}$ is linearly independent. In fact, let $\psi^{i_1\ldots i_s}_{j_1\ldots j_r}$ be numbers such that

$$\psi = \sum_{\substack{i_1,\ldots,i_s=1\\ j_1,\ldots,j_r=1}}^{n} \psi^{j_1\ldots j_r}_{i_1\ldots i_s}\theta^{i_1}\otimes\cdots\otimes\theta^{i_s}\otimes e_{j_1}\otimes\cdots\otimes e_{j_r} = 0.$$

Then $0=\psi(e_{k_1},\ldots,e_{k_s},\theta^{l_1},\ldots,\theta^{l_r}) = \psi^{l_1\ldots l_s}_{k_1\ldots k_r}$, hence the tensors are all linearly independent. Conversely, we see that for any tensor $\phi\in T^r_s(V)$ and any $v_1,\ldots v_s\in V$, $\eta^1,\ldots,\eta^r\in V^*$ we have

$$\begin{aligned}
\phi(v_1,\ldots v_s,\eta_1,\ldots,\eta^r) = \sum_{\substack{i_1,\ldots,i_s=1\\ j_1,\ldots,j_r=1}}^{n} &\phi(e_{i_1},\ldots,e_{i_s},\theta^{j_1},\ldots,\theta^{j_r})\,\theta^{i_1}\otimes\ldots\\
&\otimes\theta^{i_s}\otimes e_{j_1}\otimes\cdots\otimes e_{j_r}(v_1,\ldots v_s,\eta^1,\ldots,\eta^r).
\end{aligned}$$

The dimension of $T^r_s(V)$ is $n^r n^s$ since there are exactly $n^t$ choices of ordered $t$-tuples (with possible duplication) from a set of $n$ elements. ∎

**Definition 2.3.3.** *Let $\psi\in T^r_s(V)$ and $\{e_1,\ldots,e_n\}$ be a basis of $V$ and $\{\theta^1,\ldots,\theta^n\}$ be the associated dual basis. The numbers $\psi^{j_1\ldots j_r}_{i_1\ldots i_s}$ defined by*

$$\psi = \sum_{\substack{i_1,\ldots,i_s=1\\ j_1,\ldots,j_r=1}}^{n} \psi_{i_1\ldots i_s}^{j_1\ldots j_r}\theta^{i_1}\otimes\cdots\otimes\theta^{i_s}\otimes e_{j_1}\otimes\cdots\otimes e_{j_r}$$

*are the* components *of* $\psi$ *with respect to the basis* $\{e_1,\ldots,e_n\}$.

In the physical (and old mathematical) literature it is the standard to use for contravariant entries upper and for covariant entries lower indices. This provides a checking mechanism for the syntactical correctness of tensor formulas and also simplifies the interpretation of formulas involving tensor components. In Remark 2.3.4 below we will introduce a very effective notation (Einstein's summation convention) which is prevalent in the physics literature and has at its core the difference between upper and lower indices. Unfortunately, many modern mathematicians use lower indices for all entries on grounds of "aesthetics".

*Remark 2.3.1.* The terminology "covariant/contravariant" arose in the 19th century and refers to the transformation of tensor components under transformation of a given pair of dual bases $\{e_1,\ldots,e_n\}$, $\{\theta^1,\ldots,\theta^n\}$. Let $v=\sum_{i=1}^n v^i e_i\in V$, $\omega=\sum_{i=1}^n \omega_i\theta^i\in V^*$, and $A=\sum_{i,j}A^i_j e_i\otimes\theta^j\in T^1_1(V)$ be an invertible linear map. Then $\{\tilde e_1,\ldots,\tilde e_n\}$, $\{\tilde\theta^1,\ldots,\tilde\theta^n\}$ defined by $\tilde e_i = Ae_i$ and $\tilde\theta^i=\theta^i\circ(A^{-1})$ are also a pair of dual bases and we can write $v=\sum_{i=1}^n\tilde v^i\tilde e_i$, $\omega=\sum_{i=1}^n\tilde\omega_i\tilde\theta^i$. For any $w\in V$ we have

$$\begin{aligned}\omega(w) &= \sum_{i=1}^n \omega(\tilde w^i\tilde e_i) = \sum_{i=1}^n \tilde w^i\omega(Ae_i) = \sum_{i,j=1}^n \tilde w^i A^j_i\omega(e_j)\\ &= \sum_{i,j,k=1}^n \tilde w^i A^j_i\omega_k\theta^k(e_j) = \sum_{i,j=1}^n \tilde w^i A^j_i\omega_j\end{aligned}$$

Hence $\tilde\omega_i = \sum_{j=1}^n A^j_i\omega_j$. The components of $\omega$ transform **covariantly**, i.e., in the **same** way as the basis vectors $e_i$.

Similarly, for any $\lambda\in V^*$ we have

$$\begin{aligned}v(\lambda) &= \sum_{i=1}^n v(\tilde\lambda_i\tilde\theta^i) = \sum_{i=1}^n v(\tilde\lambda_i\theta^i\circ A^{-1}) = \sum_{i=1}^n \tilde\lambda_i\theta^i(A^{-1}v)\\ &= \sum_{i,j,k=1}^n \tilde\lambda_i(A^{-1})^j_k v^k\theta^i e_j = \sum_{i,j,k=1}^n \tilde\lambda_i(A^{-1})^i_k v^k.\end{aligned}$$

Hence $\tilde v^i = \sum_{k=1}^n (A^{-1})^i_k v^k$, the components of $v$ transform **contra**variantly with respect to the transformation $A$, i.e., **opposite** to the basis vectors $e_i$.

Another natural operation which is defined for tensors is their "contraction".

**Definition 2.3.4.** *Let $\phi \in T^r_s(V)$ and $\{e_1, \ldots, e_n; \theta^1, \ldots, \theta^n\}$ be a pair of dual bases. The* contraction *of $\phi$ with respect to the $\hat{r}$th contravariant slot and the $\hat{s}$th covariant slot of $\phi$ is defined by*

$$C^{\hat{r}}_{\hat{s}}\phi(v_1, \ldots, v_{r-1}, \omega^1, \ldots, \omega^{s-1}) = \sum_{i=1}^{n} \phi(v_1, \ldots, \overbrace{e_i}^{\hat{r}\text{th slot}}, \ldots, v_{r-1}, \omega^1, \ldots, \overbrace{\theta^i}^{\hat{q}\text{th slot}}, \ldots, \omega^{s-1}).$$

We have to show that this definition is independent of the choice of basis. In fact, if $\{\tilde{e}_1, \ldots, \tilde{e}_n\}$, $\{\tilde{\theta}^1, \ldots, \tilde{\theta}^n\}$ is another pair of dual bases then there exists a linear isomorphism $A\colon V \to V$ with $e_i = A\tilde{e}_i = \sum_{j=1}^n A^j_i \tilde{e}_j$ and $\theta^k = \tilde{\theta}^k \circ A^{-1} = \sum_{j=1}^n (A^{-1})^k_j \tilde{\theta}^j$. We calculate

$$\begin{aligned}
C^{\hat{r}}_{\hat{s}}\phi(v_1, \ldots, v_{r-1}, \omega^1, \ldots, \omega^{s-1})
&= \sum_{i=1}^{n} \phi(v_1, \ldots, e_i, \ldots, v_{r-1}, \omega^1, \ldots, \theta^i, \ldots, \omega^{s-1}) \\
&= \sum_{i,j,k=1}^{n} \overbrace{A^j_i (A^{-1})^i_k}^{=\delta^j_k} \phi(v_1, \ldots, \tilde{e}_j, \ldots, v_{r-1}, \omega^1, \ldots, \tilde{\theta}^k, \ldots, \omega^{s-1}) \\
&= \sum_{j}^{n} \phi(v_1, \ldots, \tilde{e}_j, \ldots, v_{r-1}, \omega^1, \ldots, \tilde{\theta}^j, \ldots, \omega^{s-1}).
\end{aligned}$$

**Lemma 2.3.3.** *Let $V$ be a vector space over $\mathbb{K}$ and $\phi \in T^p_q(V)$, $\psi \in T^r_s(V)$. Then*

$$\left(C^{\hat{p}}_{\hat{q}}\phi\right) \otimes \psi = C^{\hat{p}}_{\hat{q}}\left(\phi \otimes \psi\right) \qquad \text{and} \qquad \phi \otimes \left(C^{\hat{r}}_{\hat{s}}\psi\right) = C^{p+\hat{r}}_{q+\hat{s}}\left(\phi \otimes \psi\right).$$

*Proof.* This follows immediately from the definitions. ∎

Another class of natural operations on tensors are symmetry operations. We introduce below the two most important symmetry operators, *symmetrisation* sym and *anti-symmetrisation* alt of entries. First we need some technical preparation.

**Definition 2.3.5.** *A* permutation *of the numbers $(1, \ldots, p)$ is a bijection*

$$\begin{aligned}
\sigma_p\colon \{(i_1, \ldots, i_p) \mid \{i_1, \ldots, i_p\} = \{1, \ldots, p\}\} \\
\to \{(i_1, \ldots, i_p) \mid \{i_1, \ldots, i_p\} = \{1, \ldots, p\}\}.
\end{aligned}$$

*If $\sigma_p$ is a permutation we write $\sigma(i_1, \ldots, i_p) = (i_{\sigma(1)}, \ldots, i_{\sigma(p)})$. The set of all permutations of the $p$ integers $\{1, \ldots, p\}$ is denoted by $S_p$.*

*A* transposition *is a permutation which permutes only two consecutive elements and leaves all other elements fixed.*

**Lemma 2.3.4.** *The set of all permutations $S_p$ forms a group (the* permutation group*) and is generated by transpositions.*

*Proof.* That $S_p$ forms a group is clear since the collection of all bijections of a given set forms a group where the composition of maps is the group operation.

Let $\sigma(i_1, \ldots, i_p) = (i_{\sigma(1)}, \ldots, i_{\sigma(p)})$ be any permutation. Starting with the $p$-tuple $(i_1, \ldots, i_p)$ we can use successive transpositions in order to move the index $i_{\sigma(p)}$ to the last position. Assume now that $i_{\sigma(k)}, \ldots, i_{\sigma(p)}$ are at positions $k, \ldots, p$. Since $i_{\sigma(k-1)} \notin \{i_{\sigma(k)}, \ldots, i_{\sigma(p)}\}$ it must be at one of the positions $1, \ldots, k-1$. It follows that we can move $i_{\sigma(k-1)}$ to position $k-1$ by successive transpositions which all leave the last $p-k$ positions invariant. By induction we have shown that there is a finite sequence of transpositions which is equivalent to $\sigma$. ∎

**Lemma 2.3.5.** *There is a natural homomorphism*

$$\text{sign}\colon S_p \to (\{-1, 1\}, \cdot)$$

*of the permutation group into the group of two elements which is determined by* $\text{sign}(\tau_p) = -1$ *for all transpositions* $\tau_p$.

*Proof.* We prove this lemma by showing that every permutation $\sigma$ is either the product of an even number of transpositions ($\text{sign}(\sigma) = 1$) or the product of an odd number of transpositions ($\text{sign}(\sigma) = -1$).

First we show that the identity permutation id is not the product of an odd number of transpositions. Assume that id is the product of finitely many transpositions and denote by $n_{lk}$ the number of all those transpositions which interchange the numbers $l$ and $k$. The number $\eta_{lk}$ must be even since at the end $l$ must be on the same side of $k$ as at the beginning and since there are no other transpositions which interchange $k$ and $l$. If we set $n_{ll} = 0$ then the number of all transpositons is $\sum_{l=1}^{p}(\sum_{l=k}^{p} n_{lk})$ which is the sum of even numbers and therefore even.

Let now $\sigma = \tau_1 \circ \cdots \circ \tau_k = \tilde{\tau}_1 \circ \cdots \circ \tilde{\tau}_l$ where $\tau_i, \tilde{\tau}_j$ are transpositions. Since $\text{id} = (\tau_1 \circ \cdots \circ \tau_k)^{-1} \circ \tilde{\tau}_1 \circ \cdots \circ \tilde{\tau}_l = (\tau_k)^{-1} \circ \cdots \circ (\tau_1)^{-1} \circ \tilde{\tau}_1 \circ \cdots \circ \tilde{\tau}_l$ is the product of $k+l$ transpositions the number $k+l$ must be even. Hence $k$ and $l$ are both even or both odd. ∎

A permutation $\sigma_p$ acts in a natural way on a tensor $\psi \in T_p^0(V)$.

**Definition 2.3.6.** *For any $\psi \in T_p^0(V)$ and any* permutation *$\sigma_p \in S_p$ we set $\sigma_p\psi(v_1, \ldots, v_p) := \psi(v_{\sigma_p(1)}, \ldots, v_{\sigma_p(p)})$.*

**Lemma 2.3.6.** *For any permutations $\tau_p, \sigma_p \in S_p$ and any tensor $\psi \in T_p^0(V)$ we have $(\sigma_p \circ \tau_p)\psi = \tau_p(\sigma_p\psi)$.*

*Proof.* Let $v_1, \ldots, v_p \in V$. We calculate

$$\begin{aligned}(\sigma_p \circ \tau_p)\psi(v_1, \ldots, v_p) &= \psi(v_{\sigma_p \circ \tau_p(1)}, \ldots, v_{\sigma_p \circ \tau_p(p)}) \\ &= \psi(v_{\sigma_p(\tau_p(1))}, \ldots, v_{\sigma_p \circ (\tau_p(p))}) \\ &= \sigma_p \psi(v_{\tau_p(1)}, \ldots, v_{\tau_p(p)}) \\ &= \tau_p(\sigma_p \psi)(v_{\tau_p(1)}, \ldots, v_{\tau_p(p)})\end{aligned}$$

which implies the first equality. ∎

**Lemma 2.3.7.** *The maps*

$$\text{sym}\colon T_p^0(V) \to T_p^0(V), \quad \psi \mapsto \frac{1}{p!} \sum_{\sigma_p \in S_p} \sigma_p \psi \qquad \text{alt}\colon T_p^0(V) \to T_p^0(V), \quad \psi \mapsto \frac{1}{p!} \sum_{\sigma_p \in S_p} \text{sign}(\sigma_p)\sigma_p \psi$$

*are linear projections.*

*Proof.* We only prove the lemma for the operator alt since the proof for sym is completely analogous. That alt is linear is clear from the definitions. For given $\psi \in T_p^0(V)$, vectors $v_1, \ldots, v_p$ and any permutation $\tau_p$ we have

$$\begin{aligned}\text{alt}\, \psi(v_1, \ldots, v_p) &= \frac{1}{p!} \sum_{\sigma_p \in S_p} \text{sign}(\sigma_p)\, \psi(v_{\sigma_p(1)}, \ldots, v_{\sigma_p(p)}) \\ &= \frac{1}{p!} \sum_{\sigma_p \in S_p} \text{sign}(\sigma_p)\text{sign}(\tau_p)\, \psi(v_{\sigma_p \circ \tau_p(1)}, \ldots, v_{\sigma_p \circ \tau_p(p)}) \\ &= \text{sign}(\tau_p)\, \text{alt}\, \psi(v_{\tau_p(1)}, \ldots, v_{\tau_p(p)}),\end{aligned}$$

where we have used that $R_\tau : S_p \to S_p$, $\sigma_p \mapsto \sigma_p \tau_p$ is a bijection. It follows that

$$\begin{aligned}\text{alt} \circ \text{alt}\, \psi((v_1, \ldots, v_p) &= \frac{1}{p!} \sum_{\tau_p \in S_p} \text{sign}(\tau_p)\ \text{alt}\, \psi(v_{\tau_p(1)}, \ldots, v_{\tau_p(p)}) \\ &= \frac{1}{p!} \sum_{\tau_p \in S_p} \text{alt}\, \psi(v_1, \ldots, v_p) \\ &= \text{alt}\, \psi(v_1, \ldots, v_p)\end{aligned}$$

and therefore $\text{alt} \circ \text{alt} = \text{alt}$. ∎

**Definition 2.3.7.** *A covariant tensor $\psi \in T_s^0(V)$ is called* symmetric *(respectively,* anti-symmetric*) if for all $s$-tuples of vectors $(v_1, \ldots, v_s)$ and all permutations $\sigma_s$ of $(1, \ldots, s)$ the equality*

$$\psi(v_1, \ldots, v_s) = \psi(v_{\sigma_s(1)}, \ldots, v_{\sigma_s(s)})$$
$$(\textit{respectively, } \psi(v_1, \ldots, v_s) = \operatorname{sign}(\sigma_s)\psi(v_{\sigma_s(1)}, \ldots, v_{\sigma_s(s)}))$$

*holds. Symmetric and anti-symmetric contravariant tensors are defined analogously.*

[p. 61 ↓]
↓ p. 79

The set of all $r$ times contravariant and $s$ times covariant tensors on $T_xM$, where $x \in M$, form a vector bundle which generalizes the tangent bundle.

**Proposition 2.3.1.** *Let $M$ be an $n$-dimensional, smooth manifold. The set $T^r_sM = \bigcup_{x\in M} T^r_s(T_xM)$ of all $\binom{r}{s}$-tensors carries a natural vector bundle structure.*

*Proof.* Let $(\mathcal{U}_\alpha, \varphi_\alpha)$ be an atlas of $M$. Then with each chart $\mathcal{U}_\alpha$ we can associate a map

$$\psi_\alpha \colon \bigcup_{x\in\mathcal{U}_\alpha} T^r_s(T_xM) \to \phi_\alpha(\mathcal{U}) \times \mathbb{K}^{n^r n^s},$$
$$\phi_x \mapsto \Big(\varphi_\alpha(x), (\phi^{i_1\ldots i_r}_{(\alpha)\,j_1\ldots j_s})_{i_1,\ldots,i_r,j_1,\ldots,j_s\in\{1,\ldots n\}}\Big),$$

where $\phi^{i_1\ldots i_r}_{(\alpha)\,j_1\ldots j_s}$ are the components of the tensor $\phi_x$ with respect to the Gaußian basis $\partial_{x^1}, \ldots, \partial_{x^n}$. It is clear that each tensor $\psi_y \in T^r_s(T_yM)$ lies in at least one of the sets $T^r_s\mathcal{U}_\alpha := \bigcup_{x\in\mathcal{U}_\alpha} T^r_s(T_xM)$ $(\alpha \in A)$. To see that the collection $(T^r_s\mathcal{U}_\alpha, \psi_\alpha)$ forms an atlas of $T^r_sM$ we have to show that for each pair of indices $(\alpha, \beta)$ the map

$$\psi_\alpha \circ \psi_\beta^{-1} \colon \psi_\beta(T^r_s\mathcal{U}_\alpha \cap T^r_s\mathcal{U}_\beta) \to \psi_\alpha(T^r_s\mathcal{U}_\alpha \cap T^r_s\mathcal{U}_\beta)$$

is a diffeomorphism. We denote the coordinate system associated with $(\mathcal{U}_\beta, \varphi_\beta)$ by $(y^1, \ldots y^n)$ and let $x \in \mathcal{U}_\alpha \cap \mathcal{U}_\beta$. For any vector $v \in T_xM$ we can then write $v = \sum_{i=1}^n v^i_\alpha \partial_{x^i} = \sum_{i=1}^n v^i_\beta \partial_{y^i}$ where

$$v^j_\alpha = \sum_{i=1}^n v^i_\beta \frac{\partial(\varphi_\alpha \circ (\varphi_\beta)^{-1})^j}{\partial y^i}.$$

In other words, the column vectors $(v_\alpha), (v_\beta)$ consisting of the components of the vector $v$ with respect to our charts are related by $(v_\alpha) = \mathrm{D}\varphi_{\alpha\beta}(v_\beta)$, where we have set $\varphi_{\alpha\beta} = \varphi_\alpha \circ (\varphi_\beta)^{-1}$. Let $\omega$ be a 1-form. The row vectors $(\omega^\alpha), (\omega^\beta)$ consisting of the components of $\omega$ with respect to the charts must then be related by $(\omega^\alpha) = (\omega_\beta)\mathrm{D}(\varphi_{\alpha\beta})^{-1})$ since these components are defined by $\omega(v) = (\omega^\alpha) \cdot (v_\alpha) = (\omega^\beta) \cdot (v_\beta)$ for all vectors $v$. By the same argument it follows that the components $\phi^{i_1\ldots i_r}_{(\alpha)\,j_1\ldots j_s}$ and $\phi^{i_1\ldots i_r}_{(\beta)\,j_1\ldots j_s}$ are related by

$$\phi^{k_1\ldots k_r}_{(\alpha)\,l_1\ldots l_s} = \sum_{\substack{1\leq i_1\ldots i_r\leq n\\ 1\leq j_1\ldots j_s\leq n}} \phi^{i_1\ldots i_r}_{(\beta)\,j_1\ldots j_s} (\mathrm{D}\varphi_{\alpha\beta})^{k_1}_{i_1} \ldots (\mathrm{D}\varphi_{\alpha\beta})^{k_r}_{i_r}$$
$$\times (\mathrm{D}(\varphi_{\alpha\beta})^{-1})^{j_1}_{l_1} \ldots (\mathrm{D}(\varphi_{\alpha\beta})^{-1})^{j_s}_{l_s}.$$

Hence the components $(\phi_\alpha)$ and $(\phi_\beta)$ of $\phi_x$ with respect to the charts $(\mathcal{U}_\alpha, \varphi_\alpha)$, $(\mathcal{U}_\beta, \varphi_\beta)$ are related by a linear isomorphism $D_{\alpha\beta}$. Thus the map $\psi_\alpha \circ \psi_\beta^{-1}(y, (\phi_\beta)) = (\varphi_{\alpha\beta}(y), (D_{\alpha\beta}(\varphi_\beta)))$ is indeed a diffeomorphism. We have shown that $TM$ is a manifold. That $TM$ is also a vector bundle follows since for any $x \in M$ and any chart $(\mathcal{U}_\alpha, \varphi_\alpha)$ the map $T^r_s(T_xM) \to \mathbb{K}^{n^r n^s}$, $\phi_x \mapsto (\phi_\alpha)$ is a linear isomorphism. ∎

By construction of $T^r_s M$ we have $TM = T^1_0 M$.

**Definition 2.3.8.** *Let $M$ be a manifold and $r, s \in \mathbb{N} \cup \{0\}$. The vector bundle $T^r_s M$ is called the* tensor bundle *of $r$ times contravariant and $s$ times covariant tensors.*

*The tensor bundle $T^0_1 M$ is also denoted by $T^*M$ and called the* cotangent bundle *of $M$.*

*The* bundle of $s$-forms *is the subbundle $\Lambda^s M$ of $T^0_s M$ defined by*

$$\omega \in \Lambda^s M \Leftrightarrow \omega(v_1, \ldots, v_i, \ldots, v_j, \ldots v_p) = -\omega(v_1, \ldots, v_j, \ldots, v_i, \ldots v_p)$$
*for any set of vectors $\{v_1, \ldots, v_p\}$ and any pair*
*$(i, j) \subset \{1, \ldots, p\}$.*

The bundle of $s$-forms will play a fundamental rôle in Sect. 2.5 on differential forms (which may be omitted on first reading). We will give an equivalent definition in Lemma 2.5.8. See also Definition 2.3.9 for the vector space analogon $\Lambda^s(V)$.

*In the rest of this algebraic section we will introduce the space of* $p$-forms, *a special class of tensors which generalizes the notion of determinant and plays an important rôle in the analysis on manifolds. For instance, it is fundamental to the integration over a manifolds (cf. Sect. 2.5.4 below) and in the theory of differential systems (Bryant, Chern, Gardner, Goldschmidt, and Griffiths 1991). However, the usefulness of this concept will only become clear in applications and not on this technical algebraic level.*

*The material presented here will not be used before Sect. 2.5. The reader may therefore wish to postpone the study of the rest of this section.*

It turns out that the symmetry operator sym is not of particular importance. On the other hand, tensors $\psi$ which satisfy $\mathrm{alt} \circ \psi = \psi$ generalize the determinant and therefore deserve a special definition.

**Definition 2.3.9.** *A* form (of degree $p$) *(or simply* $p$-form*) is a tensor $\psi \in T^0_p(V)$ with $\mathrm{alt}\,\psi = \psi$. The vector space of all $p$-forms is denoted by $\Lambda^p(V) = \mathrm{alt}\,(T^0_p(v))$.*

The tensor product $\otimes$ induces a product $\wedge$ of forms:

**Definition 2.3.10.** *The* exterior product *(or* wedge product*)* $\wedge$ *is the bilinear map*

$$\wedge\colon \Lambda^p(V) \times \Lambda^q(V) \to \Lambda^{p+q}(V),$$
$$(\omega, \eta) \mapsto \omega \wedge \eta := \frac{(p+q)!}{p!q!}\,\mathrm{alt}(\omega \otimes \eta).$$

*Remark 2.3.2.* The normalization factor $\frac{(p+q)!}{p!q!}$ is not the only possible choice and may appear to be rather unnatural. Other factors are also common, in particular the alternative definition $\omega\tilde{\wedge}\eta = \mathrm{alt}(\omega \otimes \eta)$. We have chosen our factor in order to minimize similar combinatorial factors in formulas to come (see Remark 2.3.3).

Our choice agrees with the choice of Bryant, Chern, Gardner, Goldschmidt, and Griffiths (1991) and (Abraham and Marsden 1978), but is different from the normalization factor in (Abraham and Marsden 1967) and the normalization factor in (Kobayashi and Nomizu 1963).

**Lemma 2.3.8.** *The exterior product is associative and anti-symmetric. More concretely, let* $\omega \in \Lambda^p(V)$, $\eta \in \Lambda^q(V)$, *and* $\lambda \in \Lambda^r(V)$. *Then the formulas*

*(i)* $(\omega \wedge \eta) \wedge \lambda = \omega \wedge (\eta \wedge \lambda)$ *and*
*(ii)* $\omega \wedge \eta = (-1)^{pq}\, \eta \wedge \omega$.

*hold.*

*Proof.* (i): We calculate

$$\begin{aligned}(\omega \wedge \eta) \wedge \lambda &= \frac{((p+q)+r)!}{(p+q)!r!}\,\mathrm{alt}((\omega \wedge \eta) \otimes \lambda)\\ &= \frac{((p+q)+r)!}{(p+q)!r!}\,\frac{(p+q)!}{p!q!}\,\mathrm{alt}((\omega \otimes \eta) \otimes \lambda)\\ &= \frac{(p+q+r)!}{p!q!r!}\,\mathrm{alt}(\omega \otimes \eta \otimes \lambda).\end{aligned}$$

Analogously one shows $\omega \wedge (\eta \wedge \lambda) = \frac{(p+q+r)!}{p!q!r!}\,\mathrm{alt}(\omega \otimes \eta \otimes \lambda)$ which proves associativity.

(ii): Let $\tau_{p+q} \in S_{p+q}$ be the permutation given by $\tau_{p+q}(1, \ldots, p+q) = (q+1, \ldots, q+p, 1, \ldots, q)$. Then we have $\mathrm{sign}(\tau_{p+q}) = (-1)^{pq}$ and

$$\begin{aligned}\tau_{p+q}(\omega \otimes \eta)(v_1, \ldots, v_{p+q}) &= \omega \otimes \eta(v_{q+1}, \ldots, v_{q+p}, v_1, \ldots, v_q)\\ &= \eta \otimes \omega(v_1, \ldots, v_{p+q}).\end{aligned}$$

Using these formulas we obtain

$$
\begin{aligned}
\omega \wedge \eta &= \frac{(p+q)!}{p!q!}\,\mathrm{alt}(\omega \otimes \eta) \\
&= \frac{1}{p!q!} \sum_{\sigma_{p+q} \in S_{p+q}} \mathrm{sign}(\sigma_{p+q}) \sigma_{p+q}(\omega \otimes \eta) \\
&= \frac{1}{p!q!} \sum_{\sigma_{p+q} \in S_{p+q}} \mathrm{sign}(\sigma_{p+q}) \mathrm{sign}(\tau_{p+q}) \sigma_{p+q} \tau_{p+q}(\omega \otimes \eta) \\
&= (-1)^{pq} \frac{1}{p!q!} \sum_{\sigma_{p+q} \in S_{p+q}} \mathrm{sign}(\sigma_{p+q}) \sigma_{p+q}(\eta \otimes \omega) \\
&= (-1)^{pq} \eta \wedge \omega.
\end{aligned}
$$

∎

**Lemma 2.3.9.** *Let* $\omega^1, \dots, \omega^p \in V^*$ *and* $\pi_p \in S_p$.

*(i)* $\omega^1 \wedge \dots \wedge \omega^p = \mathrm{sign}(\pi_p) \omega^{\pi_p(1)} \wedge \dots \wedge \omega^{\pi_p(p)}$,
*(ii)* $\omega^1 \wedge \dots \wedge \omega^p = \sum_{\sigma_p \in S_p} \mathrm{sign}(\sigma_p) \omega^{\sigma_p(1)} \otimes \dots \otimes \omega^{\sigma_p(p)}$,
*(iii)* $\omega^1 \wedge \dots \wedge \omega^p = 0$ *if and only if the 1-forms* $\omega^1, \dots, \omega^p$ *are linearly dependent.*

*Proof.* (i): Since transpositions generate all permutations, it is sufficient to prove this equality for a transposition

$$\pi_p(1, \dots, i, j, \dots, n) = (1, \dots, j, i, \dots, n).$$

But in this case the assertion follows immediately from Lemma 2.3.8.

(ii): The assertion is clearly true for $p = 2$. Assume that it is true for collections of up to $q$ 1-forms $\omega^1, \dots, \omega^q$ and denote for any $\sigma_q \in S_q$ the permutation $(i_1, \dots, i_{q+1}) \mapsto (\sigma_q(i_1), \dots, \sigma_q(i_q), i_{q+1})$ by $\iota_{q+1}(\sigma_q)$. Then we have

$$
\begin{aligned}
&\omega^1 \wedge \dots \wedge \omega^q \wedge \omega^{q+1}(v_1, \dots, v_{q+1}) \\
&\quad = \frac{1}{1!q!} \sum_{\tau_{q+1} \in S_{q+1}} \mathrm{sign}(\tau_{q+1}) (\omega^1 \wedge \dots \wedge \omega^q) \\
&\qquad \otimes \omega^{q+1}(v_{\tau_{q+1}(1)}, \dots, v_{\tau_{q+1}(q+1)}) \\
&\quad = \frac{1}{q!} \sum_{\tau_{q+1} \in S_{q+1}} \sum_{\sigma_q \in S_q} \mathrm{sign}(\tau_{q+1}) \mathrm{sign}(\sigma_q) \omega^1 \otimes \dots \\
&\qquad \otimes \omega^q \otimes \omega^{q+1}(v_{\sigma_q \circ \tau_{q+1}(1)}, \dots, v_{\sigma_q \circ \tau_{q+1}(q)}, v_{\tau_{q+1}(q+1)})
\end{aligned}
$$

$$= \frac{1}{q!} \sum_{\sigma_q \in S_q} \sum_{\tau_{q+1} \in S_{q+1}} \operatorname{sign}(\iota_{q+1}(\sigma_q) \circ \tau_{q+1}) \omega^1 \otimes \dots$$
$$\otimes \omega^q \otimes \omega^{q+1} (v_{\iota_{q+1}(\sigma_q) \circ \tau_{q+1}(1)}, \dots, v_{\iota_{q+1}(\sigma_q) \circ \tau_{q+1}(q+1)})$$
$$= \sum_{\tau_{q+1} \in S_{q+1}} \operatorname{sign}(\tau_{q+1}) \omega^1 \otimes \dots$$
$$\otimes \omega^q \otimes \omega^{q+1} (v_{\tau_{q+1}(1)}, \dots, v_{\tau_{q+1}(q)}, v_{\tau_{q+1}(q+1)}).$$

(iii): If $\omega^1, \dots, \omega^p$ are linearly dependent we can assume that $\omega^1$ is a linear combinations of $\omega^2, \dots, \omega^p$ (otherwise we could renumber the $\omega^i$). There exist numbers $\alpha_i$ $(i = 2, \dots, p)$ with $\omega^1 = \sum_{i=2}^p \alpha_i \omega^i$. Hence the right hand side of

$$\omega^1 \wedge \dots \wedge \omega^p = \sum_{i=2}^p \omega^i \wedge \omega^2 \wedge \dots \wedge \omega^p$$

is the sum of products each of them containing some factor $\omega^j$ twice. Hence all summands vanish by (i).

If the forms $\omega^1, \dots, \omega^p$ are linearly independent then we can complete them to a basis $\{\omega^1, \dots, \omega^n\}$ of $V^*$. Denote the dual basis by $\{e_1, \dots, e_n\}$. If $\omega^1 \wedge \dots \wedge \omega^p$ would vanish then so would $\omega^1 \wedge \dots \wedge \omega^n$. However, we have

$$\omega^1 \wedge \dots \wedge \omega^n (e_1, \dots, e_n) = \sum_{\sigma_n \in S_n} \operatorname{sign}(\sigma_n) \omega^1(e_{\sigma_n(1)}) \cdots \omega^n(e_{\sigma_n(n)})$$
$$= \sum_{\sigma_n \in S_n} \operatorname{sign}(\sigma_n) \delta^1_{\sigma_n(1)} \cdot \dots \cdot \delta^n_{\sigma_n(n)} = 1$$

∎

**Lemma 2.3.10.** *If* $\{e_1, \dots, e_n\}$, $\{\theta^1, \dots, \theta^n\}$ *is a pair of dual bases then the set*

$$\{\theta^{i_1} \wedge \dots \wedge \theta^{i_p}\}_{1 \le i_1 < \dots < i_p \le n}$$

*forms a basis of the space* $\Lambda^p(V)$ *In particular,* $\dim(\Lambda^p(V)) = \binom{n}{p}$.

*Proof.* First we show that this set of tensors is linearly independent. Let $\eta_{i_1 \dots i_p}$ be numbers such that

$$\eta = \sum_{i_1 < \dots < i_p} \eta_{i_1 \dots i_p} \theta^{i_1} \wedge \dots \wedge \theta^{i_p} = 0$$

and let $k_1 < \dots < k_p$. Then $0 = \eta(e_{k_1}, \dots, e_{k_p}) = \eta_{k_1 \dots k_p}$, whence the $p$-forms $\{\theta^{i_1} \wedge \dots \wedge \theta^{i_p}\}_{i_1 < \dots < i_p}$ are all linearly independent.

Conversely, we know from Lemma 2.3.2 and $\operatorname{alt}(T^0_p(V)) = \Lambda^p(V)$ that the set of vectors $\{\theta^{i_1} \wedge \dots \wedge \theta^{i_p}\}_{i_1, \dots, i_p \in \{1, \dots, n\}}$ spans $\Lambda^p(V)$. Since

$$\theta^{\sigma_p(i_1)} \wedge \cdots \wedge \theta^{\sigma_p(i_p)} = \operatorname{sign}(\sigma_p)\theta^{i_1} \wedge \cdots \wedge \theta^{i_p}$$

for all permutations $\sigma_p \in S_p$ we can restrict to those $\theta^{i_1} \wedge \cdots \wedge \theta^{i_p}$ with $i_1 < \cdots < i_p$.

The dimension of $\Lambda^p(V)$ is $\binom{n}{p}$ since from a set of $n$ elements there are exactly $\binom{n}{t}$ choices of ordered $t$-tuples without duplication. ∎

**Definition 2.3.11.** *The map*

$$\begin{aligned} \lrcorner\colon V \times \Lambda^p(V) &\to \Lambda^{p-1}(V) \\ (v,\omega) &\mapsto v \lrcorner\omega\colon (w_1,\dots,w_{p-1}) \mapsto \omega(v,w_1,\dots,w_{p-1}) \end{aligned}$$

*is called the* interior product. *If $p=0$ we set $v \lrcorner\omega = 0$.*

**Lemma 2.3.11.** *Let $\omega \in \Lambda^p(V)$ and $\psi \in \Lambda^q(V)$. The interior product satisfies $v \lrcorner(\omega \wedge \eta) = (v \lrcorner\omega) \wedge \eta + (-1)^p\, \omega \wedge (v \lrcorner\eta)$ for all $p$-forms $\omega$ and all $q$-forms $\eta$.*

*Proof.* For notational purposes we set $v = w_0$. For any vectors $w_1,\dots,$ $w_{p+q-1} \in V$ we have

$$\begin{aligned} &(w_0 \lrcorner\omega) \wedge \eta(w_1,\dots,w_{p+q-1}) \\ &\quad= \frac{1}{(p-1)!q!} \sum_{\sigma_{p+q-1}\in S_{p+q-1}} \operatorname{sign}(\sigma_{p+q-1})\omega(w_0, w_{\sigma_{p+q-1}(1)},\dots \\ &\qquad\qquad \dots, w_{\sigma_{p+q-1}(p-1)}) \times \eta(w_{\sigma_{p+q-1}(p)},\dots,w_{\sigma_{p+q-1}(p+q-1)}) \end{aligned}$$

and

$$\begin{aligned} &(-1)^p\omega \wedge (w_0 \lrcorner\eta)(w_1,\dots,w_{p+q-1}) \\ &\quad= \frac{(-1)^p}{p!(q-1)!} \sum_{\tau_{p+q-1}\in S_{p+q-1}} \operatorname{sign}(\tau_{p+q-1})\omega(w_{\tau_{p+q-1}(1)},\dots,w_{\tau_{p+q-1}(p)}) \\ &\qquad\qquad \times \eta(w_0, w_{\tau_{p+q-1}(p+1)},\dots,w_{\tau_{p+q-1}(p+q-1)}). \end{aligned}$$

We consider now all possible permutationen of the ordered set $(w_0, w_1,$ $\dots, w_{p+q-1})$. We can divide them into two subgroups, the group $\overleftarrow{S^p_{p+q}}$ where $w_0$ is in one of the first $p$ positions and the subgroup $\overrightarrow{S^q_{p+q}}$ where it is in one of the last $q$ positions. Using this notation we can write

$$
\begin{aligned}
&\omega\wedge\eta(w_0, w_1, \ldots, w_{p+q-1}) \\
&= \frac{1}{p!q!}\Big( \sum_{\hat{\sigma}_{p+q}\in\overleftarrow{S^p_{p+q}}} \operatorname{sign}(\hat{\sigma}_{p+q})\omega(w_{\hat{\sigma}_{p+q}(0)}, w_{\hat{\sigma}_{p+q}(1)}, \ldots, w_{\hat{\sigma}_{p+q}(p-1)}) \\
&\qquad \times \eta(w_{\hat{\sigma}_{p+q}(p)}, \ldots, w_{\hat{\sigma}_{p+q}(p+q-1)}) \\
&\quad + \sum_{\hat{\tau}_{p+q}\in\overrightarrow{S^q_{p+q}}} \operatorname{sign}(\hat{\tau}_{p+q})\omega(w_{\hat{\tau}_{p+q}(0)}, \ldots, w_{\hat{\tau}_{p+q}(p-1)}) \\
&\qquad \times \eta(w_{\hat{\tau}_{p+q}(p)}, w_{\hat{\tau}_{p+q}(p+1)}, \ldots, w_{\tau_{p+q}(p+q-1)})\Big).
\end{aligned}
$$

Consider the first summand. For each $\hat{\sigma}_{p+q} \in \overleftarrow{S^p_{p+q}}$ we can shift $w_0$ to the first entry of $\omega$ by executing $(\hat{\sigma}_{p+q})^{-1}(0)$ transpositions. Hence there is a unique permutation $\sigma_{p+q-1} \in S_{p+q-1}$ with

$$
\begin{aligned}
&\omega(w_{\hat{\sigma}_{p+q}(0)}, \ldots, w_{\hat{\sigma}_{p+q}(p-1)})\eta(w_{\hat{\sigma}_{p+q}(p)}, \ldots, w_{\hat{\sigma}_{p+q}(p+q-1)}) \\
&\qquad = (-1)^{(\hat{\sigma}_{p+q})^{-1}(0)}\omega(w_0, w_{\sigma_{p+q}(1)}, \ldots, w_{\sigma_{p+q}(p-1)}) \\
&\qquad\qquad \times \eta(w_{\sigma_{p+q}(p)}, \ldots, w_{\sigma_{p+q}(p+q-1)}).
\end{aligned}
$$

Observe that $\hat{\sigma}_{p+q}$ is the composition of $(\hat{\sigma}_{p+q})^{-1}(0)$ transpositions and the permutation

$$
i \mapsto \begin{cases} \sigma_{p+q}(i) & \text{for } i > 0 \\ 0 & \text{for } i = 0. \end{cases}
$$

This implies $\operatorname{sign}(\hat{\sigma}_{p+q}) = (-1)^{(\hat{\sigma}_{p+q})^{-1}(0)}\operatorname{sign}(\sigma_{p+q})$. Since $w_0$ can be in each one of the $p$ possible entries of $\omega$ we obtain

$$
\begin{aligned}
&\frac{1}{p!q!} \sum_{\hat{\sigma}_{p+q}\in\overleftarrow{S^p_{p+q}}} \operatorname{sign}(\hat{\sigma}_{p+q})\omega(w_{\hat{\sigma}_{p+q}(0)}, w_{\hat{\sigma}_{p+q}(1)}, \ldots, w_{\hat{\sigma}_{p+q}(p-1)}) \\
&\qquad \times \eta(w_{\hat{\sigma}_{p+q}(p)}, \ldots, w_{\hat{\sigma}_{p+q}(p+q-1)}) \\
&= \frac{p}{p!q!} \sum_{\sigma_{p+q-1}\in S_{p+q-1}} \operatorname{sign}(\sigma_{p+q-1})\omega(w_0, w_{\sigma_{p+q-1}(1)}, \ldots, w_{\sigma_{p+q-1}(p-1)}) \\
&\qquad\qquad \times \eta(w_{\sigma_{p+q-1}(p)}, \ldots, w_{\sigma_{p+q-1}(p+q-1)}) \\
&\qquad\qquad\qquad = (w_0 \lrcorner \omega) \wedge \eta(w_1, \ldots, w_{p+q-1}).
\end{aligned}
$$

Consider now the second summand. An analogous argument for $\hat{\tau}_{p+q} \in \overleftarrow{S^q_{p+q}}$ allows us to move $w_0$ to the first entry of $\eta$. In this case we obtain an additional factor $(-1)^p$ since $\hat{\tau}_{p+q}$ is the composition of $(\hat{\tau}_{p+q})^{-1}(0)$ transpositions, a permutation

$$
i \mapsto \begin{cases} \tau_{p+q}(i) & \text{for } i > 0 \\ 0 & \text{for } i = 0, \end{cases}
$$

and the $p$ transpositions which move $w_0$ from the first entry of $\omega$ to the first entry of $\eta$. This implies $\operatorname{sign}(\hat{\tau}_{p+q}) = (-1)^p(-1)^{(\hat{\tau}_{p+q})^{-1}(0)}\tau(\sigma_{p+q})$ and we obtain

$$\begin{aligned}
\frac{1}{p!q!} & \sum_{\hat{\tau}_{p+q}\in \overrightarrow{S^q_{p+q}}} \operatorname{sign}(\hat{\tau}_{p+q})\omega(w_{\hat{\tau}_{p+q}(0)}, w_{\hat{\tau}_{p+q}(1)}, \ldots, w_{\hat{\tau}_{p+q}(p-1)}) \\
& \qquad \times \eta(w_{\hat{\tau}_{p+q}(p)}, \ldots, w_{\hat{\tau}_{p+q}(p+q-1)}) \\
&= (-1)^p \frac{q}{p!q!} \sum_{\tau_{p+q-1}\in S_{p+q-1}} \operatorname{sign}(\tau_{p+q-1})\omega(w_{\tau_{p+q-1}(1)}, \ldots, w_{\tau_{p+q-1}(p)}) \\
& \qquad \times \eta(w_0, w_{\tau_{p+q-1}(p+1)}, \ldots, w_{\tau_{p+q-1}(p+q-1)}) \\
& \qquad = (-1)^p \omega \wedge (w_0 \lrcorner \eta)(w_1, \ldots, w_{p+q-1}).
\end{aligned}$$

∎

Finally, we relate the theory of $p$-forms to the determinant of linear maps.

To motivate the definition below recall the following from linear algebra. Assume that we have an Euclidean scalar product $\langle \cdot, \cdot \rangle$ and an orthonormal basis $\{e_1, \ldots, e_n\}$. If $\{\theta^1, \ldots, \theta^n\}$ is the dual basis then one can use the $n$-form $\theta^1 \wedge \cdots \wedge \theta^n$ in order to measure the volume of parallel epipeds. For any vectors $b_1, \ldots, b_n \in V$ one defines the volume of the parallel epiped spanned by these vectors to be $\theta^1 \wedge \cdots \wedge \theta^n(b_1, \ldots, b_n)$. This number depends on the chosen scalar product but not on the orthonormal basis. The determinant of a linear map $B\colon V \to V$ is often defined as $\det(B) := \theta^1 \wedge \cdots \wedge \theta^n(b_1, \ldots, b_n)$ where $b_i = Be_i$.

This definition of a determinant obscures the fact that the determinant is independent of the choice of scalar product. The following equivalent definition is probably the most natural way to introduce the concept of a determinant.

**Definition 2.3.12.** *Let $V$ be an $n$-dimensional vector space, $W$ be a $k$-dimensional vector space over $\mathbb{K}$, and $A\colon V \to W$ a linear map.*

- *(i) The* pull-back *of $\psi$ under $A$ is the map $A^*\colon T^0_p(W) \mapsto T^0_p(V)$ defined by $A^*\psi(v_1, \ldots, v_p) = \psi(Av_1, \ldots, v_p)$.*
- *(ii) Let $\mu \in \Lambda^n(V) \setminus \{0\}$ and assume that $V = W$. Then the* determinant $\det(A)$ *of $A$ is the number defined by $A^*\mu = \det(A)\mu$.*

We have to show that the map det is well defined. First observe that $A^*$ maps $\Lambda^p(V)$ into $\Lambda^p(V)$ and recall that $\Lambda^n$ is 1-dimensional by Lemma 2.3.10. Hence $A^*\mu$ must be a multiple of $\mu$. If $\tilde{\mu}$ is any other non-vanishing $n$-form then there exists a number $\alpha \neq 0$ with $\tilde{\mu} = \alpha\mu$. Hence we have

$$\begin{aligned}
A^*\tilde{\mu}(v_1, \ldots, v_n) &= A^*(\alpha\mu)(v_1, \ldots, v_n) = \alpha\mu(Av_1, \ldots, Av_n) \\
&= \alpha \det(A)\mu(v_1, \ldots, v_n) = \det(A)\tilde{\mu}(v_1, \ldots, v_n)
\end{aligned}$$

which implies that our definition for $\det(A)$ does not depend on the chosen $n$-form.

**Lemma 2.3.12.** *Let $V$ be an $n$-dimensional vector space over $\mathbb{K}$,*

$$A : V \to V$$

*be a linear map and $\{e_1, \ldots, e_n\}$, $\{\theta^1, \ldots, \theta^n\}$ be a pair of dual bases. Let the components $A^i_j$ be defined by $Ae_i = \sum_{j=1}^n A^j_i e_j$. Then we have*

$$A^*(\theta^{i_1}\wedge\cdots\wedge\theta^{i_p}) = \sum_{j_1<\cdots<j_p} \Big( \sum_{\sigma_p\in S_p} \operatorname{sign}(\sigma_p) A^{i_1}_{\sigma_p(j_1)} \cdots A^{i_p}_{\sigma_p(j_p)} \Big) \theta^{j_1}\wedge\cdots\wedge\theta^{j_p} .$$

*In particular we have*

$$\det(A) = \sum_{\sigma_n\in S_n} \operatorname{sign}(\sigma_n) A^1_{\sigma_n(1)} \cdot \cdots \cdot A^n_{\sigma_n(n)} .$$

*Proof.* We calculate

$$\begin{aligned} A^*\theta^{i_1} \wedge \ldots \wedge \theta^{i_p}(e_{j_1}, \ldots, e_{j_p}) &= \sum_{\sigma_p\in S_p} \operatorname{sign}(\sigma_p)\theta^{i_1}(Ae_{\sigma_p(j_1)}) \cdot \cdots \cdot \theta^{i_p}(Ae_{\sigma_p(j_p)}) \\ &= \sum_{\sigma_p\in S_p} \operatorname{sign}(\sigma_p) A^{i_1}_{\sigma_p(j_1)} \cdots A^{n}_{\sigma_p(j_p)} . \end{aligned}$$

∎

*Remark 2.3.3.* We have noted in Remark 2.3.2 above that our combinatorial factors are not the only possible choice. If we had chosen $\tilde{\wedge}$ (cf. Remark 2.3.2) instead of $\wedge$, we would have had to re-define $v \tilde{\lrcorner} \omega = p\, v \lrcorner \omega$ in order to preserve Lemma 2.3.11. Another advantage of our choice is that the simple formula

$$\theta^1 \wedge \cdots \wedge \theta^n(e_1, \ldots, e_n) = 1$$

holds for any pair of dual bases $\{e_i\}_{i\in\{1,\ldots,n\}}$, $\{\theta^i\}_{i\in\{1,\ldots,n\}}$ .

### 2.3.2 Tensor fields

p. 71 ↓ [↓ p. 85]

In this section we introduce tensor fields which are tensor valued maps. All physical and geometrical structures we will encounter can be defined in terms of tensor fields or maps of tensor fields.

**Definition 2.3.13.** *The vector space of all smooth maps*

$$\phi\colon M \mapsto \bigcup_{x\in M} T^r_s M \;\; with \;\; \phi(x) \in T^r_s(T_x M) \quad \forall x \in M$$

*is denoted by $\mathcal{T}^r_s(M)$. These maps are called* $\binom{r}{s}$-tensor fields. *Often we will write $\phi_x$ instead of $\phi(x)$.*

*A $\binom{1}{0}$-tensor field is a* vector field.

*A $\binom{0}{p}$-tensor field $\omega$ with*

$$\omega(v_1, \ldots, v_i, \ldots, v_j, \ldots v_p) = -\omega(v_1, \ldots, v_j, \ldots, v_i, \ldots v_p)$$

*for any set of vectors $\{v_1, \ldots, v_p\}$ and any pair $(i, j) \subset \{1, \ldots, p\}$ is called a* differential form of degree $p$. *A differential form of degree $p$ is often simply called a* $p$-form.

*Let $\Sigma$ be another manifold and $f\colon \Sigma \to M$ be a smooth map. An* $\binom{r}{s}$-tensor field along $f$ *is a map $\psi\colon \Sigma \to \mathcal{T}^r_s(M)$ such that $\psi(x) \in T^r_s(T_{f(x)}N)$ for all $x \in \Sigma$.*

*Vector fields and differential forms along $f$ are defined analogously.*

*If $\Sigma$ is a submanifold of $M$ and $f\colon \Sigma \to M$ its canonical inclusion then we also say that $\psi$ is a* tensor field along $\Sigma$.

Let $(x^1, \ldots, x^n)$ be a coordinate system and denote by $\partial_1, \ldots, \partial_n$ its Gaußian basis. Its poinwise dual basis is given by the derivatives $(\mathrm{d}x^1, \ldots, \mathrm{d}x^n)$ (cf. Definition 2.2.7). With respect to these coordinates, any $\binom{r}{s}$-tensor field $\phi$ can be written as

$$\phi(x) = \sum_{\substack{1\leq i_1 \ldots i_r \leq n \\ 1 \leq j_1 \ldots j_s \leq n}} \phi^{i_1 \ldots i_r}_{j_1 \ldots j_s}(x)\, \partial_{i_1} \otimes \cdots \otimes \partial_{i_r} \otimes \mathrm{d}x^{j_1} \otimes \cdots \otimes \mathrm{d}x^{j_s},$$

where the $\phi^{i_1 \ldots i_r}_{j_1 \ldots j_s}$ are functions. They are the *components* (or *component functions*) of the tensor $\phi$ with respect to the coordinates $(x^1, \ldots, x^n)$. If $\phi$ is a covariant symmetric $\binom{0}{s}$ tensor field we will often drop the tensor product sign $\otimes$ in order to indicate that the symmetrisation operator sym has been applied,

$$\mathrm{sym}(\phi)(x) = \sum 1 \leq j_1 \ldots j_s \leq n \phi_{j_1 \ldots j_s}(x)\, \mathrm{d}x^{j_1} \ldots \mathrm{d}x^{j_s}.$$

If $N$ is a second manifold and $f\colon M \to N$ a smooth map then each tensor field $\phi \in \mathcal{T}^0_p(N)$ induces a unique tensor field $f^*\phi \in \mathcal{T}^0_p(M)$, its *pull-back*. This tensor field is defined by

$$f^*\phi_x(v_1, \ldots, v_p) := \phi_{f(x)}(Tf(v_1), \ldots, Tf(v_p)),$$

where $Tf\colon TM \to TN$ denotes the derivative of $f$ (Definition 2.2.8). Analogously, each tensor field $\hat{\phi} \in \mathcal{T}^p_0(M)$ induces a unique tensor field

$f_*\hat{\phi} \in \mathcal{T}^p_0(N)$, its *push-forward.* It is defined by $f_*\hat{\phi}_x(\omega^1, \ldots, \omega^p) := \phi_{f(x)}(\omega^1 \circ Tf, \ldots, \omega^p \circ Tf)$. If $\psi \in \mathcal{T}^r_s(N)$ and $g \colon M \to N$ is a local diffeomorphism we define the pull-back of $\psi$ by

$$g^*\psi(v_1, \ldots, v_s, \omega^1, \ldots, \omega^r)$$
$$:= \psi\left((g_*(v_1), \ldots, g_*(v_s), \left(g^{-1}\right)^*(\omega^1), \ldots, \left(g^{-1}\right)^*(\omega^r)\right).$$

This definition agrees with the definition of pull-back above if $\psi$ is a covariant tensor field. The push forward of a tensor field $\hat{\psi} \in \mathcal{T}^r_s(M)$ by a local diffeomoerphism $g$ is defined by $g_*\hat{\psi} = \left(g^{-1}\right)^* \hat{\psi}$. If $s = 0$ then both definitions of push-forward agree.

**Lemma 2.3.13.** *For every local diffeomorphism $g \colon M \to N$ and all tensor fields $\phi, \psi \in \mathcal{T}^r_s(M)$ we have*

$$g^*(\alpha\phi + \beta\psi) = \alpha g^*\phi + \beta g^*\psi,$$
$$g^*(\phi \otimes \psi) = g^*\phi \otimes g^*\psi,$$
$$g^* C^{\hat{r}}_{\hat{s}} \psi = C^{\hat{r}}_{\hat{s}} g^* \psi$$

*If $g$ fails to be a diffeomorphism, and $\phi, \psi$ are covariant vector fields then the first two assertions are still true. Analogous statemens also hold for the push-forward.*

*Proof.* The lemma follows directly from the definitions. ∎

*Remark 2.3.4.* It is often practical to denote a tensor field $\phi \in \mathcal{T}^r_s(M)$ by its components $\phi^{a_1 \ldots a_r}_{b_1 \ldots b_r}$ with respect to an *unspecified* Gaußian basis. The indices $a_1, \ldots a_r, b_1 \ldots b_r$ are "dummy indices" and their only purpose is to give a convenient method for explicitly denoting the entries of the tensor field. By convention, the tensor is characterized by the core symbol $\phi$, so in general we have $\phi^{a_1 \ldots a_r}_{b_1 \ldots b_r} = \phi^{c_1 \ldots c_r}_{d_1 \ldots d_r} \neq \tilde{\phi}^{c_1 \ldots c_r}_{d_1 \ldots d_r}$.[6] However, different tensor fields of different order can be denoted by the same core symbol since it is clear from the number of co- and contravariant indices that they must be different objects.

If $\alpha, \beta \in \mathbb{K}$, $\phi, \tilde{\phi}$ are two $\binom{r}{s}$-tensor field, and $\psi$ is an $\binom{\tilde{r}}{\tilde{s}}$-tensor field we write

$$\left(\alpha\phi + \beta\tilde{\phi}\right)^{a_1 \ldots a_r}_{b_1 \ldots b_s} = \alpha\phi^{a_1 \ldots a_r}_{b_1 \ldots b_s} + \beta\tilde{\phi}^{a_1 \ldots a_r}_{b_1 \ldots b_s},$$
$$(\phi \otimes \psi)^{a_1 \ldots a_{r+\tilde{r}}}_{b_1 \ldots b_{s+\tilde{s}}} = \phi^{a_1 \ldots a_r}_{b_1 \ldots b_s} \psi^{a_1 \ldots a_{\tilde{r}}}_{b_1 \ldots b_{\tilde{s}}},$$
$$\left(C^{\hat{r}}_{\hat{s}}\phi\right)^{a_1 \ldots a_{r-1}}_{b_1 \ldots b_{s-1}} = \phi^{a_1 \ldots a_{\hat{r}-1}\, c\, a_{\hat{r}+1} \ldots a_{r-1}}_{b_1 \ldots b_{\hat{s}-1}\, c\, b_{\hat{s}+1} \ldots b_{s-1}}.$$

[6] In some East European countries, some authors use conventions which are the exact opposite of ours, i.e., they have in general $\phi^{a_1 \ldots a_r}_{b_1 \ldots b_r} \neq \phi^{c_1 \ldots c_r}_{d_1 \ldots d_r} = \tilde{\phi}^{c_1 \ldots c_r}_{d_1 \ldots d_r}$. It seems likely that this variant will fade in the near future.

Since the indices determine the entries explicitly, we can write

$$\phi^{a_1\dots a_r}_{b_1\dots b_s}\psi^{a_1\dots a_{\tilde r}}_{b_1\dots b_{\tilde s}} = \psi^{a_1\dots a_{\tilde r}}_{b_1\dots b_{\tilde s}}\phi^{a_1\dots a_r}_{b_1\dots b_s}.$$

The notation for contraction is often called the *Einstein summation convention:* If in a product of tensors one index letter appears as an upper index and somewhere else as a lower index then it is understood that one has to sum over these indices in the corresponding coordinate expression. We will also use this notation in order to write a tensor with respect to a basis, for instance,

$$\sum_{a,b=1}^{n} A^a_b \partial_a \otimes \mathrm{d}x^b = A^a_b \partial_a \otimes \mathrm{d}x^b.$$

Symmetrisation and anti-symmetrisation are denoted by

$$\mathrm{sym}(\psi)_{i_1,\dots,i_s} = \psi_{(i_1,\dots,i_s)} \text{ and } \mathrm{alt}(\psi)_{i_1,\dots,i_s} = \psi_{[i_1,\dots,i_s]}.$$

An analogous notation is used for contravariant tensors and for the case that symmetrisation or anti-symmetrisation is only applied to a subset of entries. We use the delimiters | to indicate that a certain subset of entries is not symmetrised (respectively, anti-symmetrised). For instance,

$$\psi^{ab[cd|(ef)g|h]}_{i(j|k|l|(mn)|o)}$$

indicates that the entries corresponding to the indices $\{c, d, h\}$ are anti-symmetrised and that the entries corresponding to the indices $\{e, f\}$, $\{j, l, o\}$, $\{m, n\}$ are each symmetrised.

This *abstract index notation* should not be confused with the components of the tensor field with respect to a given basis. While all formulas appear to be identical, in the former case, the tensor is referred to, whereas in the latter case one simply has a collection of $\mathbb{K}$-valued functions.

The main advantages of the abstract index notation over the usual notation without indices are that tensor operations are easy to remember since tensors look like their components and that even complicated multiple contractions and tensor products can be understood at a glance. The main disadvantage is that many formulas look unnecessarily clumsy because of the jungle of indices involved. In the physical literature, the abstract index notation is usually preferred and in the mathematical literature the notation without indices is prevalent. *In this book, we will use both notations.* However, unlike in most of the physics literature here we will not reserve certain indices for special ranges. Moreover, we will sometimes write the summation sign in order to indicate the range.

## 2.4 Vector fields and ordinary differential equations

If $\gamma\colon \mathbb{K} \to M$ is a curve then for each $t \in \mathbb{K}$ the vector $\dot\gamma(t) := T_t\gamma(1)$ is a vector in $T_{\gamma(t)}M$. The following question arises naturally: given a vector field $V$ on $M$, does there exist a curve $\gamma$ with $\dot\gamma(t) = V_{\gamma(t)}$? With respect to a chart, the answer is given by the fundamental theorem for differential equations which establishes the existence and uniqueness of solutions. Because of its importance we will state it below. However this theorem should really be treated in standard courses on analysis (for mathematicians) or mathematics for physicists. See also (Dieudonné 1960, chapter 10) for a proof in a slightly more general context.

**Theorem 2.4.1 (Fundamental theorem for differential equations).** *Let $\mathcal{U} \subset \mathbb{K}^n$, $\mathcal{V} \subset \mathbb{K}^m$, $\mathcal{J} \subset \mathbb{K}$ be open subsets, and*

$$f\colon \mathcal{J} \times \tilde{\mathcal{U}} \times \mathcal{V} \to \mathbb{K}^n, \quad (t,x,y) \mapsto f(t,x,y)$$

*be $C^r$. For each $(t_0,x_0,y_0) \in \mathcal{J} \times \mathcal{U} \times \mathcal{V}$ there exists a neighbourhood $\tilde{\mathcal{J}} \times \tilde{\mathcal{V}} \subset \mathcal{J} \times \mathcal{V}$ of $(t_0,y_0)$ and a unique map $\gamma\colon (t,y) \mapsto \gamma(t,x) \in \mathbb{K}^n$ with*

$$\frac{\mathrm{d}\gamma(t,y)}{\mathrm{d}t} = f(t,\gamma(t,y),y)$$

*for all $(t,y) \in \tilde{\mathcal{J}} \times \tilde{\mathcal{V}}$ and $\gamma(t_0,y_0) = x_0$. Further, $\gamma$ is $C^r$.*

This theorem translates straightforwardly to manifolds.

**Theorem 2.4.2.** *Let $M$ be a smooth manifold and $V$ be a vector field on $M$. For every $x \in M$ there exists a subset $\mathcal{J} \subset \mathbb{K}$ and a curve $\gamma_x : \mathcal{J} \to M$ with $\gamma_x(0) = x$ and $\dot\gamma(t) = V_{\gamma(t)}$ for all $t \in \mathcal{J}$.*

*If $V_x \neq 0$ there exists a neighbourhood $\mathcal{U}$ of $\{x\} \times \{0\} \subset M \times \mathbb{R}$ such that the map $F\colon \mathcal{U} \to M$, $(t,y) \mapsto F_t(y) = \gamma_y(t)$ is well defined. Moreover, the map $y \mapsto F_{\hat t}(y)$ is a local diffeomorphism for each $\hat t$, and its inverse is given by $F_{-\hat t}$.*

*For $t, s$ small enough we have $F_t \circ F_s = F_{t+s}$.*

The curve $\gamma$ is called an *integral curve* and the map $F_t$ the *flow* of $V$.

*Proof.* Let $(\mathcal{U},\varphi)$ be a chart centered at $x$. Then the fundamental theorem for ordinary differential equations (Theorem 2.4.1) implies that there exists a solution $\beta(t)$ of the differential equation

$$\frac{\mathrm{d}}{\mathrm{d}t}\beta(t) = (\varphi_* V)_{\beta(t)}$$

with $\beta(0) = \varphi(x)$. Hence $\gamma_x = \varphi^{-1} \circ \beta$ is an integral curve of $V$ with $\gamma_x(0) = x$.

That $F$ is smooth and well defined follows from the fact that integral curves depend smoothly on parameters (cf. Theorem 2.4.1). The map

$(F_{\hat{t}})^{-1}$ is given by $(F_{\hat{t}})^{-1}(z) = \gamma_-(\hat{t})$ where $\gamma_-$ is the locally unique integral curve of $-V$ with $\gamma_-(0) = z$. Hence $F_{\hat{t}}$ is differentiable and has a differentiable inverse.

The equation $F_{t+s} = F_t \circ F_s$ follows from local uniqueness of solutions of differential equations and the fact that both $t \mapsto F_{t+s}(y)$ and $F_t \circ F_s(y)$ are integral curves of $V$ with the same initial value. ∎

An integral curve $\gamma\colon \mathcal{J} \to M$ of a vector field $V$ is called *maximal* if the existence of an integral curve $\tilde{\gamma}\colon \tilde{\mathcal{J}} \to M$ with $\mathcal{J} \subset \tilde{\mathcal{J}} \subset \mathbb{K}$ implies $\tilde{\mathcal{J}} = \mathcal{J}$. By the lemma of Zorn and local existence of integral curves each integral curve is contained in some maximal integral curve. The following Proposition shows that in the case $\mathbb{K} = \mathbb{R}$ maximal integral curves are unique.

**Proposition 2.4.1.** *If $\mathbb{K} = \mathbb{R}$ then there is a unique maximal subset $\hat{\mathcal{J}} \subset \mathbb{K}$ and a unique solution $\gamma$ of $\dot{\gamma}(t) = V_{\gamma(t)}$ defined on $\hat{\mathcal{J}}$.*

*Proof.* Let $\gamma, \tilde{\gamma}$ be integral curves of $V$ with $\gamma(0) = \tilde{\gamma}(0) = x$. We must show that these integral curves coincide on the intersection $\mathcal{J} \cap \tilde{\mathcal{J}}$ of their domains $\mathcal{J}$, $\tilde{\mathcal{J}}$. In order to do so we will prove that the set $\mathcal{K} = \{t \in \mathcal{J} \cap \tilde{\mathcal{J}} \,|\, \gamma(t) = \tilde{\gamma}(t)\}$ is both open and closed. Since in the case $\mathbb{K} = \mathbb{R}$ the set $\mathcal{J} \cap \tilde{\mathcal{J}}$ is the intersection of two open intervals and therefore connected, this set must then coincide with $\mathcal{K}$. It is clear that $\mathcal{K}$ is closed. Let $t \in \mathbb{K}$ and $(\mathcal{U}, \varphi)$ be a chart centered at $\hat{x} = \gamma(t) = \tilde{\gamma}(t)$. Given local coordinates, the problem of finding an integral curve of $V$ reduces to solving a system of ordinary differential equations. An application of the fundamental theorem for ordinary differential equation (Theorem 2.4.1) proves that there is a unique local integral curve $\beta$ through $\varphi(\hat{x})$. Hence there is a neighbourhood of $t$ such that $\gamma = \tilde{\gamma}$ on this neighbourhood. ∎

The following theorem implies that, locally, all non-vanishing vector fields are alike.

**Theorem 2.4.3.** *Let $M$ be a smooth manifold, $x \in M$, and $V$ be a vector field with $V(x) \neq 0$. Then there is a chart $(\mathcal{U}, \varphi)$ centered at $x$ such that $\varphi_* = \partial_{x^1}$ and the integral curves of $V$ are given by $t \mapsto \varphi^{-1}(t, x^2, \ldots, x^n)$.*

*Proof.* Let $N$ be an $(n-1)$-dimensional submanifold of $M$ through $x$ which is transverse to $V$ (i.e., $\mathbb{R}V(y) \oplus T_yN = T_yM$ for all $y \in N$). Let $(\mathcal{V}, \psi)$ be a chart of $N$ centered at $x$ and denote the flow of $V$ by $F_t$. There is an $\epsilon > 0$ such that the map

$$f\colon (-\epsilon, \epsilon) \times \psi(\mathcal{V}) \to M, \tag{2.4.1}$$

$$(x^1, x^2, \ldots, x^n) \mapsto F_{x^1}(\psi^{-1}(x^2, \ldots, x^n)) \tag{2.4.2}$$

is well defined. The differential of $f$ at 0 is an isomorphism since

$$F_0 = \text{id and} \left(\frac{\mathrm{d}f(x^1,\ldots,x^n)}{\mathrm{d}x^1}\right)_{(0,\ldots,0} = V(x) \neq 0.$$

Hence there exists a neighbourhood $\mathcal{W}$ of 0 where this map is an diffeomorphism. The pair $(\mathcal{U},\varphi) = (f(\mathcal{W}),(f^{-1})_{|\mathcal{U}})$ is therefore a chart centered at $x$. For any $y \in \mathcal{U}$ denote by $(y_0, \mathrm{pr}_N(y)) \in \mathbb{R} \times N \subset \mathbb{R} \times M$ the unique pair defined by $F_{y_0}(\mathrm{pr}_N(y)) = y$. Then we have $\varphi \circ F_t(y) = f^{-1} \circ F_t(y) = f^{-1} \circ F_{t+y_0}(\mathrm{pr}_N(y)) = (t+y_0, \psi(\mathrm{pr}_N(y)))$. Hence the integral curves of $V$ are indeed the curves $t \mapsto \varphi^{-1}(t, x^2, \ldots, x^n)$. It follows immediately that $\varphi_* V = \partial_{x^1}$. ∎

[p. 79 ↓]
↓ p. 118

Given a vector field $V$, one can define the derivative of a tensorfiels $\psi$ in direction $V$.

**Definition 2.4.1.** *Let $x \in M$, $\psi$ be a tensor field, $U$ be a vector field, and $F_t$ the flow of $U$. Then*

$$\pounds_U \psi(x) := \left(\left(\frac{\mathrm{d}}{\mathrm{d}t}\right)_{|t=0} {F_t}^* \psi\right)(x)$$

*is the* Lie derivative *of $\psi$.*

Here $\left(\frac{\mathrm{d}}{\mathrm{d}t}\right)_{|t=0}$ is the usual derivative in vector spaces. In fact, the expression ${F_t}^*\psi$ denotes a tensor field which is defined on a neighbourhood of $x$ if $t$ is fixed and small enough. In particular, this tensor field can be evaluated at $x$. As a function of $t$ this gives a curve in the vector space $T_s^r(T_x M)$.

The Lie derivative measures the change of $\psi$ along $V$.

**Lemma 2.4.1.** *Let $V$ be a vector field. Then the Lie derivative in direction $V$ is a derivation, i.e., for any tensor fields $\varphi, \psi$ the formulas*

$$\pounds_V(\varphi \otimes \psi) = \pounds_V \varphi \otimes \psi + \varphi \otimes \pounds_V \psi,$$
$$\pounds_V(\varphi + \psi) = \pounds_V \varphi + \pounds_V \psi.$$

*hold.*

*Proof.* These formulas follow immediately from the properties of derivatives. ∎

**Theorem 2.4.4.** *Let $U, V$ be vector fields and $f$ a smooth function. Then*

$$\pounds_U f = U \bullet f \text{ and}$$
$$\pounds_U V \bullet f = U \bullet V \bullet f - V \bullet U \bullet f.$$

*Proof.* The first equation follows immediately from

$$\pounds_U f = \frac{\mathrm{d}}{\mathrm{d}t}_{|t=0} F_t{}^* f(x) = \frac{\mathrm{d}}{\mathrm{d}t}_{|t=0} f \circ F_t(x) = \mathrm{d}f\left(\frac{\mathrm{d}}{\mathrm{d}t}_{|t=0} F_t(x)\right) = \mathrm{d}f(U_x).$$

Let $x \in M$ and $(\mathcal{U}, \varphi)$ be a chart centered at $x$. There is a neighbourhood $\mathcal{V} \subset \mathcal{U}$ of $x$ and a number $\epsilon > 0$ such that $F_t(y)$ is well defined and satisfies $F_t(y) \in \mathcal{U}$ for all for all $(t, y) \in (-\epsilon, \epsilon) \times \mathcal{V}$. An application of the Taylor formula to the map $t \mapsto f \circ F_{-t} \circ \varphi^{-1}$ implies the existence of a smooth map $\tilde{g}\colon (-\epsilon, \epsilon) \times \varphi(\mathcal{V}) \to \mathbb{R}$ with $f \circ F_{-t} \circ \varphi^{-1}(z) = f(z) + t\tilde{g}(t, z)$ for all $(t, z) \in (-\epsilon, \epsilon) \times \mathcal{V}$. The map $g_t(y) := \tilde{g}(t, \varphi(y))$ satisfies $f \circ (F_t)^{-1}(y) = f(y) + tg_t(y)$ for all $y \in \mathcal{V}$ and we obtain

$$\begin{aligned}
&\left(\frac{\mathrm{d}}{\mathrm{d}t}_{|t=0} (F_t{}^* V)_x\right) \bullet f \\
&\quad = \frac{\mathrm{d}}{\mathrm{d}t}_{|t=0} ((F_t{}^* V)_x \bullet f) = \frac{\mathrm{d}}{\mathrm{d}t}_{|t=0} (((F_{-t})_* V)_x \bullet f) \\
&\quad = \frac{\mathrm{d}}{\mathrm{d}t}_{|t=0} (((F_{-t})_* V_{F_t(x)}) \bullet f) = \frac{\mathrm{d}}{\mathrm{d}t}_{|t=0} (V_{F_t(x)} \bullet (f \circ F_{-t})) \\
&\quad = \frac{\mathrm{d}}{\mathrm{d}t}_{|t=0} (V_{F_t(x)} \bullet (f + tg_t)) = \frac{\mathrm{d}}{\mathrm{d}t}_{|t=0} (V \bullet f)_{F_t(x)} + V_x \bullet g_0 \\
&\quad = (U \bullet V \bullet f)_x - (V \bullet U \bullet f)_x,
\end{aligned}$$

where in the last step we have used

$$\begin{aligned}
(V \bullet g_0)_x &= V \bullet \left(\frac{\mathrm{d}}{\mathrm{d}t}_{|t=0} f \circ (F_t)^{-1}(\cdot)\right)_x = V_x \bullet \mathrm{d}f\left(\frac{\mathrm{d}}{\mathrm{d}t}_{|t=0} (F_t)^{-1}(\cdot)\right) \\
&= V_x \bullet \mathrm{d}f\left(\frac{\mathrm{d}}{\mathrm{d}t}_{|t=0} (F_{-t})(\cdot)\right) = -(V \bullet U \bullet f)_x.
\end{aligned}$$

∎

Theorem 2.4.4 shows that the Lie derivative of $V$ in direction $U$ is the commutator of $U$ and $V$. This motivates the following definition.

**Definition 2.4.2.** *If $U, V$ are vector fields then we call $[U, V] = \pounds_U V$ the* Lie bracket *or the* commutator *of $U$ and $V$. Vector fields* commute *if their Lie bracket vanishes.*

Commuting vector fields are of particualar interest since Gaußian vector fields $\partial_{x^k}$, $\partial_{x^l}$ are necessarily commuting. The following lemma gives the converse to this observation.

**Lemma 2.4.2 (Geometric interpretation of the Lie bracket).** *Two vector fields $U, V$ have vanishing Lie derivative, $\pounds_U V = 0$, if and only if their flows commute.*

*Proof.* Denote the flows of $U$ and $V$ by $F_t$ and $G_s$. The equation $F_t \circ G_s = G_s \circ F_t$ implies that $F_{-t} \circ G_s \circ F_t = G_s$ is the flow of $V$. Hence we have $V = \frac{\mathrm{d}}{\mathrm{d}s}(F_{-t} \circ G_s \circ F_t) = T(F_t)^{-1}\left(\frac{\mathrm{d}}{\mathrm{d}s} G_s\right) \circ F_t = F_t{}^* V$ and therefore $\mathcal{L}_U V = \left(\frac{\mathrm{d}}{\mathrm{d}t}\right)_{t=0} F_t{}^* V = \left(\frac{\mathrm{d}}{\mathrm{d}t}\right)_{t=0} V = 0$.

Conversely, assume that $\mathcal{L}_U V = 0$ which is quivalent to $\frac{\mathrm{d}}{\mathrm{d}t} F_t{}^* V = 0$. Since $F_0{}^* V = V$ an integration yields $(F_t)^* V = V$ for all $t$. This implies that $s \mapsto F_{-t} \circ G_s \circ F_t$ is an integral curve of $V$. From the uniqueness of integral curves we get $F_{-t} \circ G_s \circ F_t = G_s$ for all $t, s$. ∎

**Corollary 2.4.1.** *Let $M$ be a $n$-dimensional manifold and $\{U_1, \dots, U_n\}$ be a collection pointwise linearly independent, pairwise commuting vector fields defined on an open neighbourhood of $x \in M$. Then there exists a coordinate chart $(\mathcal{V}, \varphi)$ centred at $x$ whose Gaußian basis vector fields satisfy $\partial_{x^i} = U_i$.*

*Proof.* Denote the flow of $U_k$ by $F_t^k$ and let

$$\psi(x^1, \dots, x^n) = F^1_{x^1} \circ \dots \circ F^n_{x^n}(x)$$

for sufficiently small $(x^1, \dots, x^n) \in \mathbb{K}^n$. Since the vector fields $U_i$ are pairwise commuting so are their flows $F^i_{x^i}$ (cf. Lemma 2.4.2). Hence we have for every $i \in \{1, \dots, n\}$

$$\psi(x^1, \dots, x^n) = F^i_{x^i} \circ F^1_{x^1} \circ \dots \circ F^{i-1}_{x^{i-1}} \circ F^{i+1}_{x^{i-1}} \circ \dots \circ F^n_{x^n}(x).$$

This implies $\psi_*(E_i) = \frac{\mathrm{d}}{\mathrm{d}x^i}(x^i \mapsto \psi(x^1, \dots, x^n)) = U_i$ for the standard basis $\{E_1, \dots, E_n\}$ of $\mathbb{K}^n$. Since the vector fields $\{U_1, \dots, U_n\}$ are linear independent the map $\psi$ has maximal rank and is therefore a local diffeomorphism. Let $\mathcal{W} \subset \mathbb{K}^n$ be an open neighbourhood of 0 such that $\psi(z)$ is well defined for all $z \in \mathcal{W}$ and one-to-one on $\mathcal{W}$. We can now define $(\mathcal{U}, \varphi) = (\psi(\mathcal{W}), \psi^{-1})$. ∎

**Corollary 2.4.2.** *Let $M$ be a 2-dimensional manifold and $U$, $V$ be vector fields which are at each point linearly independent. Then $M$ admits local coordinates $(x_1, x_2)$ such that $\partial_{x^1} \parallel U$ and $\partial_{x^2} \parallel V$.*

*Proof.* By Corollary 2.4.1 we only have to show that there exist functions $f, h$ with $[fU, hV] = 0$. We calculate

$$\begin{aligned} 0 = [fU, hV] &= \nabla_{fU}(hV) - \nabla_{hV} fU \\ &= fh\nabla_U V + f\mathrm{d}h(U) - fh\nabla_V U - h\mathrm{d}f(V)U \\ &= fh[U, V] - h\mathrm{d}f(V)U + f\mathrm{d}h(U)V \\ &= fh\left([U, V] + \mathrm{d}\ln(h)(U)V - \mathrm{d}\ln(f)(U)V\right). \end{aligned}$$

Let $\omega^U$, $\omega^V$ be the 1-forms which are dual to $U, V$. Then any solution $(f, h)$ of the uncoupled system of linear ordinary differential equations

$$0 = \mathrm{d}\ln(h)(U) - \omega^V([U,V]), \qquad 0 = \mathrm{d}\ln(f)(V) + \omega^U([U,V])$$

satisfies $[fU, hV] = 0$. This system of differential equations can be solved by the fundamental theorem of ordinary differential Equations 2.4.1. ∎

In the following sections we will encounter various kinds of tensor derivatives. It is therefore practical to formalize their common properties.

**Definition 2.4.3.** *Let $D$ be a map which maps tensor fields into tensor fields. If it satisfies*

*(i) $D(\mathcal{T}^r_s(M)) \subset \mathcal{T}^r_s(M)$,*
*(ii) $D(\varphi \otimes \psi) = D(\psi) \otimes \varphi + \psi \otimes D(\varphi)$ ("product rule"),*
*(iii) $D$ commutes with contractions,*

*then $D$ is called a* derivation.

**Corollary 2.4.3.** *Let $M$ be a manifold and $V$ be a vector field on $M$. The Lie-derivative $\pounds_V$ is a derivation.*

*Proof.* We have to verify only the third property. This follows from ${F_t}^* C^{\hat{r}}_{\hat{s}} \phi = C^{\hat{r}}_{\hat{s}} {F_t}^* \phi$ and the fact that ${F_t}^*\phi \mapsto C^{\hat{r}}_{\hat{s}} {F_t}^*\phi$ is linear (so the derivative $\frac{\mathrm{d}}{\mathrm{d}t}$ can be interchanged with this operation). ∎

**Proposition 2.4.2.** *Two derivations coincide if they coincide on vector fields and functions.*

*Proof.* Writing an arbitrary tensor field $\psi$ in a coordinate representation we obtain

$$\begin{aligned} D\psi = {} & D(\psi^{i_1\ldots i_r}_{j_1\ldots j_s})\partial_{i_1} \otimes \cdots \otimes \partial_{i_r} \otimes \mathrm{d}x^{j_1} \otimes \cdots \otimes \mathrm{d}x^{j_s} \\ & + \sum_{t=1}^{r} \psi^{i_1\ldots i_r}_{j_1\ldots j_s} \partial_{i_1} \otimes \cdots \otimes D(\partial_{i_t}) \otimes \cdots \otimes \partial_{i_r} \otimes \mathrm{d}x^{j_1} \otimes \cdots \otimes \mathrm{d}x^{j_s} \\ & + \sum_{t=1}^{s} \psi^{i_1\ldots i_r}_{j_1\ldots j_s} \partial_{i_1} \otimes \cdots \otimes \partial_{i_r} \otimes \mathrm{d}x^{j_1} \otimes \cdots \otimes D(\mathrm{d}x^{j_t}) \otimes \cdots \otimes \mathrm{d}x^{j_s}. \end{aligned}$$

Hence we only have to show that $D$ is uniquely determined for tensor fields $\omega \in \mathcal{T}^0_1(M)$. But this follows from $D(\omega(V)) = D(C^1_1(\omega \otimes V)) = C^1_1(D(\omega \otimes V)) = C^1_1(D\omega \otimes V + \omega \otimes DV)) = D\omega(V) + \omega(DV)$ for arbitrary vector fields $V$ and tensor fields $\omega \in \mathcal{T}^0_1(M)$. ∎

Recall that vector fields can be considered as derivations acting on functions. We show now that the commutator of vector fields can be generalized to arbitrary derivations.

**Lemma 2.4.3.** *Let $D, \tilde{D}, \hat{D}$ be derivations. Then the commutator*

$$[D, \tilde{D}] := D \circ \tilde{D} - \tilde{D} \circ D$$

*is also a derivation. Moreover, the* Jacobi identity

$$[D, [\tilde{D}, \hat{D}]] + [\hat{D}, [D, \tilde{D}]] + [\tilde{D}, [\hat{D}, D]] = 0$$

*holds.*

*Proof.* For the first assertion we only need to check that the product rule is satisfied. This follows from

$$\begin{aligned} D \circ \tilde{D}(\varphi \otimes \psi) &= D(\tilde{D}\varphi \otimes \psi + \varphi \otimes \tilde{D}\psi) \\ &= D \circ \tilde{D}\varphi \otimes \psi + \varphi \otimes D \circ \tilde{D}\psi + \tilde{D}\varphi \otimes D\psi + D\varphi \otimes \tilde{D}\psi \end{aligned}$$

and the fact that the term $\tilde{D}\varphi \otimes D\psi + D\varphi \otimes \tilde{D}\psi$ is symmetric with respect to $D$ and $\tilde{D}$.

The second assertion is a special incident of a general property of commutators of the form $AB - BA$: The summands in

$$\begin{aligned} &[D, [\tilde{D}, \hat{D}]] + [\hat{D}, [D, \tilde{D}]], [\tilde{D}, [\hat{D}, D]] \\ &= \overbrace{D \circ \tilde{D} \circ \hat{D}}^{1} - \overbrace{D \circ \hat{D} \circ \tilde{D}}^{2} - (\overbrace{\tilde{D} \circ \hat{D} \circ D}^{3} - \overbrace{\hat{D} \circ \tilde{D} \circ D}^{4}) \\ &\quad + \overbrace{\hat{D} \circ D \circ \tilde{D}}^{5} - \overbrace{\hat{D} \circ \tilde{D} \circ D}^{4} - (\overbrace{D \circ \tilde{D} \circ \hat{D}}^{1} - \overbrace{\tilde{D} \circ D \circ \hat{D}}^{6}) \\ &\quad + \overbrace{\tilde{D} \circ \hat{D} \circ D}^{3} - \overbrace{\tilde{D} \circ D \circ \hat{D}}^{6} - (\overbrace{\hat{D} \circ D \circ \tilde{D}}^{5} - \overbrace{D \circ \hat{D} \circ \tilde{D}}^{2}) \\ &= 0 \end{aligned}$$

cancel pairwise. ∎

**Corollary 2.4.4.** *For any vector fields $U, V, W$ we have*

$$[U, [V, W] + [W, [U, V]] + [V, [W, U]] = 0.$$

*Proof.* Clear since vector fields can be considered as derivations acting on functions. ∎

**Proposition 2.4.3.** *For vector fields $U, V$, we have $\pounds_{[U,V]} = [\pounds_U, \pounds_V]$.*

*Proof.* Clearly, both $\pounds_{[U,V]}$ and $[\pounds_U, \pounds_V]$ are derivations. By Proposition 2.4.2 we only need to show that they coincide on functions and on vector fields. For any function $f$ we have $[\pounds_U, \pounds_V] \bullet f = \pounds_U \pounds_V \bullet f - \pounds_V \pounds_U \bullet f = U \bullet V \bullet f - V \bullet U \bullet f = [U,V] \bullet f = \pounds_{[U,V]} f$. Hence the formula holds for functions. Corollary 2.4.4 implies for any vector field $W$

$$\begin{aligned}[\pounds_U, \pounds_V]W &= \pounds_U(\pounds_V W) - \pounds_V(\pounds_U W) = [U,[V,W]] - [V,[U,W]] \\ &= [[U,V],W] = \pounds_{[U,V]}W.\end{aligned}$$

∎

The Lie bracket of vector fields does not only transform naturally with respect to diffeomorphism but there is also an especially simple relation if one consideres smooth maps which are not necessarily diffeomorphisms.

**Proposition 2.4.4.** *Let $f\colon M \to N$ be a smooth map and $V, W$ be vector fields on $M$. If $\tilde{V}, \tilde{W}$ are vector fields on $N$ with $T_x f(V_x) = \tilde{V}_{f(x)}$ and $T_x f(W_x) = \tilde{W}_{f(x)}$ for all $x \in M$, then the formula $T_x f([V,W]) = [\tilde{V}, \tilde{W}]_{f(x)}$ holds.*

*Proof.* Let $\varphi\colon N \to \mathbb{R}$ be a smooth function. The assertion follows from

$$\begin{aligned}V(W(\varphi \circ f)) &= V((Tf(W)(\varphi)) \circ f) = (T_x f(V)(Tf(W)(\varphi))) \circ f \\ &= (\tilde{V}(\tilde{W}(\varphi))) \circ f.\end{aligned}$$

∎

## 2.5 Differential forms

*While it is possible to avoid the usage of differential forms, they are such an important tool in analysis and mathematical physics that I have chosen to include them in this book. Differential forms will be used occasionally in the book, for instance in the treatment of electromagnetism.*

*The reader can skip this section on first reading but she or he is advised to read the motivation below.*

*This section builds on the theory of anti-symmetric tensors which is presented in Sect. 2.3.1 starting at page 72.*

Differential forms are totally anti-symmetric covariant tensors. There are areas in mathematics and physics where this anti-symmetry proves to be of great importance.

(i) *Systems of partial differential equations:* Recall that by the lemma of Schwarz the higher derivatives of a $C^\infty$ function commute. If one has a system of partial differential equations, any solution must satisfy this "integrability condition". For the existence of a solution it is often sufficient to ensure that this integrability condition holds. Since anything symmetric applied to something anti-symmetric vanishes, such conditions can be naturally expressed by the requirement that certain differential forms vanish.

(ii) *Integration:* Recall from linear algebra that the volume spanned by $n$ vectors $\{b_1, \ldots, b_n\}$ in $\mathbb{K}^n$ is given by the determinant $|\det(B)|$ where $B$ is the linear map given by $Be_i = b_i$ and $\{e_1, \ldots, e_n\}$ is the standard basis of $\mathbb{K}^n$. As the determinant is totally anti-symmetric, differential forms are its natural generalization. The lemma of Poincaré (Theorem 2.5.2) and the theorem of Stokes (Theorem 2.5.5) which unifies the classical integral theorems of Gauß and Stokes are good examples for the superiority of using differential forms.

(iii) *Physical applications:* There are also direct physical applications of differential forms. They are a prerequisitive for understanding gauge theories (cf. (Bleecker 1981)) of elementary particles and in particular the theory of electromagnetism (cf. Sect. 5.2.3).

Recall from Definition 2.3.8 that the set $\Lambda^p M = \bigcup_{x \in M} \Lambda^p(T_x M)$ of all p-forms is a vector subbundle of $T_p^0 M$.

**Definition 2.5.1.** *We denote the set of all differential forms of degree* $p$ *by* $\Omega^p(M) := \{\omega \in T_p^0(M) | \operatorname{alt} \circ \omega = \omega\}$ *(cf. Definition 2.3.13).*

If $M$ is a real manifold we will sometimes denote $\Omega^p(M)$ by $\Omega^p(M, \mathbb{R})$ (cf. Remark 2.5.2 below)

The definitions and properties of $p$-forms given in Sect. 2.3.1 carry over to differential forms in a pointwise manner.

**Lemma 2.5.1.** *Let* $\omega, \eta$ *be differential forms and* $V$ *be a vector field.*

*(i) For any smooth map* $\phi\colon M \to N$ *the exterior product satisfies* $\phi^*(\omega \wedge \eta) = \phi^*\omega \wedge \phi^*\eta$.

*(ii) If* $\phi\colon M \to N$ *is a local diffeomorphism then* $\phi^*(v \lrcorner \omega) = (\phi^* v) \lrcorner (\phi^*\omega)$ *holds.*

*(iii) The differential form* $\omega$ *can uniquley written as*

$$\omega(x) = \sum_{1 \le i_1 < \cdots < i_p \le n} \omega_{i_1 \ldots i_1}(x)\, \mathrm{d}x^{i_1} \wedge \cdots \wedge \mathrm{d}x^{i_p},$$

*where* $(x^1, \ldots, x^n)$ *is a coordinate system.*

*Proof.* The proof follows directly from the definitions. ∎

There is a simple formula which relates the Lie-bracket and the interior product.

**Lemma 2.5.2.** *Let $\omega$ be a differential form and $U, V$ vector fields. Then the formula $[U,V] \lrcorner \omega = \pounds_U(V \lrcorner \omega) - V \lrcorner \pounds_U \omega$ holds.*

*Proof.* If $\omega$ is a 0-form then we have $[U,V] \lrcorner \omega = 0$ by definition. The right hand side vanishes for the same reason. If $\omega$ is a 1-form then we have by the derivativation property of $\pounds_U$

$$\begin{aligned} [U,V] \lrcorner \omega &= \omega(\pounds_U V) = \pounds_U(\omega(V)) - (\pounds_U \omega)(V) \\ &= \pounds_U(V \lrcorner \omega) - V \lrcorner \pounds_U \omega. \end{aligned}$$

Assume that the assertion of the lemma holds for 1-forms and $p$-forms. For any $(p+1)$-form $\omega$ we find $p$-forms $\omega^i$ and 1-forms $\eta^i$ with $\omega = \sum_{i=1}^n \eta^i \wedge \omega^i$. Hence we get

$$\begin{aligned} [U,V] \lrcorner \omega &= \sum_{i=1}^n [U,V] \lrcorner (\eta^i \wedge \omega^i) \\ &= \sum_{i=1}^n ([U,V] \lrcorner \eta^i) \wedge \omega^i - \sum_{i=1}^n \eta^i \wedge [U,V] \lrcorner \omega^i \\ &= \sum_{i=1}^n \Big( (\pounds_U(V \lrcorner \eta^i) - V \lrcorner \pounds_U \eta^i) \wedge \omega^i \\ &\quad - \eta^i \wedge (\pounds_U(V \lrcorner \omega^i) - V \lrcorner \pounds_U \omega^i) \Big) \\ &= \sum_{i=1}^n \Big( (\pounds_U(V \lrcorner \eta^i)) \wedge \omega^i - (V \lrcorner \pounds_U \eta^i) \wedge \omega^i \\ &\quad - \eta^i \wedge (\pounds_U(V \lrcorner \omega^i)) + \eta^i \wedge (V \lrcorner \pounds_U \omega^i) \Big) \\ &= \sum_{i=1}^n \Big( \pounds_U \left((V \lrcorner \eta^i) \wedge \omega^i\right) - (V \lrcorner \eta^i) \wedge \pounds_U \omega^i \\ &\quad - V \lrcorner ((\pounds_U \eta^i) \wedge \omega^i) - (\pounds_U \eta^i) \wedge V \lrcorner \omega^i \\ &\quad - \pounds_U \left(\eta^i \wedge (V \lrcorner \omega^i)\right) + (\pounds_U \eta^i) \wedge V \lrcorner \omega^i \\ &\quad - V \lrcorner (\eta^i \wedge (\pounds_U \omega^i)) - (V \lrcorner \eta^i) \wedge \pounds_U \omega^i \Big) \\ &= \sum_{i=1}^n \Big( \pounds_U \left(V \lrcorner (\eta^i \wedge \omega^i)\right) - V \lrcorner (\pounds_U(\eta^i \wedge \omega^i)) \Big) \\ &= \pounds_U(V \lrcorner \omega) - V \lrcorner \pounds_U \omega. \end{aligned}$$

Hence the assertion follows for arbitrary degree by induction. ∎

By far the most important construction for differential forms is the exterior derivative which will be introduced in the following theorem.

**Theorem 2.5.1.** *For each* $p \in \mathbb{N} \cup \{0\}$ *there is a unique map*

$$\mathrm{d}\colon \Omega^p(M) \to \Omega^{p+1}(M)$$

*such that the following properties hold:*

*(i)* d *is* $\mathbb{K}$*-bilinear,*
*(ii)* $\mathrm{d} \circ \mathrm{d} = 0$
*(iii)* $\mathrm{d}(\omega \wedge \eta) = \mathrm{d}\omega \wedge \eta + (-1)^q \omega \wedge \mathrm{d}\eta$ *for all* $q$*-forms* $\omega$ *and* $r$*-forms* $\eta$,
*(iv) For* $f \in \Omega^0(M)$ *(i.e., functions* $f\colon M \to \mathbb{K}$*)* $\mathrm{d}f$ *coincides with the usual differential.*

**Definition 2.5.2.** *The operator* d *of Theorem 2.5.1 is called the* exterior derivative.

Observe that for the definition of the exterior derivative we do not need any additional structure. This fact indicates that in many applications it will play a fundamental rôle. In comparison, the Lie derivative of a tensor field is only defined with respect to a given vector field.

*Proof of Theorem 2.5.1.* First we show that d is a local operator, i.e., if $\mathcal{U}$ is an open set with compact closure and $\omega, \eta \in \Omega^p(M)$ satisfy $\omega_{|\mathcal{U}} = \eta_{|\mathcal{U}}$ then $(\mathrm{d}\omega)_{|\mathcal{U}} = (\mathrm{d}\eta)_{|\mathcal{U}}$. To see this let $\mathcal{V}$ be an open set with $\bar{\mathcal{V}} \subset \mathcal{U}$ and $h\colon M \to \mathbb{R}$ be a smooth open function with $h_{|\mathcal{V}} = 0$ and $h_{|M \setminus \bar{\mathcal{U}}} = 1$. Since $h(\omega - \eta) = \omega - \eta$ we obtain from (iii)

$$\mathrm{d}(\omega - \eta) = \mathrm{d}(h(\omega - \eta)) = \mathrm{d}h \wedge (\omega - \eta) + h\mathrm{d}(\omega - \eta)).$$

This implies $(\mathrm{d}(\omega - \eta))_{|\mathcal{V}} = 0$ since both $\mathrm{d}h$ and $h$ vanish on this set. By the arbitrariness of $\mathcal{V}$ we have therefore proved $(\mathrm{d}\omega)_{|\mathcal{U}} = (\mathrm{d}\eta)_{|\mathcal{U}}$.

Since d is a local operator we can restrict to chart neighbourhoods. We will prove the theorem by showing that for each chart $(\mathcal{U}, \varphi)$ there is a unique operator d which satisfies properties (i)–(iv) above. Let $\omega \in \Omega^p(M)$ and write $\omega = \sum_{1 \le i_1 < \cdots < i_p \le n} \omega_{i_1 \dots i_p} \mathrm{d}x^{i_1} \wedge \cdots \wedge \mathrm{d}x^{i_p}$. Properties (i)–(iv) imply

$$\begin{aligned}
\mathrm{d}\omega &= \mathrm{d}\left(\sum_{1\le i_1<\cdots<i_p\le n} \omega_{i_1\ldots i_p}\mathrm{d}x^{i_1}\wedge\cdots\wedge\mathrm{d}x^{i_p}\right)\\
&= \sum_{1\le i_1<\cdots<i_p\le n} (\mathrm{d}(\omega_{i_1\ldots i_p})\wedge\mathrm{d}x^{i_1}\wedge\cdots\wedge\mathrm{d}x^{i_p}\\
&\quad + \omega_{i_1\ldots i_p}\mathrm{d}(\mathrm{d}x^{i_1}\wedge\cdots\wedge\mathrm{d}x^{i_p}))\\
&= \sum_{1\le i_1<\cdots<i_p\le n} \mathrm{d}(\omega_{i_1\ldots i_p})\wedge\mathrm{d}x^{i_1}\wedge\cdots\wedge\mathrm{d}x^{i_p}\\
&\quad + \sum_{1\le i_1<\cdots<i_p\le n}\sum_{j=1}^{p}(-1)^{j-1}\mathrm{d}x^{i_1}\wedge\cdots\wedge\mathrm{d}x^{i_{j-1}}\\
&\quad \wedge\mathrm{d}\mathrm{d}x^{i_j}\wedge\mathrm{d}x^{i_{j+1}}\wedge\cdots\wedge\mathrm{d}x^{i_p}\\
&= \sum_{1\le i_1<\cdots<i_p\le n} \mathrm{d}(\omega_{i_1\ldots i_p})\wedge\mathrm{d}x^{i_1}\wedge\cdots\wedge\mathrm{d}x^{i_p}.
\end{aligned}$$

Thus we have shown that $\mathrm{d}\omega$ is uniquely defined if it exists. Furthermore this explicit formula also guarantees existence once we have shown that $\mathrm{d}\omega := \sum_{1\le i_1<\cdots<i_p\le n}\mathrm{d}(\omega_{i_1\ldots i_p})\wedge\mathrm{d}x^{i_1}\wedge\cdots\wedge\mathrm{d}x^{i_p}$ satisfies (i)–(iv).

Properties (i) and (iv) are clear. For (iii) we calculate for $\omega\in\Omega^p(M)$ and $\eta\in\Omega^q(M)$

$$\begin{aligned}
&\mathrm{d}(\omega\wedge\eta)\\
&\quad = \mathrm{d}\sum_{\substack{1\le i_1<\cdots<i_p\le n\\ 1\le j_1<\cdots<j_q\le n}} \omega_{i_1\ldots i_p}\eta_{j_1\ldots j_q}\wedge\mathrm{d}x^{i_1}\wedge\cdots\wedge\mathrm{d}x^{i_p}\wedge\mathrm{d}x^{j_1}\wedge\cdots\wedge\mathrm{d}x^{j_q}\\
&\quad = \sum_{\substack{1\le i_1<\cdots<i_p\le n\\ 1\le j_1<\cdots<j_q\le n}} \mathrm{d}\omega_{i_1\ldots i_p}\eta_{j_1\ldots j_q}\wedge\mathrm{d}x^{i_1}\wedge\cdots\wedge\mathrm{d}x^{i_p}\wedge\mathrm{d}x^{j_1}\wedge\cdots\wedge\mathrm{d}x^{j_q}\\
&\qquad + \sum_{\substack{1\le i_1<\cdots<i_p\le n\\ 1\le j_1<\cdots<j_q\le n}} \omega_{i_1\ldots i_p}\mathrm{d}\eta_{j_1\ldots j_q}\wedge\mathrm{d}x^{i_1}\wedge\cdots\wedge\mathrm{d}x^{i_p}\wedge\mathrm{d}x^{j_1}\wedge\cdots\wedge\mathrm{d}x^{j_q}\\
&\quad = \sum_{\substack{1\le i_1<\cdots<i_p\le n\\ 1\le j_1<\cdots<j_q\le n}} \mathrm{d}\omega_{i_1\ldots i_p}\wedge\mathrm{d}x^{i_1}\wedge\cdots\wedge\mathrm{d}x^{i_p}\wedge\eta_{j_1\ldots j_q}\mathrm{d}x^{j_1}\wedge\cdots\wedge\mathrm{d}x^{j_q}\\
&\qquad + (-1)^p\sum_{\substack{1\le i_1<\cdots<i_p\le n\\ 1\le j_1<\cdots<j_q\le n}} \omega_{i_1\ldots i_p}\wedge\mathrm{d}x^{i_1}\wedge\cdots\wedge\mathrm{d}x^{i_p}\\
&\qquad \wedge\mathrm{d}\eta_{j_1\ldots j_q}\wedge\mathrm{d}x^{j_1}\wedge\cdots\wedge\mathrm{d}x^{j_q}\\
&\quad = \mathrm{d}\omega\wedge\eta + (-1)^p\omega\wedge\mathrm{d}\eta.
\end{aligned}$$

Property (ii) is follows from

$$\mathrm{d}\mathrm{d}\omega = \sum_{1\le i_1<\cdots<i_p\le n} \mathrm{d}\mathrm{d}(\omega_{i_1\ldots i_p})\wedge\mathrm{d}x^{i_1}\wedge\cdots\wedge\mathrm{d}x^{i_p}.$$

and the fact that for any function $f \in C^\infty(M, \mathbb{R})$ we have

$$\mathrm{dd}f = \mathrm{d}\sum_{i=1}^{p} \frac{\partial f}{\partial x_i}\mathrm{d}x^i = \sum_{i,j=1}^{p} \frac{\partial^2 f}{\partial x_i \partial x_j}\mathrm{d}x^j \wedge \mathrm{d}x^i = 0,$$

where we have used that $\frac{\partial^2 f}{\partial x_i \partial x_j}$ is symmetric in $i$ and $j$ by the lemma of Schwarz. ∎

*Remark 2.5.1.* Notice that this definition of d coincides with Definition 2.2.7 in the case of 0-forms.

**Corollary 2.5.1.** *If $N$ is a second manifold and $\phi\colon M \to N$ a smooth map, then we have $\phi^*(\mathrm{d}\omega) = \mathrm{d}(\phi^*\omega)$ for all p-forms $\omega$ on $N$.*

*Proof.* This follows since properties (i)–(iv) are obviously satisfied for $\phi^*\mathrm{d}\omega$ and since the exterior derivative is unique. ∎

**Corollary 2.5.2.** *Lie derivative and exterior derivative commute: Let $\omega \in \Omega^p(M)$ and $V \in \mathcal{T}_0^1(M)$. Then the equation*

$$\pounds_V \mathrm{d}\omega = \mathrm{d}\pounds_V \omega$$

*holds.*

*Proof.* Denote the flow of $V$ by $F_t$. Then we have $\mathrm{d}(F_t)^*\omega = (F_t)^*\mathrm{d}\omega$ and therefore

$$\begin{aligned}\mathrm{d}\pounds_V\omega &= \mathrm{d}\left(\left(\frac{\mathrm{d}}{\mathrm{d}t}(F_t)^*\omega\right)_{|t=0}\right) = \left(\frac{\mathrm{d}}{\mathrm{d}t}\mathrm{d}(F_t)^*\omega\right)_{|t=0} \\ &= \left(\frac{\mathrm{d}}{\mathrm{d}t}(F_t)^*\mathrm{d}\omega\right)_{|t=0} = \pounds_V \mathrm{d}\omega.\end{aligned}$$

∎

We wish to give a formula for $\mathrm{d}\omega$ which does not depend on a chosen coordinate system. The idea is to link the exterior derivative to the Lie derivative and the interior product.

**Lemma 2.5.3.** *Let $\omega \in \Omega^p(M)$ and $V \in \mathcal{T}_0^1(M)$. Then we have*

$$\pounds_V \omega = V \lrcorner \mathrm{d}\omega + \mathrm{d}(V \lrcorner \omega)$$

*Proof.* We prove the lemma by induction. If $p = 0$, then we obviously have $V \lrcorner \omega = 0$ and $\mathcal{L}_V \omega = V \bullet \omega = \mathrm{d}\omega(V) = V \lrcorner \mathrm{d}\omega$. Assume now that the assertion has been proved for all $q \in \{0, \dots, p\}$ and let $\omega \in \Omega^{p+1}(M)$. Since the formula which we want to prove is local we can restrict to a coordinate neighbourhood $\mathcal{U}$ and write

$$\omega = \sum_{i_1 < \cdots < i_{p+1}} \omega_{i_1 \cdots i_{p+1}} \mathrm{d}x^{i_1} \wedge \cdots \wedge \mathrm{d}x^{i_{p+1}} = \omega_i \wedge \mathrm{d}x^i$$

where $\omega_i \in \Omega^p(\mathcal{U})$ are suitably chosen differential forms. Recall that $\mathcal{L}_V$ is a derivation and that therefore

$$\begin{aligned}
\mathcal{L}_V(\omega_i \wedge \mathrm{d}x^i) &= \mathcal{L}_V \left( \frac{1}{p!1!} \sum_{\sigma_{p+1} \in S_{p+1}} \mathrm{sign}(\sigma_{p+1}) \sigma_{p+1}(\omega_i \otimes \mathrm{d}x^i) \right) \\
&= \frac{1}{p!1!} \sum_{\sigma_{p+1} \in S_{p+1}} \mathrm{sign}(\sigma_{p+1}) \sigma_{p+1}(\mathcal{L}_V \omega_i \otimes \mathrm{d}x^i \\
&\quad + \omega_i \otimes \mathcal{L}_V \mathrm{d}x^i)) \\
&= \mathcal{L}_V \omega_i \wedge \mathrm{d}x^i + \omega_i \wedge \mathcal{L}_V \mathrm{d}x^i
\end{aligned}$$

holds. On the other hand, we have

$$\begin{aligned}
&V \lrcorner \mathrm{d}(\omega_i \wedge \mathrm{d}x^i) + \mathrm{d}(V \lrcorner (\omega_i \wedge \mathrm{d}x^i)) \\
&\quad = V \lrcorner (\mathrm{d}\omega_i \wedge \mathrm{d}x^i) + \mathrm{d}((V \lrcorner \omega_i) \wedge \mathrm{d}x^i + (-1)^p \omega_i \wedge (V \lrcorner \mathrm{d}x^i)) \\
&\quad = \overbrace{(V \lrcorner \mathrm{d}\omega_i) \wedge \mathrm{d}x^i}^{\text{induction}} + (-1)^{p+1} \mathrm{d}\omega_i \wedge (V \lrcorner \mathrm{d}x^i) \\
&\quad\quad + \overbrace{\mathrm{d}(V \lrcorner \omega_i) \wedge \mathrm{d}x^i}^{\text{induction}} + (-1)^p \mathrm{d}\omega_i \wedge (V \lrcorner \mathrm{d}x^i) \\
&\quad\quad + \overbrace{(-1)^p (-1)^p \omega_i \wedge \mathrm{d}(V \lrcorner \mathrm{d}x^i)}^{*} \\
&\quad = \overbrace{\mathcal{L}_V \omega_i \wedge \mathrm{d}x^i}^{\text{induction}} + \overbrace{\omega_i \wedge \mathcal{L}_V \mathrm{d}x^i}^{*}
\end{aligned}$$

where we have used $\mathrm{d}(V \lrcorner \mathrm{d}x^i) = \mathrm{d}(\mathrm{d}x^i(V)) = \mathrm{d}(\mathcal{L}_V x^i) = \mathcal{L}_V \mathrm{d}x^i$ (cf. Corollary 2.5.2). ∎

We can now use the preceding lemma in order to prove an invariant formula for the exterior derivative.

**Proposition 2.5.1.** *Let $\omega \in \Omega^p(M)$ and $V_0, \dots, V_p$ be vector fields. Then the exterior derivative of $\omega$ is given by*

$$\begin{aligned}
\mathrm{d}\omega(V_0,\ldots,V_p) &= \sum_{i=0}^{p}(-1)^i \pounds_{V_i}\big(\omega(V_0,\ldots,\widehat{V_i},\ldots,V_p)\big) \\
&\quad + \sum_{0\le i<j\le p}(-1)^{i+j}\omega(\pounds_{V_i}V_j, V_0,\ldots,\widehat{V_i},\ldots,\widehat{V_j},\ldots,V_p) \\
&= \sum_{i=0}^{p}(-1)^i V_i \bullet \big(\omega(V_0,\ldots,\widehat{V_i},\ldots,V_p)\big) \\
&\quad + \sum_{0\le i<j\le p}(-1)^{i+j}\omega([V_i,V_j], V_0,\ldots,\widehat{V_i},\ldots,\widehat{V_j},\ldots,V_p)
\end{aligned}$$

*where* $\widehat{\cdot}$ *means that the corresponding vector field is left out.*

*Proof.* The second equality follows trivially from the first equality. We will prove the proposition by induction. If $\omega \in \Omega^0(M)$ then $\mathrm{d}\omega(V_0) = V_0 \bullet \omega = \pounds_{V_0}(\omega)$ which implies the equality in the case $p = 0$. Assume now that the assertion has been proven for $q \in \{0,\ldots,p\}$ and let $\omega \in \Omega^{p+1}(M)$. Lemma 2.5.3 implies

$$\begin{aligned}
&\mathrm{d}\omega(V_0,\ldots,V_{p+1}) = (V_0 \lrcorner \mathrm{d}\omega)(V_1,\ldots,V_{p+1}) \\
&\quad = \pounds_{V_0}\omega(V_1,\ldots,V_{p+1}) - \overbrace{\mathrm{d}(V_0 \lrcorner \omega)(V_1,\ldots,V_{p+1})}^{\text{induction}} \\
&\quad = \pounds_{V_0}\big(\omega(V_1,\ldots,V_{p+1})\big) - \sum_{i=1}^{p+1}\omega(V_1,\ldots,V_{i-1},\pounds_{V_0}V_i,V_{i+1},\ldots,V_{p+1}) \\
&\qquad - \overbrace{\sum_{j=1}^{p+1}(-1)^{j-1}\pounds_{V_j}\big((V_0 \lrcorner \omega)(V_1,\ldots,V_{j-1},V_{j+1},\ldots,V_{p+1})\big)}^{\text{induction}} \\
&\qquad - \overbrace{\sum_{1\le j<k\le p+1}(-1)^{j-1+k-1}(V_0 \lrcorner \omega)(\pounds_{V_j}V_k, V_1,\ldots,}^{\text{induction}} \\
&\qquad \overbrace{\widehat{V_j},\ldots,\widehat{V_k},\ldots,V_{p+1}))}^{\text{induction}} \\
&\quad = \pounds_{V_0}\big(\omega(V_1,\ldots,V_{p+1})\big) \\
&\qquad - \sum_{i=1}^{p+1}(-1)^{i-1}\omega(\pounds_{V_0}V_i, V_1,\ldots,V_{i-1},V_{i+1},\ldots,V_{p+1}) \\
&\qquad + \sum_{j=1}^{p+1}(-1)^j \pounds_{V_j}\big(\omega(V_0,V_1,\ldots,V_{j-1},V_{j+1},\ldots,V_{p+1})\big) \\
&\qquad + \sum_{1\le j<k\le p+1}(-1)^{j+k}\omega(\pounds_{V_j}V_k, V_0,V_1,\ldots,\widehat{V_j},\ldots,\widehat{V_k},\ldots,V_{p+1})).
\end{aligned}$$

$$= \sum_{j=0}^{p+1} (-1)^j \pounds_{V_j} \omega(V_0, \ldots, \widehat{V_j}, \ldots, V_p)$$

$$+ \sum_{0 \le j < k \le p+1} (-1)^{j+k} \omega(\pounds_{V_j} V_k, V_0, \ldots, \widehat{V_j}, \ldots, \widehat{V_k}, \ldots, V_p)$$

∎

### 2.5.1 The lemma of Poincaré

*The lemma of Poincare is the generalization of the following two facts familiar to physicists.*
*(i) Every curl-free vector field has a local scalar potential and*
*(ii) every divergence-free vector field has a local vector potential.*
*The lemma of Poincaré is a good example for the elegance of differential forms.*

To give the reader a better idea how the lemma of Poincaré arises we will briefly recall the introduction of scalar and vector potentials and then translate this discussion into the language of differential forms.

(i) Let $F: \mathcal{U} \subset \mathbb{R}^3 \to \mathbb{R}^3$ be a smooth map which satisfies $\mathrm{rot}(F) = 0$. Then for every $x \in \mathcal{U}$ there is a neighbourhood $\mathcal{V}$ and a function $f: \mathcal{V} \to \mathbb{R}$ with $F_{|V} = \mathrm{grad}(f)$. In Mechanics this mathematical fact is applied to conservative force fields $F$. The function $f$ is called the associated *scalar potential.*

(ii) Consider now a map $V: \mathcal{U} \subset \mathbb{R}^3 \to \mathbb{R}^3$ whose divergence vanishes, $\frac{\partial V^1}{\partial x^1} + \frac{\partial V^2}{\partial x^2} + \frac{\partial V^3}{\partial x^3} = 0$. Then for every $x \in \mathcal{U}$ there is a neighbourhood $\mathcal{V}$ and another map $W: \mathcal{V} \to \mathbb{R}^3$ such that $\mathrm{rot}(W) = V$. The map $W$ is called the *vector potential* of $V$. Vector potentials arise for instance in the elementary theory of electromagnetism.

In the language of differential forms, these (and similar) results can be treated in a simple, unified manner. To see this, we will briefly sketch two classes of isomorphisms which will be studied in more generality and more detail in Sect. 4.2.

The standard scalar product $\langle \cdot, \cdot \rangle_{\mathbb{R}^3}$ of $\mathbb{R}^3$ induces an isomorphism which maps vector fields to the 1-forms, $V \mapsto V^\flat := \langle V, \cdot \rangle_{\mathbb{R}^3}$. The inverse map is denoted by $\sharp: \omega \mapsto \omega^\sharp$, where $\langle \omega^\sharp, \cdot \rangle_{\mathbb{R}^3} = \omega$. These isomorphisms are will defined since the scalar product is non-degenerate. With respect to the standard orthonormal basis this isomorphism just interchanges column and row vectors.

The scalar product also induces an isomorphism between 0-forms and 3-forms and an isomorphism between 1-forms and 2-forms. Let $(x^1, x^2, x^3)$ be the standard coordinate system of $\mathbb{R}^3$ and set

$$\begin{aligned}\star 1 &= \mathrm{d}x^1 \wedge \mathrm{d}x^2 \wedge \mathrm{d}x^3,\\ \star \mathrm{d}x^1 &= \mathrm{d}x^2 \wedge \mathrm{d}x^3,\\ \star \mathrm{d}x^2 &= \mathrm{d}x^3 \wedge \mathrm{d}x^1,\\ \star \mathrm{d}x^3 &= \mathrm{d}x^1 \wedge \mathrm{d}x^2.\end{aligned}$$

This defines linear isomorphisms $\star\colon \Lambda^0(\mathbb{R}^3) \to \Lambda^3(\mathbb{R}^3)$ and $\star\colon \Lambda^1(\mathbb{R}^3) \to \Lambda^2(\mathbb{R}^3)$. They can be extended to 2- and 3-forms by demanding $\star\star\omega = \omega$ for all forms in $\mathbb{R}^3$.

Observe that $\mathrm{grad}(f) = \mathrm{d}f^\sharp$, $\mathrm{rot}(F) = (\star \mathrm{d}F^\flat)^\sharp$, $\mathrm{div}(V) = \star(\mathrm{d} \star V^\flat)$. Hence the (i) and (ii) above relating $F$ and $f$, $V$ and $W$ translate into the assertions

(i) If the 1-form $F^\flat$ satisfies $\mathrm{d}F^\flat = 0$, then locally there exists a 0-form $f$ with $\mathrm{d}f = F^\flat$;
(ii) If the 2-form $\star V^\flat$ satisfies $\mathrm{d} \star V^\flat = 0$, then locally there exists a 1-form $W^\flat$ with $\mathrm{d}W^\flat = \star V^\flat$.

The lemma of Poincaré generalizes these facts.

**Theorem 2.5.2 (Lemma of Poincaré).** *Let $M$ be a manifold and $\omega \in \Omega^p(M)$ with $p \geq 1$. If $\mathrm{d}\omega = 0$, then for every $x \in M$ there is a neighbourhood $\mathcal{U}$ of $x$ and a $(p-1)$-form $\theta \in \Omega^{p-1}(\mathcal{U})$ such that $\omega_{|\mathcal{U}} = \mathrm{d}\theta$.*

The proof of the theorem will be a corollary to the following lemma.

**Lemma 2.5.4.** *Let $M$ be a smooth manifold and define for any $t \in [0,1]$ the map*

$$i_t\colon M \to [0,1] \times M, \quad x \mapsto (t,x).$$

*There exists a linear map $K\colon \Omega^{q+1}([0,1] \times M) \to \Omega^q(M)$ $(q \geq 0)$ with $\mathrm{d} \circ K + K \circ \mathrm{d} = (i_1)^* - (i_0)^*$.*

*Proof.* For $\omega \in \Omega^{q+1}([0,1] \times M)$ and $v_1, \ldots, v_q \in T^x M$ we define

$$K\omega_x(v_1,\ldots,v_q) = \int_0^1 \left(i_t{}^*(\partial_t \lrcorner \omega_{(t,x)})\right)(v_1,\ldots,v_q)\,\mathrm{d}t.$$

Let $V_1, \ldots, V_{q+1}$ be vector fields on $M$. Then, using Proposition 2.5.1, we obtain

$$\begin{aligned}&\mathrm{d}K\omega(V_1,\ldots,V_{q+1})\\ &\quad= \sum_{a=1}^{q+1}(-1)^{a-1} V_a \bullet \int_0^1 i_t{}^*(\partial_t \lrcorner \omega_{(t,x)})(V_1,\ldots,\widehat{V_a},\ldots,V_{q+1})\,\mathrm{d}t\\ &\qquad+ \sum_{1\leq a<b\leq q+1} (-1)^{(a-1)+(b-1)} \int_0^1 i_t{}^*(\partial_t \lrcorner \omega_{(t,x)})([V_a,V_b],V_1,\ldots\\ &\qquad\ldots,\widehat{V_a},\ldots,\widehat{V_b},\ldots,V_{q+1})\,\mathrm{d}t.\end{aligned}$$

For any vector field $V$ on $M$ we will denote the canonical lift of $V$ to $\mathbb{R} \times M$, given by $(t,x) \mapsto (0,V) \in T_t\mathbb{R} \oplus T_xM$, by the same symbol $V$. In the following it is notationally advantageous to write $V_0$ for $\partial_t$. With these notations and using $[V_0, V_a] = [\partial_t, V_a] = [(\partial_t, 0), (0, V_a)] = 0$ the $(p+1)$-form $\mathrm{d}K\omega$ is given by

$$\begin{aligned}
&\mathrm{d}K\omega(V_1, \ldots, V_{q+1}) \\
&\quad = \sum_{a=1}^{q+1} (-1)^{a-1} V_a \bullet \int_0^1 \omega_{(t,x)}(V_0, V_1, \ldots, \widehat{V_a}, \ldots, V_{q+1})\,\mathrm{d}t \\
&\qquad + \sum_{1 \le a < b \le q+1} (-1)^{a+b} \int_0^1 \omega_{(t,x)}(V_0, [V_a, V_b], V_1, \ldots \\
&\qquad\qquad \ldots, \widehat{V_a}, \ldots, \widehat{V_b}, \ldots, V_{q+1})\,\mathrm{d}t \\
&\quad = -\sum_{a=1}^{q+1} (-1)^{a} \int_0^1 V_a \bullet \Big(\omega_{(t,x)}(V_0, V_1, \ldots, \widehat{V_a}, \ldots, V_{q+1})\Big)\,\mathrm{d}t \\
&\qquad - \sum_{0 \le a < b \le q+1} (-1)^{a+b} \int_0^1 \omega_{(t,x)}([V_a, V_b], V_0, V_1, \ldots \\
&\qquad\qquad \ldots, \widehat{V_a}, \ldots, \widehat{V_b}, \ldots, V_{q+1})\,\mathrm{d}t.
\end{aligned}$$

Analogously, $K\mathrm{d}\omega$ is given by

$$\begin{aligned}
K\mathrm{d}\omega(V_1, \ldots, V_{q+1}) &= \int_0^1 i_t{}^*(\partial_t \lrcorner \mathrm{d}\omega_{(t,x)})(V_1, \ldots, V_{q+1})\,\mathrm{d}t \\
&= \int_0^1 \mathrm{d}\omega_{(t,x)}(V_0, V_1, \ldots, V_{q+1})\,\mathrm{d}t \\
&= \int_0^1 \sum_{a=0}^{q+1} (-1)^a V_a \bullet (\omega_{(t,x)}(V_0, \ldots, \widehat{V^a}, \ldots, V_{q+1}))\mathrm{d}t \\
&\quad + \sum_{0 \le a < b \le q+1} (-1)^{a+b} \int_0^1 \omega_{(t,x)}([V_a, V_b], V_1, \ldots \\
&\quad \ldots, \widehat{V_a}, \ldots, \widehat{V_b}, \ldots, V_{q+1})\,\mathrm{d}t,
\end{aligned}$$

Taking the sum, $\mathrm{d} \circ K + K \circ \mathrm{d}$, all terms with $a \ge 1$ cancel and we arrive at

$$\begin{aligned}
(\mathrm{d} \circ K + K \circ \mathrm{d})\omega(V_1, \ldots, V_{q+1}) &= \int_0^1 \partial_t \bullet (\omega_{(t,x)}(V_1, \ldots, V_{q+1})\,\mathrm{d}t \\
&= (i_1)^*\omega(V_1, \ldots, V_{q+1}) - (i_0)^*\omega(V_1, \ldots, V_{q+1}).
\end{aligned}$$

∎

*Proof of Theorem 2.5.2.* Let $(\mathcal{V}, \varphi)$ be a chart with $\varphi(x) = 0$ and $\varphi(\mathcal{V}) \supset \{(x^1, \ldots, x^n) | \sum_{a=1}^n (x^a)^2 \leq 1\}$. Setting

$$\mathcal{U} := \varphi^{-1}(\{(x^1, \ldots, x^n) | \sum_{a=1}^{n} (x^a)^2 < 1\})$$

and $F(t, y) := \varphi^{-1}(t^2 \varphi(y))$ we have constructed a smooth map $F \colon [0,1] \times \mathcal{U} \to \mathcal{U}$ which satisfies $F(0, y) = x$ and $F(1, y) = y$ for all $y \in \mathcal{U}$.

Since $F \circ i_1$ is the identity on $\mathcal{U}$ and $F \circ i_0$ the constant map $y \mapsto x$, we have for any $p$-form $\omega$, $(F \circ i_1)^* \omega = \omega$ and $(F \circ i_0)^* \omega = 0$. The assertion follows now from

$$\mathrm{d} \circ K F^* \omega - K \circ \mathrm{d} F^* \omega = i_1^* F^* \omega - i_0^* F^* \omega = \omega$$

and $\mathrm{d}F^* \omega = F^* \mathrm{d}\omega = 0$. ∎

Observe that the proof of the lemma of Poincaré allows us to calculate $\theta = K F^* \omega$ explicitly.

**Definition 2.5.3.** *A differential form* $\omega \in \Omega^p(M)$ *is* closed *if* $\mathrm{d}\omega = 0$ *and is* exact *if there exists a differential from* $\tilde{\omega} \in \Omega^{p-1}(M)$ *with* $\omega = \mathrm{d}\tilde{\omega}$.

In this terminology the lemma of Poincaré simply states that every closed differential form is locally exact. In general, this is not true globally. Corollary 2.5.3 below is a global version of the lemma of Poincaré.

**Definition 2.5.4.** *Let* $M, N$ *be manifolds and* $f, \tilde{f} \colon M \to N$ *be smooth maps.* $f, \tilde{f}$ *are* homotopic *if there is a differentiable map* $F \colon [0,1] \times M \to N$ *such that* $F(0, x) = f(x)$ *and* $F(1, x) = \tilde{f}(x)$ *for all* $x \in M$. *The map* $F$ *is called an* homotopy.

*A connected manifold* $M$ *is called* contractible *if the maps* $\mathrm{id} \colon M \to M$, $x \mapsto x$ *and the constant map* $c_{x_0} \colon M \to M$, $x \mapsto x_0$ *are homotopic.*

If $x_0, x_1$ are in $M$ and $\gamma \colon [0,1] \mapsto M$ is a curve from $x_0$ to $x_1$ then $F(t, x) = \gamma(t)$ is homotopy between the two constant maps $c_{x_0}$ and $c_{x_1}$. Hence the following lemma implies that contractibility does not depend on the choice of $x_0$.

**Lemma 2.5.5.** *Homotopy is an equivalence relation which we denote by* $\simeq$

*Proof.* The relation $f \simeq f$ is clear since we can choose $F(t, x) = x$ for all $t$. If $f \simeq g$ and $F$ is a homotopy between $f$ and $g$ then $\tilde{F}(t, x) := F(1-t, x)$ is a homotopy between $g$ and $f$ whence $g \simeq f$. Assume that $f \simeq g$ and $g \simeq h$ and let $F$, (respectively, $\tilde{F}$) be homotopies between $f$ and $g$ (respectively, $g$ and $h$). Let $\phi \colon [0, 1/2] \to [0,1]$ be a smooth map which satisfies

(i) $\phi(0) = 0$,
(ii) $\phi(1/2) = 1$,
(iii) $\frac{d^k\phi(1/2)}{dt^k} = 0$ for all $k \geq 1$

and $\psi(t) = 1 - \phi(1-t)$. Then the map

$$\hat{F}\colon [0,1] \to M, \qquad (t,x) \mapsto \begin{cases} F(\phi(t), x) & \text{if } t \in [0, 1/2) \\ \tilde{F}(\psi(t), x) & \text{if } t \in [1/2, 1] \end{cases}$$

is a homotopy between $f$ and $h$, ∎

**Lemma 2.5.6.** *Let $f, \tilde{f}\colon M \to N$ be smooth homotopic maps and $\omega \in \Omega^p(N)$ be a closed differential form of degree $p \geq 1$. Then $f^*\omega - \tilde{f}^*\omega$ is exact.*

*Proof.* Let $F$ be a homotopy between $f$ and $\tilde{f}$ and $K\colon \Omega^{p+1}(M) \to \Omega^p(M)$ be the operator defined in Lemma 2.5.4. Then we have

$$dKF^*\omega = dKF^*\omega + KdF^*\omega = (i^1)^*F^*\omega - (i^0)^*F^*\omega = \tilde{f}^*\omega - f^*\omega.$$

∎

**Corollary 2.5.3.** *Let $M$ be a contractible manifold, $p \geq 1$, and $\omega \in \Omega^p(M)$ be a closed differential form. Then $\omega$ is exact.*

*Proof.* Let $c_{x_0}\colon M \to M$ the constant map $x \mapsto x_0$. The differential form $\mathrm{id}^*\omega - (c_{x_0})^*\omega$ is exact by Lemma 2.5.6 and the assertion follows from $\mathrm{id}^*\omega = \omega$, $(c_{x_0})^*\omega = 0$. ∎

### 2.5.2 The theorem of Frobenius

The guiding idea in analysis is that it is much simpler to work with a linearization of a function than with the function itself. Analogously, it is often much simpler to specify properties of the tangent bundle of a manifold than to describe the manifold itself. We are therefore interested in the following problem.

> Let $E$ be a vector subbundle of $TM$. What are the necessary and sufficient conditions for the (local) existence of a submanifold $N \subset M$ with $TN = E_N$?

If $N$ is a submanifold of $M$ then $TN$ is a subbundle of $TM$ which has the property that for any two sections $U$, $V$ of $TN$ (i.e. any two vector fields along $N$ which are at each point tangent) the commutator $[U, V]$ is again a section of $N$. This follows since $N$ is a manifold in its own right.

On the other hand, not every subbundle $E$ of $TM$ has this property, for instance, take $M = \mathbb{R}^3$ and $E = \{a(z\partial_x + \partial_y) + b\partial_z \mid a, b \in \mathbb{R}\}$. For this vector bundle we have $[z\partial_x + \partial_y, \partial_z] = -\partial_x \notin E_{(x,y,z)}$. Hence there cannot exist a submanifold $N$ of $M$ with $TN = E_N$. This motivates the following definition.

**Definition 2.5.5.** *A vector subbundle $E$ of $TM$ is called* integrable *if for any two sections $U, V$ of $E$ the Lie derivatie $[U, V]$ is also a section of $E$.*

*An* integral manifold *of $E$ is a submanifold $N$ of $M$ with $TN \subset E$. An integral manifold $N$ is called* maximal *if $T_xN = E_x$ for all $x \in N$.*

The theorem of Frobenius (cf. Theorem 2.5.3 below) asserts that integrability is also sufficient for the local existence integral manifolds. This justifies the terminology. In order to verify integrability it is sufficient to consider a single frame for $E$, i.e, $k$ linearly independent locally defined vector fields $V_1, \ldots, V_k$ which at each point $x$ span $E_x$.

**Lemma 2.5.7.** *A $k$-vector subbundle $E$ of $TM$ is integrable if and only if there exists a local frame $\{V_1, \ldots, V_k\}$ of $E$ such that for all $i, j$ the commutator $[V_i, V_j]$ is a section of $E$.*

*Proof.* It is clear that the condition is necessary. Let $U, V$ be any sections of $E$. Then there are functions $\alpha^i$, $\beta^i$ with $U = \sum_{i=1}^k \alpha^i V_i$ and $V = \sum_{i=1}^k \beta^i V_i$ and

$$[U, V] = (U \bullet \beta^i - V \bullet \alpha^i)V_i + \alpha^i \beta^j [V_i, V_j] \in E.$$

∎

**Theorem 2.5.3 (Theorem of Frobenius, contravariant form).** *Let $E$ be a smooth subbundle of $TM$. Then through every $x \in M$ there is a locally unique maximal integral manifold $N_x$ of $M$ if and only if $E$ is integrable. Moreover, $N_x$ depends smoothly on $x$.*

The basic idea of proof is as follows. Let $V_1, \ldots V_k$ be a frame of $E$ and $N^1$ be the submanifold swept out by the integral curve of $V_1$ through $x$. At each $y \in N^1$ we can consider the integral curve of $V_2$ through $y$. All these integral curves together form a set subset $N^2$ of $M$. Now for any $z \in N^2$ we take the integral curve of $V_3$ through $z$. These integral curves form a subset $N^3$ of $M$ and so on. One then has to check that $N^k$ really is a submanifold. This is not entirely straightforward because in general, the intermediate sets $N^2, \ldots, N^{k-1}$ are *not* submanifolds. It can be shown, however, that the subsets $N^2, \ldots, N^{k_1}$ are submanifolds of $M$ if the frame $\{V_1, \ldots, V_k\}$ is carefully chosen.

We will prove the theorem for an equivalent covariant form (cf. Theorem 2.5.4 below) using an analogous strategy.

A $k$-dimensional vector subbundle of $M$ can be described as the intersection $\bigcap_{i=1}^{n-k} \text{kern}(\omega^i)$, where $\omega^i$ are some suitably chosen, at each point linearly independent 1-forms. In fact, let $\{V_{n-k}, \ldots, V_n\}$ be a frame for $E$ and $\{V_1, \ldots, V_{n-k}\}$ be a completion to a frame of $TM$. If $\{\omega^1, \ldots, \omega^n\}$ is the dual basis, then $E = \bigcap_{i=1}^{n-k} \text{kern}(\omega^i)$.

**Definition 2.5.6.** *A collection of finitely many pointwise linearly independent 1-forms $\{\omega_i\}$ is called a* Pfaffian system. *The Pfaffian system is* integrable *if the vector subbundle $E = \bigcap_{i=1} \text{kern}(\omega^i)$ is integrable.*

If a 1-form $\omega^i$ vanishes on a submanifold $N$, i.e., $\omega^i(v) = 0$ for all $v \in TN$, then $\mathrm{d}\omega$, being intrinsic to $N$, should also vanish: $\mathrm{d}\omega^i(u, v) = 0$ for all $u, v \in T_xN$. This is equivalent to $\left(\mathrm{d}\omega^i \wedge \omega^1 \wedge \cdots \wedge \omega^{n-k}\right)_x = 0$ which is the same integrability condition as before:

**Lemma 2.5.8.** *Let $\omega^1, \ldots, \omega^{n-k}$ be 1-forms which are linearly independent at each point. Then the vector subbundle $E := \bigcap_{i=1}^{n-k} \text{kern}(\omega^i)$ is integrable if and only if $\mathrm{d}\omega^i \wedge \omega^1 \wedge \cdots \wedge \omega^{n-k} = 0$ for all $i \in \{1, \ldots, n-k\}$.*

*Proof.* We extend $\{\omega^1, \ldots, \omega^{n-k}\}$ to a coframe $\{\omega^1, \ldots, \omega^n\}$ of $T^*M$ and let $\{V_1, \ldots, V_n\}$ be the dual basis. $E$ is then spanned by $\{V_{n-k+1}, \ldots, V_n\}$. We calculate

$$\mathrm{d}\omega^i(V_k, V_l) = V_k \bullet \omega^i(V_l) - V_l \bullet \omega^i(V_k) - \omega^i([V_k, V_l]) = -\omega^i([V_k, V_l]).$$

It follows immediately that $\mathrm{d}\omega^i(V_k, V_l)$ ($i \in \{1, \ldots, n-k\}$) vanishes for all $V_k, V_l \in E$ if and only if $[V_k, V_l] \in E$ for all $V_k, V_l \in E$. ∎

**Theorem 2.5.4 (Theorem of Frobenius, covariant form).** *Let $\omega^1, \ldots, \omega^{n-k}$ be a Pfaffian system which satisfies $\mathrm{d}\omega^i \wedge \omega^1 \wedge \cdots \wedge \omega^{n-k} = 0$ for all $i \in \{1, \ldots, n-k\}$. Then there exist coordinates $(x^1, \ldots, x^n)$ and functions $\omega^i_a$ with*

$$\omega^i = \omega^i_j(x^1, \ldots, x^n)\mathrm{d}x^j \quad (j = 1, \ldots, n-k).$$

The manifold $N = \{y \in M \,|\, x^j(y) = x^j(x) \forall j \in \{1, \ldots, n-k\}\}$ is a maximal integral manifold of $E := \bigcap_{i=1}^{n-k} \text{kern}(\omega^i)$. In particular, Theorems 2.5.3 and 2.5.4 are equivalent.

*Proof of Theorem 2.5.4.* As outlined above, we will prove the theorem by induction over $k$.

Let $k = 1$. Then $E := \bigcap_{i=1}^{n-1} \text{kern}(\omega^i)$ is a one dimensional vector subbundle spanned by a single, non-vanishing vector field $V$. By Theorem 2.4.3 there exist coordinates $(x^1, \ldots, x^n)$ such that $V = \partial_{x^n}$. Since $\omega^i(V) = 0$ ($i = 1, \ldots, n-1$), there must exist functions $\omega^i_j$ with $\omega^i = \sum_{j=1}^{n-1} \omega^i_j(x^1, \ldots, x^n)\mathrm{d}x^j$.

We assume now that the theorem has been proven for $k = 1, \dots, \hat{k}-1$.

Let $f$ be any function such that $\mathrm{d}f, \omega^1, \dots, \omega^{n-\hat{k}}$ are linearly independent. Clearly, this system of differential forms satisfies the integrability conditions

$$\begin{aligned} \mathrm{dd}f \wedge \mathrm{d}f \wedge \omega^1 \wedge \cdots \wedge \omega^{n-\hat{k}} &= 0, \\ \mathrm{d}\omega^i \wedge \mathrm{d}f \wedge \omega^1 \wedge \cdots \wedge \omega^{n-\hat{k}} &= 0. \end{aligned}$$

By our induction assumption there exist coordinates $(\hat{x}^1, \dots, \hat{x}^n)$ with

$$\begin{aligned} \mathrm{d}f &= f_1(\hat{x})\mathrm{d}\hat{x}^1 + \cdots + f_{n-(\hat{k}-1)}\mathrm{d}\hat{x}^{n-(\hat{k}-1)}, \\ \omega^i &= \omega^i_1(\hat{x})\mathrm{d}\hat{x}^1 + \cdots + \omega^i_{n-(\hat{k}-1)}\mathrm{d}\hat{x}^{n-(\hat{k}-1)}. \end{aligned} \tag{2.5.3}$$

We will now show that there is a Pfaffian system $\{\Omega^1, \dots, \Omega^{n-\hat{k}}\}$ which defines the same vector bundle as the original Pfaffian system but does not depend on $\hat{x}^{n-\hat{k}+2}, \dots, \hat{x}^n$. Equation (2.5.3)implies that $f$ must be a function of $\hat{x}^1, \dots, \hat{x}^{n-(\hat{k}-1)}$ only. We can therefore substitute one of the coordinate functions by $f$, say $f = \hat{x}^{n-(\hat{k}-1)}$. At each $x \in M$ the two sets of 1-forms $\{\mathrm{d}\hat{x}^1, \dots, \mathrm{d}\hat{x}^{n-\hat{k}}, \mathrm{d}f\}$ and $\{\mathrm{d}f, \omega^1, \dots, \omega^{n-\hat{k}}\}$ are each pointwise linearly independent and span at each $x \in M$ the same subspace of $T^*_x M$. Hence there are functions $h^i_j$ with $\omega^i = h^i_j \mathrm{d}\hat{x}^j + h^i_{n-(\hat{k}-1)}\mathrm{d}f$, where $(h^i_j)_{i,j=1,\dots,n-\hat{k}}$ is an invertible matrix at each point. This implies that there are functions $h^i$ such that the differential forms $\Omega^i = \mathrm{d}\hat{x}^i + h^i(\hat{x})\mathrm{d}f$ span the same space as $\{\omega^1, \dots, \omega^{n-\hat{k}}\}$. Since the vector bundle $E := \bigcap_{i=1}^{n-k} \mathrm{kern}(\omega^i) = \bigcap_{i=1}^{n-k} \mathrm{kern}(\Omega^i)$ is integrable we obtain from Lemma 2.5.8

$$\begin{aligned} 0 &= \mathrm{d}\Omega^i \wedge \Omega^1 \wedge \cdots \wedge \Omega^{n-\hat{k}} \\ &= \mathrm{d}h^i \wedge \mathrm{d}f \wedge \Omega^1 \wedge \cdots \wedge \Omega^{n-\hat{k}} \\ &= \mathrm{d}h^i \wedge \mathrm{d}f \wedge \mathrm{d}\hat{x}^1 \wedge \cdots \wedge \mathrm{d}\hat{x}^{n-\hat{k}}. \end{aligned}$$

It follows that the functions $h^i$ (and therefore also the 1-forms $\Omega^i$) depend only on $\hat{x}^1, \dots, \hat{x}^{n-\hat{k}+1}$ (where we have used $f = \hat{x}^{n-\hat{k}+1}$).

Since the 1-forms $\Omega^1, \dots, \Omega^{n-\hat{k}}$ do not depend on $\hat{x}^{n-\hat{k}+2}, \dots, \hat{x}^n$ the Pfaffian system $\{\Omega^1, \dots, \Omega^{n-\hat{k}}\}$ can also be considered as a Pfaffian system of the space $M^{n-\hat{k}+1}$ parametrized by $\hat{x}^1, \dots, \hat{x}^{n-(\hat{k}-1)}$. These forms define a 1-vector subbundle of $TM^{n-\hat{k}+1}$. At the beginning of the proof we have already established the assertion in this case, hence there are new coordinates $(x^1, \dots, x^{n-\hat{k}+1})$ of $M^{n-\hat{k}+1}$ and functions $\Omega^i_j$ such that $\Omega^i = \sum_{j=1}^{n-\hat{k}} \Omega^i_j(x)\mathrm{d}x^j$. These equalities also hold for the 1-forms $\Omega^i$ considered on $M$ since they only depend on $\hat{x}^1, \dots, \hat{x}^{n-(\hat{k}-1)}$.

The assertion follows now from the fact that the 1-forms $\omega^i$ are linear combinations of the 1-forms $\Omega^i$. ∎

### 2.5.3 Orientable real manifolds

*This section is a prerequisitive for the following section on integration.*

We were led to the definition of a manifold by the localization of the global concept of a vector space and have seen that a manifold can be thought of as a collection of local $\mathbb{R}^n$s which have been patched together. The example of a sphere shows that, globally, a manifold may be very different from $\mathbb{R}^n$. In this section we give a (very primitive) global classification of manifolds by dividing the collection of all manifolds into two classes. More sophisticated global classifications can be found in books on differential topology such as (Guillemin and Pollack 1974) or (Bott and Tu 1982).

**Definition 2.5.7.** *A real manifold $M$ is* orientable *if there exists a nowhere vanishing $n$-form $\nu$ on $M$. An orientation of $M$ is the choice of one of the two equivalence classes $\{f\nu|\, f \in C^\infty(M, \mathbb{R}^+\backslash\{0\})\}$, $\{-f\nu|\, f \in C^\infty(M, \mathbb{R}^+ \setminus \{0\})\}$. An* oriented *manifold is an orientable manifold together with an orientation.*

**Proposition 2.5.2.** *A real manifold $M$ is orientable if and only if it has an atlas $\{(\mathcal{U}_k, \varphi_k)\}_{k\in\mathbb{N}}$ such that for all $a, b \in \mathbb{N}$ and all $x \in \mathcal{U}_a \cap \mathcal{U}_b$ the differential $\mathrm{D}\left(\varphi_a \circ (\varphi_b)^{-1}\right)_x : \mathbb{R}^n \to \mathbb{R}^n$ has positive determinant.*

*Proof.* "$\Rightarrow$": Assume that $\nu$ is a nowhere vanishing $n$-form and let

$$\{(\mathcal{V}_j, \psi_j)\}_{j\in\mathbb{N}}$$

be a countable atlas. In order to simplify notation we renumber these charts such that for each $k > 2$ there is a $j < k$ with $\mathcal{V}_j \cap \mathcal{V}_k \neq \emptyset$.

We set $(\mathcal{U}_1, \varphi_1) = (\mathcal{V}_1, \psi_1)$ thereby trivially defining an atlas for $\mathcal{V}_1$ which satisfies the positive determinant condition This atlas can be extended by an induction argument. Assume that we have defined an atlas

$$\{(\mathcal{U}_1, \varphi_1), \ldots, (\mathcal{U}_k, \varphi_k)\}$$

for the set $\mathcal{V}_1 \cup \cdots \cup \mathcal{V}_k$ which satisfies the positive determinant condition and let $j \in \{1, \ldots, k\}$ be an index with $\mathcal{V}_{k+1} \cap \mathcal{V}_j \neq \emptyset$.

There are a nowhere vanishing function $f_j : \varphi(\mathcal{U}_j) \to \mathbb{R}$ with $(\varphi_j)_*\nu = f_j \mathrm{d}x^1 \wedge \cdots \wedge \mathrm{d}x^n$ and a nowhere vanishing function $\tilde{f}_{k+1} : \varphi(\mathcal{V}_{k+1}) \to \mathbb{R}$ satisfying $(\psi_{k+1})_*\nu = \tilde{f}_{k+1}\mathrm{d}x^1 \wedge \cdots \wedge \mathrm{d}x^n$. Since neither $f_j$ nor $f_{k+1}$ vanish on $\varphi_j(\mathcal{U}_j) \cap \psi_{k+1}(\mathcal{V}_{k+1})$, we have either $f_j \cdot \tilde{f}_{k+1}(y) > 0$ or $f_j \cdot \tilde{f}_{k+1}(y) < 0$ for all $y \in \varphi_j(\mathcal{U}_j) \cap \psi_{k+1}(\mathcal{V}_{k+1})$. In the first case we set

$$(\mathcal{U}_{k+1}, \varphi_{k+1}) = (\mathcal{V}_{k+1}, \psi_{k+1}) \text{ and } f_{k+1} = \tilde{f}_{k+1}$$

whereas in in the second case we set

$$(\mathcal{U}_{k+1}, \varphi_{k+1}) = (\mathcal{V}_{k+1}, \chi_{1\leftrightarrow 2} \circ \psi_{k+1}) \text{ and } f_{k+1} = -\tilde{f}_{k+1},$$

where $\chi_{1\leftrightarrow 2}$ is the reflection defined by

$$\chi_{1\leftrightarrow 2}(y^1, y^2, y^3, \ldots, y^n) = (y^2, y^1, y^3, \ldots, y^n).$$

In either case we have then $(\varphi_{k+1})_*\nu = f_{k+1}\mathrm{d}x^1\wedge\cdots\wedge\mathrm{d}x^n$ and $f_j \cdot f_{k+1} > 0$. This implies

$$\begin{aligned}\det(D\left(\varphi_j \circ (\varphi_{k+1})^{-1}\right) \mathrm{d}x^1\wedge\cdots\wedge\mathrm{d}x^n &\\ &= \left(\varphi_j \circ (\varphi_{k+1})^{-1}\right)^* \mathrm{d}x^1 \wedge\cdots\wedge \mathrm{d}x^n \\ &= \frac{f_j}{f_{k+1} \circ \left(\varphi_j \circ (\varphi_{k+1})^{-1}\right)} \mathrm{d}x^1 \wedge\cdots\wedge \mathrm{d}x^n\end{aligned}$$

and therefore $\det(D\left(\varphi_j \circ (\varphi_{k+1})^{-1}\right) > 0$ in either case. We still have to show that $\det(D\left(\varphi_i \circ (\varphi_{k+1})^{-1}\right) > 0$ for any $i \in \{1, \ldots, k\}$ with $\mathcal{U}_i \cap \mathcal{U}_{k+1} \neq 0$. For $l \in \{1, \ldots, k+1\}$ let $f_l$ be defined by $(\varphi_l)_*\nu = f_l \mathrm{d}x^1 \wedge\cdots\wedge \mathrm{d}x^n$ $(l \in \{p, q\})$. Since all $f_l$ $(l \in \{1, \ldots, k\}$ have the same sign and also $f_{k+1}$ and $f_j$ have the same sign it follows that $\mathrm{sign}(f_i) = \mathrm{sign}(f_j) = \mathrm{sign}(f_{k+1})$. Hence $\det(D\left(\varphi_i \circ (\varphi_{k+1})^{-1}\right) > 0$

"$\Leftarrow$": Let $\{g_k : M \to [0,1]\}_{k\in\mathbb{N}}$ be a partition of unity subordinate to $\{\mathcal{U}_k\}_{k\in\mathbb{N}}$ and let

$$\nu = \sum_{k=1}^{n} g_k (\varphi_k)^*(\mathrm{d}x^1 \ldots \mathrm{d}x^n).$$

This is a smooth, well defined $n$-form since at each $x$ only finitely many $g_k(x)$ are non-zero and the support of each $g_k$ is contained in $\mathcal{U}_k$. Let $x \in \mathcal{U}_i$ and $i_1, \ldots i_p$ all indices with $g_{i_j}(x) \neq 0$. Then

$$(\varphi_i)_*\nu_x = \sum_{j=1}^{p} g_{i_j}(x) \mathrm{D}\left(\varphi_{i_j} \circ (\varphi_i)^{-1}\right)_{|\varphi_i(x)} \mathrm{d}x^1 \wedge\cdots\wedge \mathrm{d}x^n$$

does not vanish since all $g_{i_j}$ are strictly positive and all

$$\mathrm{D}\left(\varphi_{i_j} \circ (\varphi_i)^{-1}\right)_{|\varphi_i(x)}$$

have the same sign. ∎

**Definition 2.5.8.** *Let $M$ be an oriented manifold and $\nu \in \Omega^n(M)$ be a representative of the orientation. An* oriented atlas *is an atlas $\{(\mathcal{U}_a, \varphi_a)\}_{a\in A}$ such that for each $a \in A$ there exists a strictly positive function $\nu_{a,1\ldots n} : \mathcal{U}_a \to \mathbb{R}^+$ with*

$$(\varphi_a)_* \nu = \nu_{a,1\ldots n} \mathrm{d}x^1 \ldots \mathrm{d}x^n,$$

*where $(x^1, \ldots, x^n)$ are the standard coordinates of $\mathbb{R}^n$. A* positively oriented chart *is a chart which belongs to an oriented atlas.*

*Example 2.5.1 (Möbius band, continued from page 49).* The Möbius band $M$ defined in Example 2.1.2 is not orientable. Using the same notation as in its definition, the set $\mathcal{U}_1$ intersects $\mathcal{U}_2$ in two subsets, $\mathcal{W}_+ = \pi^{-1}((a, 2a) \times (-b, b))$ and and $\mathcal{W}_- = \pi^{-1}((0, a) \times (-b, b))$. The map $\mathrm{D}\left(\varphi_1 \circ (\varphi_2)^{-1}\right)_{|x}$ has positive determinant for all $x \in \mathcal{W}_+$ and negative determinant for all $x \in \mathcal{W}_-$. Assume now that $M$ is orientable, i.e., that there is a nowhere vanishing 2-form $\nu$ on $M$. Then there are nowhere vanishing functions

$$f_1, : \mathcal{V}_1 \to \mathbb{R}, \qquad f_2 : \mathcal{V}_2 \to \mathbb{R}$$

with

$$\left((\varphi_1)^{-1}\right)^* \nu = f_1 \mathrm{d}x^1 \wedge \mathrm{d}x^2 \text{ and } \left((\varphi_2)^{-1}\right)^* \nu = f_2 \mathrm{d}x^1 \wedge \mathrm{d}x^2.$$

From $\left((\varphi_2)^{-1}\right)^* \nu = f_2 \mathrm{d}x^1 \wedge \mathrm{d}x^2 = \left((\varphi_1 \circ (\varphi_2)^{-1}\right)^* (f_1 \mathrm{d}x^1 \wedge \mathrm{d}x^2)$ we obtain that $f_2(x) = \det(\mathrm{D}\left(\varphi_1 \circ (\varphi_2)^{-1}\right) f_1$ for all $x \in \mathcal{W}_+ \cup \mathcal{W}_-$. Since the determinant $\det(\mathrm{D}\left(\varphi_1 \circ (\varphi_2)^{-1}\right)$ changes sign we get a contradiction. Thus the Möbius band is not orientable.

It is a simple but good exercise to actually build a Möbius band from paper and to verify using this model that there are closed curves along which there does not exist any continuous frame.

### 2.5.4 Integration on real manifolds

*In this section, we restrict to $\mathbb{K} = \mathbb{R}$. This is necessary since we need to employ partitions of unity. See Remark 2.5.2 below for the integration of complex valued functions.*

One usually introduces integration as a method to determine the volume of an open, bounded region $B \subset \mathbb{R}^n$. The main idea is as follows. We divide $B$ into small parallel epipeds $B_i$. Each $B_i$ carries a number $f(B_i)$ representing the volume of $B_i$. Summing up all these numbers gives an approximation for the volume of $B$. Clearly, the function $f$ which maps parallel epipeds into real numbers must satisfy certain properties. The most obvious property is that if we divide $B_i$ into two disjunct parallel epipeds $A_i, C_i$ with $B_i = A_i \cup C_i$, we have $f(B_i) \approx f(A_i) + f(C_i)$, at least if $B_i$ is sufficiently small. Then, choosing an infinite sequence $\{\{B_{i,a}\}_{i\in I(a)}\}_a$ ($a \in \mathbb{N}$) of such divisions we obtain a sequence of numbers $\{\sum_{i\in I(a)} f(B_{i,a})\}_{a\in\mathbb{N}}$ which in most cases of interest has a well defined limit, the volume vol($B$) of $B$. Linear algebra indicates the following choice for $f$. Let $\{e_1, \ldots, e_n\}$ the standard basis of $\mathbb{R}^n$, $\{\theta^1, \ldots, \theta^n\}$ its dual basis, and $b_{i,1}, \ldots, b_{i,n}$ those vectors which span the parallel epiped $B_i$. The number $f(B_i) = \theta^1 \wedge \cdots \wedge \theta^n(b_{i,1}, \ldots, b_{i,n})$ is then the Euclidian volume of $B_i$ with respect to the standard Euclidean scalar product. This function clearly satisfies the the additivity condition above. If one knows how to determine volumes one can also integrate continuous functions $\psi\colon B \to \mathbb{R}$ (for instance mass densities) by replacing the differential form $\theta^1 \wedge \cdots \wedge \theta^n$ with the differential form $\psi\theta^1 \wedge \cdots \wedge \theta^n$.

Let us now turn to manifolds. The main problem here is that we do not have a linear space in which to embed the cubes. However, by now we are familiar with the idea of translating concepts to their infinitesimal counterparts. Since the tangent space was introduced as the linear approximation of the manifold, it is natural to place our parallel epipeds which use the linear structure of $\mathbb{R}^n$ into the tangent spaces rather than into the manifold itself. In other words, we divide $M$ into small sets $\mathcal{V}_i$ such that each of these sets corresponds to a parallel epiped in $T_{x_i}M$, where $x_i$ is a point in $\mathcal{V}_i$. We cannot define a canonical volume because a general manifold does not have a preferred frame $\{E_1, \ldots, E_n\}$. This indicates that it is more natural to integrate $n$-forms directly than to define the volume of a set first. We will later recover the volume as aspecial case (cf. Definition 4.2.1).

To simplify part of our discussions we will define integration for $n$-forms which are not necessarily smooth.

**Definition 2.5.9.** *Let $M$ be a real manifold. We denote by $\Omega^n_{\mathrm{c}}(M)$ the set of all continuous $n$-forms.*

We clearly have $\Omega^n(M) \subset \Omega^n_{\mathrm{c}}(M)$. Let $(\mathcal{U}, \varphi)$ be a chart and $\omega \in \Omega^n_{\mathrm{c}}(M)$ be an $n$-form with compact supp($\omega$) $\subset \mathcal{U}$. Writing $\varphi = (x^1, \ldots, x^n)$ there is a unique smooth function $\omega_{1\ldots n}$ with $\omega = \omega_{1\ldots n}\mathrm{d}x^1 \wedge \cdots \wedge \mathrm{d}x^n$. We define

$$\int_{(\mathcal{U},\varphi)} \omega = \int_{(\mathcal{U},\varphi)} \omega_{1\ldots n}\mathrm{d}x^1 \wedge \cdots \wedge \mathrm{d}x^n := \int_{\varphi(\mathcal{U})} \omega_{1\ldots n} \circ \varphi^{-1}\mathrm{d}x^1 \ldots \mathrm{d}x^n,$$

where the last expression is the usual integration in $\mathbb{R}^n$. We still have to show

(i) that $\int_{\mathcal{U},\varphi} \omega$ does not depend on the chosen chart,
(ii) how to extend this local definition to manifolds which may not be covered by a single chart,
(iii) how to extend this local definition to $n$-forms which do not have compact support.

(i): Let $(\mathcal{U}, \psi)$ be another chart, denote by $(y^1, \ldots, y^n)$ the corresponding coordinate functions and set

$$\eta = \operatorname{sign}\left((\det(\{\frac{\partial x^a}{\partial y^b}\}))\right).$$

Then we have

$$\begin{aligned}\omega = \omega_{i\ldots n} \mathrm{d}x^1 \wedge \cdots \wedge \mathrm{d}x^n &= \omega_{i\ldots n} \det(\{\frac{\partial x^a}{\partial y^b}\}) \circ \psi \, \mathrm{d}y^1 \wedge \cdots \wedge \mathrm{d}y^n \\ &= \eta \omega_{i\ldots n} \left|\det(\{\frac{\partial x^a}{\partial y^b}\})\right| \circ \psi \, \mathrm{d}y^1 \wedge \cdots \wedge \mathrm{d}y^n\end{aligned}$$

and therefore

$$\begin{aligned}\int_{\psi(\mathcal{V})} & \omega_{1\ldots n} \circ \psi^{-1} \mathrm{d}y^1 \ldots \mathrm{d}y^n \\ &= \int_{\psi(\mathcal{U})} \omega_{i\ldots n} \circ \psi^{-1} \left|\det(\{\frac{\partial x^a}{\partial y^b}\})\right| \mathrm{d}y^1 \ldots \mathrm{d}y^n \\ &= \eta \int_{\varphi\circ\psi(\psi(\mathcal{U}))} \omega_{i\ldots n} \circ \psi^{-1} \circ (\varphi \circ \psi)^{-1} \mathrm{d}x^1 \ldots \mathrm{d}x^n \\ &= \eta \int_{\varphi(\mathcal{U})} \omega_{1\ldots n} \circ \varphi^{-1} \mathrm{d}x^1 \ldots \mathrm{d}x^n.\end{aligned}$$

Hence the definition is indeed coordinate independent if one fixes an orientation in advance and restricts to an oriented atlas.

In order to address (ii) and to define integration globally we will employ a partition of unity.

**Definition 2.5.10.** *Let $M$ be an oriented $n$-dimensional real manifold and $\omega \in \Omega^n_c(M)$ be a $n$-form with compact support. Then*

$$\int_M \omega := \sum_{a\in A} \int_{(\mathcal{U}_{b(a)}, \varphi_{b(a)})} f_a \, \omega,$$

*where $\{(\mathcal{U}_b, \varphi_b)\}_{\beta\in B}$ is an oriented atlas such that each $\mathcal{U}_b$ has compact closure, $f_a$ a partition of unity subordinate to $\{(\mathcal{U}_b\}_{b\in B}$ and $b(a)$ is an index with* $\operatorname{supp}(f_a) \subset \mathcal{U}_{b(a)}$.

Since each $x \in M$ has a neighbourhood which is intersected by only finitely many $\operatorname{supp}(f_a)$ and $\operatorname{supp}(\omega)$ is compact the sum in the definition above is finite.

Let $\{g_c\}_{c\in C}$ be another partition of unity subordinate to $\{(\mathcal{U}_b\}_{b\in B}$ and $\tilde{b}(c)$ be an index with $\operatorname{supp}(g_c) \subset \mathcal{U}_{\tilde{b}(c)}$. Since all sums involved are finite we can calculate

$$\begin{aligned}
\sum_{a\in A}\int_{(\mathcal{U}_{b(a)},\varphi_{b(a)})} f_a\,\omega &= \sum_{a\in A}\int_{(\mathcal{U}_{b(a)},\varphi_{b(a)})}\sum_{c\in C} g_c\,f_a\,\omega\\
&=\sum_{a\in A}\sum_{c\in C}\int_{(\mathcal{U}_{b(a)},\varphi_{b(a)})} g_c\,f_a\,\omega\\
&=\sum_{c\in C}\sum_{a\in A}\int_{(\mathcal{U}_{\tilde b(c)},\varphi_{\tilde b(c)})} g_c\,f_a\,\omega\\
&=\sum_{c\in C}\int_{(\mathcal{U}_{\tilde b(c)},\varphi_{\tilde b(c)})} g_c\sum_{a\in A} f_a\,\omega\\
&=\sum_{c\in C}\int_{(\mathcal{U}_{\tilde b(c)},\varphi_{\tilde b(c)})} g_c\,\omega.
\end{aligned}$$

This implies that the definition is independent of the chosen partition of unity. Since by coordinate invariance it is also independent of the chosen oriented atlas our definition of integration over $n$-forms with compact support is well defined.

We will now address point (iii). If $\omega \in \Omega^n_c(M)$ does not have compact support, its integral may not exist. This is completely analogous to the integration of functions $f\colon \mathbb{R}\to\mathbb{R}$. For our purposes the following extension is sufficient.

**Definition 2.5.11.** *Let $M$ be an oriented $n$-dimensional real manifold and $\omega \in \Omega^n_c(M)$. Let $\{\mathcal{U}_a,\varphi_a)\}_{a\in A}$ be an oriented atlas and for each $a$ let $\omega^{(a)}_{1\ldots n}\colon \mathcal{U}_a \to \mathbb{R}$ be defined by $(\varphi_a)_*\omega = \omega^{(a)}_{1\ldots n}\mathrm{d}x^1\wedge\cdots\wedge\mathrm{d}x^n$. The* modulus *of $\omega$ is the continuos $n$-form $|\omega|$ locally defined by $(\varphi_a)_*|\omega| = |\omega^{(a)}_{1\ldots n}|\mathrm{d}x^1\wedge\cdots\wedge\mathrm{d}x^n$.*

Let $M$ be an oriented $n$-dimensional real manifold and $(\mathcal{U}_a,\varphi_a)_{a\in\mathbb{N}}$ be a countable oriented atlas such that each $\mathcal{U}_a$ has compact closure. As a preparation to the following definition we first need to give a meaning to the expression $\{\int_{\mathcal{U}_1\cup\cdots\cup\mathcal{U}_k}|\omega|\}_{k\in\mathbb{N}}$ for arbitrary $\omega\in\Omega^n_c(M)$ which do not necessarily have compact support. For each $j\in\{1,\ldots,k\}$ let $I_j=\{l\in\{j+1,\ldots,k\}\,|\,U_j\cap U_l\neq\emptyset\}$. if $\eta\in\Omega^n_c(\mathcal{U}_1\cup\cdots\cup\mathcal{U}_k)$ has compact support then we have $\int_{\mathcal{U}_1\cup\cdots\cup\mathcal{U}_k}\eta=\sum_{j=1}^k\left(\int_{(\mathcal{U}_j,\varphi_j}\eta-\sum_{l\in I_j}\int_{(\mathcal{U}_j\cap U_l,\varphi_j}\eta\right)$. If $\omega\in\Omega^n_c(M)$ is a general differential form we *define*

$$\int_{\mathcal{U}_1\cup\cdots\cup\mathcal{U}_k}\omega=\sum_{j=1}^k\left(\int_{(\mathcal{U}_j,\varphi_j}\omega-\sum_{l\in I_j}\int_{(\mathcal{U}_j\cap U_l,\varphi_j}\omega\right).$$

**Definition 2.5.12.** *Let $M$ be an oriented $n$-dimensional real manifold and $(\mathcal{U}_a,\varphi_a)_{a\in\mathbb{N}}$ be a countable oriented atlas such that each $\mathcal{U}_a$ has compact closure. The $n$-form $\omega\in\Omega^n_c(M)$ is* integrable *if the (monotonically increasing) sequence*

$$\{\int_{\mathcal{U}_1\cup\cdots\cup\mathcal{U}_k} |\omega|\}_{k\in\mathbb{N}} = \sum_{a\in A_k} \int_{(\mathcal{U}_{b(a)},\varphi_{b(a)})} f_a^k\, \omega,$$

*where* $\{f_a^k : \mathcal{U}_1 \cup \cdots \cup \mathcal{U}_k\}$ *a partition of unity subordinate to* $\{(\mathcal{U}_b\}_{b\in B}$ *and* $b(a)$ *is an index with* $\operatorname{supp}(f_a) \subset \mathcal{U}_{b(a)}$. *is bounded.*

Clearly, any $\omega \in \Omega^n_c(M)$ with compact support is integrable. The definition is independent of the chosen atlas. To see this let $(\mathcal{V}_b, \psi_b)_{b\in\mathbb{N}}$ be a second countable oriented atlas such that each $\mathcal{V}_b$ has compact closure. Then for each $k \in \mathbb{N}$ there is a $j(k) \in \mathbb{N}$ such that $\mathcal{U}_1 \cup \ldots \mathcal{U}_k \subset \mathcal{V}_1 \cup \ldots \mathcal{V}_{j(k)}$. Hence $\int_{\mathcal{U}_1\cup\cdots\cup\mathcal{U}_k} |\omega| \le \int_{\mathcal{V}_1\cup\cdots\cup\mathcal{V}_{j(k)}} |\omega|$ which implies that $\{\int_{\mathcal{U}_1\cup\cdots\cup\mathcal{U}_k} |\omega|\}_{k\in\mathbb{N}}$ is bounded if $\{\int_{\mathcal{V}_1\cup\cdots\cup\mathcal{V}_j} |\omega|\}_{j\in\mathbb{N}}$ is bounded.

We can now define integration for differential forms which may not have compact support.

**Definition 2.5.13.** *Let* $M$ *be an oriented* $n$*-dimensional real manifold and* $(\mathcal{U}_a, \varphi_a)_{a\in\mathbb{N}}$ *be a countable oriented atlas such that each* $\mathcal{U}_a$ *has compact closure. For any integrable* $\omega \in \Omega^n(\mathbb{R}^n)$ *we define its integral by*

$$\int_M \omega = \lim_{k\to\infty} \int_{\mathcal{U}_1\cup\cdots\cup\mathcal{U}_k} \omega.$$

This limit is well defined since integrability implies that

$$|\int_{\mathcal{U}_1\cup\cdots\cup\mathcal{U}_k} \omega - \int_{\mathcal{U}_1\cup\cdots\cup\mathcal{U}_{k+l}} \omega| \le \int_{\mathcal{U}_k\cup\cdots\cup\mathcal{U}_{k+l}} |\omega| \le \int_{\bigcup_{j=k}^\infty} |\omega| \to \infty$$

for $k \to \infty$ and therefore that $\{\int_{\mathcal{U}_1\cup\cdots\cup\mathcal{U}_k} \omega\}_{k\in\mathbb{N}}$ is a Cauchy sequence.

Observe that we can integrate any integrable $n$-form over any subset open subset $\mathcal{U}$ and any (not necessarily integrable) $n$-form over any open subset $\mathcal{V}$ with compact closure since these forms are clearly integrable with respect to the (open) submanifolds $\mathcal{U}$ and $\mathcal{V}$.

The following two lemmas are direct consequences of our definitions.

**Lemma 2.5.9 (Linearity of integration).** *Let* $M$ *be a real manifold and* $\omega, \tilde{\omega} \in \Omega^n_c(M)$ *differential forms which are integrable. Then we have for any real numbers* $a, b$

$$\int_M (a\omega + b\tilde{\omega}) = a\int_M \omega + b\int_M \tilde{\omega}.$$

**Lemma 2.5.10.** *Let* $M, N$ *be oriented real manifolds and* $f\colon M \to N$ *be a diffeomorphism such that* $f_*\nu$ *is a representative of the orientation of* $N$ *for each representative* $\nu$ *of the orientation of* $M$. *Then we have for any* $\omega \in \omega \in \Omega^n_c(M)$

$$\int_M f^*\omega = \int_N \omega.$$

*Complex valued functions play a very important rôle in functional analysis and quantum mechanics and vector valued differential forms are used in both physical and mathematical gauge theory. (For instance, the analogue of the electromagnetic field strength is described by a Lie algebra valued differential form.) Readers interested in integration over such objects can find some elementary definitions in the remark below. In this book, however, we will use these concepts only in order to motivate the definition of mean curvature vector field (cf. Definition 4.4.2).*

*Remark 2.5.2.* We have restricted to $\mathbb{K} = \mathbb{R}$ since partitions of unity do not exist in general over complex manifolds. It is possible however to integrate complex valued differential forms over a real manifold $M$. Let $M$ be a real manifold. A *complex valued differential form* is a map $\omega \colon x \mapsto \omega_x$ where $\omega_x \colon \overbrace{T_xM \times \cdots \times T_xM}^{p \text{ entries}} \to \mathbb{C}$ is multi-linear and anti-symmetric. Denote by $\Omega^p(M, \mathbb{C})$ the space of all complex valued differential form of degree $p$. For each $\omega \in \Omega^p(M, \mathbb{C})$ there exist two uniqely defined differential forms $\omega^{\mathrm{re}}, \omega^{\mathrm{im}} \in \Omega^p(M, \mathbb{R})$ such that $\omega = \omega^{\mathrm{re}} + \mathbf{i}\omega^{\mathrm{im}}$. We call $\omega \in \Omega^n(M, \mathbb{C})$ *integrable* if both $\omega^{\mathrm{re}}$ and $\omega^{\mathrm{im}}$ are integrable. The *integral* over an integrable complex valued $n$ form $\omega$ is defined by

$$\int_M \omega = \int_M \omega^{\mathrm{re}} + \mathbf{i} \int_M \omega^{\mathrm{im}}$$

We wish to show that integration over complex valued differential forms is $\mathbb{C}$-linear. Additivity is clear. With $\omega \in \Omega^n(M, \mathbb{C})$ and $a = a^{\mathrm{re}} + \mathbf{i}a^{\mathrm{im}} \in \mathbb{C}$ we have

$$\begin{aligned}
\int_M a\omega &= \int_M (a^{\mathrm{re}} + \mathbf{i}a^{\mathrm{im}})(\omega^{\mathrm{re}} + \mathbf{i}\omega^{\mathrm{im}}) \\
&= \int_M ((a^{\mathrm{re}}\omega^{\mathrm{re}} - a^{\mathrm{im}}\omega^{\mathrm{im}}) + \mathbf{i}(a^{\mathrm{re}}\omega^{\mathrm{im}} + a^{\mathrm{im}}\omega^{\mathrm{re}})) \\
&= (a^{\mathrm{re}} \int_M \omega^{\mathrm{re}} - a^{\mathrm{im}} \int_M \omega^{\mathrm{im}}) + \mathbf{i}(a^{\mathrm{re}} \int_M \omega^{\mathrm{im}} + a^{\mathrm{im}} \int_M \omega^{\mathrm{re}})) \\
&= (a^{\mathrm{re}} + \mathbf{i}a^{\mathrm{im}}) \int_M (\omega^{\mathrm{re}} + \mathbf{i}\omega^{\mathrm{im}}) \\
&= a \int_M \omega.
\end{aligned}$$

This definition can be further extended to vector spaces. Let $\mathbb{K} = \mathbb{R}$ or $\mathbb{K} = \mathbb{C}$ and $\mathfrak{V}$ be a $k$-dimensional vector space over $\mathbb{K}$. A *vector valued differential form of degree* $p$ is a smooth map $\omega \colon x \mapsto \omega_x$ where

$$\omega_x \colon \overbrace{T_xM \times \cdots \times T_xM}^{p \text{ entries}} \to \mathfrak{V}$$

is multi-linear and anti-symmetric. Denote by $\Omega^p(M, \mathfrak{V})$ the space of all vector valued differential form of degree $p$. If $\omega \in \Omega^p(M, \mathfrak{V})$ and

$\{\mathfrak{v}_1, \ldots, \mathfrak{v}_k$ a fixed basis of $\mathfrak{V}$, there exist $k$ uniquely defined differential forms $\omega^\alpha \in \Omega^p(M)$ with $\omega = \sum_{\alpha=1}^k \omega^\alpha \mathfrak{v}_\alpha$. A vector valued differential form $\omega$ is called *integrable* if all $\omega^\alpha$ are integrable. It is clear that the definition is independent of the chosen basis. We can now define the *integral* of a vector valued differential form $\omega$ via

$$\int_M \omega = \sum_{\alpha=1}^k \left(\int_M \omega^\alpha\right) \mathfrak{v}_\alpha .$$

This definition is independent of the chosen basis since for any other basis $\tilde{\mathfrak{v}}_1, \ldots, \tilde{\mathfrak{v}}_k$ with $\mathfrak{v}_\alpha = A_\alpha^\beta \tilde{\mathfrak{v}}_\beta$ we have $\omega = \sum_{\beta=1}^k \tilde{\omega}^\beta \tilde{\mathfrak{v}}_\beta$ where $\tilde{\omega}^\beta = \sum_{\alpha=1}^k A_\alpha^\beta \omega^\alpha$ and therefore

$$\begin{aligned}\sum_{\beta=1}^k \left(\int_M \tilde{\omega}^\beta\right) \tilde{\mathfrak{v}}_\beta &= \sum_{\alpha=1,\beta=1}^k \left(\int_M A_\alpha^\beta \omega^\alpha\right) \tilde{\mathfrak{v}}_\alpha = \sum_{\alpha=1,\beta=1}^k A_\alpha^\beta \left(\int_M \omega^\alpha\right) \tilde{\mathfrak{v}}_\alpha \\ &= \sum_{alpha=1}^k \left(\int_M \omega^\alpha\right) \mathfrak{v}_\alpha .\end{aligned}$$

We turn now to the theorem of Stokes which generalizes the classical integration formulas of Gauß and Stokes which in turn generalize fundamental theorem of calculus, $\int_a^b f'(x)\mathrm{d}x = f(b) - f(a)$.

First we need to define the concept of a manifold with boundary.

**Definition 2.5.14.** *The pair $(M, \partial M)$ is an $n$-dimensional* oriented manifold with boundary *if there exists an $n$-dimensional oriented manifold $\tilde{M}$ and an embedding $\iota\colon M \to \tilde{M}$ such that*

- *(i) the topological boundary $\partial\iota(M)$ is an $(n-1)$-dimensional submanifold of $\tilde{M}$ which is diffeomorphic to $\partial M$,*
- *(ii) for each $x \in \partial\iota(M)$ there is a positively oriented chart $(\mathcal{U}, \varphi)$ of $\tilde{M}$ centered at $x$ with $\varphi(\mathcal{U} \cap \iota(M)) = \{y \in \varphi(\mathcal{U}) \,|\, y^1 < 0\}$.*

*An oriented manifold with boundary is* compact *if $\iota(M) \cup \partial\iota(M) \subset \tilde{M}$ is compact.*

We will usually identify $M$ with $\iota(M)$ and $\partial M$ with $\partial\iota(M)$.

Let $\tilde{M}$ be a manifold and $N$ a hypersurface in $\tilde{M}$ such that $M = \tilde{M} \setminus N$ is connected. Then $(M, N)$ is not an oriented manifold with boundary, even if $\tilde{M}$ is oriented and compact. The following theorem of Stokes would not hold for $(M, N)$.

**Theorem 2.5.5 (Theorem of Stokes).** *Let $(M, \partial M)$ be an oriented compact real manifold with boundary and $\omega \in \Omega^{n-1}(M)$. Assume that*

$d\omega$ *and* $\omega$ *are integrable over* $M$ *and* $\partial M$*, respectively. Then Stokes' formula*

$$\int_{\partial M} \omega = \int_M d\omega$$

*holds.*

*Proof.* Let $\{\mathcal{U}_j, \varphi_j\}_{j\in\mathbb{N}}$ be an oriented atlas of $\tilde{M}$ such that for each $j$ there are intervals $(a_j^1, b_j^1), \dots (a_j^n, b_j^n)$ with $\varphi_j(\mathcal{U}_j) = (a_j^1, b_j^1) \times \cdots \times (a_j^n, b_j^n)$. The definition of a manifold with boundary implies that we can also assume that for each $\mathcal{U}_j$ which intersects $\partial M$ the equality $\varphi_j(\mathcal{U}_j \cap M) = \{y \in \varphi(\mathcal{U}) | \, y^1 < 0\}$ holds. Let $\{f_j\}_j \in \mathbb{N}$ be a partition of unity subordinate to $\{\mathcal{U}_j, \varphi_j\}_{j\in\mathbb{N}}$

We show first that

$$\int_{M\cap\mathcal{U}_j} d(f_j\omega) = \int_{\partial M\cap\mathcal{U}_j} f_j\omega \tag{2.5.4}$$

holds for each $j \in \mathbb{N}$. We may write $f_j\omega = \sum_{i=1}^n \omega_k^j dx^1 \wedge \cdots \wedge dx^{i-1} \wedge dx^{i+1} \wedge \cdots \wedge dx^n$ which implies $d(f_j\omega) = \sum_{i=1}^n (-1)^{i+1} \partial_i \omega_k^j dx^1 \wedge \cdots \wedge dx^n$. We set

$$[a_j, b_j]^i = (a_j^1, b_j^1) \times \dots \overbrace{(a_j^i, b_j^i)}^{\wedge} \cdots \times (a_j^n, b_j^n) \subset \mathbb{R}^{n-1},$$

where $\overbrace{(a_j^i, b_j^i)}^{\wedge}$ indicates that the $i$th interval is omitted. Since $\omega_k^j$ has compact support in $(a_j^1, b_j^1) \times \cdots \times (a_j^n, b_j^n)$ the left hand side of Equation 2.5.4 is given by

$$\begin{aligned}
&\int_{M\cap\mathcal{U}_j} d(f_j\omega) \\
&\quad= \int_{M\cap\mathcal{U}_j} \sum_{i=1}^n (-1)^{i+1} \frac{\partial\omega_k^j}{\partial x^i} dx^1 \wedge \cdots \wedge dx^n \\
&\quad= \int_{(a_j^1, b_j^1)\times\cdots\times(a_j^n, b_j^n)} \sum_{i=1}^n (-1)^{i+1} \frac{\partial\omega_k^j}{\partial x^i} dx^1 \dots dx^n \\
&\quad= \int_{(a_j^1, b_j^1)\times\cdots\times(a_j^n, b_j^n)} \frac{\partial\omega_k^j}{\partial x^1} dx^1 \dots dx^n \\
&\qquad+ \sum_{i=2}^n (-1)^{i+1} \int_{[a_j, b_j]^i} \Big( \overbrace{\omega_i^j(x^1, \dots, b_j^i, \dots, x^n)}^{=0} \\
&\qquad - \overbrace{\omega_i^j(x^1, \dots, a_j^i, \dots, x^n)}^{=0} \Big) dx^1 \dots \overbrace{dx^i}^{\wedge} \dots dx^n
\end{aligned}$$

$$= \begin{cases} \int_{[a_j,b_j]^1} \left(\omega_1^j(0,x^2,\dots,x^n) - \overbrace{\omega_1^j(a_j^1,x^2,\dots,x^n)}^{=0}\right) \mathrm{d}x^2 \dots \mathrm{d}x^n & \text{if } \mathcal{U}_j \text{ intersects } \partial M, \\ \int_{[a_j,b_j]^1} \left(\overbrace{\omega_1^j(b_j^1,x^2\dots,x^n)}^{=0} - \overbrace{\omega_1^j(a_j^1,x^2,\dots,x^n)}^{=0}\right) \mathrm{d}x^2 \dots \mathrm{d}x^n & \text{otherwise.} \end{cases}$$

There are three possible cases.

(i) If $M \cap \mathcal{U}_j = \emptyset$ then $\partial M \cap \mathcal{U}_j = \emptyset$ and both integrals vanish.

(ii) If $M \cap \mathcal{U}_j \neq \emptyset$ but $\partial M \cap \mathcal{U}_j = \emptyset$ the right hand side vanishes because we integrate over an empty set. The left hand side vanishes by our calculation

(iii) Assume that $M \cap \mathcal{U}_j \neq \emptyset$ and $\partial M \cap \mathcal{U}_j \neq \emptyset$. Then

$$\begin{aligned} \int_{M\cap\mathcal{U}_j} \mathrm{d}(f_j\omega) &= \int_{(a_j^2,b_j^2)\times\cdots\times(a_j^1,b_j^n)} \omega_1^j(0,x^2,\dots,x^n \,\mathrm{d}x^2 \dots \mathrm{d}x^n \\ &= \int_{\partial M\cap\mathcal{U}_j} \omega_1^j(0,x^2,\dots,x^n)\mathrm{d}x^2 \wedge \dots \wedge \mathrm{d}x^n \\ &= \int_{\partial M\cap\mathcal{U}_j} f_j\omega \end{aligned}$$

since the pull back of $\mathrm{d}x^1$ to $\partial M$ vanishes and therefore also the $n-1$-form

$$\omega_j^i(0,x^2,\dots,x^n)\mathrm{d}x^1 \wedge \dots \wedge \widehat{\mathrm{d}x^i} \wedge \dots \wedge \mathrm{d}x^n \text{ pulled back to } \partial M (i \geq 2).$$

We conclude the proof by summing over all local integrations:

$$\begin{aligned} \int_M \mathrm{d}\omega &= \int_M \left(\sum_{j=1}^{\infty} \mathrm{d}(f_j\omega)\right) = \sum_{j=1}^{\infty} \int_{M\cap U_j} \mathrm{d}(f_j\omega) \\ &= \sum_{j=1}^{\infty} \int_{\partial M\cap U_j} f_j\omega = \int_{\partial M} \omega. \end{aligned}$$

∎

**Corollary 2.5.4.** *Let $M$ be an $n$-dimensional, oriented, compact, real manifold and $\omega \in \Omega^{n-1}(M)$. Then $\int_M \mathrm{d}\omega = 0$.*

*Proof.* Since $M$ is a manifold without boundary, $\partial M = \emptyset$ and $\int_M \mathrm{d}\omega = \int_{\partial M} \omega = \int_{\emptyset} \omega = 0$. ∎

As an application of Stokes' theorem we prove that for every even-dimensional unit sphere every vector field must have a zero.[7] This theorem is also known as the "theorem of the hedgehog" since it shows that it is impossible to perfectly comb an "ideal" hedgehog.

**Lemma 2.5.11.** *Let $M$ be an $n$-dimensional, oriented, compact, real manifold and $f, \tilde{f}\colon M \to M$ homotopic maps. Then*

$$\int_M f^*\omega = \int_M \tilde{f}^*\omega$$

*for all $\omega \in \Omega^n(M)$.*

*Proof.* Lemma 2.5.6 implies that there exists a differential form $\tilde{\omega} \in \Omega^{n-1}(M)$ with $f^*\omega - \tilde{f}^*\omega = \mathrm{d}\tilde{\omega}$. Hence the assertion follows from Corollary 2.5.4. ∎

**Theorem 2.5.6.** *Let $S^n$ be the unit sphere in $\mathbb{R}^{n+1}$ and $V$ be a vector field on $S^n$. If $n$ is even then there is a point $x_0 \in S^n$ with $V(x_0) = 0$.*

*Proof.* If $V(x) \neq 0$ for all $x \in S^n$ we can normalize $V$ with respect to the Euclidean scalar product $\langle \cdot, \cdot \rangle_{\mathbb{R}^{n+1}}$ of $\mathbb{R}^{n+1}$, i.e., we can assume without loss of generality that $\langle V, V \rangle_{\mathbb{R}^{n+1}} = 1$. For each $x \in S^n$ we can identify $V(x)$ with a tangent vector of $\mathbb{R}^{n+1}$ and, since $T_x\mathbb{R}^{n+1}$ and $\mathbb{R}^{n+1}$ are canonically isomorphic further with a point in $\mathbb{R}^{n+1}$. Our normalization implies then that we have defined a map $V\colon S^n \to S^n$. Let $\gamma$ be an integral curve of $V$ with $\gamma(0) = x$. Then $\langle \gamma(t), \gamma(t) \rangle_{\mathbb{R}^{n+1}} = 1$ implies $0 = \left(\frac{\mathrm{d}}{\mathrm{d}t} \langle \gamma(t), \gamma(t) \rangle_{\mathbb{R}^{n+1}}\right)_{|t=0} = 2\langle x, V(x) \rangle_{\mathbb{R}^{n+1}}$. Hence $V(x)$ is perpendicular to $x$ and the homotopy

$$\begin{aligned} F\colon [0,1] \times S^n &\to S^n \\ x &\mapsto \cos(\pi t)x + \sin(\pi t)V(x) \end{aligned}$$

is well defined. It satisfies $F(0,x) = x$ and $F(1,x) = -x$ for all $x \in S^n$. Since $-\mathrm{id}$ is homotopic to id and the diffeomorphism $-\mathrm{id}$ changes the orientation of $S^n$ (for $n$ even) Lemma 2.5.11 implies that for every $\omega \in \Omega^n(S^n)$

$$\int_{S^n} \omega = -\int_{S^n} (-\mathrm{id})^*\omega = -\int_{S^n} \omega.$$

holds. This in turn implies $\int_{S^n} \omega = 0$ for all $n$-forms $\omega$ which is certainly not true. Hence our initial assumption $V(x) \neq 0$ for all $x \in S^n$ must be wrong. ∎

## 2.6 Connections and projective structures

p. [↓ F

There is one feature of $\mathbb{A}^n$ which we have ignored so far in our efforts to localize spacetime. Given any two different points in $\mathbb{A}^n$ there is a unique line passing through them. This global structure has an infinitesimal counterpart given by the directional derivative of vector fields. In fact, these lines are exactly those curves $\gamma\colon [a,b] \to \mathbb{A}^n$ which satisfy $\mathrm{D}\dot\gamma(\dot\gamma) = 0$. (Observe that this expression is well defined, i.e. for any vector field $V$ with $V(\gamma(t)) = \dot\gamma(t)$ we have $\mathrm{D}V(\dot\gamma) = 0$.)

Recall that in Chap. 1 we have relied on this *affine structure* in order to introduce inertial observers. Here we will introduce a generalization of it to general manifolds. This will be done by generalizing the directional derivative D.

In the affine space $\mathbb{A}^n$ the difference of a derivative of a map $\psi\colon \mathbb{A}^n \to \mathbb{A}^n$ and a vector field $V\colon \mathbb{A}^n \to \mathbb{K}^n$ is blurred. In fact, let $\{e_1,\ldots,e_n\}$ be the standard basis of $\mathbb{K}^n$. Since

- the derivative of $\psi$ in direction $w$ at $x$ is given by $x + \frac{\partial\psi^i}{\partial x^j} w^j e_i$ and
- the derivative of $V$ in direction $w$ at $x$ is given by $\frac{\partial V^i}{\partial x^j} w^j e_i$

it is difficult to see the difference between both kinds of derivatives. Accordingly, they are commonly both denoted by D. Consider now a map $\phi\colon M \to M$ and a vector field $U$ on $M$. The derivative of $\phi$ at $x$ in direction $w_x$ is now given by $T_x\phi(w_x) \in T_{\phi(x)}M$ whereas the analogous derivative of $V$ is given by $T_xV(w_x) \in T_{V(x)}TM$. We do not obtain a vector but an element in the tangent bundle of the tangent bundle. It follows that this derivative cannot be used for defining straight lines.

In order to obtain an analogue of the directional derivative with values in $TM$ we will need an additional structure, a *connection*. The following definition is an expression of the idea that infinitesimally a connection $\nabla$ should be the same as the usual derivative D.

**Definition 2.6.1.** *A* covariant derivative *or* connection $\nabla$ *is a map*

$$\nabla\colon \mathcal{T}^1_0(M) \to \mathcal{T}^1_1(M), \quad V \to \nabla V, \quad \nabla V(W) =: \nabla_W V =: \nabla_a V^b W^a \partial_{x^b}$$

*such that for all vector fields $U, V, W$, all functions $f, h$, and all $\alpha, \beta \in \mathbb{R}$ the following holds.*

*(i)* $\nabla_{fV + hW} U = f\nabla_V U + h\nabla_W U$,
*(ii)* $\nabla_W(\alpha U + \beta V) = \alpha\nabla_W U + \beta\nabla_W V$,
*(iii)* $\nabla_W fU = (W \bullet f)U + f\nabla_W U$.

*The* torsion *of $\nabla$ is the tensor field*

$$(U,V) \mapsto \mathrm{Tor}(U,V) = \nabla_U V - \nabla_V U - [U,V].$$

[7] Recall that we only consider smooth vector fields

*A covariant derivative* $\nabla$ *is called* torsion-free *if in addition to (i)–(iii) the equation*

*(iv)* $\mathrm{Tor} = 0.$

*holds.*

This definition is justified by the following theorem.

**Theorem 2.6.1.** *A map* $\nabla\colon \mathcal{T}^1_0(M) \to \mathcal{T}^1_1(M)$ *is a torsion-free covariant derivative if and only if for each* $x \in M$ *there exist coordinates* $(x^0, \ldots, x^{n-1})$ *centered at* $x$ *such that*

$$\left(\frac{\partial V^a}{\partial x^b}\right)_{|x} = (\nabla_b V^a)_{|x} \, . \tag{2.6.5}$$

*holds at* $x$ *for all vector fields* $V$.

*Proof.* Assume first that there exist coordinates such that the Equation 2.6.5 is satisfied. Then it is clear that properties (i)–(iv) of a torsion-free covariant derivative are satisfied.

Let $\nabla$ be a covariant derivative and $(U,\varphi)$, $(\tilde{U},\tilde{\varphi})$ be two charts centered at $x$. We denote the coordinates with respect to these two charts by $x^i(y) = \varphi^i(y)$, $\tilde{x}^i(y) = \tilde{\varphi}^i(y)$. Observe that $(\mathrm{D}(\varphi \circ \tilde{\varphi}^{-1}))^a_b = \frac{\partial x^a}{\partial \tilde{x}^b}$ and $(\mathrm{D}(\tilde{\varphi} \circ \varphi^{-1}))^a_b = \frac{\partial \tilde{x}^a}{\partial x^b}$.

Setting $\nabla_{\partial_{x^a}} \partial_{x^b} = \Gamma^c_{ab} \partial_{x^c}$ and $\nabla_{\partial_{\tilde{x}^a}} \partial_{\tilde{x}^b} = \tilde{\Gamma}^c_{ab} \partial_{\tilde{x}^c}$ we have

$$\begin{aligned} \nabla_W V &= W^a \partial_{x^a} V^c \partial_{x^c} + \Gamma^c_{ab} \partial_{x^c} W^a V^b \partial_{x^c} \\ &= \tilde{W}^a \partial_{\tilde{x}^a} \tilde{V}^c \partial_{x^c} + \tilde{\Gamma}^c_{ab} \partial_{\tilde{x}^c} \tilde{W}^a \tilde{V}^b \partial_{x^c}, \end{aligned}$$

where $\tilde{V}^a = \frac{\partial \tilde{x}^a}{\partial x^b} V^b$, $\tilde{W}^a = \frac{\partial \tilde{x}^a}{\partial x^b} W^b$, and $\partial_{\tilde{x}^a} = \frac{\partial x^b}{\partial \tilde{x}^a} \partial_{x^b}$. Hence we obtain

$$\begin{aligned} \widetilde{(\nabla_W V)}^c &= \tilde{W}^a \partial_{\tilde{x}^a} \tilde{V}^c + \tilde{\Gamma}^c_{ab} \tilde{W}^a \tilde{V}^b \\ &= \frac{\partial \tilde{x}^a}{\partial x^d} W^d \frac{\partial x^e}{\partial \tilde{x}^a} \partial_{x^e} \left( \frac{\partial \tilde{x}^c}{\partial x^f} V^f \right) + \tilde{\Gamma}^c_{ab} \frac{\partial \tilde{x}^a}{\partial x^d} W^d \frac{\partial \tilde{x}^b}{\partial x^f} V^f \\ &= \frac{\partial \tilde{x}^a}{\partial x^d} \frac{\partial x^e}{\partial \tilde{x}^a} \frac{\partial \tilde{x}^c}{\partial x^f} W^d \partial_{x^e} V^f + \frac{\partial \tilde{x}^a}{\partial x^d} \frac{\partial x^e}{\partial \tilde{x}^a} \frac{\partial^2 \tilde{x}^c}{\partial x^f \partial x^e} W^d V^f \\ &\quad + \frac{\partial \tilde{x}^a}{\partial x^d} \frac{\partial \tilde{x}^b}{\partial x^f} \tilde{\Gamma}^c_{ab} W^d V^f \\ &= \frac{\partial \tilde{x}^c}{\partial x^f} W^d \partial_{x^d} V^f + \left( \frac{\partial^2 \tilde{x}^c}{\partial x^f \partial x^d} + \frac{\partial \tilde{x}^a}{\partial x^d} \frac{\partial \tilde{x}^b}{\partial x^f} \tilde{\Gamma}^c_{ab} \right) W^d V^f \\ &= \frac{\partial \tilde{x}^c}{\partial x^e} \left( W^d \partial_{x^d} V^e + \frac{\partial x^e}{\partial \tilde{x}^h} \left( \frac{\partial^2 \tilde{x}^h}{\partial x^f \partial x^d} + \frac{\partial \tilde{x}^a}{\partial x^d} \frac{\partial \tilde{x}^b}{\partial x^f} \tilde{\Gamma}^h_{ab} \right) W^d V^f \right) \end{aligned}$$

From $\widetilde{(\nabla_W V)}^c = \frac{\partial \tilde{x}^c}{\partial x^e} (\nabla_W V)^e = \frac{\partial \tilde{x}^c}{\partial x^e} W^d \partial_{x^d} V^e + \Gamma^e_{df} W^d V^f$ we get therefore

$$\Gamma^e_{df} = \frac{\partial x^e}{\partial \tilde{x}^h} \frac{\partial^2 \tilde{x}^h}{\partial x^f \partial x^d} + \frac{\partial x^e}{\partial \tilde{x}^h} \frac{\partial \tilde{x}^a}{\partial x^d} \frac{\partial \tilde{x}^b}{\partial x^f} \tilde{\Gamma}^h_{ab}.$$

Now we can show that a covariant derivative is torsion-free if and only if there exist coordinates such that the equation in the theorem holds. Expressing the condition of being torsion-free in coordinates we see that it is equivalent to $\Gamma^c_{ab} = \Gamma^c_{ba}$ in any coordinate system $(x^0, \ldots, x^{n-1})$. If the covariant derivative is induced by coordinates in which the $\tilde{\Gamma}^c_{ab}$ vanish at $x$, then in any other any coordinate system $(x^0, \ldots, x^{n-1})$ we have $\Gamma^e_{fd} = -\frac{\partial x^e}{\partial \tilde{x}^h} \frac{\partial^2 \tilde{x}^h}{\partial x^f \partial x^d}$ which is clearly symmetric in $e$ and $f$. For the converse, assume that the $\tilde{\Gamma}^c_{ab}$ are symmetric in $a, b$ and (without loss of generality) that $\tilde{x}^a(x) = 0$. We will now consider a quadratic coordinate transformation of the form $\tilde{x}^a = x^a + \frac{1}{2} A^a_{bc} x^b x^c$, where $A^a_{bc}$ is symmetric in $b$ and $c$. At $\tilde{x}^a = 0$ we have then

$$\frac{\partial \tilde{x}^a}{\partial x^b} = \delta^a_b, \quad \frac{\partial x^a}{\partial \tilde{x}^b} = \delta^a_b, \quad \frac{\partial^2 \tilde{x}^c}{\partial x^a \partial x^b} = A^c_{ab},$$

whence $\Gamma^c_{ab} = \tilde{\Gamma}^c_{ab} + A^c_{ab}$ at $\tilde{x}^a = 0$. Our assertion follows by choosing $A^c_{ab} = -\left(\tilde{\Gamma}^c_{ab}\right)_{|x}$. ∎

For later reference we collect the coordinate expressions derived in the proof of Theorem 2.6.1 in the following corollary.

**Corollary 2.6.1.** *Let $(x^0, \ldots, x^{n-1})$ be a local coordinate system on a manifold with connection $(M, \nabla)$. Then there exist functions $\Gamma^a_{bc} = \Gamma^a_{(bc)}$ such that for each vector field $V$ the covariant derivative $\nabla V$ is given by*

$$\nabla_b V^a = \partial_b V^a + \Gamma^a_{bc} V^c.$$

*and the function $\Gamma^a_{bc}$ transform under a coordinate transformation*

$$(x^0, \ldots, x^{n-1}) \mapsto (\tilde{x}^0, \ldots, \tilde{x}^{n-1})$$

*according to*

$$\tilde{\Gamma}^a_{bc} = \frac{\partial \tilde{x}^a}{\partial x^h} \frac{\partial^2 x^h}{\partial \tilde{x}^b \partial \tilde{x}^c} + \frac{\partial \tilde{x}^a}{\partial x^h} \frac{\partial x^d}{\partial \tilde{x}^b} \frac{\partial x^e}{\partial \tilde{x}^c} \Gamma^h_{de}.$$

**Lemma 2.6.1.** *Let $(M, \nabla)$ be a manifold with connection. There exists a unique extension of $\nabla$ to general tensor fields,*

$$\nabla : T^r_s(M) \to T^r_{s+1}(M), \quad \psi \mapsto \nabla \psi$$

*such that $\nabla_V : T^r_s(M) \to T^r_s(M)$ is a derviation for every tensor field $V$.*

*Proof.* It is clear that we can extend $\nabla_V$ to tensor fields as a derviation. Uniqueness follows directly from Proposition 2.4.2. ∎

In special relativity, timelike straight lines represent freely falling particles and lightlike straight lines represent light signals. In Euclidean space, straight lines are the shortest curves between any two points. One way to define a straight line $\gamma\colon (a,b) \mapsto \mathbb{A}^n$ is to require that the acceleration $\ddot{\gamma} = \mathrm{D}\dot{\gamma}(\dot{\gamma})$ vanishes. This definition carries over to manifolds with connections as follows.

Let $t \mapsto V(t)$ be a vector field along a curve $t \mapsto \gamma(t)$. Then we can extend $V$ to a vector field $\tilde{V}$ which is defined in a neighbourhood of the path of $\gamma$. It follows from Corollary 2.6.1 that the covariant derivative of $\tilde{V}$ restricted to $\gamma$ in direction $\dot{\gamma}$ is independent of the extension $\tilde{V}$. Hence the following definition makes sense.

**Definition 2.6.2.** *Let $\nabla$ be a covariant derivative, $t \mapsto \gamma(t)$ a curve, and $t \mapsto V(t)$ a vector field along $\gamma$.*

*(i) Then $\dot{V}(t) = \nabla_{\dot{\gamma}(t)} V(t) = \left(\frac{\mathrm{d}}{\mathrm{d}t} V^a(t) + \Gamma^a_{bc} V^c(t) \dot{\gamma}^b(t)\right) \partial_a$ is the* covariant derivative of $V$ along $\gamma$.

*(ii) A* pregeodesic *is a curve $\gamma$ satisfying $\nabla_{\dot{\gamma}}\dot{\gamma} \parallel \dot{\gamma}$.*

*(iii) A pregeodesic $\gamma$ is a* geodesic *if $\nabla_{\dot{\gamma}}\dot{\gamma} = 0$.*

*(iv) A geodesic is* complete *if it is defined for all $t \in \mathbb{K}$.*

*(v) A* projective structure *is a maximal equivalence class of to connections which all have the same pregeodesics.*

The notation $\nabla_{\dot{\gamma}(t)} V(t)$ can be justified as follows. Let $W$ be any vector field with $W(\gamma(t)) = V(t)$. Then we have

$$\begin{aligned}
(\nabla_{\dot{\gamma}(t)} W) \circ \gamma &= \left((\dot{\gamma}^b \partial_b W^a + \Gamma^a_{bc} \dot{\gamma}^b W^c)\partial_a\right) \circ \gamma \\
&= \left(\frac{\mathrm{d}}{\mathrm{d}t}(W \circ \gamma)^a + \Gamma^a_{bc} \dot{\gamma}^b W^c \circ \gamma\right)\partial_a \\
&= \left(\frac{\mathrm{d}}{\mathrm{d}t} V^a + \Gamma^a_{bc} \dot{\gamma}^b V^c\right)\partial_a .
\end{aligned}$$

Hence the notation comes from identifying $V$ and $W$ which are different maps but assign to each point $\gamma(t)$ the same vector. [p. 118 ↓] ↓ p. 128

We will now introduce the notion of *parallel transport*. The following motivation may seem to be mathematically imprecise. Nevertheless, mathematicians used such arguments (before the advent of the French Bourbaki school which introduced a new level of precision in mathematics) in order to introduce connections. These arguments capture very well some of the intuition which leads to the notion of a covariant derivative and are therefore worth knowing, even though they have to be taken with a grain of salt: One may view a covariant derivative as a *connection* between infinitesimally neighbouring tangent spaces. Let $V$ be a vector field, $x \in M$ and $\xi \in T_xM$. The idea that $T_xM$ is the infinitesimal approximation of the manifold $M$ near $x$ is sometimes expressed by stating that $x+\xi$ is a point infinitesimally close to $x$. (Strictly speaking, this "addition" of points and vectors in general manifolds does not make sense. Assume that $M$ is a submanifold of $\mathbb{K}^{n+k}$. Then we can identify $T_xM$ with a linear subspace of $\mathbb{K}^{k+m}$ and $x+\xi$ does make sense. This point will in general not lie on $M$ but still be close to $M$ if $\xi$ is small.) Let $v_x \in T_xM$ which we want to "parallelly" translate to the point $x+\xi$. We will then have $v_{x+\xi} = v_x + \delta v$ where $\delta v$ is small. Since this is an infinitesimal process $\delta v$ should depend linearly on $v$ and the difference vector $\xi$. This bilinear map *connects* the tangent spaces of our infinitesimally neighbouring points. In coordinates, we have $\delta v^a = \Gamma^a_{bc} v^b \xi^c$. Consider now a curve $\gamma$ between two distant points $x, y \in M$. We can divide the curve into infinitesimal segments and parallelly translate a vector $v_x \in T_xM$ successively along these infinitesimal curve segments. The concatanation of these infinitesimal parallel translations gives a linear map $\mathbf{P}_\gamma \colon T_xM \to T_yM$ which in general will depend on the intermediate tangent spaces and therefore on the curve $\gamma$. These ideas will be made precise in Definition 2.6.3 and Proposition 2.6.1 below.

**Definition 2.6.3.** *Let $(M, \nabla)$ be a manifold with connection and*

$$\gamma \colon (a,b) \to M$$

*be a curve. A vector field $V$ along $\gamma$ is called* parallelly transported *along $\gamma$ if $\nabla_{\dot\gamma} V(t) = 0$ for all $t \in (a,b)$. A parallelly transported vector field is often simply called* parallel.

In particular, a geodesic is a curve whose tangent vector is parallelly transported. For this reason, geodesics are sometimes called *auto-parallel* curves.

**Proposition 2.6.1.** *Let $(M, \nabla)$ be a manifold with connection and let $\gamma \colon [a,b] \to M$ be a curve which can be smoothly extended into both directions. The map*

$$\mathbf{P}_\gamma \colon T_{\gamma(a)}M \to T_{\gamma(b)}M, \quad v \mapsto V(b),$$

*where $V \colon [a,b] \to TM$ is the unique parallel vector field along $\gamma$ with $V(a) = v$, is a linear isomorphism.*

*Proof.* Since $\gamma([a,b])$ is compact, there exist finitely many charts which cover the curve $\gamma$. Without loss of generality we can assume that $\gamma$ is contained in a single chart. (Otherwise we could divide $\gamma$ in segments which are contained in single charts. A successive application of the proposition to these segments would imply the proposition in the general case.) The equation $\nabla_{\dot\gamma} V(t) = 0 \quad \forall t \in (a,b)$ reduces to a system of first order differential equations for the coefficients $V^a(t)$: $\dot{V}^a + \Gamma^a_{bc}\dot{\gamma}^b V^c = 0$. It follows from the fundamental theorem for ordinary differential equations (cf. Theorem 2.4.1) that the map $\mathbf{P}_\gamma$ is well defined. Letting $\tilde{\gamma}(t) \mapsto \gamma(a+b-t)$ and $\tilde{V}(t) = V(a+b-t)$ we clearly have $\nabla_{\dot{\gamma}} V(t) = 0$ if and only if $\nabla_{\dot{\tilde{\gamma}}} \tilde{V}(t) = 0$. This implies that $\left(\mathbf{P}_\gamma\right)^{-1} = \mathbf{P}_{\tilde{\gamma}}$ and a fortiori that $\mathbf{P}_\gamma$ is an isomorphism. ∎

It is possible to recover the covariant derivative from parallel transport. The relation between these concepts is similar to the relation between the Lie-derivative and the Lie-transport. But note that there is no Lie-transport along a single curve.

**Proposition 2.6.2.** *Let $(M, \nabla)$ be a manifold with connection and let $\gamma\colon [a,b] \to M$ be a curve which can be smoothly extended into both directions. For each $s \in [0, b-t)$ let $\tilde{\gamma}_{t,s}\colon [0,s] \to M$, $\sigma \mapsto \gamma(t+s-\sigma)$. Then for any vector field $V(t)$ along $\gamma$ we have*

$$\nabla_{\dot{\gamma}(t)} V(t) = \lim_{s \to 0} \frac{1}{s} \left( \mathbf{P}_{\tilde{\gamma}_{t,s}} (V(t+s)) - V(t) \right).$$

*Proof.* Let $W(\sigma) = \mathbf{P}_{\tilde{\gamma}_{t,\sigma}}(V(t+s))$. Then $\nabla_{\dot{\tilde{\gamma}}_{t,\sigma}} W = 0$ implies $\frac{\mathrm{d}}{\mathrm{d}\sigma} W^a = -\Gamma^a_{bc} W^b \frac{\mathrm{d}}{\mathrm{d}\sigma}(\tilde{\gamma}_{t,\sigma})^c$. Using a Taylor expansion it is easy to see that there exists a smooth vector field $U$ along $\tilde{\gamma}_{t,\sigma}$ with

$$W(\sigma) = W(0) - \sigma \Gamma^a_{bc} W^b(0) \left( \left( \frac{\mathrm{d}}{\mathrm{d}\sigma} \tilde{\gamma}_{t,\sigma} \right)_{|\sigma=0} \right)^c \partial_a + U(\sigma)\sigma^2.$$

Setting $\sigma = s$ we obtain from $W(0) = V(t+s)$ and $\left(\frac{\mathrm{d}}{\mathrm{d}\sigma}\tilde{\gamma}_{t,\sigma}\right)_{|\sigma=0} = -\dot{\gamma}(t)$ the equation

$$\mathbf{P}_{\tilde{\gamma}_{t,s}} V(t+s) - V(t) = V(t+s) - V(t) + s\Gamma^a_{bc} V^b(t)\dot{\gamma}^c(t)\partial_a - U(s)s^2$$

which implies the assertion. ∎

**Corollary 2.6.2.** *A connection is uniquely determined by its parallel transport.*

Since inertial observers played an important rôle in Chap. 1, a thorough understanding of geodesics and projective classes should be important for the globalization of the results in Chap. 1 (cf. Chap. 3). But they are also of independent geometric interest. In fact, the classical developement of Euclidean geometry is built on the concept of straight lines (and therefore on the concept of projective classes).

Theorem 2.6.1 implies that torsion-free connections are a straightforward globalization of the usual derivative of vector fields in vector spaces. Proposition 2.6.3 shows that also from the viewpoint of geodesics it is sufficient to consider only torsion-free connections. p. 1 [↓ p.

**Lemma 2.6.2.** *Let* $\nabla$, $\tilde{\nabla}$ *be connections on a manifold* $M$. *Then their difference* $\nabla - \tilde{\nabla}$ *is a tensor field.*

*Proof.* We have to show that $(U,V) \mapsto S(U,V) = \nabla_U V - \tilde{\nabla}_U V$ is function-linear. The only non-trivial part of this assertion follows from

$$\begin{aligned} S(U, fV) &= \nabla_U(fV) - \tilde{\nabla}_U(fV) \\ &= (U \bullet f)V + f\nabla_U V - (U \bullet f)V - f\tilde{\nabla}_U V \\ &= f\nabla_U V - f\tilde{\nabla}_U V = fS(U,V). \end{aligned}$$

∎

**Proposition 2.6.3.** *Let* $\tilde{\nabla}$ *be a connection. Then there exists a unique torsion-free connection* $\nabla$ *which has the same geodesics as* $\tilde{\nabla}$.

*Proof.* We define $\nabla_V W = \tilde{\nabla}_V W - \frac{1}{2}\widetilde{\mathrm{Tor}}(V,W)$, where $\widetilde{\mathrm{Tor}}(V,W) = \tilde{\nabla}_V W - \tilde{\nabla}_W V - [V,W]$ is the torsion tensor of $\tilde{\nabla}$. Since $\widetilde{\mathrm{Tor}}(W,W) = 0$ for all $W$, both connections have the same geodesics. Further, $\nabla$ is torsion-free by construction. For uniqueness note that if we add any any additional, non-vanishing, in the covariant entries skew symmetric tensor field $S \in \mathcal{T}_2^1(M)$ to $\nabla$ we loose the property $\mathrm{Tor} = 0$. On the other hand, for any non-vanishing, in the covariant entries symmetric tensor field $S \in \mathcal{T}_2^1(M)$ there exists a vector $v_x$ with $S(v_x, v_x) \neq 0$. This implies that the geodesics with respect to the connections $\nabla$ and $\nabla + S$ which have initial velocity $v_x$ do not coincide. ∎

However, there exists infinitely many torsion-free connections with the same pregeodesics. This means that each projective class contains infinitely many torsion-free connections.

**Lemma 2.6.3.** *Let* $\nabla$ *and* $\tilde{\nabla}$ *be torsion-free connections and* $\mathfrak{P}$ *be the projective structure generated by* $\nabla$. *Then* $\tilde{\nabla} \in \mathfrak{P}$ *if and only if there exists a one-form* $\theta$ *such that* $\tilde{\nabla} - \nabla = \theta \otimes \mathrm{id} + \mathrm{id} \otimes \theta$ *(or, in coordinates,* $\tilde{\Gamma}^a_{bc} - \Gamma^a_{bc} = 2\delta^a_{(b}\theta_{c)}$*).*

*Proof.* Let $\Sigma^a_{bc} = \tilde{\Gamma}^a_{bc} - \Gamma^a_{bc}$. Then their pregeodesics can coincide only if $\Sigma^a_{bc} v^b v^c \parallel v^a$ for all vectors $v^a$. This implies $\Sigma^a_{bc} v^b v^c v^d - \Sigma^d_{bc} v^b v^c v^a = 0$ for all vectors $v$ which is equivalent to $\delta^e_{(d}\Sigma^a_{bc)} - \delta^a_{(d}\Sigma^e_{bc)} = 0$. Since $\Sigma$ is symmetric in the lower indices $b$, $c$ we get

$$\begin{aligned} S^{ea}_{bcd} &:= \delta^e_{(d}\Sigma^a_{bc)} - \delta^a_{(d}\Sigma^e_{bc)} \\ &= \frac{1}{3}\Big( (\delta^e_b\Sigma^a_{cd} - \delta^a_b\Sigma^e_{cd}) + (\delta^e_c\Sigma^a_{db} - \delta^a_c\Sigma^e_{db}) + (\delta^e_d\Sigma^a_{bc} - \delta^a_d\Sigma^e_{bc}) \Big). \end{aligned}$$

Contracting the indices $e$ and $b$ gives $S^{ba}_{bcd} = n\Sigma^a_{cd} - \Sigma^a_{cd} + \Sigma^a_{dc} - \delta^a_c\Sigma^b_{db} + \Sigma^a_{dc} - \delta^a_d\Sigma^b_{bc} = (n+1)\Sigma^a_{cd} - 2\delta^a_{(c}\theta_{d)}$, where $\theta_d := (n+1)^{-1}\Sigma^b_{bd}$. It follows that $\Sigma^a_{cd} = 2\delta^a_{(c}\theta_{d)}$.

For the converse notice that $\tilde{\nabla}_V V = \nabla_V V + \theta(V)V$, whence both connections have the same pregeodesics. ∎

The following corollary will be used in Sect. 3.2

**Corollary 2.6.3.** *Let $(M, \nabla)$ be a manifold with connection. Let $\tilde{\nabla}$ be a connection such that for every $x \in M$ there is an open set $\mathfrak{U}_x \subset T_xM$ such that the pregeodesics with initial velocity $v_x \in \mathfrak{U}_x$ with respect to $\nabla$ and $\tilde{\nabla}$ coincide. Then $\nabla$ and $\tilde{\nabla}$ generate the same projective structure*

*Proof.* We use the same notation as in the proof of Lemma 2.6.3. It is sufficient to note that $\Sigma^a_{bc}v^bv^cv^d - \Sigma^d_{bc}v^bv^cv^a = 0$ for all vectors $v \in \mathfrak{U}_x$ already implies $\delta^e_{(d}\Sigma^a_{bc)} - \delta^a_{(d}\Sigma^e_{bc)} = 0$. ∎ [p. 124 ↓] ↓ p. 155

Written down in coordinates, the geodesic equation $\ddot{\gamma}^a + \Gamma^a_{bc}\dot{\gamma}^b\dot{\gamma}^c = 0$ reduces to a system of second order differential equations on $M$. Alternatively, it can be considered as a system of first order equations on the tangent bundle $TM$. In coordinates, this system of differential equations is given by

$$\frac{\mathrm{d}}{\mathrm{d}t}\gamma^a = v^a, \quad \frac{\mathrm{d}}{\mathrm{d}t}v^a = -\Gamma^a_{bc}v^bv^c.$$

The corresponding vector field on $TM$ is called the *geodesic spray* $\Gamma \in \mathcal{T}^1_0(TM)$, and can be invariantly defined by

$$\Gamma(v_x) := T_0\dot{\gamma}_{v_x}(\partial_t),$$

where $\gamma_{v_x}$ is the unique maximal geodesic with $\gamma_{v_x}(0) = x$ and $\dot{\gamma}_{v_x}(0) = v_x$. The following proposition justifies the definition.

**Proposition 2.6.4.** *Let $(M, \nabla)$ be a manifold with connection and $\Gamma$ be the geodesic spray. If $\lambda\colon (a,b) \to TM$ is an integral curve of $\Gamma$, then $\pi_{TM} \circ \lambda$ is a geodesic. Conversely, for every geodesic $\gamma$ there exists a unique integral curve $\lambda$ of $\Gamma$ with $\pi_{TM} \circ \lambda = \gamma$.*

*Proof.* Let $v_x \in T_xM$ and $\gamma_{v_x}$ be a geodesic with $\dot{\gamma}(t_0) = v_x$. This geodesic defines a curve $\lambda_{v_x}(t) := \dot{\gamma}_{v_x}(t)$ in $TM$. Clearly, $\pi_{TM} \circ \lambda_{v_x} = \gamma_{v_x}$ and $\frac{\mathrm{d}}{\mathrm{d}t}\lambda_{v_x}(t) = T_t\lambda_{v_x}(\partial_t) = T_t\dot{\gamma}_{v_x}(\partial_t) = T_0\frac{\mathrm{d}}{\mathrm{d}t}(\gamma_{v_x} \circ \tau_t)(\partial_t)$, where $\tau_t$ is the translation $s \mapsto t+s$. Since $\frac{\mathrm{d}}{\mathrm{d}t}(\gamma_{v_x} \circ \tau_t)(0)$ is the velocity vector of $\gamma_{v_x}$ at $t$, $\lambda_{v_x}$ is an integral curve of $\Gamma$.

Conversely, let $\lambda$ be an integral curve of $\Gamma$ and consider the geodesic $\gamma_{\lambda(t_0)}$ with $\dot{\gamma}_{\lambda(t_0)}(t_0) = \lambda(t_0)$. By the construction above, $\dot{\gamma}_{\lambda(t_0)}$ is an integral curve of $\Gamma$. Since its initial point in $TM$ is $\lambda(t_0)$, it must coincide with $\lambda(t_0)$ by the uniqueness part of the fundamental theorem for ordinary differential equations (Theorem 2.4.1). ∎

We have seen that a manifold is locally isomorphic to $\mathbb{A}^n$ (considered as a set with differentiable structure). It is not true, however, that a manifold with connection is isomorphic to $\mathbb{A}^n$ with its affine structure. In fact, for a manifold with connection there do not generally exist charts which map geodesics into straight lines. For this to be the case a necessary condition would be that the connection is in the same projective class as the canonical connection of $\mathbb{A}^n$. We will now show that one can still identify the geodesics which pass through a *given* point with all straight lines in $\mathbb{A}^n$ which pass throug a given intersection point (cf. Lemma 2.6.4).

As a consequence of Proposition 2.6.4 and the fundamental theorem for ordinary differential equations, each $x \in M$ has a neighbourhood $\mathcal{W}$ of 0 in $T_xM$ and there is a $\delta > 0$ such that for all $v \in \mathcal{W}$ the geodesic $\gamma_v\colon (-\delta, \delta) \to M$ with $\dot{\gamma}(0) = v$ is defined. By choosing $\mathcal{W}$ small enough we can normalize the interval $[-\delta, \delta]$. In fact, observe that $\dot{\gamma}_{av}(t) = a\dot{\gamma}_v(at)$ for any $a \in \mathbb{K}$. Hence for every $x \in M$ the zero vector $0 \in T_xM$ has a neighbourhood $\mathcal{U} \subset T_xM$ such that for all $v \in \mathcal{U}$ the geodesics $t \mapsto \gamma_v(t)$ with initial velocity $v$ is defined on the interval $[-1, 1]$.

**Definition 2.6.4.** *Let $(M, \nabla)$ be a manifold with connection. The map*

$$\exp\colon \{v \in TM \,|\, \gamma_v(1) \text{ is defined}\} \to M$$
$$v \mapsto \exp(v) := \exp_x(v) := \gamma_v(1),$$

*where $x = \pi_{TM}(v)$, is called the* exponential map *of* $\nabla$.

It follows from the fundamental theorem for ordinary differential equations that the set of all $v \in TM$, for which the integral curve $\lambda$ of $\Gamma$ with $\lambda(0) = v$ is defined up to (including) parameter value 1, is open. Hence the domain of exp is open in $TM$. This also implies that for any $x$ the intersection of this domain with $T_xM$ is open in $T_xM$. This set is also star-shaped as a consequence of the equation $\dot{\gamma}_{av}(t) = a\dot{\gamma}_v(at)$ for all $a \in \mathbb{R}$.

**Proposition 2.6.5.** *For each point $x \in M$ there exists a neighbourhood $\mathcal{U}$ of $0 \in T_xM$ such that $\exp_x$ is a diffeomorphism from $\mathcal{U}$ onto a neighbourhood $U$ of $x \in M$.*

*Proof.* Let $v \in T_xM$ and $\tilde{v} = \left(\frac{\mathrm{d}}{\mathrm{d}t}(tv)\right)_{|t=0} \in T_0T_xM$. Then

$$T\exp_x(\tilde{v}) = \left(\frac{\mathrm{d}}{\mathrm{d}t}\exp_x(tv)\right)_{|t=0} = \left(\frac{\mathrm{d}}{\mathrm{d}t}\gamma_{tv}(1)\right)_{|t=0} = \left(\frac{\mathrm{d}}{\mathrm{d}t}\gamma_v(t)\right)_{|t=0} = v$$

which implies that $T\exp_x$ is an isomorphism. Now the assertion follows from the inverse function theorem. ∎

**Corollary 2.6.4.** *There is a neighbourhood $\tilde{U}$ of $0_x \in TM$ such that*

$$\mathrm{Exp}\colon \tilde{U} \to M \times M, \qquad v_y \mapsto (y, \exp(v_y))$$

*is a diffeomorphism onto its image.*

*Proof.* The corollary follows from the fact that $T\mathrm{Exp} = T\pi_{TM} \oplus T\exp$ is non-singular whenever $T\exp$ is non-singular. ∎

Proposition 2.6.5 provides especially practical coordinates. Let $\{e_1, \ldots, e_n\}$ be a basis of $T_xM$ and write $v_x = v^i e_i$ for each vector $v_x \in T_xM$. Then $\tilde{B}_r(0_x) = \{v_x \,|\, \sqrt{\sum_{a=1}^n (v^a)^2} < r\}$ is a neighbourhood of $0_x \in T_xM$. The map $\exp_x\colon \tilde{B}_r(0) \to B_r(x) := \exp(\tilde{B}_r(0_x))$ is a diffeomorphism for any sufficiently small $r > 0$. Hence we can define a coordinate system by $x^a(y) := \left({\exp_x}^{-1}(y)\right)^a$. These coordinates are called *normal coordinates* and the corresponding chart is called a *normal chart*.

**Lemma 2.6.4.** *Let $(M, \nabla)$ be a manifold with connection, $x \in M$, and $(\mathcal{U}, \varphi)$ be a normal coordinate chart centered at $x$. Then $\varphi$ maps the geodesics through $x$ onto the straight lines through $0 \in \mathbb{K}^n$.*

*Furthermore, the Christoffel symbols with respect to the chart $(\mathcal{U}, \varphi)$ satisfy $\Gamma^a_{(bc)}(x) = 0$.*

*Proof.* The first assertion follows immediately from the construction of normal coordinates.

To prove the second assertion we must only show that $\Gamma^a_{bc} v^b v^c$ vanishes for all vectors $v \in T_xM$. Let $\gamma$ be the geodesic through $x$ with $\dot{\gamma}(0) = v$. Since its coordinate expression is a straight line the coordinate components satisfy $\ddot{\gamma}^a = 0$ and therefore $0 = \ddot{\gamma}^a(t) + \Gamma^a_{bc}(\gamma(t))\dot{\gamma}^b(t)\dot{\gamma}^c(t) = \Gamma^a_{bc}(\gamma(t))\dot{\gamma}^b(t)\dot{\gamma}^c(t)$. At $t = 0$ this equation reduces to $\Gamma^a_{bc}(x)v^b v^c = 0$. ∎

In Euclidean space, a convex $\mathcal{U}$ set is characterized by the requirement that any two points $x, y \in \mathcal{U}$ can be joined by a straight line which is contained in $\mathcal{U}$. For a manifold with connection we call a set $\mathcal{U}$ *convex* if any two points $x, y \in \mathcal{U}$ can be joined by a unique geodesic which is contained in $\mathcal{U}$. We will now show that each point has a *convex neighbourhood*.

**Theorem 2.6.2.** *For each $x \in M$ there is a sequence of convex neighbourhoods $\mathcal{U}_n$ with $\bigcap_{n=1}^{\infty} \mathcal{U}_n = \{x\}$.*

*Proof.* Let $(x^1, \ldots, x^n)$ be a normal coordinate system and $(\mathcal{U}, \varphi)$ the corresponding chart. For $y \in \text{Image}(\exp_x) \cap \mathcal{U}$ we define the distance function $d(y) := \sqrt{(x^a(y))^2}$. Let $B_r(x) = \{y \in \text{Image}(\exp_x) \cap \mathcal{U} \,|\, d(y) < r\}$.

Choosing $r$ small enough, there exists a neighbourhood $\tilde{W}(r)$ of $0_x \in TM$ such that Exp maps $\tilde{W}(\mathrm{r})$ diffeomorphically onto $B_r(x) \times B_r(x)$. For $\tilde{r} \to 0$ the neighbourhood $\tilde{W}(\tilde{r})$ shrinks to the set $\{0_x\} \subset T_x M$. Because of the continuity of exp and $\exp(0_x) = x$, there is an $\tilde{r} \in (0, r)$ such that $\exp(tw_y) \in B_x(r)$ for all $t \in [0,1]$ and $w_y \in \tilde{W}(\tilde{r})$.

We will show that (for $\tilde{r}$ sufficiently small), any two points $y, z \in B_{\tilde{r}}(x)$ are joined by a geodesic which does not leave $B_{\tilde{r}}(x)$. Let $\gamma$ be the geodesic starting at $y$ with velocity vector $\text{Exp}^{-1}(y, z)$. This geodesic joins $y$ with $z$ by the definition of Exp.

We have to show that for $\tilde{r}$ small enough the curve $\gamma$ cannot leave $B_{\tilde{r}}(x)$. The idea of the the proof is as follows. Since this geodesic would have to re-enter $B_{\tilde{r}}(x)$, in the given coordinates it would have to be curved at least of order $1/\tilde{r}$ somewhere outside $B_{\tilde{r}}(x)$. On the other hand, geodesics are generalizations of straight lines. This should give a contradiction for $\tilde{r}$ small enough. While this is the geometrical idea of proof, it may not be entirely clear from the analytical implementation which we will present now.

If $\gamma$ would leave $B_{\tilde{r}}(x)$, then the map $t \to d(\gamma(t))$ would have a maximum which is bigger than $\tilde{r}$. At this maximum we have $\frac{\mathrm{d}}{\mathrm{d}t} d \circ \gamma = 0$ and $\frac{\mathrm{d}^2}{\mathrm{d}t^2} d \circ \gamma \leq 0$. We can directly compute

$$\frac{\mathrm{d}^2}{\mathrm{d}t^2} d \circ \gamma = \frac{\sum_{a=1}^{n} \left( (\dot{\gamma}^a)^2 + \gamma^a \ddot{\gamma}^a \right)}{d(\gamma(t))} = \frac{\sum_{a=1}^{n} \left( \delta_{ab} - \gamma^a \Gamma^a_{bc} \right) \dot{\gamma}^a \dot{\gamma}^b}{d(\gamma(t))},$$

where in the last equality we have used the geodesic equation. Since

$$\Gamma^a_{bc}(x) \dot{\gamma}^a \dot{\gamma}^b = \Gamma^a_{(bc)}(x) \dot{\gamma}^a \dot{\gamma}^b \text{ and } \Gamma^a_{(bc)}(x) = 0,$$

we can choose the original $r$ so small that $\delta_{ab} - x^a(y) \Gamma^a_{(bc)}(y)$ is positive definite for all $y \in B_r(x)$. The curve $\gamma$ does not leave $B_x(r)$ by the constructions of $\tilde{r}$. Hence we must have $\frac{\mathrm{d}^2}{\mathrm{d}t^2} d \circ \gamma > 0$ at the maximum which gives a contradiction. ∎

## 2.7 Examples of connections

p. 121 ↓
[↓ p. 133]

*In this section, we will introduce two examples of connections which will both become important in Chap. 3.*

### 2.7.1 The Levi-Cività connection

The following definition is a generalization of Euclidean space $(\mathbb{A}^n, \langle\cdot,\cdot\rangle_{\mathbb{R}^n})$ and Minkowski space $(\mathbb{A}^n, \eta)$ to real manifolds which are not necessarily affine spaces.

**Definition 2.7.1.** *A* pseudo-Riemannian manifold $(M, g)$ *is a real manifold $M$ together with a symmetric $\binom{0}{2}$-tensor field $g$ which is everywhere non-degenerate. We will often simply write $\langle u, v\rangle$ instead of $g(u, v)$.*

The *norm* of a vector $u$ with respect to $g$ is defined by $||u|| = \sqrt{|g(u,u)|}$, but in general it is *not* a norm in the sense of linear algebra (the triangle inequality only holds if $g$ is positive or negative definite). Let $i, j, \nu \in \{1, \ldots, n\}$ and define

$$(\eta_\nu)_{ij} := \begin{cases} -1 & \text{if } i = j \leq \nu, \\ 1 & \text{if } i = j > \nu, \\ 0 & \text{otherwise.} \end{cases}$$

If $(M, g)$ is a pseudo-Riemannian manifold then there is a $\nu \in \{1, \ldots, n\}$ such that for each point $x \in M$ there is a basis $\{e_1, \ldots, e_n\}$ of $T_xM$ which satisfies

$$g(e_i, e_j) = (\eta_\nu)_{ij}\,. \tag{2.7.6}$$

We say that $g$ has *signature* $(\overbrace{-,\ldots,-}^{\nu \text{ times}}, \overbrace{+,\ldots,+}^{(n-\nu)\text{ times}})$ and call $\nu$ the *index* of $g$. A basis $\{e_1, \ldots, e_n\}$ satisfying Equation (2.7.6) is called an *orthonormal basis* and a local frame $\{E_1, \ldots E_n\}$ such that for each $x$ in its domain of definition $\{E_1(x), \ldots E_n(x)\}$ is an orthonormal basis is called an *orthonormal frame*.

**Definition 2.7.2.** *A pseudo-Riemannian manifold is called a* Riemannian manifold *if $g$ is positive definite and is called a* Lorentzian manifold *if $g$ has signature $(-,+,\ldots,+)$.*

**Lemma 2.7.1.** *Let $(M, g)$ be a pseudo-Riemannian manifold. Then each $x \in M$ has a neighbourhood $\mathcal{U}$ which is the domain of an orthonormal frame.*

*Proof.* Let $\mathcal{U}$ be a coordinate neighbourhood of $x$ and denote the induced Gaußian frame by $\{\partial_1, \ldots \partial_n\}$. We apply a variant of the Schmidt orthogonalisation procedure to this frame. There exist functions $A_1^j$ on $\mathcal{U}$ such that $U_1(x) := \sum_{j=1}^n A_1^j(x)\partial_j$ satisfies $g(U_1, U_1) \neq 0$ at all points $x \in \mathcal{U}$. We will use an inductuion argument in order to define a frame $\{\tilde{E}_1, \ldots, \tilde{E}_n\}$ which is orthonormal up to a permutation of the frame vector fields. Let $\tilde{E}_1 = U_1/||U_1||$. Now assume that we have constructed pointwise linearly independent vector fields $\{\tilde{E}_1, \ldots, \tilde{E}_{i-1}\}$ such that $g(\tilde{E}_k, \tilde{E}_l) = 0$ for $k \neq l$ and $g(\tilde{E}_k, \tilde{E}_k) = \pm 1$. There exist functions $A_i^j$ on $\mathcal{U}$ such that $U_i(x) := \sum_{j=1}^n A_i^j(x)\partial_j$ does not lie in span$\{\tilde{E}_1, \ldots, \tilde{E}_{i-1}\}$ and

$$\tilde{U}_i = U_i - \sum_{k=1}^{i-1} \frac{g(U_i, \tilde{E}_k)}{g(\tilde{E}_k, \tilde{E}_k)} \tilde{E}_k$$

satisfies $g(\tilde{U}_i, \tilde{U}_i) \neq 0$. The vector field $\tilde{E}_i = \tilde{U}_i / ||\tilde{U}_i||$ is well defined and satisfies $g(\tilde{E}_i, \tilde{E}_i) = \pm 1$, $g(\tilde{E}_k, \tilde{E}_i) = 0$ for $k \in \{1, \ldots, i-1\}$. Finally, let $\{E_1, \ldots, E_n\}$ be a suitable permutation of $\{\tilde{E}_1, \ldots, \tilde{E}_n\}$. ∎

Minkowski spacetime has a metric $\eta$ which is constant with respect to the usual derivative D, i.e. $\mathrm{D}\eta = 0$. This is equivalent to the fact that parallel transport of vectors is an isometry. Furthermore, D is the only torsion-free connection which satisfies this requirement. For a general pseudo-Riemannian manifold, there exists a unique connection for which an analogous statement holds.

**Theorem 2.7.1.** *Let $(M, g)$ be a pseudo-Riemannian real manifold. Then there exists a unique torsion-free connection $\nabla$ which satisfies $\nabla g = 0$.*

This connection is called the *Levi-Cività connection.*

*Proof of Theorem 2.7.1.* Recall that the connection $\nabla$ is torsion-free if and only if $\nabla_U V - \nabla_V U = [U, V]$ for all vector fields $U, V$. A short calculation shows that the condition $\nabla g = 0$ is equivalent to $U\langle V, W\rangle = \langle \nabla_U V, W\rangle + \langle V, \nabla_U W\rangle$. The strategy for the proof is to use these two conditions in order to calculate the only possible candidate for the Levi-Cività connection. It is then easy to verify that this candidate satisfies all the relevant equations.

To exploit that $\nabla$ is assumed to be torsion-free consider the difference

$$\begin{aligned} U\langle V, W\rangle - W\langle U, V\rangle &= \langle \nabla_U V, W\rangle + \langle V, \nabla_U W\rangle \\ &\quad - \langle \nabla_W U, V\rangle - \langle U, \nabla_W V\rangle \\ &= \langle \nabla_U V, W\rangle - \langle U, \nabla_W V\rangle + \langle [U, W], V\rangle . \end{aligned}$$

If we add $V\langle U, W\rangle = \langle \nabla_V U, W\rangle + \langle U, \nabla_V W\rangle$ to this equation the right-hand side becomes $\langle \nabla_U V, W\rangle + \langle \nabla_V U, W\rangle + \langle U, [V, W]\rangle + \langle [U, W], V\rangle$. Here we can eliminate $\langle \nabla_V U, W\rangle$ using that $\nabla$ is assumed to be torsion-free: $\langle \nabla_V U, W\rangle = \langle [V, U], W\rangle + \langle \nabla_U V, W\rangle$. Putting everything together we finally obtain the *Koszul Formula*

$$\begin{aligned} \langle \nabla_U V, W\rangle = \frac{1}{2}\Big( & U\langle V, W\rangle + V\langle U, W\rangle - W\langle U, V\rangle - \langle U, [V, W]\rangle \\ & + \langle V, [W, U]\rangle - \langle W, [V, U]\rangle \Big). \end{aligned} \tag{2.7.7}$$

∎

The following proposition gives a slightly more geometric characterization of the Levi-Cività connection.

**Proposition 2.7.1.** *Let $(M, g)$ be a pseudo-Riemannian manifold and $\nabla$ be a torsion-free connection. Then $\nabla g = 0$ if and only if parallel transport is an isometry.*

*Proof.* Assume first that $\nabla$ is the Levi-Cività connection. Let $t \mapsto \gamma(t)$ be a curve and $t \mapsto U(t)$, $t \mapsto V(t)$ be two parallel vector fields along $\gamma$. Then we calculate $\frac{\mathrm{d}}{\mathrm{d}t}\langle U(t), V(t)\rangle = \nabla_{\dot\gamma(t)}\langle U(t), V(t)\rangle = \left\langle \nabla_{\dot\gamma(t)}U, V(t)\right\rangle + \left\langle U(t), \nabla_{\dot\gamma(t)}V\right\rangle = 0$ which implies that $\langle U, V\rangle$ is independent of $t$. Consequently, parallel transport is an isometry.

Conversely, assume that parallel transport is an isometry and let $u, v, w$ be vectors. We choose a curve $\gamma$ with $\dot\gamma(0) = w$ and parallelly transported vector fields $U, V$ with $U(0) = u$, $V(0) = v$. Then we obtain $0 = \nabla_{\dot\gamma(t)}\langle U, V\rangle = (\nabla_{\dot\gamma(t)}g)(U, V) + \left\langle \nabla_{\dot\gamma(t)}U, V(t)\right\rangle + \left\langle U(t), \nabla_{\dot\gamma(t)}V\right\rangle = (\nabla_{\dot\gamma(t)}g)(U, V)$. At $t = 0$ this implies $0 = (\nabla_w g)(u, v)$ which proves the assertion. ∎

### 2.7.2 The Weyl connection

Let $M$ be a real manifold and $g$ be a metric on $M$. A second metric $\tilde g$ is *conformal* to $g$ is there is a positive function $\Omega\colon M \to \mathbb{R}^+ \setminus \{0\}$ with $g = \Omega^2 \tilde g$. A *conformal structure* $\mathfrak{C}$ is an equivalence class of conformal metrics. In the next chapter we will see that the Michelson Morley experiment directly leads to a conformal structure rather than a Lorenzian metric.

Given a conformal structure $\mathfrak{C}$ there is a class of adapted connections. This generalizes the Levi-Cività connection of the previous section.

**Definition 2.7.3.** *A triple $(M, \mathfrak{C}, \nabla)$, where $M$ is a $n$-dimensional manifold, $\mathfrak{C}$ a conformal structure on $M$, and $\nabla$ a torsion-free connection is called a* Weyl structure *if for every $g \in \mathfrak{C}$ there exists a one-form $\varphi$ such that $\nabla g = \varphi \otimes g$. The connection $\nabla$ is called a* Weyl connection

In the following we will use the exterior derivative $\mathrm{d}\omega$ of a p-form $\omega \in \mathcal{T}_1^0(M)$ (cf. Theorem 2.5.1). In Theorem 2.7.2 below we will also use the lemma of Poincaré (Theorem 2.5.5).

Readers who have omitted Sect. 2.5 can replace $\mathrm{d}\omega_{ab}$ by $2!\partial_{[a}\omega_{b]}$. Using this equality the lemma of Poincaré can be understood in our special case.

**Lemma 2.7.2.** *Let $(M, \mathfrak{C}, \nabla)$ be a Weyl structure. Then the 2-form $F = -\frac{1}{2}\mathrm{d}\varphi$ is independent of $g \in \mathfrak{C}$.*

*Proof.* Let $g \in \mathfrak{C}$ and $\tilde{g} = \Omega^2 g$. Then we have

$$\nabla \tilde{g} = 2\Omega \mathrm{d}\Omega \otimes g + \Omega^2 \nabla g = (\varphi + 2\mathrm{d}\ln \Omega) \otimes \tilde{g} = \tilde{\varphi} \otimes \tilde{g}.$$

Hence $F = -\frac{1}{2}\mathrm{d}\varphi = -\frac{1}{2}\mathrm{d}\tilde{\varphi}$ does not depend on the choice of $g$. ∎

The 2-form $F$ is called the *length curvature* of the Weyl structure. We will motivate this term in Sect. 2.8.1 below.

**Theorem 2.7.2.** *Let $(M, \mathfrak{C}, \nabla)$ be a manifold with Weyl structure and $x \in M$. Then $x$ has a neighbourhood $\mathcal{U}$ such that for the induced Weyl structure $(\mathcal{U}, \mathfrak{C}, \nabla)$ the following statements are equivalent.*

*(i)* $F = 0$,
*(ii) There exists a $\tilde{g} \in \mathfrak{C}$ which has Levi-Cività connection $\nabla$.*

*Proof.* We first show that (ii) implies (i). Let $g \in \mathfrak{C}$ be any metric and $\tilde{g} = \Omega^2 g$ such that $\nabla \tilde{g} = 0$. Then we have $\nabla g = \varphi \otimes g = \nabla(\Omega^2 \tilde{g}) = 2\Omega \mathrm{d}\Omega \otimes \tilde{g} = \Omega^3 \mathrm{d}\Omega \otimes g$. Hence $\varphi = \Omega^3 \mathrm{d}\Omega$ and $\mathrm{d}\varphi = 0$.

For "(i) $\Rightarrow$ (ii)" note that $F = \frac{1}{2}\mathrm{d}\varphi = 0$. Hence an application of the lemma of Poincaré (Theorem 2.5.2) implies the existence of a neighbourhood $\mathcal{U}$ of $x$ and of a function $f\colon \mathcal{U} \to \mathbb{R}$ with $\mathrm{d}f = \varphi$. Consequently, $\nabla(\mathrm{e}^{-f} g) = -\mathrm{e}^{-f}\mathrm{d}f \otimes g + \mathrm{e}^{-f}\mathrm{d}f \otimes g = 0$. ∎

**Corollary 2.7.1.** *Let $(M, \mathfrak{C}, \nabla)$ be a Weyl structure and assume that there exists a parallel, non-vanishing $n$-form $\mu$. Then there is an (up to sign) uniqe metric $g \in \mathfrak{C}$ such that $\nabla$ is the Levi-Cività connection of $g$ and $|\Omega(E_1, \ldots, E_n)| = 1$ for every $g$-orthonormal basis $\{E_1, \ldots, E_n\}$.*

*Proof.* For any metric $g \in \mathfrak{C}$ we define an $n$-form $\tilde{\mu}$ as follows. We let $\{E_1, \ldots, E_n\}$ be an orthonormal basis with dual basis $\{\theta^1, \ldots, \theta^n\}$ and denote $g(E_i, E_i) \in \{-1, 1\}$ by $\epsilon_i$. Then $\tilde{\mu}$ is defined by $\tilde{\mu} = \theta^1 \wedge \cdots \wedge \theta^n$.

Since $\Lambda^n(T_x M)$ is 1-dimensional there is a unique $g \in \mathfrak{C}$ such that $\tilde{\mu} = \mu$. For this metric and any vector $v$ we calculate

$$\begin{aligned}
0 &= \nabla_v \left(\mu(E_1, \ldots, E_n)\right) \\
&= \left(\nabla_v \mu\right)(E_1, \ldots, E_n) + \sum_{i=1}^n \mu(E_1, \ldots, E_{i-1}, \nabla_v E_i, E_{i+1}, \ldots, E_n) \\
&= \sum_{i=1}^n \mu(E_1, \ldots, E_{i-1}, g(\nabla_v E_i, , E_i)\epsilon_i E_i, E_{i+1}, \ldots, E_n) \\
&= \sum_{i=1}^n g(\nabla_v E_i, E_i)\epsilon_i
\end{aligned}$$

$$= \sum_{i=1}^{n} \frac{1}{2}\epsilon_i \left(\nabla_v \left(g(E_i, E_i)\right) - \left(\nabla_v g\right)(E_i, E_i)\right)$$

$$= -\sum_{i=1}^{n} \frac{1}{2}\epsilon_i \varphi(v) g(E_i, E_i) = -\frac{n}{2}\varphi(v).$$

∎

**Definition 2.7.4.** *Let $(M, \mathfrak{C}, \nabla)$ be a Weyl structure and $\gamma$ be a smooth timelike (or spacelike) curve. We call $\gamma$ affinely parametrized if*

$$\nabla_{\dot\gamma(t)}\dot\gamma(t) \perp \dot\gamma(t)$$

*for all $t$ in the domain of definition of $\gamma$.*

It is clear that timelike pregeodesics are affinely parametrized if and only if they are geodesics.

**Lemma 2.7.3.** *Let $(M, \mathfrak{C}, \nabla)$ be a Weyl structure and $t \mapsto \gamma$ be a smooth curve with $g(\dot\gamma(t), \dot\gamma(t)) \neq 0$ for all $t$ and all $g \in \mathfrak{C}$. Then there exists a reparametrization $\tilde\gamma(s) = \gamma(t(s))$ such that $\tilde\gamma$ is affinely parametrized. If $t \mapsto \gamma(t)$ is affinely parametrized then $s \mapsto \gamma(t(s))$ is affinely parametrized if and only if there exist $a, b \in \mathbb{R}$ such that $s = at + b$.*

*Proof.* We denote $\frac{\mathrm{d}}{\mathrm{d}t}$ by a dot, $\dot{(\cdot)}$, and $\frac{\mathrm{d}}{\mathrm{d}s}$ by a prime, $(\cdot)'$. Let $\tilde\gamma(s) = \gamma(t(s))$. Then

$$\begin{aligned}
&g(\nabla_{\dot\gamma(t)}\dot\gamma(t), \dot\gamma(t)) \\
&= g(\nabla_{\dot{\tilde\gamma}(s(t))}\dot{\tilde\gamma}(s(t)), \dot{\tilde\gamma}(s(t))) \\
&= g(\nabla_{\tilde\gamma'(s)\frac{\mathrm{d}s}{\mathrm{d}t}}\left(\tilde\gamma'(s)\frac{\mathrm{d}s}{\mathrm{d}t}\right), \tilde\gamma'(s)\frac{\mathrm{d}s}{\mathrm{d}t}) \\
&= \left(\frac{\mathrm{d}s}{\mathrm{d}t}\right)^2 \left(g(\nabla_{\tilde\gamma'(s)}\left(\tilde\gamma'(s)\frac{\mathrm{d}s}{\mathrm{d}t}\right), \tilde\gamma'(s))\right) \\
&= \left(\frac{\mathrm{d}s}{\mathrm{d}t}\right)^2 \left(\frac{\mathrm{d}t}{\mathrm{d}s}\frac{\mathrm{d}^2 s}{\mathrm{d}t^2} g(\tilde\gamma'(s), \tilde\gamma'(s)) + \frac{\mathrm{d}s}{\mathrm{d}t} g(\nabla_{\tilde\gamma'(s)}\tilde\gamma'(s), \tilde\gamma'(s))\right)
\end{aligned}$$

implies that $\tilde\gamma$ is affinely parametrized if and only if

$$\frac{\mathrm{d}^2 s}{\mathrm{d}t^2} = \frac{\mathrm{d}s}{\mathrm{d}t}\frac{g(\nabla_{\dot\gamma(t)}\dot\gamma(t), \dot\gamma(t))}{g(\dot\gamma(t), \dot\gamma(t))}$$

holds. The first assertion follows immediately from the fundamental theorem for ODEs. If $t \mapsto \gamma(t)$ is already affinely parametrized, the differential equation reduces to $\frac{\mathrm{d}^2 s}{\mathrm{d}t^2} = 0$ and the second assertion follows. ∎

[p. 128 ↓]
→8
↓ p. 147

[8] The following section will be of great importance in Chap. 5. However, at this point in time it is better first to skip forward.

## 2.8 Curvature

p. 175 ↓
[↓ p. 138]

In Sect. 2.6 we have seen that the covariant derivative defines a notion of parallel transport along curves. Given a small loop $\gamma\colon [0,1] \mapsto M$ with $\gamma(0) = \gamma(1) = x$, this parallel transport defines a map

$$R_\gamma\colon T_xM \to T_xM, \quad v_x \mapsto \mathbf{P}_\gamma v_x$$

While in Minkowski spacetime and in Euclidean space we always have $\mathbf{P}_\gamma v_x = v_x$, in general the vector $\mathbf{P}_\gamma v_x$ depends on the loop $\gamma$. The following theorem shows that if one restricts to rectangular loops then there exists a well defined limit where $\gamma \to \{x\}$.

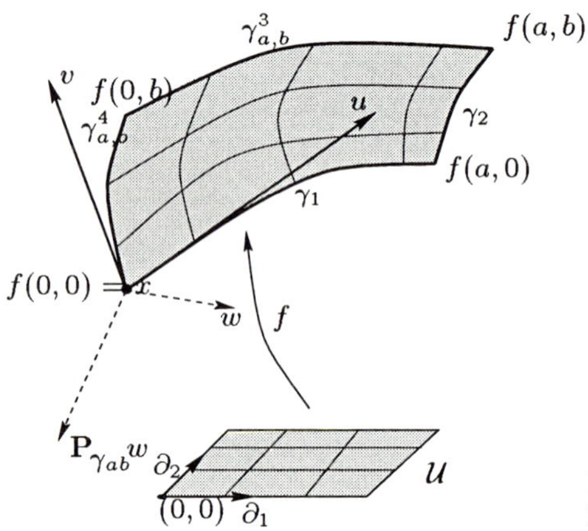


**Fig. 2.8.1.** The immersed surface in Theorem 2.8.1

**Theorem 2.8.1.** *Let $(M,\nabla)$ be a manifold with connection and $x \in M$, $u,v,w \in T_xM$. Then there exists a well defined tensor field*

$$R\colon T_xM \times T_xM \times T_xM \to T_xM$$
$$(u,v,w) \mapsto R(u,v)w$$

*with the following property.*

*Let $\mathcal{U} \subset \mathbb{R}^2$ be an open neighbourhood of $(0,0)$ and $f\colon \mathcal{U} \to M$ be an immersed 2-surface such that $f_*\partial_1 = u$, $f_*\partial_2 = v$. For any $a > 0$, $b > 0$ with $[0,a] \times [0,b] \subset \mathcal{U}$ let $\gamma_{a,b}$ be the closed curve*

$$\gamma_{a,b}\colon [0, 2a+2b] \to f(\mathcal{U}) \subset M$$
$$t \mapsto \begin{cases} f(t,0) & \textit{for} \quad 0 \le t < a \\ f(a, t-a) & \textit{for} \quad a \le t < a+b \\ f(2a+b-t, b) & \textit{for} \quad a+b \le t < 2a+b \\ f(0, 2a+2b-t) & \textit{for} \quad 0 \le t < a. \end{cases}$$

*Then*

$$\lim_{(a,b)\to(0,0)} \frac{1}{ab}\left(\mathbf{P}_{\gamma_{ab}} w - w\right) = -R(u,v)w$$

*holds.*

*If $U, V, W$ are vector fields with $U_x = u$, $V_x = v$, and $W_x = w$ then $R(u,v)w$ is given by*

$$R(u,v)w = \left(\nabla_U \nabla_V W - \nabla_V \nabla_U W - \nabla_{[U,V]} W\right)_{|x}.$$

*Proof.* It is easy to check that

$$R(u,v)w = \left(\nabla_U \nabla_V W - \nabla_V \nabla_U W - \nabla_{[U,V]} W\right)_{|x}$$

does not depend on the extensions $U, V, W$ and is therefore a tensor field.

We denote the four segments of the curve $\gamma_{a,b}$ by $\gamma^1_{a,b}, \ldots \gamma^4_{a,b}$, respectively. Let $W(a,b) = \mathbf{P}_{\gamma^2_{a,b}} \circ \mathbf{P}_{\gamma^1_{a,b}}(w)$ and observe that $W$ depends smoothly on $(a,b)$. From the definition of parallel transpor we have

$$\nabla_{f_*\partial_2} W(a,b) = \nabla_{\dot\gamma^2_{a,b}} W(a,b) = 0.$$

This and $W(0,0) = w$ immediately imply

$$R(u,v)w = -\left(\nabla_{\partial_2} \nabla_{\partial_1} W\right)_{(0,0)}.$$

Observe that the curve segment $\gamma^4_{a,b}$ is independent of $a$. In particular, $\gamma^4_{a,b} = \gamma^4_{0,b}$ which implies $\left(\mathbf{P}_{\gamma^4_{a,b}}\right)^{-1} w = W(0,b)$. Hence we obtain

$$\begin{aligned}
&\frac{1}{ab}\left(\mathbf{P}_{\gamma_{ab}} w - w\right) \\
&\quad = \frac{1}{ab}\left(\mathbf{P}_{\gamma^4_{a,b}} \circ \mathbf{P}_{\gamma^3_{a,b}} \circ \mathbf{P}_{\gamma^2_{a,b}} \circ \mathbf{P}_{\gamma^1_{a,b}}(w) - w\right) \\
&\quad = \frac{1}{ab}\left(\mathbf{P}_{\gamma^4_{a,b}} \circ \left(\mathbf{P}_{\gamma^3_{a,b}} \circ \left(\mathbf{P}_{\gamma^2_{a,b}} \circ \mathbf{P}_{\gamma^1_{a,b}}(w)\right) - (\mathbf{P}_{\gamma^4_{a,b}})^{-1}\right)(w)\right) \\
&\quad = \frac{1}{b}\mathbf{P}_{\gamma^4_{a,b}}\left(\frac{1}{a}\left(\mathbf{P}_{\gamma^3_{a,b}}(W(a,b)) - W(0,b)\right)\right).
\end{aligned}$$

If $b = 0$ then $W$ is simply the parallel transport of $w$ along $\dot\gamma^1_{a,b}$ which implies

$$\nabla_{f_*\partial_1} W(a,0) = \nabla_{\dot\gamma^1_{a,b}} W(a,0) = 0, \qquad (2.8.8)$$

and therefore

$$\frac{1}{ab}\left(\mathbf{P}_{\gamma_{ab}}w - w\right) = \frac{1}{b}\left(\mathbf{P}_{\gamma^4_{a,b}}\left(\nabla_{f_*\partial_1}W(0,b)\right) - \nabla_{f_*\partial_1}W(0,0)\right)$$
$$+\frac{1}{b}\mathbf{P}_{\gamma^4_{a,b}}\Big(\frac{1}{a}(\mathbf{P}_{\gamma^3_{a,b}}(W(a,b)) - W(0,b))$$
$$- \nabla_{f_*\partial_1}W(0,b)\Big). \tag{2.8.9}$$

Since $\gamma^4_{a,b}$ does not depend on $a$ the parallel transport along $\gamma^4_{a,b}$, $\mathbf{P}_{\gamma^4_{a,b}}$, does not depend on $a$ either. By Proposition 2.6.2 the limit of the first summand is therefore given by

$$\lim_{a,b\to 0}\frac{1}{b}\left(\mathbf{P}_{\gamma^4_{a,b}}(\nabla_{f_*\partial_1}W(0,b)) - \nabla_{f_*\partial_1}W(0,0)\right)$$
$$= \lim_{b\to 0}\frac{1}{b}\left(\mathbf{P}_{\gamma^4_{0,b}}\left(\nabla_{f_*\partial_1}W(0,b)\right) - \nabla_{f_*\partial_1}W(0,0)\right)$$
$$= \nabla_{f_*\partial_2}\nabla_{f_*\partial_1}W(0,0).$$

In order to complete the proof we need only to show that the second summand in Equation (2.8.9),

$$\tilde{\varphi}(a,b) = \frac{1}{b}\mathbf{P}_{\gamma^4_{a,b}}\Big(\frac{1}{a}(\mathbf{P}_{\gamma^3_{a,b}}(W(a,b)) - W(0,b)) - \nabla_{f_*\partial_1}W(0,b)\Big),$$

has the limit 0 for $(a,b)\to 0$.

We will first show that this summand is continuous in $(a,b)$. Since

$$\varphi(a,b) = \mathbf{P}_{\gamma^4_{a,b}}\left((\mathbf{P}_{\gamma^3_{a,b}}(W(a,b)) - W(0,b)) - a\nabla_{f_*\partial_1}W(0,b)\right)$$

depends smoothly on $(a,b)$, an application of the Taylor formula yields

$$\varphi(a,b) = \psi(a) + \theta(b) + ab\tilde{\psi}(a,b),$$

where $\psi, \theta, \tilde{\psi}$ are smooth functions. The curve $\gamma^3_{0,b}$ is constant which implies $\mathbf{P}_{\gamma^3_{0,b}} = \mathrm{id}$ and therefore $\varphi(0,b) = 0$ for all $b$. The curve $\gamma^3_{a,0}$ is inverse to the curve $\gamma^1_{a,0}$. In analogy to Equation (2.8.8) we obtain $\nabla_{f_*\partial_1}W(0,0) = 0$ and therefore

$$\varphi(a,0) = \mathbf{P}_{\gamma^3_{a,0}}W(a,0) - W(0,0) - a\nabla_{f_*\partial_1}W(0,0)$$
$$= \mathbf{P}_{\gamma^3_{a,0}}W(a,0) - W(0,0) = 0.$$

From $\varphi(0,b) = \varphi(a,0) = 0$ we conclude that both $\psi$ and $\theta$ are constant and that their sum vanishes. Hence we have $\varphi(a,b) = ab\tilde{\psi}(a,b) = ab\tilde{\varphi}(a,b)$ and

$$\begin{aligned}\tilde{\varphi}(a,b) &= \frac{1}{ab}\varphi(a,b)\\ &= \frac{1}{b}\left(\frac{1}{a}\left(\mathbf{P}_{\gamma^3_{a,b}}(W(a,b)) - W(0,b)\right) - \nabla_{f_*\partial_1}W(0,b)\right)\end{aligned}$$

is smooth. In particular, $\tilde{\varphi}$ is continuous. The equation $\tilde{\varphi}(0,0) = 0$ follows now from

$$\lim_{a\to 0}\frac{1}{a}\left(\mathbf{P}_{\gamma^3_{a,b}}(W(a,b)) - W(0,b)\right) = \nabla_{f_*\partial_1}W(0,b),$$

for any $b > 0$. ∎

**Definition 2.8.1.** *The* $\binom{1}{3}$*-tensor* $R$ *is called the* curvature tensor *(or* Riemann tensor*) of* $(M,\nabla)$.

**Lemma 2.8.1 (Bianchi identities).** *Let* $(M,\nabla)$ *be a manifold with connection. Then* $R$ *satisfies the* second Bianchi identity,

$$(\nabla_u R)(v,w) + (\nabla_v R)(w,u) + (\nabla_w R)(u,v) = 0.$$

*If* $\nabla$ *is torsion-free, then* $R$ *also satisfies the* first Bianchi identity,

$$R(u,v)w + R(v,w)u + R(w,u)v = 0.$$

*Proof.* Consider a normal coordinate system $(x^1,\ldots,x^n)$ centered at $x \in M$ and let $U,V,W$ be extensions of $u,v,w \in T_xM$ such that these vector fields have constant components with respect to our coordinate system. We have then $[U,V]_y = [U,W]_y = [V,W]_y = 0$ for all $y$ in this coordinate neighbourhood and $\nabla_{U_x}V = \nabla_{V_x}W = \nabla_{W_x}U = 0_x$. For any vector field $X$ we get

$$\begin{aligned}\left(\nabla_{U_x}R\right)(V,W)X &= \nabla_{U_x}(R(V,W)X) - R(\nabla_{U_x}V,W)X\\ &\quad - R(V,\nabla_{U_x}W)X - R(V,W)\nabla_{U_x}X\\ &= \nabla_{U_x}(R(V,W)X) - R(V,W)\nabla_{U_x}X\\ &= \nabla_{U_x}(\nabla_V\nabla_W X - \nabla_W\nabla_V X) - \nabla_{V_x}\nabla_W\nabla_U X\\ &\quad + \nabla_{W_x}\nabla_V\nabla_U X\\ &= ([\nabla_U,[\nabla_V,\nabla_W]])_x X.\end{aligned}$$

The second Bianchi identity follows now immediately from Lemma 2.4.3.

For the first Bianchi identify we assume in addition that $\nabla$ is torsion free. We calculate

$$\begin{aligned}R(u,v)w + R(v,w)u + R(w,u)v \\ &= \nabla_U\nabla_V W - \nabla_V\nabla_U W + \nabla_V\nabla_W U - \nabla_W\nabla_V U \\ &\quad + \nabla_W\nabla_U V - \nabla_U\nabla_W V \\ &= \nabla_U\left(\nabla_V W - \nabla_W V\right) + \nabla_V\left(\nabla_W U - \nabla_U W\right) \\ &\quad + \nabla_W\left(\nabla_U V - \nabla_V U\right) \\ &= 0 + 0 + 0 = 0\end{aligned}$$

[p. 134 ↓]
↓ p. 180

∎

**Definition 2.8.2.** *Let $(M, \nabla)$ be a manifold with connection. The tensor* $\mathrm{Ric}(u, v) = \mathrm{tr}(R(\cdot, u)v)$ *is called the* Ricci tensor. *We denote by $F \in \Omega^2(M)$ the $2/n$-fold multiple of the anti-symmetric part of* Ric,

$$F(u, v) = \frac{1}{n}\left(\mathrm{Ric}(u, v) - \mathrm{Ric}(v, u)\right).$$

**Lemma 2.8.2.** *Let $(M, \nabla)$ be a manifold with torsion-free connection. Then $F = -\frac{1}{n}\mathrm{tr}(R(\cdot, \cdot))$ holds.*

*Proof.* Let $\{E_1, \ldots, E_n\}$ be a basis, $\{\omega^1, \ldots, \omega^n\}$ be dual basis and $u, v \in T_x M$. From the first Bianchi identity and the antisymmetry of $R(\cdot, \cdot)$ we obtain $R(u, v)E_a = -R(E_a, u)v + R(E_a, v)u$ and therefore

$$\begin{aligned}nF(u, v) &= \omega^a\left(R(E_a, u)v - R(E_a, v)u\right) \\ &= -\omega^a\left(R(u, v)E_a\right) = -\mathrm{tr}(R(u, v)).\end{aligned}$$

∎

### 2.8.1 Applications to Weyl structures

The differential form $F$ has a particularly geometric interpretation if $\nabla$ is a Weyl connection, justifying the following definition

**Definition 2.8.3.** *Let $(M, \mathfrak{C}, \nabla)$ be a manifold with Weyl structure. Then we call $F$ the* length curvature *and $K$, defined by $R(u, v)w = K(u, v)w + F(u, v)w$ the* directional curvature.

To motivate these terms coined by Weyl[9] we will need the following lemma.

**Lemma 2.8.3.** *Let $(M, \mathfrak{C}, \nabla)$ be a manifold with Weyl structure. Then $F = \frac{1}{2}\mathrm{d}\varphi$ and $g(K(u, v)w, w) = 0$ for all vectors $u, v, w$ and all $g \in \mathfrak{C}$.*

---

[9] In the German original, they are called *Streckenkrümmung* and *Richtungskrümmung*.

*Proof.* Let $U, V, W$ be tensor fields with $U_x = u$, $V_x = v$, and $W_x = w$. We can also assume that $[U, V] = 0$.

$$\begin{aligned}
&g(\nabla_U \nabla_V W, W) \\
&\quad = -(\nabla_U g)(\nabla_V W, W) - g(\nabla_V W, \nabla_U W) + \nabla_U \left(g(\nabla_V W, W)\right) \\
&\quad = -\varphi(U) g(\nabla_V W, W) - g(\nabla_V W, \nabla_U W) + \nabla_U \Big(\frac{1}{2} V \bullet g(W, W) \\
&\qquad - \frac{1}{2}(\nabla_V g)(W, W)\Big) \\
&\quad = -\varphi(U)\Big(\frac{1}{2} V \bullet g(W, W) - \frac{1}{2}(\nabla_V g)(W, W)\Big) - g(\nabla_V W, \nabla_U W) \\
&\qquad + \frac{1}{2} U \bullet V \bullet g(W, W) - \frac{1}{2}\nabla_U \Big(\varphi(V) g(W, W)\Big) \\
&\quad = -\frac{1}{2}\Big(\varphi(U) V \bullet g(W, W) + \varphi(V) U \bullet g(W, W)\Big) \\
&\qquad + \frac{1}{2}\varphi(U)\varphi(V) g(W, W) - g(\nabla_V W, \nabla_U W) \\
&\qquad + \frac{1}{2} U \bullet V \bullet g(W, W) - \frac{1}{2}(\nabla_U \varphi)(V) g(W, W) \\
&\qquad - \frac{1}{2}\varphi(\nabla_U V) g(W, W).
\end{aligned}$$

This implies (using $[U, V] = 0$)

$$\begin{aligned}
g(R(U, V)W, W) &= g(\nabla_U \nabla_V W - \nabla_V \nabla_U W, W) \\
&= -\frac{1}{2}\left((\nabla_U \varphi)(V) - (\nabla_V \varphi)(U)\right) g(W, W) \\
&= -\frac{1}{2} \mathrm{d}\varphi(U, V) g(W, W).
\end{aligned}$$

Since $W$ is arbitrary, we obtain $\operatorname{tr}(R(U, V)) = -\frac{n}{2}\mathrm{d}\varphi(U, V)$ which proves the first claim. The second claim follows from

$$\begin{aligned}
g(K(U, V)W, W) &= g(R(U, V)W, W) - g(F(U, V)W, W) \\
&= -\frac{1}{2}\mathrm{d}\varphi(U, V) g(W, W) - g(-\frac{1}{2}\mathrm{d}\varphi(U, V)W, W) = 0.
\end{aligned}$$

∎

Let $w \in T_x M$ and $\gamma$ be a small loop of the type given in Theorem 2.8.1. Then the parallel transport $\mathbf{P}_\gamma w$ is approximately

$$\begin{aligned}
\mathbf{P}_\gamma w &\approx w + abR(u, v)w \\
&= w + abF(u, v)w + \underbrace{abK(u, v)w}_{\in w^\perp}.
\end{aligned}$$

It follows that $1+abF(u,v)$ is the factor by which the parallel transport stretches the (relative) length of $w$ and that $abK(u,v)w$ is the change of direction of $w$ due to the parallel transport. Since the parallel transport of a vector does not leave its relative length invariant it is impossible to compare lengths at different points. There is an important consequence to this fact, the so-called *clock paradox of second kind* (cf. Sect. 3.3).

## 2.9 Variation of geodesics

*In this section we investigate the infinitesimal analogue of 1-parameter families of geodesics.*

*This section is technical and can be omitted on first reading. It is a prerequisitive for Sects. 4.5 and 4.6 in the chapter on pseudo-Riemannian manifolds. Section 4.5 is concerned with metric preserving diffeomorphisms and used in the discussion of cosmological models (Chap. 6). Sect. 4.6 is necessary for understanding the complete proofs of the singularity theorems which are presented in Chap. 9.*

In Chap. 5 we will see that freely falling particles can mathematically be described by (certain) geodesics. A cloud of such particles moving in spacetime corresponds then to a smooth 3-parameter family of geodesics. It is therefore of physical interest to study families of geodesics. Here we will study the slightly simpler sub-case of 1-parameter families and its infinitesimal analogue.

Let $f\colon \Sigma \to M$ be a differentiable map and recall that a *(smooth) vector field along* $f$ is a smooth map $X\colon \Sigma \to TM$ with $\pi_{TM} \circ X(x) = f(x)$ for all $x \in \Sigma$. We denote the space of vector fields along $f$ by $\mathcal{T}_0^1(f)$. Any smooth vector field $X$ on $M$ induces a vector field $\hat{X}\colon x \mapsto \hat{X}_x := X_{f(x)}$. A vector field $U$ on $\Sigma$ also induces a natural vector field along $f$ via $x \mapsto f_* U_x$. Important examples of this construction are given by vector fields along curves and by vector fields along canonical immersions of submanifolds (cf. Sect. 4.4).

**Lemma 2.9.1.** *Let* $f\colon \Sigma \to M$, $U, V \in \mathcal{T}_0^1(\Sigma)$, *and* $X, Y \in \mathcal{T}_0^1(f)$. *Then* $\overset{f}{\nabla}_U X := \left(U^\beta \partial_\beta X^a + \Gamma^a_{bc}(f_*U)^b X^c\right)\partial_a$ *is a well defined vector field along* $f$ *and satisfies the following properties.*

*(i)* $\overset{f}{\nabla}_U X$ *is function-linear in* $U$ *and* $\mathbb{R}$*-linear in* $X$*;*

*(ii)* $\overset{f}{\nabla}_U(\varphi X) = \mathrm{d}\varphi(U)X + \varphi \overset{f}{\nabla}_U X$ *for all functions* $\varphi \in C^\infty(\Sigma)$*;*

*(iii)* $\overset{f}{\nabla}_U f_*V - \overset{f}{\nabla}_V f_*U - f_*[U,V] = \mathrm{Tor}(f_*U, f_*V)$.

*Proof.* We have to show that the definition is invariant under coordinate transformations. Let $\psi$ be a diffeomorphism of $\Sigma$ and $\phi$ be a diffeomorphism of $M$. Let $\psi\colon \Sigma \to \Sigma$ a diffeomorphism. Then we obtain

$$\begin{aligned}
\psi_*\left(\overset{f}{\nabla}_U X\right) &= \overset{f\circ\psi^{-1}}{\nabla}_{\psi_* U}(X\circ\psi^{-1}) \\
&= \Big(\partial_\gamma \psi^\beta U^\gamma (\partial_\delta X^a \partial_\beta (\psi^{-1})^\delta) \circ \psi^{-1} \\
&\quad + \Gamma^a_{bc}(\partial_\gamma f^b \partial_\beta (\psi^{-1})^\gamma)(\partial_\delta \psi^\beta U^\delta) X^c \circ \psi^{-1}\Big)\partial_a \\
&= \Big((U^\beta \partial_\beta X^a + \Gamma^a_{bc}(\partial_\beta f^b U^\beta) X^c)\partial_a\Big) \circ \psi^{-1}.
\end{aligned}$$

Hence the formula does not depend on the coordinates chosen for $\Sigma$.

Let $\phi\colon M \to M$ be a diffeomeorphism. We obtain

$$\begin{aligned}
\overset{\phi\circ f}{\nabla}_U \phi_* X &= \Big(U^\beta \partial_\beta ((\partial_d \phi^a)\circ f\, X^d) \\
&\quad + \Gamma^a_{bc}(\partial_\beta(\phi\circ f)^b U^\beta)(\partial_d\phi^c)\circ f\, X^d\Big)\partial_a \\
&= \Big(U^\beta ((\partial_e\partial_d\phi^a)\circ f\, \partial_\beta f^e X^d + (\partial_d\phi^a)\circ f\, \partial_\beta X^d) \\
&\quad + \Gamma^a_{bc}(\partial_e\phi^b)\circ f\, \partial_\beta f^e U^\beta (\partial_d\phi^c)\circ f\, X^d\Big)\partial_a \\
&= U^\beta \Big((\partial_d\phi^a)\circ f\, \partial_\beta X^d + ((\partial_e\partial_d\phi^a)\circ f \\
&\quad + \Gamma^a_{bc}(\partial_e\phi^b)\circ f\, (\partial_d\phi^c)\circ f\,)\partial_\beta f^e X^d\Big)\partial_a
\end{aligned}$$

The Gaußian basis vector field with respect to the coordinates induced by $\phi$ are given by $\phi_*\partial_a$. Taking this into account we see that the Christoffel symbols transform as given in Corollary 2.6.1. This implies that our coordinate formula defines a well defined vector field along $f$. Equations (i)–(iii) follow directly from our coordinate expression. ∎

If $f$ is an immersion then the covariant derivative along $f$ can be calculated entirely in $M$.

**Lemma 2.9.2.** *Let $f\colon \Sigma \to M$ be an immersion, $U, V \in \mathcal{T}^1_0(\Sigma)$, and $X, Y \in \mathcal{T}^1_0(f)$. Let $\tilde{U}, \tilde{X}$ be vector fields on $M$ which conicide with $f_* U$ and $X$ at all points $y = f(x)$. Then we have $\overset{f}{\nabla}_U X = \left(\nabla_{\tilde{U}}\tilde{X}\right)\circ f$ at all $x \in \Sigma$.*

This lemma justifies writing $\nabla_{f_* U} X$ instead of $\left(\nabla_{\tilde{U}}\tilde{X}\right)\circ f$. We will use this notation extensively in Sect. 4.4.

*Proof of Lemma 2.9.2.* Let $x \in \Sigma$. Since $f$ is an immersion there exists a neighbourhood of $\mathcal{U}$ of $x$, a neighbourhood $\mathcal{V} \subset \mathbb{K}^{n-\dim(\Sigma)}$ of 0, and a local diffeomorphsim $F\colon \mathcal{U}\times\mathcal{V} \to M$ with $F(x,0) = f(x)$ for all $x \in \mathcal{U}$. We may extend $U$, $X$ to $\mathcal{U}\times\mathcal{V}$ such that $\tilde{U} = F_* U$ and $\tilde{X} = X \circ F^{-1}$. Then we obtain

$$\begin{aligned}\left(\nabla_{\tilde U}\tilde X\right)^b \circ F &= \tilde U^a \partial_a \tilde X^b \circ F - (\Gamma^b_{ac}\tilde U^a \tilde X^c)\circ F\\ &= \partial_i F^a U^i \partial_a (X^b \circ F^{-1})\circ F - (\Gamma^b_{ac}\circ F)\,(F_*U)^a X^c\\ &= \partial_i F^a U^i \partial_j X^b \partial_a (F^{-1})^j \circ F - (\Gamma^b_{ac}\circ F)\,(F_*U)^a X^c\\ &= U^j \partial_j X^b - (\Gamma^b_{ac}\circ F)\,(F_*U)^a X^c.\end{aligned}$$

Restricting the last expression to $\mathcal{U} \times \{0\}$ gives

$$\left(\nabla_{\tilde U}\tilde X\right)\circ F(x,0) = \left(\overset{f}{\nabla}_U X\right)_x$$

for all $x \in \mathcal{U} \subset \Sigma$. ∎

**Lemma 2.9.3.** *Let* $f\colon \Sigma \to M$, $U, V \in \mathcal{T}^1_0(\Sigma)$, *and* $X \in \mathcal{T}^1_0(f)$. *Then the equation*

$$R(f_*U, f_*V)X = \overset{f}{\nabla}_U \overset{f}{\nabla}_V X - \overset{f}{\nabla}_V \overset{f}{\nabla}_U X - \overset{f}{\nabla}_{[U,V]} X$$

*holds.*

*Proof.* The equation follows directly from the definition of the Riemann tensor. ∎

Observe that $f$ does not need to be an immersion and therefore $f(\Sigma)$ may not be an immersed submanifold. This is important for the following application.

**Definition 2.9.1.** *A* geodesic variation *is a map* $f\colon (-\delta,\delta)\times(a,b) \to M$, $(s,t) \mapsto f(s,t) \in M$ *such that for each* $s$ *the curve* $t \mapsto f(s,t)$ *is a geodesic. We denote the velocity of the geodesics by* $f_t := T_{(s,t)}f(\partial_t)$ *and the* deviation vector field *along the geodesic* $f(s,\cdot)$ *by* $f_s := T_{(s,t)}f(\partial_s)$ .

**Proposition 2.9.1.** *Let* $f\colon (s,t) \mapsto f(s,t) \in M$ *be a geodesic variation. Then* $f_t$ *satisfies the geodesic equation* $\overset{f}{\nabla}_{\partial_t} f_t = 0$ *and the deviation vector field* $f_s$ *satisfies the equation*

$$\overset{f}{\nabla}_{\partial_t}\overset{f}{\nabla}_{\partial_t} f_s + R(f_s,f_t)f_t - (\overset{f}{\nabla}_{\partial_t}\mathrm{Tor})(f_t f_s) - \mathrm{Tor}(f_t, \overset{f}{\nabla}_{\partial_t} f_s) = 0.$$

*Proof.* The geodesic equation follows directly from the definition of a geodesic variation. Observe that $[f_s, f_t] = [f_*(\partial_s), f_*(\partial_t)] = f_*[\partial_s,\partial_t] = 0$ and that therefore

$$\begin{aligned}\overset{f}{\nabla}_{\partial_t}\overset{f}{\nabla}_{\partial_t} f_s &= \overset{f}{\nabla}_{\partial_t}\overset{f}{\nabla}_{\partial_s} f_t + \overset{f}{\nabla}_{\partial_t}\overbrace{[f_t,f_s]}^{=0} + \overset{f}{\nabla}_{\partial_t}(\mathrm{Tor}(f_t,f_s))\\ &= R(f_t,f_s)f_t + \overset{f}{\nabla}_{\partial_s}\overbrace{\overset{f}{\nabla}_{\partial_t} f_t}^{=0} + (\overset{f}{\nabla}_{\partial_t}\mathrm{Tor})(f_t,f_s)\\ &\quad + \mathrm{Tor}(f_t, \overset{f}{\nabla}_{\partial_t} f_s).\end{aligned}$$

∎

It is often sufficient to consider the infinitesimal analogue of geodesic variations. This justifies the following definition.

**Definition 2.9.2.** *A* Jacobi field *is a vector field $J$ along a geodesic $\gamma$ which satisfies the* Jacobi equation

$$\nabla_{\dot\gamma}\nabla_{\dot\gamma}J + R(J,\dot\gamma)\dot\gamma - (\nabla_{\dot\gamma}\mathrm{Tor})(\dot\gamma, J) - \mathrm{Tor}(\dot\gamma, \nabla_{\dot\gamma}J) = 0.$$

**Proposition 2.9.2.** *Let $\gamma\colon [a,b) \to M$ be a geodesic. The Jacobi fields along $\gamma$ span a $2n$-dimensional linear space and any Jacobi field $J$ along $\gamma$ is uniquely determined by $J(a), \nabla_{\dot\gamma(a)}J$.*

*Proof.* Without loss of generality we may consider a single chart which contains the geodesic. The Jacobi equation reduces then to a system of $n$ ordinary second order differential equations, or, equivalently, to a system of $2n$ ordinary first order differential equations. Hence the assertion follows from the fundamental theorem for ordinary differential equations (cf. Theorem 2.4.1). ∎

**Corollary 2.9.1.** *Let $\gamma\colon [a,b] \to M, t \mapsto \exp_x((t-a)u_x)$ be a geodesic. A vector field $J$ along $\gamma$ is a Jacobi field which vanishes at $x = \gamma(a)$ if and only if there is a vector $v_x \in T_xM$ with $J = \exp((t-a)(u_x + sv_x))_*\,\partial_s$.*

*Proof.* It is clear that, given such a vector $v_x$, $J$ is a Jacobi field along $\gamma$ which vanishes at $x$. Proposition 2.9.2 implies that the Jacobi fields along $\gamma$ which vanish at $x$ span an $n$-dimensional vector space and are characterized by their velocity vector $\nabla_{\dot\gamma(a)}J$. The assertion follows since

$$\nabla_{\dot\gamma(a)} \exp((t-a)(u_x + sv_x))_*\,\partial_s = v_x$$

for all $v_x \in T_xM$ and $T_xM$ is an $n$-dimensional vector space. ∎

**Definition 2.9.3.** *Two points $x, y \in M$ are* conjugate *if there is a geodesic $\gamma$ joining $x$ and $y$ and a non-zero Jacobi field $J$ along $\gamma$ which vanishes both at $x$ and $y$.*

**Proposition 2.9.3.** *Two points $x, y \in M$ are conjugate if and only if there exists an $u_x \in T_xM$ in the domain of* exp *such that* $\exp(u_x) = y$ *and $T_{u_x}\exp_x\colon T_{u_x}T_xM \to T_yM$ fails to have maximal rank.*

*Proof.* If $x, y$ are conjugate, there is a geodesic $\gamma\colon [0,1] \to M$ joining these two points and a Jacobi field $J$ along $\gamma$ which is non-zero but vanishes at $x$ and $y$. Let $u_x$ be the uniquely determined vector which satisfies $\exp(tu_x) = \gamma(t)$. By Corollary 2.9.1 there exists a vector $v_x \in T_xM \setminus \{0\}$ such that $J(t) = T_{(0,t)}\exp(t(u_x + sw_x))(\partial_s)$. The assumption $J(1) = 0$

implies that the linear map $T_{(0,1)}\exp(t(u_x + sw_x))\colon \mathbb{K}^2 \to T_yM$ does not have maximal rank which in turn implies that $T_{u_x}\exp_x : T_{u_x}T_xM \to T_yM$ does not have maximal rank.

To prove the converse assertion, we just choose the vectors $u_x$, $v_x \neq 0$ by the requirement that $\exp_x(u_x) = 0$ and $T_{u_x}\exp\left((\frac{\mathrm{d}}{\mathrm{d}s}{}_{|s=0}(u_x + sv_x)\right) = 0$. The vector field $J(t) = T_{(0,t)}\exp(t(u_x + sw_x))(\partial_s)$ along $\gamma(t) = \exp_x(tu_x)$ is then a non-zero Jacobi field which vanishes at $x$ and $y$. ∎

**Proposition 2.9.4.** *Let $\gamma\colon [a,b] \to M$ be a geodesic without conjugate points. For every pair of vectors $w_{\gamma(a)} \in T_{\gamma(a)}M$, $\tilde{w}_{\gamma(b)} \in T_{\gamma(b)}M$ there is a unique Jacobi field $J$ along $\gamma$ with $J(a) = w_{\gamma(a)}$ and $J(b) = \tilde{w}_{\gamma(b)}$.*

*Proof.* There are vectors $u_{\gamma(a)} \in T_{\gamma(a)}M$ with $\gamma(t) = \exp_{\gamma(a)}((t-a)u_{\gamma(a)})$ and $\tilde{u}_{\gamma(b)} \in T_{\gamma(b)}M$ with $\gamma(t) = \exp_{\gamma(b)}((b-t)\tilde{u}_{\gamma(b)})$. Since $\gamma$ does not have conjugate points the linear maps

$$T_{(b-a)u_{\gamma(a)}}\exp_{\gamma(a)} : T_{(b-a)u_{\gamma(a)}}T_{\gamma(a)}M \to T_{\gamma(b)}M$$

and

$$T_{(b-a)\tilde{u}_{\gamma(b)}}\exp_{\gamma(b)} : T_{(b-a)\tilde{u}_{\gamma(b)}}T_{\gamma(b)}M \to T_{\gamma(a)}M$$

are both isomorphisms. Hence there are vectors $v_{\gamma(a)}$ and $\tilde{v}_{\gamma(b)}$ such that

$$\tilde{w}_{\gamma(b)} = T_{(b-a)u_{\gamma(a)}}\exp_{\gamma(a)}\left((b-a)\frac{\mathrm{d}}{\mathrm{d}s}_{|s=0}\left(u_{\gamma(a)} + sv_{\gamma(a)}\right)\right),$$

$$w_{\gamma(a)} = T_{(b-a)\tilde{u}_{\gamma(b)}}\exp_{\gamma(b)}\left((b-a)\frac{\mathrm{d}}{\mathrm{d}s}_{|s=0}\left(\tilde{u}_{\gamma(b)} + s\tilde{v}_{\gamma(b)}\right)\right).$$

Let $J_1$ , $J_2$ be the Jacobi fields defined by

$$J_1 = T_{(0,t)}\exp((t-a)(u_{\gamma(a)} + sv_{\gamma(a)}))(\partial_s),$$
$$J_2 = T_{(0,t)}\exp((b-t)(\tilde{u}_{\gamma(b)} + s\tilde{v}_{\gamma(b)}))(\partial_s).$$

Then $J_1$ vanishes at $\gamma(a)$ and has the value $\tilde{w}_{\gamma(b)}$ at $\gamma(b)$ whereas $J_2$ has the value $w_{\gamma(a)}$ at $\gamma(a)$ and vanishes at $\gamma(b)$. Observe that the sum $J_1(t)+J_2(t)$ is well defined since $\pi_{TM}\circ J_1(t) = \pi_{TM}\circ J_2(t) = \gamma(t)$. By the linearity of the Jacobi equation the vector field $J = J_1 + J_2$ along $\gamma$ is also a Jacobi field which has correct values at both $\gamma(a)$ and $\gamma(b)$. This proves existence. Uniqueness is clear since the space of solutions has dimension $2n$ which is just the dimension of the vector space $T_{\gamma(a)}M \oplus T_{\gamma(b)}M$. ∎

Jacobi fields can also be used to calculate the differential of the exponential map restricted to an $n$-dimensional "vertical" subspace of $T_{u_x}TM$ which consists of all $v_{[u]} = \frac{\mathrm{d}}{\mathrm{d}t}(u_x + tv_x)_{|t=0} \in T_{u_x}TM$, where $v_x \in T_xM$.

**Proposition 2.9.5.** *Let $x \in M$, $u_x, v_x \in T_xM$, and $J$ be the Jacobi field along $t \mapsto \gamma(t) = \exp(tu_x)$ with $J(0) = 0$ and $\nabla_{\dot\gamma} J(0) = v_x$. Then the differential of the exponential map in direction $v_{[u]}$ is given by*

$$T_{u_x} \exp_x(v_{[u]}) = J(1).$$

*Proof.* Let $f\colon (s,t) \mapsto \exp_x(t(u+sv)) \in M$ and

$$f_t = T_{(s,t)}f(\partial_t) = T_{t(u+sv)} \exp_x((u+sv)_{[t(u+sv)]}),$$
$$f_s = T_{(s,t)}f(\partial_s) = T_{t(u+sv)} \exp_x((tv)_{[t(u+sv)]}).$$

Since $f$ is a geodesic variation with $f(s,0) = x \quad \forall s$ the vector field $(f_s)_{|s=0}$ is a Jacobi vector field along $\gamma$ which vanishes at 0. From $[f_s, f_t] = 0$ we obtain $\nabla_{\dot\gamma} J = \overset{f}{\nabla}_{\partial_t} f_s = \overset{f}{\nabla}_{\partial_s} f_t + \mathrm{Tor}(f_s, f_t)$. At $t = 0$, we have $f_s = 0$ and $f_t = T_0 \exp_x((u+sv)_{[0]})$), whence $\nabla_{\dot\gamma} J(0) = (\frac{\partial}{\partial s})_{|s=0} T_0 \exp_x((u+sv)_{[0]}) = T_0 \exp_x((v)_{[0]}) = v$. The Jacobi vector field $J$ also satisfies $J(1) = T_{(0,1)}f(\partial_s) = T_{(u)} \exp_x(v_{[u]})$. The assertion follows since initial value and derivative characterize Jacobie fields uniquely. ∎

# 3. Space and time from a global point of view

p. 133 ↓
[↓ p. 152]

*The content of this chapter is mainly physical. In Sect. 3.1 we show that the experiment of Michelson and Morley indicates that spacetime admits a conformal structure. A conformal structure is not sufficient to describe spacetime adequately. In Sect. 3.2 we generalize the notion of inertial observers which leads to the existence of a projective structure. (One of Einstein's key observations was that this projective structure is closely linked to the phenomenon of gravity. This will be persued in Chap. 5).*

*In Sect. 3.3 we use our physical postulates in order to show that the conformal and the projective structures of spacetime form a Weyl structure. Here we closely follow (Ehlers, Pirani, and Schild 1972). The proofs in this section are technical and can be omitted without loss of continuity. In Sect. 2.8 we introduce a further physical postulate which restricts the Weyl structure to a Lorentzian manifold.*

## 3.1 Light rays: the conformal structure

In Sect. 1.4 we have seen that spacetime is endowed with an invariant field of light cones (cf. Postulate 1.4.1). In analogy to the discussion in the previous chapter we will define these light cones infinitesimally, i.e., in the tangent spaces rather than in spacetime itself.

The discussion in Chap. 1 may seem to indicate that for each $x \in M$ the tangential space $T_x M$ can be identified with $(\mathbb{R}^n, \eta)$. Since for each non-degenerate bilinear form $g_x$ of signature $(-,+\ldots,+)$ there exist linear coordinates $(x^0, x^1, \ldots, x^{n-1})$ such that $g_x = -\mathrm{d}\left(x^0\right)^2 + \delta_{ij}\mathrm{d}x^i\mathrm{d}x^j$ $(i, j \in \{1, \ldots, n-1\})$, one may be tempted to simply replace Minkowski spacetime $(\mathbb{A}^n, \eta)$ by a general Lorentzian manifold $(M, g)$. However, the Michelson-Morley experiment only determines the paths of light rays.[1] In other words, from the Michelson-Morley experiment alone one can only infer the existence of a *conformal structure* $(\mathbb{A}^n, \mathfrak{C}_\eta)$, where $\mathfrak{C}_\eta = \{\Omega^2\eta |\ \Omega \in C^\infty(\mathbb{A}^n, \mathbb{R}^+ \setminus \{0\})\}$. (In Sect. 1.4 we used the affine structure

[1] For the definition of wave length we needed the Euclidean structure of space. This seems to indicate that implicitly we used a Lorentzian metric rather than a conformal structure in order to interpret the Michelson-Morley experiment. However, the outcome of this experiment is a null-effect, i.e., $\Delta Z \approx 0$ which is *independent* of the Euclidean structure chosen.

of $\mathbb{A}^n$ to single out a constant representative $\eta \in \mathfrak{C}_\eta$. This is not possible for general manifolds.)

Hence in a global setting the Michelson-Morley experiment leads to the following postulate.

**Postulate 3.1.1 (Existence of a conformal structure).** *(Conformal) spacetime is a pair $(M, \mathfrak{C})$, where $M$ is a n-dimensional manifold and $\mathfrak{C}$ a conformal structure of signature $(-, + \ldots, +)$. This conformal structure is given by the paths of light rays.*

We will call a conformal structure of signature $(-, +, \ldots, +)$ *Lorentzian*.

We will see that, given a Lorentzian conformal structure, it is possible to recover light rays (cf. Postulate 3.1.2 below and the discussion leading to it).

**Definition 3.1.1.** *Let $(M, \mathfrak{C})$ be a manifold with Lorentzian conformal structure. A* null (or lightlike) hypersurface *$N$ is a hypersurface such that (for any $g \in \mathfrak{C}$) the induced metric on $N$ is positive semi-definite but not positive definite.*

Let $(M, \mathfrak{C})$ be a manifold with a Lorentzian conformal structure and $N \subset M$ be a null hypersurface. At each point $x \in N$ there exists a *unique* 1-dimensional subspace $l_x \subset T_xM$ which is tangent to $N$ and satisfies $g(v, v) = 0$ for all $v \in l_x$, $g \in \mathfrak{C}$. For, if there where two such vectors $v_1, v_2 \in T_xN$ which were not collinear then there would exist a vector $w \in \operatorname{span}\{v_1, v_2\}$ with $g(w, w) < 0$ for all $w \in \mathfrak{C}$. But this is impossible since $g$ restricted to $T_xN$ is positive semidefinite. If $U, V$ are vector fields with $U_x \in l_x \setminus \{0_x\}$, $V_x \in l_x \setminus \{0_x\}$ for all $x \in N$ then the integral curves of $V$ and $U$ in $N$ are reparametrizations of each other.

**Definition 3.1.2.** *Let $(M, \mathfrak{C})$ be a manifold with Lorentzian conformal structure and $N \subset M$ a null hypersurface. A curve $\gamma$ in $N$ whose velocity vector satisfies $g(\dot\gamma, \dot\gamma) = 0$ for every metric $g \in \mathfrak{C}$ is called a* conformal null geodesic

**Lemma 3.1.1.** *Let $\gamma$ be a conformal null geodesic and $g \in \mathfrak{C}$. Then $\gamma$ satisfies the differential equation*

$$\ddot\gamma^a + g^{ad}\left(\partial_b g_{dc} - \frac{1}{2}\partial_d g_{bc}\right)\dot\gamma^b\dot\gamma^c \quad \| \, \dot\gamma^a .$$

*Proof.* First we will show that the coordinate expression is invariant with respect to coordinate transformations and with respect to changes of the representative $g \in \mathfrak{C}$. If $\nabla$ is the Levi-Cività connection of $g$ then we have

$$(\nabla_{\dot\gamma}\dot\gamma)^a = \ddot\gamma^a + g^{ad}\left(\partial_b g_{dc} - \frac{1}{2}\partial_d g_{bc}\right)\dot\gamma^b\dot\gamma^c,$$

whence we have only to show that for any other metric $e^{2f}g \in \mathfrak{C}$ with Levi-Cività connection $\hat{\nabla}$ we have $\nabla_{\dot{\gamma}}\dot{\gamma} \parallel \hat{\nabla}_{\dot{\gamma}}\dot{\gamma}$. But this follows from

$$e^{-2f}g^{ad}\left(\partial_b(e^{2f}g_{dc}) - \frac{1}{2}\partial_d(e^{2f}g_{bc})\right)\dot{\gamma}^b\dot{\gamma}^c = g^{ad}\left(\partial_b g_{dc} - \frac{1}{2}\partial_d g_{bc}\right)\dot{\gamma}^b\dot{\gamma}^c$$
$$+ 2\partial_b f\dot{\gamma}^a - g^{ad}\partial_d f\,\overbrace{g_{bc}\dot{\gamma}^b\dot{\gamma}^c}^{=0}.$$

Let $N$ be a null hypersurface containing $\gamma$. We may chose coordinates $(x^1, \ldots, x^{n-1})$ for $N$ such that $\partial_1$ spans $l_x$ for each $x$. Since the bilinear form $g_{|\mathrm{span}\{\partial_2,\ldots,\partial_{n-1}\}}$ is positive definite and $g_{|TN}$ does not have full rank, we have $g_{1i} = 0 \forall i \in \{1, \ldots, n-1\}$. We may extend the coordinate system to $(x^0, \ldots, x^{n-1})$ such that $g(\partial_0, \partial_1) = -1$ at $N$. In these coordinates we have (after normalizing $\dot{\gamma}$) $\dot{\gamma} = \partial_1$ and we obtain $g^{ad}\left(\partial_b g_{dc} - \frac{1}{2}\partial_d g_{bc}\right)\dot{\gamma}^b\dot{\gamma}^c = \frac{1}{2}g^{ad}\left(-\partial_d g_{11}\right)$. Since in our coordinates we have $\ddot{\gamma} = 0$, the lemma is proved once we have seen that $g^{ad}\left(-\partial_d g_{11}\right) \parallel (\partial_1)^a$ or, equivalently, $\partial_d g_{11} \parallel \dot{\gamma}^a g_{ad} = -\delta^0_d$. This equation follows from $\partial_i g_{11} = 0 \,\forall i \in \{1, \ldots, n-1\}$. ∎

**Corollary 3.1.1.** *Let $v \in T_xM \setminus \{0\}$ be a vector with $g(v, v) = 0$. Then there exists an (up to re-parametrization) unique conformal null geodesic $\gamma$ through $x = \gamma(0)$ with $\dot{\gamma}(0) = v$.*

*Proof.* By the fundamental theorem for ODEs, given any function $\lambda(t)$, the differential equation

$$\ddot{\gamma}^a + g^{ad}\left(\partial_b g_{dc} - \frac{1}{2}\partial_d g_{bc}\right)\dot{\gamma}^b\dot{\gamma}^c - \lambda\dot{\gamma}^a = 0$$

has a unique solution for any $v \in T_xM$. It is easy to see that for any two functions $\lambda, \tilde{\lambda}$ the solution curves are identical up to a reparametrization. ∎

The preceding corollary shows that at each point there is a unique conformal null geodesic in any direction $\mathbb{R}v$ where $g(v, v) = 0$ for all $g \in \mathfrak{C}$. This implies that there are exactly as many conformal null geodesics as there are light rays. In addition, it is easy to see that in the case of the Lorentzian conformal structure induced by Minkowski spacetime the light rays defined in Sect. 1.4 and the conformal null geodesics coincide. Hence we feel justified to link our infinitesimal Lorentzian conformal structure to light rays in spacetime by identifying them with conformal null geodesics.

**Postulate 3.1.2 (Light rays).** *The light rays of spacetime are the conformal null geodesics of its Lorentzian conformal structure.*

A Lorentzian conformal structure is all we need in order to investigate causality. The following definition is a straight forward generalization of Definition 1.4.6.

**Definition 3.1.3.** *Let $\mathfrak{C}$ be a Lorentzian conformal structure of signature $(-,+,\ldots,+)$, $g \in \mathfrak{C}$, and $\mathcal{A},\mathcal{U} \subset M$.*

*(i) A vector $w$ is called* spacelike, *if* $g(w,w) > 0$, timelike *if* $g(w,w) < 0$, *and* lightlike *(or* null*) if* $g(w,w) = 0$. *The vector $w$ is called* causal *if it is timelike or lightlike.*
*A vector field $V$ is* timelike *(respectively,* lightlike *or* null, causal, spacelike*) if for each $x \in M$ the vector $V_x$ is timelike (respectively, lightlike or null, causal, spacelike).*

*(ii) The Lorentzian conformal structure $\mathfrak{C}$ is* time orientable *if there is a global timelike vector field $V$.*
*Assume that $\mathfrak{C}$ is time orientable. A* time orientation *is an equivalence class of timelike vector fields $V$ where $V \sim W$ if $g(V_x, W_x) < 0$ at some point $x \in M$.*
*Let $[V]$ be a time orientation of $\mathfrak{C}$. A causal vector $u$ is called* future directed *(respectively,* past directed*) if $g(u,V) < 0$ (respectively, $g(u,V) > 0$).*

*(iii) A curve $\gamma$ is called* spacelike *(resp.,* timelike, lightlike, causal, future directed, past directed*) if all its velocity vectors $\dot\gamma$ are spacelike (resp., timelike, lightlike, causal, future directed, past directed). A timelike curve is often called a* world line *when one whishes to emphasize that it can represent the history of a (small) material object.*

*(iv) The* chronological future *of a set $\mathcal{A}$ relative to $\mathcal{U}$ is*

$$I^+(\mathcal{A},\mathcal{U}) = \{x \in M \,|\, \exists \text{ a future directed, timelike curve } \gamma \subset \mathcal{U} \text{ from } \mathcal{A} \text{ to } x\}$$

*The* causal future *of a set $\mathcal{A}$ relative to $\mathcal{U}$ is*

$$J^+(\mathcal{A},\mathcal{U}) = \{x \in M \,|\, \exists \text{ a future directed, causal curve } \gamma \subset \mathcal{U} \text{ from } \mathcal{A} \text{ to } x\}$$

*There are analogous definitions for the* chronological past *$I^-(\mathcal{A},\mathcal{U})$ and the* causal past *$J^-(\mathcal{A},\mathcal{U})$ of $\mathcal{A}$ relative to $\mathcal{U}$. If $\mathcal{U} = M$ we omit the term "relative to $M$" and write $I^+(\mathcal{A})$, etc. If $\mathcal{A} = \{x\}$ is a single point we write $I^+(x,\mathcal{U})$ etc.*

Definition 3.1.3 is independent of the chosen representative $g \in \mathfrak{C}$.

**Lemma 3.1.2.** *Let $x \in M$ and $\mathcal{U}$ be an open neighbourhood of $x$. Then $I^+(x,\mathcal{U})$ is open.*

*Proof.* Let $y \in I^+(x,\mathcal{U})$ and $\gamma \subset$ be a timelike curve from $x$ to $y$. Choose a coordinate system $(x^0,\ldots,x^{n-1})$, let $\mathcal{V} \subset \mathcal{U}$ be a (small enough) compact neighbourhood of $y$, and let $z \in \mathcal{V} \cap \gamma$. Then the (coordinate) straight line $\ell$ from $z$ to $y$ is timelike. By compactness of $\mathcal{V}$ there exists an $\alpha > 0$ such that any straight line $\ell'$ from $z$ with $\angle(\ell,\ell') < \alpha$ satisfies $\sup\{g(\dot{\ell}'(t),\dot{\ell}'(t)) | \ell'(t) \in \mathcal{V}\} < \frac{1}{2}\sup\{g(\dot{\ell}(t),\dot{\ell}(t)) | \ell(t) \in \mathcal{V}\} < 0$. Hence these lines are all timelike in $\mathcal{V}$. Since they fill a whole neighbourhood of $y$ the assertion is proven. ∎

The set $J^+(x,\mathcal{U})$ does not need to be closed relative to $\mathcal{U}$. However, as we will see later, $J^+(x,\mathcal{U})$ is closed if $\mathcal{U}$ is chosen small enough.

**Corollary 3.1.2.** *Let $\mathcal{U}$ be open and $\mathcal{A}$ be any subset of $\mathcal{U}$.*

$$\begin{aligned} I^+(\mathcal{A},\mathcal{U}) &= I^+(I^+(\mathcal{A},\mathcal{U}),\mathcal{U}) = I^+(J^+(\mathcal{A},\mathcal{U}),\mathcal{U}) = J^+(I^+(\mathcal{A},\mathcal{U}),\mathcal{U}) \\ &\subset J^+(\mathcal{A},\mathcal{U}) = J^+(J^+(\mathcal{A},\mathcal{U}),\mathcal{U}). \end{aligned}$$

*Proof.* The only inclusions which are not obvious are $I^+(J^+(\mathcal{A},\mathcal{U}),\mathcal{U}) \subset I^+(\mathcal{A},\mathcal{U})$ and $J^+(I^+(\mathcal{A},\mathcal{U}),\mathcal{U}) \subset I^+(\mathcal{A},\mathcal{U})$.

Let $x \in I^+(J^+(\mathcal{A},\mathcal{U}),\mathcal{U})$. Then there exist a $y \in \mathcal{A}$ and a $z \in M$, a causal curve $\mu\colon [0,1/2] \to M$ from $y$ to $z$ and a timelike curve $\lambda\colon [1/2,1] \to M$ from $z$ to $x$. The concatenation $\gamma\colon [0,1] \to M$ of $\mu$ and $\lambda$ is a (piecewise) causal curve from $y$ to $x$ which is timelike near $x$. Fix a metrik $g \in \mathfrak{C}$ and let $U$ be a timelike vector field along $\gamma$ which satisfies $U(1) = \dot{\gamma}(1)$. There is an $\epsilon > 0$ such that $g(\dot{\gamma}(t),\dot{\gamma}(t)) < -\epsilon$ for all $t \in (1-\epsilon,1)$. Let $\phi\colon [0,1] \to \mathbb{R}^+$ be a smooth function which satisfies $\phi(0) = \phi(1) = 0$ and $\phi(t) > 0$ for all $t \in [0,1-\epsilon]$. For any (small enough) $s > 0$ let $f(s,t) = \exp_{\gamma(t)}(s\phi(t)V(t))$. The curves $t \mapsto f(s,t)$ all connect $y$ with $x$ and $f(0,t) = \gamma(t)$. We denote derivatives with respect to $t$ by a dot and with respect to $s$ by a prime. We calculate

$$\begin{aligned} (g(\dot{f}(s,t),\dot{f}(s,t)))' &= \nabla_{f_*\partial_s}(g(\dot{f}(s,t),\dot{f}(s,t))) = 2g(\nabla_{f'}\dot{f},\dot{f}) \\ &= 2g(\nabla_{\dot{f}}f',\dot{f}) = 2g(\nabla_{\dot{\gamma}}(\phi V),\dot{\gamma}) = 2\dot{\phi}g(V,\dot{\gamma}). \end{aligned}$$

Hence we have $(g(\dot{f}(s,t),\dot{f}(s,t)))'_{(0,t)} < 0$ for all $t \in [0,1-\epsilon]$ and the curves $t \mapsto f(s,t)$ are timelike on $[0,1-\epsilon]$ for $s > 0$ small enough. Since we have $g(\dot{\gamma}(t),\dot{\gamma}(t)) < -\epsilon < 0$ for all $t \in (1-\epsilon,1]$ the curves also satisfy $g(\dot{f}(s,t),\dot{f}(s,t)) < -\epsilon/2 < 0$ for sufficiently small $|s|$. This proves that we have obtained a timelike curve $t \mapsto f(s_0,t)$ from $y$ to $x$. This curve may have a kink at the parameter value $1/2$ where the original curve $\gamma$ passes through $z$. Using Lemma 2.1.7 and a coordinate chart it is not difficult to smooth out $t \mapsto f(s_0,t)$ near $t = 1/2$ while preserving that this curve is timelike.

The inclusion $J^+(I^+(\mathcal{A},\mathcal{U}),\mathcal{U}) \subset I^+(\mathcal{A},\mathcal{U})$ can be shown analogously. ∎

**Lemma 3.1.3.** $I^+(\mathcal{A}) = \operatorname{int}(J^+(\mathcal{A})),\ J^+(\mathcal{A}) \subset \overline{I^+(\mathcal{A})}$.

*Proof.* By Lemma 3.1.2, $I^+(\mathcal{A}) = \bigcup_{x \in \mathcal{A}} I^+(x)$ is open. The inclusion $I^+(\mathcal{A}) \subset \operatorname{int}(J^+(\mathcal{A}))$ is clear. Let $x \in \operatorname{int}(J^+(\mathcal{A}))$. Since $\operatorname{int}(J^+(\mathcal{A}))$ is open, there is an $y \in I^-(x) \cap \operatorname{int}(J^+(\mathcal{A}))$. Hence $x \in I^+(y) \subset I^+(J^+(\mathcal{A})) = I^+(\mathcal{A})$.

Let $x \in J^+(\mathcal{A})$, $\gamma$ be a causal curve from $\mathcal{A}$ to $x$, and $\mathcal{U}$ be a neighbourhood of $x$. Since any small enough deformation of $\gamma$ has future endpoint in $\mathcal{U}$, we can deform $\gamma$ thereby obtaining a timelike curve from $\mathcal{A}$ to $\mathcal{U}$. ∎

**Definition 3.1.4.** *Let $(M, \mathfrak{C})$ be a Lorentzian conformal structure, $x \in M$, and $\mathcal{U}$ be an open neighbourhood of $x$. We call*

$$C_x^+(\mathcal{U}) = \{y \in M \mid \ \exists \text{ a future directed conformal null geodesic } \gamma \subset \mathcal{U} \text{ from } x \text{ to } y\}$$

*the* integrated future light cone of $x$ relative to $\mathcal{U}$. *There are analogous definitions for the* integrated past light cone *and the* integrated light cone. *If $\mathcal{U} = M$, we omit the term "relative to $M$" and write $C_x^+$, $C_x^-$, and $C_x$.*

**Proposition 3.1.1.** *Let $(M, \mathfrak{C})$ be a manifold with Lorentzian conformal strucute. Then each $x \in M$ has an open neighbourhood $\mathcal{U}$ diffeomeorphic to $\mathbb{R}^n$ such that*

*(i) $C_x^+(\mathcal{U}) \setminus \{x\}$ is a smooth hypersurface which is diffeomorphic to $S^{n-2} \times \mathbb{R}$,*
*(ii) $C_x^+(\mathcal{U}) \cap I^+(x, U) = \emptyset$,*
*(iii) $C_x^+(\mathcal{U}) \subset \overline{I^+(x, \mathcal{U})} = J^+(x, \mathcal{U})$.*

[p. 147 ↓] →2 ↓ p. 154

Proposition 3.1.1 will be a corollary to Lemma 3.1.4 below which is a result from Lorentzian geometry. Choose a representative $g \in \mathfrak{C}$ and denote the associated Levi-Cività connectionn by $\nabla$. The exponential map allows us to identify the causal structure of a *convex neighbourhood* of $x \in M$ with the causal structure of an open, convex set of the tangent space at $x$.

**Lemma 3.1.4.** *Let $(M, \mathfrak{C})$ be Manifold with Lorentzan conformal structure, $g \in \mathfrak{C}$, and $\mathcal{U}$ be a convex neighbourhood of $x \in M$ with respect to the Levi-Cività connection of $g$.*

*(i) $y \in I^+(x, \mathcal{U})$ (respectively $J^+(x, \mathcal{U})$) if and only if $y = \exp_x(v)$ where $v$ a future pointing timelike (resp., causal) vector.*
*(ii) $J^+(x, \mathcal{U}) = \overline{I^+(x, \mathcal{U})}$,*

[2] The complete proof of Proposition 3.1.1 requires the material from page 126 immediately after the proof of Proposition 2.6.4 up to the end of Sect. 2.6.

*Proof of Proposition 3.1.1.* Choose any $g \in \mathfrak{C}$ and let $u \in T_xM$ be a timelike vector with $g(u,u) = -1$. Each vector $v \in T_xM$ can be uniquely decomposed as $v = v^0 u + v_\perp$ here $v_\perp \in u^\perp = \{v \in T_xM \,|\, g(u,v) = 0\}$ and $v^0 \in \mathbb{R}$. The bilinear form $h(v,w) = g(v,w) + 2v^0 w^0$ is a Euclidean scalar product on $T_xM$. For every $\epsilon > 0$ let $B_\epsilon(0_x) = \{v \in T_xM \,|\, h(v,v) < \epsilon\}$. This set is obviously a neigbourhood of $0_x$ in $T_xM$. A similar argument as in the proof of Theorem 2.6.2 shows that $\mathcal{U}_\epsilon = \exp_{0_x}(B_\epsilon(0_x))$ is a convex neighbourhood of $x$ for small enough $\epsilon$.

The set $S_\epsilon = \{v \in T_xM \,|\, g(v,v) = 0, g(u,v) = -1\sqrt{\epsilon/2}\}$ is a submanifold of $T_xM$ which is diffeomorphic to the $(n-2)$-dimensional round sphere $S^{n-2} = \{z \in \mathbb{R}^{n-1} \,|\, \sum_{i=1}^{n-1}(z^i)^2 = 1\}$ and which lies in the boundary of $B_\epsilon(0_x)$. The map $(0,1) \times S_\epsilon \to M$, $(t,v) \mapsto \exp_x(tv)$ is a parameterisation of $C_x^+(\mathcal{U}_\epsilon) \setminus \{x\}$ which proves the first claim (i).

Since the map $\exp_x : B_\epsilon(0_x) \to \mathcal{U}_\epsilon$ is a diffeomorphism assertion (ii) follows from Lemma 3.1.4 and the fact that every causal vector $w \in B_\epsilon(0_x)q$ is either timelike or null. Assertion (iii) is a trivial consequence of Lemma 3.1.4 (ii). ∎

*Proof of Lemma 3.1.4.* (i): We prove the statement for $I^+(x,\mathcal{U})$ (the proof for $J^+(x,\mathcal{U})$ is analogous). The exponential map is a diffeomorphism of a neighbourhood $\tilde{\mathcal{U}}$ of $0_x \in T_xM$ onto $\mathcal{U}$. For any geodesic $\gamma$ we obtain $\nabla_{\dot\gamma}(g(\dot\gamma,\dot\gamma) = 2g(\nabla_{\dot\gamma}\dot\gamma,\dot\gamma) = 0$ whence the velocity vectors of geodesics do not change their causal class. It follows that $\exp_x$ maps timelike vectors into $I^+(x,\mathcal{U})$. We have to show that for each point $y \in I^+(x,\mathcal{U})$ the vector $(\exp_x)^{-1}(y) = v \in T_xM$ is necessarily timelike.

The double cone $\tilde{C}_x = \{v \in \tilde{\mathcal{U}} \,|\, g(v,v) = 0\}$ divides $\tilde{\mathcal{U}} \setminus \tilde{C}_x$ into 3 connected components: the future and past full cones of timelike vectors $(\tilde{C}_x^{\circ,+}, \tilde{C}_x^{\circ,-})$ and the set of spacelike vectors $(\tilde{C}_x^{\circ,s})$. Applying the diffeomeorphism $\exp_x : \tilde{\mathcal{U}} \to \mathcal{U}$ we see that the set $C_x(\mathcal{U}) = \{z \in \mathcal{U} \,|\, \exists v \in \tilde{C}_x$ with $z = \exp_x(v)\}$ divides $\mathcal{U}$ into the sets $C_x^{\circ,+}, C_x^{\circ,-}, C_x^{\circ,s}$, respectively. For every $y \in I^+(x,\mathcal{U})$ there is a timelike curve $\gamma\colon [a,b] \to \mathcal{U}, t \mapsto \gamma$ which connects $x$ and $y$. From $g(\dot\gamma(a),\dot\gamma(a)) < 0$ we know that $\gamma$ must initially enter $C_x^{\circ,+}$. If the assertion does not hold then $y \in C_x^+(\mathcal{U}) \cup C_x^{\circ,s}$. Since $I^+(x,\mathcal{U})$ is open we can assume without loss of generality that $y \in C_x^{\circ,s}$. Hence $\gamma$ must intersect $C_x$ at some point $\gamma(t_0)$ where $\gamma$ leaves $C_x^{\circ,+}$. Since $\dot\gamma(t_0)$ is timelike and future directed it is transverse to $C_x$ at $\gamma(t_0)$ and points into $C^{\circ,+}$. But this is a contradiction to the construction of the point $\gamma(t_0)$.

For (ii) it is sufficient to note that the set of causal vectors in $T_xM$ is the closure of the set of timelike vectors in $T_xM$. ∎

p. 152 ↓
[↓ p. 155]
Since a Lorentzian conformal structure allows to measure angles and relative lengths (at a given point), it is sufficient for spatial geometry at one point. We just loose an absolute calibration.

## 3.2 Inertial observers: the projective structure

So far we only have considered light propagation. In order to specialize our structure further we must take into account other fundamental properties of nature. In Chap. 1 inertial observers have played an important theoretical rôle. Since they are not subject to any physical forces one can physically implement inertial observers (or particles) by freely falling observers (or particles). We will use freely falling or inertial observers as the other input into our theory besides the Lorentzian conformal structure induced by light rays.

The following postulate reflects that the movement of an inertial observer depends on his/her initial velocity and initial position. This is the main content of Galilei's law of inertia.

**Postulate 3.2.1 (Existence of inertial observers).** *Through any point $x \in (M, g)$ and for any timelike direction $\mathbb{R}u$ there exists (up to parametrization and extension) exactly one* inertial observer $\gamma\colon \mathbb{R} \to M$ *which passes through $x$ with velocity $\dot\gamma \parallel \mathbb{R}u$.*

Postulate 3.2.1 singles out a collection of paths in spacetime. In Minkowski space, inertial observers move along straight, timelike lines. This is again a global characterization which we need to overcome by formulating it infinitesimally, i.e. in the tangent bundle rather than in spacetime itself. An infinitesimal description of (unparameterized) inertial observers in Minkowski space is that their spatial acceleration vanishes. This property can be generalized as follows.

**Postulate 3.2.2 (Law of inertia).** *For each $x \in M$ there exists a chart $(\mathcal{U}_x \varphi_x)$ centered at $x$ such that with respect to the corresponding coordinate system we have for all inertial observers passing through $x$*

$$\frac{\mathrm{d}^2\gamma^a}{\mathrm{d}t^2} \parallel \dot\gamma^a .$$

*at $x$. The chart maps $\varphi_x : \mathcal{U}_x \to \mathbb{R}^n$ depend smoothly on the parameter $x$.*

While by its very formulation Postulate 3.2.2 is independent of the chosen charts, we need to express it in an arbitrary coordinate system. In Sect. 2.6 we have seen that the derivative of vector fields is coordinate dependent and therefore not well defined in a general manifold. However, we can use Theorem 2.6.1 and Corollary 2.6.1 in order to define a connection with respect to which Postulate 3.2.2 can be formulated in a manifestly coordinate free manner.

Let $(\tilde{\mathcal{U}}, \tilde{\varphi})$ be any chart and $x \in \tilde{\mathcal{U}}$. We define the Christoffel symbols $\tilde{\Gamma}^a_{bc}$ with respect to this chart at $x$ by

$$\tilde{\Gamma}^a_{bc} = \frac{\partial \tilde{x}^a}{\partial x^h} \frac{\partial^2 x^h}{\partial \tilde{x}^b \partial \tilde{x}^c}.$$

where $(x^1, \ldots, x^n)$ are the coordinates with respect to the chart $(\mathcal{U}_x, \tilde{\varphi}_x)$ provided by Postulate 3.2.2. We will now prove that this construction gives a well defined connection $\nabla$. Let $(\hat{\mathcal{U}}, \hat{\varphi})$ be a second chart with $x \in \hat{\mathcal{U}}$. Then we have $\hat{\Gamma}^a_{bc} = \frac{\partial htx^a}{\partial x^h} \frac{\partial^2 x^h}{\partial \hat{x}^b \partial \hat{x}^c}$. We need to show that these two definitions give the same connection. Indeed, we calculate

$$\begin{aligned}
\tilde{\Gamma}^a_{bc} &= \frac{\partial \tilde{x}^a}{\partial x^h} \frac{\partial^2 x^h}{\partial \tilde{x}^b \partial \tilde{x}^c} \\
&= \frac{\partial \tilde{x}^a}{\partial \hat{x}^k} \frac{\partial \hat{x}^k}{\partial x^h} \frac{\partial}{\partial \tilde{x}^b} \left( \frac{\partial x^h}{\partial \hat{x}^l} \frac{\partial \hat{x}^l}{\partial \tilde{x}^c} \right) \\
&= \frac{\partial \tilde{x}^a}{\partial \hat{x}^k} \frac{\partial \hat{x}^k}{\partial x^h} \left( \frac{\partial^2 x^h}{\partial \hat{x}^l \partial \hat{x}^m} \frac{\partial \hat{x}^l}{\partial \tilde{x}^c} \frac{\partial \hat{x}^m}{\partial \tilde{x}^b} + \frac{\partial x^h}{\partial \hat{x}^l} \frac{\partial^2 \hat{x}^l}{\partial \tilde{x}^c \partial \tilde{x}^b} \right) \\
&= \frac{\partial \tilde{x}^a}{\partial \hat{x}^k} \frac{\partial \hat{x}^l}{\partial \tilde{x}^c} \frac{\partial \hat{x}^m}{\partial \tilde{x}^b} \left( \frac{\partial \hat{x}^k}{\partial x^h} \frac{\partial^2 x^h}{\partial \hat{x}^l \partial \hat{x}^m} \right) + \frac{\partial \tilde{x}^a}{\partial \hat{x}^l} \frac{\partial^2 \hat{x}^l}{\partial \tilde{x}^c \partial \tilde{x}^b} \\
&= \frac{\partial \tilde{x}^a}{\partial \hat{x}^l} \frac{\partial^2 \hat{x}^l}{\partial \tilde{x}^c \partial \tilde{x}^b} + \frac{\partial \tilde{x}^a}{\partial \hat{x}^k} \frac{\partial \hat{x}^l}{\partial \tilde{x}^c} \frac{\partial \hat{x}^m}{\partial \tilde{x}^b} \hat{\Gamma}^k_{lm},
\end{aligned}$$

which is exactly the transformation formula provided by Corollary 2.6.1. Thus we have a well defined torsion-free connection $\Gamma$ and Postulate 3.2.2 can be restated as follows. There is a connection $\nabla$ such that inertial observers are pregeodesics.

[p. 154 ↓] →3 ↓ p. 124

Given our collection of inertial observers, $\{(\mathcal{U}_x, \varphi_x)\}_{x \in M}$ is not the only collection of compatible charts such that the formula in Postulate 3.2.2 holds. In fact, Corollary 2.6.3 implies that exactly those collections $\{(\mathcal{V}_x, \psi_x)\}_{x \in M}$ which induce torsion-free connections that have the same pre-geodesics as $\nabla$ are also possible choices. This implies the following corollary.

p. 125 ↓ [↓ p. 157]

**Corollary 3.2.1.** *Postulates 3.2.1, 3.2.2 determine a projective structure $\mathfrak{P}$ such that each particle is a pregeodesics with respect to any $\nabla \in \mathfrak{P}$.*

Weyl characterized the connection as a field which forces a particle to be transported parallelly with itself in space and time. Thus we have arrived at a geometrical explanation for the law of inertia postulated by Galilei.

---

[3] In order to understand the corollary below we need to know a little bit more about projective structures.

## 3.3 Compatibility: Weyl structure

We have obtained a Lorentzian conformal and a projective structure but the relationship of these two geometrical structures is still unspecified. We will now introduce a further postulate which links observers and light rays and therefore these two structures. It is an experimental fact that one can chase light rays with material observers arbitrarily closely, provided one uses enough energy. The following is a formalization of this idea.

**Postulate 3.3.1 (Compatibility with the causal structure).** *Each $x \in M$ has a neighbourhood $\mathcal{U}$ such that for all $y \in \mathcal{U} \setminus \{x\}$ we have*

$y = \gamma(t)$ *for an inertial observer* $\gamma$ *through* $x \Leftrightarrow y \in I^+(x,\mathcal{U}) \cup I^-(x,\mathcal{U})$.

As a first consequence of this compatibility axiom we can determine light rays using the connection instead of the Lorentzian conformal structure.

**Lemma 3.3.1.** *The conformal null geodesics coincide with those pregeodesics which are somewhere lightlike.*

*Proof.* Let $x \in M$ and let $\mathcal{U}$ be the intersection of the neighbourhoods of $x$ which are provided by Proposition 3.1.1 and Postulate 3.3.1. Let $\mu$ be a conformal null geodesic from $x$ to some fixed point $y \in \mathcal{U}$. Proposition 3.1.1 implies that $\mu$ lies in the boundary of $I^+(x,\mathcal{U})$. Hence there is a sequence of points $y_i \in I^+(x,\mathcal{U})$ which converges to $y$. For each $i$ let $\gamma_i : [0,1] \mapsto \mathcal{U}$ be the pregeodesic which corresponds to the inertial observer which moves from $x$ to $y_i$ (cf. Postulate 3.3.1). Let $v \setminus \{0\}$ be an accumulation point of the (bounded) sequence $\dot{\gamma}_i(0) \in T_x M$ and let $\gamma$ be the pregeodesic with $\dot{\gamma}(0) = v$. By the continuous dependence of solutions of differential equations on initial conditions and parameters (cf. Theorem 2.4.1) there are for each point $\gamma(s)$ ($s \in [0,1]$) and each neighbourhood $\mathcal{V}$ of $\gamma(s)$ infinitely many pregeodesics $\gamma_i$ which intersect $\mathcal{V}$ and whose velocity vectors at $s$ converge to $\dot{\gamma}(s)$. This implies that $\gamma$ is causal and that $\gamma \subset \overline{I^+(x,\mathcal{U})}$. Since $y_i \to y$ and $\overline{\mathcal{U}}$ is compact the pregeodesic $\gamma$ reaches $y$. The inclusion $J^+(I^+(x,\mathcal{U}),\mathcal{U}) = I^+(x,\mathcal{U})$ and $y \in \gamma \cap C_x^+(\mathcal{U})$ imply $\gamma \subset \overline{I^+(x,\mathcal{U})} \setminus I^+(x,\mathcal{U}) = C_x^+(\mathcal{U})$. Since $C_x^+$ is a null hypersurface and $\gamma$ is lightlike (causal but not timelike) $\gamma$ must be a conformal null geodesic. That $\gamma$ coincides with $\mu$ follows now from the uniqueness of conformal null geodesics.

For the converse we simply need to note that both pregeodesics which have an initial lightlike velocity vector and conformal null geodesics are uniquely determined by initial point $x$ and initial velocity direction $\mathbb{R}\dot{\gamma}(0)$. ∎

In the rest of this section we will show that our postulates imply the existence of a natural Weyl structure.

**Theorem 3.3.1.** *Let $\mathfrak{C}$ be a Lorentzian conformal structure and $\mathfrak{P}$ be a projective structure such that the Postulates 3.1.1, 3.2.1, 3.2.2, and 3.3.1 are satisfied. Then there exists a unique $\nabla \in \mathfrak{P}$ such that for all $g \in \mathfrak{C}$ there is a one-form $\varphi$ with $\nabla g = \varphi \otimes g$.*

[p. 155 ↓] →4 ↓ p. 162

The proof of this theorem will be split into several lemmas.

**Lemma 3.3.2.** *Let $g$ be a Lorentzian metric and $\Delta_{abc} = \Delta_{(abc)}$ be a totally symmetric tensor such that $\Delta(v, v, v) = 0$ for all null vectors $v$. Then there exists a one-form $\vartheta$ such that $\Delta_{(abc)} = \vartheta_{(a}g_{bc)}$.*

*Proof.* sym($\vartheta \otimes g$) clearly satisfies the condition of the lemma. We will now verify that this is the only possible choice.

Let $\mathrm{t}$ be a vector with $g(\mathrm{t}, \mathrm{t}) = -1$ and $e \in \mathrm{t}^\perp$ with $g(e, e) = 1$. Then $\mathrm{t} \pm e$ are null vectors and from

$$\Delta^b(\mathrm{t} \pm e, \mathrm{t} \pm e, \mathrm{t} \pm e) = \Delta_{(abc)} \left(\mathrm{t}^a\mathrm{t}^b\mathrm{t}^c \pm 3\mathrm{t}^a\mathrm{t}^b e^c + 3\mathrm{t}^a e^b e^c \pm e^a e^b e^c\right) = 0$$

we obtain

$$0 = \Delta_{(abc)} \left(\mathrm{t}^a\mathrm{t}^b\mathrm{t}^c + 3\mathrm{t}^a e^b e^c\right), \qquad (3.3.1)$$

$$0 = \Delta_{(abc)} \left(3\mathrm{t}^a\mathrm{t}^b e^c + e^a e^b e^c\right). \qquad (3.3.2)$$

Setting

$$\vartheta_c = -\left(3\Delta_{(abc)}\mathrm{t}^a\mathrm{t}^b + 2\Delta(\mathrm{t}, \mathrm{t}, \mathrm{t})\mathrm{t}^d g_{dc}\right).$$

Equation (3.3.2) is equivalent to

$$\Delta(e, e, e) = -3(\Delta_{(abc)}\mathrm{t}^a\mathrm{t}^b e^c) = \vartheta(e) = g(e, e)\vartheta(e) \quad \forall e \in \mathrm{t}^\perp.$$

Analogously, Equation (3.3.1) is equivalent to

$$\Delta_{(abc)}e^a e^b \mathrm{t}^c = -\frac{1}{3}\Delta_{(abc)}\mathrm{t}^a\mathrm{t}^b\mathrm{t}^c = \frac{1}{3}\vartheta(\mathrm{t}) = g_{(ab}\vartheta_{c)}e^a e^b \mathrm{t}^c.$$

Finally, the definition of $\vartheta$ implies

$$\Delta_{(abc)}\mathrm{t}^a\mathrm{t}^b e^c = -\frac{1}{3}\vartheta(e) = \frac{1}{3}g(\mathrm{t}, \mathrm{t})\vartheta(e) = g_{(ab}\vartheta_{c)}\mathrm{t}^a\mathrm{t}^b e^c.$$

and

$$\Delta_{(abc)}\mathrm{t}^a\mathrm{t}^b\mathrm{t}^c = g(\mathrm{t}, \mathrm{t})\vartheta(\mathrm{t}).$$

By the polarization identity for symmetric 3-tensors, we know that $\Delta_{(abc)}$ coincides with $g_{(ab}\vartheta_{c)}$ on a basis and our claim is proved. ∎

[4] The proof of Theorem 3.3.1 is rather technical. It can be omitted without loss of continuity.

**Lemma 3.3.3.** *Let $g \in \mathfrak{C}$ and let*

$$G^a_{bc} = \frac{1}{2} g^{ad} \left( \partial_b g_{dc} + \partial_c g_{bd} - \partial_d g_{bc} \right).$$

*Then $\Delta^a{}_{bc} = \Gamma^a_{bc} - G^a_{bc}$ are the components of a tensor field which satisfies $\Delta_{a[bc]} = 0$. There is a one-form $\sigma$ with*

$$\Delta_{abc} = g_{(ab}\sigma_{c)} + \frac{2}{3}\Delta_{[ab]c} + \frac{2}{3}\Delta_{[ac]b}.$$

*Proof.* The numbers $G^a_{bc}$ are just the Christoffel symbols of the Levi-Cività connection associates with $g$ and the difference of two connections is always a tensor field. The symmetry property $\Delta_{a[bc]} = 0$ is immediate since both connections are torsion-free.

In order to prove the decomposition formula for $\Delta$ we will first show that $\Delta_{abc} = \Delta_{(abc)} + \frac{2}{3}\Delta_{[ab]c} - \frac{2}{3}\Delta_{[ca]b}$. Since $\Delta$ is symmetric in the last two indices, the total symmetrization

$$\Delta_{(abc)} = \frac{1}{6} \left( \Delta_{abc} + \Delta_{bca} + \Delta_{cab} + \Delta_{bac} + \Delta_{bac} + \Delta_{bac} \right)$$

can be simplified to $\Delta_{(abc)} = \frac{1}{3} \left( \Delta_{abc} + \Delta_{bca} + \Delta_{cab} \right)$. From

$$\Delta_{[ab]c} = \frac{1}{2} \left( \Delta_{abc} - \Delta_{bac} \right) = \frac{1}{2} \left( \Delta_{abc} - \Delta_{bca} \right) \text{ and}$$
$$\Delta_{[ca]b} = \frac{1}{2} \left( \Delta_{cab} - \Delta_{acb} \right) = \frac{1}{2} \left( \Delta_{cab} - \Delta_{abc} \right)$$

we obtain $\Delta_{[ab]c} - \Delta_{[ca]b} = \Delta_{abc} - \frac{1}{2} \left( \Delta_{bac} + \Delta_{cab} \right)$, and therefore $\Delta_{(abc)} + \frac{2}{3}\Delta_{[ab]c} - \frac{2}{3}\Delta_{[ca]b} = \frac{1}{3} \left( \Delta_{abc} + \Delta_{bca} + \Delta_{cab} + 2\Delta_{abc} - \Delta_{bac} - \Delta_{cab} \right) = \Delta_{abc}$.

In order to finish the proof we must show $\Delta_{(abc)} = g_{(ab}\sigma_{c)}$ for some suitably chosen 1-form $\sigma$. Let $\gamma$ be a pregeodesic which is lightlike at $x = \gamma(0)$. By Lemma 3.3.1, $\gamma$ is a conformal null geodesics which implies $\nabla_{\dot\gamma} \left( g(\dot\gamma, \dot\gamma) \right) = 0$. For any pregeodesic $\mu$ there exists a function $\lambda(t)$ such that $\nabla_{\dot\mu(t)}\dot\mu(t) = \mu(t)\dot\mu(t)$. This implies

$$\begin{aligned}
\nabla_{\dot\gamma} \left( g(\dot\gamma, \dot\gamma) \right) &= (\nabla g)(\dot\gamma, \dot\gamma, \dot\gamma) + 2g(\dot\gamma, \nabla_{\dot\gamma}\dot\gamma) \\
&= (\nabla g)(\dot\gamma, \dot\gamma, \dot\gamma) + 2\lambda(t) g(\dot\gamma, \dot\gamma) \\
&= \partial_a g_{bc} \dot\gamma^a \dot\gamma^b \dot\gamma^c - 2 g_{bd} \Gamma^d_{ac} \dot\gamma^a \dot\gamma^b \dot\gamma^c + 2\lambda(t) g(\dot\gamma, \dot\gamma) \\
&= 2 \left( \frac{1}{2} \left( \partial_a g_{bc} + \partial_c g_{ac} - \partial_b g_{ac} \right) - g_{bd} \Gamma^d_{ac} \right) \dot\gamma^a \dot\gamma^b \dot\gamma^c \\
&\quad + 2\lambda(t) g(\dot\gamma, \dot\gamma) \\
&= -2\Delta_{abc} \dot\gamma^a \dot\gamma^b \dot\gamma^c + 2\lambda(t) g(\dot\gamma, \dot\gamma).
\end{aligned} \tag{3.3.3}$$

From $\nabla_{\dot\gamma}(g(\dot\gamma,\dot\gamma)) = 0$ we obtain $\Delta_{abc}v^a v^b v^c = 0$ for every null vector $v$. The assertion follows now from Lemma 3.3.2. ∎

**Lemma 3.3.4.** *The causal character of each pregeodesic is constant.*

*Proof.* Let $\gamma$ be a pregeodesic with $\nabla_{\dot\gamma(t)}\dot\gamma(t) = \lambda(t)\dot\gamma(t)$. Lemma 3.3.3 and Equation (3.3.3) in its proof imply

$$\nabla_{\dot\gamma}(g(\dot\gamma,\dot\gamma)) = 2\left((\lambda(t) - \sigma(\dot\gamma)\right)g(\dot\gamma,\dot\gamma).$$

By the uniqueness theorem of ordinary differential equations, $g(\dot\gamma,\dot\gamma)$ cannot change its sign. ∎

**Lemma 3.3.5.** *There exists 1-forms $\varphi,\theta$ and a symmetric tensor field $\tilde L^a_{bc} = \tilde L^a_{(bc)}$ with $\tilde L^a_{ba} = 0$, $\tilde L_{(abc)} = 0$ such that*

$$\Delta_{abc} = \frac{1}{2}\varphi_a g_{bc} - g_{a(b}\varphi_{c)} + g_{a(b}\theta_{c)} + \tilde L_{a(bc)}.$$

*Proof.* Let $p_a = -\frac{2}{1-n}\frac{4}{3}\Delta_{[ab]}{}^b$. This implies

$$g^{ac}p_{[a}g_{b]c} = \frac{1}{2}g^{ac}\left(p_a g_{bc} - p_b g_{ac}\right) = -\frac{n-1}{2}p_a = \frac{4}{3}\Delta_{[ab]}{}^b.$$

Hence $\hat L_{abc} = \frac{4}{3}\Delta_{[ab]c} - p_{[a}g_{b]c}$ satisfies $\hat L^a_{ba} = 0$ and $\hat L_{(abc)} = 0$. Lemma 3.3.3 implies the equation

$$\Delta_{abc} = g_{(ab}\sigma_{c)} + \frac{1}{2}p_{[a}g_{b]c} + \frac{1}{2}p_{[a}g_{c]b} + \hat L_{a(bc)}.$$

The tensor $\tilde L^a_{bc} = \hat L^a_{(bc)}$ also satisfies $\tilde L^a_{ba} = 0$ because $\hat L_{acb}$ is antisymmetric in $a$ and $c$ and has therefore vanishing trace. Hence we can replace $\hat L$ by $\tilde L$.

Since $\frac{1}{2}p_{[a}g_{b]c} + \frac{1}{2}p_{[a}g_{c]b} = \frac{1}{2}p_a g_{bc} - \frac{1}{2}g_{a(b}p_{c)}$ and $g_{(ab}\sigma_{c)} = \frac{1}{3}(\sigma_a g_{bc} + 2g_{a(b}\sigma_{c)})$, the assertion follows with $\varphi_a = p_a + \frac{2}{3}\sigma_a$ and $\theta_a = \frac{1}{2}p_a + \frac{4}{3}\sigma_a$. ∎

Consider now a null geodesic $\gamma$. Since it is both a geodesic with respect to the Lorentzian conformal structure $\mathfrak{C}$ and with respect to the projective structure $\mathfrak{P}$ we have $\Delta^a{}_{bc}\dot\gamma^b\dot\gamma^c \parallel \dot\gamma^a$. From Lemma 3.3.5 and $g(\dot\gamma,\dot\gamma) = 0$ we get $\tilde L^a_{bc}\dot\gamma^b\dot\gamma^c \parallel \dot\gamma^a$. Hence the tensor $\tilde L$ must also satisfy $\tilde L(N,N) \parallel N$ for all lightlike vectors $N$. This property allows us to simplify $\tilde L$ further.

**Lemma 3.3.6.** *Let $n \geq 3$ and $L(\cdot,\cdot)\colon T_xM \times T_xM \to T_xM$ be a symmetric tensor such that $L(N,N) \parallel N$ for all null vectors $N$. For each $g \in \mathfrak{C}$ there exist 1-forms $\lambda,\mu$ such that*

$$L(v,w) = \frac{1}{2}\left(\lambda(v)w + \lambda(w)v\right) + g(v,w)\mu^\sharp \quad \forall v,w \in T_xM$$

*Proof.* Let $\{E_0, \ldots, E_{n-1}\}$ be an orthonormal basis and $\{\omega^0, \ldots, \omega^{n-1}\}$ be the dual basis. We can write $L = L^a_{bc} \otimes \omega^b \otimes \omega^c \otimes E_a$, where $L^a_{bc} = L^a_{cb}$ for all indices $a, b, c$. We consider lightlike vectors given by $N = E_0 + \cos(\theta)E_A + \sin(\theta)E_B$, where $A \neq B$ and $\theta \in [0, 2\pi]$. In order to exploit the condition $L(N,N) = c(N)N$, we expand $L(N,N)$ as a Fourier polynomial in $\theta$. Using $\cos^2\theta = \frac{1}{2}(1+\cos(2\theta))$, $\sin\theta\cos\theta = \frac{1}{2}\sin(2\theta)$, and $\sin^2\theta = \frac{1}{2}(1-\cos(2\theta))$, we calculate

$$\begin{aligned} L(N,N) &= L(E_0,E_0) + \cos^2(\theta)L(E_A,E_A) + \sin^2(\theta)L(E_B,E_B) \\ &\quad + 2\cos(\theta)L(E_0,E_A) + 2\sin(\theta)L(E_0,E_B) \\ &\quad + 2\sin(\theta)\cos(\theta)L(E_A,E_B) \\ &= L(E_0,E_0) + \frac{1}{2}\left(L(E_A,E_A) + L(E_B,E_B)\right) \\ &\quad + 2L(E_0,E_A)\cos(\theta) + 2L(E_0,E_B)\sin(\theta) \\ &\quad + \frac{1}{2}\left(L(E_A,E_A) - L(E_B,E_B)\right)\cos(2\theta) + L(E_A,E_B)\sin(2\theta) \\ &= c(\theta)\left(E_0 + \cos(\theta)E_A + \sin(\theta)E_B\right) \end{aligned}$$

for some function $c(\theta)$. The left hand side is a Fourier polynomial of order 2 which implies that $c$ must be a Fourier polnomial of order $\leq 1$ since otherwise the right hand side would be a Fourier polynomial of order $\geq 3$. Hence we can write $c(\theta) = \alpha_{AB} + \beta_{AB}\cos\theta + \gamma_{AB}\sin\theta$. The right hand side is then given by

$$\begin{aligned} &\alpha_{AB}E_0 + \frac{1}{2}\beta_{AB}E_A + \frac{1}{2}\gamma_{AB}E_B + (\beta_{AB}E_0 + \alpha_{AB}E_A)\cos\theta \\ &\quad + (\gamma_{AB}E_0 + \alpha_{AB}E_B)\sin\theta + \left(\frac{1}{2}\beta_{AB}E_A - \frac{1}{2}\gamma_{AB}E_B\right)\cos(2\theta) \\ &\quad + \left(\frac{1}{2}\gamma_{AB}E_A + \frac{1}{2}\beta_{AB}E_B\right)\sin(2\theta). \end{aligned}$$

A comparison of coefficients gives

$$\begin{aligned} L(E_0,E_0) &+ \frac{1}{2}\left(L(E_A,E_A) + L(E_B,E_B)\right) \\ &= \alpha_{AB}E_0 + \frac{1}{2}\beta_{AB}E_A + \frac{1}{2}\gamma_{AB}E_B, \end{aligned} \tag{3.3.4}$$

$$2L(E_0,E_A) = \beta_{AB}E_0 + \alpha_{AB}E_A, \tag{3.3.5}$$

$$2L(E_0,E_B) = \gamma_{AB}E_0 + \alpha_{AB}E_B, \tag{3.3.6}$$

$$L(E_A,E_A) - L(E_B,E_B) = \beta_{AB}E_A - \gamma_{AB}E_B, \tag{3.3.7}$$

$$L(E_A,E_B) = \frac{1}{2}\gamma_{AB}E_A + \frac{1}{2}\beta_{AB}E_B. \tag{3.3.8}$$

We obtain a linear system of equations for the components $L^a_{bc}$ $(b \leq c)$ of $L = L^a_{(bc)} \otimes \omega^b \otimes \omega^c \otimes E_a$. From Equation 3.3.5 we obtain immediately $L^B_{0A} = 0$ for $B \notin \{0, A\}$,

$$L^0_{0A} = \frac{1}{2}\beta_{AB}, \quad L^A_{0A} = \frac{1}{2}\alpha_{AB}. \tag{3.3.9}$$

Further, neither $\beta_{AB}$ nor $\alpha_{AB}$ can depend on $B$ since the left hand sides of Equations (3.3.9) are independent of $E_B$. Equation 3.3.6 implies in addition that $L^B_{0B} = \frac{1}{2}\alpha_{AB}$. Hence $\alpha_{AB}$ cannot depend on $A$ either. We will therefore write $\lambda_0 := \alpha_{AB}$ and $\lambda_A := \beta_{AB}$. Equation 3.3.8 implies $L^0_{AB} = 0$ for $A \neq B$ and $L^C_{AB} = 0$ for pairwise different $A, B, C$. We also obtain $L^A_{AB} = \frac{1}{2}\gamma_{AB}$ and $L^B_{AB} = \frac{1}{2}\lambda_A$. These two equations are only consistent if $\gamma_{AB} = \lambda_B$. Equation 3.3.7 can be used to eliminate $L(E_B, E_B)$ in Equation 3.3.4 resulting in

$$\begin{aligned} L(E_0, E_0) + L(E_A, E_A) = \alpha_{AB}E_0 + \frac{1}{2}\beta_{AB}E_A + \frac{1}{2}\gamma_{AB}E_B + \frac{1}{2}\beta_{AB}E_A \\ - \frac{1}{2}\gamma_{AB}E_B = \lambda_0 E_0 + \lambda_A E_A. \end{aligned}$$

This equation implies $L^0_{00} + L^0_{AA} = \lambda_0$ which is independent of $A$. Hence $L^0_{AA} = L^0_{BB}$ for $A \neq B$ we can set $\mu^0 := -L^0_{AA}$. We also have $L^B_{00} = -L^B_{AA}$ for $B \neq A$ and $L^A_{00} + L^A_{AA} = \lambda_A$. If we set $\mu^A = L^A_{00}$, all coefficients are determined. It is now straightforward to check that $L^a_{(bc)} = \delta^a_{(b}\lambda_{c)} + \mu^a g_{bc}$. Conversely, this tensor indeed has all the properties listed in the lemma. ∎

**Corollary 3.3.1.** *Let $L(\cdot,\cdot)\colon T_xM \times T_xM \to T_xM$ be a symmetric tensor such that $L(N, N) \parallel N$ for all null vectors $N$ and $g(L(v, v), v) = 0$ for all vectors $v$. Then for each $g \in \mathfrak{C}$ there exists a 1-form $\lambda$ such that*

$$L(v, w) = \frac{1}{2}\left(\lambda(v)w + \lambda(w)v\right) - g(v, w)\lambda^\sharp \quad \forall v, w \in T_xM,$$

*where $\lambda^\sharp$ is defined by $\lambda(v) = g(\lambda^\sharp, v)$ for all $v \in T_xM$.*

*Proof.* This follows immediately from Lemma 3.3.6 and the additional condition $g(L(v, v)v) = 0$ for all vectors $v$. ∎

We are now ready to prove the main result of this section.

*Proof of Theorem 3.3.1.* We will first show that there exists a unique $\nabla \in \mathfrak{P}$ such that for each representative $g \in \mathfrak{C}$ there is a one-form $\varphi$ with

$$\begin{aligned} (\nabla_V W)^a = V^b\partial_b W^a + g^{ad}\Big(\frac{1}{2}\left(\partial_b g_{dc} + \partial_c g_{bd} - \partial_d g_{bc}\right) \\ + \frac{1}{2}\varphi_d g_{bc} - g_{d(b}\varphi_{c)}\Big)V^b W^c. \end{aligned} \tag{3.3.10}$$

Recall the formula for $\Delta$ provided by Lemma 3.3.5. Since $\tilde{L}^a_{bc}$ satisfies the assumptions of Corollary 3.3.1 there exists a one-form $\lambda$ such that $\tilde{L}_{a(bc)} = g_{a(b}\lambda_{c)} - \lambda_a g_{bc}$. The property $\tilde{L}^a_{ba} = 0$ implies $0 = g^{ac}\tilde{L}_{a(bc)} = \frac{1}{2}(n+1)\lambda_b - \lambda_b = \frac{1}{2}(n-1)\lambda_b$, whence $\lambda = 0$ and therefore $\tilde{L} = 0$. Hence we have

$$\begin{aligned}\Gamma_{abc} - \frac{1}{2}\left(\partial_b g_{ac} + \partial_c g_{ba} - \partial_a g_{bc}\right) &= \Delta^a_{bc} \\ &= \frac{1}{2}\varphi_d g_{bc} - g_{d(b}\varphi_{c)} + g_{a(b}\theta_{c)}.\end{aligned}$$

An application of Lemma 2.6.3 implies that there is a unique $\nabla \in \mathfrak{P}$ such that $\theta = 0$. This proves Equation (3.3.10).

From Equation (3.3.10) we obtain

$$\begin{aligned}\nabla_a g_{bc} &= \partial_a g_{bc} - \Gamma^d_{ab} g_{dc} - \Gamma^d_{ac} g_{db} \\ &= \partial_a g_{bc} - \frac{1}{2}\left(\partial_a g_{cb} + \partial_b g_{ac} - \partial_c g_{ab}\right) - \frac{1}{2}\varphi_c g_{ab} + g_{c(a}\varphi_{b)} \\ &\quad - \frac{1}{2}\left(\partial_a g_{bc} + \partial_c g_{ab} - \partial_b g_{ac}\right) - \frac{1}{2}\varphi_b g_{ac} + g_{b(a}\varphi_{c)} \\ &= \varphi_a g_{bc}.\end{aligned}$$

Hence $(M, \mathfrak{C}, \nabla)$ is a Weyl structure and the theorem is proved. ∎

## 3.4 Reduction to the Lorentzian structure

p. 157 ↓
[↓ p. 165]

There are experimental facts which indicate that a general Weyl structure has features which have no counterpart in our actual universe. It seems therefore necessary to specify the geometrical structure of space-time further.

It is plausible to identify an affine parametrization (cf. Definition 2.7.4) of an inertial observer with a *standard clock* carried by the observer. The freedom $t \mapsto at + b$ corresponds to the freedom to choose the zero on the time axis and to choose the unit in which time is measured. An *atomic clock* roughly works at follows (detail of this mechanism can be found in textbooks on Quantum mechanics). Each atom has a characteristic minimal energy $E$ which it can absorb. A very short while after the absorbtion of such a package of energy the atom will emit a photon whose frequency $\nu$ with respect to the rest frame of the atom is given by $E = h\nu$. Since this frequency is characteristic for each sort of atom it can be used to build a clock.

In Sect. 1.4.3 we have seen how to calculate $E$ from the world lines of the atom and the photon in the context of special relativity. Unfortunately, we cannot simply apply this calculation here since the number $E$ did depend on the Minkowski metric $\eta$ and not merely on its conformal class. But it is very suggestive to identify this atomic clock with the standard clock $t$ given by the affine parameter of the world line of the atom.

We will now see that this identification gives rise to a global effect which has not been observed. Let $x, y \in M$, $y \in I^+(x)$ be two events and consider two atoms of the same kind which are moving from $x$ to $y$ along different paths $\gamma_1 : [0, \alpha_1] \to M$, $\gamma_2 : [0, \alpha_2] \to M$ in spacetime (cf. Fig. 3.4.1). We will assume that they move initially along the same

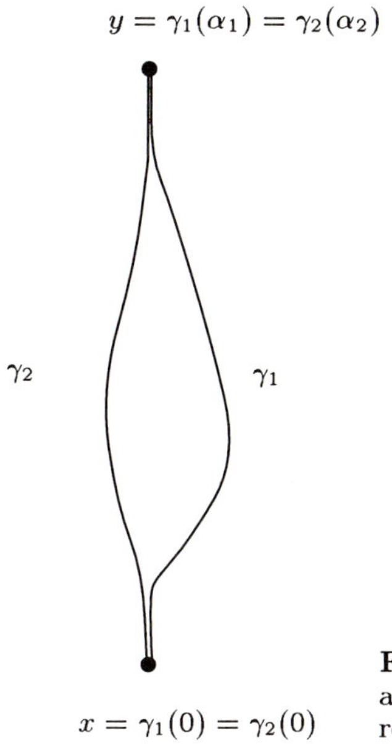


**Fig. 3.4.1.** The worldlines from $x$ to $y$ of two atoms which are initially and finally at rest with respect to each other

path in spacetime and that their clocks are initially calibrated at $x$, i.e., there is an $\epsilon > 0$ such that $\dot{\gamma}_1(t) = \dot{\gamma}_2(t)$ for all $t \in [0, \epsilon)$. We will also assume that just before reaching $y$ they are again moving side by side. For these observers we obtain

$$\begin{aligned}\int_0^{\alpha_A} \frac{\nabla_{\dot{\gamma}_A(t)} g(\dot{\gamma}_A(t), \dot{\gamma}_A(t))}{g(\dot{\gamma}_A(t), \dot{\gamma}_A(t))} dt &= \int_0^{\alpha_A} \frac{\left(\nabla_{\dot{\gamma}_A(t)} g\right)(\dot{\gamma}_A(t), \dot{\gamma}_A(t))}{g(\dot{\gamma}_A(t), \dot{\gamma}_A(t))} \mathrm{d}t \\ &\quad + 2\int_0^{\alpha_A} \frac{g(\nabla_{\dot{\gamma}_A(t)} \dot{\gamma}_A(t), \dot{\gamma}_A(t))}{g(\dot{\gamma}_A(t), \dot{\gamma}_A(t))} \mathrm{d}t \\ &= \int_0^{\alpha_A} \varphi(\dot{\gamma}_A(t)) \mathrm{d}t.\end{aligned}$$

Since the frequencies were initially equal we obtain

$$\begin{aligned}&\ln\left(g(\dot{\gamma}_1(\alpha_1), \dot{\gamma}_1(\alpha_1))\right) - \ln\left(g(\dot{\gamma}_2(\alpha_2), \dot{\gamma}_2(\alpha_2))\right) \\ &\quad = \ln\left(g(\dot{\gamma}_1(\alpha_1), \dot{\gamma}_1(\alpha_1))\right) - \ln\left(g(\dot{\gamma}_1(0), \dot{\gamma}_1(0))\right) \\ &\qquad + \ln\left(g(\dot{\gamma}_2(0), \dot{\gamma}_2(0))\right) - \ln\left(g(\dot{\gamma}_2(\alpha_2), \dot{\gamma}_2(\alpha_2))\right).\end{aligned}$$

Let $Q$ be any 2-surface which is bounded by the curves $\gamma_1$, $\gamma_2$. An application of the theorem of Stokes (Theorem 2.5.5)[5] gives

$$\begin{aligned}\ln\left(g(\dot\gamma_1(\alpha_1),\dot\gamma_1(\alpha_1))\right) - \ln\left(g(\dot\gamma_2(\alpha_2),\dot\gamma_2(\alpha_2))\right) \\ = \int_0^{\alpha_1}\varphi(\dot\gamma_1(t))\mathrm{d}t - \int_0^{\alpha_2}\varphi(\dot\gamma_2(t))\mathrm{d}t \\ = \int_Q \mathrm{d}\varphi = -2\int_Q F.\end{aligned}$$

Hence the parameterization of both curves and therefore — by our identification — the frequencies of both atoms are different at $y$, even though they were the same at $x$. As a consequence, the frequency of an atom clock would depend of the history of the atoms constituting it. This does not seem to be the case. Moreover, the spectrum of far away stars is apparently independent of the history of the atoms which constitute these stars.[6] Hence we conclude that $\mathrm{d}\varphi = 0$.

**Postulate 3.4.1 (No second clock effect).** *The length curvature $F = -\frac{1}{2}\mathrm{d}\varphi$ of spacetime vanishes identically.*

Notice that the justification of Postulate 3.4.1 requires more interpretation than the justification of our other axioms. One may argue that it is the weakest link in the chain of arguments which leads to general relativity.

**Corollary 3.4.1.** *Assume that Postulates 3.1.1, 3.2.1, 3.2.2, 3.3.1, and 3.4.1 hold. Then spacetime is a Lorentzian manifold $(M, g)$.*

*Proof.* This follows immediately from Theorem 2.7.2. ∎

We have now arrived at a geometrical structure which gives the framework for a description of space and time. The arguments which lead to Corollary 3.4.1 may seem so compelling that the reader could ask herself or himself why we started with Newton's theory of spacetime instead of motivating our postulates directly. In fact, from a purely conceptional point of view it is advantageous to analyze the measurement of space and time relations and to use this analysis in order to arrive at a Lorentz structure. This program has been carried out by Ehlers, Pirani, and Schild (1972) who also arrive at a Weyl structure and reduce it to a Lorentz structure via Postulate 3.4.1. With our preparation this

[5] The gist of our argument is that the difference of the frequences at $\gamma_1(\alpha_1) = \gamma_2(\alpha_2)$ is non-zero. That this is the case for suitable paths $\gamma_1$, $\gamma_2$ unless $\varphi$ can be chosen to vanish should be plausible even without appealing to the theorem of Stokes. It is needed for a strict proof though.

[6] This argument does not depend on the identification of atomic clocks with the affine parameter of their world lines.

article is highly readible and certainly recommended to physicists who are interested in the operational approach.

We have used a more historic approach for two reasons.

Firstly, most readers are familiar with the classical description of spacetime (albeit less formalized, perhaps).

Secondly, Newton's theory is also very compelling on first sight. So are Galilei's theory and the special theory of relativity. What is more, when these theories where still young and generally accepted it was very difficult to see how to improve them. In fact, most physicists and philosophers would have claimed that these theories are correct in an absolute way. There is no doubt that the Lorentzian description of spactime will not be the last improvement either. It is even a prominent topic of current research to try to incorporate general relativity into a new general quantum theory of spacetime. It is generally believed that this new theory will be qualitatively very different from the geometrical theory we have presented here. The reader should recall that we started with macroscopic properties of rays of light which do not take into account the quantum nature of light. Also, we have always assumed that space and time are continuous rather than discrete. Hence there are several points where our theory of spacetime may prove inadedquate. It must also be said, however, that a conceptionally satisfying theory of quantum gravity does not yet exist.

Our description of spacetime is much better than previous theories and *to date* there is no other theory which describes the global properties of spacetime better. [p. 162 ↓] ↓ p. 167

# 4. Pseudo-Riemannian manifolds

p. 165 ↓
→1
[↓ p. 170]

*We have learned that spacetime can be described by a Lorentzian manifold. In this section we will investigate the slightly more general case of pseudo-Riemannian manifolds in detail. The development of the theory of spacetime will be continued in Chap. 5 where we motivate Einstein's equation, the central equation in general relativity which links matter to gravitation.*

*Readers who wish to get to Einstein's equation quickly may skip most of Chap. 4. They only need to read Definition 4.2.2 and Sect. 4.3 up to and including corollary 4.3.1.*

*For mathematicians, this chapter contains the essentials of (pseudo)-Riemannian differential geometry. Almost everything we present here will be used in the following physically motivated sections.*

*Prerequisitives of this chapter: Sect. 2.7.1 and Sect. 2.8 (up to but not including Lemma 2.8.2).*

Recall that a pseudo-Riemannian manifold $(M, g)$ consists of an $n$-dimensional manifold $M$ and a symmetric, everywhere non-degenerate $\binom{0}{2}$-tensor field $g$. We will often denote $g$ by $\langle \cdot, \cdot \rangle$. A pseudo-Riemannian $(M, g)$ is called a *Riemannian manifold* (respectively, *Lorentzian manifold*) if $g$ has signature is $(+, \ldots, +)$ (respectively, $(-, +, \ldots, +)$). The simplest example of a Riemannian manifold is *Euclidean Space*, $(\mathbb{R}^n, \mathrm{d}(x^1)^2 + \cdots + \mathrm{d}(x^n)^2)$, and the simplest example of a Lorentzian manifold is *Minkowski spacetime*, $(\mathbb{R}^n, -\mathrm{d}(x^0)^2 + \mathrm{d}(x^1)^2 + \cdots + \mathrm{d}(x^{n-1})^2)$.

In this book, we are especially interested in Lorentzian manifolds as mathematical models of spacetime. Riemannian submanifolds (cf. Sect. 4.4) of codimension 1 can be thought of as instants of time. They will play an important rôle when we discuss the initial value problem in Sect. 5.4. Pseudo-Riemannian manifolds which are neither Lorentzian nor Riemannian are rarely applied in physics. However, it does not come at any additional cost to widen the discussion to this more general case.

*Unless explicitly stated otherwise all geometrical objects are understood to be derived from the metric and the Levi-Cività connection.*

*Remark 4.0.1.* The investigation of hypersurfaces in Euclidean space has a very long tradition in mathematics and it has led to many important (mathematical) developments. (Pseudo)-Riemannian manifolds are the

[1] We only collect those facts which are essential to an understanding of Einstein's equation which will be presented in the next chapter.

natural generalization of these hypersurfaces and therefore of independent interest to mathematicians. While we will not push this angle, readers primarily interested in learning differential geometry should keep in mind the following example of a Riemannian manifold.

Let $M \subset \mathbb{R}^n$ be a hypersurface and consider for each $x \in M$ the tangent space $T_xM$ as a subspace of $\mathbb{R}^n$. To be concrete, let $\iota\colon M \to \mathbb{R}^n$ be the natural inclusion and identify $T_xM$ with $\iota_* T_xM \subset T_{\iota(x)}\mathbb{R}^n \approx \mathbb{R}^n$. We denote the standard scalar product of $\mathbb{R}^n$ by $\langle\cdot,\cdot\rangle_{\mathbb{R}^n}$ and define the Riemannian metric $g$ of $M$ by

$$g(v,w) = \langle \iota_* v, \iota_* w\rangle_{\mathbb{R}^n}$$

for all vectors $v, w \in T_xM$.

While this class of examples is rather simple, it is sufficient for visualizing most important features of Riemannian manifolds. Whereas Euclidean space is trivial in the sense that $\mathrm{d}(x^1)^2 + \cdots + \mathrm{d}(x^n)^2$ is a constant tensor field with respect to appropriate coordinates, $g$ is nonconstant in general and the curvature of its Levi-Cività connection does not vanish.

*Example 4.0.1.* The simplest, non-trivial example is given by the sphere $S^2 = \{x \in \mathbb{R}^3 |\, (x^1)^2 + (x^2)^2 + (x^3)^2 = 1\} \subset \mathbb{R}^3$. Denote by $\iota\colon S^2 \mapsto \mathbb{R}^3$, $x \mapsto x$ the canonical inclusion. In this case we have $T_xM = \{y \in \mathbb{R}^n |\, \langle x, y\rangle_{\mathbb{R}^3} = 0\}$. We can parametrize (a dense open subset of) the sphere using the chart $(\mathcal{U}, \varphi)$ where

$$\varphi^{-1}(\theta,\phi) = \begin{pmatrix} \cos\phi\cos\theta \\ \sin\phi\cos\theta \\ \sin\theta \end{pmatrix}.$$

Let $\{E_1, E_2\}$ be the standard orthonormal basis of $\mathbb{R}^2$. From

$$(\varphi^{-1})_* E_1 = \frac{\partial\varphi^{-1}}{\partial\theta} = \partial_\theta = \begin{pmatrix} -\cos\phi\sin\theta \\ -\sin\phi\sin\theta \\ \cos\theta \end{pmatrix},$$

$$(\varphi^{-1})_* E_2 = \frac{\partial\varphi^{-1}}{\partial\phi} = \partial_\phi = \begin{pmatrix} -\sin\phi\cos\theta \\ \cos\phi\cos\theta \\ 0 \end{pmatrix}$$

we obtain $g_{\theta\theta} = g(\partial_\theta, \partial_\theta) = 1$, $g_{\theta\varphi} = 0$, $g_{\phi\phi} = \cos^2\theta$. We could now use the Koszul formula (Equation (2.7.7)) to calculate the Levi-Cività connection and we could determine the Riemann tensor through the formula given in Theorem 2.8.1. In Sect. 4.4 we will study general submanifolds and present better techniques for calculating these quantities.

A submanifold of Minkowski space does not necessarily inherit a Lorentzian metric — in fact, the example $C_x^+(\mathcal{U})\setminus\{x\}$ shows that a hypersurface in Minkowski space may not inherit any pseudo-Riemannian metric. In contrast to the Riemannian case, it is probably not wise to try gaining intuition for Lorentzian manifolds from studying submanifolds of Minkowski space. We should instead use the intuition we gained in the previous chapters. Lorentzian manifolds serve as models for space and time as a unit — and in this physical way their geometry can be understood best.

Since the metric of a pseudo-Riemannian manifold is everwhere non-degenerate we have a canonical isomorphism of vectors and one-forms.

**Lemma 4.0.1.** *Let $(M, g)$ be a pseudo-Riemannian manifold. The metric induces an isomorphism $(\cdot)^\flat : T_xM \mapsto T_x^*M$, $v \mapsto v^\flat$, where $v^\flat(w) = g_x(v, w)$ for all $w \in T_xM$.*

*Proof.* This follows immediately from the fact that $g$ is non-degenerate. ∎

We denote the inverse isomorphism by $(\cdot)^\sharp : T_x^*M \mapsto T_xM$. This isomorphism can be naturally extended to tensor fields.

**Definition 4.0.1.** . *Let $\psi \in T_s^r(T_xM)$. Then we define the completely covariant tensor $\psi^\flat$ by*

$$\psi^\flat(v_1, \ldots, v_{s+r}) = \psi(v_1, \ldots, v_s, (v_{s+1})^\flat, \ldots, (v_{s+r})^\flat)$$

*and the completely contravariant tensor $\psi^\sharp$ by*

$$\psi^\sharp(\omega^1, \ldots, \omega^{r+s}) = \psi((\omega_1)^\sharp, \ldots, (\omega^s)^\sharp, \omega^{s+1}, \ldots, \omega^{r+s1}).$$

*The components of $\psi^\flat$ are often simply denoted by $\psi_{i_1 \ldots i_s j_1 \ldots j_r}$ and the components of $\psi^\sharp$ by $\psi^{i_1 \ldots i_s j_1 \ldots j_r}$.*

The isomorphisms $(\cdot)^\sharp$, $(\cdot)^\flat$ are often referred to as "raising and lowering of indices". This terminology is motivated by their expression in the abstract index notation. We write $(g^\sharp)_{ab} = g^{ab}$, $(v^\flat)_a = g_{ab}v^b =: v_a$, and $(\omega^\sharp)^a = g^{ab}\omega_b =: \omega^a$. The symbols "♯" and "♭" should be easy to remember since there is an analogous notation in music.

One of the most important inequalities in linear algebra is the Cauchy-Schwarz inequality for positive definite scalar products $\langle \cdot, \cdot \rangle_{\mathbb{R}^n}$. It states that for every pair of vectors $v, w$ the inequality

$$|\langle v, w \rangle_{\mathbb{R}^n}| \leq \sqrt{\langle v, v \rangle_{\mathbb{R}^n}} \sqrt{\langle w, w \rangle_{\mathbb{R}^n}}$$

holds. This inequality clearly generalises to Riemannian manifolds. To us, an analoguous inequality for Lorentzian manifolds is much more important.

**Lemma 4.0.2 (Inverse triangle inequality).** *Let $(M, g)$ be a Lorentzian manifold and let $v, w$ be causal vectors. Then the inequality*

$$|\langle v, w\rangle| \geq \sqrt{|\langle v, v\rangle|}\sqrt{|\langle w, w\rangle|}$$

*holds.*

*Proof.* The inequality holds trivially for null vectors. Hence we can assume that both $w$ and $v$ are timelike. There is a vector $e \perp v$ and a number $\alpha$ such that $w = \alpha v + e$. This implies $\langle e, e\rangle \geq 0$, $0 < \langle w, w\rangle = \alpha^2 \langle v, v\rangle + \langle e, e\rangle$ and therefore

$$\langle v, w\rangle^2 = \alpha^2 \langle v, v\rangle^2 = (\langle w, w\rangle - \langle e, e\rangle)\langle v, v\rangle \geq \langle w, w\rangle \langle v, v\rangle .$$

[p. 167 ↓]
↓ p. 172

∎

*In the rest of this section we prove a somewhat technical lemma which may be omitted on first reading.*

For any $u \in T_xM$, the $n$-dimensional "vertical" subspace $T_{u_x}T_xM$ of $T_{u_x}TM$ is given by the image of the map $T_xM \to T_uTM$, $v \mapsto \tilde{v}_{[u]} := \frac{\mathrm{d}}{\mathrm{d}t}(u + tv)_{|t=0}$. We can equip $T_{u_x}T_xM$ with a pseudo-scalar product by defining $\left\langle \tilde{v}_{[u]}, \tilde{w}_{[u]}\right\rangle = \langle v, w\rangle$. In general, the push forward of the exponential map fails to be a linear isometry of the spaces $T_{u_x}T_xM$ and $T_xM$. However, if one of the vectors $\tilde{v}_{[u]}$, $\tilde{w}_{[u]}$ is aligned with $u_{[u]}$, we have the following invariance.

**Lemma 4.0.3 (Gauß lemma).** *Let $x \in M$, $u \in T_xM \setminus \{0\}$, and $\tilde{v}_{[u]}$, $\tilde{w}_{[u]} \in T_{u_x}T_xM$. If there are $v \in \mathbb{R}u \subset T_xM$ and $w \in T_xM$ with $\tilde{v}_{[u]} = \frac{\mathrm{d}}{\mathrm{d}t}(u + tv)_{|t=0}$ and $\tilde{w}_{[u]} = \frac{\mathrm{d}}{\mathrm{d}t}(u + tw)_{|t=0}$, then*

$$\left\langle T_x \exp(\tilde{v}_{[u]}), T_x \exp(\tilde{w}_{[u]})\right\rangle = \langle v, w\rangle .$$

*Proof.* The assertion is a formula which is linear in $v_{[u]}$ and $w_{[u]}$. Since $u$ and $v$ are parallel we can therfore replace $v$ by $u$ in this formula. Consider the map $f\colon (s,t) \mapsto \exp_x(t(u + sw)) \in M$ and let

$$f_t := T_{(s,t)}f(\partial_t) = T_{t(u+sw)} \exp_x(\widetilde{(u + sw)}_{[t(u+sw)]})$$
$$f_s := T_{(s,t)}f(\partial_s) = T_{t(u+sw)} \exp_x(\widetilde{tw}_{[t(u+sw)]}).$$

Observe that we have $\overset{f}{\nabla}_{\partial_s} f_t = \overset{f}{\nabla}_{\partial_t} f_s$ for all $s$ since $\nabla$ is torsion free. The curve $t \mapsto \exp_x(t(u + sw))$ is a geodesic, which implies

$$\overset{f}{\nabla}_{\partial_t} \langle f_t, f_s\rangle = \Big\langle \overbrace{\overset{f}{\nabla}_{\partial_t} f_t}^{=0}, f_s \Big\rangle + \Big\langle f_t, \overset{f}{\nabla}_{\partial_t} f_s \Big\rangle = \Big\langle f_t, \overset{f}{\nabla}_{\partial_s} f_t \Big\rangle$$
$$= \frac{1}{2} \overset{f}{\nabla}_{\partial_s} \langle f_t, f_t\rangle = \frac{1}{2} \overset{f}{\nabla}_{\partial_s} \langle u + sw, u + sw\rangle = \langle u, w\rangle .$$

This equation can be easily integrated yielding $\langle f_t, f_s\rangle = t\langle u, w\rangle$. Hence the assertion follows by setting $(s,t) = (0,1)$. ∎

## 4.1 Existence of Lorentzian and Riemannian manifolds

*This section is included for its theoretical interest and can be omitted on first reading.*

**Theorem 4.1.1.** *Every manifold $M$ carries a positive definite metric $g$.*

*Proof.* Let $\{(\mathcal{U}_\alpha, \varphi_\alpha)\}_{\alpha\in A}\}$ be a collection of charts which covers $M$ and $\{f_\alpha\}_{\alpha\in A}$ be a partition of unity subordinate to $\{\mathcal{U}_\alpha\}_{\alpha\in A}$. The bilinear form

$$g_x := \sum_{\alpha\in A} f_\alpha(x)(\varphi_\alpha)^*(\mathrm{d}x^1 \otimes \mathrm{d}x^1 + \cdots + \mathrm{d}x^n \otimes \mathrm{d}x^n.$$

is evidently well defined and symmetric. Let $j$ be an index with $x \in \operatorname{supp}(f_{\alpha_j})$ and $v \in T_xM$ be any non-zero vector. The estimate

$$g_x(v,v) = \sum_{\alpha\in A} f_\alpha(x) \sum_{i=1}^n \left(((\varphi_\alpha)_* v)^i\right)^2 > f_{\alpha_j}(x) \sum_{i=1}^n \left(((\varphi_{\alpha_j})_* v)^i\right)^2 > 0$$

implies then that $g_x$ is positive definite. ∎

A *line field* (or *line bundle*) is a 1-vector subbundle with base manifold $M$ of the tangent bundle $TM$. A nowhere vanishing vector field $U$ generates a line bundle but it is possible to have line bundles which are not generated by vector fields (cf. the example of a Möbius band).

**Theorem 4.1.2.** *A Manifold $M$ carries a Lorentzian metric if and only if it admits a (non-oriented) line field $l$.*

*Proof.* Let $h$ be a Riemannian metric of $M$ and assume that there exists a line field $l$ on $M$. Each point $x \in M$ has a neighbourhood $\mathcal{U}$ such that for all $y \in \mathcal{U}$ the line field $l$ can be represented by $l_y = \mathbb{R}U_y$, where $U$ is a vector field on $\mathcal{U}$. We can assume without loss of generality that $h(U,U) = 1$. Then $U$ is determined up to a factor $\pm 1$. Hence the $\binom{0}{2}$-tensor field $g := h - 2U^\flat \otimes U^\flat$ is globally well defined. We restrict now attention to $\mathcal{U}$ again. Let $\{U_2, \ldots, U_n\}$ be a completion of $U$ to a local orthonormal frame with respect to $h$. That the tensor field $g$ has signature $(-,+,\ldots,+)$ and is therefore a Lorentzian metric follows from $g(U_i,U_j) = h(U_i,U_j) = \delta_{ij}$, $g(U,U_i) = 0$, $g(U,U) = 1 - 2 = -1$,

Conversely, assume that $M$ admits a Lorentzian metric $g$ and let $h$ be a Riemannian metric on $M$. Then at each point $x \in M$ we have a linear map $A\colon T_xM \to T_xM$ defined by $A^i_j = g^{ik}h_{jk}$, where we use the Lorentzian metric $g$ to raise or lower indices. The tensor $A$ is symmetric with respect to $h$ since

$$h(v,Aw) = h_{ki}v^k A^i_j w^j = h_{ki}g^{il}h_{jl}v^k w^j = h_{jl}A^l_k v^k w^j = h(w,Av)$$

for all vectors $v, w$. Hence there is an $h$-orthonormal basis of eigenvectors of $A$. Since $g$ has signature $(-1, 1, \ldots, 1)$ the linear map has one negative and $n-1$ positive eigenvalues. It follows that a each point $x$ there is a uniquely determined 1-dimensional subspace $l_x$ of $T_xM$ which is spanned by the eigenvectors corresponding to the negative eigenvalue. Clearly, $l$ defines a line field of $M$. ∎

**Corollary 4.1.1.** *The Sphere* $S^2 = \{x \in \mathbb{R}^3 | \, (x^1)^2 + (x^2)^2 + (x^3)^2 = 1\}$ *does not admit a Lorentzian metric.*

*Proof.* If there is a Lorentzian metric on $S^2$ then there is also a line field $l$ on $S^2$. Let $h$ be a Riemannian metric. Then there are at each $x \in S^2$ exactly two vectors $\pm U_x \in l_x$ with $h(U_x, U_x) = 1$. We will now construct a non-vanishing vector field $V$ on $S^2$. At $x$ we choose $V_x = U_x$. and consider all great circles through $x$. These great circles cover all of $S^2$ and each point but $\{\pm x\}$ is intersected by exactly one such great circle. The points $x$ and $-x$ are intersected by all great circles. We define now $V$ on $S^2 \setminus \{-x\}$ as follows. Along each great circle $\gamma$ we let $V_{\gamma(t)} \in \{U_{\gamma(t)}, -U_{\gamma(t)}\}$ be uniquely determined vector such that $t \mapsto U_{\gamma(t)}$ is smooth. These vector fields along great circles have each a limit vector in $\{U_{-x}, -U_{-x}\}$. Let now $\gamma_1 : [0, \pi] \to S^2$, $\gamma_2 : [0, \pi] \to S^2$ be two great arcs from $x$ to $-x$. Since we can smoothly rotate one arc into the other and $V$ depends smoothly on parameters it is clear that both limits $\lim_{t\to\pi} V_{\gamma_1(t)} \in \{U_{-x}, -U_{-x}\}$ and $\lim_{t\to\pi} V_{\gamma_2(t)} \in \{U_{-x}, -U_{-x}\}$ must coincide. But this implies that all arcs have the same limit at $\{-x\}$ and that therefore we have defined a continuous, non-vanishing vector field $V$. This contradicts Theorem 2.5.6 and therefore there cannot be a Lorentzian metric on $S^2$. ∎

We will see in Proposition 8.1.1 that there are other reasons why one may want to discard compact models of spacetime.

## 4.2 The volume form and the Hodge star operator

p. 170 ↓
[↓ p. 175]

*In Sect. 2.5.4 we have seen that for a general manifold without additional structures it is possible it to define an integration of n-forms but not an integration of functions. In an oriented pseudo-Riemannian manifold there is a canonical n-form, the volume form. This allows us to define the integral over a function (cf. Definition 4.2.2).*

*The Hodge star operator is a canonical isomorphism of p-forms and n − p forms. It can be used to put Maxwell's Equations which describe electromagnetism and light (cf. Sect. 5.2.3) into an especially simple form. A further motivation is given in the introduction to Sect. 2.5.1.*

*The discussion of the Hodge star operator can be omitted on first reading and will not be needed in the rest of this book.*

*This section draws on Sect. 2.5*

The isomorphisms $\flat$, $\sharp$ (cf. Definition 4.0.1) induce a pseudo-scalar product on $T^r_s M$.

**Lemma 4.2.1.** *The bilinear form*

$$g[{}^r_s]: T^r_s(T_x M) \times T^r_s(T_x M) \to \mathbb{R},$$
$$(\psi, \phi) \mapsto C^1_1 \dots C^{r+s}_{r+s} \psi^\sharp \otimes \phi^\flat$$

*is a non-degenerate, symmetric bilinear form. The pseudo-scalar product $g[{}^r_s]$ is positive definite if $g$ is positive definite.*[2]

*Proof.* For $\psi, \phi \in T^r_s(T_x M)$ we calculate

$$\begin{aligned}(\psi^\sharp \otimes \phi^\flat)^{a_1 \dots a_{s+r}}_{b_1 \dots b_{s+r}} &= \psi^{a_1 \dots a_s}_{c_{s+1} \dots c_{s+r}} \phi^{d_1 \dots d_s}_{b_{s+1} \dots b_{s+r}} g^{a_{s+1} c_{s+1}} \dots g^{a_{s+r} c_{s+r}} \\ &\quad \times g_{b_1 d_1} \cdots g_{b_s d_s} \\ &= \psi^{a_1 \dots a_s a_{s+1} \dots a_{s+r}} \phi^{d_1 \dots d_s d_{s+1} \dots d_{s+r}} g_{b_1 d_1} \cdots g_{b_{s+r} d_{s+r}}.\end{aligned}$$

It follows that the total contraction of this tensor is symmetric in its constituents $\psi^{a_1 \dots a_s a_{s+1} \dots a_{s+r}}$ and $\phi^{d_1 \dots d_s d_{s+1} \dots d_{s+r}}$. Let

$$\{e_1, \dots, e_n, \omega^1, \dots, \omega^n\}$$

be a pair of dual orthonormal bases and $\psi \in T^r_s(T_x M)$. Since $(\omega^i)^\sharp = \pm e_i$ and $(e_i)^\flat = \pm \omega^i$ (the sign depending on the signature and $i$) we have

$$g[{}^r_s](\psi, e_{a_1} \otimes \dots \otimes e_{a_s} \otimes \omega^{b_1} \otimes \dots \otimes \omega^{n_r}) = \pm \psi(e_{a_1}, \dots, e_{a_s}, \omega^{b_1}, \dots, \omega^{n_r})$$

which in turn implies that $g[{}^r_s]$ is non-degenerate. The assertion for Riemannian metrics follows from

$$\begin{aligned}g[{}^r_s](\psi, \psi) = \sum_{\substack{a_1, \dots, a_s, \\ b_r, \dots, b_r, \\ c_1, \dots, c_s, \\ d_1, \dots, d_r}} &\psi(e_{a_1}, \dots, e_{a_s}, \omega^{b_1}, \dots, \omega^{n_r}) \\ &\times \psi(e_{c_1}, \dots, e_{c_s}, \omega^{d_1}, \dots, \omega^{d_r}) \delta_{a_1 c_1} \dots \delta_{a_s c_s} \delta_{b_1 d_1} \dots \delta_{b_r d_r}.\end{aligned}$$

∎

The metric measures not only the length and the angle of vectors but also the volume of a parallel-epiped spanned by a collection of linearly independent vectors.

**Lemma 4.2.2.** *Let $(M, g)$ be a pseudo-Riemannian manifold and $x \in M$. Then $x$ has a neighbourhood $\mathcal{U}$ such that there exist exactly two $n$-forms $\mu$, $-\mu$ with $|g[{}^0_n](\mu_\pm, \mu_\pm)| = n!$. Furthermore, if $\{E_1, \dots, E_n\}$ is an orthonormal frame then $\mu_x(E_1, \dots, E_n) \in \{-1, 1\}$.*

[2] The converse is not true since $g[{}^0_n]$ is always definite since it is a non-vanishing bilinear form on a one-dimensional space.

*Proof.* The first claim follows from the fact that $\Lambda^n(T_xM)$ is a 1-dimensional vector space and $g[^0_n]_x$ a non-vanishing bilinear form on it.

Let $\{\tilde{E}_1, \ldots, \tilde{E}_n\}$ be an orthonormal frame and $\{\omega^1, \ldots \omega^n\}$ be the dual frame. Then the tensor fields $\varphi_\pm = \pm\omega^1 \otimes \cdots \otimes \omega^n$ satisfy

$$|g[^0_n](\varphi_\pm, \varphi_\pm)| = 1.$$

Hence $\mu_\pm = \pm\omega^1 \wedge \cdots \wedge \omega^n = \pm n! \operatorname{alt}(\varphi)$ satisfies $|g[^0_n](\mu_\pm, \mu_\pm)| = n!$. If $\{E_1, \ldots, E_n\}$ is any other orthonormal frame and $A \in \mathcal{T}^1_1(\mathcal{U})$ is defined by $\tilde{E}_i = AE_i = A^j_i E_j$ then $\mu_\pm(\tilde{E}_1, \ldots, \tilde{E}_n) = \det(A^i_j)\mu_\pm(E_1, \ldots, E_n) = \det(A^i_j)$. The second claim follows since the determinant of a linear map which transforms orthonormal bases into orthonormal bases is always equal to either 1 or $-1$. ∎

The preceeding lemma implies (by continuity) that on a connected pseudo-Riemannian manifold there are at most 2 $n$-forms $\pm\mu$ which satisfy the normalization condition $|g[^0_n](\mu_\pm, \pm)| = n!$. Evidently, this is the case if and only if there exists a non-vanishing $n$-form on $M$, i.e., if $M$ is orientable.

**Definition 4.2.1.** *If $(M, g)$ is an oriented pseudo-Riemannian manifold with orientation $\mathcal{O}$ then the* volume form *is the unique $n$-form $\mu_M \in \mathcal{O}$ with $|g[^0_n](\mu_M, \mu_M)| = 1$. If $\mathcal{U} \subset M$ is an open set with compact closure we define the volume of $\mathcal{U}$ by* $\operatorname{vol}(\mathcal{U}) = \int_\mathcal{U} \mu_M$.[3]

The following definition reduces to the volume integral in Euclidean geometry

**Definition 4.2.2.** *Let $(M, g)$ be an oriented pseudo-Riemannian manifold with volume form $\mu$. Then the* Integral of a function $f$ *over an open set $\mathcal{U}$ is defined by $\int_\mathcal{U} f\mu$.*[4]

**Lemma 4.2.3.** *Let $(M, g)$ be an oriented pseudo-Riemannian manifold and $(x^1, \ldots, x^n)$ a positively oriented coordinate system. Then the coordinate expression of $\mu_M$ is given by*

$$\mu_M = \sqrt{|\det\left((g_{ij})_{1\le i,j\le n}\right)|}\,\mathrm{d}x^1 \wedge \cdots \wedge \mathrm{d}x^n.$$

*Proof.* If $(y^1, \ldots, y^n)$ is another positively oriented coordinate system then the number $\det\left(\left(\frac{\partial x^a}{\partial y^b}\right)_{a,b}\right)$ is positive. We denote the metric components with respect to $(y^1, \ldots, y^n)$ by $\tilde{g}_{ab}$ and obtain

[3] Readers who have omitted Sect. 2.5 can replace $\int_\mathcal{U} \mu_M$ by $\int_{\varphi(\mathcal{U})} \sqrt{|\det((g_{ij})_{i,j})}\mathrm{d}x^1 \ldots \mathrm{d}x^n$, where $\varphi$ is chart map whose domain contains $\mathcal{U}$.

[4] As in the preceding footnote, $\int_\mathcal{U} f\mu$ may be replaced by $\int_\mathcal{U} \mu_M$ by $\int_{\varphi(\mathcal{U})} f \circ \varphi^{-1} \sqrt{|\det((g_{ij})_{i,j})}\mathrm{d}x^1 \ldots \mathrm{d}x^n$.

$$\tilde{g}_{ab} = \frac{\partial x^c}{\partial y^a}\frac{\partial x^d}{\partial y^b} g_{cd}$$

which in turn implies

$$\begin{aligned}\sqrt{\det\left((\tilde{g}_{ab})_{1\le a,b\le n}\right)} &= \sqrt{\det\left(\left(\frac{\partial x^a}{\partial y^b}\right)_{1\le a,b\le n}\right)^2 \det\left((\tilde{g}_{ab})_{1\le a,b\le n}\right)} \\ &= \det\left(\left(\frac{\partial x^a}{\partial y^b}\right)_{1\le a,b\le n}\right)\sqrt{\det\left((\tilde{g}_{ab})_{1\le a,b\le n}\right)}.\end{aligned}$$

Hence the $n$-form given by $\sqrt{|\det\left((g_{ij})_{1\le i,j\le n}\right)|}\mathrm{d}x^1 \wedge \cdots \wedge \mathrm{d}x^n$ for any positively oriented coordinate system is globally well defined. At $x$ we can choose a positively coordinate system with $g_{ab}(x) = \pm\delta_{ab}$. Calculating

$$\sqrt{|\det\left((g_{ij})_{1\le i,j\le n}\right)|}\mathrm{d}x^1 \wedge \cdots \wedge \mathrm{d}x^n$$

in this coordinate system implies that this n-form coincides with $\mu_M$. ∎

**Definition 4.2.3.** *Let $\varphi \in \mathcal{T}^r_s(M)$ be a tensor field on $(M, g)$. Then the* divergence *of $\varphi$, $\operatorname{div}(\varphi)$, is a tensor field in $\mathcal{T}^{r-1}_s(M)$ and given by*

$$\begin{aligned}&\operatorname{div}\varphi(\lambda^1, \ldots, \lambda^{r-1}, v_1, \ldots, v_s) \\ &\qquad = \sum_{a=1}^{n} \left(\nabla_{E_a}\varphi\right)(\theta^a, \lambda^1, \ldots, \lambda^{r-1}, v_1, \ldots, v_s),\end{aligned}$$

*where $\{E_1, \ldots, E_n; \theta^1, \ldots \theta^n\}$ is (any) pair of dual, orthonormal bases.*

This definition is independent of the chosen orthonormal basis — a fact which can be seen best in coordinates: the equation $\operatorname{div}(\varphi)^{i_1\ldots i_{r-1}}_{j_1\ldots j_s}(x) = \nabla_k\varphi^{ki_1\ldots i_{r-1}}_{j_1\ldots j_s}(x)$ is clearly invariant and agrees with the definition above if the Gaußian frame is chosen to be orthonormal at $x$.

[p. 172 ↓]
↓ p. 134

**Lemma 4.2.4.** *Let $(M, g)$ be an oriented pseudo-Riemannian manifold and $U$ be a vector field on $M$. Then the formula*

$$\pounds_U \mu_M = (\operatorname{div}(U))\mu_M$$

*holds.*

*Proof.* Let $x \in M$ and choose normal coordinates centred at $x$. Since the Christoffel symbols of $\nabla$ vanish at $x$ we have

$$\nabla_a = \partial_a \text{ and } \partial_a\left(\sqrt{\det\left((\tilde{g}_{ab})_{1\le a,b\le n}\right)}\right) = 0.$$

Hence using Lemma 2.5.3 we obtain at $x$

$$
\begin{aligned}
\pounds_U \mu_M &= \pounds_U \left( \sqrt{\det\left( (\tilde{g}_{ab})_{1 \le a,b \le n} \right)} \mathrm{d}x^1 \wedge \cdots \wedge \mathrm{d}x^n \right) \\
&= \left( U \bullet \sqrt{\det\left( (\tilde{g}_{ab})_{1 \le a,b \le n} \right)} \right) \mathrm{d}x^1 \wedge \cdots \wedge \mathrm{d}x^n \\
&\quad + \sqrt{\det\left( (\tilde{g}_{ab})_{1 \le a,b \le n} \right)} \pounds_U \left( \mathrm{d}x^1 \wedge \cdots \wedge \mathrm{d}x^n \right) \\
&= \sqrt{\det\left( (\tilde{g}_{ab})_{1 \le a,b \le n} \right)} \sum_{a=1}^{n} \mathrm{d}x^1 \wedge \cdots \wedge \pounds_U \mathrm{d}x^a \wedge \cdots \wedge \mathrm{d}x^n \\
&= \sqrt{\det\left( (\tilde{g}_{ab})_{1 \le a,b \le n} \right)} \sum_{a=1}^{n} \mathrm{d}x^1 \wedge \cdots \wedge \mathrm{d}(U \lrcorner \mathrm{d}x^a) \wedge \cdots \wedge \mathrm{d}x^n \\
&= \sqrt{\det\left( (\tilde{g}_{ab})_{1 \le a,b \le n} \right)} \sum_{a=1}^{n} \mathrm{d}x^1 \wedge \cdots \wedge \mathrm{d}U^a \wedge \cdots \wedge \mathrm{d}x^n \\
&= \sqrt{\det\left( (\tilde{g}_{ab})_{1 \le a,b \le n} \right)} \sum_{a=1}^{n} \frac{\partial U^a}{\partial x^a} \mathrm{d}x^1 \wedge \cdots \wedge \mathrm{d}x^n \\
&= \left( \sum_{a=1}^{n} \frac{\partial U^a}{\partial x^a} \right) \mu_M . = (\mathrm{div}(U))\, \mu_M .
\end{aligned}
$$

Since $x$ was arbitrary we have $\pounds_U \mu_M = (\mathrm{div}(U))\mu$ everywhere. ∎

**Corollary 4.2.1 (Theorem of Gauß).** *Let $(M, g)$ be an oriented pseudo-Riemannian manifold, $U$ be a vector field on $M$, and $\mathcal{V} \subset M$ an open subset of $M$ with smooth boundary $\partial\mathcal{V}$. Assume that $\partial\mathcal{V}$ is a smooth submanifold and that there is a vector field $\mathfrak{n}$ along $\partial\mathcal{V}$ with $g(v, \mathfrak{n}) = 0$ for all $v \in T\partial\mathcal{V}$ and $g(\mathfrak{n}, \mathfrak{n}) = \eta \in \{-1, 1\}$. Then we have*

$$
\int_{\mathcal{V}} \mathrm{div}(U) \mu_M = \eta \int_{\partial\mathcal{V}} \langle \mathfrak{n}, U \rangle \, \mu_{\partial\mathcal{V}} .
$$

*Proof.* Lemmas 4.2.4 and 2.5.3 imply $(\mathrm{div}(U))\mu_M = \pounds_U \mu_M = U \lrcorner \mathrm{d}\mu_M + \mathrm{d}(U \lrcorner \mu_M) = \mathrm{d}(U \lrcorner \mu_M)$. Denote by $\iota \colon \partial\mathcal{V} \to M$, $x \mapsto x$ the canonical inclusion. Then we have $\iota^*(U \lrcorner \mu_M) = \eta \langle U, \mathfrak{n} \rangle \, \mathfrak{n} \lrcorner \mu_M = \eta \langle U, \mathfrak{n} \rangle \, \mu_{\partial\mathcal{V}}$. Hence the Theorem of Stokes (Theorem 2.5.5) implies

$$
\int_{\mathcal{V}} \mathrm{div}(U) \mu_M = \int_{\mathcal{V}} \mathrm{d}(U \lrcorner \mu_M) = \eta \int_{\partial\mathcal{V}} \langle \mathfrak{n}, U \rangle \, \mu_{\partial\mathcal{V}} .
$$

∎

We will now introduce a canonical isomorphism between the space of $p$-forms $\Lambda^p(T_x M)$ and the space of $(n-p)$-forms $\Lambda^{n-p}(T_x M)$. See Sect. 2.5.1 for the motivation of this isomorphism.

**Proposition 4.2.1.** *Let $(M,g)$ be an oriented pseudo-Riemannian manifold with volume form $\mu$. For each $p \in \{0,\dots,n\}$ there is a unique linear isomorphism*

$$\star\colon \Lambda^p(T_xM) \to \Lambda^{n-p}(T_xM),\ \psi \mapsto \star\psi$$

*such that $g[{}^{0}_{n-p}](\star\psi,\phi) = g[{}^{0}_{n}](\psi\wedge\phi,\mu)$ for all $\psi \in \Lambda^p(T_xM)$, $\phi \in \Lambda^{n-p}(T_xM)$.*

The operator $\star$ is called the *Hodge star operator* and the isomorphism induced by $\star$ the *Hodge star isomorphism*.

*Proof.* Let $\{e_1,\dots,e_n\}$, $\{\omega^1,\dots,\omega^n\}$ be a dual pair of orthonormal bases and $\eta_i = \langle e_i, e_i\rangle$. Then the set $\{\omega^{a_1}\wedge\cdots\wedge\omega^{a_{n-p}}\}_{a_1<\cdots<a_{n-p}}$ is an orthonormal basis of $\Lambda^{n-p}(T_xM)$ with

$$g[{}^{0}_{n-p}](\omega^{a_1}\wedge\omega^{a_{n-p}},\omega^{b_1}\wedge\cdots\wedge\omega^{b_{n-p}}) = \eta_{a_1}\delta_{a_1b_1}\cdots\eta_{a_{n-p}}\delta_{a_{n-p}b_{n-p}},$$

where we have assumed $a_1<\cdots<a_{n-p}$, $b_1<\cdots<b_{n-p}$. We can therefore write

$$\begin{aligned}
\star\psi &= \sum_{a_1<\cdots<a_{n-p}} \star\psi(e_{a_1},\dots,e_{a_{n-p}})\omega^{a_1}\wedge\cdots\wedge\omega^{a_{n-p}} \\
&= \sum_{a_1<\cdots<a_{n-p}} g[{}^{0}_{n-p}](\star\psi,\omega^{a_1}\wedge\cdots\wedge\omega^{a_{n-p}}) \\
&\qquad \times \eta_{a_1}\cdots\eta_{a_{n-p}}\omega^{a_1}\wedge\cdots\wedge\omega^{a_{n-p}} \\
&= \sum_{a_1<\cdots<a_{n-p}} g[{}^{0}_{n}](\psi\wedge\omega^{a_1}\wedge\cdots\wedge\omega^{a_{n-p}},\mu) \\
&\qquad \times \eta_{a_1}\cdots\eta_{a_{n-p}}\omega^{a_1}\wedge\cdots\wedge\omega^{a_{n-p}}.
\end{aligned}$$

The last formula proves that $\star\psi$ is uniquely defined by the defining property of $\star$. Further, our explicit basis representation shows that $\star\psi$ exists. Since $g[{}^{0}_{n-p}](\star\psi,\phi) = g[{}^{0}_{n}](\psi\wedge\phi,\mu)$ is an invariant equation, our representation of $\star\psi$ does not depend on the chosen basis. ∎

**Corollary 4.2.2.** *Let $\psi \in \Lambda^p(T_xM)$ and $\{e_1,\dots,e_n\}$, $\{\omega^1,\dots,\omega^n\}$ be a pair of orthonormal dual bases. Writing $\eta_i := \langle e_i,e_i\rangle$ we have*

$$\star\psi = \sum_{a_1<\cdots<a_{n-p}} g[{}^{0}_{n}](\psi\wedge\omega^{a_1}\wedge\cdots\wedge\omega^{a_{n-p}},\mu)\eta_{a_1}\cdots\eta_{a_{n-p}}\omega^{a_1}\wedge\cdots\wedge\omega^{a_{n-p}}.$$

**Proposition 4.2.2.** *The operator $\star\colon \Lambda^p(T_xM)\to\Lambda^{n-p}(T_xM)$ is an isometry for even index $\nu$ and an anti-isometry for odd index $\nu$.*

*Proof.* Let $\{e_1,\dots,e_n\}$, $\{\omega^1,\dots,\omega^n\}$ be a pair of orthonormal dual bases and assume $a_1<\cdots<a_p$, $a_{p+1}<\cdots<a_n$, $b_1<\cdots<b_p$, $b_{p+1}<\cdots<b_n$, $\{a_1,\dots a_n\} = \{b_1,\dots,b_n\} = \{1,\dots,n\}$. By Corollary 4.2.2 we have

$$\star\omega^{a_1}\wedge\cdots\wedge\omega^{a_p}=\mathrm{sign}(a_1,\ldots,a_n)\eta_{a_{p+1}}\cdots\eta_{a_n}\omega^{a_{p+1}}\wedge\cdots\wedge\omega^{a_n}$$

and therefore

$$\begin{aligned}g[^0_{n-p}](\star\omega^{a_1}\wedge\cdots\wedge\omega^{a_n},\star\omega^{b_1}\wedge\cdots\wedge\omega^{b_n})\\&=\mathrm{sign}(a_1,\ldots,a_n)\mathrm{sign}(b_1,\ldots,b_n)\eta_{a_{p+1}}\cdots\eta_{a_n}\eta_{b_{p+1}}\cdots\eta_{b_n}\\&\quad\times g[^0_{n-p}](\omega^{a_{p+1}}\wedge\cdots\wedge\omega^{a_n},\omega^{b_{p+1}}\wedge\cdots\wedge\omega^{b_n})\\&=g[^0_{n-p}](\omega^{a_{p+1}}\wedge\cdots\wedge\omega^{a_n},\omega^{b_{p+1}}\wedge\cdots\wedge\omega^{b_n})\\&=\eta_{a_{p+1}}\cdots\eta_{a_n}\delta_{a_{p+1}b_{p+1}}\cdots\delta_{a_nb_n},\end{aligned}$$

where we have used that $g[^0_{n-p}](\omega^{a_{p+1}}\wedge\cdots\wedge\omega^{a_n},\omega^{b_{p+1}}\wedge\cdots\wedge\omega^{b_n})$ vanishes unless $a_{p+1}=b_{p+1},\ldots,a_n=b_n$. Since

$$g[^0_p](\omega^{a_1}\wedge\cdots\wedge\omega^{a_p},\omega^{b_1}\wedge\cdots\wedge\omega^{b_p})=\eta_{a_1}\cdots\eta_{a_p}\delta_{a_1b_1}\cdots\delta_{a_pb_p}$$

and (by our selection and ordering of indices)

$$\delta_{a_{p+1}b_{p+1}}\cdots\delta_{a_nb_n}=\delta_{a_1b_1}\cdots\delta_{a_pb_p}$$

holds, we have

$$\begin{aligned}g[^0_{n-p}](\star\omega^{a_1}\wedge\cdots\wedge\omega^{a_n},\star\omega^{b_1}\wedge\cdots\wedge\omega^{b_n})\\&=\frac{\eta_{a_{p+1}}\cdots\eta_{a_n}}{\eta_{a_1}\cdots\eta_{a_p}}g[^0_p](\omega^{a_1}\wedge\cdots\wedge\omega^{a_p},\omega^{b_1}\wedge\cdots\wedge\omega^{b_p})\\&=(-1)^\nu g[^0_p](\omega^{a_1}\wedge\cdots\wedge\omega^{a_p},\omega^{b_1}\wedge\cdots\wedge\omega^{b_p}).\end{aligned}$$

∎

**Lemma 4.2.5.** *Let $(M,g)$ be an oriented pseudo-Riemannian manifold of index $\nu$ with volume form $\mu$. Then*

*(i)* $\star 1=\mu,\quad \star\mu=(-1)^\nu$;
*(ii)* $g[^0_{n-p}](\star\phi,\psi)=(-1)^{(n-p)p}g[^0_p](\phi,\star\psi)$ *for all* $\phi\in\Lambda^p(T_xM)$, $\psi\in\Lambda^{n-p}(T_xM)$;
*(iii)* $\phi\wedge\star\psi=g[^0_p](\phi,\psi)\,\mu$ *for all* $\phi,\psi\in\Lambda^p(T_xM)$;
*(iv)* $\star\star\phi=(-1)^{\nu+p(n-p)}\phi$.

*Proof.* (i): By definition we have $g[^0_n](\star 1,\mu)=g[^0_n](1\wedge\mu,\mu)=g[^0_n](\mu,\mu)$, and the first claim follows since $g[^0_n]$ is non-degenerate and $\Lambda^n(T_xM)$ is a one dimensional vector space. The other claim can be proved analogously but it also follows from (iv).

(ii): Using Lemma 2.3.8 we can directly calculate

$$\begin{aligned}g[^0_{n-p}](\star\phi,\psi)&=g[^0_n](\phi\wedge\psi,\mu)=(-1)^{p(n-p)}g[^0_n](\psi\wedge\phi,\mu)\\&=(-1)^{p(n-p)}g[^0_{n-p}](\star\psi,\phi).\end{aligned}$$

(iii): From Proposition 4.2.2 we get

$$g[{}^0_n](\phi \wedge \star\psi, \mu) = g[{}^{\,0}_{n-p}](\star\phi, \star\psi) = (-1)^\nu g[{}^0_p](\phi, \psi).$$

The assertion follows now from $g[{}^0_n](\mu, \mu) = \eta_1 \dots \eta_n = (-1)^\nu$.

(iv): For every $\psi \in \Lambda^p(T_x M)$ we have

$$\begin{aligned} g[{}^0_p](\star\star\phi, \psi) &= g[{}^0_n](\star\phi \wedge \psi, \mu) \\ &= g[{}^0_n]((-1)^{p(n-p)}\psi \wedge \star\phi, \mu) \\ &= (-1)^{p(n-p)} g[{}^0_n](g[{}^0_p](\psi, \phi)\mu, \mu) \\ &= (-1)^{p(n-p)}(-1)^\nu g[{}^0_p](\psi, \phi), \end{aligned}$$

whence the claim follows from the non-degeneracy of $g[{}^0_p]$. ∎

**Lemma 4.2.6.** *Let $(M, g)$ be an oriented pseudo-Riemannian manifold and $\omega \in \Omega^p(M)$. Then we have* $\mathrm{div}(\omega) = (-1)^{p-1} \star \mathrm{d} \star \omega$.

*Proof.* Let $x \in M$ and choose a normal coordinate system centered at $x$ such that $\{\partial_1, \dots, \partial_n\}$ is orthonormal at $x$. Since the equation to be proved is linear in $\omega$ we can assume without loss of generatlity that $\omega = f \mathrm{d}x^{i_1} \wedge \dots \wedge \mathrm{d}x^{i_p}$ where $i_1 < \dots < i_p$. We obtain at $x$

$$\star\omega = f \mathrm{sign}(i_1, \dots, i_n)\eta_{i_{p+1}} \dots \eta_{i_n} \mathrm{d}x^{i_{p+1}} \wedge \dots\dots \mathrm{d}x^{i_n}.$$

Since the first derivative of the metric components at $x$ vanish we can ignore them in the calculation of $\mathrm{d} \star \omega$ and get

$$(\mathrm{d} \star \omega)_x = \sum_{j=1}^n \partial_j f \mathrm{sign}(i_1, \dots, i_n)\eta_{i_{p+1}} \dots \eta_{i_n} \mathrm{d}x^j \wedge \mathrm{d}x^{i_{p+1}} \wedge \dots \wedge \mathrm{d}x^{i_n}$$

Let $\tilde{\eta} \in \{-1, 1\}$ be the number which satisfis $\mathrm{d}x^j \wedge \mathrm{d}x^{i_{p+1}} \wedge \dots \wedge \mathrm{d}x^{i_n} = \tilde{\eta} \mathrm{d}x^{j_p} \wedge \dots \wedge \mathrm{d}x^{j_n}$ where $j_p < \dots < j_n$ and let $\sigma_{[j]}$ be the permutation of $i_1, \dots, i_p$ which moves $j$ into the last entry and which leaves the relative order of all other entries fixed. Then we get $\mathrm{sign}(j_1, \dots, j_n) = \mathrm{sign}(i_1, \dots, i_p)\tilde{\eta}\mathrm{sign}(\sigma_{[j]})$. We get then

$$\begin{aligned} &\star(\mathrm{d}x^j \wedge \mathrm{d}x^{i_{p+1}} \wedge \dots \wedge \mathrm{d}x^{i_n}) \\ &\quad = \tilde{\eta}\mathrm{sign}(j_1, \dots, j_n)\eta_{j_1} \dots \eta_{j_{p-1}} \mathrm{d}x^{j_1} \wedge \dots\dots \mathrm{d}x^{j_{p-1}} \\ &\quad = \mathrm{sign}(i_1, \dots, i_n)\mathrm{sign}(\sigma_{[j]})\eta_{j_1} \dots \eta_{j_{p-1}} \mathrm{d}x^{j_1} \wedge \dots\dots \mathrm{d}x^{j_{p-1}} \\ &\quad = \mathrm{sign}(i_1, \dots, i_n)\mathrm{sign}(\sigma_{[j]})\eta_{j_1} \dots \eta_{j_{p-1}} \\ &\qquad \times \partial_j \lrcorner \left(\mathrm{d}x^j \wedge \mathrm{d}x^{i_1} \wedge \dots \wedge \widehat{\mathrm{d}x^j} \wedge \dots \wedge \mathrm{d}x^{i_p}\right) \\ &\quad = (-1)^{p-1}\mathrm{sign}(i_1, \dots, i_n)\eta_{j_1} \dots \eta_{j_{p-1}} \\ &\qquad \times \partial_j \lrcorner \left(\mathrm{d}x^j \wedge \mathrm{d}x^{i_1} \wedge \dots \wedge \mathrm{d}x^{i_p}\right). \end{aligned}$$

Hence we get

$$\begin{aligned}(\star \mathrm{d} \star \omega)_x &= \sum_{j=1}^{n} \partial_j f \operatorname{sign}(i_1, \ldots, i_n) \eta_{i_{p+1}} \cdots \eta_{i_n} (-1)^{p-1} \\ &\quad \times \operatorname{sign}(i_1, \ldots, i_n) \eta_{j_1} \cdots \eta_{j_{p-1}} \partial_j \lrcorner \left( \mathrm{d}x^j \wedge \mathrm{d}x^{i_1} \wedge \cdots \wedge \mathrm{d}x^{i_p} \right) \\ &= \sum_{j=1}^{n} \partial_j f \eta_j (-1)^{p-1} \partial_j \lrcorner \left( \mathrm{d}x^j \wedge \mathrm{d}x^{i_1} \wedge \cdots \wedge \mathrm{d}x^{i_p} \right) \\ &= (-1)^{p-1} \operatorname{div}(\omega)_x .\end{aligned}$$

∎

## 4.3 Curvature of pseudo-Riemannian manifolds

p. 138 ↓
[↓ p. 184]

*The material presented in this section up to and including Corollary 4.3.1 is fundamental for Sect. 5 on relativity.*

*The rest of this section can be omitted on first reading.*

The Riemann (or curvature) tensor $R(U,V)W = \nabla_U \nabla_V W - \nabla_V \nabla_U W - \nabla_{[U,W]} W$ of a connection has been introduced geometrically in Sect. 2.8. Since it arises in such a natural way, one would suspect that the Riemann tensor of the Levi-Cività connection plays also an important rôle in the physical theory of spacetime. This is indeed so as we will see in Chap. 5. Here we will merely investigate some of its mathematical properties.

The Riemann tensor satisfies a collection of useful algebraic identities.

**Proposition 4.3.1.** *Let $(M, g)$ be a pseudo-Riemannian manifold, $x \in M$, and $t, u, v, w \in T_x M$. Then*

$$\begin{aligned} R(u,v) &= -R(v,u), \\ R(u,v)w + R(v,w)u + R(w,u)v &= 0 \quad \textit{(First Bianchi identity)}, \\ \langle R(u,v)w, t \rangle &= -\langle R(u,v)t, w \rangle , \\ \langle R(u,v)w, t \rangle &= \langle R(w,t)u, v \rangle . \end{aligned}$$

*Proof.* The first identity is clear from the definition of the curvature tensor. The second identity is just the first Bianchi identity which holds for any torsion-free connection (cf. Lemma 2.8.1).

Since

$$\begin{aligned}&\langle R(u,v)(w+t), w+t \rangle \\ &\quad = \langle R(u,v)w, w \rangle + \langle R(u,v)w, t \rangle + \langle R(u,v)t, w \rangle + \langle R(u,v)t, t \rangle ,\end{aligned}$$

the third identity is proved once we have shown that $\langle R(u,v)w, w \rangle = 0$ for all vectors $u, v, w \in T_x M$. We can assume that there are vector fields $U, V, W$ which are linear combinations of Gaußian vector fields with constant coefficients and satisfy $U_x = u$, $V_x = v$, $W_x = w$. Then we have $[U, V] = 0$ and calculate

$$\begin{aligned}
\langle R(u,v)w,w\rangle &= \langle \nabla_U \nabla_V W, W\rangle - \langle \nabla_V \nabla_W, W\rangle \\
&= \nabla_U \langle \nabla_V W, W\rangle - \langle \nabla_V W, \nabla_U W\rangle \\
&\quad - \nabla_V \langle \nabla_U W, W\rangle + \langle \nabla_U W, \nabla_V W\rangle \\
&= \nabla_U \langle \nabla_V W, W\rangle - \nabla_V \langle \nabla_U W, W\rangle \\
&= \frac{1}{2} U \bullet V \bullet \langle W, W\rangle - \frac{1}{2} V \bullet U \bullet \langle W, W\rangle \\
&= 0,
\end{aligned}$$

where in the last equality ew have used that the vector fields have constant coefficients.

The first Bianchi identity implies

$$\begin{aligned}
0 = 0 + 0 + 0 + 0 = \\
&= \overbrace{\langle R(u,v)w,t\rangle}^{1\,(-)} + \overbrace{\langle R(v,w)u,t\rangle}^{2\,(-)} + \overbrace{\langle R(w,u)v,t\rangle}^{3\,(+)} \\
&+ \overbrace{\langle R(t,u)v,w\rangle}^{4\,(-)} + \overbrace{\langle R(u,v)t,w\rangle}^{1\,(-)} + \overbrace{\langle R(v,t)u,w\rangle}^{5\,(+)} \\
&+ \overbrace{\langle R(w,t)u,v\rangle}^{6\,(-)} + \overbrace{\langle R(t,u)w,v\rangle}^{4\,(-)} + \overbrace{\langle R(u,w)t,v\rangle}^{3\,(+)} \\
&+ \overbrace{\langle R(v,w)t,u\rangle}^{2\,(-)} + \overbrace{\langle R(w,t)v,u\rangle}^{6\,(-)} + \overbrace{\langle R(t,v)w,u\rangle}^{5\,(+)} \\
&\qquad = 2\langle R(w,u)v,t\rangle + 2\langle R(v,t)u,w\rangle .
\end{aligned}$$

Here we have used the first and third identity to show that the $(-)$ terms cancel and the $(+)$ terms are equal. ∎

Recall that by taking the trace we can construct a $\binom{0}{2}$ tensor Ric from the curvature tensor. In the pseudo-Riemannian case we can contract Ric with the metric to obtain a function.

**Definition 4.3.1.** *Let $(M, g)$ be a pseudo-Riemannian manifold. Then the function* Scal $:= \mathrm{tr}(\mathrm{Ric}) = g^{ab}\mathrm{Ric}_{ab}$ *is called the* scalar curvature *of $g$.*

The second Bianchi identity (Lemma 2.8.1) has the following consequence which will be of importance in Sect. 5.3.

**Lemma 4.3.1.** *Let $(M, g)$ be a pseudo-Riemannian manifold. Then*

$$\mathrm{div}(\mathrm{Ric} - \frac{1}{2}\mathrm{Scal}\, g) = 0.$$

*Proof.* From the second Bianchi identity we infer

$$0 = \sum_{a=1}^{n} \omega^a \Big( (\nabla_{E_a} R)(u,v)w + (\nabla_v R)(E_a,u)w + (\nabla_u R)(v,E_a)w \Big)$$
$$= \operatorname{div}(R)(u,v)w + \nabla_v \operatorname{Ric}(u,w) - \nabla_u \operatorname{Ric}(v,w),$$

where we have used the property $\nabla g = 0$ of the Levi-Cività connection. Replacing $v$ and $w$ with a contraction gives $0 = 2\operatorname{div}(\operatorname{Ric})(u) - \nabla_u \operatorname{Scal}$. ∎

**Proposition 4.3.2.** *Let $(M,g)$ be a pseudo-Riemannian manifold and let $(x^1,\ldots,x^n)$ be normal coordinates centred at $x_0 \in M$. In these coordinates, the metric tensor $g$ satisfies*

$$g_{ab}(x) = g_{ab}(x_0) - \frac{1}{3} R_{acbd}(x_0) x^c x^d + o(|x^c|^2).$$

*Proof.* Since $(x^1,\ldots,x^n)$ are normal coordinates centred at $x_0$ The Christoffel symbols satisfy $\Gamma^k_{ij}(x_0) = 0$. The Koszul formula (2.7.7) implies that

$$\Gamma^k_{ij} = \frac{1}{2} g^{kl} \big(\partial_i g_{lj} + \partial_j g_{il} - \partial_l g_{ij}\big).$$

Hence we obtain

$$2(g_{kl}\Gamma^k_{ij} + g_{ki}\Gamma^k_{lj}) = \partial_i g_{lj} + \partial_j g_{il} - \partial_l g_{ij} + \partial_l g_{ij} + \partial_j g_{il} - \partial_i g_{lj} = 2\partial_j g_{il}$$

which implies that the first partial derivatives of the metric components vanish at $x_0$. It follows immediately that then also al $\partial_a g^{bc}$ vanish at $x_0$. At $x_0$ we can calculate for the Riemann tensor

$$\begin{aligned}
R^a_{bcd} &= (R(\partial_c,\partial_d)\partial_b)^a \\
&= \Big(\nabla_{\partial_c}(\Gamma^i_{bd}\partial_i) - \nabla_{\partial_d}(\Gamma^i_{bc}\partial_i)\Big)^a \\
&= \partial_c \Gamma^a_{bd} - \partial_d \Gamma^a_{bc} \\
&= \frac{1}{2} g^{ae} \Big(\partial_c\partial_b g_{ed} + \partial_c\partial_d g_{be} - \partial_c\partial_e g_{bd} - \partial_d\partial_b g_{ec} - \partial_d\partial_c g_{be} + \partial_d\partial_e g_{bc}\Big) \\
&= \frac{1}{2} g^{ae} \Big(\partial_c\partial_b g_{ed} - \partial_c\partial_e g_{bd} - \partial_d\partial_b g_{ec} + \partial_d\partial_e g_{bc}\Big).
\end{aligned} \tag{4.3.1}$$

We will now show that $\partial_a\partial_b g_{cd} = \partial_c\partial_d g_{ab}$ at $x_0$. Recall that in normal coordinates centred at $x_0$ all rays $t \to (tx^1,\ldots tx^n)$ which pass through $x_0$ are geodesics. Hence we get $\Gamma^a_{bc}(tx)x^b x^c = 0$ for all $x = (x^1,\ldots,x^n)$ small enough. This implies

$$\begin{aligned}
0 &= \frac{\partial}{\partial x^d}(2g_{ae}\Gamma^e_{bc}(tx)x^b x^c) \\
&= \frac{\partial}{\partial x^d}\left((\partial_b g_{ac}(tx) + \partial_c g_{ba}(tx) - \partial_a g_{bc}(tx))\, x^b x^c\right) \\
&= t\,(\partial_d\partial_b g_{ac}(tx) + \partial_d\partial_c g_{ba}(tx) - \partial_d\partial_a g_{bc}(tx))\, x^b x^c \\
&\quad + 2\,(\partial_d g_{ac}(tx) + \partial_c g_{da}(tx) - \partial_a g_{dc}(tx))\, x^c.
\end{aligned}$$

Observe that $\partial_d g_{ac}(0) = 0$ and that therefore

$$\lim_{t\to 0}\frac{1}{t}\partial_d g_{ac}(tx) = \mathrm{d}(\partial_d g_{ac}(0))(x) = \partial_e\partial_d g_{ac}(0)x^e.$$

Hence dividing the equation above by $t$ and taking the limit $t \to 0$ we get

$$\begin{aligned}
0 &= (\partial_d\partial_b g_{ac}(0) + \partial_d\partial_c g_{ba}(0) - \partial_d\partial_a g_{bc}(0))\, x^b x^c \\
&\quad + 2\,(\partial_e\partial_d g_{ac}(0) + \partial_e\partial_c g_{da}(0) - \partial_e\partial_a g_{dc}(0))\, x^c x^e. \\
&= x^b x^c(4\partial_b\partial_d g_{ac}(0) - \partial_d\partial_a g_{bc}(0) + 2\partial_b\partial_c g_{da}(0) - 2\partial_b\partial_a g_{dc}(0)).
\end{aligned}$$

Contracting this equation with $x^d$ we obtain

$$0 = (2\partial_b\partial_d g_{ac}(0) - \partial_d\partial_a g_{bc}(0))x^b x^c x^d$$

Since $(x^1, \ldots, x^n)$ is arbitrary (for small enough values) this equation implies that $G_{abcd} := 2\partial_{(d}\partial_b g_{|a|c)}(0) - \partial_{(d}\partial_{|a|} g_{bc)}(0) = 0$. We obtain

$$\begin{aligned}
0 &= G_{abcd} + G_{bcad} + G_{cabd} \\
&= \frac{1}{3}(2\partial_d\partial_b g_{ac}(0) - \partial_d\partial_a g_{bc}(0) + 2\partial_b\partial_c g_{ad}(0) - \partial_b\partial_a g_{cd}(0) \\
&\quad + 2\partial_c\partial_d g_{ab}(0) - \partial_c\partial_a g_{db}(0)) \\
&\quad + \frac{1}{3}(2\partial_d\partial_c g_{ba}(0) - \partial_d\partial_b g_{ca}(0) + 2\partial_c\partial_a g_{bd}(0) - \partial_c\partial_b g_{ad}(0) \\
&\quad + 2\partial_a\partial_d g_{bc}(0) - \partial_a\partial_b g_{dc}(0)) \\
&\quad + \frac{1}{3}(2\partial_d\partial_a g_{cb}(0) - \partial_d\partial_c g_{ab}(0) + 2\partial_a\partial_b g_{cd}(0) - \partial_a\partial_c g_{bd}(0) \\
&\quad + 2\partial_b\partial_d g_{ca}(0) - \partial_b\partial_c g_{da}(0)) \\
&= \frac{1}{3}(3\partial_d\partial_b g_{ac}(0) + 3\partial_d\partial_a g_{bc}(0) + 3\partial_c\partial_d g_{ab}(0)) \\
&= \partial_d\partial_a g_{bc}(0) + \partial_d\partial_b g_{ca}(0) + \partial_d\partial_c g_{ab}(0).
\end{aligned} \tag{4.3.2}$$

This last equation implies now

$$\begin{aligned}
0 &= G_{dbca} \\
&\frac{1}{3}(2\partial_a\partial_b g_{dc}(0) - \partial_a\partial_d g_{bc}(0) + 2\partial_b\partial_c g_{da}(0) - \partial_b\partial_d g_{ca}(0) \\
&\quad + 2\partial_c\partial_a g_{db}(0) - \partial_c\partial_d g_{ab}(0)) \\
&= \frac{2}{3}(\partial_a\partial_b g_{dc}(0) + \partial_b\partial_c g_{da}(0) + \partial_c\partial_a g_{db}(0)).
\end{aligned}$$

We interchange $b$ and $d$ in this equation and get $\partial_a\partial_d g_{bc}(0)+\partial_d\partial_c g_{ba}(0)+\partial_c\partial_a g_{bd}(0)=0$. A comparison with Equation (4.3.2) gives $\partial_c\partial_a g_{bd}(0)=\partial_d\partial_b g_{ca}(0)$ which is equivalent to our assertion $\partial_a\partial_b g_{cd}=\partial_c\partial_d g_{ab}$.

It follows that Equation (4.3.1) implies

$$R^a_{bcd}(0)=g^{ae}(0)\Big(\partial_c\partial_b g_{ed}(0)-\partial_c\partial_e g_{bd}(0)\Big)$$

and therefore

$$\begin{aligned}R_{acbd}(0)+R_{adbc}(0)&=\partial_b\partial_c g_{ad}(0)-\partial_b\partial_a g_{cd}(0)+\partial_b\partial_d g_{ac}(0)-\partial_b\partial_a g_{dc}(0)\\&=\partial_b\partial_c g_{ad}(0)+\partial_b\partial_d g_{ac}(0)-\partial_b\partial_a g_{cd}(0)\\&=-3\partial_b\partial_a g_{cd}(0),\end{aligned}$$

where we have used Equation (4.3.2). The assertion of the proposition follows now from a Taylor expansion of $g_{ab}(x)$ around the point $x_0$. ∎

**Corollary 4.3.1.** *Let $A$ be a tensor field which is pointwise defined as an algebraic expression of $g$ and its first two derivatives. Then $A$ is an algebraic expression of $g$ and the Riemann tensor $R$.*

*Proof.* Proposition 4.3.2 implies that any tensor field $A$ which is a pointwise invariant function of $g$ and its first two derivatives only depends on [p. 180 ↓] $g$ and $R$. ∎

↓ p. 204

For 2-dimensional manifolds, the Riemann tensor reduces to a function, a much simpler geometric quantity. To see this, let $\{e_1,e_2\}$ be an orthonormal basis of $T_xM$. Since $\langle R(\cdot,\cdot)\cdot,\cdot\rangle$ is anti-symmetric in the first two and in the second two entries the expression $\langle R(e_1,e_2)e_2,e_1\rangle$ does not depend on the chosen orthonormal basis and defines a function $K\colon M\to\mathbb{R}$. It is easy to see that this function describes the curvature tensor uniquely. (For explicit formulas cf. Sect. 4.3.1)

For higher dimensional pseudo-Riemannian manifolds such a simple relationship does not exist. However, it is possible to define a function $K$ which maps the space of non-degenerate 2-dimensional subspaces of $T_xM$ into the real numbers.

Let

$$G_2^{\text{nondeg}}(TM)=\{\text{span}\{u_x,v_x\}\,\big|\,u_x\not\parallel v_x,\,g_{|\text{span}\{u_x,v_x\}}\text{ is non-degenerate}\}$$

be the set of all two-dimensional subspaces in $TM$ which are either spacelike or timelike.

**Definition 4.3.2.** *The function*

$$K\colon G_2^{\text{nondeg}}(TM)\mapsto\mathbb{R},\quad \text{span}\{u_x,v_x\}\mapsto\frac{-\langle R(u_x,v_x)u_x,v_x\rangle}{\langle u_x,u_x\rangle\langle v_x,v_x\rangle-\langle u_x,v_x\rangle^2}$$

*is called the* sectional curvature *of $M$.*

Observe that this expression is well defined since the denominator does not vanish for any $\Pi \in G_2^{\text{nondeg}}(TM)$ and both the nominator and the denominator transform by a factor $\det(A)^2$ under a change of basis $\tilde{u}_x = A_1^1 u_x + A_1^2 v_x$, $\tilde{v}_x = A_2^1 u_x + A_2^2 v_x$.

Those pseudo-Riemannian manifolds for which the sectional curvature reduces to a function on $M$ should be especially interesting.

**Proposition 4.3.3.** *Let $(M, g)$ be a pseudo-Riemannian manifold and $x \in M$. If the sectional curvature satisfies $K(\Pi_x) = K(\tilde{\Pi}_x)$ for all $\Pi_x, \tilde{\Pi}_x \in T_xM$ then there exists a number $c \in \mathbb{R}$ with $R_x(u,v)w = c\left(\langle v, w\rangle u - \langle u, w\rangle v\right)$ for all $u, v, w \in T_xM$.*

*Proof.* Assume that $K(\Pi_x)$ does not depend on the choice of plane in $T_xM$. Then the definition of the sectional curvature implies that given an orthonormal basis $\{e_1, \ldots, e_n\}$ we have

$$\langle R(e_i, e_j)e_j, e_i\rangle = c\left(\langle e_i, e_i\rangle \langle e_j, e_j\rangle - \langle e_i, e_j\rangle^2\right)$$

for some constant $c$. From the tensorial property of $\langle R(\cdot,\cdot)\cdot,\cdot\rangle$ we conclude that

$$\langle R(u,v)v, u\rangle = c\left(\langle u, u\rangle \langle v, v\rangle - \langle u, v\rangle^2\right)$$

for all $u, v \in T_xM$. For every $w \in T_xM$, we obtain

$$\begin{aligned}
0 &= \langle R(u+w, v)v, u+w\rangle - c\left(\langle u+w, u+w\rangle \langle v, v\rangle - \langle u+w, v\rangle^2\right) \\
&= \langle R(u,v)v, u\rangle + \langle R(w,v)v, u\rangle + \langle R(u,v)v, w\rangle + \langle R(w,v)v, w\rangle \\
&\quad - c\big(\langle u, u\rangle \langle v, v\rangle + 2\langle u, w\rangle \langle v, v\rangle + \langle w, w\rangle \langle v, v\rangle \\
&\quad - \langle u, v\rangle^2 - 2\langle u, v\rangle \langle w, v\rangle - \langle v, w\rangle^2\big) \\
&= 2\langle R(u,v)v, w\rangle + c\big(\langle u, u\rangle \langle v, v\rangle - \langle u, v\rangle^2 + \langle w, w\rangle \langle v, v\rangle - \langle w, v\rangle^2 \\
&\quad - \langle u, u\rangle \langle v, v\rangle - 2\langle u, w\rangle \langle v, v\rangle - \langle w, w\rangle \langle v, v\rangle \\
&\quad + \langle u, v\rangle^2 + 2\langle u, v\rangle \langle w, v\rangle + \langle v, w\rangle^2 \\
&\ 2\langle R(u,v)v - c(\langle v, v\rangle u - \langle u, v\rangle v), w\rangle
\end{aligned}$$

which implies $R(u,v)v = c(\langle v, v\rangle u - \langle u, v\rangle v)$ for all $u, v \in T_xM$. We can now polarize again and get

$$\begin{aligned}
0 &= R(u, v+w)(v+w) - c\big(\langle v+w, v+w\rangle u - \langle u, v+w\rangle (v+w)) \\
&= R(u,v)v - c(\langle v, v\rangle u - \langle u, v\rangle v) + R(u,v)w - c(\langle v, w\rangle u - \langle u, w\rangle v) \\
&\quad + R(u,w)v - c(\langle w, v\rangle u - \langle v, u\rangle w) + R(u,w)w \\
&\quad - c(\langle w, w\rangle u - \langle w, u\rangle w) \\
&= R(u,v)w - c(\langle v, w\rangle u - \langle u, w\rangle v) + R(u,w)v - c(\langle w, v\rangle u - \langle v, u\rangle w).
\end{aligned}$$

This implies $R(u,v)w - c(\langle v, w\rangle u - \langle u, w\rangle v) = R(w,u)v - c(\langle u, v\rangle w - \langle w, v\rangle u)$ for all $u, v, w \in T_xM$,i.e the expression $R(u,v)w - c(\langle v, w\rangle u -$

$\langle u,w\rangle v)$ is invariant with respect to cyclic permutation. Since the cyclic $R(u,v)w + R(w,u)v + R(v,w)_u$ vanishes (cf. Proposition 4.3.1) and

$$\langle u,w\rangle v - \langle v,w\rangle u + \langle v,u\rangle w - \langle u,v\rangle w + \langle w,v\rangle u - \langle w,u\rangle v = 0,$$

the cyclic sum of $R(u,v)w - c(\langle v,w\rangle u - \langle u,w\rangle v)$ gives $0 = 3\big(R(u,v)w - c(\langle v,w\rangle u - \langle u,w\rangle v)\big)$. ∎

**Proposition 4.3.4 (Lemma of Schur).** *If* $\dim(M) \geq 3$ *and the sectional curvature* $K\colon G_2^{\mathrm{nondeg}}(TM) \mapsto \mathbb{R}$ *reduces to a function on* $M$ *then it is constant.*

*Proof.* Proposition 4.3.3 implies that there exists a function $x \mapsto c(x)$ with $R_x(u,v)w = c(x)\,(\langle v,w\rangle u - \langle u,w\rangle v)$ for all $u,v,w \in T_xM$. Let $U, V, W$ be vector fields whose covariant derivatives vanish at $x$ and which satisfy $U_x = u$, $V_x = v$, $W_x = w$. Then we obtain for any vector $t \in T_xM$

$$(\nabla_t R)\,(u,v)w = \mathrm{d}c(t)\,(\langle v,w\rangle u - \langle u,w\rangle v)\,.$$

Hence the the second Bianchi identity (Lemma 2.8.1) implies

$$\begin{aligned} 0 = \mathrm{d}c(t)\,(\langle v,w\rangle u - \langle u,w\rangle v) + \mathrm{d}c(v)\,(\langle u,w\rangle t - \langle t,w\rangle u) \\ + \mathrm{d}c(u)\,(\langle t,w\rangle v - \langle v,w\rangle t)\,. \end{aligned}$$

Let $u$ be a vector which is not lightlike and $w \neq 0$ be orthogonal to $u$. We can choose $t = w$ and $v$ to be orthogonal to $w$. Since $\dim(M) \geq 3$ we can also assume that $u, v, w$ are linearly independent. The equation above reduces then to

$$0 = (-\mathrm{d}c(v)u + \mathrm{d}c(u)v)\,\langle w,w\rangle\,.$$

which in turn implies $\mathrm{d}c(u) = 0$. Hence $\mathrm{d}c$ vanishes on all vectors which are not lightlike and therefore must vanish identically. ∎

Observe that the preceding proposition is false for two-dimensional pseudo-Riemannian manifolds.

Proposition 4.3.4 motivates the following definition.

**Definition 4.3.3.** *A pseudo-Riemannian manifold* $(M,g)$ *has* constant curvature *if the sectional curvature* $K\colon G_2^{\mathrm{nondeg}}(TM) \mapsto \mathbb{R}$ *is constant.*

In Proposition 4.5.2 below we will locally classify all pseudo-Riemannian manifolds of constant curvature.

### 4.3.1 2-dimensional pseudo-Riemannian manifolds

*We collect formulas for the Levi-Cività connection and the curvature tensor of 2-dimensional pseudo-Riemannian manifolds. These formulas will be used in Sect. 7.4.*

2-dimensional pseudo-Riemannian manifolds are the lowest dimensional pseudo-Riemannian manifolds which are not trivial.

Let $\eta_1, \eta_2 \in \{-1, 1\}$ and let $\{E_1, E_2\}$ be an orthonormal frame with $g(E_1, E_1) = \eta_1$, $g(E_1, E_2) = 0$, $g(E_1, E_1) = \eta_2$. By Corollary 2.4.2 there are coordinates $(t, q)$ and functions $\lambda(t,q)$, $\nu(t,q)$ such that $E_1 = \mathrm{e}^{-\nu(t,q)}\partial_t$ and $E_2 = \mathrm{e}^{-\lambda(t,q)}\partial_q$.

**Proposition 4.3.5.** *Let* $(M, g)$ *be a 2-dimensional pseudo-Riemannian manifold. Then* $[E_1, E_2] = \mathrm{d}\nu(E_2)E_1 - \mathrm{d}\lambda(E_1)E_2$ *and the Levi-Cività connection is given by*

$$\nabla_{E_1}E_1 = -\eta_1\eta_2\mathrm{d}\nu(E_2)E_2, \quad \nabla_{E_1}E_2 = \mathrm{d}\nu(E_2)E_1,$$
$$\nabla_{E_2}E_1 = \mathrm{d}\lambda(E_1)E_2, \quad \nabla_{E_2}E_2 = -\eta_1\eta_2\mathrm{d}\lambda(E_1)E_1.$$

*The curvature is completely determined by the scalar curvature* Scal,

$$R(E_1, E_2)E_1 = -\frac{\eta_2}{2}\mathrm{Scal}\, E_2, \quad R(E_1, E_2)E_2 = \frac{\eta_1}{2}\mathrm{Scal}\, E_1,$$
$$\mathrm{Ric} = \frac{1}{2}\mathrm{Scal}\, g, \quad K(T_xM) = \frac{1}{2}\mathrm{Scal},$$

*where*

$$\mathrm{Scal} = -2\Big(\eta_1\left(E_1 \bullet E_1 \bullet \lambda + (\mathrm{d}\lambda(E_1))^2\right) + \eta_2\left(E_2 \bullet E_2\nu + (\mathrm{d}\nu(E_2))^2\right)\Big).$$

*Proof.* We calculate first the commutator of the vector field $E_1$, $E_2$, using the ordinary derivative of $\mathbb{R}^2$ with respect to the coordinates $(t, q)$.

$$\begin{aligned}
[E_1, E_2] &= \mathrm{D}(\mathrm{e}^{-\lambda}\partial_q)(\mathrm{e}^{-\nu}\partial_t) - \mathrm{D}(\mathrm{e}^{-\nu}\partial_t)(\mathrm{e}^{-\lambda}\partial_q) \\
&= -\mathrm{e}^{-\nu}\mathrm{e}^{-\lambda}\mathrm{d}\lambda(\partial_t)\partial_q + \mathrm{e}^{-\nu}\mathrm{e}^{-\lambda}\mathrm{D}\partial_q(\partial_t) + \mathrm{e}^{-\nu}\mathrm{e}^{-\lambda}\mathrm{d}\nu(\partial_q)\partial_t \\
&\quad - \mathrm{e}^{-\nu}\mathrm{e}^{-\lambda}\mathrm{D}\partial_t(\partial_q) \\
&= -\mathrm{d}\lambda(E_1)E_2 + \mathrm{d}\nu(E_2)E_1 - \mathrm{e}^{-\nu}\mathrm{e}^{-\lambda}\overbrace{[\partial_t, \partial_q]}^{=0}.
\end{aligned}$$

Our formulas for the Levi-Cività connection follow directly from

$$
\begin{aligned}
\left\langle \nabla_{E_1} E_1, E_1 \right\rangle &= \frac{1}{2} E_1 \bullet \eta_1 = 0, \\
\left\langle \nabla_{E_2} E_1, E_1 \right\rangle &= \frac{1}{2} E_2 \bullet \eta_1 = 0, \\
\left\langle \nabla_{E_1} E_2, E_2 \right\rangle &= \frac{1}{2} E_1 \bullet \eta_2 = 0, \\
\left\langle \nabla_{E_2} E_2, E_2 \right\rangle &= \frac{1}{2} E_2 \bullet \eta_2 = 0, \\
\left\langle \nabla_{E_1} E_2, E_1 \right\rangle &= \left\langle \nabla_{E_1} E_2 - \nabla_{E_2} E_1, E_1 \right\rangle \\
&= \langle \mathrm{d}\nu(E_2) E_1 - \mathrm{d}\lambda(E_1) E_2, E_1 \rangle = \eta_1 \mathrm{d}\nu(E_2), \\
\left\langle \nabla_{E_1} E_1, E_2 \right\rangle &= -\left\langle E_1, \nabla_{E_1} E_2 \right\rangle = -\eta_1 \mathrm{d}\nu(E_2), \\
\left\langle \nabla_{E_2} E_1, E_2 \right\rangle &= -\left\langle \nabla_{E_1} E_2 - \nabla_{E_2} E_1, E_2 \right\rangle \\
&= -\langle \mathrm{d}\nu(E_2) E_1 - \mathrm{d}\lambda(E_1) E_2, E_2 \rangle = \eta_2 \mathrm{d}\lambda(E_1) E_2 \\
\left\langle \nabla_{E_2} E_2, E_1 \right\rangle &= -\left\langle E_2, \nabla_{E_2} E_1 \right\rangle = -\eta_2 \mathrm{d}\lambda(E_1)
\end{aligned}
$$

The curvature can now be calculated straightforwardly.

$$
\begin{aligned}
R(E_1, E_2) E_1 &= \nabla_{E_1} \nabla_{E_2} E_1 - \nabla_{E_2} \nabla_{E_1} E_1 - \nabla_{[E_1, E_2]} E_1 \\
&= \nabla_{E_1} (\mathrm{d}\lambda(E_1) E_2) + \eta_1 \eta_2 \nabla_{E_2} (\mathrm{d}\nu(E_2) E_2) \\
&\quad - \mathrm{d}\nu(E_2) \nabla_{E_1} E_1 + \mathrm{d}\lambda(E_1) \nabla_{E_2} E_1 \\
&= E_1 \bullet E_1 \bullet \lambda \, E_2 + \mathrm{d}\lambda(E_1) \mathrm{d}\nu(E_2) E_1 + \eta_1 \eta_2 E_2 \bullet E_2 \bullet \nu \, E_2 \\
&\quad - \mathrm{d}\nu(E_2) \mathrm{d}\lambda(E_1) E_1 + \eta_1 \eta_2 (\mathrm{d}\nu(E_2))^2 E_2 + (\mathrm{d}\lambda(E_1))^2 E_2 \\
&= \Big( E_1 \bullet E_1 \bullet \lambda + (\mathrm{d}\lambda(E_1))^2 \\
&\qquad + \eta_1 \eta_2 \left( E_2 \bullet E_2 \bullet \nu + (\mathrm{d}\nu(E_2))^2 \right) \Big) E_2.
\end{aligned}
$$

Since we have

$$
\begin{aligned}
\mathrm{Ric}(E_1, E_1) &= \mathrm{tr}(R(\cdot, E_1) E_1) = \theta^2 (R(E_2, E_1) E_1) \\
&= \eta_2 \langle R(E_2, E_1) E_1, E_2 \rangle = -\eta_2 \langle R(E_1, E_2) E_1, E_2 \rangle, \\
\mathrm{Ric}(E_1, E_2) &= \mathrm{Ric}(E_2, E_1) = \mathrm{tr}(R(\cdot, E_2) E_1) \\
&= \theta^2 (R(E_2, E_2) E_1) + \theta^1 (R(E_1, E_2) E_1) = 0, \\
\mathrm{Ric}(E_2, E_2) &= \mathrm{tr}(R(\cdot, E_2) E_2) = \eta_1 \langle R(E_1, E_2) E_2, E_1 \rangle \\
&= -\eta_1 \langle R(E_1, E_2) E_1, E_2 \rangle,
\end{aligned}
$$

we obtain $\mathrm{Ric} = \frac{1}{2} \mathrm{Scal}\, g$, where

$$\begin{aligned}\mathrm{Scal} &= -2\eta_1\eta_2 \langle R(E_1, E_2)E_1, E_2\rangle \\ &= -2\Big(\eta_1 \left(E_1 \bullet E_1 \bullet \lambda + (\mathrm{d}\lambda(E_1))^2\right) \\ &\quad + \eta_2 \left(E_2 \bullet E_2 \bullet \nu + (\mathrm{d}\nu(E_2))^2\right)\Big).\end{aligned}$$

Finally, the sectional curvature is given by

$$K(T_x M) = -\frac{\langle R(E_1, E_2)E_1, E_2\rangle}{\langle E_1, E_1\rangle \langle E_2, E_2\rangle - \langle E_1, E_2\rangle^2} = \frac{1}{2}\mathrm{Scal}.$$

∎

## 4.4 Submanifolds

Given a pseudo-Riemannian manifold $(M, g)$ and an (immersed) submanifold $f\colon \Sigma \to M$ it is of interest which geometrical structure $\Sigma$ inherits from $(M, g)$. Examples for physically especially interesting submanifolds are spacelike hypersurfaces (describing an instant of time, cf. Sect. 5.4) or the integrated light cone. In Chap. 9 *closed trapped surfaces*, a class of submanifolds of codimension 2, will play a central rôle.

*Remark 4.4.1.* In this section we will explicitly refer to the immersion $f$. Very often the immersed submanifold is a subset of $M$ and $f$ is the canonical injection. In this case it is more convenient to omit any references to $f$. We will often speak of tensor fields *along* $\Sigma$ instead of tensor fields along $f$.

We denote the space of vector fields along $\Sigma$ by $\mathcal{T}_0^1(f)$ and will use the notation introduced following Lemma 2.9.2. In this notation, Lemma 2.9.1 is fundamental for the following.

**Lemma 4.4.1.** *Let $(M, g)$ be a pseudo-Riemannian manifold and $f\colon \Sigma \to M$ be an immersed submanifold of $M$. Then the following holds for all vector fields $U, V \in \mathcal{T}_0^1(\Sigma)$, $X, Y \in \mathcal{T}_0^1(f)$.*

*(i)* $\nabla_{f_* U} f_* V - \nabla_{f_* V} f_* U = f_*[U, V]$,

*(ii)* $\mathrm{d}\langle X, Y\rangle (U) = \left\langle \nabla_{f_* U} X, Y\right\rangle + \left\langle X, \nabla_{f_* U} Y\right\rangle$.

*Proof.* These properties follow immediately from the definition of $\nabla_{f_* U} X$ (cf. Lemma 2.9.1) and the fact that $\nabla$ is a Levi-Cività connection. ∎

**Definition 4.4.1.** *Let $f\colon \Sigma \to M$ be an immersed pseudo-Riemannian manifold and $g$ be the metric of $M$.*

*The map $f$ (or the immersed submanifold $\Sigma$) is called* non-degenerate *if $f^*g$ is a non-degenerate $\binom{0}{2}$-tensor field. A non-degenerate submanifold is also called a* pseudo-Riemannian submanifold, *and a non-degenerate hypersurface is also called a* pseudo-Riemannian hypersurface.

*Let $\Sigma$ be a non-degenerate immersed submanifold of $(M,g)$ and denote by $(T_x\Sigma)^\perp$ the set of all $v \in T_{f(x)}M$ with $g(v,w) = 0$ for all $w \in f_*T_x\Sigma$. The decomposition $T_{f(x)}M = f_*T_x\Sigma \oplus (T_x\Sigma)^\perp$ induces projections*

$$v \mapsto v^\top \in f_*T_x\Sigma \text{ and } v \mapsto v^\perp \in (T_x\Sigma)^\perp$$

*such that $v = v^\top + v^\perp$ for every $v \in T_{f(x)}M$.*

**Lemma 4.4.2.** *Let $\Sigma$ be a non-degenerate immersed submanifold of $(M,g)$ and $U, V \in \mathcal{T}_0^1(\Sigma)$. Then $\boldsymbol{\nabla}_U V$ defined by*

$$f_*\boldsymbol{\nabla}_U V := \left(\nabla_{f_*U} f_*V\right)^\top$$

*is the Levi-Cività connection of $(\Sigma, f^*g)$.*

*Proof.* Observe that $\boldsymbol{\nabla}_U V$ is well defined since $f$ is an immersion. Properties (i)–(iv) in Lemma 2.9.1 imply that the Koszul equation (cf. Equation 2.7.7 in Theorem 2.7.1) is satisfied for $\boldsymbol{\nabla}$. ∎

**Lemma 4.4.3.** *Let $\Sigma$ be a non-degenerate immersed submanifold of $(M,g)$. The map*

$$\mathrm{I\!I}\colon \mathcal{T}_0^1(\Sigma) \times \mathcal{T}_0^1(\Sigma) \to \left(\mathcal{T}_0^1(\Sigma)\right)^\perp,$$
$$(U,V) \mapsto \mathrm{I\!I}(U,V) := \left(\nabla_{f_*U} f_*V\right)^\perp$$

*is symmetric in $U$ and $V$ and function-linear. $\mathrm{I\!I}$ is called the* shape tensor.

*Proof.* The map $\mathrm{I\!I}$ is clearly function-linear in its first entry. Recall that the Lie bracket of any two vector fields tangent to $\Sigma$ is itself tangent to $\Sigma$. Hence $\mathrm{I\!I}(U,V) - \mathrm{I\!I}(V,U) = \left(\nabla_{f_*U} f_*V - \nabla_{f_*V} f_*U\right)^\perp = (f_*[U,V])^\perp = 0$, which in turn implies that $\mathrm{I\!I}$ is symmetric. But then $\mathrm{I\!I}$ must also be function-linear in its second entry. ∎

So far we have considered vector fields tangent to $\Sigma$. We obtain similar relationships for vector fields normal to $\Sigma$.

**Lemma 4.4.4.** *Let $\Sigma$ be a non-degenerate immersed submanifold of $(M,g)$. Let $U$ be a vector field on $\Sigma$ and $N$ be a vector field along $\Sigma$ such that $N(x) \in (T_{f(x)} f_* \Sigma)^\perp$ for all $x \in \Sigma$. Then $\nabla_{f_* V} N$ is well defined and we have*

$$\nabla_{f_* V} N = \left(\nabla_{f_* V} N\right)^\perp - \langle I\!I(V,\cdot), N\rangle^\sharp ,$$

*where $\sharp$ denotes the lift of indices with respect to the induced metric $f^* g$.*

*Proof.* It follows immediately from the coordinate expression that $\nabla_{f_* V} N$ does not depend on the extensions of $V$, $N$ off $\Sigma$. Hence it is well defined. Let $X$ be a vector field along $\Sigma$ and $X^\top$, $X^\perp$ its tangent and normal part. Then we have

$$\begin{aligned}
\left\langle \nabla_{f_* V} N, X\right\rangle &= \left\langle \nabla_{f_* V} N, X^\perp\right\rangle + \left\langle \nabla_{f_* V} N, X^\top\right\rangle \\
&= \left\langle \left(\nabla_{f_* V} N\right)^\perp, X^\perp\right\rangle + \nabla_{f_* V} \overbrace{\langle N, X^\top\rangle}^{=0} \\
&\quad - \left\langle N, \nabla_{f_* V} X^\top\right\rangle \\
&= \left\langle \left(\nabla_{f_* V} N\right)^\perp, X^\perp\right\rangle - \left\langle N, \left(\nabla_{f_* V} X^\top\right)^\perp\right\rangle \\
&= \left\langle \left(\nabla_{f_* V} N\right)^\perp, X^\perp\right\rangle - \langle N, I\!I(V, X^\top)\rangle .
\end{aligned}$$

∎

**Lemma 4.4.5.** *Let $\Sigma$ be a non-degenerate immersed submanifold of $(M,g)$ and let $\gamma\colon [a,b] \to \Sigma$ be a smooth curve. For every vector $n \in (T_{\gamma(a)}\Sigma)^\perp$ there is a unique vector field $N$ along $f \circ \gamma$ with*

*(i)* $N(a) = n$,
*(ii)* $N(t) \in (f_* T_{\gamma(t)}\Sigma)^\perp$ *for all* $t \in [a,b]$,
*(iii) and* $(\nabla_{f_* \dot\gamma} N)^\perp = 0$.

*We will write* $N(t) = \mathbf{P}^\perp_{\gamma|[a,t]} n$ *and refer to it as the* normal parallel transport *of* $n$.

*Proof.* Let $E_1, \dots, E_{n-\dim(\Sigma)}$ be orthonormal vector fields along $f \circ \gamma$ which span $(T_{\gamma(t)}\Sigma)^\perp$ at every point. We can decompose any vector field $N$ along $f \circ \gamma$ with values in $(T\Sigma)^\perp$ as $N(t) = N^i(t) E_i(t)$. Similarly, there are functions $\Gamma^i_j\colon [a,b] \to \mathbb{R}$ such that $(\nabla_{f_* \dot\gamma} E_j)^\perp = \Gamma^i_j E_i$. With respect to these decompositions we obtain $(\nabla_{f_* \dot\gamma} N)^\perp = (\frac{\mathrm{d}}{\mathrm{d}t} N^i + N^j \Gamma^i_j) E_i$. Hence the equation $(\nabla_{f_* \dot\gamma} N)^\perp = 0$ reduces to a first order system of ordinary differential equations and each solution is uniquely determined by its initial values $N^i(a) = n^i$. ∎

**Proposition 4.4.1 (Gauß equation).** *Let $\Sigma$ be an immersed non-degenerate submanifold of $(M, g)$. Denote by ${}^{\Sigma}R$ the Riemann tensor of $(\Sigma, g)$ and let $U, V, W, X \in \mathcal{T}_0^1(\Sigma)$. Then we have*

$$\begin{aligned} &f^*g({}^{\Sigma}R(U,V)W, X) \\ &\quad = \langle R(U,V)W, X\rangle + \langle I\!I(U,X), I\!I(V,W)\rangle - \langle I\!I(U,W), I\!I(V,X)\rangle . \end{aligned}$$

*Proof.* Since this is a tensor equation we can assume that $[U, V] = 0$. We calculate

$$\begin{aligned} &\langle R(f_*U, f_*V)f_*W, f_*X\rangle \\ &\quad = \left\langle \nabla_{f_*U}\nabla_{f_*V} f_*W, f_*X\right\rangle - \left\langle \nabla_{f_*V}\nabla_{f_*U} f_*W, f_*X\right\rangle \\ &\quad = \left\langle \nabla_{f_*U} f_*\left(\boldsymbol{\nabla}_V W\right), f_*X\right\rangle + \left\langle \nabla_{f_*U} I\!I(V,W), f_*X\right\rangle \\ &\qquad - \left\langle \nabla_{f_*V} f_*\left(\boldsymbol{\nabla}_U W\right), f_*X\right\rangle - \left\langle \nabla_{f_*V} I\!I(U,W), f_*X\right\rangle \\ &\quad = \langle f_*\left(\boldsymbol{\nabla}_U\boldsymbol{\nabla}_V W\right), f_*X\rangle - \langle f_*\left(\boldsymbol{\nabla}_V\boldsymbol{\nabla}_U W\right), f_*X\rangle \\ &\qquad + \nabla_{f_*U}\overbrace{\langle I\!I(V,W), f_*X\rangle}^{=0} - \left\langle I\!I(V,W), \nabla_{f_*U} f_*X\right\rangle \\ &\qquad - \nabla_{f_*V}\overbrace{\langle I\!I(U,W), f_*X\rangle}^{=0} + \left\langle I\!I(U,W), \nabla_{f_*V} f_*X\right\rangle \\ &\quad = \langle f_*\left({}^{\Sigma}R(U,V)W\right), f_*X\rangle - \langle I\!I(V,W), I\!I(U,X)\rangle \\ &\qquad + \langle I\!I(U,W), I\!I(V,X)\rangle . \end{aligned}$$

∎

Since $I\!I$ is neither a tensor field on $\Sigma$ nor a tensor field of $M$, the covariant derivative of $I\!I$ is not defined a priori. However, it is easy to check that for any vectors $u, v, w \in T_x\Sigma$ and any vector fields $U, V, W$ with $U_x = u, V_x = v, W_x = w$ the the expression

$$\left(\nabla_{f_*W} I\!I\right)(U,V) = \nabla_{f_*W}(I\!I(U,V)) - I\!I(\boldsymbol{\nabla}_W U, V) - I\!I(U, \boldsymbol{\nabla}_W V)$$

depends only on the values of $u, v, w$. This justifies to call $\nabla I\!I$ as defined above the *covariant derivative of* $I\!I$.

**Proposition 4.4.2 (Codazzi equation).** *Let $\Sigma$ be a non-degenerate immersed submanifold of $(M, g)$ and let $U, V, W \in \mathcal{T}_0^1(\Sigma)$. Then we have*

$$(R(f_*U, f_*V)f_*W)^{\perp} = \left(\nabla_{f_*U} I\!I\right)^{\perp}(V,W) - \left(\nabla_{f_*V} I\!I\right)^{\perp}(U,W).$$

*Proof.* Let $N$ be a vector field along $\Sigma$ which is orthogonal to $T\Sigma$.

$$
\begin{aligned}
&\langle R(f_*U, f_*V)f_*W, N\rangle \\
&\quad= \left\langle \nabla_{f_*U}\nabla_{f_*V} f_*W, N\right\rangle - \left\langle \nabla_{f_*V}\nabla_{f_*U} f_*W, N\right\rangle \\
&\qquad - \left\langle \nabla_{f_*[U,V]} f_*W, N\right\rangle \\
&\quad= \left\langle \nabla_{f_*U} f_*\left(\boldsymbol{\nabla}_V W\right), N\right\rangle + \left\langle \nabla_{f_*U} I\!I(V,W), N\right\rangle \\
&\qquad - \left\langle \nabla_{f_*V} f_*\left(\boldsymbol{\nabla}_U W\right), N\right\rangle - \left\langle \nabla_{f_*V} I\!I(U,W), N\right\rangle \\
&\qquad - \langle I\!I([U,V],W), N\rangle \\
&\quad= \langle I\!I(U, \boldsymbol{\nabla}_V W), N\rangle + \left\langle \nabla_{f_*U} I\!I(V,W), N\right\rangle \\
&\qquad - \langle I\!I(V, \boldsymbol{\nabla}_U W), N\rangle - \left\langle \nabla_{f_*V} I\!I(U,W), N\right\rangle \\
&\qquad - \langle I\!I(\boldsymbol{\nabla}_U V, W), N\rangle + \langle I\!I(\boldsymbol{\nabla}_V U, W), N\rangle \\
&\quad= \left\langle \left(\nabla_{f_*U} I\!I\right)(V,W), N\right\rangle - \left\langle \left(\nabla_{f_*V} I\!I\right)(U,W), N\right\rangle .
\end{aligned}
$$

The assertion follows since $N$ was an arbitrary vector field with values in $(T\Sigma)^{\perp}$. ∎

For some purposes the shape tensor is too complex and also too rich in information. In order to obtain a simpler geometrical quantity one can average at a given point $x \in \Sigma$ the shape tensor over all direction in the submanifold.

Consider a $k$-dimensional Riemannian submanifold $\Sigma \subset M$. We can then identify the set of directions in $T_x\Sigma$ with the unit sphere $S^{k-1} = \{v \in T_x\Sigma \,|\, (f^*g)_x(v,v) = 1\}$. This set is a compact Riemannian submanifold of the Euclidean space $(T_x\Sigma, (f^*g)_x)$. We denote by $\mu_{S^{k-1}}$ the volume form of $S^{k-1}$ considered as submanifold of $(T_x\Sigma, (f^*g)_x)$. A good definition for the average of $I\!I$ over the directions in $T_x\Sigma$ is then

$$
\text{Average} = \frac{1}{\operatorname{vol}(S^{k-1})} \int_{S^2} I\!I_x(\cdot,\cdot)\mu_{S^2},
$$

where the vector space valued integral is defined as in Remark 2.5.2. This method would not work for pseudo-Riemannian manifolds which are not Riemannian since in this case the set of directions does not correspond to a compact pseudo-Riemannian submanifold of $T_x\Sigma$.

We will now calculate the average in the Riemannian case. This leads to a formula which can be straightforwardly generalized to arbitrary pseudo-Riemannian submanifolds. For simplicity we only consider a 3-dimensional submanifold $\Sigma \subset M$. The general case is analogous but calculationally more elaborate. Recall from Example 4.0.1 that we can parametrize a dense open subset of $S^2$ using the chart map $\varphi$ given by

$$\varphi^{-1}\colon (-\pi,\pi)\times(0,2\pi)\to S^2, \quad (\theta,\phi)\mapsto \begin{pmatrix}\cos\phi\cos\theta\\ \sin\phi\cos\theta\\ \sin\theta\end{pmatrix},$$

where $T_x\Sigma$ is identified with $\mathbb{R}^3$ via an orthonormal basis $\{e_1,e_2,e_3\}$. Denoting the induced metric on $S^2$ by $g_{S^2}$ we have $(g_{S^2})_{\theta\theta}=(g_{S^2})(\partial_\theta,\partial_\theta)=1$, $(g_{S^2})_{\theta_\varphi}=0$, $(g_{S^2})_{\phi\phi}=\cos^2\theta$. Hence Lemma 4.2.3 implies $\mu_{S^2}=|\cos\theta|\mathrm{d}\theta\wedge\mathrm{d}\phi$. Let $e_4,\dots,e_n$ be an orthonormal basis of $(T_x\Sigma)^\perp$. There are bilinear forms $I\!I^i$ with

$$I\!I_x(v,w)=\sum_{i=4}^{n} I\!I^i(v,w)e_i$$

for all $v,w\in T_xM$. Since the volume of $S^2$ is given by $\int_{S^2}\mu_{S^2}=\frac{4}{3}\pi$ we obtain for the average of $I\!I_x$

$$\begin{aligned}
&\frac{3}{4\pi}\sum_{i=4}^{n}\int_{S^2} I\!I^i(\cdot,\cdot)\mu_{S^2}\,e_i\\
&\quad=\frac{3}{4\pi}\sum_{i=4}^{n}\int_{-\pi/2}^{\pi/2}\int_0^{2\pi} I\!I^i\left(\begin{pmatrix}\cos\phi\cos\theta\\ \sin\phi\cos\theta\\ \sin\theta\end{pmatrix},\begin{pmatrix}\cos\phi\cos\theta\\ \sin\phi\cos\theta\\ \sin\theta\end{pmatrix}\right)\cos(\theta)\mathrm{d}\phi\mathrm{d}\theta\,e_i\\
&\quad=\frac{3}{4\pi}\sum_{i=4}^{n}\int_{-\pi/2}^{\pi/2}\int_0^{2\pi}\Big(\cos^2\phi\cos^2\theta I\!I^i_{11}+2\cos\phi\sin\phi\cos^2\theta I\!I^i_{12}\\
&\qquad+2\cos\phi\cos\theta\sin\theta I\!I^i_{13}+\sin^2\phi\cos^2\theta I\!I^i_{22}\\
&\qquad+2\sin\phi\cos\theta\sin\theta I\!I^i_{23}+\sin^2\theta I\!I^i_{33}\Big)\cos(\theta)\mathrm{d}\phi\mathrm{d}\theta\,e_i\\
&\frac{3}{4\pi}\sum_{i=4}^{n}\int_{-\pi/2}^{\pi/2}\left(\cos^3\theta\pi I\!I^i_{11}+\cos^3\theta\pi I\!I^i_{22}+2\cos\theta\sin^2\theta\pi I\!I^i_{33}\right)\cos\theta\mathrm{d}\theta\,e_i\\
&\quad=\sum_{i=4}^{n}\left(I\!I^i_{11}+I\!I^i_{22}+I\!I^i_{33})\right)e_i\\
&\quad=\sum_{j=1}^{3} f^*g(e_j,e_j)I\!I(e_j,e_j).
\end{aligned}$$

This justifies the following definition.

**Definition 4.4.2.** *Let $f\colon\Sigma\to M$ be a non-degenerate immersed submanifold of the pseudo-Riemannian manifold $(M,g)$. The* mean curvature vector field *$H$ is defined by*

$$H_x=\frac{1}{\dim(\Sigma)}\sum_{i=1}^{\dim(\Sigma)} f^*g(e_i,e_i)I\!I(e_i,e_i),$$

*where $\{e_1,\dots,e_{\dim(\Sigma)}\}$ is an orthonormal basis of $T_xM$.*

It is easy to see that $H_x$ does not depend on the choice of orthonormal basis. The mean curvature vector field plays a prominent role in the investigation of black holes (cf. Chap. 9) and is closely linked to the theory of minimal surfaces (cf. Lemma 4.4.8 below). The normalisation factor $\frac{1}{\dim(\Sigma)}$ is more common in the mathematical literature than the more logical alternative 1. Since unusual normalisations are even worse then bad normalisations we have retained this factor.

**Definition 4.4.3.** *Let $\Sigma$ be a non-degenerate immersed submanifold of $(M, g)$. A (local) vector field $\mathfrak{n}\colon \Sigma \to T\Sigma^{\perp}$ along $\Sigma$ with $\langle \mathfrak{n}, \mathfrak{n}\rangle = \pm 1$ is called a* normal vector field *or simply a* normal

*The* shape operator $S_{\mathfrak{n}} : T_x\Sigma \to T_x\Sigma$ *of $\Sigma$ associated with $\mathfrak{n}$ is defined by $\langle S_{\mathfrak{n}} u, v\rangle = \langle I\!I(u,v), \mathfrak{n}\rangle$.*

*Denote by $\pi_{\mathfrak{n}}\colon (T_x\Sigma)^{\perp} \to \mathbb{R}\mathfrak{n}$ the orthogonal projection. Then the* second fundamental form $k_{\mathfrak{n}}$ *corresponding to $\mathfrak{n}$ is defined by $\pi_{\mathfrak{n}}(I\!I(u,v)) = k(u,v)\mathfrak{n}$.*

**Definition 4.4.4.** *Let $\Sigma$ be a non-degenerate immersed hypersurface of $(M, g)$. The shape operator and second fundamental form of a non-degenerate hypersurface are simply denoted by $S$ and $k$.*

The term *second fundamental form* comes from the theory of hypersurfaces in Euclidean space $(\mathbb{R}^3, \langle\cdot,\cdot\rangle_{\mathbb{R}^3})$ which predates Riemannian geometry. The normal space of a non-degenerate hypersurface $\Sigma$ is 1-dimensional. Therefore there exists an (up to sign) unique normal vector field $\mathfrak{n}\colon \Sigma \to TM$ along $\Sigma$. Hence even without choosing a normal, $k$ and $S$ are uniquely determined up to sign. In Euclidean geometry, the induced metric $g = f^* \langle\cdot,\cdot\rangle_{\mathbb{R}^3}$ is called *first fundamental form* since it allows to determine fundamental geometrical quantities such as angles and lengths. The second fundamental form $k$ is another $\binom{0}{2}$-tensor field. Since the Gauß equation reduces to

$$\left\langle {}^{\Sigma}R(u,v)w, t\right\rangle = k(u,t)k(v,w) - k(u,w)k(v,t),$$

the curvature of the surface $\Sigma$ is completely determined by $k$. Moreover, $k$ also determines the shape tensor and therefore how a hypersurface is curved in space. Hence $k$ is indeed a geometrically fundamental quantity which justifies the name *second fundamental form.*

**Lemma 4.4.6.** *Let $\Sigma$ be a non-degenerate hypersurface with normal $\mathfrak{n}$. Then the shape operator is given by $Su = -\nabla_{f_* u}\mathfrak{n}$ and the second fundamenal form by $k(u,v) = -\left\langle \nabla_{f_* u}\mathfrak{n}, f_* v\right\rangle \langle \mathfrak{n}, \mathfrak{n}\rangle$. The shape operator $S$ is self-adjoint and the second fundamental form $k$ is symmetric.*

*Proof.* Let $V$ be a vector field with $V_x = v$. Then we have $\langle Su, f_* v\rangle = \langle I\!I(u,v), \mathfrak{n}\rangle = \left\langle \nabla_{f_* u} f_* V, \mathfrak{n}\right\rangle = -\left\langle f_* v, \nabla_{f_* u}\mathfrak{n}\right\rangle$. The normalization

$$\langle \mathfrak{n}, \mathfrak{n} \rangle = \pm 1$$

implies $\left\langle \nabla_{f_* u} \mathfrak{n}, \mathfrak{n} \right\rangle = 0$ and therefore the first assertion. For the second fundamental form we calculate

$$k(u, v) = \langle I\!I(u, v), \mathfrak{n} \rangle \langle \mathfrak{n}, \mathfrak{n} \rangle = -\left\langle \nabla_{f_* u} \mathfrak{n}, f_* v \right\rangle \langle \mathfrak{n}, \mathfrak{n} \rangle .$$

The self-adjointness of $S$ and the symmetry of $k$ follow from $I\!I(u, v) = I\!I(v, u)$. ∎

In analogy to the mean curvature vector field one can introduce the average of the shape operator.

**Definition 4.4.5.** *Let $f \colon \Sigma \to M$ be a non-degenerate immersed hypersurface. Then the* mean curvature *is given by $H(x) = \frac{1}{n-1} \operatorname{tr}(k_x)$.*

The most important mathematical application of this concept arises in the theory of minimal submanifolds.

**Lemma 4.4.7.** *Let $(M, g)$ be an oriented pseudo-Riemannian manifold and $\Sigma \subset M$ a pseudo-Riemannian hypersurface in $M$ with normal $\mathfrak{n}$. Then the volume form of $\Sigma$ is given by*

$$\mu_\Sigma = \mathfrak{n} \lrcorner \mu_M$$

*(which implicitely defines an orientation of $\Sigma$).*

*Proof.* It follows directly from Definition 4.2.1 that either $\mathfrak{n} \lrcorner \mu_M$ or $-\mathfrak{n} \lrcorner \mu_M$ is the volume form of $\Sigma$. ∎

*The rest of this section is an illustration of the concept of mean curvature primarily directed to mathematicians.*

Let $(M, g)$ be an oriented, Riemannian manifold and consider a compact submanifold $B \subset M$ of co-dimension 2. We denote the space of all smooth hypersurfaces $\Sigma \subset M$ whose boundary is $B$ by $\mathfrak{M}(B)$. Then the question naturally arises which $\Sigma \in \mathfrak{M}(B)$ has minimal volume $\operatorname{vol}(\Sigma) = \int_\Sigma \mu_\Sigma$. In general, there may not be a minimizing hypersurface or it may not be unique. Nevertheless, it is relatively easy to derive a necessary condition any minimizing hypersurface must satisfy.

Assume that $\Sigma$ is a minimizing hypersurface and let $U$ be a vector field such that $U(x) = h(x)\mathfrak{n}(x)$ for some function $h \colon \Sigma \to \mathbb{R}$. If $h(x) = 0$ for all $x \in B$ then the flow $F_t$ of $U$ generates a smooth 1-parameter family $\Sigma_t = F_t(\Sigma)$ of hypersurfaces in $\mathfrak{M}(B)$. Hence a necessary condition that $\Sigma$ has minimal volume is given by $\left(\frac{\mathrm{d}}{\mathrm{d}t}\right)_{t=0} \int_{F_t(\Sigma)} \mu_{F_t(\Sigma)} = 0$.

**Lemma 4.4.8.** *Let $(M,g)$ be an oriented pseudo-Riemannian manifold and $\Sigma$ be an oriented non-degenerate hypersurface with normal $\mathfrak{n}$. Let $\tilde{\mathfrak{n}}$ be a vector field in a neighbourhood of $\Sigma$ which satisfies*

*(i)* $\mathfrak{n}(x) = \tilde{\mathfrak{n}}(x)$ *for all* $x \in \Sigma$ *and*
*(ii)* $\langle \tilde{\mathfrak{n}}, \tilde{\mathfrak{n}} \rangle = \pm 1$

*and $h$ be a function and $U = h\tilde{\mathfrak{n}}$. For the flow $F_t$ of $U$ we have*

$$\left(\frac{\mathrm{d}}{\mathrm{d}t}\right)_{t=0} \int_{F_t(\Sigma)} \mu_{F_t(\Sigma)} = (n-1) \int_\Sigma \eta_\Sigma h H \mu_\Sigma,$$

*where $H$ is the mean curvature of $\Sigma$ and $\eta_\Sigma = \langle \mathfrak{n}, \mathfrak{n} \rangle \in \{-1, 1\}$.*

*Proof.* Since all objects are smooth we can interchange differentiation with respect to $t$ and integration over $\Sigma$ and obtain, using Lemma 2.5.2,

$$\begin{aligned}
\left(\frac{\mathrm{d}}{\mathrm{d}t}\right)_{t=0} \int_{F_t(\Sigma)} \mu_{F_t(\Sigma)} &= \left(\frac{\mathrm{d}}{\mathrm{d}t}\right)_{t=0} \int_\Sigma {F_t}^*(\mu_\Sigma) = \int_\Sigma \left(\frac{\mathrm{d}}{\mathrm{d}t}\right)_{t=0} {F_t}^* \mu_\Sigma \\
&= \int_\Sigma \pounds_U(\mu_\Sigma) = \int_\Sigma \pounds_U(\mathfrak{n} \lrcorner \mu_M) \\
&= \int_\Sigma [U, \mathfrak{n}] \lrcorner \mu_M + \mathfrak{n} \lrcorner \pounds_U \mu_M \\
&= \int_\Sigma ([h\tilde{\mathfrak{n}}, \mathfrak{n}] \lrcorner \mu_M + \operatorname{div}(h\tilde{\mathfrak{n}}) \mathfrak{n} \lrcorner \mu_M) \\
&= \int_\Sigma (-\mathrm{d}h(\mathfrak{n}) + \operatorname{div}(h\tilde{\mathfrak{n}})) \mathfrak{n} \lrcorner \mu_M = \int_\Sigma h \operatorname{div}(\tilde{\mathfrak{n}}) \mu_\Sigma.
\end{aligned}$$

Let $\{E_1, \ldots E_n\}$ be an orthonormal frame with $E_n = \tilde{\mathfrak{n}}$ and $\theta^1, \ldots, \theta^n$ be the dual frame. Then we get

$$\begin{aligned}
\operatorname{div}(\tilde{\mathfrak{n}}) &= \sum_{a=1}^{n} \theta^a (\nabla_{E_a} \tilde{\mathfrak{n}}) = \sum_{a=1}^{n-1} \theta^a (\nabla_{E_a} \tilde{\mathfrak{n}}) + \theta^n (\nabla_{E_n} E_n) \\
&= -\sum_{a=1}^{n-1} \theta^a (S E_a) \pm \overbrace{\left\langle E_n, \nabla_{E_n} E_n \right\rangle}^{=0} \\
&= (n-1) \langle \mathfrak{n}, \mathfrak{n} \rangle H.
\end{aligned}$$

∎

Since for each function $h$ with $h_{|B} = 0$ we obtain a 1-parameter family of hypersurfaces $\Sigma_t \in \mathfrak{M}(B)$ with $\Sigma_0 = \Sigma$ a smooth hypersurface can only extremize volume if $H = 0$. For if there would be a point $x \in \Sigma \setminus \{B\}$ with $H(x) \neq 0$ then there would exist a neighbourhood $\mathcal{U}_x$ of $x$ in $\Sigma$ and a function $h$ with

(i) $h(y) = 0$ for all $y \in B \cup (\Sigma \setminus \mathcal{U}_x)$,
(ii) $h(y)H(y) \geq 0$ for all $y \in B \cup \Sigma$,
(iii) $h(x)H(x) > 0$.

This would imply $\int_\Sigma \eta_\Sigma h H \mu_\Sigma \neq 0$ in contradiction to the extremality of $\mathrm{vol}(\Sigma_0)$. The equation $H = 0$ can be understood as a differential equation for a function $f$ which describes the hypersurface $\Sigma$. It is a classical problem in differential geometry to solve this equation. A comprehensivive monograph on this subject (for $M = \mathbb{R}^3$) is (Nitsche 1975).

### 4.4.1 Hyperquadrics

*In this section we study the simplest non-trivial class of hypersurfaces of $\mathbb{R}^n$. These examples will be used in Chaps. 7 and 6.*

Let

$$\eta_\nu : \mathbb{R}^n \times \mathbb{R}^n \to \mathbb{R}, \qquad (x, y) \mapsto -\sum_{j=1}^{\nu} x^j y^j + \sum_{k=\nu+1}^{n} x^k y^k.$$

be the standard pseudo-scalar product of index $\nu$.

**Proposition 4.4.3.** *The pseudo-Riemannian manifold $(\mathbb{R}^n \setminus \{0\}, \eta_\nu)$ is foliated by hypersurfaces $\mathrm{Quad}_\nu^{n-1}(c) := \{x \in \mathbb{R}^n \setminus \{0\} \,|\, \eta_\nu(x,x) = c\}$, where $c \in \mathbb{R}$. The hypersurface $\mathrm{Quad}_\nu^{n-1}(c)$ is non-degenerate if and only if $c \neq 0$.*

These hypersurfaces $\mathrm{Quad}_\nu^{n-1}(c)$ are called *hyperquadrics.*

*Proof.* It is clear that every $x \in \mathbb{R}^n \setminus \{0\}$ lies in exactly one subset $\mathrm{Quad}_\nu^{n-1}(c)$. The set $\mathrm{Quad}_\nu^{n-1}(c)$ is the zero set of $x \mapsto f_c(x) = \eta_\nu(x,x) - c$. Since $\mathrm{d}f = -\sum_{j=1}^{\nu} x^j \mathrm{d}x^j + \sum_{k=\nu+1}^{n} x^k \mathrm{d}x^k$ does not vanish unless $x = 0$ the map $f$ has constant rank in $\{x \in \mathbb{R}^n \,|\, x \neq 0\}$. Hence Proposition 2.1.1 implies that $\mathrm{Quad}_\nu^{n-1}(c)$ is a hypersurface of $\mathbb{R}^n \setminus \{0\}$. The tangent space at $x$ is given by $\{v \in \mathbb{R}^n \,|\, \eta_\nu(x,v) = 0\}$.

Assume that $c = 0$. Then we have $\eta_\nu(x,x) = 0$ and therefore $x \in T_x\mathrm{Quad}_\nu^{n-1}(c)$. Since for any tangent vector $v$ we have $\eta_\nu(x,v) = 0$ the induced metric is degenerate.

Assume now that $c \neq 0$. Suppose there was a vector $w \in T_x\mathrm{Quad}_\nu^{n-1}(c)$ with $\eta_\nu(w,v) = 0$ for all $v \in T_x\mathrm{Quad}_\nu^{n-1}(c)$. Since we have also $\eta_\nu(x,w) = 0$, this would imply $\eta_\nu(w,y) = 0$ for all $y \in \mathbb{R}^n$. This is impossible since $\eta_\nu$ is non-degenerate. ∎

**Lemma 4.4.9.** *Let $c \neq 0$. The map*

$$\iota : \mathrm{Quad}_\nu^{n-1}(c) \to \mathrm{Quad}_{n-\nu}^{n-1}(-c),$$
$$(x^1, \ldots, x^\nu, x^{\nu+1}, \ldots, x^n) \to (x^{\nu+1}, \ldots, x^n, x^1, \ldots, x^\nu)$$

*is an anti-isometry.*

*Proof.* The hypersurfaces $\mathrm{Quad}_\nu^{n-1}(c)$ and $\mathrm{Quad}_{n-\nu}^{n-1}(-c)$ coincide since

$$\eta_\nu(x,x) = -\eta_{n-\nu}(\iota(x),\iota(x))$$

for all $x \in \mathbb{R}^n$. The map $\iota$ is an anti-isometry since for each pair of vectors $u, v \in T_x\mathrm{Quad}_\nu^{n-1}(c) = T_x\mathrm{Quad}_{n-\nu}^{n-1}(-c)$ we have $\eta_\nu(v,w) = -\eta_{n-\nu}(\iota_* v, \iota_* w)$. ∎

Lemma 4.4.9 implies that it is possible to restrict attention to those hypersurfaces $\mathrm{Quad}_\nu^{n-1}(c)$ with $c \geq 0$.

**Definition 4.4.6.** *The* pseudo-sphere *of dimension $n$, index $\nu$, and radius $r$ is the hyperquadric $S_\nu^{n-1}(r) = \mathrm{Quad}_\nu^{n-1}(r^2)$. If $\nu = 0$ we write $S^{n-1}(r)$ instead of $S_0^{n-1}(r)$ and $S^{n-1}$ instead of $S^{n-1}(1)$.*

**Lemma 4.4.10.** *The pseudo-sphere $S_\nu^{n-1}(r)$ is diffeomorphic to $\mathbb{R}^\nu \times S^{n-1-\nu}$, where $S^k$ denotes the $k$-dimensional unit sphere.*

*Proof.* The diffeomorphism $m_r : \mathbb{R}^n \to \mathbb{R}^n$, $x \mapsto \frac{1}{r}x$ maps $S_\nu^{n-1}(r)$ onto $S_\nu^{n-1}(1)$. It is therefore sufficient to prove the lemma for $r = 1$. Consider the map $f\colon \mathbb{R}^\nu \times S^{n-1-\nu} \to \mathbb{R}^n$, $(y,z) \mapsto (y, \sqrt{1+|y|^2}\, z)$. The equation $\eta_\nu(f(y,z), f(y,z)) = 1$ implies that the image of $f$ is a subset of $S_\nu^{n-1}(1)$. The map $f$ has also a smooth inverse map, $f^{-1}(y,\tilde{z}) = (y, \sqrt{1+|y|^2}^{-1}\, \tilde{z})$, hence it is in fact a diffeomorphism onto $S_\nu^{n-1}(1)$. ∎

**Lemma 4.4.11.** *Let $c \neq 0$. The hyperquadric $\mathrm{Quad}_\nu^{n-1}(c)$ is a manifold of constant curvature $1/c$.*

*Proof.* The shape operator is given by

$$Sv = -\mathrm{D}\left(\frac{x}{\sqrt{|c|}}\right)(v) = -\frac{v}{\sqrt{|c|}},$$

whence the shape tensor reads

$$I\!I(u,v) = -\eta_\nu\left(\frac{u}{\sqrt{|c|}}, v\right)\frac{c}{|c|}\frac{x}{\sqrt{|c|}}$$

and the Gauß equation (Proposition 4.4.1) reduces to

$$\eta_\nu(R(u,v)v,u) = \frac{c}{|c|^2}\left(\eta_\nu(u,u)\eta_\nu(v,v) - \eta_\nu(u,v)^2\right).$$

∎

### 4.4.2 Umbilic and totally geodesic submanifolds

*The investigation of submanifolds is a classical field in differential geometry. Naturally, one concentrates on submanifolds whose shape tensor is of especially simple form since only for these classes one has a chance of at least a partial classification. In this section we collect those elementary definitions and results which should be covered in a course of differential geometry.*

*Readers who are primarily interested in physical aspects can skip this section.*

**Definition 4.4.7.** *Let $f\colon \Sigma \to M$ be an immersed pseudo-Riemannian submanifold of a pseudo-Riemannian manifold $(M, g)$.*

*A point $x \in \Sigma$ is called* umbilic *if there exists a vector $\mathfrak{n} \in T_x M$ with $I\!I(u,v) = f^*g(u,v)\mathfrak{n}$ for all $u, v \in T_x\Sigma$. The submanifold $\Sigma$ is called* (totally) umbilic *if all points in $M$ are umbilic.*

*The submanifold $\Sigma$ is called* totally geodesic *if $I\!I = 0$.*

**Lemma 4.4.12.** *Let $f\colon \Sigma \to M$ be an immersed non-degenerate submanifold of the pseudo-Riemannian manifold $(M, g)$. Then the following statements are equivalent.*

*(i) $\Sigma$ is totally geodesic;*
*(ii) For every curve $\gamma \subset \Sigma$ we have: $\gamma$ is a geodesic of $(\Sigma, f^*g)$ if and only if $f \circ \gamma$ is a geodesic with respect to $(M, g)$.*
*(iii) For every $v \in T_x\Sigma$ we have: If $\gamma$ is the maximal geodesic in $M$ with $\dot\gamma(0) = f_*v$ then there is a $\delta > 0$ such that $\gamma([-\delta, \delta]) \subset f(\Sigma)$*
*(iv) For every curve $\gamma\colon [a,b] \to \Sigma$ and every $v \in T_{\gamma[a]}\Sigma$ we have: The parallel transport of $v$ along $\gamma$ satisfies $f_*\mathbf{P}_\gamma(v) = \mathbf{P}_{f\circ\gamma}(f_*v)$.*

*Proof.* (i) $\Leftrightarrow$ (ii): Recall that $\nabla_{f_*\dot\gamma} f_*\dot\gamma = f_*\boldsymbol{\nabla}_{\dot\gamma}\dot\gamma + I\!I(\dot\gamma,\dot\gamma)$. Clearly, $I\!I = 0$ implies that the curve $\gamma$ is a geodesic if and only if $f \circ \gamma$ is. Conversely, from the symmetry of $I\!I$ we infer that

$$I\!I = 0 \quad \Leftrightarrow \quad I\!I(v,v) = 0 \ \forall v.$$

The assertion follows from the decomposition above since for each $v \in T_x\Sigma$ there is a geodesic $\gamma$ with $\dot\gamma(0) = v$.

(ii) $\Leftrightarrow$ (iii): Trivial.

(i) $\Leftrightarrow$ (iv): Let $\gamma$ be a curve and $v \in T_x\Sigma$. The parallel transport $\mathbf{P}_\gamma(v)$ of $v$ along $\gamma$ satisfies

$$0 = f_*\boldsymbol{\nabla}_{\dot\gamma}\mathbf{P}_\gamma(v) = \nabla_{f_*\dot\gamma} f_*\mathbf{P}_\gamma(v) - I\!I(\dot\gamma, \mathbf{P}_\gamma(v)).$$

Hence the equivalence "(i) $\Leftrightarrow$ (iv)" follows in the same way as "(i) $\Leftrightarrow$ (ii)". ∎

The following lemma shows that totally geodesic submanifolds are characterized by the infinitesimal neighbourhood of a single point.

**Proposition 4.4.4.** *Let $f\colon \Sigma \to M$ and $\tilde{f}\colon \tilde{\Sigma} \to M$ be two totally geodesic submanifolds of the pseudo-Riemannian manifold $(M, g)$. If there are points $x \in \Sigma$, $\tilde{x} \in \tilde{\Sigma}$ with $f_* T_x \Sigma = \tilde{f}_* T_{\tilde{x}} \tilde{\Sigma}$ then there are neighbourhoods $\mathcal{U}$ of $x$ and $\tilde{\mathcal{U}}$ of $\tilde{x}$ with $f(\mathcal{U}) = \tilde{f}(\tilde{\mathcal{U}})$.*

*Proof.* Since $\exp_x$ is a local diffeomeorphism near $0_x \in T_x\Sigma$ there are neighbourhoods $\mathcal{W}$, $\tilde{\mathcal{W}}$ of $x$, $\tilde{x}$ which are swept out by geodesics through $x$, $\tilde{x}$. The image of these geodesics under $f$, $\tilde{f}$ are geodesics of $M$ by Lemma 4.4.12. Since the tangent vectors of these two sets of geodesics at $f(x) = \tilde{f}(\tilde{x})$ each form a neighbourhood of $0 \in f_* T_x \Sigma = \tilde{f}_* T_{\tilde{x}} \tilde{\Sigma}$, there is a neighbourhood $\mathcal{U}$ of $f(x) = \tilde{f}(\tilde{x})$ with $\mathcal{U} \cap f(\mathcal{W}) = \mathcal{U} \cap \tilde{f}(\tilde{\mathcal{W}})$. ∎

### 4.4.3 Warped products

*Many standard spacetimes have a generalized product structure, the warped product structure. To write down curvature expressions for this class in general will save work in Chaps. 6 and 8.*

**Definition 4.4.8.** *Let $(\Sigma, g_\Sigma)$, $(F, g_F)$ be pseudo-Riemannian manifolds of dimensions $n_\Sigma$, $n_F$ and $r\colon \Sigma \to \mathbb{R}^+ \setminus \{0\}$ be a smooth function. Then the pseudo-Riemannian manifold*

$$\left(\Sigma \times F, \pi_\Sigma{}^* g_\Sigma + (r \circ \pi_\Sigma)^2 \, \pi_F{}^* g_\Sigma\right),$$

*where $\pi_\Sigma\colon \Sigma \times F \to \Sigma$ and $\pi_F\colon \Sigma \times F \to F$) are the canonical projections, is called the* warped product *of $\Sigma$ and $F$ with* warping function *$r$.*

We can identify vector fields $X$ on $\Sigma$ (respectively, $V$ on F) with the vector field $\tilde{X}$ satisfying $\pi_{\Sigma*}\tilde{X} = X$, $\pi_{F*}\tilde{X} = 0$ (respectively with $\tilde{V}$ satisfying $\pi_{\Sigma*}\tilde{V} = 0$, $\pi_{F*}\tilde{V} = V$). We call $\tilde{X}$ (respectively, $\tilde{V}$) the *lift* of $X$ (respectively, $V$). In the following, we will make use of this identification and denote both vector fields by the same letter. Further, for any vector field $\xi$ on $\Sigma \times F$ there are unique vector fields $X$ on $\Sigma$ and $V$ on $F$ with $\xi = X + V$.

For every $x \in M$ we denote the submanifold $\Sigma \times \{\pi_F(x)\}$ of $\Sigma \times F$ by $\Sigma_x$ and the submanifold $\{\pi_\Sigma(x)\} \times F$ by $F_x$.

**Lemma 4.4.13.** *Let $X, Y$ be vector fields on $\Sigma \times F$ which are lifts of vector fields on $\Sigma$ and $U, V$ vector fields on $\Sigma \times F$ which are lifts of vector fields on $F$. Then*

*(i)* $\pi_{\Sigma*}\nabla_Y X = \nabla_{\pi_{\Sigma*}Y}\pi_{\Sigma*}X$,
*(ii)* $\pi_{F*}\nabla_Y X = 0$,
*(iii)* $\nabla_X U = \nabla_U X = \mathrm{d}(\ln r)(X)U$,
*(iv)* $\pi_{\Sigma*}\nabla_U V = -\langle U, V\rangle \operatorname{grad}(\ln r)$,

*(v)* $\pi_{F*}\nabla_U V = \nabla_{\pi_{F*}U}\pi_{F*}V$.

*Proof.* These equations can be verified using the Koszul formula (Equation (2.7.7)).

(i), (ii): Since $\pi_\Sigma : \Sigma_x \to \Sigma$ is an isometry, we only have to show that $\langle \nabla_X Y, V\rangle = 0$ for all vector fields $V$ which are tangent to the fibre $F$. From Proposition 2.4.4 we get $(\pi_\Sigma)_*[X,V] = 0$ and $(\pi_F)_*[X,V] = 0$ since $(\pi_\Sigma)_*V = 0$ and $(\pi_F)_*X = 0$. The Koszul formula (2.7.7) reduces therefore to $2\langle \nabla_X Y, V\rangle = -V\langle X,Y\rangle - \langle V,[X,Y]\rangle$. Since $\langle X,Y\rangle$ is a function on $\Sigma$ the first summand vanishes. The second summand vanishes since $[X,Y]$ is tangent to $\Sigma_x$.

(iii): From $[X,U] = 0$ we get $\nabla_U X = \nabla_X U$. The covariant derivative $\nabla_X U$ is tangent to the fibres $F_x$ since $\langle Y, \nabla_X U\rangle = -\langle \nabla_X Y, U\rangle = 0$ for all vector fields $Y$ which are lifts of vector fields on $\Sigma$. The Koszul formula implies

$$\begin{aligned} 2\langle \nabla_X U, V\rangle &= X\langle U,V\rangle = X(r^2 g_F(U,V)) = 2r\mathrm{d}r(X)g_F(U,V) \\ &= \frac{2}{r}\mathrm{d}r(X)\langle U,V\rangle . \end{aligned}$$

(iv): The equation follows from

$$\begin{aligned} \langle \nabla_U V, X\rangle &= -\langle V, \nabla_U X\rangle = -\left\langle V, \frac{1}{r}\mathrm{d}r(X)U\right\rangle \\ &= -\frac{1}{r}\langle U,V\rangle \langle \mathrm{grad}(r), X\rangle . \end{aligned}$$

(v): For each $x \in M$ the fibre $F_x$ is a submanifold of $M$ whose induced metric $g_{|F}$ is a constant multiple of the metric on $F$, $g_{|F_x} = r^2(x)g_F$. Hence their Levi-Cività connections coincide and the assertion follows. ∎

**Corollary 4.4.1.** *Let $\gamma = (\gamma_\Sigma, \gamma_F)$ be a curve in $\Sigma \times F$. A curve $\gamma$ is a geodesic if and only if*

*(i)* $\nabla_{\dot\gamma_\Sigma}\dot\gamma_\Sigma = \langle \dot\gamma_F, \dot\gamma_F\rangle\, \mathrm{grad}(\ln r)$,

*(ii)* $\nabla_{\dot\gamma_F}\dot\gamma_F = -2\mathrm{d}(\ln r)(\dot\gamma_\Sigma)\dot\gamma_F$.

*Proof.* We can write $\dot\gamma(t) = X(\gamma(t)) + V(\gamma(t))$, where $X, V$ are vector fields such that $X(\gamma(t)$ is tangent to $\Sigma_{\gamma(t)}$ and $V(\gamma(t))$ is tangent to $F_{\gamma(t)}$ at all $t$. Then we have $\nabla_{\dot\gamma}\dot\gamma = \nabla_X X + \nabla_X V + \nabla_V X + \nabla_V V$ and the assertion follows if we project this vector to $T\Sigma_x$ and $TF_x$. ∎

**Lemma 4.4.14.** *Let $X, Y, Z$ be vector fields on $\Sigma \times F$ which are lifts of vector fields on $\Sigma$ and $U, V, W$ be vector fields on $\Sigma \times F$ which are lifts of vector fields on $F$. Then*

*(i)* $\pi_{\Sigma *}R(X,Y)Z = R_\Sigma(\pi_{\Sigma *}X, \pi_{\Sigma *}Y)\pi_{\Sigma *}Z$,
*(ii)* $\pi_{F *}R(X,Y)Z = 0$,
*(iii)* $R(X,Y)U = 0$,
*(iv)* $R(X,U)Y = \frac{1}{r}\nabla\nabla r(X,Y)U$,
*(v)* $R(X,U)V = -\frac{1}{r}\langle U,V\rangle \nabla_X \mathrm{grad}(r)$,
*(vi)* $R(U,V)X = 0$,
*(vii)* $\pi_{\Sigma *}R(U,V)W = 0$,
*(viii)* $\pi_{F *}R(U,V)W = R_F(U,V)W + \frac{1}{r^2}\langle \mathrm{grad}(r), \mathrm{grad}(r)\rangle (\langle U,W\rangle V - \langle V,W\rangle U)$.

*Proof.* Assertions (i) and (ii) follow directly from Lemma 4.4.13 (i) and (ii).

(iii): We may choose $X$, $Y$ such that $[X,Y]=0$. Then

$$\begin{aligned} R(X,Y)U &= \nabla_X\nabla_Y U - \nabla_Y\nabla_X U \\ &= \nabla_X\left(\mathrm{d}\ln r(Y)U\right) - \nabla_Y\left(\mathrm{d}\ln r(X)U\right) \\ &= \left(\nabla\nabla\ln(r)(X,Y) - \mathrm{d}\ln r(\nabla_X Y) + \mathrm{d}\ln(r)(Y)\mathrm{d}\ln(r)(X)\right)U \\ &\quad - \left(\nabla\nabla\ln(r)(Y,X) - \mathrm{d}\ln r(\nabla_Y X) + \right. \\ &\quad \left.\mathrm{d}\ln(r)(X)\mathrm{d}\ln(r)(Y)\right)U \\ &= \mathrm{d}\ln r([X,Y])U = 0. \end{aligned}$$

(iv): Since $[X,U]=0$ we have

$$\begin{aligned} R(X,U)Y &= \nabla_X\nabla_U Y - \nabla_U\nabla_X Y \\ &= \nabla_X\left(\mathrm{d}\ln r(Y)U\right) - \mathrm{d}\ln r(\nabla_X Y)U \\ &= \nabla\nabla\ln r(X,Y)U + \mathrm{d}\ln r(\nabla_X Y)U \\ &\quad + \mathrm{d}\ln r(Y)\nabla_X U - \mathrm{d}\ln r(\nabla_X Y)U \\ &= \left(\nabla\nabla\ln r(X,Y) + \mathrm{d}\ln r(X)\mathrm{d}\ln r(Y)\right)U \\ &= \frac{1}{r}\nabla\nabla r(X,Y)U. \end{aligned}$$

(vi): We can directly calculate

$$\begin{aligned} R(U,V)X &= \nabla_U\nabla_V X - \nabla_V\nabla_U X - \nabla_{[U,V]}X \\ &= \nabla_U(\mathrm{d}\ln r(X)V) - \nabla_V(\mathrm{d}\ln r(X)U) - \mathrm{d}\ln r(X)[U,V] \\ &= \overbrace{\nabla\nabla(\ln r(X))(U,V)}^{1} + \overbrace{\mathrm{d}\ln r(X)\nabla_U V}^{2} - \overbrace{\nabla\nabla(\ln r(X))(V,U)}^{1} \\ &\quad - \overbrace{\mathrm{d}\ln r(X)\nabla_V U}^{2} - \overbrace{\mathrm{d}\ln r(X)[U,V]}^{2} \\ &= 0 \end{aligned}$$

where we have used that $[U,V] = \nabla_U V - \nabla_V U$.

(v): Since $\langle R(X,U)V,W\rangle = -\langle R(V,W)X,U\rangle = 0$, the vector

$$R(X,U)V$$

must be tangent to $\Sigma_x$. The assertion follows from

$$\langle R(X,U)V,Y\rangle = -\langle R(X,U)Y,V\rangle = -\frac{1}{r}\nabla\nabla r(X,Y)\,\langle U,V\rangle\,.$$

(vii): This follows from $\langle R(U,V)W,X\rangle = -\langle R(U,V)X,W\rangle = 0$.

(viii) Observe first that the Levi-Cività connection induced on the fibre $F_x$ equals the Levi-Cività connection of $g_F$ since both metrics differ only by a constant factor $r^2(x)$. The result follows from the Gauß equation (Proposition 4.4.1) since by Lemma 4.4.13 (iv) the shape tensor is given by $I\!I(U,V) = -\langle U,V\rangle\,\mathrm{grad}(\ln r)$. ∎

**Lemma 4.4.15.** *Let $X,Y$ be vector fields on $\Sigma\times F$ which are lifts of vector fields on $\Sigma$ and $U,V$ be vector fields on $\Sigma\times F$ which are lifts of vector fields on $F$. Then*

*(i)* $\mathrm{Ric}(X,Y) = \mathrm{Ric}_\Sigma(\pi_{\Sigma *}X,\pi_{\Sigma *}Y) - \frac{n_F}{r}\nabla\nabla r(X,Y)$,
*(ii)* $\mathrm{Ric}(X,U) = 0$,
*(iii)* $\mathrm{Ric}(U,V) = \mathrm{Ric}_F(U,V) - \left(\frac{\Delta r}{r} + \frac{(n_F-1)}{r^2}\langle \mathrm{grad}(r),\mathrm{grad}(r)\rangle\right)$ $\times\langle U,V\rangle$,
*(iv)* $\mathrm{Scal} = \mathrm{Scal}_\Sigma + \frac{1}{r^2}\mathrm{Scal}_F - \frac{2n_F}{r}\Delta r - \frac{n_F(n_F-1)}{r^2}\langle\mathrm{grad}(r),\mathrm{grad}(r)\rangle$.

*Proof.* Since $\mathrm{Ric}(X,Y) = \mathrm{tr}(R(\cdot,X)Y) = \mathrm{tr}_\Sigma(R(\cdot,X)Y)+\mathrm{tr}_F(R(X,\cdot)Y)$, (i) follows directly from Lemma 4.4.14 (i), (iv) while assertion (ii) is a consequence of Lemma 4.4.14 (iii) and (v). Formula (iii) is implied by Lemma 4.4.14 (v) and (viii) and assertion (iv) is just the metric trace of (i) and (iii). ∎

## 4.5 Isometries and Killing vector fields

p. 184 ↓
[↓ p. 205]

*An isometry is a diffeomorphism which preserves the metric. pseudo-Riemannian manifolds with many isometries are especially simple. The relevance to the theory of space and time comes from the fact that observations indicate that our universe is well approximated by Lorentzian manifolds with many isometries (cf. Chap. 6).*

**Definition 4.5.1.** *Let $(M,g)$ and $(\tilde M,\tilde g)$ be pseudo-Riemannian manifolds. An* isometry *is a diffeomorphism $\phi\colon M\to\tilde M$ which preserves the metric, $(\phi^*\tilde g) = g$. A* local isometry *is a local diffeomorphism $\phi$ such that $(\phi^*\tilde g)_x = g_x$ at all points $x\in M$.*

**Lemma 4.5.1.** *Let $(M,g)$ and $(\tilde M,\tilde g)$ be pseudo-Riemannian manifolds and $\mathcal{U}\subset\tilde M$ be a connected open set. If $\phi,\psi\colon\mathcal{U}\to\tilde M$ are local isometries, then $\phi=\psi$ if and only if there is a point $x\in\mathcal{U}$ with $T_x\phi = T_x\psi$.*

*Proof.* The two isometries clearly coincide on the closed set $\mathcal{V} = \{y \in \mathcal{U} \mid T_y\phi = T_y\psi\}$. Since $\mathcal{V}$ is non-empty and $\mathcal{U}$ is connected, we only need to show that $\mathcal{V}$ is open. Let $y \in \mathcal{V}$ and $\mathcal{W}$ be a normal neighbourhood of $y$. Then for every $z \in \mathcal{W}$ there is a vector $v[z] \in T_yM$ with $z = \exp_y(v[z])$. But this implies $\phi(z) = \phi(\exp_y(v[z])) = \exp_{\phi(y)}(T_y\phi(y[z])) = \exp_{\psi(y)}(T_y\psi(y[z])) = \psi(\exp_y(v[z])) = \psi(z)$. Hence $\phi_{|\mathcal{W}} = \psi_{|\mathcal{W}}$ and therefore $\mathcal{W} \subset \mathcal{V}$. ∎

**Definition 4.5.2.** *A* Killing vector field *is a vector field $\xi$ whose flow defines local isometries.*

*Lorentzian manifold $(M, g)$ is* stationary *in a region $\mathcal{U} \subset M$ if there is a timelike Killing vector field in $\mathcal{U}$. It is* static *in $\mathcal{U}$ if this Killing vector field is orthogonal to spacelike hypersurfaces.*

Clearly, only very special pseudo-Riemannian manifolds can have non-zero Killing vector fields. A simple example is given by a metric which does not depend on one of the coordinates. Then the corresponding Gaußian vector field is a Killing vector field.

**Lemma 4.5.2.** *A vector field $\xi$ is Killing if and only if $\mathcal{L}_\xi g = 0$ if and only if $\nabla\xi^\flat$ is antisymmetric. In this case we have $\nabla\xi^\flat = \frac{1}{2}\mathrm{d}\xi^\flat$.*

*Proof.* The first equivalence is clear since the Lie derivative $\mathcal{L}_\xi$ is the derivative along the integral curves of $\xi$. In order to prove the second equivalence we calculate

$$\begin{aligned}
(\mathcal{L}_\xi g)(U,V) &= \mathcal{L}_\xi \langle U,V\rangle - \left\langle \mathcal{L}_\xi U, V\right\rangle - \left\langle U, \mathcal{L}_\xi V\right\rangle \\
&= \overbrace{\nabla_\xi \langle U,V\rangle}^{\star} - \overbrace{\left\langle \nabla_\xi U, V\right\rangle}^{\star} + \langle \nabla_U \xi, V\rangle - \overbrace{\left\langle U, \nabla_\xi V\right\rangle}^{\star} \\
&\quad + \langle U, \nabla_V \xi\rangle \\
&= \nabla\xi^\flat(U,V) + \nabla\xi^\flat(V,U).
\end{aligned}$$

Here we have used that for a Levi-Cività connection the terms marked with a $\star$ add to zero. It follows that $\xi$ is a Killing vector field if and only if $\nabla\xi^\flat$ is anti-symmetric. Now the assertion follows from $(\mathrm{d}\xi^\flat)_{ab} = 2\nabla_{[a}\xi_{b]}$. ∎ [p. 204 ↓] ↓ p. 251

**Proposition 4.5.1.** *Let $\xi_1, \xi_2$ be Killing vector fields. Then $[\xi_1, \xi_2]$ is also a Killing vector field, i.e., the Killing vector field on a pseudo-Riemannian manifold form a Lie algebra.*

*Proof.* We have only to show that the commutator of two Killing vector fields is a Killing vector field. From Proposition 2.4.3 we know that $\mathcal{L}_{[\xi,\eta]}\psi = [\mathcal{L}_\xi, \mathcal{L}_\eta]\psi$ for any tensor $\psi$. In particular we obtain $\mathcal{L}_{[\xi,\eta]}g = \mathcal{L}_\xi\mathcal{L}_\eta g - \mathcal{L}_\eta\mathcal{L}_\xi g = 0 - 0 = 0$. ∎

**Lemma 4.5.3.** *Let $\xi$ be a Killing vector field and $\gamma$ be a geodesic. Then $\xi_{|\gamma}$ is a Jacobi field and $s \mapsto \langle \xi_{\gamma(s)}, \dot\gamma_{\gamma(s)} \rangle$ is constant.*

*Proof.* Denote the flow of $\xi$ by $F_t$. Since $F_t$ is an isometry for each $t$ and $s \mapsto \gamma(s)$ is a geodesic, the curve $s \mapsto F_t(\gamma(s))$ is also a geodesic. Hence $(s,t) \mapsto F_t(\gamma(s))$ is a variation of geodesics and its deviation vector field, $\frac{\mathrm{d}}{\mathrm{d}t} F_t \gamma(s) = \xi_{\gamma(s)}$, is a Jacobi field. For the second property note that $\nabla_\xi{}^\flat$ is anti-symmetric by Lemma 4.5.2. Hence $\nabla_{\dot\gamma} \langle \xi, \gamma \rangle = \left\langle \nabla_{\dot\gamma} \xi, \dot\gamma \right\rangle + \left\langle \xi, \nabla_{\dot\gamma} \dot\gamma \right\rangle = \nabla \xi^\flat (\dot\gamma, \dot\gamma) = 0.$ ∎

*In the rest of this section we will investigate highly symmetric pseudo-Riemannian manifolds. These results are of independent mathematical interest and will be used in Chaps. 7, 6.*

**Definition 4.5.3.** *A pseudo-Riemannian manifold $(M,g)$ is called* locally symmetric *if* $\nabla R = 0$.

This definition implies that the components of $R$ with respect to a parallelly propagated frame are constant functions.

**Lemma 4.5.4.** *Let $(M,g)$ be a pseudo-Riemannian manifold. It is locally symmetric if and only if for every curve $\gamma$ and all vector fields $U, V, W$ wich are parallelly propagated along $\gamma$ the vector field $R(U,V)W$ is also paralley propagated along $\gamma$.*

*Proof.* The equation $\nabla R = 0$ implies for parallelly transported vector fields $U, V, W$

$$\begin{aligned} \nabla_{\dot\gamma}(R(U,V)W) &= \overbrace{\left(\nabla_{\dot\gamma} R\right)}^{=0}(U,V)W + R(\overbrace{\nabla_{\dot\gamma} U}^{=0}, V)W \\ &\quad + R(U, \overbrace{\nabla_{\dot\gamma} V}^{=0})W + R(U,V)\overbrace{\nabla_{\dot\gamma} W}^{=0} \\ &= 0 \end{aligned}$$

along $\gamma$. Hence $R(U,V)W$ is also parallel along $\gamma$.

Conversely, let $\xi, u, v, w \in T_x M$ and $\gamma$ be a curve with $\dot\gamma(0) = \xi$. Let $U, V, W$ be the parallel propagation of $u, v, w$ along $\gamma$ and assume that the vector field $R(U,V)W$ along $\gamma$ is also parallel. Then the assertion follows from

$$\begin{aligned} \left(\nabla_\xi R\right)(u,v)w &= \overbrace{\nabla_{\dot\gamma(0)}(R(U,V)W)}^{=0} - R(\overbrace{\nabla_{\dot\gamma(0)} U}^{=0}, v)w \\ &\quad - R(u, \overbrace{\nabla_{\dot\gamma(0)} V}^{=0})w - R(u,v)\overbrace{\nabla_{\dot\gamma(0)} W}^{=0} \\ &= 0. \end{aligned}$$

∎

**Theorem 4.5.1.** *Let $(M, g)$ and $(\tilde{M}, \tilde{g})$ be locally symmetric manifolds and $x \in M$, $\tilde{x} \in \tilde{M}$. If there exists a linear isometry $A: T_xM \to T_{\tilde{x}}\tilde{M}$ with $AR(u,v)w = \tilde{R}(Au, Av)Aw$ for all $u, v, w \in T_xM$, then there are neighbourhoods $\mathcal{U}, \tilde{\mathcal{U}}$ of $x, \tilde{x}$ and a unique isometry $\phi: \mathcal{U} \to \tilde{\mathcal{U}}$ with $T_x\phi = A$.*

*Proof.* We only need to prove existence since uniqueness follows from Lemma 4.5.1. We will show that for some normal neighbourhood $\mathcal{U}$ of $x$ the map $\phi: \mathcal{U} \to \tilde{M}$, $y \mapsto \exp_{\tilde{x}} \circ A \circ \exp_x{}^{-1}$ is a local isometry. First note that $\phi$ is well defined if $\mathcal{U}$ is sufficiently small. By Proposition 2.6.5 there is for every $y \in \mathcal{U}$ a unique $w_x \in T_xM$ with $\exp(w_x) = y$. Further, for every $u_y \in T_yM$ there is a unique $\bar{u}_{w_x} \in T_{w_x}(T_xM)$ with $T_{w_x}\exp_x(\bar{u}_{w_x}) = u_y$. Since $\bar{u}_{w_x} \in T_{w_x}T_xM \subset T_{w_x}TM$ there is a vector $\bar{u}_x$ such that $\bar{u}_{w_x} = \left(\frac{\mathrm{d}}{\mathrm{d}t}\right)_{|t=0}(w_x + \bar{u}_x)$. It follows therefore from Proposition 2.9.5 that $\langle u_y, u_y\rangle = \langle J(1), J(1)\rangle$, where $J$ is the unique Jacobi vector field along the curve $\gamma: t \mapsto f(0,t)$ with $J(0) = 0$ and $\nabla_{\dot{\gamma}}J(0) = \bar{u}_x$. From the definition of $\phi$ we get

$$\phi_*(u_y) = T\exp_{\tilde{x}} TAT\exp_x{}^{-1}(u_y) = T\exp_{\tilde{x}} TA(\bar{u}_{w_x}) = T\exp_{\tilde{x}}(A\bar{u}_{Aw_x})$$

and by the same argument as before it follows that there is a Jacobi field $\tilde{J}$ along the geodesic $\tilde{\gamma}: t \mapsto \exp(tAw_x)$ which satisfies $\tilde{J}(0) = 0$, $\nabla_{\dot{\tilde{\gamma}}}J(0) = A\bar{u}_x$, and $\tilde{g}(\phi_*(u_y), \phi_*(u_y)) = \tilde{g}(\tilde{J}(1), \tilde{J}(1))$.

Let $\{E_1, \ldots, E_n\}$ be a parallelly propagated frame along $\gamma$ with $E_1 = \dot{\gamma}$ and let $\{\tilde{E}_1, \ldots, \tilde{E}_n\}$ be the unique orthonormal, parallelly propagated frame along $\tilde{\gamma}$ with $\tilde{E}_i(0) = AE_i$ $(i \in \{1, \ldots, n\})$. With respect to these frames the Jacobi equations for $J$ and $\tilde{J}$ are given by

$$\frac{\mathrm{d}^2 J^i}{\mathrm{d}t^2} + \sum_{k=1}^{n} R^i{}_{1k1}J^k \quad \text{and} \quad \frac{\mathrm{d}^2 \tilde{J}^i}{\mathrm{d}t^2} + \sum_{k=1}^{n} \tilde{R}^i_{1k1}\tilde{J}^k.$$

Here we have used that $\tilde{\gamma}$ is the unique geodesic with $\dot{\tilde{\gamma}}(0) = A\dot{\gamma}(0)$ which implies that $\tilde{E}_1 = \dot{\tilde{\gamma}}$. The functions $R^i{}_{1k1}$ and $\tilde{R}^i_{1k1}$ are each constant by Lemma 4.5.4. Since we assume $AR(u,v)w = \tilde{R}(Au, Av)Aw$ for all $u, v, w \in T_xM$, the definition of our parallel frames implies $R^i{}_{1k1} = \tilde{R}^i_{1k1}$ for all $i, k$. Further, the functions $J^i, \tilde{J}^k$ satisfy $J^k(0) = \tilde{J}^k(0) = 0$ and (by the definition of our frames) $\frac{\mathrm{d}}{\mathrm{d}t}J^k(0) = \frac{\mathrm{d}}{\mathrm{d}t}\tilde{J}^k(0)$. Hence the fundamental theorem for differential equations 2.4.1 implies $J^k(t) = \tilde{J}^k(t)$ for all $k$ and we get

$$\begin{aligned}
\tilde g(\phi_* u_y, \phi_* u_y) = \tilde g(\tilde J(1), \tilde J(1)) &= \sum_{i,k=1}^{n} \tilde J^i(1)\tilde J^k(1)\tilde g(\tilde E_i(1), \tilde E_k(1)) \\
&= \sum_{i,k=1}^{n} \tilde J^i(1)\tilde J^k(1)\tilde g(\tilde E_i(0), \tilde E_k(0)) \\
&= \sum_{i,k=1}^{n} J^i(1)J^k(1)\tilde g(AE_i(0), AE_k(0)) \\
&= \sum_{i,k=1}^{n} J^i(1)J^k(1)g(E_i(0), E_k(0)) = g(J(1), J(1) \\
&= g(u_y, u_y).
\end{aligned}$$

and the assertion follows from the polarization identity

$$g(u,v) = \frac{1}{2}(g(u+v, u+v) - g(u,u) - g(v,v)).$$

∎

**Proposition 4.5.2.** *Let $(M,g)$ and $(\tilde M, \tilde g)$ be pseudo-Riemannian manifolds with constant curvature $c$ and $\tilde c$. They are locally isometric if and only if they have the same dimension and signature and satisfy $c = \tilde c$.*

*Proof.* Observe first that the conditions are necessary.

We show now that a pseudo-Riemannian manifold with constant curvature is necessarily locally symmetric. Let $t, u, v, w \in T_xM$ and $U, V, W$ vector fields which satisfy $U_x = u, V_x = v, W_x = w$ and whose covariant derivatives vanish at $x$. From Proposition 4.3.3 we get

$$\begin{aligned}
(\nabla_t R)(u,v)w &= \nabla_t (R(U,V)W) - R(\nabla_t U, v)w \\
&\quad - R(u, \nabla_t V)w - R(u,v)\nabla_t W \\
&= c\nabla_t(\langle V, W\rangle U - \langle U, W\rangle V) = 0.
\end{aligned}$$

If $(M,g)$ and $(\tilde M, \tilde g)$ have the same dimension and signature then for any $x \in T_xM, \tilde x \in T_{\tilde x}\tilde M$ there exists a linear isometry $A\colon T_xM \to T_{\tilde x}\tilde M$. If $(M,g)$ and $(\tilde M, \tilde g)$ have the same constant curvature then this isometry satisfies $AR(u,v)w = \tilde R(Au, Av)Aw$ for all $u, v, w \in T_xM$ and the assertion follows from Theorem 4.5.1. ∎

**Corollary 4.5.1.** *Let $(M,g)$ be a pseudo-Riemannian manifold with non-zero constant curvature. Then there is a hyperquadric $\mathrm{Quad}_\nu^{n-1}(c)$ $(c \neq 0)$ which is locally isometric to $(M,g)$.*

A global classification of pseudo-Riemannian manifolds with constant curvature is much more difficult (Wolf 1977). The following Lemma indicates that Hyperquadrics have a very large isometry group.

**Lemma 4.5.5.** *Let $c \neq 0$. and $x, y \in \mathrm{Quad}_\nu^{n-1}(c)$. For any pair of orthonormal bases $\{e_1, \ldots, e_n\} \subset T_x\mathrm{Quad}_\nu^{n-1}(c)$ and $\{f_1, \ldots, f_n\} \subset T_y\mathrm{Quad}_\nu^{n-1}(c)$ there is an isometry $\phi\colon \mathrm{Quad}_\nu^{n-1}(c) \to \mathrm{Quad}_\nu^{n-1}(c)$ with $\phi_*(e_i) = f_i$ $(i \in \{1, \ldots, n\})$.*

*Proof.* Let $\bar{\phi}\colon \mathbb{R}^n \to \mathbb{R}^n$ be the linear map which maps $x, e_1, \ldots, e_n$ to $y$, $f_1, \ldots, f_n$. It is an isometry of $(\mathbb{R}^n, \eta_\nu)$ onto itself and therefore also maps $\mathrm{Quad}_\nu^{n-1}(c)$ onto itself. Since $\bar{\phi}_*(e_i) = \phi(e_i) = f_i$, the restriction $\phi$ of $\bar{\phi}$ to $\mathrm{Quad}_\nu^{n-1}(c)$ is the desired isometry. ∎

## 4.6 Length and energy functionals

*In this section we will study the problem of extremizing the length functional and find necessary and sufficient conditions in the Riemannian and the Lorentzian case. Here we will lay the foundation for many surprisingly strong global theorems in differential geometry (e.g. the Theorem of Myers,cf. (O'Neill 1983, theorem 10.24)) and for the singularity theorems in general relativity (cf. Chap. 9).*

*This section is mathematically more involved than the other sections in this chapter and can be skipped on first reading.*

*This section uses material from Sect. 2.9.*

In Riemannian geometry the length of a curve measures how much wire one would need to model the curve in space. It is a fundamental geometrical experience in Euclidean geometry that for any given (not too distant) pair of points there is a curve of shortest length which connects them.

In Lorentzian geometry the length of a causal curve can be interpreted as the proper time an observer needs in order to traverse this world line. Since in special relativity moving clocks are slower (twin paradoxon) one expects that for any two (not too distant), causally related point there is a longest causal curve which connects them.

For other signatures the problem of extremizing length does not lead to non-trivial results (cf. Lemma 4.6.9)

For the discussion in this section it is technically advantageous to widen the class of admissable curves to the *continuous, piecewise smooth* curves. The advantage lies in the fact that in many situations it is much easier to construct a continuous, piecewise smooth curve with certain properties than a smooth curve.

**Definition 4.6.1.** *Let $(M, g)$ be a pseudo-Riemannian manifold and $\gamma\colon [a, b] \to M$ be a continuous, piecewise smooth curve in $M$.*

*Then the length of* $\gamma$ *is defined by* $L(\gamma) := \int_a^b \sqrt{|g(\dot\gamma(t),\dot\gamma(t))|}\,dt$.

This definition makes sense since a piecewise smooth curve has a well defined derivative everywhere but on a set of measure zero. It is independent of the chosen parametrization. In the case of Euclidean space it coincides with the length one would define through the approximation of $\gamma$ by polygons. The following lemma guarantees that there are no repercussions in considering piecewise smooth curves instead of smooth curves.

**Lemma 4.6.1.** *Let* $\gamma\colon [a,c] \to M$ *be a piecewise smooth curve. Then there is a sequence of smoothly immersed curves* $\gamma_i\colon [a,c] \to M$ *which converge pointwise to* $\gamma$ *and satisfy* $\dot\gamma_i(t) \to \dot\gamma(t)$, $\lim_{i\to\infty} L(\gamma_i) = L(\gamma)$.

*Proof.* Assume that $\gamma$ is the concatenation of two smooth curves $\mu\colon [a,b] \to M$ and $\lambda\colon [b,c] \to M$ where $\mu(b) = \lambda(b)$. We choose a coordinate system $(x^1,\dots,x^n)$ such that $\lambda$ is given by $t \mapsto (t,0,\dots,0)$. Let $i_0 \in \mathbb{N}$ such that $b - 2^{-i_0} > a$ and let $t_i = b - 2^{-i}$ where $i > i_0$ and $i \in \mathbb{N}$. By Lemma 2.1.7 there are smooth functions $\varphi_i, \psi_i\colon [a,c] \to [0,1]$ such that

$$\varphi_i(t)\begin{cases} = 1 & \text{for all } t \in [a,t_{i-1}],\\ > \frac12 & \text{for all } t < t_{i+1}],\\ = 0 & \text{for all } t \in [b,c],\end{cases},$$

$$\psi_i(t)\begin{cases} = 0 & \text{for all } t \in [a,t_{i-1}],\\ > \frac12 & \text{for all } t > t_i],\\ = 1 & \text{for all } t \in [b,c],\end{cases}.$$

We define the curve $\mu_i$ with respect to our coordinate system by

$$(\gamma_i)^k(t) = \mu^k(t_{i-1}) + \int_{t_{i-1}}^t \left(\varphi_i(s)\left(\dot\mu^k(s) + c_i^k\psi_i(s)\right) + \delta_1^k\psi_i(s)\right) ds,$$

where the constants $c_i^k$ are determined by the condition $\lambda_(b) = \gamma_i(b)$. Notice that $\mu(t) = \gamma_i(t)$ for $t \le t_{i-1}$ and $\gamma_i(t) = \lambda(t)$ for $t \ge b$. Since $\mu$ is smooth there is a number $c > 0$ such that $|\dot\mu^k(t)| < c$ and $|\mu^k(t) - \lambda(b)| = |\mu^k(t) - \mu(b)| \le c||b-t|$ for all $t \in [a,b]$. Hence using $\varphi_(t)\psi_i(t) > 1/4$ for $t \in [t_i, t_{i+1}]$ we obtain

$$\begin{aligned} |c_i^k|\frac{t_{i+1}-t_i}{4} &\le \left|\int_{t_{i-1}}^b c_i^k\varphi_i(s)\psi_i(s)ds\right| \\ &\le |\lambda^k(b) - \mu^k(t_{i-1}| + \left|\int_{t_{i-1}}^b \varphi_i(s)\dot\mu^k(s)ds\right| \\ &\quad + \left|\int_{t_{i-1}}^b \delta_1^k\psi_i(s)ds\right| \\ &\le (2c+1)(b - t_{i+1}). \end{aligned}$$

From $t_i = b - 2^{-1}$ we get $|c_i^k| \leq 2(2c+1)$. Since $c_i^k$, $\varphi_i(t)$, $\psi_i(t)$ are uniformly bounded with respect to $i$, the curves $\gamma_i$ converge pointwise to $\gamma$ and the lengths of $\gamma_i$ converge to the length of $\gamma$. ∎

**Corollary 4.6.1.** *Assume that $(M, g)$ is a Lorentzian metric and that $\gamma\colon [a, c] \to M$ is piecewise smooth future directed causal curve. Then there is sequence of smoothly immersed timelike curves $\gamma_i\colon [a, c] \to M$ which satisfies $\dot\gamma_i(t) \to \dot\gamma(t)$, $\gamma_i(t) \to \gamma(t)$, and $L(\gamma_i) \to L(\gamma)$.*

*Proof.* We can assume without loss of generality that $\gamma$ is the concatenatuoin of two smooth curves $mu\colon [a, b] \to M$ and $\lambda\colon [b, c] \to M$ with $\mu(b) = \lambda(b)$ such that both $\dot\mu(b)$ and $\dot\lambda(b)$ point into the same future cone. There are sequences of timelike curves $\mu_i$ and $\lambda_i$ which converge to $\mu$ and $\lambda$ and satisfy $\mu_i(b) = \lambda_i(b) = \mu(b)$. For eachs such pair of curves Lemma 4.6.1 provides a sequence $\gamma_{i,j}$ of curves sucht that $\gamma_{i,j}$ converges to the concatenation of $\mu_i$ and $\lambda_i$. Since both, $\lambda_i$ and $\mu_i$ are timelike and future directed so is $\gamma_{i,j}$ for $j$ enough. We can assume without loss of generality that all $\gamma_{i,j}$ are timelike. It follows that the sequence $\{\gamma_{i,i}\}_{i\in\mathbb{N}}$ consists of timelike curves and converges to $\gamma$. ∎

### 4.6.1 Variation of length and energy

In Euclidean space, the shortest curve between two points is the straight line connecting them. In Minkowski space, the longest causal curve between two points $x, y \in I^+(x)$, is also the straight line connecting them.

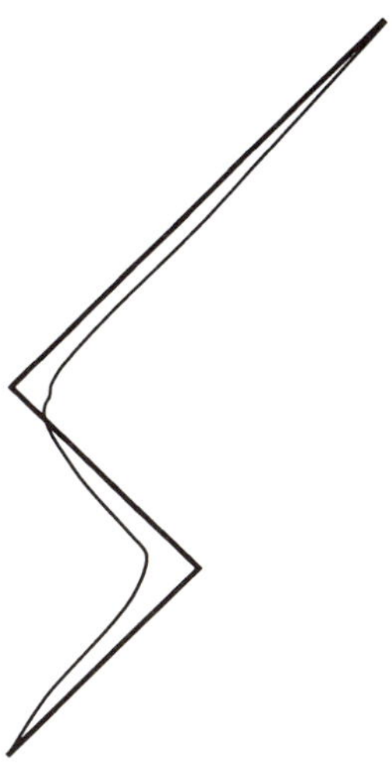

**Fig. 4.6.1.** A broken lightlike geodesic can be smoothed out by a curve of arbitrarily small length

In a general Riemannian manifold "without holes" it is intuitively clear that any two points can be joined by at least one shortest curve.

In a Lorentzian manifold, the infimum over the length of all curves which connect $x$ and $y$ *is always zero* since we can join any two points by a broken lightlike geodesic which then can be smoothed out to give a smooth curve of arbitrarily small length (cf. Fig. 4.6.1 and Corollary 4.6.1). It is also clear that there does not exist a curve of maximal length connecting $x$ and $y$ since we can always choose a spiralling spacelike curve of arbitrarily large length (cf. Fig. 4.6.1). However, we will see below that in many situations there exist curves connecting causally related curves $x$ and $y$ which maximize $L$ in the class of all causal curves.[5]

pseudo-Riemannian manifolds which are neither Riemanian nor Lorentzian do not admit any non-trivial solutions to the length extremizing problem, even if one restricts to spacelike or timelike curves. These arguments will be made precise in Lemma 4.6.9 below.

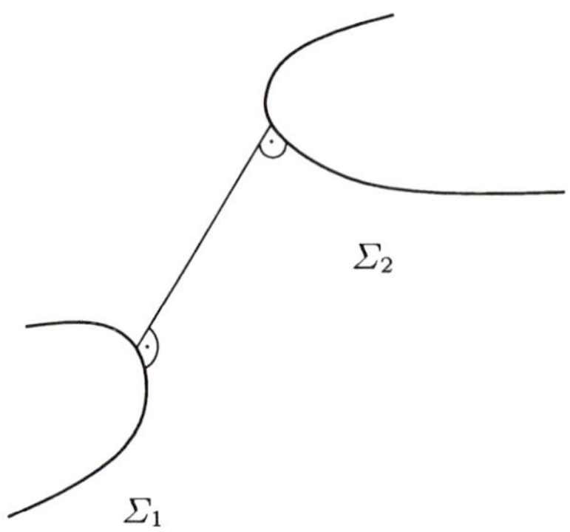


**Fig. 4.6.2.** A curve minimizing the distance between two spacelike submanifolds $\Sigma_1$ and $\Sigma_2$

We will actually investigate the slightly more general problem where the endpoints $x$, $y$ are replaced by submanifolds without boundary $\Sigma_1$, $\Sigma_2$ (cf. Fig. 4.6.2). In order to solve the length extremizing problem in the Riemannian and in the Lorentzian case we will study 1-parameter families of curves $f\colon [a,b] \times (-\epsilon, \epsilon) \to M$, $(s,t) \mapsto f(s,t)$ such that $f(s,a) =\in \Sigma_1$ and $f(s,b) \in \Sigma_2$ for all $s$. If $\gamma$ extremizes the length functional $L$ for all smooth curves[6] which connect $\Sigma_1$ with $\Sigma_2$, then we have $\frac{\mathrm{d}}{\mathrm{d}s}_{|s=0} L(f(s,\cdot))$ for all such 1-parameter families with $\gamma(t) = f(0,t)$. Through the investigation of $\frac{\mathrm{d}^2}{\mathrm{d}s^2}_{|s=0} L(f(s,\cdot))$ we will arrive at sufficient conditions.

**Definition 4.6.2.** *Let $\Sigma_1$, $\Sigma_2$ be submanifolds of $M$ and $\gamma\colon [a,b] \to M$ be a curve which connects $\Sigma_1$ with $\Sigma_2$.*

*A continuous variation $f\colon (-\epsilon,\epsilon) \times [a,b] \to M$, $(s,t) \mapsto f(s,t)$ of $\gamma$ is called* piecewise smooth *if there are numbers $t_1, \ldots, t_k \in (a,b)$ such that $f_{|(-\epsilon,\epsilon)\times[t_i,t_i+1]}$ is smooth, where $t_0 := a, t_{k+1} := b$ and $i \in \{0,\ldots,k\}$.*

[5] Our examples also imply that the length extremizing problem does not have a solution if one restricts to spacelike curves instead of causal curves.

[6] In the Lorentzian case: all smooth, causal curves.

*A (continuous, piecewise smooth) variation $f$ of $\gamma$* connects $\Sigma_1$ with $\Sigma_2$ *if $f(s,a) \in \Sigma_1$ and $f(b,s) \in \Sigma_2$ for all $s \in (-\epsilon, \epsilon)$.*

*We denote the vector field $T_{(s,t)}f(\partial_s)$ along $f$ by $f_s$, the vector field $T_{(s,t)}f(\partial_t)$ along $f$ by $f_t$ (where defined), and call the (piecewise smooth) vector field $\xi(t) := (f_s))_{|s=0}$ along $\gamma$ the* variation vector field.

**Lemma 4.6.2.** *Let $\gamma\colon [a,b] \to M$ be a smooth curve which connects two submanifolds $\Sigma_1$, $\Sigma_2$. For any vector field $\xi$ along $\gamma$ with $\xi(a) \in T_{\gamma(a)}\Sigma_1$, $\xi(b) \in T_{\gamma(b)}\Sigma^2$ there exists a variation $f$ of $\gamma$ which connects $\Sigma_1$ with $\Sigma_2$ and which has variation vector field $\xi$.*

*Proof.* Let $\mu_1 \subset \Sigma_1$ and $\mu_2 \subset \Sigma_2$ be smooth curves with $\mu_1(0) = \gamma(a)$, $\mu_2(0) = \gamma(b)$, $\dot{\mu}_1(0) = \xi(a)$, and $\dot{\mu}_2(0) = \xi(b)$. We can now extend $\xi$ to a vector field $\Xi$ such that $\mu_i$ $(i \in \{1,2\})$ are integral curves of $\Xi$. If $F$ denotes the flow of $\Xi$ we set $f(s,t) = F_s(\gamma(t))$. ∎

If $\gamma\colon [a,b] \to M$ is piecewise smooth then $\dot{\gamma}$ is discontinous at those points where $\gamma$ fails to be smooth. We will therefore need the following technical definition.

**Definition 4.6.3.** *Let $\gamma$ be a continuous, piecewise smooth curve and $V$ be a piecewise smooth vector field along $\gamma$. For each $t_0 \in [a,b]$ we set*

$$\Delta V(t_0) := \lim_{t \to t_0, t > t_0} V(t) - \lim_{t \to t_0, t < t_0} V(t).$$

It is clear that $\Delta V(t_0) = 0$ if and only if $V$ is continuous at $t_0$.

**Lemma 4.6.3 (First variation of arc length).** *Let $\gamma\colon [a,b] \to M$ be a spacelike or timelike, continuous, piecewise smooth curve, let $\eta =$ $\operatorname{sign}(\langle\dot{\gamma},\dot{\gamma}\rangle)$, and $f\colon [-\epsilon,\epsilon] \times [a,b] \to M, (s,t) \mapsto f(s,t)$ be a continuous, piecewise smooth variation of $\gamma$ with variation vector field $\xi$. Denote by $t_1,\ldots,t_k \in (a,b)$ the points where $\gamma$ fails to be smooth. Then the derivative of $L$ with respect to $s$ is given by*

$$\begin{aligned}\left(\frac{\mathrm{d}}{\mathrm{d}s} L(f(s,\cdot))\right)_{|s=0} &= -\eta \int_a^b \left\langle \nabla_{\dot{\gamma}} \left( \frac{\dot{\gamma}}{\sqrt{|\langle \dot{\gamma},\dot{\gamma}\rangle|}} \right), \xi \right\rangle \mathrm{d}t \\ &\quad - \sum_{i=1}^{k} \left\langle \delta \frac{\dot{\gamma}(t_i)}{\sqrt{|\langle \dot{\gamma}(t_i), \dot{\gamma}(t_i)\rangle|}}, \xi(t_i) \right\rangle \\ &\quad - \eta \left\langle \frac{\dot{\gamma}}{\sqrt{|\langle \dot{\gamma},\dot{\gamma}\rangle|}}, \xi \right\rangle \Bigg|_a^b .\end{aligned}$$

*Proof.* If $\epsilon$ is small enough, all curves $f(s,\cdot)$ are either timelike or spacelike, and the integrand $\sqrt{|\langle\langle f_t, f_t\rangle\rangle|}$ is differentiable for all $(s,t)$ where $f$ is differentiable. We can therefore exchange on every smooth piece of $\gamma$ the differentiation and the integration. Now the assertion follows from

$$\frac{\mathrm{d}}{\mathrm{d}s}\sqrt{\eta\langle f_t, f_t\rangle} = \eta\left\langle \frac{f_t}{\sqrt{\eta\langle f_t, f_t\rangle}}, \overset{f}{\nabla}_{\partial_s} f_t \right\rangle = \eta\left\langle \frac{f_t}{\sqrt{\eta\langle f_t, f_t\rangle}}, \overset{f}{\nabla}_{\partial_t} f_s \right\rangle$$
$$= \eta\left\langle \frac{f_t}{\sqrt{\eta\langle f_t, f_t\rangle}}, f_s \right\rangle^{\cdot} - \eta\left\langle \overset{f}{\nabla}_{\partial_t}\left(\frac{f_t}{\sqrt{\eta\langle f_t, f_t\rangle}}\right), f_s \right\rangle.$$

∎

**Corollary 4.6.2.** *Let $(M,g)$ be a Riemannian or Lorentzian manifold, $\Sigma_1$, $\Sigma_2$ be two submanifolds without boundary, and $\gamma$ be a curve connecting $\Sigma_1$ with $\Sigma_2$.*

*(i) If $(M,g)$ is Riemannian and $\gamma$ is shorter than any other (neighbouring) curve connecting $\Sigma_1$ and $\Sigma_2$ then $\gamma$ is a pregeodesic which intersects both $\Sigma_1$, and $\Sigma_2$ orthogonally.*

*(ii) Assume that $\Sigma_1$, $\Sigma_2$ are Riemannian submanifolds or degenerate to a point. If $(M,g)$ is Lorentzian and $\gamma$ a causal curve which is longer than any other causal curve connecting $\Sigma_1$ and $\Sigma_2$ then $\gamma$ is a causal pregeodesic which intersects both $\Sigma_1$, and $\Sigma_2$ orthogonally.*

*Proof.* We prove only (i) since the other case is completely analogous.

We assume first that $\gamma$ is not a pregeodesic. Observe that it is a pregeodesic if and only if $V(t) := \nabla_{\dot\gamma}\left(\frac{\dot\gamma}{\sqrt{|\langle\dot\gamma,\dot\gamma\rangle|}}\right) = 0$ for all $t$. Hence by our assumption there would be a point $t_0 \in (a,b)$ with $V(t_0) \neq 0$. Let $\xi_1$ be a vector field along $\gamma$ such that $\langle V(t_0), \xi_1(t_0)\rangle < 0$. BY continuity there is a neighbourhood $(t_-, t_+)$ of $t_0$ such that $\langle V(t), \xi_1(t)\rangle < 0$ for all $t \in (t_-, t_+)$. Let now $\varphi$ be a smooth, positive function with support in $(t_0, t_+)$ and $\varphi(t_0) \neq 0$. (Such a function exists by Lemma 2.1.7). Then $\langle V(t), \varphi(t)\xi_1(t)\rangle \geq 0$ for all $t \in [a,b]$ and does not vanish in a neighbourhood of $t_0$. Taking a variation with variation vector field $\xi = \varphi\xi_1$ we obtain therefore $\left(\frac{\mathrm{d}}{\mathrm{d}s}L(f(s,\cdot))\right)_{|s=0} < 0$. This implies that there are shorter curves than $\gamma$ in contradiction to our assumption.

We assume now that $\gamma$ is a pregeodesic but does not intersect $\Sigma_1$ orthogonally. Then there is a vector $v \in T_{\gamma(a)}\Sigma_1$ with $\langle\dot\gamma(a), v\rangle < 0$. Let $\xi$ be a variation vector field with $\xi(a) = v$ and $\xi(b) = 0$. Since $\xi(a)$ is tangential to $\Sigma_1$ there is a variation $f$ of $\gamma$ with variation vector field $\xi$ such that $f(s,\cdot)$ connects $\Sigma_1$ with $\Sigma_2$ for all $s$. Again we obtain $\left(\frac{\mathrm{d}}{\mathrm{d}s}L(f(s,\cdot))\right)_{|s=0} < 0$ in contradition to our assumption that $\gamma$ is length minimizing. It follows that $\gamma$ intersects $\Sigma_1$ orthogonally.

Finally observe that the same argument holds equally well for $\Sigma_2$. ∎

The discussion above does not apply to null curves since $L$ involves the square root of $\langle\dot\gamma,\dot\gamma\rangle$. This problem can be avoided if one consideres the *energy* of the curve $\gamma$, $E(\gamma) := \int_a^b \frac{1}{2}\langle\dot\gamma(t)\dot\gamma(t)\rangle\,\mathrm{d}t$.[7]

Unlike the length functional this integral expression *does depend* on the parametrization of the curve. While it is not true that spacelike or timelike curves which extremize $L$ also extremize the energy integral, we will see below that this property almost holds.

**Lemma 4.6.4 (First variation of energy).** *Let $\gamma\colon [a,b]\to M$ be a continuous, piecewise smooth curve and $f\colon[-\epsilon,\epsilon]\times[a,b]\to M, (s,t)\mapsto f(s,t)$ be a continuous, piecewise smooth variation with variation vector field $\xi$. Denote by $t_1,\dots,t_k\in(a,b)$ the points where $\gamma$ fails to be smooth. Then the derivative of $E$ with respect to $s$ is given by*

$$\left(\frac{\mathrm{d}}{\mathrm{d}s}E(f(s,\cdot))\right)_{|s=0} = \int_a^b \left\langle\nabla_{\dot\gamma}\dot\gamma,\xi\right\rangle\mathrm{d}t + \sum_{i=1}^k \langle\delta\dot\gamma(t_i),\xi(t_i)\rangle + \langle\dot\gamma,\xi\rangle\Big|_a^b .$$

*Proof.* Consider a piece of $\gamma$ where it is smooth. The assertion follows from

$$\frac{1}{2}\frac{\mathrm{d}}{\mathrm{d}s}\langle f_t,f_t\rangle = \left\langle f_t,\overset{f}{\nabla}_{\partial_s}f_t\right\rangle = \left\langle f_t,\overset{f}{\nabla}_{\partial_t}f_s\right\rangle = \langle f_t,f_s\rangle^{\cdot} - \left\langle\overset{f}{\nabla}_{\partial_t}f_t,f_s\right\rangle .$$

∎

It follows that a curve which extremizes energy is a geodesic (and not merely a pregeodesic).

We will now derive sufficient conditions for curves to extremize the length between submanifolds without boundary. But first we need a technical lemma.

**Lemma 4.6.5.** *Let $\gamma$ be a spacelike or timelike pregeodesic and denote the orthogonal projection to the orthogonal complement of $\gamma$ by $(\cdot)^\perp$. Then for every vector field $V$ along $\gamma$ the formula $(\nabla_{\dot\gamma}V)^\perp = \nabla_{\dot\gamma}(V^\perp)$ holds.*

*Proof.* The vector field $V$ can be decomposed into its part orthogonal to $\dot\gamma$, $W$, and its part tangent to $\dot\gamma$, $\varphi\dot\gamma$, where $\varphi$ is a smooth function. From $V=\varphi\dot\gamma+W$ and $\langle\dot\gamma,W\rangle=0$ we get

$$(\nabla_{\dot\gamma}V)^\perp = \mathrm{d}\varphi(\dot\gamma)(\dot\gamma)^\perp + \varphi\left(\nabla_{\dot\gamma}\dot\gamma\right)^\perp + \left(\nabla_{\dot\gamma}W\right)^\perp = \left(\nabla_{\dot\gamma}W\right)^\perp ,$$

[7] The name "energy" comes from the fact that in the Riemannian case the integrand is just the kinetic energy of a mass point of mass 1. In the Lorentzian case the integrand has nothing to do with energy.

where we have used that a curve is a pregeodesic if and only if $\nabla_{\dot\gamma}\dot\gamma \parallel \dot\gamma$. The assertion follows now since

$$\left\langle \left(\nabla_{\dot\gamma} W\right), \dot\gamma\right\rangle = \langle W, \dot\gamma\rangle^{\cdot} - \left\langle W, \nabla_{\dot\gamma}\dot\gamma\right\rangle = 0 - 0 = 0$$

implies $\left(\nabla_{\dot\gamma} W\right)^{\perp} = \nabla_{\dot\gamma} W = \nabla_{\dot\gamma}\left(V^{\perp}\right)$. ∎

In the following we will freely interchange $\perp$ and $\nabla_{\dot\gamma}$ when $\gamma$ is a geodesic.

**Lemma 4.6.6 (second variation of arc length).** *Let $\gamma$ be a spacelike or timelike geodesic with $\eta = \langle\dot\gamma, \dot\gamma\rangle \in \{-1, 1\}$ and let $f\colon [-\epsilon, \epsilon] \times [a, b] \to M, (s, t) \mapsto f(s, t)$ be a continuous, piecewise smooth variation of $\gamma$. If we denote by $t_1, \ldots, t_k \in (a, b)$ the points where $f(s, \cdot)$ fails to be smooth then the second derivative of $L \circ f$ with respect to $s$ is given by*

$$\begin{aligned}
\left(\frac{\mathrm{d}^2}{\mathrm{d}s^2} L(f(s,\cdot))\right)_{|s=0} &= \eta \int_a^b \left(\left\langle (\nabla_{\dot\gamma}\xi)^{\perp}, (\nabla_{\dot\gamma}\xi)^{\perp}\right\rangle + \langle R(\xi, \dot\gamma)\xi, \dot\gamma\rangle\right) \mathrm{d}t \\
&\quad + \eta \left\langle \left(\overset{f}{\nabla}_{\partial_s} f_s\right)_{|s=0}, \dot\gamma\right\rangle\Big|_a^b \\
&= -\eta \int_a^b \left\langle \nabla_{\dot\gamma}\nabla_{\dot\gamma}\xi^{\perp} + R(\xi^{\perp}, \dot\gamma)\dot\gamma, \xi^{\perp}\right\rangle \mathrm{d}t \\
&\quad - \eta \sum_{i=1}^{k} \left\langle \Delta(\nabla_{\dot\gamma}\xi)^{\perp}(t_i), \xi^{\perp}(t_i)\right\rangle \\
&\quad + \eta \left\langle \left(\overset{f}{\nabla}_{\partial_s} f_s\right)_{|s=0}, \dot\gamma\right\rangle\Big|_a^b
\end{aligned}$$

*where $\xi$ denotes the variation vector field and $(\cdot)^{\perp}$ the orthogonal projection to $(\dot\gamma)^{\perp}$.*

*Proof.* Using the formula for the first derivative in the proof of Lemma 4.6.3 we obtain

$$\begin{aligned}
&\frac{\mathrm{d}^2}{\mathrm{d}s^2}\sqrt{\eta\langle f_t, f_t\rangle} \\
&= \eta \frac{\mathrm{d}}{\mathrm{d}s}\left\langle \frac{f_t}{\sqrt{\eta\langle f_t, f_t\rangle}}, \overset{f}{\nabla}_{\partial_s} f_t\right\rangle \\
&= -\frac{\eta\cdot\eta}{\sqrt{\eta\langle f_t, f_t\rangle}^{3}}\left\langle f_t, \overset{f}{\nabla}_{\partial_s} f_t\right\rangle\left\langle f_t, \overset{f}{\nabla}_{\partial_s} f_t\right\rangle \\
&\quad + \frac{\eta}{\sqrt{\eta\langle f_t, f_t\rangle}}\left(\left\langle \overset{f}{\nabla}_{\partial_s} f_t, \overset{f}{\nabla}_{\partial_s} f_t\right\rangle + \left\langle f_t, \overset{f}{\nabla}_{\partial_s}\overset{f}{\nabla}_{\partial_s} f_t\right\rangle\right).
\end{aligned} \tag{4.6.3}$$

From $\overset{f}{\nabla}_{\partial_s} f_t = \overset{f}{\nabla}_{\partial_t} f_s$ and

$$\overset{f}{\nabla}_{\partial_s} \overset{f}{\nabla}_{\partial_s} f_t = \overset{f}{\nabla}_{\partial_s} \overset{f}{\nabla}_{\partial_t} f_s = \overset{f}{\nabla}_{\partial_t} \overset{f}{\nabla}_{\partial_s} f_s + R(f_s, f_t) f_s$$

we get $\left\langle \overset{f}{\nabla}_{\partial_s} f_t, \overset{f}{\nabla}_{\partial_s} f_t \right\rangle = \left\langle \overset{f}{\nabla}_{\partial_t} f_s, \overset{f}{\nabla}_{\partial_t} f_s \right\rangle$ and

$$\left\langle f_t, \overset{f}{\nabla}_{\partial_s} \overset{f}{\nabla}_{\partial_s} f_t \right\rangle = \left\langle f_t, \overset{f}{\nabla}_{\partial_t} \overset{f}{\nabla}_{\partial_s} f_s \right\rangle + \langle f_t, R(f_s, f_t) f_s \rangle .$$

Since for every vector field $V$ along $\gamma$ we have $V = V^\perp + \eta \frac{\langle V, f_t \rangle}{\langle f_t, f_t \rangle} f_t$ and therefore $\langle V, V \rangle = \langle V^\perp, V^\perp \rangle + \frac{\langle V, \dot\gamma \rangle^2}{\langle \dot\gamma, \dot\gamma \rangle}$ the second and third summand in Equation (4.6.3) simplify to

$$\frac{\eta}{\sqrt{\eta \langle f_t, f_t \rangle}} \left\langle \left( \overset{f}{\nabla}_{\partial_t} f_s \right)^\perp, \left( \overset{f}{\nabla}_{\partial_t} f_s \right)^\perp \right\rangle .$$

Using the product formula for the term $\left\langle f_t, \overset{f}{\nabla}_{\partial_t} \overset{f}{\nabla}_{\partial_s} f_s \right\rangle$ we finally obtain

$$\begin{aligned} \left( \frac{\mathrm{d}^2}{\mathrm{d}s^2} L(f(s,\cdot)) \right)_{|s=0} = \int_a^b \frac{\eta}{\sqrt{\eta \langle f_t, f_t \rangle}} \Bigg( & \left\langle \left( \overset{f}{\nabla}_{\partial_t} f_s \right)^\perp, \left( \overset{f}{\nabla}_{\partial_t} f_s \right)^\perp \right\rangle \\ & + \left\langle f_t, \overset{f}{\nabla}_{\partial_s} f_s \right\rangle^{\cdot} - \left\langle \overset{f}{\nabla}_{\partial_t} f_t, \overset{f}{\nabla}_{\partial_s} f_s \right\rangle \\ & + \langle f_t, R(f_s, f_t) f_s \rangle \Bigg) \mathrm{d}t \end{aligned}$$

and the first equality in the assertion follows since $f(0,\cdot) = \gamma$ is a geodesic with $\langle \dot\gamma, \dot\gamma \rangle = \eta$.

Since for every smooth piece of the variation $f$ the equation

$$\left\langle \nabla_{\dot\gamma} \xi^\perp, \nabla_{\dot\gamma} \xi^\perp \right\rangle = \left\langle \nabla_{\dot\gamma} \xi^\perp, \xi^\perp \right\rangle^{\cdot} - \left\langle \nabla_{\dot\gamma} \nabla_{\dot\gamma} \xi^\perp, \xi^\perp \right\rangle$$

holds, the second equality follows from an integration by parts using

$$\langle R(\xi, \dot\gamma)\xi, \dot\gamma \rangle = \langle R(\xi^\perp, \dot\gamma)\xi^\perp, \dot\gamma \rangle = \langle R(\xi^\perp, \dot\gamma)\dot\gamma, \xi^\perp \rangle$$

∎

There is an analogous formula for the second variation of the energy integral.

**Lemma 4.6.7 (second variation of energy).** *Let $\gamma$ be a geodesic and $f\colon [-\epsilon,\epsilon]\times[a,b]\to M, (s,t)\mapsto f(s,t)$ be a continuous, piecewise smooth variation of $\gamma$. Denote by $t_1,\ldots,t_k \in (a,b)$ the points where $f(s,\cdot)$ fails to be smooth. Then the second derivative of $E$ with respect to $s$ is given by*

$$\begin{aligned}\left(\frac{\mathrm{d}^2}{\mathrm{d}s^2}E(f(s,\cdot))\right)_{|s=0} &= \int_a^b \left(\left\langle \nabla_{\dot\gamma}\xi, \nabla_{\dot\gamma}\xi\right\rangle + \langle R(\xi,\dot\gamma)\xi,\dot\gamma\rangle\right)\mathrm{d}t \\ &\quad + \left\langle \left(\overset{f}{\nabla}_{\partial_s} f_s\right)_{|s=0}, \dot\gamma \right\rangle \\ &= -\int_a^b \left\langle \nabla_{\dot\gamma}\nabla_{\dot\gamma}\xi + R(\xi,\dot\gamma)\dot\gamma, \xi\right\rangle \\ &\quad - \sum_{i=1}^{k}\left\langle \Delta\nabla_{\dot\gamma}\xi, \xi(t_i)\right\rangle \\ &\quad + \eta\left\langle \left(\overset{f}{\nabla}_{\partial_s} f_s\right)_{|s=0}, \dot\gamma \right\rangle\end{aligned}$$

*Proof.* The first equation follows from

$$\begin{aligned}\frac{1}{2}\frac{\mathrm{d}^2}{\mathrm{d}s^2}\langle f_t(s,\cdot), f_t(s,\cdot)\rangle &= \left\langle \overset{f}{\nabla}_{\partial_s} f_t, \overset{f}{\nabla}_{\partial_s} f_t\right\rangle + \left\langle \overset{f}{\nabla}_{\partial_s}\overset{f}{\nabla}_{\partial_s} f_t, f_t\right\rangle \\ &= \left\langle \overset{f}{\nabla}_{\partial_t} f_s, \overset{f}{\nabla}_{\partial_t} f_s\right\rangle + \left\langle \overset{f}{\nabla}_{\partial_s}\overset{f}{\nabla}_{\partial_t} f_s, f_t\right\rangle \\ &= \left\langle \overset{f}{\nabla}_{\partial_t} f_s, \overset{f}{\nabla}_{\partial_t} f_s\right\rangle + \left\langle \overset{f}{\nabla}_{\partial_t}\overset{f}{\nabla}_{\partial_s} f_s, f_t\right\rangle \\ &\quad + \langle R(f_s,f_t)f_s, f_t\rangle \\ &= \left\langle \overset{f}{\nabla}_{\partial_t} f_s, \overset{f}{\nabla}_{\partial_t} f_s\right\rangle + \left\langle \overset{f}{\nabla}_{\partial_s} f_s, f_t\right\rangle^{\cdot} \\ &\quad - \left\langle \overset{f}{\nabla}_{\partial_s} f_s, \overset{f}{\nabla}_{\partial_t} f_t\right\rangle + \langle R(f_s,f_t)f_s, f_t\rangle .\end{aligned}$$

The second equality follows from an integration by parts exactly as in the proof of Lemma 4.6.6. ∎

**Lemma 4.6.8.** *Let $\Sigma_1$, $\Sigma_2$ be submanifolds of $M$ and $\gamma\colon [a,b]\to M$ be a curve from $\Sigma_1$ to $\Sigma_2$ which intersects both submanifolds orthogonally. Assume that $f$ is a (continuous, piecewise smooth) variation of $\gamma$ which connects $\Sigma_1$ with $\Sigma_2$. Then*

$$\left\langle \left(\overset{f}{\nabla}_{\partial_s} f_s\right)_{|s=0}, \dot\gamma\right\rangle\Bigg|_a^b = \langle I\!I_{\Sigma_2}(\xi(b),\xi(b)), \dot\gamma(b)\rangle - \langle I\!I_{\Sigma_1}(\xi(a),\xi(a)), \dot\gamma(a)\rangle$$

*holds, where* $I\!I_{\Sigma_i}$ *denotes the shape tensor of* $\Sigma_i$.

*Proof.* The assertion follows immediately from the definition of the shape tensor and our assumption $\gamma(a) \perp \Sigma_1$, $\gamma(b) \perp \Sigma_2$. ∎

Lemma 4.6.8 implies that for a geodesic variation the second derivative $\left(\frac{d^2}{ds^2}L(f(s,\cdot))\right)_{|s=0}$ (respectively, $\left(\frac{d^2}{ds^2}E(f(s,\cdot))\right)_{|s=0}$) is a quadratic form on the space of all variation vector fields along the central geodesic $\gamma$. The associated bilinear form is called the *index form* $I_\gamma^L$ of $\gamma$ (respectively, $I_\gamma^E$ of $\gamma$). It is an infinite dimensional analogue to the Hessian of a function.

**Definition 4.6.4.** *Let* $\Sigma_1$, $\Sigma_2$ *be submanifolds of* $M$ *or points in* $M$ *and* $\gamma\colon [a,b] \to M$ *be a geodesic which connects* $\Sigma_1$ *and* $\Sigma_2$ *and intersects both submanifolds orthogonally. Denote by* $\mathcal{T}^1_{\Sigma_1,\Sigma_2}\gamma$ *be the space of piecewise smooth vector fields along* $\gamma$ *which are tangent to* $\Sigma_1$ *at* $a$ *and to* $\Sigma_2$ *at* $b$.

*The* energy index form *is the bilinear form*

$$
\begin{aligned}
I^{E,\gamma}_{\Sigma_1,\Sigma_2}\colon & \mathcal{T}^1_{\Sigma_1,\Sigma_2}\gamma \times \mathcal{T}^1_{\Sigma_1,\Sigma_2}\gamma \to \mathbb{R},\\
&(\xi_1,\xi_2) \mapsto \int_a^b \left( \left\langle \nabla_{\dot\gamma}\xi_1, \nabla_{\dot\gamma}\xi_2 \right\rangle + \langle R(\xi_1,\dot\gamma)\xi_2,\dot\gamma\rangle \right) dt\\
&\qquad + \langle I\!I_{\Sigma_2}(\xi_1(b),\xi_2(b)),\dot\gamma(b)\rangle - \langle I\!I_{\Sigma_1}(\xi_1(a),\xi_2(a)),\dot\gamma(a)\rangle .
\end{aligned}
$$

*If* $\gamma$ *is either spacelike or timelike and satisfies* $\langle\dot\gamma,\dot\gamma\rangle = \eta \in \{-1,1\}$, *the* length index form *is defined by*

$$
\begin{aligned}
I^{L,\gamma}_{\Sigma_1,\Sigma_2}\colon & \mathcal{T}^1_{\Sigma_1,\Sigma_2}\gamma \times \mathcal{T}^1_{\Sigma_1,\Sigma_2}\gamma \to \mathbb{R},\\
&(\xi_1,\xi_2) \mapsto \eta \int_a^b \left( \left\langle \nabla_{\dot\gamma}\xi_1^\perp, \nabla_{\dot\gamma}\xi_2^\perp \right\rangle + \langle R(\xi_1,\dot\gamma)\xi_2,\dot\gamma\rangle \right) dt\\
&\qquad + \eta\,\langle I\!I_{\Sigma_2}(\xi_1(b),\xi_2(b)),\dot\gamma(b)\rangle\\
&\qquad - \eta\,\langle I\!I_{\Sigma_1}(\xi_1(a),\xi_2(a)),\dot\gamma(a)\rangle
\end{aligned}
$$

**Corollary 4.6.3.** *Let* $\gamma$ *be a geodesic from* $\Sigma_1$ *to* $\Sigma_2$ *which intersects these submanifolds orthogonally. The index form* $I^{L,\gamma}_{\Sigma_1,\Sigma_2}$ *is positive semi-definite if* $\gamma$ *minimizes length and negative negative semi-definite if* $\gamma$ *maximizes length.*

*Proof.* For $L$ the assertion follows from the Taylor expansion

$$
L(f(s,\cdot)) = L(\gamma) + s\overbrace{\left(\frac{d}{ds}L(f(s,\cdot)\right)_{s=0}}^{=0} + \frac{s^2}{2}\left(\frac{d^2}{ds^2}L(f(s,\cdot)\right)_{s=0} + O(s^3).
$$

The proof for the energy integral is exactly the same. ∎

The following lemma summarizes in which cases there is a non-trivial extremizing problem for $E$ and $L$. In particular, it implies that the extremizing problem has only in the Riemannian and Lorentzian case non-trivial solutions.

**Lemma 4.6.9.** *Let $(M, g)$ be a pseudo-Riemannian manifold, $\Sigma_1$, $\Sigma_2$ be submanifolds of $M$ without boundary, and $\gamma\colon [a,b] \to M$ be a geodesic which connects $\Sigma_1$ with $\Sigma_2$ and intersects both submanifolds orthogonally. If $\gamma$ is a null geodesic assume in addition that $\dot\gamma(a) \notin T_{\gamma(a)}\Sigma_1$ and $\dot\gamma(b) \notin T_{\gamma(b)}\Sigma_2$.*

*(i) If $I^{E,\gamma}_{\Sigma_1,\Sigma_2}$ is positive (respectively, negative) semi-definite then $g$ has signature $(+\cdots+)$ (respectively, $(-\cdots-)$).*

*(ii) Let $I^{E,\gamma,\perp}_{\Sigma_1,\Sigma_2}$ be the bilinear form $I^{E,\gamma}_{\Sigma_1,\Sigma_2}$ restricted to $\dot\gamma^\perp$. If $I^{E,\gamma,\perp}_{\Sigma_1,\Sigma_2}$ is positive (respectively, negative) definite then either*

*– $g$ has signature $(+\cdots+)$ (respectively, $(-\cdots-)$)*

*or*

*– $g$ has signature $(-+\cdots+)$ (respectively, $(-\cdots-+)$)*

*– $\gamma$ is causal (respectively, $\gamma$ is spacelike or null)*

*– $\Sigma_1$, $\Sigma_2$ are spacelike (respectively, $\Sigma_1$, $\Sigma_2$ are timelike) at $\gamma(a)$, $\gamma(b)$*

*(iii) If the index form $I^{L,\gamma}_{\Sigma_1,\Sigma_2}$ is positive semi-definite then $g$ has signature $(+\cdots+)$ or $(-\cdots-)$.*

*(iv) If the index form $I^{L,\gamma}_{\Sigma_1,\Sigma_2}$ is negative semi-definite then either*

*– $g$ has signature $(-+\cdots+)$*

*– $\gamma$ is timelike*

*– $\Sigma_1$, $\Sigma_2$ are spacelike at $\gamma(a)$, $\gamma(b)$*

*or*

*– $g$ has signature $(-\cdots-+)$*

*– $\gamma$ is spacelike*

*– $\Sigma_1$, $\Sigma_2$ are timelike at $\gamma(a)$, $\gamma(b)$*

*Proof.* Let $\delta \in \mathbb{R}\setminus\{0\}$ and $v \in T_{\gamma(a)}M$ be a vector with $\langle v, v\rangle = \delta$. For every $k \in \mathbb{N}$ we consider the variation vector field $W_k(t) = \frac{1}{k}V(t)\sin((t-a)\frac{k\pi}{b-a})$, where $V$ is the parallel translation of $v$ along $\gamma$. This variation vector field vanishes at the endpoints of $\gamma$ and is therefore in $\mathcal{T}^1_{\Sigma_1,\Sigma_2}\gamma$. The equation

$$
\begin{aligned}
I^{E,\gamma}_{\Sigma_1,\Sigma_2}(W_k, W_k) &= \int_a^b \left(\left\langle \nabla_{\dot\gamma} W_k, \nabla_{\dot\gamma} W_k\right\rangle + \langle R(W_k, \dot\gamma)W_k, \dot\gamma\rangle\right) \\
&= \int_a^b \left(\left(\frac{\pi}{b-a}\right)^2 \delta\cos^2\left((t-a)\frac{k\pi}{b-a}\right)\right. \\
&\quad \left. + \frac{1}{k^2}\langle R(V,\dot\gamma)V,\dot\gamma\rangle \sin^2\left((t-a)\frac{k\pi}{b-a}\right)\right)
\end{aligned}
$$

implies that for sufficiently large $k$ the integrand has the same sign as $\delta$, $\operatorname{sign}(I^{E,\gamma}_{\Sigma_1,\Sigma_2}(W_k, W_k)) = \operatorname{sign}(\delta)$.

Assertion (i) follows immediately from $\delta = \langle v, v\rangle$.

For the rest of the prove we will assume in addition that $\langle\dot\gamma(a), v\rangle = 0$. Consequently, we have $\langle W_k(t), \dot\gamma(t)\rangle = 0$ for all $t \in [a, b]$.

(ii): Assume that $I^{E,\gamma,\perp}_{\Sigma_1,\Sigma_2}$ is positive semi-definite and suppose either that

- there is a 2-dimensional subspace of $T_xM$ restricted to which $g$ is negative definite or that
- there is a 1-dimensional subspace restricted to which $g$ is negative definite and that $\gamma$ is spacelike.

In both cases there is a vector $v \in T_{\gamma(a)}M$ with $\delta := \langle v, v\rangle < 0$ and $v \perp \dot\gamma_a$. Hence $\operatorname{sign}(I^{E,\gamma,\perp}_{\Sigma_1,\Sigma_2}(W_k, W_k)) = -1$ and $I^{E,\gamma,\perp}_{\Sigma_1,\Sigma_2}$ cannot be positive semi-definite. This contradicts our assumption, and therefore it follows that either

- $g$ is Riemannian or
- $g$ is Lorentzian and $\gamma$ is causal.

Since $\dot\gamma(a) \notin T_{\gamma(a)}\Sigma_1$ ($\dot\gamma(b) \notin T_{\gamma(b)}\Sigma_2$) and $\Sigma_1$ ($\Sigma_b$) is orthogonal to $\dot\gamma$ the submanifold must be spacelike at $\gamma(a)$ ($\gamma(b)$). The proof for negative semi-definite $I^{E,\gamma,\perp}_{\Sigma_1,\Sigma_2}$ is completely analogous.

The index form $I^{L,\gamma}_{\Sigma_1,\Sigma_2}$ is only defined for spacelike or timelike geodesics. For large enough $k$ the relation $W_k \perp \dot\gamma$ implies

$$I^{L,\gamma}_{\Sigma_1,\Sigma_2}(W_k, W_k) = \eta I^{E,\gamma}_{\Sigma_1,\Sigma_2}(W_k, W_k)$$

and therefore $\operatorname{sign}(I^{L,\gamma}_{\Sigma_1,\Sigma_2}(W_k, W_k)) = \operatorname{sign}(\eta\delta)$.

(iii): Assume that $g$ is not definite. Then $v$ can be chosen such that $\delta = -\eta$. This implies that $I^{L,\gamma}_{\Sigma_1,\Sigma_2}$ is not positive semi definite either.

(iv): There is nothing to prove if $(M, g)$ is a 2-dimensional Lorentzian manifold. Suppose that either

- $g$ has not signature $(- + \cdots +)$ (respectively $(- \cdots - +)$) or that
- $\dim(M) \geq 3$ and $\gamma$ is spacelike (respectively, timelike).

Then $v$ can be chosen such that $\delta = \eta$ and $\operatorname{sign}(I^{L,\gamma}_{\Sigma_1,\Sigma_2}(W_k, W_k)) = \operatorname{sign}(\eta\delta) = 1$ for large enough $k$ implies that $I^{L,\gamma}_{\Sigma_1,\Sigma_2}$ is not negative semi definite. ∎

**Corollary 4.6.4.** *Let $(M, g)$ be a Riemannian or a Lorentzian manifold and $\gamma\colon [a, b] \to M$ be a spacelike or timelike geodesic which connects $\Sigma_1$ with $\Sigma_2$ and intersects both submanifolds orthogonally.*

*(i) If $I^{L,\gamma}_{\Sigma_1,\Sigma_2}$ is definite then for all variations $f\colon [-\epsilon,\epsilon]\times[a,b]\to M$ with non-vanishing variation vector field there is a $\delta\in(0,\epsilon)$ such that*
- $L(\gamma) < L(f(s,\cdot)) \quad \forall s\in[-\delta,\delta]$ *in the Riemannian case, and*
- $L(\gamma) > L(f(s,\cdot)) \quad \forall s\in[-\delta,\delta]$ *in the Lorentzian case.*

*(ii) If $I^{L,\gamma}_{\Sigma_1,\Sigma_2}$ is not semi-definite, there exist variations $f\colon [-\epsilon,\epsilon]\times[a,b]\to M$ of $\gamma$ such that*
- $L(\gamma) > L(f(s,\cdot)) \quad \forall s\in[-\epsilon,\epsilon]$ *in the Riemannian case, and*
- $L(\gamma) < L(f(s,\cdot)) \quad \forall s\in[-\epsilon,\epsilon]$ *in the Lorentzian case.*

*Proof.* To prove (i) let $f$ be a variation of $\gamma$ with non-vanishing variation vector field $\xi$. Since $L(f(s,\cdot)) = L(\gamma) + \frac{1}{2}s^2 I^{L,\gamma}_{\Sigma_1,\Sigma_2}(\xi,\xi) + o(s^2)$. The assertion follows from $I^{L,\gamma}_{\Sigma_1,\Sigma_2}(\xi,\xi)\neq 0$ and Lemma 4.6.9.

For the proof of (ii) let $\xi_+,\xi_-$ variation fields with $I^{L,\gamma}_{\Sigma_1,\Sigma_2}(\xi_+,\xi_+) > 0$ and $I^{L,\gamma}_{\Sigma_1,\Sigma_2}(\xi_-,\xi_-) < 0$. By Lemma 4.6.2 there are variations $f_\pm$ of $\gamma$ with variation vector fields $\xi_\pm$. Now the assertion follows from $L(f_\pm(s,\cdot)) = L(\gamma) + \frac{1}{2}s^2 I^{L,\gamma}_{\Sigma_1,\Sigma_2}(\xi_\pm,\xi_\pm) + o(s^2)$. ∎

In Lemma 4.6.15 below we will extend this result to null geodesics.

### 4.6.2 Conjugate and focal points

The existence of conjugate points is closely linked to semi-definiteness of the index form in the case that $\Sigma_1$, $\Sigma_2$ are points. Before investigating this relationship we will need to collect a few important facts about Jacobi vector fields in pseudo-Riemannian manifolds.

**Lemma 4.6.10.** *Let $\gamma\colon [a,b]\to M$ be a geodesic and $J,\tilde J$ be Jacobi fields which vanish at some point $t_0\in[a,b]$. Then we have $\left\langle J,\nabla_{\dot\gamma}\tilde J\right\rangle = \left\langle \nabla_{\dot\gamma}J,\tilde J\right\rangle$.*

*Proof.* It follows from

$$\begin{aligned}\left\langle \nabla_{\dot\gamma}J,\tilde J\right\rangle^{\cdot} &= \left\langle R(J,\dot\gamma)\tilde J,\dot\gamma\right\rangle + \left\langle \nabla_{\dot\gamma}J,\nabla_{\dot\gamma}\tilde J\right\rangle\\ &= \left\langle R(\tilde J,\dot\gamma)J,\dot\gamma\right\rangle + \left\langle \nabla_{\dot\gamma}\tilde J,\nabla_{\dot\gamma}J\right\rangle = \left\langle J,\nabla_{\dot\gamma}\tilde J\right\rangle^{\cdot}\end{aligned}$$

that $\left\langle J,\nabla_{\dot\gamma}\tilde J\right\rangle - \left\langle \nabla_{\dot\gamma}J,\tilde J\right\rangle$ is constant. Hence the assertion is an immediate consequence of $J(t_0)=\tilde J(t_0)=0$ . ∎

We show now that it is possible to split Jacobi fields into a tangential and into an orthogonal part unless $\gamma$ is a null geodesic. Moreover, the tangential part is always trivial.

**Lemma 4.6.11.** *Let $\gamma\colon [a,b] \to M$ be a geodesic, $\xi$ be a vector field along $\gamma$ with $\xi(t) \parallel \dot\gamma(t)$ for all $t \in [a,b]$, and $J$ be a Jacobi field along $\gamma$.*

*(i) The vector field $\xi$ along $\gamma$ is a Jacobi field if and only if there are numbers $\alpha, \beta$ with $\xi(t) = (\alpha t + \beta)\dot\gamma(t)$.*

*(ii) The following statements are equivalent.*

*(a) $\langle J(t), \dot\gamma(t)\rangle = 0$ for all $t \in [a,b]$,*

*(b) there are two different numbers $c, d \in [a,b]$ with $\langle J(c), \dot\gamma(c)\rangle = 0$ and $\langle J(d), \dot\gamma(d)\rangle = 0$,*

*(c) there is a number $c \in [a,b]$ with $\langle J(c), \dot\gamma(c)\rangle = 0$ and $\left\langle \nabla_{\dot\gamma(c)} J, \dot\gamma(c) \right\rangle = 0$.*

*Proof.* (i): If $\xi \parallel \dot\gamma$ we can write $\xi(t) = \phi(t)\dot\gamma(t)$. Since $R(\phi(t)\dot\gamma, \dot\gamma)\dot\gamma = 0$ the Jacobi equation reduces to $\nabla_{\dot\gamma}\nabla_{\dot\gamma}(\phi(t)\dot\gamma(t)) = 0$ which in turn is equivalent to $\ddot\phi(t) = 0$.

(ii): The assertion follows once we have shown that

$$\varphi(t) := \langle J(t), \dot\gamma(t)\rangle$$

satisfies $\ddot\varphi = 0$. But this follows from

$$\begin{aligned}\ddot\varphi &= \nabla_{\dot\gamma}\nabla_{\dot\gamma}\langle J, \dot\gamma\rangle = \nabla_{\dot\gamma}\left(\left\langle \nabla_{\dot\gamma} J, \dot\gamma\right\rangle + \left\langle J, \overbrace{\nabla_{\dot\gamma}\dot\gamma}^{=0}\right\rangle\right) \\ &= \left\langle \nabla_{\dot\gamma}\nabla_{\dot\gamma} J, \dot\gamma\right\rangle = \langle -R(J,\dot\gamma)\dot\gamma, \dot\gamma\rangle = 0.\end{aligned}$$

∎

**Corollary 4.6.5.** *Let $\gamma\colon [a,b] \to M$ be a geodesic which is timelike or spacelike and $J$ be a Jacobi field along $\gamma$. Then the orthogonal projections $J^\top$ to $\dot\gamma$ and $J^\perp$ to $\dot\gamma^\perp$ are also Jacobi fields along $\gamma$.*

*Proof.* Without loss of generality assume that $\langle\dot\gamma, \dot\gamma\rangle = \eta \in \{-1, 1\}$. Then $J^\top$ is given by $\eta\langle J, \dot\gamma\rangle\dot\gamma$. Since $\gamma$ is a geodesic we obtain $\nabla_{\dot\gamma} J^\top = \eta\left\langle \nabla_{\dot\gamma} J, \dot\gamma\right\rangle\dot\gamma = \left(\nabla_{\dot\gamma} J\right)^\top$. In the same way we get

$$\nabla_{\dot\gamma}\nabla_{\dot\gamma} J^\top = \left(\nabla_{\dot\gamma}\nabla_{\dot\gamma} J\right)^\top.$$

From $J^\top \parallel \dot\gamma$ we obtain $R(J^\top, \dot\gamma)\dot\gamma = 0$ and therefore

$$\begin{aligned}\nabla_{\dot\gamma}\nabla_{\dot\gamma} J^\top + R((J^\top, \dot\gamma)\dot\gamma &= \nabla_{\dot\gamma}\nabla_{\dot\gamma} J^\top = \left(\nabla_{\dot\gamma}\nabla_{\dot\gamma} J\right)^\top \\ &= \left(\nabla_{\dot\gamma}\nabla_{\dot\gamma} J\right)^\top + (R(J,\dot\gamma)\dot\gamma)^\top = 0,\end{aligned}$$

where we have used $R(\cdot,\dot\gamma)\dot\gamma)^\top = 0$ (cf. Proposition 4.3.1). From $J^\perp = J - J^\top$ and $\nabla_{\dot\gamma}\nabla_{\dot\gamma}J^\top = 0$ we get $\nabla_{\dot\gamma}\nabla_{\dot\gamma}J^\perp = \nabla_{\dot\gamma}\nabla_{\dot\gamma}J$. The second assertion follows now from $R(J^\perp,\dot\gamma)\dot\gamma = R(J,\dot\gamma)\dot\gamma$. ∎

Let $\gamma\colon [a,b] \to M$ be a geodesic and $J$ be a Jacobi field which vanishes at $a$ and $c \in (a,b)$. There is a geodesic variation $f\colon (s,t) \mapsto f(s,t)$ of $\gamma$ with variation vector field $J$ (cf. Proposition 2.9.1 and Corollary 2.9.1). These geodesics pass through $\gamma(a)$ and "intersect $\gamma(c)$ to first order" though they *may not* actually meet this point. Up to second order the geodesics segments $f(s,\cdot)_{[a,c]}$ have all the same length as $\gamma_{|[a,c]}$. It is therefore plausible to expect that $\gamma$ will not extremize length beyond $\gamma(c)$.

A typical example where the geodesics meet in both points $\gamma(a)$ and $\gamma(c)$ is given by those great circles of the unit sphere sphere $S^2$ which intersect both the south pole at $\gamma(a)$ and the north pole at $\gamma(c)$. Since in this example all neighbouring curves intersect $\gamma(c)$ at an angle different from zero, they can be deformed so that they meet $\gamma(b)$ and have length shorter than $\gamma$ (cf. Fig. 4.6.3). This is also true in the general case, even if the neighbouring geodesics do not actually intersect $\gamma(c)$ (cf. Theorem 4.6.1).

For arbitrary submanifolds $\Sigma_1$, $\Sigma_2$ an analogous statement cannot be formulated. There is simply no canonical way to compare different submanifolds along a curve. Hence we cannot speak of a "first point where a given normal geodesic fails to minimize length between $\Sigma_1$ and $\Sigma_2$". However, we can ask at which point a geodesic $\gamma$ orthogonal to a submanifold $\Sigma$ fails to minimize distance from $\Sigma$. In order to answer this question we will need to generalize the concept of a pair of conjugate points to a pair which consists of a submanifold and a point.

**Definition 4.6.5.** *Let $\Sigma \subset M$ be a submanifold and $\gamma\colon [a,b] \to M$ be a geodesic with $\gamma(a) \in \Sigma$, $\dot\gamma(a) \in (T_{\gamma(a)}\Sigma)^\perp \setminus T_{\gamma(a)}\Sigma$. The point $\gamma(c)$ is called a* focal point *of $\Sigma$ along $\gamma$ if there is a Jacobi field $J$ along $\gamma$ with*

*(i) $J(a) \in T_{\gamma(a)}\Sigma$, $J(c) = 0$,*
*(ii) $\left\langle \nabla_{\dot\gamma}J(a), v\right\rangle + \langle I\!I(J(a),v),\dot\gamma(a)\rangle = 0$ for all $v \in T_{\gamma(a)}\Sigma$.*

Observe that in the case that $\Sigma$ is just a single point, condition (ii) is empty and the definition reduces to the definition of a pair of conjugate points. The following lemma explains why we also demand condition (ii).

**Lemma 4.6.12.** *Let $\Sigma$ be a submanifold of $M$ and $\gamma\colon [a,b] \to M$ be a geodesic with $\gamma(a) \in \Sigma$, $\dot\gamma(a) \in (T_{\gamma(a)}\Sigma)^\perp \setminus T_{\gamma(a)}\Sigma$.*

*If $f\colon (-\epsilon,\epsilon) \times [a,b] \to M$ is a variation of $\gamma$ through geodesics orthogonal to $\Sigma$ then the variation vector field $\xi$ satisfies*

*(i) $\xi$ is a Jacobi field,*

*(ii)* $\xi(a) \in T_{\gamma(a)}\Sigma$

*(iii)* $\left\langle \nabla_{\dot\gamma}\xi(a), v \right\rangle + \langle I\!I(\xi(a), v), \dot\gamma(a)\rangle = 0$ *for all* $v \in T_{\gamma(a)}\Sigma$.

*Conversely, let* $\xi$ *be a vector field along* $\gamma$ *which satifies (i)–(iii). Then there is a variation* $f\colon (-\epsilon, \epsilon) \times [a, b] \to M$ *of* $\gamma$ *through geodesics orthogonal to* $\Sigma$ *which has variation vector field* $\xi$.

*Proof.* "⇒": Assume that $f$ is a variation through normal geodesics. That $\xi$ is a Jacobi field follows from Proposition 2.9.1 and property (ii) follows from $f(s, a) \in \Sigma$ for all $s$. The equation

$$\nabla_{\dot\gamma}\xi = \overset{f}{\nabla}_{\partial_t} f_s = \overset{f}{\nabla}_{\partial_s} f_t$$

implies that for every vector field $V$ tangent to $\Sigma$

$$\begin{aligned}\left\langle \nabla_{\dot\gamma}\xi, V \right\rangle &= \left\langle \overset{f}{\nabla}_{\partial_s} f_t, V \right\rangle = \overset{f}{\nabla}_{\partial_s} \overbrace{\langle f_t, V\rangle}^{=0} - \left\langle f_t, \overset{f}{\nabla}_{\partial_s} V \right\rangle \\ &= -\langle f_t, I\!I(f_s, V)\rangle .\end{aligned}$$

holds. This proves (iii).

"⇐": Let $\mu\colon (-\epsilon, \epsilon) \to \Sigma$ be a curve with $\dot\mu(0) = \xi(a)$ and $V$ be a vector field along $\mu$ with $V(s) \perp T_{\mu(s)}\Sigma$ and $V(0) = \dot\gamma(0)$. We define $f(s, t) = \exp(tV(s))$ thereby obtaining a variation of $\gamma$ through geodesics normal to $\Sigma$. This variation has variation vector field $\xi$ if and only if $f_s(0, a) = \xi(a)$ and $\nabla_{\dot\gamma(a)} f_s = \nabla_{\dot\gamma(a)}\xi$. We clearly have $f_s(0, a) = \dot\mu(0) = \xi(a)$ for any choice of $V$. Since $\nabla_{\dot\gamma(a)} f_s = \nabla_{f_s} f_t = \nabla_{\dot\mu(0)} V$ we have to choose $V$ such that $\nabla_{\dot\mu(0)} V = \nabla_{\dot\gamma(a)}\xi$. To see that this is always possible, let $V(s) = \mathbf{P}^{\perp}_{\mu|[0,s]}\dot\gamma(a) + s\mathbf{P}^{\perp}_{\mu|[0,s]}\left(\nabla_{\dot\gamma(a)}\xi\right)^{\perp}$ (cf. Lemma 4.4.5). Then we have $V(0) = \dot\gamma(a)$ and, using Lemmas 4.4.4 and 4.4.5,

$$\begin{aligned}\nabla_{\dot\mu(0)} V &= \nabla_{\dot\mu(0)} \mathbf{P}^{\perp}_{\mu|[0,s]}\dot\gamma(a) + \mathbf{P}^{\perp}_{\mu|[0,0]}\left(\nabla_{\dot\gamma(a)}\xi\right)^{\perp} \\ &= \overbrace{\left(\nabla_{\dot\mu(0)} \mathbf{P}^{\perp}_{\mu|[0,s]}\dot\gamma(a)\right)^{\perp}}^{=0} - \langle I\!I(\dot\mu, \cdot), \dot\gamma(a)\rangle^{\sharp} + \left(\nabla_{\dot\gamma(a)}\xi\right)^{\perp} \\ &= -\langle I\!I(f_s, \cdot), f_t\rangle^{\sharp} + \left(\nabla_{\dot\gamma(a)}\xi\right)^{\perp} \\ &= \nabla_{\dot\gamma(a)}\xi,\end{aligned}$$

where the last inequality follows from

$$\begin{aligned}-\langle I\!I(f_s, W), f_t\rangle &= -\left\langle \overset{f}{\nabla}_{\partial_s} W, f_t \right\rangle = \left\langle W \overset{f}{\nabla}_{\partial_s} f_t \right\rangle \\ &= \left\langle W, \overset{f}{\nabla}_{\partial_t} f_s \right\rangle = \left\langle W, \nabla_{\dot\gamma(a)}\xi \right\rangle\end{aligned}$$

for all vector fields $W$ which are tangent to $\Sigma$. ∎

If $\gamma$ is not a null geodesic then the neighbouring geodesics provided by Lemma 4.6.12 are of the same causal type as $\gamma$. However, if $\gamma$ is null then this property is not guaranteed. The following lemma clarifies the situation for null geodesics.

**Lemma 4.6.13.** *Let $\Sigma$ be a submanifold of $M$ and $\gamma\colon [a,b] \to M$ be a null geodesic with $\gamma(a) \in \Sigma$, $\dot\gamma(a) \in (T_{\gamma(a)}\Sigma)^\perp \setminus T_{\gamma(a)}\Sigma$.*

*Let $\xi$ be a vector field along $\gamma$ which satisfies properties (i)–(iii) in Lemma 4.6.12. Then there is a variation $f\colon (-\epsilon,\epsilon) \times [a,b] \to M$ of $\gamma$ through null geodesics orthogonal to $\Sigma$ and with variation vector field $\xi$ if and only if $\langle \xi(t), \dot\gamma(t)\rangle = 0$ for all $t \in [a,b]$.*

*Proof.* Assume that $f$ is such a variation through null geodesics. The equation $\langle f_t, f_t\rangle = 0$ implies $0 = \left\langle \overset{f}{\nabla}_{\partial_s} f_t, f_t \right\rangle = \left\langle \overset{f}{\nabla}_{\partial_t} f_s, f_t \right\rangle$ and therefore $\left\langle \nabla_{\dot\gamma(a)}\xi, \dot\gamma(a)\right\rangle = 0$. Lemma 4.6.12 (ii) and Lemma 4.6.11 (ii) imply $\langle \xi(t), \dot\gamma(t)\rangle = 0$ for all $t \in [a,b]$.

Conversely, assume that $\xi$ is a vector field along $\gamma$ which satisfies (i)–(iii) of Lemma 4.6.12 and is orthogonal to $\gamma$. We construct the variation $f$ as in the proof of Lemma 4.6.12 but now we will choose $V$ so that it is null at every point of $\mu$. To do so, let $s \mapsto W(s) \in T_{\gamma(a)}M$ be a curve with $W(0) = \dot\gamma(a)$ and $\langle W(s), W(s)\rangle = 0$ for all $s$. As in the proof of Lemma 4.6.12 let $\mu\colon s \mapsto \mu(s) \in \Sigma$ be a curve with $\mu(0) = \gamma(a)$ and $\dot\mu(0) = \xi(a)$. We set $V(s) = \mathbf{P}^\perp_{\mu|[0,s]} W(s)$. Clearly, $V(s)$ is normal to $\Sigma$ at all $s$ and $\nabla_{\dot\mu}\langle V(s), V(s)\rangle = 2\left\langle \nabla_{\dot\mu}V(s), V(s)\right\rangle = 2\left\langle \left(\nabla_{\dot\mu}V(s)\right)^\perp, V(s)\right\rangle = 0$ implies $\langle V(s), V(s)\rangle = 0$ for all $s$. Hence $f(s,t) = \exp(tV(s))$ is a variation of $\gamma$ through null geodesics normal to $\Sigma$. We have to choose $W$ such that the variation vector field of $f$ coincides with $\xi$. Since we have $f_s(0,a) = \dot\mu(0) = \xi(a)$ we only have to arrange $W$ such that

$$\overset{f}{\nabla}_{\partial_t} f_s(0,a) = \nabla_{\dot\gamma}\xi.$$

From $-\langle I\!I(f_s,\cdot), f_t\rangle^\sharp = \left(\nabla_{\dot\gamma(a)}\xi\right)^\top$ (cf. proof of Lemma 4.6.12) we obtain

$$\begin{aligned}\left(\overset{f}{\nabla}_{\partial_t} f_s\right)_{(0,a)} &= \left(\overset{f}{\nabla}_{\partial_s} f_t\right)_{(0,a)} = \left(\nabla_{\dot\mu}V(s)\right)_{(0,a)} \\ &= \left(\nabla_{\dot\mu}V(s)\right)^\perp_{(0,a)} - \langle I\!I(\dot\mu(0),\cdot), V(a)\rangle\end{aligned}$$

$$
\begin{aligned}
&= (\nabla_{\dot{\mu}} V(s))^{\perp}{}_{(0,a)} + \left(\nabla_{\dot{\gamma}(a)}\xi\right)^{\top} \\
&= \left(\nabla_{\dot{\mu}(0)}\left(\mathbf{P}^{\perp}_{\mu|[0,s]} W(s)\right)\right)^{\perp} + \left(\nabla_{\dot{\gamma}(a)}\xi\right)^{\top} \\
&= \left(\nabla_{\dot{\mu}}\left(\mathbf{P}^{\perp}_{\mu|[0,s]} W(0)\right)_{|s=0} + \mathbf{P}^{\perp}_{\mu|[0,s]}\left(\frac{\mathrm{d}}{\mathrm{d}s} W(s)\right)_{|s=0}\right)^{\perp} \\
&\quad + \left(\nabla_{\dot{\gamma}(a)}\xi\right)^{\top} \\
&= \left(\frac{\mathrm{d}}{\mathrm{d}s} W(s)\right)_{|s=0} + \left(\nabla_{\dot{\gamma}(a)}\xi\right)^{\top}.
\end{aligned}
$$

We have $\left\langle \nabla_{\dot{\gamma}}\xi, \dot{\gamma}(a)\right\rangle = 0$ since $\xi$ is orthogonal to $\gamma$ (Lemma 4.6.11). Since $\dot{\gamma}(a) \perp T_{\gamma(a)}\Sigma$ we get therefore $\left\langle \left(\nabla_{\dot{\gamma}}\xi\right)^{\perp}, \dot{\gamma}(a)\right\rangle = 0$. The tangent space of the null cone $C_{\gamma(a)} \subset T_{\gamma(a)}M$ at the point $\dot{\gamma}(a)$ is just $\{X \in T_{\gamma(a)} \mid \langle \dot{\gamma}(a), X\rangle = 0\}$. Hence $\left(\nabla_{\dot{\gamma}}\xi\right)^{\perp}$ is a tangent vector to $C_{\gamma(a)}$ at $\dot{\gamma}(a)$ and we can choose $W$ such that $(\frac{\mathrm{d}}{\mathrm{d}s}W)_{|s=0} = \left(\nabla_{\dot{\gamma}}\xi\right)^{\perp}$. But this implies $\overset{f}{\nabla}_{\partial_t} f_s(0,a) = \nabla_{\dot{\gamma}}\xi$ and we are done. ∎

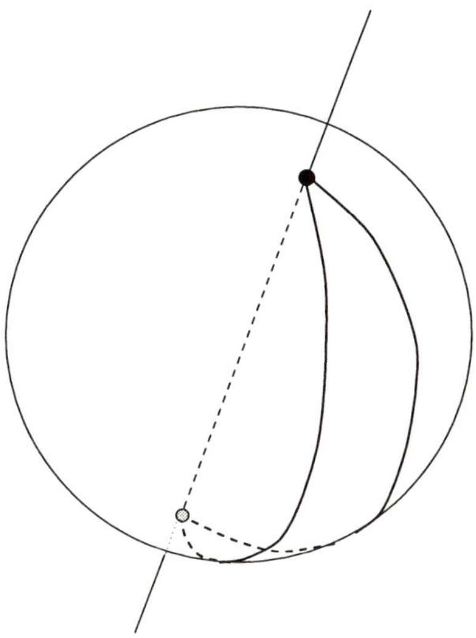

**Fig. 4.6.3.** Conjugate points on the sphere

We need another technical lemma analogous to Lemma 4.6.10

**Lemma 4.6.14.** *Let $\Sigma$ be a pseudo-Riemannian submanifold of $M$ and $\gamma\colon [a,b] \to M$ be a geodesic which intersects $\Sigma$ orthogonally at $\gamma(a)$. If*

*$J_1, J_2$ are Jacobi fields which satisfy $J_i(a) \in T_{\gamma(a)}\Sigma$ and $\left\langle \nabla_{\dot\gamma} J_i(a), v\right\rangle + \langle I\!I(J_i(a), v), \dot\gamma(a)\rangle = 0$ for all $v \in T_{\gamma(a)}\Sigma$, then $J_1, J_2$ satisfy*

$$\left\langle J_1(t), \nabla_{\dot\gamma(t)} J_2\right\rangle = \left\langle \nabla_{\dot\gamma(t)} J_1, J_2(t)\right\rangle$$

*for all $t \in [a, b]$.*

*Proof.* From the proof of Lemma 4.6.10 we know

$$\nabla_{\dot\gamma}\left(\left\langle \nabla_{\dot\gamma} J_1, J_2\right\rangle - \left\langle J_1, \nabla_{\dot\gamma} J_2\right\rangle\right) = 0.$$

Hence the assertion follows from

$$\begin{aligned}\left\langle \nabla_{\dot\gamma(a)} J_1, J_2(a)\right\rangle &- \left\langle J_1(a), \nabla_{\dot\gamma(a)} J_2\right\rangle \\ &= -\langle I\!I(J_1(a), J_2(a)), \dot\gamma(a)\rangle + \langle I\!I(J_2(a), J_1(a)), \dot\gamma(a)\rangle = 0.\end{aligned}$$

∎

We can now present a theorem which links focal points to length extremizing geodesics.

**Theorem 4.6.1.** *Let $(M, g)$ be a Riemannian or a Lorentzian manifold, $\Sigma$ be a Riemannian submanifold, and $\gamma\colon [a, b] \to M$ be a geodesic which intersects $\Sigma$ orthogonally at $\gamma(a)$. If $(M, g)$ is Lorentzian we also assume that $\gamma$ is timelike*

*(i) The submanifold $\Sigma$ does not have focal points along $\gamma$ if and only if*
- *the index form $I^{L,\gamma}_{\Sigma,\{\gamma(b)\}}$ is positive semi-definite in the Riemannian and negative semi-definite in the Lorentzian case.*
- *If $I^{L,\gamma}_{\Sigma,\{\gamma(b)\}}(V, V) = 0$, then there exists a function $v_0$ with $V = v_0\dot\gamma$, i.e., $V$ corresponds to a reparametrization of $\dot\gamma$.*

*(ii) The point $\gamma(b)$ is the only focal point of $\Sigma$ along $\gamma$ if and only if*
- *the index form is positive semi-definite in the Riemannian and negative semi-definite in the Lorentzian case.*
- *there is a non-vanishing vector field $V\colon [a, b] \to M$ along $\gamma$ which satisfies $I^{L,\gamma}_{\Sigma,\{\gamma(b)\}}(V, V) = 0$ and has values in $\dot\gamma^\perp$.*

*(iii) There is a focal point $\gamma(c)$ of $\Sigma$ along $\gamma$ with $c < b$ if and only if the index form is not semi-definite.*

*Proof.* Observe that the cases (i)–(iii) are mutually exclusive and comprise all possibilities. Hence it is sufficient to prove the "if"-parts only.

(i): Since there are no focal points of $\Sigma$ along $\gamma$ there are $n - 1$ linearly independen Jacobi fields $J_i\colon [a, b] \to TM$ which

(a) are everywhere perpendicular to $\dot\gamma$,

(b) satisfy $J_i(a) \in T_{\gamma(a)}\Sigma$ and $\left\langle \nabla_{\dot\gamma} J_i(a), v \right\rangle + \langle I\!I(J_i(a), v), \dot\gamma(a)\rangle = 0$ for all $v \in T_{\gamma(a)}\Sigma$.
(c) form a basis of $\gamma(t)^\perp$ for every parameter value $t \in (a, b]$.

Let $V$ be a variation vector field which is tangent to $\Sigma$ at $a$ and vanishes at $b$. Because of (c) there are smooth functions $v^i : (a, b] \to \mathbb{R}$ $(i = 0, \ldots, n-1)$ with $V = v^0\dot\gamma + \sum_{i=1}^{n-1} v^i J_i$. We will now calculate $\left\langle \nabla_{\dot\gamma} V^\perp, \nabla_{\dot\gamma} V^\perp \right\rangle + \langle R(V^\perp, \dot\gamma)V^\perp, \dot\gamma\rangle$.

$$
\begin{aligned}
\left\langle \nabla_{\dot\gamma} V^\perp, \nabla_{\dot\gamma} V^\perp \right\rangle &= \left\langle \nabla_{\dot\gamma} V^\perp, \dot v^k J_k + v^k \nabla_{\dot\gamma} J_k \right\rangle \\
&= \left\langle \nabla_{\dot\gamma} V^\perp, v^k \nabla_{\dot\gamma} J_k \right\rangle + \left\langle \dot v^i J_i + v^i \nabla_{\dot\gamma} J_i, \dot v_k J_k \right\rangle \\
&= \left\langle \nabla_{\dot\gamma} V^\perp, v^k \nabla_{\dot\gamma} J_k \right\rangle + \left\langle v^i \nabla_{\dot\gamma} J_i, \dot v_k J_k \right\rangle + \left\langle \dot v^i J_i, \dot v_k J_k \right\rangle \\
&\overset{\text{Lemma 4.6.14}}{=} \left\langle \nabla_{\dot\gamma} V^\perp, v^k \nabla_{\dot\gamma} J_k \right\rangle + \left\langle v^i J_i, \dot v_k \nabla_{\dot\gamma} J_k \right\rangle \\
&\quad + \left\langle \dot v^i J_i, \dot v_k J_k \right\rangle \\
&= \left\langle \nabla_{\dot\gamma} V^\perp, v^k \nabla_{\dot\gamma} J_k \right\rangle + \left\langle v^i J_i, \nabla_{\dot\gamma} \left( v_k \nabla_{\dot\gamma} J_k \right) \right\rangle \\
&\quad - \left\langle v^i J_i, v_k \nabla_{\dot\gamma} \nabla_{\dot\gamma} J_k \right\rangle + \left\langle \dot v^i J_i, \dot v_k J_k \right\rangle \\
&= \nabla_{\dot\gamma} \left\langle V^\perp, v^k \nabla_{\dot\gamma} J_k \right\rangle + \left\langle V^\perp, R(V^\perp, \dot\gamma)\dot\gamma \right\rangle \\
&\quad + \left\langle \dot v^i J_i, \dot v_k J_k \right\rangle,
\end{aligned}
$$

where in the last equation we have used that $J_k$ are Jacobi fields. Hence we get

$$
\left\langle \nabla_{\dot\gamma} V^\perp, \nabla_{\dot\gamma} V^\perp \right\rangle + \langle R(V^\perp, \dot\gamma)V^\perp, \dot\gamma\rangle = \langle \dot v^i J_i, \dot v^i J_i\rangle + \left\langle V^\perp, v^k \nabla_{\dot\gamma} J_k \right\rangle^{\cdot}.
$$

Assume for the moment that the functions $v^k$ have continuous extensions to $a$. Then (b), $V(b) = 0$, $V^\perp(a) = V(a) \in T_{\gamma(a)}\Sigma$, and the fact that all $v^i$ have bounded extensions to $a$ imply

$$
\begin{aligned}
\int_a^b \left\langle V^\perp, v^k \nabla_{\dot\gamma} J_k \right\rangle^{\cdot} \mathrm{d}t &= - \left\langle V^\perp, v^k \nabla_{\dot\gamma} J_k \right\rangle_{t=a} \\
&= \left\langle I\!I(V(a), v^k(a) J_k), \dot\gamma(a) \right\rangle .
\end{aligned}
$$

Writing $\eta = \langle \dot\gamma, \dot\gamma\rangle$ we obtain

$$\begin{aligned}
I^{L,\gamma}_{\Sigma,\{\gamma(b)\}}(V,V) &= \eta \int_a^b \Big( \big\langle \nabla_{\dot\gamma} V^\perp, \nabla_{\dot\gamma} V^\perp \big\rangle + \langle R(V^\perp,\dot\gamma)V^\perp,\dot\gamma\rangle \Big)\,\mathrm{d}t \\
&\quad - \eta \,\langle I\!I(V(a),V(a)),\dot\gamma(a)\rangle \\
&= \eta \int_a^b \langle \dot v^i J_i, \dot v^i J_i\rangle\,\mathrm{d}t + \eta\,\langle I\!I(V(a), v^k(a)J_k),\dot\gamma(a)\rangle \\
&\quad - \eta\,\langle I\!I(V(a),V(a)),\dot\gamma(a)\rangle \\
&= \eta \int_a^b \langle \dot v^i J_i, \dot v^i J_i\rangle\,\mathrm{d}t.
\end{aligned}$$

Since $\gamma$ is timelike in the Lorentzian case the vector field $V$ is always spacelike and we can conclude that $\eta I^{L,\gamma}_{\Sigma,\{\gamma(b)\}}(V,V)$ is semi-definite. Furthermore, $\eta I^{L,\gamma}_{\Sigma,\{\gamma(b)\}}(V,V) = 0$ if and only if all $v^i$ are constant, which in turn implies that $V^\perp$ vanishes since it vanishes at $b$.

We still have to show that the functions $v^k$ have continuous extensions to $a$. There are $\dim(\Sigma)$ vectors $\{J_{i_1}(a),\dots,J_{i_{\dim(\Sigma)}}(a)\}$ which form a basis of $T_{\gamma(a)}\Sigma$. This follows from Lemma 4.6.12 since it is possible to construct normal geodesic variations of $\dot\gamma$ in any direction tangential to $\Sigma$. We can also assume that $J_k(a) = 0$ for all $k \notin \{i_1,\dots,i_{\dim(\Sigma)}\}$ because otherwise we could subtract a suitable linear combination of the $J_{i_l}$ from $J_k$. For each $t$ we decompose $V(t)$ into a part $U$ tangential to $\operatorname{span}\{J_{i_1}(t),\dots,J_{i_{\dim(\Sigma)}}(t)\}$ and a part $\tilde U$ tangential to $\operatorname{span}\{J_k(t) \,|\, k \notin \{i_1,\dots,i_{\dim(\Sigma)}\}\}$. It is clear that all $v^{i_l}$ have smooth extensions to $a$ since $\{J_{i_1}(t),\dots,J_{i_{\dim(\Sigma)}}(t)\}$ are linearly independent for all $t\in[a,b]$. Since $\tilde U = V - v^{i_l}J_{i_l}$ and $J_k$ ($k\notin\{i_1,\dots,i_{\dim(\Sigma)}\}$) are smooth and vanish at $a$ there are vector fields $W$, $K_k$ such that $\tilde U = (t-a)W$ and $J_k = (t-a)K_k$. These vector fields satisfy $W(a) = \nabla_{\dot\gamma}\tilde U(a)$ and $K_k(a) = \nabla_{\dot\gamma}J_k(a)$. From $\tilde U^\perp = v^k J_k$ we get $W^\perp = v^k K_k$, and from the linear independence of $\{J_k(t)\}$ ($t\in(a,b]$) and the fact that $J_k(a) = 0$ we infer that the vector fields $K_k$ are linearly independent near $a$. Hence there are smooth one forms $\omega^i$ along $\gamma$ with $\omega^i(K_k) = \delta^i_k$ near $a$. Consequently, the function $v^k = \omega^k(W^\perp)$ (near $a$) has a smooth extension to $a$.

(ii): Let $\{b_i\}\to b$ be a strictly monotonically increasing sequence and $V$ be a vector field along $\gamma$ which is tangent to $\Sigma$ at $a$ and vanishes at $b$. Since $V$ vanishes at $b$, there exists a well defined vector field $W$ such that $V(t) = (b-t)W(t)$. Let $V_i$ be the piecewise differentiable vector field given by $V_i(t) = (b_i - t)W(t)$ for $t\in[a,b_i)$ and $V_i(t) = 0$ for $t\in[b_i,b]$. This gives a sequence of piecewise smooth vector fields $\{V_i\}$ with $V_{i[a,b_i]} = V_{|[a,b_i]}$ and $V_{|[b_i,b]} = 0$. Since $b$ is the first conjugate point part (i) of the theorem implies that $I^{L,\gamma}_{\Sigma,\{\gamma(b)\}}(V_i,V_i) \geq 0$. Hence $I^{L,\gamma}_{\Sigma,\{\gamma(b)\}}(V_i,V_i) \to I^{L,\gamma}_{\Sigma,\{\gamma(b)\}}(V,V)$ ($i\to\infty$) implies that $\eta I^{L,\gamma}_{\Sigma,\{\gamma(b)\}}$ is positive semi-definite. To see that there exist non-trivial variations with vanishing index form let $J$ be a Jacobi field orthogonal to $\gamma$ according to Definition 4.6.5. Since $J = J^\perp$ we have

$$\begin{aligned}
I^{L,\gamma}_{\Sigma,\{\gamma(b)\}}(J,J) &= \eta \int_a^b \left( \left\langle \nabla_{\dot\gamma} J, \nabla_{\dot\gamma} J \right\rangle + \langle R(J,\dot\gamma)J, \dot\gamma\rangle \right) \mathrm{d}t \\
&\quad - \eta \left\langle I\!I(J(a),J(a)), \dot\gamma(a)\right\rangle \\
&= \eta \int_a^b \left( \left\langle \nabla_{\dot\gamma} J, J \right\rangle^{\cdot} \overbrace{- \left\langle \nabla_{\dot\gamma}\nabla_{\dot\gamma} J, J \right\rangle - \langle R(J,\dot\gamma)\dot\gamma, J\rangle}^{=0 \text{ since } J \text{ is a Jacobi field}} \right) \mathrm{d}t \\
&\quad - \eta \left\langle I\!I(J(a),J(a)), \dot\gamma(a)\right\rangle \\
&= \eta \left\langle I\!I(J(a),J(a)), \dot\gamma(a)\right\rangle - \eta \left\langle I\!I(J(a),J(a)), \dot\gamma(a)\right\rangle = 0
\end{aligned}$$

(iii): Let $c \in (a,b)$ be the first focal point of $\Sigma$ along $\gamma$ and t $J$ be a non-vanishing Jacobi field according to Definition 4.6.5. Then $\lim_{t\to c} \nabla_{\dot\gamma} J(t) \neq 0$ and the piecewise smooth vector field

$$V(t) = \begin{cases} J(t) & \text{for } t \in [a,c], \\ 0 & \text{for } t \in (c,b] \end{cases}$$

satisfies $\Delta\nabla_{\dot\gamma} V(c) = -\lim_{t\to c} \nabla_{\dot\gamma} J(t) \neq 0$. Let $\delta > 0$ and $W$ be a vector field along $\gamma$ which satisfies

$$W(a) = W(b) = 0, \quad \langle W(t), \dot\gamma(t)\rangle = 0, \quad \left\langle W(c), \Delta\nabla_{\dot\gamma} V(c) \right\rangle > 0.$$

By the definition of $J$ we have $I^{L,\gamma}_{\Sigma,\{\gamma(b)\}}(V,V) = 0$ and $I^{L,\gamma}_{\Sigma,\{\gamma(b)\}}(V,W) = -2\eta \langle \Delta V(c), W(c)\rangle$. This in turn implies

$$\begin{aligned}
&I^{L,\gamma}_{\Sigma,\{\gamma(b)\}}(V+\delta W, V+\delta W) \\
&\quad = I^{L,\gamma}_{\Sigma,\{\gamma(b)\}}(V,V) + 2\delta I^{L,\gamma}_{\Sigma,\{\gamma(b)\}}(V,W) + \delta^2 I^{L,\gamma}_{\Sigma,\{\gamma(b)\}}(W,W) \\
&\quad = -2\eta\delta \langle \Delta V(c), W(c)\rangle + \delta^2 I^{L,\gamma}_{\Sigma,\{\gamma(b)\}}(W,W).
\end{aligned}$$

Hence for $\delta$ sufficiently small we obtain $\operatorname{sign}(I^{L,\gamma}_{\Sigma,\{\gamma(b)\}}(V+\delta W, V+\delta W)) = \operatorname{sign}(-\eta\delta)$. Since we can replace $W$ by $-W$ this equation implies that $I^{L,\gamma}_{\Sigma,\{\gamma(b)\}}$ fails to be semi-definite. ∎

In order to have an analogue of Theorem 4.6.1 for null geodesics we need to use the energy index form.

**Theorem 4.6.2.** *Let $(M,g)$ be a Riemannian or a Lorentzian manifold, $\Sigma$ be a Riemannian submanifold, and $\gamma\colon [a,b] \to M$ be a spacelike (Riemannian case) or causal (Lorentzian case) geodesic with $\dot\gamma(a) \in (T_{\gamma(a)}\Sigma)^\perp$.*

*(i) The submanifold $\Sigma$ does not have focal points along $\gamma$ if and only if*
- *the index form $I^{E,\gamma,\perp}_{\Sigma,\{\gamma(b)\}}$ is positive semi-definite,*
- *$I^{E,\gamma,\perp}_{\Sigma,\{\gamma(b)\}}(V,V) = 0$ implies taht there exists a function $v_0$ with $V = v_0\dot\gamma$, i.e., $V$ corresponds to a reparametrization of $\dot\gamma$.*

*(ii) The point $\gamma(b)$ is the only focal point of $\Sigma$ along $\gamma$ if and only if*
- *the index form $I^{E,\gamma,\perp}_{\Sigma,\{\gamma(b)\}}$ is semi-definite and*
- *there is a non-vanishing vector field $V\colon [a,b] \to M$ along $\gamma$ which satisfies $I^{E,\gamma,\perp}_{\Sigma,\{\gamma(b)\}}(V,V) = 0$ and has values in $\dot\gamma^\perp$.*

*(iii) There is a focal point $\gamma(c)$ of $\Sigma$ along $\gamma$ with $c < b$ if and only if the index form $I^{E,\gamma,\perp}_{\Sigma,\{\gamma(b)\}}$ is not semi-definite.*

*Proof.* If $\gamma$ is timelike (in the Lorentzian case) or spacelike (in the Riemannian case) then the proof is completely analogous to the proof of Theorem 4.6.1. We can therefore restrict to the case that $\gamma$ is a null geodesic and $\Sigma$ is spacelike.

(i): Again, there are Jacobi vector fields which satisfy assertions (a)–(c) in the proof of Theorem 4.6.1 (i). Since every Jacobi field satisfying (a)–(c) must be a linear combination of the $J_i$ we can assume without loss of generality that $J_1(t) = t\dot\gamma(t)$. Exactly as in the proof of Theorem 4.6.1 we obtain for every vector field $V$ along $\gamma$ with $V(a) \in T_{\gamma(a)}\Sigma$ and $V(b) = 0$

$$I^{E,\gamma,\perp}_{\Sigma,\{\gamma(b)\}}(V,V) = \int_a^b \langle \dot v^i J_i, \dot v^i J_i\rangle \, \mathrm{d}t,$$

where $t \mapsto v^i(t)$ are functions defined by $V(t) = \sum_{i=1}^{n-1} v^i(t) J_i(t)$. The equation $\langle J_i, \dot\gamma\rangle = 0$ implies that each $J_i$ must be spacelike or null, whence $I^{E,\gamma,\perp}_{\Sigma,\{\gamma(b)\}}(V,V) \geq 0$. Moreover, the Jacobi field $J_i$ is spacelike for $i \geq 2$. By the linear independence of the Jacobi fields at every point $\neq a$ we have $I^{E,\gamma,\perp}_{\Sigma,\{\gamma(b)\}}(V,V) > 0$ unless $\dot v^2 = \ldots \dot v^{n-1} = 0$. Since $V(b) = 0$ the equation $I^{E,\gamma,\perp}_{\Sigma,\{\gamma(b)\}}(V,V) = 0$ implies that $v^2 = \cdots = v^{n-1} = 0$ and therefore $V = v^1 J_1$ which is parallel to $\dot\gamma$.

The proof of assertions (ii) and (iii) is completely analogous to the proof of Theorem 4.6.1 (ii), (iii). ∎

We can now extend Corollary 4.6.4 to null geodesics in the case that one of the submanifolds $\Sigma_1$, $\Sigma_2$ degenerates to a point. This is achieved by showing that there is a variation of $\gamma$ through timelike curves from the spacelike submanifold $\Sigma$ to $\gamma(b)$ if $I^{E,\gamma,\perp}_{\Sigma,\{\gamma(b)\}}$ is not semi-definite. Observe that we cannot use the same argument as in Corollary 4.6.4. While we would obtain the existence of a variation $f$ with $E(f(s,\cdot)) < 0$ for all $s$, it would not be clear that these varied curves are everywhere timelike.

**Lemma 4.6.15.** *Let $(M,g)$ be a Lorentz manifold, $\Sigma$ be a spacelike submanifold, and $\gamma\colon [a,b] \to M$ be a null geodesics which intersects $\Sigma$ orthogonally. If $I^{E,\gamma,\perp}_{\Sigma,\{\gamma(b)\}}$ is not semi-definite then there is a timelike curve from $\Sigma$ to $\gamma(b)$ arbitrarily close to $\gamma$.*

*Proof.* The strategy of proof is as follows. Theorem 4.6.2 implies that there is a first focal point $\gamma(c)$ $(c \in (a,b))$ of $\Sigma$ along $\gamma$. For some small $\delta \in (0, b-c)$ we will construct two vector fields $\xi(t)$ and $A(t)$ along $\gamma_{|[a,c+\delta]}$ such that for every variation $f$ of $\gamma$ with $f_s(0,t) = \xi(t)$ and $\overset{f}{\nabla}_{\partial_s(0,t)} f_s = A(t)$ we have $\langle f_t(t,s), f_t(t,s)\rangle < 0$ for $s > 0$ sufficiently small. We will show that there is such a variation which, in addition, satisfies $f(s,a) \in \Sigma$ and $f(s,c+\delta) = \gamma(c+\delta)$ for all $s$. It is then possible to join $\Sigma$ and $\gamma(c+\delta)$ by a timelike curve arbitrarily close to $\gamma$. This curve can in turn be slightly deformed in order to arrive at a timelike curve from $\Sigma$ to $\gamma(b)$.

Observe that $A$ cannot be chosen completely independently of $\xi$. In fact, at $\gamma(a) \in \Sigma$ we have must have $A^\perp = \left(\overset{f}{\nabla}_{\partial_s} f_s\right)^\perp = I\!I(f_s, f_s) = I\!I(\xi,\xi)$. From the proof Lemma 4.6.7 we see that

$$\begin{aligned}
\frac{1}{2}\left(\frac{\mathrm{d}^2}{\mathrm{d}s^2}\langle f_t, f_t\rangle\right)_{|s=0} &= \left\langle \overset{f}{\nabla}_{\partial_t} f_s, \overset{f}{\nabla}_{\partial_t} f_s \right\rangle + \left\langle \overset{f}{\nabla}_{\partial_t} \overset{f}{\nabla}_{\partial_s} f_s, f_t \right\rangle \\
&\quad + \langle R(f_s, f_t) f_s, f_t\rangle . \\
&= -\left\langle \overset{f}{\nabla}_{\partial_t} \overset{f}{\nabla}_{\partial_t} f_s + R(f_s, f_t) f_t, f_s \right\rangle \\
&\quad + \left\langle \overset{f}{\nabla}_{\partial_t} f_s, f_s \right\rangle^{\cdot} + \left\langle \overset{f}{\nabla}_{\partial_t} \overset{f}{\nabla}_{\partial_s} f_s, f_t \right\rangle
\end{aligned}$$

holds. If we can construct vector fields $\xi$, $A$ with

$$\left\langle \nabla_{\dot\gamma}\nabla_{\dot\gamma}\xi + R(\xi,\dot\gamma)\dot\gamma, \xi\right\rangle > 0 - \left\langle \nabla_{\dot\gamma}\xi, \xi\right\rangle^{\cdot} - \left\langle \nabla_{\dot\gamma}A, \dot\gamma\right\rangle > 0$$

then any corresponding variation $f$ satisfies $\langle f_t, f_t\rangle < 0$ for small $s > 0$.

Let $\gamma(c)$ $(c \in (a,b))$ be the first focal point of $\Sigma$ along $\gamma$ and let $J$ be a Jacobi field according to Definition 4.6.5. It follows from Lemma 4.6.11 (ii) that this Jacobi field is everywhere orthogonal to $\gamma$.

If there would be a point $d \in (a,c)$ with $J(d) = \alpha\dot\gamma(d)$ then the Jacobi field $J(t) - t\frac{\alpha}{d}\dot\gamma(t)$ would have a zero at $d$ and satisfy all the conditions of Definition 4.6.5. Hence there would be a focal point $\gamma(d)$ of $\Sigma$ along $\gamma$ before $\gamma(c)$ in contradiction to the definition of $c$. We have therefore shown that $J(t) \in (\dot\gamma(t))^\perp \setminus \mathbb{R}\dot\gamma(t)$ for all $t \in (a,c)$.

The derivative of $J$ at $c$ satisfies $\nabla_{\dot\gamma} J(c) \in T_{\gamma(c)}M \setminus \mathbb{R}\dot\gamma(c)$ since $J$ is non-trivial and not parallel to $\dot\gamma$. This implies that there is a $\delta > 0$

such that $c$ is the only point in $(a, c+\delta]$ where $J$ is tangent to $\gamma$. Hence there exists a spacelike vector field $U$ along $\gamma$ with value in $(\dot{\gamma})^{\perp}$ and a function $\varphi\colon [a,b] \to \mathbb{R}$ such that

- $\langle U(t), U(t)\rangle = 1$ for all $t \in [a, c+\delta]$,
- $J(t) = \varphi(t)U(t)$ for all $t \in [a, c+\delta]$,
- $\varphi(t) > 0$ for all $t \in (a, c)$.
- $\varphi(t) < 0$ for all $t \in (c, c+\delta)$.

We will now construct $\xi$ by slighly stretching $U$. Let $\psi\colon [a, c+\delta] \to \mathbb{R}$ be a function and consider $\xi = (\varphi + \psi)U$. From

$$\begin{aligned}\nabla_{\dot{\gamma}}\nabla_{\dot{\gamma}}\xi + R(\xi,\dot{\gamma})\dot{\gamma} &= \nabla_{\dot{\gamma}}\nabla_{\dot{\gamma}}(\psi U) + \psi R(U,\dot{\gamma})\dot{\gamma}\\ &= \ddot{\psi}U + 2\dot{\psi}\nabla_{\dot{\gamma}}U + \psi(\nabla_{\dot{\gamma}}\nabla_{\dot{\gamma}}U + R(U,\dot{\gamma})\dot{\gamma})\end{aligned}$$

We get

$$\left\langle \nabla_{\dot{\gamma}}\nabla_{\dot{\gamma}}\xi + R(\xi,\dot{\gamma})\dot{\gamma}, \xi\right\rangle = (\varphi+\psi)\left(\ddot{\psi} + \psi\left\langle \nabla_{\dot{\gamma}}\nabla_{\dot{\gamma}}U + R(U,\dot{\gamma})\dot{\gamma}, U\right\rangle\right).$$

There is a number $\lambda_1 > 0$ such that $-(\lambda_1)^2 < \left\langle \nabla_{\dot{\gamma}}\nabla_{\dot{\gamma}}U + R(U,\dot{\gamma})\dot{\gamma}, U\right\rangle_{|t}$ for all $t \in [a, c+\delta]$. Let $\lambda_2 > 0$ and $\psi(t) = \lambda_2(\mathrm{e}^{\lambda_1 t} - \mathrm{e}^{\lambda_1 a})$. Then we obtain

$$\begin{aligned}&\ddot{\psi} + \psi\langle \nabla_{\dot{\gamma}}\nabla_{\dot{\gamma}}U + R(U,\dot{\gamma})\dot{\gamma}, U\rangle\\ &\qquad = \psi\left((\lambda_1)^2 + \left\langle \nabla_{\dot{\gamma}}\nabla_{\dot{\gamma}}U + R(U,\dot{\gamma})\dot{\gamma}, U\right\rangle\right) + \lambda_2(\lambda_1)^2\mathrm{e}^{\lambda_1 a}\\ &\qquad \geq \lambda_2(\lambda_1)^2\mathrm{e}^{\lambda_1 a} > 0\end{aligned}$$

for all $t \in (a, c+\delta)$. We set $\lambda_2 = -\varphi(c+\delta)/(\mathrm{e}^{\lambda_1(c+\delta)} - \mathrm{e}^{\lambda_1 a})$ which is positive. Observe that $\psi(a) = 0$, $\varphi(c+\delta) + \psi(c+\delta) = 0$, and $\varphi(t) + \psi(t) > 0$ for all $t \in (a, c]$. We can assume that $c+\delta$ is the first zero after $c$ since otherwise we could replace $\delta$ by a smaller number. To summarize, the vector field $\xi$ satisfies $\xi(a) = J(a)$, $\xi(c+\delta) = 0$, and $\left\langle \nabla_{\dot{\gamma}}\nabla_{\dot{\gamma}}\xi + R(\xi,\dot{\gamma})\dot{\gamma}, \xi\right\rangle > 0$ for all $t \in (a, c+\delta)$. We will now construct the acceleration vector field $A$.

There is a basis $e_1, \ldots, e_{n-1}$ of $(\dot{\gamma}(a))^{\perp}$ such that

(i) $e_{n-1} = \dot{\gamma}(a)$
(ii) $\operatorname{span}\{e_1, \ldots, e_{\dim(\Sigma)}\} = T_{\gamma(a)}\Sigma$,
(iii) $\langle e_i, e_k\rangle = \delta_{ik}$ for all $i \in \{1, \ldots, n-2\}$, $k \in \{1, \ldots, n-1\}$.

Let $e_n \in T_{\gamma(a)}M$ be the null vector with $\langle e_n, e_{n-1}\rangle = -1$ and $\langle e_n, e_i\rangle = 0$ for all $i \in \{1, \ldots, n-2\}$. We denote by $e_i(t)$ $(i \in \{1, \ldots, n\})$ the parallel transport of $e_i$ along $\gamma$ $(i \in \{1, \ldots, n\})$. There are numbers $\Xi^k$ $(k \in \{\dim(\Sigma)+1, \ldots, n\})$ such that $I\!I(\xi(a), \xi(a)) = \sum_{k=\dim(\Sigma)+1}^{n} \Xi^k e_k$. Let

$$A(t) = \sum_{k=\dim(\Sigma)+1}^{n-1} \frac{c+\delta-t}{c+\delta-a} \Xi^k e_k - \mu(t) e_n,$$

where

$$\mu(t) = \left( \langle I\!I(\xi_a, \xi_a), \dot\gamma(a) \rangle + \left\langle \xi_a, \nabla_{\dot\gamma(a)} \xi \right\rangle \right) \frac{c+\delta-t}{c+\delta-a} - \left\langle \xi_t, \nabla_{\dot\gamma(t)} \xi \right\rangle.$$

We have $\mu(a) = \langle I\!I(\xi_a, \xi_a), \dot\gamma(a) \rangle = -\Xi^n$ and

$$\begin{aligned}
\mu(c+\delta) &= \left\langle \xi_{c+\delta}, \nabla_{\dot\gamma(c+\delta)} \xi \right\rangle = \frac{1}{2} \nabla_{\dot\gamma(c+\delta)} \langle \xi, \xi \rangle \\
&= \overbrace{(\varphi + \psi)_{|c+\delta}}^{=0} (\dot\varphi + \dot\psi)_{|c+\delta} = 0.
\end{aligned}$$

Hence the vector field $A$ satisfies $A(a) = A(a)^\perp = I\!I(\xi_a, \xi_a)$ and $A(c+\delta) = 0$. We will now show that $\left\langle \nabla_{\dot\gamma} A, \dot\gamma \right\rangle + \nabla_{\dot\gamma} \left\langle \nabla_{\dot\gamma} \xi, \xi \right\rangle \leq 0$. We calculate

$$\begin{aligned}
\left\langle \nabla_{\dot\gamma} A, \dot\gamma \right\rangle &= \left\langle \sum_{k=\dim(\Sigma)+1}^{n-1} \nabla_{\dot\gamma} \left( \frac{c+\delta-t}{c+\delta-a} \Xi^k e_k \right) - \nabla_{\dot\gamma} (\mu e_n), e_{n-1} \right\rangle \\
&= \left\langle \sum_{k=\dim(\Sigma)+1}^{n-1} \left( \frac{c+\delta-t}{c+\delta-a} \right)^{\cdot} \Xi^k e_k - \dot\mu e_n, e_{n-1} \right\rangle = \dot\mu \\
&= -\frac{\langle I\!I(\xi_a, \xi_a), \dot\gamma(a) \rangle + \left\langle \xi_a, \nabla_{\dot\gamma(a)} \xi \right\rangle}{c+\delta-a} - \nabla_{\dot\gamma} \left\langle \xi_t, \nabla_{\dot\gamma(t)} \xi \right\rangle
\end{aligned}$$

and our claim follows from

$$\begin{aligned}
&\langle I\!I(\xi_a, \xi_a), \dot\gamma(a) \rangle + \left\langle \xi_a, \nabla_{\dot\gamma(a)} \xi \right\rangle \\
&\qquad = \langle I\!I(J_a, J_a), \dot\gamma(a) \rangle + \frac{1}{2} \nabla_{\dot\gamma(a)} \langle \xi, \xi \rangle \\
&\qquad = \left\langle \overset{f}{\nabla}_{\partial_s} f_s, f_t \right\rangle_{|(a,0)} + \phi(a)(\dot\psi(a) + \dot\phi(a)) \\
&\qquad = -\left\langle f_s, \nabla_{f_t} f_s \right\rangle_{|(a,0)} + \phi(a)(\dot\psi(a) + \dot\phi(a)) \\
&\qquad = -\phi(a)\dot\phi(a) + \phi(a)(\dot\psi(a) + \dot\phi(a)) = \phi(a)\dot\psi(a) \geq 0.
\end{aligned}$$

We will now show that there is a variation $f\colon (-\epsilon, \epsilon) \times [a, c+\delta] \to M$ such that

1. $f_s(0,t) = J(t)$, $\nabla_{f_s(0,t)} f_s = A(t)$ for all $t \in [a, c+\delta]$,
2. $f(s,a) \in \Sigma$ for all $s \in (-\epsilon, \epsilon)$,

3. $f(s,c+\delta)=\gamma(c+\delta)$ for all $s\in(-\epsilon,\epsilon)$.

By construction of $\xi$ and $A$ such a variation must satisfy $\langle f_t, f_t\rangle < 0$ for all $s\neq 0$. Hence we will obtain timelike curves from $\Sigma$ to $\gamma(c+\delta)$ which are arbitrarily close to $\gamma$ which in turn implies that there also exist timelike curves from $\Sigma$ to $\gamma(b)$.

Let $\mu\colon(-\epsilon,\epsilon)$ be a curve in $\Sigma$ with $\mu(0)=\gamma(a)$, $\dot\mu(0)=\xi(a)$, and $\boldsymbol{\nabla}_{\dot\mu(0)}\dot\mu=0$ where $\boldsymbol{\nabla}$ is the Levi-Cività connection of $\Sigma$. Let $\hat f$ be a variation such that $\hat f(s,a)=\mu(s)$ and $\hat f(s,c+\delta)=\gamma(c+\delta)$ for all $s\in(-\epsilon,\epsilon)$. For $\epsilon$ sufficiently small there is a smooth map $V\colon(s,t)\mapsto V(s,t)\in T_{\gamma(t)}M$ with $\exp_{\gamma(t)}(Z(s,t))=\hat f(s,t)$, where $Z(0,t)=0_{\gamma(t)}$. We will now modify this variation so that the modified variation has variation vector field $\xi$ and acceleration vector field $A$. Restricting the equation $\partial_s\exp(Z(s,t))=T_{Z(s,t)}\exp_{\gamma(t)}\frac{\mathrm{d}}{\mathrm{d}s}Z(s,t)$ to $s=0$ implies that the variation vector field $\hat\xi$ of $\hat f$ is given by $\left(\frac{\mathrm{d}}{\mathrm{d}s}Z\right)_{(0,t)}$. In order to calculate the acceleration vector field $\hat A$ let $\{x^1,\dots,x^n,y^1,\dots,y^n\}$ be a coordinate system of $TM$ such that $y^k(v)=v^k$ for any vector $v$. For any given $v\in T_xM$ let $\lambda$ be the geodesic given by $\lambda(s)=\exp_x(sv)$. Then the calculation

$$
\begin{aligned}
0=\left(\nabla_{\dot\lambda}\dot\lambda\right)^i &= \frac{\mathrm{d}^2}{\mathrm{d}s^2}\left(\lambda^i(s)\right)+\Gamma^i_{jk}\dot\lambda^j(s)\dot\lambda^k(s)\\
&=\frac{\mathrm{d}^2}{\mathrm{d}s^2}\left(\exp_x(sv)\right)+\Gamma^i_{jk}\dot\lambda^j(s)\dot\lambda^k(s)\\
&=\frac{\partial^2}{\partial y^j y^k}\left(\exp_x\right)^i v^j v^k+\Gamma^i_{jk}\dot\lambda^j(s)\dot\lambda^k(s)
\end{aligned}
$$

implies $\frac{\partial^2}{\partial y^j y^k}\left(\exp_x\right)^i(0)=-\Gamma^i_{jk}(x)$. If follows that the equation

$$
\hat A=\left(\nabla_{\hat\xi}\partial_s\exp_{\gamma(t)}(Z)\right)_{|(0,t)}=\left(\frac{\mathrm{d}^2}{\mathrm{d}s^2}Z\right)_{(0,t)}
$$

holds and completely analogously that the variation

$$
f(s,t)=\exp_{\gamma(t)}\left(Z(s,t)+s(\xi(t)-\hat\xi(t))+\frac{s^2}{2}(A(t)-\hat A(t))\right)
$$

has variation vector field $\xi$ and acceleration vector field $A$. By construction, each of the curves $t\mapsto f(s,t)$ starts in $\Sigma$ and ends in $\gamma(c+\delta)$. ∎

### 4.6.3 Existence of focal points

The existence of focal points depends on three factors, the curvature of the pseudo-Riemannian manifold near $\gamma$, the length of $\gamma$, and the shape of the submanifold $\Sigma$. A typical result is the following.

**Proposition 4.6.1.** *Let $\Sigma$ be a spacelike hypersurface and $\gamma\colon [a,b] \to M$ be a geodesic with $\langle\dot\gamma,\dot\gamma\rangle = \eta \in \{-1,1\}$ and $\dot\gamma(a) \in \left(T_{\gamma(a)}\Sigma\right)^{\perp}$. If $\mathrm{Ric}(\dot\gamma(t),\dot\gamma(t)) \geq 0$ for all $t$ and the mean curvature vector field $H$ of $\Sigma$ satisfies $\left\langle H_{\gamma(a)},\dot\gamma(a)\right\rangle =: c > 0$, then there is a focal point of $\Sigma$ along $\gamma$ before $\gamma(a+1/c)$, provided $c > 1/(b-a)$.*

*Proof.* Let $\{e_1,\ldots,e_{n-1}\}$ be a basis of $T_{\gamma(a)}\Sigma$ and $E_i$ the parallel translation of $e_i$ along $\gamma$. The vector field $\xi_i(t) := (1+ca-ct)E_i(t)$ vanishes at $a+1/c$ and is tangent to $\Sigma$ at $t=a$. Since

$$
\begin{aligned}
\sum_{i=1}^{n-1} I^{L,\gamma}_{\Sigma,\{\gamma(a+1/c)\}}(\xi_i,\xi_i)
&= \eta \sum_{i=1}^{n-1}\Big(c + \int_a^{a+1/c} (1+ca-ct)^2 \langle R(E_i,\dot\gamma)E_i,\dot\gamma\rangle\,\mathrm{d}t \\
&\quad - \langle\dot\gamma(a), I\!I(E_i(a),E_i(a))\rangle\Big) \\
&= \eta\Big((n-1)c - \int_a^{a+1/c} (1+ca-ct)^2 \mathrm{Ric}(\dot\gamma(t),\dot\gamma(t))\mathrm{d}t \\
&\quad - \left\langle\dot\gamma(a),(n-1)H_{\gamma(a)}\right\rangle\Big) \\
&= -\eta \int_a^{a+1/c} (1+ca-ct)^2 \mathrm{Ric}(\dot\gamma(t),\dot\gamma(t))\mathrm{d}t
\end{aligned}
$$

we have found a vector field $\xi$ whith negative index in the Riemannian and positive index in the Lorentzian case. If follows that $I^{L,\gamma}_{\Sigma,\{\gamma(a+1/c)\}}$ is semi-definite or indefinite and therefore the existence of a focal point before $\gamma(a+1/c)$ ∎

The inequality given in Proposition 4.6.1 is sharp. Consider a sphere of radius $r$ and with inner normal $n$. Then its mean curvature vector field is given by $H_x = \frac{1}{r}n$. The geodesic $\gamma$ with $\dot\gamma(a) = n_x$ satisfies $\langle H_x,\dot\gamma(a)\rangle = \frac{1}{r}$ and intersects the centre of the sphere at $\gamma(a+r)$. Clearly, the centre is the first focal point of the sphere along $\gamma$.

There is an analogue of Proposition 4.6.1 for null geodesics and spacelike submanifolds of co-dimension 2. This analogue will become important in Chap. 9 on singularities in general relativity.

**Proposition 4.6.2.** *Let $\Sigma$ be a spacelike submanifold of dimension $n-2$ and $\gamma\colon [a,b] \to b$ be a null geodesic with $\dot\gamma(a) \in \left(T_{\gamma(a)}\Sigma\right)^{\perp}$. If*

$$\mathrm{Ric}(\dot\gamma(t),\dot\gamma(t)) \geq 0$$

*for all $t$ and the mean curvature vector field $H$ of $\Sigma$ satisfies*

$$\left\langle H_{\gamma(a)},\dot\gamma(a)\right\rangle =: c > 0,$$

*then there is a focal point of $\Sigma$ along $\gamma$ before $\gamma(a+1/c)$, provided $c > 1/(b-a)$.*

*Proof.* The proof is analogous to the proof of Proposition 4.6.1. We choose a basis $\{e_1, \ldots, e_{n-2}\}$ of $T_{\gamma(a)}\Sigma$. There are a spacelike unit vector $e_{n-1}$ and a timelike unit vector $e_n$, both orthogonal to $\Sigma$ such that $\dot\gamma(a) = e_{n-1} + e_n$. We denote the parallel translation of $e_k$ along $\gamma$ by $E_k$. Then we have

$$\begin{aligned}
\mathrm{Ric}(\dot\gamma, \dot\gamma) &= \sum_{i=1}^{n-2} \langle R(E_i, \dot\gamma)\dot\gamma, E_i\rangle + \langle R(E_{n-1}, \dot\gamma)\dot\gamma, E_{n-1}\rangle - \langle R(E_n, \dot\gamma)\dot\gamma, E_n\rangle \\
&= \sum_{i=1}^{n-2} \langle R(E_i, \dot\gamma)\dot\gamma, E_i\rangle + \langle R(E_{n-1}, E_n)E_n, E_{n-1}\rangle \\
&\quad - \langle R(E_n, E_{n-1})E_{n-1}, E_n\rangle \\
&= \sum_{i=1}^{n-2} \langle R(E_i, \dot\gamma)\dot\gamma, E_i\rangle .
\end{aligned}$$

This implies

$$\begin{aligned}
\sum_{i=1}^{n-3} & I^{E,\gamma,\perp}_{\Sigma,\{\gamma(a+1/c)\}}((1+ca-ct)E_i, (1+ca-ct)E_i) \\
&= \sum_{i=1}^{n-1} \left( c + \int_a^{a+1/c} (1+ca-ct)^2 \langle R(E_i, \dot\gamma)E_i, \dot\gamma\rangle \,\mathrm{d}t \right. \\
&\quad \left. - \langle \dot\gamma(a), I\!I(E_i(a), E_i(a))\rangle \right) \\
&= (n-1)c - \int_a^{a+1/c} (1+ca-ct)^2 \mathrm{Ric}(\dot\gamma(t), \dot\gamma(t))\mathrm{d}t \\
&\quad - \langle \dot\gamma(a), (n-1)H_{\gamma(a)}\rangle \\
&= -\int_a^{a+1/c} (1+ca-ct)^2 \mathrm{Ric}(\dot\gamma(t), \dot\gamma(t))\mathrm{d}t .
\end{aligned}$$

Hence $I^{E,\gamma,\perp}_{\Sigma,\{\gamma(a+1/c)\}}$ is not positive semi-definite and therefore there is a focal point before $\gamma(a+1/c)$. ∎

It is also possible to prove the existence of conjugate points of geodesics $\gamma$, if they are sufficiently long and if suitable curvature conditions hold along $\gamma$ (cf. Proposition 4.6.3 below). This result is central to the singularity theorem which is presented in Chap. 9. Since in the case of a single geodesic we do not have initial conditions which guarantee focussing in one direction, the proof of Proposition 4.6.3 will be much more delicate than the proof of Propositions 4.6.1 and 4.6.2. The rest of this section will be devoted to proving the following proposition:[8]

**Proposition 4.6.3.** *Let $(M, g)$ be a Riemannian or Lorentzian manifold and $\gamma$ be a complete geodesic which is spacelike in the Riemannian and causal in the Lorentzian case. If* $\mathrm{Ric}(\dot\gamma(t), \dot\gamma(t)) \geq 0$ *for all $t$ and if there is a $t_0$ such that the map*

$$R\colon (\dot\gamma(t_0))^\perp \to (\dot\gamma(t_0))^\perp, \quad v \mapsto Rv := R(v, \dot\gamma)\dot\gamma$$

*is not identically zero. Then $\gamma$ has a pair of conjugate points.*

The proof of this proposition requires more results from the theory of conjugate points.

We have seen before that there is an $(n-1)$-dimensional subspace of Jacobi fields along $\gamma$ which have values in $(\dot\gamma)^\perp$ and vanish at a given point. In the case that $\gamma$ is a null geodesic, there is always a linear combinations of these Jacobi fields which is equal to the trivial Jacobi field $t \mapsto (\alpha t + \beta)\dot\gamma(t)$. This uninteresting subspace disappears if one consideres the factor space $(\dot\gamma(t))^\perp / \mathbb{R}\dot\gamma(t) = \{[v] \,|\, v \in (\dot\gamma(t))^\perp, v \sim w \Leftrightarrow v - w \parallel \dot\gamma(t)\}$ instead of the orthogonal complement $(\dot\gamma(t))^\perp$. This space coincides with the orthogonal complement if $\gamma$ is timelike or spacelike. If $\gamma$ is a null geodesic then working with the factor space has two advantages: As we have indicated above, $[(\alpha t + \beta)\dot\gamma(t)] = [0]$. More importantly, the metric $g$ induces a metric $[g]$ on $(\dot\gamma(t))^\perp / \dot\gamma(t)$ which is positive definite instead of degenerate.

**Definition 4.6.6.** *Let $\gamma\colon [a, b] \to M$ be a geodesic and $t \in [a, b]$. Two vectors $v, w \in (\dot\gamma(t))^\perp$ are called equivalent, $v \sim w$, if there is a number $\alpha \in \mathbb{R}$ with $v = w + \alpha\dot\gamma(t)$. We denote the space of equivalence classes $\{[v] \,|\, v \in (\dot\gamma(t))^\perp\}$ by $[\dot\gamma(t)]^\perp$ and set $[\dot\gamma]^\perp = \bigcup_{t \in [a,b]} [\dot\gamma(t)]^\perp$.*

*A map $[A](t)\colon [\dot\gamma(t)]^\perp \times \cdots \times [\dot\gamma(t)]^\perp \times \left([\dot\gamma(t)]^\perp\right)^* \times \cdots \times \left([\dot\gamma(t)]^\perp\right)^* \to \mathbb{R}$ along $\gamma$ which is linear in each of its entries is called a* tensor class *along $\gamma$.*

From the definition it is clear that any tensor $A$ of $(\dot\gamma(t))^\perp$ induces a tensor class $[A]$ at $t$ via

$$[A]([v_1], \ldots, [v_s], [\varphi^1], \ldots, [\varphi^r]) = A(v_1, \ldots, v_s, \varphi^1, \ldots, \varphi^r)$$

where $\varphi^i$ is defined by $[\varphi^i]([v]) = \varphi^i(v)$ for all $v \in (\dot\gamma(t))^\perp$ (In particular, $\varphi^i(\dot\gamma) = 0$). Conversely, any tensor class is induced by a tensor in this way.

The metric and the covariant derivative in direction $\dot\gamma$ induce analogous objects for tensor classes.

[8] We follow closely the presentation in (Beem and Ehrlich 1981). There is also a much shorter proof in (Hawking and Ellis 1973) which I failed to understand.

**Lemma 4.6.16.** *Let $\gamma$ be a geodesic and $[A]$ be a tensor class along $\dot\gamma$ and $\hat A$ be any tensor field along $\gamma$ with $[\hat A] = [A]$. Then $[\dot A] := \nabla_{\dot\gamma}\hat A$ is well defined.*

*If $\gamma$ is a null geodesic then the metric $[g]\colon [\dot\gamma]^\perp \times [\dot\gamma]^\perp \to \mathbb{R}$, $[v,w] \mapsto [g]([v],[w]) := g(v,w)$ is well defined and positive definite.*

*The operator $[R]\colon [\dot\gamma(t)]^\perp \to [\dot\gamma(t)]^\perp$, $[v] \mapsto [R(v,\dot\gamma)\dot\gamma]$ is well defined.*

*Proof.* For any one-form $\varphi$ satisfying $\varphi(\dot\gamma) = 0$ we have

$$\begin{aligned}
&\left(\nabla_{\dot\gamma(t)}\varphi\right)(V(t) + f(t)\dot\gamma(t))\\
&\qquad = \nabla_{\dot\gamma(t)}\left(\varphi(V(t) + f(t)\dot\gamma(t))\right) - \varphi\left(\nabla_{\dot\gamma(t)}(V(t) + f(t)\dot\gamma(t))\right)\\
&\qquad = \nabla_{\dot\gamma(t)}(\varphi(V(t)))\\
&\qquad\quad - \varphi\left(\nabla_{\dot\gamma(t)}V(t) + \mathrm{d}f\dot\gamma(t)\dot\gamma(t) + f\nabla_{\dot\gamma(t)}\dot\gamma(t)\right)\\
&\qquad = \nabla_{\dot\gamma(t)}(\varphi(V(t)) - \varphi\left(\nabla_{\dot\gamma(t)}V(t)\right) = \nabla_{\dot\gamma}\varphi(V(t)),
\end{aligned}$$

where we have used that $\nabla_{\dot\gamma}\dot\gamma = 0$. For any vector field $V$ along $\gamma$ we have $[\nabla_{\dot\gamma}(V(t) + f(t)\dot\gamma(t))] = [\nabla_{\dot\gamma}V(t) + \mathrm{d}f(\dot\gamma)\dot\gamma + f\nabla_{\dot\gamma}\dot\gamma] = [\nabla_{\dot\gamma}V(t)]$. Hence the first assertion holds for 1-forms, vector fields, and (trivially) for functions along $\gamma$. Since $\nabla_{\dot\gamma}$ is a derivation this proves the first claim.

The second claim follows since there is a basis $e_1, \ldots, e_{n-1}$ of $(\dot\gamma(t))^\perp$ such that $g(e_i, e_j) = \delta_{ij}$ for $i,j \in \{1,\ldots,n-2\}$ and $g(e_{n-1}, e_k) = 0$ for $k \in \{1,\ldots,n-1\}$.

The third assertion is a consequence of $R(\dot\gamma,\dot\gamma)\dot\gamma = 0$. ∎

**Definition 4.6.7.** *A* Jacobi tensor class *is a tensor class $[A]\colon [\dot\gamma]^\perp \to [\dot\gamma]^\perp$ along $\gamma$ for which $[\ddot A] + [R][A] = [0]$ holds.*

**Lemma 4.6.17.** *A tensor class $[A]\colon [\dot\gamma]^\perp \to [\dot\gamma]^\perp$ along $\gamma$ is a Jacobi tensor class if and only there is a tensor field $A$ along $\gamma$ which induces $[A]$ and has the property that $t \mapsto AV(t)$ is a Jacobi field for every parallel vector field $V$ with values in $(\dot\gamma)^\perp$.*

*Proof.* Suppose that $A$ is a tensor field along $\gamma$ such that $AV \in (\dot\gamma)^\perp$ is a Jacobi field for any parallel vector field $V$ with values in $(\dot\gamma)^\perp$. It follows immediately from

$$\begin{aligned}
\left(\nabla_{\dot\gamma}\nabla_{\dot\gamma}[A] + [R][A]\right)[V] &= \left[\left(\nabla_{\dot\gamma}\nabla_{\dot\gamma}A\right)V\right] + [R(AV,\dot\gamma)\dot\gamma]\\
&= \left[\nabla_{\dot\gamma}\nabla_{\dot\gamma}(AV) + R(AV,\dot\gamma)\dot\gamma\right]
\end{aligned}$$

that $[A]$ is then a Jacobi tensor class. Conversely observe that there are exactly $(2n-2)$ linearly independent Jacobi vector fields $J_i$ along

$\gamma$ which have values in $(\dot\gamma)^\perp$. Let $\{E_1, \ldots, E_{n-1}\}$ be a parallel frame of $(\dot\gamma)^\perp$ and define the tensor field $A\colon (\dot\gamma)^\perp \to (\dot\gamma)^\perp$ by $Av = \sum_{i,j=1}^{n-1} A_i^j v^i J_j$ where $v = \sum_{i=1}^{n-1} v^i E_i$ and the $A_i^j$ are (constant) real numbers. It is clear that for every parallel vector field $V \in (\dot\gamma)^\perp$ the vector field $AV$ is a Jacobi field with values in $(\dot\gamma)^\perp$. Hence $[A]$ is a Jacobi tensor class. Since the differential equation $[\ddot A] + [R][A] = 0$ implies that the space of Jacobi tensor fields is $4(n-1)^2$-dimensional if $\gamma$ is timelike or spacelike and $4(n-2)^2$-dimensional if $\gamma$ is null every Jacobi tensor field can be generated by some tensor field $A$ as constructed above. ∎

It is clear from the proof of Lemma 4.6.17 that the columns of a Jacobi tensor class with respect to a parallel basis of $(\dot\gamma(t))^\perp$ are just Jacobi fields expressed in this basis.

**Corollary 4.6.6.** *Let $\gamma\colon [a,b] \to M$ be a geodesic and $t_0, t_1 \in [a,b]$ $(t_0 \neq t_1)$.*

*(i) For any pair of tensor classes*

$$[A_0]\colon [\dot\gamma(t_0)]^\perp \to [\dot\gamma(t_0)]^\perp, \quad [\dot A_0]\colon [\dot\gamma(t_0)]^\perp \to [\dot\gamma(t_0)]^\perp$$

*there is a unique Jacobi tensor class $[A]$ with $[A](t_0) = [A_0]$ and $[\dot A](t_0) = \dot A_0$;*

*(ii) Assume that $\gamma$ does not have conjugate points and let*

$$[A_0]\colon [\dot\gamma(t_0)]^\perp \to [\dot\gamma(t_0)]^\perp, \quad [A_1]\colon [\dot\gamma(t_1)]^\perp \to [\dot\gamma(t_1)]^\perp$$

*be a given pair of tensor classes. Then there is a unique Jacobi tensor class $[A]$ with $[A](t_0) = [A_0]$ and $[A](t_1) = [A_1]$.*

*Proof.* The assertions follow immediately from Lemma 4.6.11 and Propositions 2.9.2, 2.9.4. ∎

The following lemma is clear from the definitions and the fact that a non-vanishing Jacobi field which is parallel to $\dot\gamma$ has at most one zero:

**Lemma 4.6.18.** *Let $\gamma$ be a geodesic. Two points $\gamma(c)$, $\gamma(d)$ are conjugate if and only if the Jacobi tensor class $[A]$ which satisfies $[A](c) = [0]$, $[\dot A]](c) = \mathrm{id}$ is singular at $d$.*

**Definition 4.6.8.** *Let $\gamma$ be a geodesic and $[B](t)\colon [\dot\gamma(t)]^\perp \to [\dot\gamma(t)]^\perp$ be a tensor class along $\gamma$. Then the adjoint of $[B]$ with respect to $[g]_{\gamma(t)}$ is denoted by $[B]^*$.*

**Lemma 4.6.19.** *Let $[A]$ be a Jacobi tensor class along a geodesic $\gamma\colon [a,b] \to M$ and assume that there is a number $t_0 \in [a,b]$ with $[A](t_0) = [0]$. If $[A]$ is non-singular at $t$ then the tensor class $[\dot A][A]^{-1}$ is self-adjoint at all $t$.*

*Proof.* Let $V, W$ be parallel vector fields along $\gamma$ with values in $(\dot\gamma)^\perp$. The equations

$$\begin{aligned}\nabla_{\dot\gamma}\Big(\big\langle AV, \nabla_{\dot\gamma}AW\big\rangle - \big\langle \nabla_{\dot\gamma}AV, AW\big\rangle\Big)&\\ &= \big\langle \nabla_{\dot\gamma}AV, \nabla_{\dot\gamma}AW\big\rangle + \big\langle AV, \nabla_{\dot\gamma}\nabla_{\dot\gamma}AW\big\rangle\\ &\quad - \big\langle \nabla_{\dot\gamma}\nabla_{\dot\gamma}AV, AW\big\rangle - \big\langle \nabla_{\dot\gamma}AV, \nabla_{\dot\gamma}AW\big\rangle\\ &= \langle -R(AV,\dot\gamma)\dot\gamma, AW\rangle - \langle AV, -R(AW,\dot\gamma)\dot\gamma\rangle = 0\end{aligned}$$

and $A(t_0) = 0$ imply that $\big\langle Av, \nabla_{\dot\gamma}Aw\big\rangle = \big\langle \nabla_{\dot\gamma}Av, Aw\big\rangle$ for all vectors $v, w \in (\dot\gamma)^\perp$. It follows that

$$\begin{aligned}\big\langle \nabla_{\dot\gamma}AA^{-1}v, w\big\rangle &= \big\langle \nabla_{\dot\gamma}A(A^{-1}v), A(A^{-1}w)\big\rangle\\ &= \big\langle A(A^{-1}v), \nabla_{\dot\gamma}A(A^{-1}w)\big\rangle\\ &= \big\langle v, \nabla_{\dot\gamma}AA^{-1}w\big\rangle.\end{aligned}$$

∎

The self-adjoint tensor class $[\dot A][A]^{-1}$ has a direct geometrical interpretation in terms of congruences of geodesics. Let $\gamma\colon [a,b] \to M$ be a spacelike (in the Riemannian case) or timelike (if $M$ is Lorentzian) geodesic[9] and $f\colon \mathbb{R}^{n-1}\times\mathbb{R} \to M$, $(s^1,\dots,s^{n-1},t) \mapsto f(s^1,\dots,s^{n-1},t)$ be a smooth $(n-1)$-parameter geodesic variation of $\gamma$. We may assume that each geodesic $t \mapsto f(s^1,\dots,s^{n-1},t)$ satisfies $\langle f_t, f_t\rangle \in \{-1,1\}$ and that at $t = a$ the vectors $\{f_t, f_{s^1},\dots,f_{s^{n-1}}\}$ are linearly independent. Then $U = f_t(s^1,\dots,s^{n-1},t)$ is a well defined vector field near $\gamma(a)$. The covariant derivative of $U$ geometric properties of our congruence of geodesics. The function $\theta = \operatorname{div}(U)$ measures the divergence of neighbouring geodesics. Analogously, $\omega^\flat = dU^\flat$ is the infinitesimal rotation and $\sigma$, the traceless, symmetric part of $\nabla U$, the infinitesimal, volume preserving distortion of neighbouring geodesics. At $(s^1,\dots,s^{n-1}) = 0$ we can recover this information in terms of Jacobi classes.

The Jacobi field $f_{s^i}$ satisfies

$$\nabla_{f_t}\langle f_s, f_t\rangle = \big\langle \nabla_{f_t}f_s, f_t\big\rangle = \big\langle \nabla_{f_s}f_t, f_t\big\rangle = 0.$$

Hence $\langle f_{s^i}(0,\dots,0,a), \dot\gamma(a)\rangle = 0$ implies $\langle f_{s^i}(0,\dots,0,t), \dot\gamma(t)\rangle = 0$ for all $t \in [a,b]$. Since we can replace the parameter $t$ by $t + h(s^1,\dots,s^{n-1})$, where $h$ is an arbitrary function, we can always parameterize our geodesics such that $f_{s^k}(0,\dots,0,t) \in (\dot\gamma(t))^\perp$ for all $k$ and $t$. Let $\{E_i(t)\}_{i=1,\dots,n-1}$

[9] The interpretation in for lightlike geodesics is slightly less direct.

be a orthonormal basis of $(\dot\gamma)^\perp$ which is parallel along $\gamma$ and denote by $A\colon (\dot\gamma)^\perp \to (\dot\gamma)^\perp$ the tensor field along $\gamma$ that maps $E_i$ into $f_{s^i}(0,\dots,0,t)$. It follows that $[A]$ is a Jacobi tensor class and that

$$(\nabla_{\dot\gamma}A)A^{-1}f_{s^i} = (\nabla_{\dot\gamma}A)E_i = \nabla_{\dot\gamma}(AE_i) = \nabla_{f_t}f_{s^i} = \nabla_{f_{s^i}}f_t = \nabla_{f_{s^i}}U.$$

Since the vector fields $\{f_{s^1},\dots,f_{s^n}\}$ span $(\dot\gamma)^\perp$ we conclude that $\nabla_v U = (\nabla_{\dot\gamma}A)A^{-1}v$ for all $v \in \dot\gamma^\perp$. This motivates the following definition.

**Definition 4.6.9.** *Let $[A]$ be a Jacobi tensor class along the geodesic $\gamma$. Then the* expansion $\theta$ *of $[A]$ is defined by*

$$\theta = \mathrm{tr}([\dot A][A]^{-1}),$$

*the* vorticity tensor $\omega$ *of $[A]$ by*

$$\omega(t) = \frac{1}{2}\left([\dot A][A]^{-1} - \left([\dot A][A]^{-1}\right)^*\right),$$

*and its* shear tensor $\sigma$ *by*

$$\sigma(t) = \begin{cases} \frac{1}{2}\left([\dot A][A]^{-1} + \left([\dot A][A]^{-1}\right)^*\right) - \frac{\theta(t)}{n-1}\mathrm{id} & \text{if } \langle\dot\gamma,\dot\gamma\rangle \in \{-1,1\}, \\ \frac{1}{2}\left([\dot A][A]^{-1} + \left([\dot A][A]^{-1}\right)^*\right) - \frac{\theta}{n-2}\mathrm{id} & \text{if } \langle\dot\gamma,\dot\gamma\rangle = 0. \end{cases}$$

The following lemma implies that $\theta(t) = \mathrm{tr}([\dot A][A]^{-1})$ diverges where $A$ is singular.

**Lemma 4.6.20.** *For any a Jacobi vector class $[A]$ we have*

$$\theta = \frac{1}{\det([A])}(\det([A]))^{\cdot},$$

*where* det *is any parallel determinant function. (In particular, one can choose the determinant function induced by the metric $[g]$).*

*Proof.* Let $r = n-1$ if $\gamma$ is spacelike or timelike and $r = n-2$ if $\gamma$ is null.

Since the space of parallel determinant functions along $\gamma$ is 1-dimensional, it is clear that the the formula in the assertion is independent of the choice of det.

Assume that $[A](t_0)$ is non-singular. There is a parallel linear tensor class $[B]$ such that $[A](t_0)[B](t_0) = \mathrm{id}$. Let $[C] = [A][B]$ and $\{[E_i]\}_{i=1,\dots,r}$ be a parallel orthonormal basis of $[\dot\gamma]^\perp$. We choose the determinant function defined by $\det([D]) = \mathrm{Det}(([D]^i_k)_{i,k})$, where Det is the standard determinant in $\mathbb{R}^r$ and $[D][E_k] = [D]^i_k[E_i]$. At $t = t_0$ we have $[C][E_k] = [E_k]$ and and $\mathrm{Det}([E_1],\dots,[E_r]) = 1$. This implies

$$
\begin{aligned}
(\det[C])^{\cdot}{}_{|t=t_0} &= (\det([C][E_1], \ldots, [C][E_{n-1}])^{\cdot}{}_{|t=t_0} \\
&= \sum_{i=1}^{r} \det([E_1], \ldots, [E_{i-1}], [\dot{C}][E_i], [E_{i+1}], \ldots, [E_r])_{|t=t_0} \\
&= \sum_{i=1}^{n-1} [\dot{C}]_i^i \det([E_1], \ldots, [E_{i-1}], [E_i], [E_{i+1}], \ldots, [E_r])_{|t=t_0} \\
&= \operatorname{tr}([\dot{C}])_{|t=t_0}.
\end{aligned}
$$

Since $[A] = [C][B]^{-1}$ we obtain therefore at $t = t_0$

$$
\begin{aligned}
(\det[A])^{\cdot} &= (\det[C] \det[B]^{-1})^{\cdot} = (\det[C])^{\cdot} \det[B]^{-1} \\
&= \operatorname{tr}([\dot{C}]) \det[B]^{-1} = \operatorname{tr}([\dot{C}][B]^{-1}[B]) \det[C]^{-1} \det([C][B]^{-1}) \\
&= \operatorname{tr}([\dot{A}][B]) \det[C]^{-1} \det[A] = \operatorname{tr}([\dot{A}][A]^{-1}[C]) \det[C]^{-1} \det[A].
\end{aligned}
$$

At $t_0$ we have $[C] = \mathrm{id}$ and therefore $(\det[A])^{\cdot} = \operatorname{tr}([\dot{A}][A]^{-1}) \det[A]$. ∎

**Lemma 4.6.21 (Raychaudhury equation).** *Let $(M, g)$ be a Lorentzian or Riemannian manifold and $\gamma$ be a causal geodesic if $(M, g)$ is Lorentzian and a spacelike geodesic otherwise. If $[A]$ is a Jacobi tensor class then its expansion satisfies*

$$
\dot{\theta} = \begin{cases} -\operatorname{Ric}(\dot{\gamma}, \dot{\gamma}) - \operatorname{tr}(\omega^2) - \operatorname{tr}(\sigma^2) - \frac{1}{n-1}\theta^2 & \text{if } \langle \dot{\gamma}, \dot{\gamma} \rangle \in \{-1, 1\}, \\ -\operatorname{Ric}(\dot{\gamma}, \dot{\gamma}) - \operatorname{tr}(\omega^2) - \operatorname{tr}(\sigma^2) - \frac{1}{n-2}\theta^2 & \text{if } \langle \dot{\gamma}, \dot{\gamma} \rangle = 0. \end{cases}
$$

*Proof.* Let $r = n-1$ if $\gamma$ is spacelike or timelike and $r = n-2$ if $\gamma$ is null. Let $\{E_i\}_{i=1,\ldots,r}$ be a parallel orthonormal frame of $(\dot{\gamma})^{\perp}$. Since $[A]$ is a Jacobi class we have $([\dot{A}][A]^{-1})^{\cdot} = [\ddot{A}][A]^{-1} + [\dot{A}](-[A]^{-1}[\dot{A}][A]^{-1}) = -[R] - ([\dot{A}][A]^{-1})^2$ and therefore

$$
\begin{aligned}
\dot{\theta} &= \left(\operatorname{tr}\left([\dot{A}][A]^{-1}\right)\right)^{\cdot} = \operatorname{tr}\left(\left([\dot{A}][A]^{-1}\right)^{\cdot}\right) \\
&= -\operatorname{tr}([R]) - \operatorname{tr}([\dot{A}][A]^{-1})^2 \\
&= -\sum_{i=1}^{r} g\left(R(E_i, \dot{\gamma})\dot{\gamma}, E_i\right) - \operatorname{tr}\left(\left(\omega + \sigma + \frac{\theta}{r}\,\mathrm{id}\right)^2\right).
\end{aligned}
$$

If $r = n-1$ it is clear that $\sum_{i=1}^{r} g\left(R(E_i, \dot{\gamma})\dot{\gamma}, E_i\right) = -\operatorname{Ric}(\dot{\gamma}, \dot{\gamma})$. If $\gamma$ is null we can find a parallel spacelike vector field $E_{n-1}$ and a timelike vector field $E_n$ such that $\{E_1, \ldots, E_n\}$ are orthonormal and $\dot{\gamma} = E_{n-1} + E_n$. Then we have

$$\begin{aligned}
\mathrm{Ric}(\dot\gamma,\dot\gamma) &= \sum_{i=1}^{n} g\left(R(\dot\gamma,E_i)\dot\gamma,E_i\right)\\
&= \sum_{i=1}^{n-2} g\left(R(\dot\gamma,E_i)\dot\gamma,E_i\right)\\
&\quad + g\left(R\left(E_{n-1}+E_n,E_{n-1}\right)\left(E_{n-1}+E_n\right),E_{n-1}\right)\\
&\quad - g\left(R\left(E_{n-1}+E_n,E_n\right)\left(E_{n-1}+E_n\right),E_n\right)\\
&= \sum_{i=1}^{n-2} g\left(R(\dot\gamma,E_i)\dot\gamma,E_i\right) + g\left(R\left(E_n,E_{n-1}\right)E_n,E_{n-1}\right)\\
&\quad - g\left(R\left(E_{n-1},E_n\right)E_{n-1},E_n\right)\\
&= \sum_{i=1}^{n-2} g\left(R(\dot\gamma,E_i)\dot\gamma,E_i\right),
\end{aligned}$$

where we have used the symmetries of $R$ and the fact that

$$\mathrm{tr}(B) = \sum_{i=1}^{n-1} g(BE_i,E_i) - g(BE_n,E_n)$$

for every linear map $B$. Hence in either case, $r = n-1$ or $r = n-2$, we get

$$\dot\theta = \mathrm{Ric}(\dot\gamma,\dot\gamma) - \mathrm{tr}\left(\omega^2+\sigma^2+\frac{\theta^2}{r^2}\mathrm{id}+\frac{2}{r}(\omega+\sigma)+\omega\sigma+\sigma\omega\right)$$

By definition we have $\mathrm{tr}(\omega) = \mathrm{tr}(\sigma) = 0$. For any tensor $[B]$ we have $\mathrm{tr}(([B]+[B]^*)([B]-[B]^*)) = \mathrm{tr}([B]^2) - \mathrm{tr}(([B]^*)^2) + \mathrm{tr}([B]^*[B]) - \mathrm{tr}([B][B]^*)$. Since the definition of the trace implies

$$\begin{aligned}
\mathrm{tr}([B]^2) &= \sum_{i=1}^{n-1}[g]\left([B]^2[E_i],[E_i]\right) = \sum_{i=1}^{n-1}[g]\left([E_i],([B]^*)^2)[E_i]\right)\\
&= \mathrm{tr}(([B]^*)^2)
\end{aligned}$$

and

$$\mathrm{tr}([B]^*[B]) = \mathrm{tr}([B]([B]^*[B])[B]^{-1}) = \mathrm{tr}(B[B]^*)$$

we conclude that $\mathrm{tr}(\omega\sigma) = \mathrm{tr}(\sigma\omega) = 0$. ∎

**Lemma 4.6.22.** *Let $\gamma$ be a timelike or spacelike geodesic and $[A]$, $[B]$ be Jacobi tensor classes along $\gamma$. Then the tensor class $([A]^*)^{\cdot}[B] - [A]^*[\dot B]$ is parallel along $\gamma$.*

*Proof.* First observe that $[R]$ is self-adjoint. In fact,

$$[g]\,([R][v],[w]) = \langle R(v,\dot\gamma)\dot\gamma, w\rangle = \langle R(w,\dot\gamma)\dot\gamma, v\rangle = [g]\,([v],[R][w])$$

for all vectors $[v],[w]$ implies $[R]=[R]^*$. Hence we obtain

$$\begin{aligned}\Big(([A]^*)^{\cdot}[B]-[A]^*[\dot B]\Big)^{\cdot} &= ([A]^*)^{\cdot\cdot}[B]+([A]^*)^{\cdot}[\dot B]-([A]^*)^{\cdot}[\dot B]\\ &\quad -[A]^*[\ddot B]\\ &= ([\ddot A])^*[B]-[A]^*[\ddot B]\\ &= -([R][A])^*[B]+[A]^*[R][B]\\ &= -[A]^*[R]^*[B]+[A]^*[R][B]=0.\end{aligned}$$

∎

We are now in a position to prove Proposition 4.6.3

*Proof of Proposition 4.6.3.* Let $\gamma\colon \mathbb{R}\to M$ be a complete geodesic and $r=n-1$ if $\gamma$ is spacelike or timelike and $r=n-2$ if $\gamma$ is null. We choose $t_0\in\mathbb{R}$ such that $R(\cdot,\dot\gamma(t_0))\dot\gamma(t_0)\neq 0$. The symmetries of $R$ imply then that the induced operator $[R]\colon [\dot\gamma(t_0)]^\perp\to[\dot\gamma(t_0)]^\perp$, $[v]\mapsto [R(v,\dot\gamma(t_0))\dot\gamma(t_0)]$ does not vanish. We need to show that $\gamma$ has a pair of conjugate points. Let $J_\pm$ be the space of all Jacobi tensor classes $[A]$ which satisfy $\omega=0$, $[A](t_0)=\mathrm{id}$, and $\mathrm{tr}([\dot A](t_0))\gtreqless 0$.

We will first show that each $[A]\in J_\pm$ satisfies $\det[A](t)=0$ for some $t\lessgtr t_0$. Suppose (without loss of generality) that $[A]\in J_-$. Since the shear $\sigma$ is self-adjoint, $\mathrm{tr}(\sigma^2)\geq 0$ and the Raychaudhury equation implies $\dot\theta\leq -\frac{1}{r}\theta^2$. If there is a $t_1>t_0$ with $\theta(t_1)<0$ then an integration implies $\frac{1}{\theta(t)}\geq\frac{1}{\theta(t_1)}+\frac{t-t_1}{n-1}$ for all $t\geq t_1$. Since the right hand side vanishes for some $t=t_2>t_1$ the expansion $\theta(t)$ must diverge at $t_2$. Consequently, $\det([A])$ vanishes at $t_2$. If there is not any $t_1>t_0$ with $\theta(t_1)<0$ then the inequalities $\theta(t_0)=\mathrm{tr}([\dot A](t_0))\leq 0$ and $\dot\theta\leq-\frac{1}{r}\theta^2$ imply $\theta(t)=0$ for all $t\geq t_0$. From the Raychaudhury equation we see that $\sigma=0$ and therefore also $[\dot A][A]^{-1}=0$ for all $t\geq t_0$. Because of $([\dot A][A]^{-1})^{\cdot}=-[R]-([\dot A][A]^{-1})^2$ this would imply $R(\cdot,\dot\gamma,\dot\gamma)=0$ for all $t\geq t_0$ in contradiction to our assumption on $t_0$. The proof for $[A]\in J_+$ is completely analogous.

For each $\hat t>t_0$ let $[B_{\hat t}]$ the unique Jacobi tensor class which satisfies

$$[B_{\hat t}](\hat t)=0 \text{ and } [B_{\hat t}](t_0)=\mathrm{id}.$$

Assume for the moment that we have proved the existence of a Jacobi tensor class $[B]$ with $[B](t)=\lim_{\hat t\to\infty}[B_{\hat t}](t)$ and $\det([B](t_1))\neq 0\quad\forall t_1>t_0$. Since all Jacobi tensor classes $[B_{\hat t}]$ have vanishing vorticity $\omega$ so has

$[B]$. Moreover, $[B](t_0) = \mathrm{id}$ implies that the Jacobi tensor class $[B]$ must lie in either $J_-$ or $J_+$. From the (still to be proven) fact that $[B](t)$ is non-singular for $t > t_0$ we infer that $[B] \in J_+ \setminus J^-$. It follows that $\mathrm{tr}([\dot{B}](t_0)) > 0$ and therefore that there is a $\hat{t} > t_0$ with $\mathrm{tr}([\dot{B}_{\hat{t}}](t_0)) > 0$. Since $[B_{\hat{t}}](t_0) = \mathrm{id}$ this implies that the expansion $\theta_{[B_{\hat{t}}]}$ of $[B_{\hat{t}}]$ at $t_0$ is strictly positive. From the inequality $\dot{\theta}_{[B_{\hat{t}}]} \leq -\frac{1}{n-1}(\theta_{B_{\hat{t}}})^2$ we obtain

$$\frac{1}{\theta_{[B_{\hat{t}}]}(t_0)} \geq \frac{1}{\theta_{[B_{\hat{t}}]}(t)} + \frac{t_0 - t}{n-1} \text{ for all } t < t_0.$$

Since $\frac{t_0 - t}{n-1} \to \infty$ $(t \to -\infty)$ this implies the existence of a $t_1$ which satisfies $\det([B_{\hat{t}}](t_1)) = 0$. Hence there is a non-vanishing, parallel vector field $V$ such that

$$V(t) \in (\dot{\gamma}(t))^{\perp} \text{ for all } t \quad \text{and} \quad B_{\hat{t}}V(t_1) = 0.$$

Since the non-trivial Jacobi vector field defined by $J := B_{\hat{t}}V$ vanishes at both $t_1$ and $\hat{t}$ our geodesic $\gamma$ has a pair of conjugate points.

We still have to show that $[B]$ does exist and that $[B](t)$ is non-singular for all $t > t_0$. In order to do so we will first obtain a formula for $[B_{\hat{t}}]$ which depends only on a single, given Jacobi tensor class.

Let $[A]$ be the Jacobi tensor class which is uniquely determined by $[A](t_0) = 0$ and $[\dot{A}](t_0) = \mathrm{id}$. This tensor class is non-singular for all $t > t_0$ since $[A](t_0) = 0$ and $\gamma_{|[t_0,\infty)}$ does not have conjugate points. Let $\{[E_i]\}_{i=1,\dots,r}$ be a parallel orthonormal frame of $[\dot{\gamma}]^{\perp}$ along $\gamma$. With respect to this frame we define a tensor class $[C]$ by

$$([C][v])^i = [A]^i_j(t) \int_t^{\hat{t}} \left(([A]^*[A])^{-1}\right)^j_k (s) ds [v]^k.$$

Observe that this definition is independent of the chosen frame since any two parallel orthonormal frames are related by a constant orthonormal matrix $D$ and since such matrices satisfy $D^* D = \mathrm{id}$. We will show below that $[C] = [B_{\hat{t}}]$.

But first we need to check that $[C]$ is a Jacobi tensor class.

$$\begin{aligned}
([\dot{C}](t)[v])^i &= [\dot{A}]^i_j(t) \int_t^{\hat{t}} \left(([A]^*[A])^{-1}\right)^j_k (s) ds [v]^k \\
&\quad - [A]^i_j(t) \left(([A]^*[A])^{-1}\right)^j_k (t)[v]^k \\
&= [\dot{A}]^i_j(t)([A]^{-1})^j_l([C][v])^l - (([A]^*)^{-1}[v])^i
\end{aligned}$$

and therefore

$$\begin{aligned}[\ddot{C}](t)[v] &= [\ddot{A}][A]^{-1}[C][v] - [\dot{A}][A]^{-1}[\dot{A}][A]^{-1}[C][v] \\ &\quad + [\dot{A}][A]^{-1}\left([\dot{A}][A]^{-1}[C] - ([A]^*)^{-1}\right)[v] \\ &\quad + ([A]^*)^{-1}[\dot{A}]^*([A]^*)^{-1}[v] \\ &= [\ddot{A}][A]^{-1}[C][v] + ([\dot{A}][A]^{-1} - ([\dot{A}][A]^{-1})^*)([A]^*)^{-1}[v] \\ &= [\ddot{A}][A]^{-1}[C][v],\end{aligned}$$

where in the last equation we have used that $[\dot{A}][A]^{-1}$ is self adjoint by Lemma 4.6.19. Hence we get

$$\begin{aligned}[\ddot{C}] + [R][C] &= [\ddot{A}][A]^{-1}[C] + [R][A][A]^{-1}[C] \\ &= ([\ddot{A}] + [R][A])[A]^{-1}[C] = 0\end{aligned}$$

and $[C]$ is indeed a Jacobi tensor class.

Now we show $[C] = [B_{\hat{t}}]$. Since $[C](\hat{t}) = 0 = [B_{\hat{t}}](\hat{t})$ the equality $[B_{\hat{t}}] = [C]$ follows once we have shown $[\dot{C}](\hat{t}) = [\dot{B}_{\hat{t}}](\hat{t})$. Lemma 4.6.22 and $(([A]^*)^{\cdot}[B_{\hat{t}}] - [A]^*[\dot{B}_{\hat{t}}])_{|t_0} = \mathrm{id}$ imply

$$([A]^*)^{\cdot}[B_{\hat{t}}] - [A]^*[\dot{B}_{\hat{t}}] = \mathrm{id}$$

at all $t \in [a, b]$. In particular, we get $\mathrm{id} = -[A]^*(\hat{t})[\dot{B}_{\hat{t}}](\hat{t})$ since $[B_{\hat{t}}](\hat{t}) = 0$. This in turn is equivalent to $[\dot{B}_{\hat{t}}](\hat{t}) = -([A]^*)^{-1}(\hat{t}))$. On the other hand, $[C](\hat{t}) = 0$ implies $[\dot{C}](\hat{t}) = [\dot{A}](\hat{t})[A]^{-1}(\hat{t})[C](\hat{t}) - ([A]^*)^{-1}(\hat{t}) = -([A]^*)^{-1}(\hat{t}) = [\dot{B}_{\hat{t}}](\hat{t})$. This completes the proof of $[C] = [B_{\hat{t}}]$.

We can now employ the formula

$$([B_{\hat{t}}])^i_k = [A]^i_j(t) \int_t^{\hat{t}} \left(([A]^*[A])^{-1}\right)^j_k (s) \mathrm{d}s$$

in order to show that $[B] = \lim_{\hat{t}\to\infty}[B_{\hat{t}}]$ exists if for some $a < t_0$ the geodesic segment $\gamma\colon [a, \infty) \to M$ does not have any conjugate points. We will prove this by showing that both $\lim_{\hat{t}\to\infty}[B_{\hat{t}}](t_0)$ and $\lim_{\hat{t}\to\infty}[\dot{B}_{\hat{t}}](t_0)$ exist (cf. Corollary 4.6.6). The existence of the first limit is trivial since $[B_{\hat{t}}](t_0) = \mathrm{id}$ for all $\hat{t}$. By Lemma 4.6.19 and $[B_{\hat{t}}](t_0) = \mathrm{id}$ it follows that $([\dot{B}_{\hat{t}}])^*(t_0) = [\dot{B}_{\hat{t}}](t_0)$ for all $\hat{t}$. Polarization implies then that $[\dot{B}_{\hat{t}}](t_0)$ is uniquely determined by the numbers $[g]\left([\dot{B}_{\hat{t}}][v], [v]\right)$ where $[v] \in [\dot{\gamma}(t_0)]^{\perp}$. We will now show that $[g]\left([\dot{B}_{\hat{t}}][v], [v]\right)$ have a well defined limit for every $[v] \in [\dot{\gamma}(t_0)]^{\perp}$. Since we use an (orthonormal) parallel frame along $\gamma$ and $[\dot{A}](t_0) = \mathrm{id}$ we have

$$[g]\left([\dot{B}_{\hat{t}}](t_0)[v],[v]\right) = \delta_{il}[\dot{A}]^i_j(t_0)\int_t^{\hat{t}}\left(([A]^*[A])^{-1}\right)^j_k(s)\mathrm{d}s[v]^k[v]^l$$
$$- [g]\left(([A]^*)^{-1}(t)[v],[v]\right)$$
$$= \int_t^{\hat{t}}\delta_{il}\left(([A]^*[A])^{-1}\right)^i_k(s)\mathrm{d}s[v]^k[v]^l$$
$$- [g]\left(([A]^*)^{-1}(t)[v],[v]\right)$$

and therfore for all $\hat{t}_+ > \hat{t}_-$

$$[g]([\dot{B}_{\hat{t}_+}](t_0)[v],[v]) - [g]\left([\dot{B}_{\hat{t}_-}](t_0)[v],[v]\right)$$
$$= \int_{\hat{t}_-}^{\hat{t}_+}\delta_{il}\left(([A]^*[A])^{-1}\right)^i_k(s)[v]^k[v]^l\mathrm{d}s.$$
$$= \int_{\hat{t}_-}^{\hat{t}_+}[g]\left(([A]^*[A])^{-1}(s)[v]^k[E_k](s),[v]^i[E_i](s)\right)\mathrm{d}s.$$

The last expression is non-negative since

$$[g](([A]^*[A])^{-1}[v],[v]) = [g](([A]^*[A])^{-1}[v],[A]^*[A](([A]^*[A])^{-1}[v])$$
$$= [g]([A](([A]^*[A])^{-1}[v],[A](([A]^*[A])^{-1}[v])$$

and $[g]$ is positive definite. Hence the function $\hat{t}\mapsto[g]\left([\dot{B}_{\hat{t}}](t_0)[v],[v]\right)$ is monotonically increasing for every $[v]\in[\dot{\gamma}(t_0)]^\perp$. We will now show that $[g]\left([\dot{B}_{\hat{t}}](t_0)[v],[v]\right) < [g]\left([\dot{B}_a(t_0)][v],[v]\right)$ for all $[v]\in[\gamma(t_0)]^\perp$. This will give an upper bound for the monotonically increasing function

$$\hat{t}\mapsto[g]\left([\dot{B}_{\hat{t}}](t_0)[v],[v]\right)$$

thereby ensuring that the limit exists. Theorem 4.6.2 implies that the index form $I^{E,\gamma,\perp}_{\{\gamma(a)\},\{\gamma(t_0)\}}$ is positive definite in all the cases we consider. Hence applying the piecewise smooth Jacobi vector field

$$J(t) = \begin{cases} B_a(t)\mathbf{P}_{\gamma|[t,t_0]}v & \text{for } t\in[a,t_0], \\ B_{\hat{t}}(t)\mathbf{P}_{\gamma|[t_0,t]}v & \text{for } t\in(t_0,\hat{t}]. \end{cases}$$

to $I^{E,\gamma,\perp}_{\{\gamma(a)\},\{\gamma(t_0)\}}$ we obtain

$$\begin{aligned}
0 &\le I^{E,\gamma,\perp}_{\{\gamma(a)\},\{\gamma(t_0)\}}(J,J) \\
&= \int_a^{\hat t} \left(\left\langle \nabla_{\dot\gamma} J, \nabla_{\dot\gamma} J\right\rangle + \langle R(J,\dot\gamma)J,\dot\gamma\rangle\right)\mathrm{d}t \\
&= -\int_a^{\hat t} \left\langle \nabla_{\dot\gamma}\nabla_{\dot\gamma} J + R(J,\dot\gamma)\dot\gamma, J\right\rangle - \left\langle \Delta(\nabla_{\dot\gamma(t_0)} J), J(t_0)\right\rangle \\
&= -\left\langle \dot B_{\hat t}(t_0)v, B_{\hat t}(t_0)v\right\rangle + \left\langle \dot B_a(t_0)v, B_a(t_0)v\right\rangle \\
&= -[g]\left([\dot B_{\hat t}](t_0)[v], [B_{\hat t}](t_0)[v]\right) + [g]\left([\dot B_a(t_0)][v], [B_a](t_0)[v]\right) \\
&= -[g]\left([\dot B_{\hat t}](t_0)[v], [v]\right) + [g]\left([\dot B_a](t_0)[v], [v]\right)
\end{aligned}$$

which implies the desired bound. Let $[\dot B^0]\colon (\dot\gamma(t_0)^\perp \to (\dot\gamma(t_0)^\perp$ be the unique selfadjoint tensor class defined by

$$[g]\left([\dot B^0][v],[v]\right) = \lim_{\hat t\to\infty} [g]\left([\dot B_{\hat t}](t_0)[v],[v]\right).$$

Then $[\dot B_{\hat t}](t_0) \to [\dot B^0]$ and, consequently, $[B_{\hat t}] \to [B]$, where $[B]$ is the Jacobi tensor class defined by $[B](t_0) = \mathrm{id}$, $[\dot B](t_0) = [\dot B^0]$. This proves the existence of $[B]$.

We have still to show that $[B](t)$ is non-singular for $t > t_0$. From our construction it is clear that $[B]$ is given by

$$[B]^i_k(t) = [A]^i_j(t) \int_t^\infty \left(([A]^*[A])^{-1}\right)^j_k(s)\mathrm{d}s.$$

Let $[v] \in [\dot\gamma(t)]^\perp \setminus \{0\}$ and $V$ be the parallel vector field along $\gamma$ with $V(t) = v$. Then

$$\begin{aligned}
&[g]([A]^{-1}[B][v],[v]) \\
&\quad= \int_t^\infty [g]\left(([A]^*[A])^{-1}[V](s),[V](s)\right)\mathrm{d}s \\
&\quad= \int_t^\infty [g]\left([A]([A]^*[A])^{-1}[V](s),[A]([A]^*[A])^{-1}[V](s)\right)\mathrm{d}s \\
&\quad> 0
\end{aligned}$$

implies that the operator $([A]^{-1}[B])(t)$ is is non-singular. Thus $B$ is the composition of non-singular operators and therefore also non-singular. ∎

# 5. General relativity

p. 205 ↓
[↓ p. 266]

*Einstein's equation is of the form $\mathcal{D}g = T$, where $\mathcal{D}$ is an operator acting on the Lorentzian metric $g$ and $T$ an expression which describes the distribution of matter in the universe. In Sect. 5.1 we motivate that $T$ should be a symmetric $\binom{0}{2}$ tensor field which is divergence-free, and in Sect. 5.3 we find an expression for $\mathcal{D}g$.*

## 5.1 Matter

In Chaps. 1 and 3 we did not explicitly consider gravity. However, one of the main insights of Einstein was that gravity and the geometry of spacetime are closely linked. His argument is very simple and runs roughly as follows.

The movement of a particle which is subjected to a fixed external "force field" depends on its initial location, its initial velocity, its mass, and its *charge* (i.e. its "sensitivity" to the force field). For instance, a particle in an electric field which is initially at rest will move to one side if it is positively charged, to the opposite side if it is negatively charged and not at all if it is neutral. To be more concrete, consider a reference frame $(\tau, \mathfrak{t})$ in a Galilei spacetime and suppose that there is a non-relativistic particle $(m, \gamma)$ which is located in an electric field $\vec{\mathfrak{E}}$ and has the electric charge $\mathfrak{e}$. Then the formula

$$m\ddot{\vec{\gamma}} = \mathfrak{e} \cdot \vec{\mathfrak{E}}(t, \vec{\gamma})$$

holds. Similarly, let $\vec{\mathfrak{G}}$ be a gravitational field and $\mathfrak{g}$ be the "gravitational charge"[1] of the particle. Then

$$m \cdot \ddot{\vec{\gamma}} = \mathfrak{g} \cdot \vec{\mathfrak{G}} \tag{5.1.1}$$

holds. It is an experimental fact that the quotient $\frac{\mathfrak{e}}{m}$ depends on the particle whereas the analogous quotient $\frac{\mathfrak{g}}{m}$ is a *universal constant* and can be set $= 1$ (Eötvös 1896). Einstein concluded that this fact is not a mere coincidence but reveals that gravitation is an acceleration (rather than a force) and therefore something geometrical. He therefore replaced the equation $\ddot{\vec{\gamma}} = \vec{\mathfrak{G}}$ by the geodesic equation $\nabla_{\dot{\gamma}}\dot{\gamma} = 0$ and the "force field" $\mathfrak{G}$ by the connection $\nabla$ of spacetime. This point of view leads

[1] It is usually called the *passive gravitational mass.*

to a physical interpretation of inertial observers: They are simply those observers which are freely falling.

It is an experimental fact that the matter distribution[2] spacetime determines gravity. Hence we have to look for an equation of the form

$$\mathcal{D}g = T, \tag{5.1.2}$$

where $(M, g)$ is an $n$-dimensional Lorentzian manifold[3], $\mathcal{D}$ is some kind of operator acting on the metric $g$, and $T$ contains the information on the matter distribution. The "correct" form of $T$ cannot be derived. First of all, it is beyond doubt that matter cannot be described by a smooth object in spacetime but instead demands a quantum description. This implies that we can hardly expect a description from fundamental, physically suggestive principles. $T$ will therefore be a classical approximation, i.e. something phenomenological. Consequently, our final form for Equation (5.1.2), Equation (5.3.11) will appear to be grounded less firmly than the spacetime structure. However, the reader should recall that in the derivation of the Lorentzian structure of spacetime we already assumed that light can be described in an entirely classical (i.e. non-quantum) way.

The only matter models we had considered so far where special-relativistic point particles (cf. p. 40) which admit a straight-forward generalization.

**Definition 5.1.1.** *A* particle *is a pair* $(m, \gamma)$, *where* $m \geq 0$ *is the* mass *of the particle and* $\gamma$ *is a curve in* $M$ *with* $g(\dot\gamma(t), \dot\gamma(t)) = -1$ *for all* $t \in M$, *representing the* history *of the particle.*

Exactly as in the special-relativistic analogy, an infinitesimal observer $v$ at $x = \gamma(0)$ measures the energy $E_v = -mg(\dot\gamma(0), v)$ and the spatial momentum $\dot\gamma(0)^\perp = \dot\gamma(0) - \frac{E_v}{m}v$. The following simple observation will serve as a guidance for defining $T$.

**Lemma 5.1.1.** *Let* $x = \gamma(t_0) \in M$ *and* $\{w_1, \ldots, w_n\}$ *be* $n$ *linearly independent timelike vectors with* $\langle w_i, w_i \rangle = -1$. *Then* $m\dot\gamma(t_0)$ *and* $m$ *are determined by the numbers* $E_{w_1}, \ldots, E_{w_n}$.

*Proof.* Since $\{g(w_i, \cdot)\}_{i=1,\ldots,n}$ is a basis of $T_x^*M$, $m\dot\gamma$ is uniquely determined by $E_{w_1}, \ldots, E_{w_n}$ and $m$ can be calculated from $-m^2 = g(m\dot\gamma, m\dot\gamma)$. ∎

---

[2] Here we use the term "matter" in a rather wide sense encompassing all forms of energy. This is motivated by the special-relativistic equation $E = mc^2$ ($c = 1$: velocity of light) which asserts that (rest) mass is simply a form of energy (cf. Sect. 1.4.3).

[3] The spacetime we live in appears to be a 4-dimensional Lorentzian manifold. However, in this book we will not specialize to $n = 4$.

In other words, we only need to know the energy function

$$E\colon \{v \in T_xM \,|\, g(v,v) = -1\} \to \mathbb{R}, \quad w \mapsto E_w$$

in order to recover the complete information about a single particle.

Since $g$ is a smooth object, we would expect $T$ to be smooth also. This indicates that point particles which are not depending smoothly on the coordinates of $M$ cannot be used to constitute $T$. The simplest way to obtain a smooth matter distribution from a collection of particles is to consider averages instead of individual particles.

**Definition 5.1.2.** *A* congruence of particles *is a pair* $(\epsilon, U)$, *where* $\epsilon\colon M \to \mathbb{R}$ *is a function and* $U$ *is a future directed vector field with* $g(U,U) = -1$.

The integral curves of $U$ are identified with the world lines of the particles and the *energy density function* $\epsilon$ with the energy density, measured by comoving observers. To keep the presentation simple we will restrict to a special case and assume that $\mathrm{d}U^\flat \wedge U^\flat = 0$, i.e. that there exists locally a spacelike hypersurface $\Sigma$ which is orthogonal to $U$ (cf. Theorem 2.5.4). If $B \subset \Sigma$ is a compact region then an observer flowing with the particles measures for the energy of those particles which pass through $B$ the quantity

$$E = \int_B \epsilon \mu_\Sigma,$$

where $\mu_\Sigma$ is the induced volume form (cf. Definition 4.2.2[4]). Since a single observers must be identified with a timelike curve rather than a congruence of curves this expression should be understood as an approximation for small $B$. It is clear that we recover the definition of a point particle if the compact set $\mathrm{supp}(\epsilon) \cap \Sigma$ shrinks to a point and the energy density $\epsilon$ increases adequately.

A different congruence of observers, represented by a vector field $V$ with $g(V,V) = -1$ and $\mathrm{d}V^\flat \wedge V^\flat = 0$, will measure a different energy content,

$$E_V = \int_{B_V} \epsilon_V \mu_{\Sigma_V},$$

where

- $\Sigma_V$ is a spacelike hypersurface orthogonal to $V$,
- $\mu_{\Sigma_V}$ the volume form induced on $V$,
- $B_V = \{x \in \Sigma_V \,|\, \exists$ a particle through $x$ which intersects $\Sigma\}$, and

[4] Readers who have not read Sect. 2.5.4 may wish to do so now. Alternatively, they may (for the time being) refer to the footnote in Definition 4.2.2. In the following we will make use of calculus for differential forms (Sect. 2.5) in order to avoid clumsy but straightforward calculations.

– $\epsilon_V$ is a function which depends on the congruence of particles $(\epsilon, U)$ and the observer field $V$.

We will now *motivate* a transformation law $\epsilon \mapsto \epsilon_V$ through comparison with special relativity

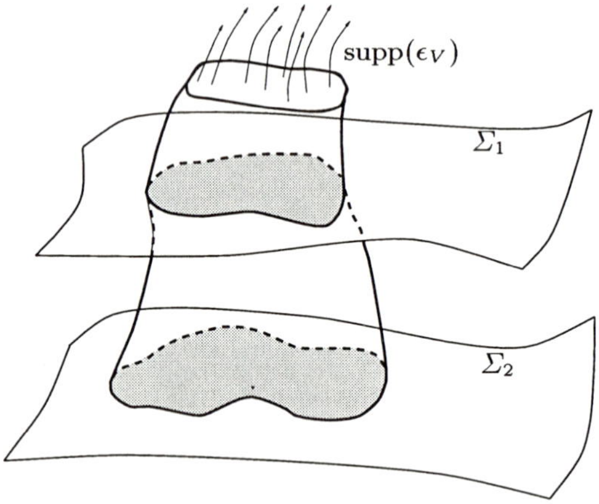


**Fig. 5.1.1.** A localized congruence

Consider Minkowski space $(\mathbb{A}^n, \eta)$ and the inertial observer field

$$V : x \mapsto v$$

where $v$ is a vector with $\eta(v, v) = -1$. Assume also that the congruence is localized, i.e. restricted to any spacelike hypersurface $\Sigma$, $\epsilon_V$ has compact support (cf. Fig. 5.1.1) The inertial (or freely falling) observers $t \mapsto x + tv$ with common rest space $x + v^{\perp}$ measure the energy

$$E_V = \int_{x+v^{\perp}} \epsilon_V \, \mu_{x+v^{\perp}} = \int_{x+v^{\perp}} \epsilon_V \, (V \lrcorner \mu_{\mathbb{A}^n}),$$

where $\mu_{\mathbb{A}^n}$ is the volume form of $\eta$. We assume now that the congruence of particles is freely falling, i.e. the field $U$ satisfies the geodesic equation $\nabla_U U = 0$. In the limit that $\mathrm{supp}(\epsilon_V) \cap \Sigma$ shrinks to a point we would recover the energy associated with a single freely falling particle.

Since $\nabla_U U = 0$ we can choose linear coordinates $(x^0, x^1, \ldots, x^{n-1})$ such that

$$U = \partial_0, \quad V = \frac{1}{\sqrt{1 - ||\vec{V}||^2}} \left( \partial_0 + ||\vec{V}|| \partial_1 \right).$$

This gives

$$\begin{aligned}
E_V &= \int_{x+V^\perp} \epsilon_V \frac{1}{\sqrt{1-\|\vec{V}\|^2}} \left(\partial_0 + \|\vec{V}\|\partial_1\right) \lrcorner \mu_\eta \\
&= \int_{x+V^\perp} \epsilon_V \frac{1}{\sqrt{1-\|\vec{V}\|^2}} \Big(\mathrm{d}x^1 \wedge \cdots \wedge \mathrm{d}x^{n-1} \\
&\qquad - \|\vec{V}\| \mathrm{d}x^0 \wedge \mathrm{d}x^2 \cdots \wedge \mathrm{d}x^{n-1}\Big) \\
&= \frac{1}{\sqrt{1-\|\vec{V}\|^2}} \int_{x+V^\perp} \epsilon_V \mathrm{d}x^1 \wedge \cdots \wedge \mathrm{d}x^{n-1} \\
&\quad - \frac{\|\vec{V}\|}{\sqrt{1-\|\vec{V}\|^2}} \int_{x+V^\perp} \epsilon_V \mathrm{d}x^0 \wedge \mathrm{d}x^2 \cdots \wedge \mathrm{d}x^{n-1} \\
&= \frac{1}{\sqrt{1-\|\vec{V}\|^2}} \int_{x^1_-}^{x^1_+} \tilde{\epsilon}_V \mathrm{d}x^1 - \frac{\|\vec{V}\|}{\sqrt{1-\|\vec{V}\|^2}} \int_{x^0_-}^{x^0_+} \tilde{\epsilon}_V \mathrm{d}x^0,
\end{aligned}$$

where $\tilde{\epsilon}_V(x^0, x^1) = \int \epsilon_V \mathrm{d}x^2 \wedge \cdots \wedge \mathrm{d}x^{n-1}$ and $x^i_\pm$ denote the maximal (minimal) values of $x^i$ restricted to the support of $\epsilon_V$ in $x + V^\perp$. In this hyperplane, we have $x^0 = \|\vec{V}\| x^1$ and therefore

$$E_V = \sqrt{1-\|\vec{V}\|^2} \int_{x^1_-}^{x^1_+} \tilde{\epsilon}_V \mathrm{d}x^1.$$

Let $\gamma\colon t \mapsto (t, x^1, \ldots, x^{n-1})$ and let $(m, \gamma)$ be the corresponding freely falling particle with rest mass $m$. Then its energy — measured in its own rest frame — is $\tilde{E}_0 = -\langle\dot{\gamma}, m\dot{\gamma}\rangle = -m\langle\partial_0, \partial_0\rangle = m$. The energy measured by the infinitesimal observer $v = \left(\sqrt{1-\|\vec{V}\|^2}\right)^{-1} (\partial_0 + \|\vec{V}\|\partial_1)$, is given by

$$\tilde{E}_v = -\langle V, m\dot{\gamma}\rangle = \frac{m}{\sqrt{1-\|\vec{V}\|^2}} = \frac{\tilde{E}_0}{\sqrt{1-\|\vec{V}\|^2}}.$$

An analogous relationship should also hold for our smooth congruence $U$ since this congruence can be used to smoothly model a point particle. Hence we should have

$$E_v = \frac{1}{\sqrt{1-\|\vec{V}\|^2}} \int_{x^1_-}^{x^1_+} \tilde{\epsilon} \mathrm{d}x^1.$$

Since $x \mapsto \epsilon_V(x)$ was arbitrary this equation implies $\epsilon_V = \frac{1}{1-\|\vec{V}\|^2}\epsilon$. This transformation law indicates that $\epsilon_V$ depends quadratically on $V$.

**Postulate 5.1.1 (Tensorial character of energy momentum).** *The map $T$ is a symmetric $\binom{0}{2}$-tensor field and the energy density measured by an infinitesimal observer $v$ is given by $T(v,v)$.*

In the special case of our congruence of (non-interacting) particles there is a simple, well defined tensor field $T_U$ , namely

$$T_U = \epsilon\, U^\flat \otimes U^\flat .$$

Observe that $\epsilon_V = T_U(V,V)$ is in accordance with the transformation law derived above.

The following lemma indicates that it is enough to know all possible energy densities $\epsilon_v$ in order to recover the tensor $T$ (cf. Lemma 5.1.1).

**Lemma 5.1.2.** *Let $T$ be a symmetric $\binom{0}{2}$-tensor. Then $T$ is uniquely determined by the values $T(u,u)$ for all vectors $u$ with $g(u,u) = -1$.*

*Proof.* Let $T, S$ be two symmetric tensors with $T(u,u) = S(u,u)$ for all $u$ with $\langle u,u\rangle = -1$ and let $v$ be a timelike vector. Then the vector $v/\sqrt{-\langle v,v,\rangle}$ satisfies $g(v/\sqrt{-\langle v,v,\rangle}, v/\sqrt{-\langle v,v,\rangle}) = -1$ which yields $T(v,v) = S(v,v)$. Since the space of all timelike vectors is open there is for every vector $w$ a $\delta > 0$ such that $v + tw$ is timelike for all $t \in [-\delta,\delta]$. Hence we obtain

$$\begin{aligned} T(w,w) &= \frac{1}{2}\left(\frac{\mathrm{d}^2}{\mathrm{d}t^2} T(v+tw, v+tw)\right)_{|t=0} \\ &= \frac{1}{2}\left(\frac{\mathrm{d}^2}{\mathrm{d}t^2} S(v+tw, v+tw)\right)_{|t=0} = S(w,w) \end{aligned}$$

and the claim follows from the polarization identity. ∎

Conservation of energy and momentum is another fundamental property of matter which we wish to encode in our theory. We will find an infinitesimal formulation which (in special cases) recovers conservation of momentum (cf. Equation (1.4.12)). In Sects. and 1.2.1 and 1.4.3 we have simply stated conservation of momentum. These conservation laws can actually be derived within the theory of point particle mechanics. This is the content of the Noether Theorem which is covered in textbooks on mechanics. The main non-mechanical input for the Noether theorem is the Galilei group (in the non-relativistic case) and the Poincaré group (in the relativistic case). Recall that the Poincaré group is the set of all isometries of Minkowski spacetime. In order to find an infinitesimal formulation of conservation of momentum we will therefore have to employ Killing vector fields which can be regarded as infinitesimal analogues of 1-parameter groups of isometries.

**Lemma 5.1.3.** *Let $T$ be a symmetric $\binom{0}{2}$-tensor with* $\mathrm{div}(T) = 0$ *and $\xi$ be a Killing field. Then* $\mathrm{div}\left(T(\xi, \cdot)^\sharp\right) = 0$.

*Proof.* since $T$ is divergence-free we have

$$\mathrm{div}(T(\xi,\cdot)^\sharp) = \nabla_a\left(T^{ab}\xi_b\right) = \left(\nabla_a T^{ab}\right)\xi_b + T^{ab}\left(\nabla_a \xi_b\right) = T^{ab}\left(\nabla_a \xi_b\right).$$

Now the symmetry of $T$ and the anti-symmetry of $\nabla\xi^\flat$ (cf. Lemma 4.5.2) imply that the second summand also vanishes. ∎

Let $\{\Sigma_t\}_{t\in\mathbb{R}}$ be a foliation of $M$ into spacelike hypersurfaces with future normals $\mathfrak{n}_t$. A *world tube with respect to* $\{\Sigma_t\}_{t\in\mathbb{R}}$ is an open subset $\mathcal{W}$ of $M$ with piecewise smooth, timelike boundary such that the intersection $\mathcal{W}\cap\Sigma_t$ is connected for all $t$. If $\mathcal{W}$ is a world tube with respect to $\{\Sigma_t\}_{t\in\mathbb{R}}$ then we denote the subset $\bigcup_{t\in[t_1,t_2]}\mathcal{W}\cap\Sigma_t$ by $\mathcal{W}_{t_1,t_2}$ and the part of the boundary which is not contained in $\Sigma_{t_1}\cup\Sigma_{t_2}$ by $\mathcal{W}_{\mathrm{time}}$.

**Corollary 5.1.1.** *Let $T$ be a symmetric $\binom{0}{2}$-tensor field with* $\mathrm{div}(T) = 0$ *and $\xi$ be a Killing field. Let $t_1 < t_2$ and $\mathcal{W}$ be a world tube with respect to $\{\Sigma_t\}_{t\in\mathbb{R}}$ such that* $\mathrm{supp}(T)\cap\mathcal{W}_{\mathrm{time}} = \emptyset$. *Then the following conservation law holds.*

$$\int_{\Sigma_{t_2}\cap\mathcal{W}_{t_1,t_2}} \left\langle \mathfrak{n}_{t_2}, T(\xi,\cdot)^\sharp\right\rangle \mathfrak{n}_{t_1} \lrcorner \mu_M = \int_{\Sigma_{t_1}\cap\mathcal{W}_{t_1,t_2}} \left\langle \mathfrak{n}_{t_1}, T(\xi,\cdot)^\sharp\right\rangle, \mathfrak{n}_{t_2} \lrcorner \mu_M .$$

*Proof.* We have

$$\begin{aligned}
&\left\langle \mathfrak{n}_t, T(\xi,\cdot)^\sharp\right\rangle (\mathfrak{n}_t \lrcorner \mu_M)(V_1,\dots,V_{n-1}) \\
&\qquad = \left\langle \mathfrak{n}_t, T(\xi,\cdot)^\sharp\right\rangle \mu_M(\mathfrak{n}_t, V_1,\dots,V_{n-1}) \\
&\qquad = -(T(\xi,\cdot)^\sharp \lrcorner \mu_M)(V_1,\dots,V_{n-1})
\end{aligned}$$

for any $(n-1)$-tuple of vector fields tangent to $\Sigma_t$. Hence pulled back to $\Sigma_{t_{1/2}}$ we get $\left\langle \mathfrak{n}_t, T(\xi,\cdot)^\sharp\right\rangle \mathfrak{n}_t \lrcorner \mu_M = -T(\xi,\cdot)^\sharp \lrcorner \mu_M$. This (and $\mathrm{supp}(T)\cap\mathcal{W}_{\mathrm{time}} = \emptyset$) imply

$$\begin{aligned}
\int_{\partial\mathcal{W}} \left\langle \mathfrak{n}_t, T(\xi,\cdot)^\sharp\right\rangle \mathfrak{n}_t \lrcorner \mu_M &= \int_{\Sigma_{t_2}\cap\mathcal{W}_{t_1,t_2}} \left\langle \mathfrak{n}_{t_2}, T(\xi,\cdot)^\sharp\right\rangle \mathfrak{n}_{t_1} \lrcorner \mu_M \\
&\quad - \int_{\Sigma_{t_1}\cap\mathcal{W}_{t_1,t_2}} \left\langle \mathfrak{n}_{t_1}, T(\xi,\cdot)^\sharp\right\rangle \mathfrak{n}_{t_2} \lrcorner \mu_M ,
\end{aligned}$$

where we have used that the future and past boundaries $\Sigma_{t_2}$ and $\Sigma_{t_1}$ have opposite induced orientations. The assertion follows now from the Theorem of Stokes 2.5.5 since $\mathrm{d}(\left\langle \mathfrak{n}_t, T(\xi,\cdot)^\sharp\right\rangle (\mathfrak{n}_t \lrcorner \mu_M)) = -\mathrm{div}(T(\xi,\cdot)^\sharp)\mu_M = 0$.[5] ∎

[5] Readers who have skipped Sect. 2.5 may instead apply the integral theorem of Gauß.

Hence the quantity $\int_{\Sigma_t} \langle \mathfrak{n}_t, T(\xi, \cdot)^\sharp \rangle (\mathfrak{n}_t \lrcorner \mu_M)$ is independent of the time parameter $t$ defined by the foliation if $\operatorname{div}(T) = 0$. We will now identify this quantity with a component of special-relativistic momentum in the context of Sect. 1.4.3.

We will assume that spacetime is isometric to Minkowski spacetime before a time $t_1$ represented by a spacelike hypersurface $\Sigma_{t_1}$ and after a time $t_2$ represented by a spacelike hypersurface $\Sigma_{t_2}$. We will study a matter model which consists of $k$ freely falling congruences of particles in the region before $\Sigma_{t_1}$ and after $\Sigma_{t_2}$. In between these hypersurfaces interactions or collisions may take place. Hence in this region we will (at this point) neither make an assumption on the matter model nor on the metric.

To be concrete, consider the set $\mathbb{A}^n$, a point $o \in \mathbb{A}^n$, and a non-vanishing constant 1-form $\tau$. This 1-form defines a foliation of $\mathbb{A}$ with affine hypersurfaces $\Sigma_t = \{x \in \mathbb{A} \,|\, \pi(x-o) = t\}$. Let $\eta$ be a Minkowski metric such that $\Sigma_t$ are spacelike hypersurfaces and let $V$ be the timelike, future directed constant vector field which is orthogonal to all $\Sigma_t$ and satisfies $\eta(V, V) = -1$. Assume that the spacetime $(M, g)$ satisfies $M = \mathbb{A}^n$ and $g_{|\{x \in M \,|\, \pi(x) \notin [t_1, t_2]\}} = \eta_{|\{x \in M \,|\, \pi(x) \notin [t_1, t_2]\}}$. Let $(\epsilon_i, U_i)_{i=1,\dots,k}$ be the congruences of particles defined at all points $x$ with $\pi(x) \notin (t_1, t_2)$. Assume that $\nabla_{U_i} U_i = 0$ and that the energy densities $\epsilon_i$ satisfy $\operatorname{supp}(\epsilon_i) \cap \{x \in M \,|\, \pi(x) \in \{t_1, t_2\}\}$ is compact.

Let $T$ be a symmetric $\binom{0}{2}$ tensor field with $\operatorname{div}(T) = 0$ and

$$T_x = \sum_{i=1}^{k} \epsilon_i(x) (U_i)_x^\flat \otimes (U_i)_x^\flat \text{ for } \pi(x) \notin (t_1, t_2)$$

and assume that $(M, g)$ admits a Killing vector field $\xi$. Corollary 5.1.1 implies that

$$\int_{\Sigma_{t_1}} \langle V, , T(\xi, \cdot)^\sharp \rangle \, V \lrcorner \mu_{\mathbb{A}^n} = \int_{\Sigma_{t_2}} \langle V, , T(\xi, \cdot)^\sharp \rangle \, V \lrcorner \mu_{\mathbb{A}^n}$$

Conversely, it is clear that $\operatorname{div}(T(\xi)) = 0$ must hold if the integral equality is valid for all such particle flows.

Since the vector fields $U_i$ are constant for $x \in \Sigma_t$ ($t \notin [t_1, t_2]$) we get $\operatorname{div}(T) = \sum_{i=1}^{k} \mathrm{d}\epsilon_i(U_i) U_i = 0$. If the vector fields $U_i$ are at each point linearly independent then this equation implies $\mathrm{d}\epsilon_i(U_i) = 0$,i.e. the energy density of each particle flow is constant along its flow lines. In the following we will *assume* that this is also the case if these vector fields are (pointwise) linearly dependent. Then the in-going and out-going rest masses $(m_i)_1$, $(m_i)_2$ defined by

$$(m_i)_a = \int_{(z_i)_a + (U_i)^\perp} \epsilon_i U_i \lrcorner \mu_{\mathbb{A}^n}$$

where $(z_i)_a$ is chosen such that

– $\operatorname{supp}(\epsilon_i) \cap \{x \in M \,|\, \pi(x) \in [t_1, t_2]\} = \emptyset$ and
– $\pi((z_i)_1) < t_1$, $\pi((z_i)_2) > t_2$

are well defined constants.

Assume now that the Killing vector field $\xi$ is constant for $t > t_2$, $t < t_1$, and denote these constant vector fields by $\xi_2$, $\xi_1$. For each vector field $U_i$ and $a \in \{1, 2\}$ there are vectors $(e_i)_a$ with $\langle V, (e_i)_a \rangle = 0$, $\langle (e_i)_a, (e_i)_a \rangle = 1$, and

$$(U_i)_{|\Sigma_{t_a}} = \frac{1}{\sqrt{1 - \|\vec{U}_i\|^2}} (V + \|\vec{U}_i\| (e_i)_a).$$

The integrals in the formula above reduce to

$$\begin{aligned}
\int_{\Sigma_{t_a}} \langle V, T(\xi, \cdot)^\sharp \rangle \, V \lrcorner \mu_{\mathbb{A}^n} &= \sum_{i=1}^k \int_{\Sigma_{t_a}} \epsilon_i \, \langle V, U_i \rangle \, \langle U, \xi_a \rangle \, V \lrcorner \mu_{\mathbb{A}^n} \\
&= -\sum_{i=1}^k \langle U_i, \xi_a \rangle \int_{\Sigma_{t_a}} \frac{\epsilon_i}{\sqrt{1 - \|(\vec{U}_i)_{|\Sigma_{t_a}}\|^2}} V \lrcorner \mu_{\mathbb{A}^n} \\
&= -\sum_{i=1}^k \langle U_i, \xi_a \rangle \int_{(z_i)_a + (U_i)^\perp} \epsilon_i \, U_i \lrcorner \mu_{\mathbb{A}^n} \\
&= -\sum_{i=1}^k (m_i)_a \, \langle U_i, \xi_a \rangle
\end{aligned}$$

where we have taken the length contraction into account (cf. Fig. 5.1.2). Hence we recover conservation of special-relativistic momentum. This motivates to demand the second matter postulate

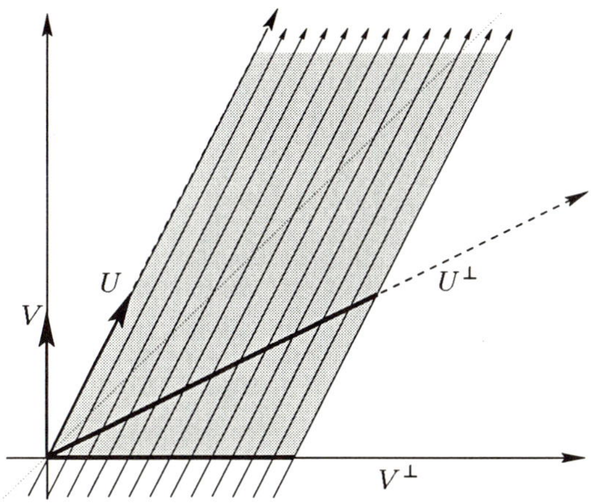


**Fig. 5.1.2.** Transformation of the mass density in special relativity

**Postulate 5.1.2 (Infinitesimal conservation law).** *The tensor field $T$ has vanishing divergence,* $\mathrm{div}(T) = 0$.

Postulate 5.1.2 is interpreted as an infinitesimal formulation of conservation of energy and momentum. That these quantities are conserved is intuitively clear from the absence of a perpetuum mobile. However, the infinitesimal formulation implies a *true* conservation law only if space-time is endowed with a Killing vector field. In general, this is not the case. It follows that conservation of energy can only hold infinitesimally.

**Definition 5.1.3.** *A symmetric* $\binom{0}{2}$*-tensor field with* $\mathrm{div}(T) = 0$ *is called an* energy momentum tensor. *It is sometimes called* stress energy momentum tensor *or* stress energy tensor.

## 5.2 Some specific matter models

If $T$ and $g$ are simultaneously diagonalizable,

$$g = \begin{pmatrix} -1 & 0 & \dots & 0 \\ 0 & 1 & \ddots & \vdots \\ \vdots & \ddots & \ddots & 0 \\ 0 & \dots & 0 & 1 \end{pmatrix}, \quad T = \begin{pmatrix} \epsilon & 0 & \dots & 0 \\ 0 & p_1 & \ddots & \vdots \\ \vdots & \ddots & \ddots & 0 \\ 0 & \dots & 0 & p_{n-1} \end{pmatrix}, \tag{5.2.3}$$

then $\epsilon$ is interpreted as the *energy density* with respect to the flow of matter and $p_i$ ($i \in \{1, \dots, n-1\}$) are interpreted as *principal pressures*. To motivate this interpretation we will simplify to a perfect fluid, i.e. a matter distribution for which all principal pressures are equal.

### 5.2.1 The perfect fluid

**Definition 5.2.1.** *Let $\epsilon, p \colon M \to \mathbb{R}$ be smooth functions and $U$ be a vector field with $g(U,U) = -1$. Then*

$$T = (\epsilon + p)U^\flat \otimes U^\flat + p\, g$$

*is called the energy momentum tensor of a* perfect fluid. *A perfect fluid with $p = 0$ is called* dust.

Observe that the energy momentum tensor $T = \epsilon U^\flat \otimes U^\flat$ considered in the motivation of Postulates 5.1.1 and 5.1.2 describes dust.

**Lemma 5.2.1.** *Let $T$ be the energy momentum tensor of a perfect fluid. Then* $\mathrm{div}(T) = 0$ *is equivalent to*

$$d\epsilon(U) = -(\epsilon + p)\mathrm{div}(U), \quad (\epsilon + p)\nabla_U U = -\pi_{U^\perp}\mathrm{grad}(p),$$

*where $\pi_{U^\perp}$ denotes the projection to the orthogonal complement of $U$.*

*Proof.* It is straightforward to calculate

$$\begin{aligned}
(\mathrm{div}T)_a &= g^{cb}\nabla_c T_{ab} \\
&= g^{cb}(\partial_c(\epsilon+p)U_aU_b + (\epsilon+p)\big((\nabla_c U_a)U_b + U_a(\nabla_c U_b)\big) + \partial_c p g_{ab}) \\
&= \mathrm{d}(\epsilon+p)(U)U_a + (\epsilon+p)(\nabla_U U)_a + (\epsilon+p)\mathrm{div}(U)U_a + \mathrm{grad}(p)_a \\
&= (\mathrm{d}\epsilon(U) + (\epsilon+p)\mathrm{div}U)U_a \\
&\quad + (\epsilon+p)(\nabla_U U)_a + \big((U^\flat\otimes U^\flat + g)(\mathrm{grad}(p),\cdot)\big)_a .
\end{aligned}$$

The assertion follows since $\nabla_U U \perp U$ and $(U^\flat \otimes U^\flat + g)$ is the metric projected to $U^\perp$. ∎

The vector field $U$ is the velocity of the fluid particles and $\epsilon$ the energy a comoving observer would measure. If the divegence of this vector field is negative then the particles are getting closer together and consequently the energy density should increase. This is expressed in the first equation. The second equation states that the spatial acceleration of the fluid particles is proportional to grad$(p)$. This indicates that $p$ should be interpreted as the pressure exerted on the fluid.

Perfect fluids are phenomenological models and the equations implied by $\mathrm{div}(T) = 0$ tend to develop shock waves. It is therefore often argued that perfect fluids exhibit properties which are not shared by real matter. However, perfect fluids are prevalent in cosmological models of the universe.

### 5.2.2 The collisionless gas

An attempt to arrive at a more realistic matter model is to consider a relativistic gas. The idea is that we do not have a congruence of particles but that each individual particle can move in any direction. The energy momentum tensor is then obtained by averaging over all particle velocities. Let $(x^0, \dots x^{n-1})$ be a coordinate system of $M$ and choose the *canonical coordinates* $(x^0, \dots, x^{n-1}, p_0, \dots p_{n-1})$ of $T^*M$ which are defined by $\alpha = p_a(\alpha)\mathrm{d}x^a$ for every 1-form $\alpha \in T_xM$. A *relativistic gas* can be described by an energy momentum tensor

$$T_{ab}(x) = \int_{\mathsf{P}^+(x)} p_a p_b f(x,p)\left(-\det((g_{cd})_{c,d=0,\dots,n-1})\right)^{-\frac{1}{2}} \mathrm{d}p_1 \wedge \dots \wedge \mathrm{d}p_n,$$

where $\mathsf{P}^+(x) \subset T_x^*M$ denotes the set of future causal 1-forms and

$$f \colon \mathsf{P}^+(x) \to \mathbb{R}^+$$

is a density function. We assume that for $|p_a| \to \infty$ the function $f(x,\cdot)$ is sufficiently rapidly decreasing so that the integral is well defined. Observe that the $n$-form

$$(-\det((g_{cd})_{c,d=0,\ldots,n-1}))^{-\frac{1}{2}}\, \mathrm{d}p_1 \wedge \cdots \wedge \mathrm{d}p_n$$

does not depend on the choice of coordinates $(x^1, \ldots, x^{n-1})$. The relativistic gas is *collisionless* if the *Liouville equation* $\mathrm{d}f(X_H) = 0$ holds, where

$$X_H = g^{ab} p_a \partial_{x^b} - \frac{1}{2}\partial_{x^c} g^{ab} p_a p_b \partial_{p_c}.^6$$

Using a system of normal coordinates it is easy to see that $\mathrm{d}f(X_H) = 0$ implies $\mathrm{div}(T) = 0$.

If $U$ is a vector field and $f$ was replaced by the delta distribution $\delta(\sqrt{\epsilon}U^a - p^a)$ one would obtain dust. Hence dust may be viewed as a "gas" whose molecules are all aligned and move into a preferred direction determined by the vector field $U$.

Analogously, a *relativistic photon gas* is given by an energy momentum tensor of the form

$$T_{ab}(x) = \int_{\mathsf{P}_0^+} p_a p_b f_0(x,p)\, \nu_{\mathsf{P}_0^+(x)},$$

where $\mathsf{P}_0^+(x) \subset T_x^* M$ denotes the submanifold of non-vanishing future null 1-forms at $x$, $\nu_{\mathsf{P}_0^+(x)}$ is a non-vanishing, oriented $n-1$-form on $\mathsf{P}_0^+(x)$, and $f_0\colon \mathsf{P}_0^+(x) \to \mathbb{R}^+$ is the photon density function with respect to $\nu_{\mathsf{P}_0^+(x)}$. We assume that for $|p_a| \to \infty$ the function $f_0$ is sufficiently rapidly decreasing so that the integral is well defined.

The following lemma implies that the energy density associated with a relativistic gas is always positive.

**Lemma 5.2.2.** *Let $T$ be the energy momentum tensor of a relativistic gas (respectively, photon gas) with $f \geq 0$ (respectively, $f_0 \geq 0$). Then $T(u,u) > 0$ for all timelike vectors $u$ unless the density function $f$ (respectively, $f_0$) vanishes.*

*Proof.* This is clear since for each vector $u$ the integrand in the definition for $T(u,u)$ is positive unless $f$ (respectively, $f_0$) vanishes. ∎

**Lemma 5.2.3.** *Let $T$ be the energy momentum tensor of a photon gas. Then $\mathrm{tr}(T) = 0$.*

*Proof.* The assertion follows from

$$\mathrm{tr}(T) = g^{ab} \int_{\mathsf{P}_0^+} p_a p_b f_0(x,p)\, \nu_{\mathsf{P}_0^+(x)} = \int_{\mathsf{P}_0^+} g^\sharp(p,p) f_0(x,p)\, \nu_{\mathsf{P}_0^+(x)} = 0$$

since the 1-forms $p$ are null. ∎

---

[6] Readers who have knowledge of mechanics will notice that $X_H$ is just the Hamilton vector field to the Hamiltonian function $H(x,p) = \frac{1}{2}g^{ab}p_a p_b$. The equation $\mathrm{d}f(X_H) = 0$ expresses then conservation of mechanical energy (cf. (Ehlers 1973) for details).

### 5.2.3 The electromagnetic field

An electromagnetic field can be described by a 2-form $F$ which satisfies *Maxwell's equations*,

$$\mathrm{d}F = 0, \tag{5.2.4}$$

$$\mathrm{div}(F) = J, \tag{5.2.5}$$

where $J$ is interpreted as an electromagnetic current one form. The first equation can be geometrically explained within gauge theory (A small volume which contains the essentials of gauge theory is (Bleecker 1981)[7]). The second equation does not have any content without a prior interpretation of $J$. For our purposes it is sufficient to note that $J$ is linked to other forms of matter.

*Remark 5.2.1.* Using the Hodge star operator we can write $\star \mathrm{d} \star F = J$ instead of $\mathrm{div}(F) = J$.

If there is no interaction between electromagnetism and the other matter fields, i.e. if matter is neutral, then we have the set of equations

$$\mathrm{d}F = 0, \tag{5.2.6}$$

$$\mathrm{div}(F) = 0. \tag{5.2.7}$$

These equations are called the *source-free Maxwell equations.*

The electromagnetic part of the energy momentum tensor is given by

$$(T_{\mathrm{el}})_{ab} = \frac{1}{4\pi}\left(g^{cd}F_{ac}F_{bd} - \frac{1}{4}\langle F, F\rangle g_{ab}\right). \tag{5.2.8}$$

We will sketch in Sect. 5.3.1 below how one may justify these formulas.

[7] There are many mathematical texts on "gauge theory" which are very misleading. For "mathematical convenience" (or lack of physical knowledge) the Lorentzian metric of spacetime is replaced by a Riemannian metric. This leads to equations which are of a very different nature from those which describe physics. Only in very special cases (a prerequisite is that all functions are analytic) is it possible to convert results of the Riemannian theory to the Lorentzian theory using an analytic extension argument according to which one can "rotate" a Riemannian theory into a corresponding Lorentzian theory, where both theories are embedded in a complex theory. In the literature on quantum field theory this rotation is known as the *Wick rotation.* The Riemannian analogue of gauge theory is mathematically (but not necessarily physically) of interest because it is linked to the well developed theory of *elliptic* partial differential equations. Gauge theory, on the other hand, is linked to *hyperbolic* partial differential equations. To sell the Riemannian analogue as gauge theory has presumably the advantage that it makes it easier to get funds for research in pure mathematics. On the other hand, it does confuse people. A pure Mathematician who worked in a field closely related to this Riemannian analogue and who saw work using the Lorentzian metric instead of a Riemannian metric once even asked me whether this Lorentzian approach would also be useful to physics!

**Lemma 5.2.4.** *Let $F$ be a closed 2-form and assume that $T_{el}$ is given by Equation 5.2.8. Then we have* $\operatorname{div}(T_{el}) = F(\cdot, \frac{1}{4\pi}\operatorname{div}(F^\sharp)) = F(\cdot, J^\sharp)$

*Proof.* Since $dF = 0$ we have $\nabla_a F_{bc} + \nabla_b F_{ca} + \nabla_c F_{ab} = 0$ which implies $\frac{1}{2}F^{ac}\nabla_b F_{ca} = -F^{ac}\nabla_a F_{bc}$. This gives

$$\begin{aligned}
4\pi \operatorname{div}(T)_b &= \nabla^a \left( g^{cd} F_{ac} F_{bd} - \frac{1}{4} g^{ef} g^{cd} F_{ec} F_{fd} g_{ab} \right) \\
&= g^{cd} F_{bd} \nabla^a F_{ac} + g^{cd} F_{ac} \nabla^a F_{bd} - \frac{1}{4} g^{ef} g^{cd} F_{fd} g_{ab} \nabla^a F_{ec} \\
&\quad - \frac{1}{4} g^{ef} g^{cd} F_{ec} g_{ab} \nabla^a F_{fd} \\
&= F(\cdot, \operatorname{div}(F^\sharp)) + g^{cd} F_{ac} \nabla^a F_{bd} - \frac{1}{2} F^{ec} \nabla_b F_{ec} \\
&= F(\cdot, \operatorname{div}(F^\sharp)) + F^{ac} \nabla_a F_{bc} + \frac{1}{2} F^{ec} \nabla_b F_{ce} \\
&= F(\cdot, \operatorname{div}(F^\sharp)).
\end{aligned}$$

∎

**Corollary 5.2.1.** *Assume that the source-free Maxwell equations hold. Then*

$$\operatorname{div}(T_{el}) = 0.$$

*Remark 5.2.2.* Recall that in the derivation of the Lorentzian structure of spacetime we assumed that light rays can be described by null geodesics. Since light is electromagnetic radiation we should now check that this identification is consistent with the description of electromagnetism in this section. However, this would require a proper discussion of electromagnetism which is beyond the scope of this book. Readers with knowledge of electromagnetism may consult (De Felice and Clarke 1990, section 7.8) for the identification of light with lightlike geodesics. Here we can only say that null geodesics can be taken as a description of light rays in an (observer-dependent) limit.

## 5.3 Einstein's equation

Recall that the equation which links geometry and matter should be of the form

$$\mathcal{D}g = T.$$

In the preceeding two sections we have motivated that the the right-hand side of this equation should be a symmetric, divergence-free $\binom{0}{2}$ tensor field. Now we will find an expression for the left-hand side.

In the Newtonian theory of gravitation, gravity is described by the equations

$$\ddot{\vec{\gamma}} = \vec{\mathfrak{G}} = \mathrm{grad}(\phi), \tag{5.3.9}$$
$$\Delta\phi = k\varrho, \tag{5.3.10}$$

where $\phi$ is the Newtonian potential for the gravitational field. Equation 5.3.10 is a second order partial differential equation for the Newtonian potential $\phi$ and describes how it is related to the mass density *varrho* of the universe.

Recall that we have replaced Equation 5.3.9 by the geodesic equation $\nabla_{\dot{\gamma}}\dot{\gamma} = 0$ which is equivalent to

$$\ddot{\gamma}^a = -\Gamma^a_{bc}\dot{\gamma}\dot{\gamma}.$$

It follows that the Christoffel symbols $\Gamma^a_{bc}$ have a rôle similar to the gravitational force field $\vec{\mathfrak{G}}$. One obtains the Christoffel symbols from $g$ via differentiation, just as one obtains the graviational force field $\mathfrak{G}$ from the Newtonian potential $\phi$ through differentiation. This indicates that $\phi$ corresponds to the metric $g$. Since the Newtonian potential is related to the matter distribution via a second order partial differential equation, we expect that $g \mapsto \mathcal{D}g$ is likewise a second order operator.

**Postulate 5.3.1 (Gravitation is determined by a 2nd-order pde).** *In any given coordinate system, $\mathcal{D}\colon g \mapsto \mathcal{D}g$ is a pointwise smooth function of $g_{cd}$, $\partial_a g_{cd}$, and $\partial_a\partial_b g_{cd}$.*

**Theorem 5.3.1.** *Let $(M,g)$ be a Lorentzian manifold such that* $\mathrm{dScal} \neq 0$ *and $\mathcal{D}g$ be a $\binom{0}{2}$ tensor field which satisfies Postulate 5.3.1 and*

$$\mathrm{div}(\mathcal{D}g) = 0$$

*(cf. Postulate 5.1.2).*

*If, in addition, $\mathcal{D}g$ is linear in $\partial_a\partial_b g$ then there exist constants $\Lambda, \mu \in \mathbb{R}$ such that*

$$\mathcal{D}g = \mu(\mathrm{Ric} - \frac{1}{2}\mathrm{Scal}\, g) + \Lambda g.$$

*Proof.* By Corollary 4.3.1 and the linearity assumption $\mathcal{D}g$ must be of the form

$$\mathcal{D}g = c_1\mathrm{Ric} + c_2\mathrm{Scal}\, g + c_3 g.$$

Lemma 4.3.1 implies now

$$0 = \mathrm{div}(\mathcal{D}g) = \left(\frac{c_1}{2} + c_2\right)\mathrm{dScal}.$$

Hence the result follows by our assumption that there is an $x \in M$ with $\mathrm{dScal}_{|x} \neq 0$. ∎

*Remark 5.3.1.* The assumption that $\mathcal{D}g$ is linear in its highest derivatives is rather awkward. Lovelock (1972) has shown that in 4-dimensional (but not in higher dimensional (!)) Lorentzian manifolds this assumption is not needed. Unfortunately, his proof is far too involved to be reproduced here.

*Remark 5.3.2.* Observe that we did not even need to assume symmetry of $\mathcal{D}g$, i.e. Postulate 5.1.1 is superfluous. However, the symmetry assumption was important to prove the conservation property Corollary 5.1.1 which motivates the requirement $\operatorname{div}(T) = 0$.

In conclusion, our postulates imply that gravity is governed by Einstein's equation as defined below.

**Definition 5.3.1.** Einstein's equation *(with* cosmological constant*)* $\Lambda \in \mathbb{R}$ *is given by*

$$\mathrm{Ric} - \frac{1}{2}\mathrm{Scal}\, g + \Lambda g = 8\pi T, \tag{5.3.11}$$

*where $T$ is the energy momentum tensor describing the matter distribution.*

In the above form, Einstein's equation is valid in *geometrical units* where the Gravitational constant and the velocity of light are set to 1 (cf. (Wald 1984, appendix F) for explicit translation rules to other units).

Astronomical observations seem to imply that $|\Lambda|$ is very small, if not zero. It should also be noted that the Newtonian theory of gravitation arises as a limit for $c \to \infty$ ($c$: velocity of light) if and only if $\Lambda = 0$. We will therefore often set $\Lambda = 0$ and obtain

[p. 251 ↓]
→8
↓ p. 266

$$\mathrm{Ric} - \frac{1}{2}\mathrm{Scal}\, g = 8\pi T. \tag{5.3.12}$$

### 5.3.1 The Lagrangian formulation of Einstein's equation

*In this section an alternative way which leads to Einstein's equation is sketched. This approach also aids in finding an appropriate energy momentum tensor. Unlike the rest of this book, this chapter rests on an underlying principle which is difficult to verify directly.*

*This section can be omitted on first reading and is not required for any other part of this book.*

[8] Our guide ends with Einstein's equation. For what follows we will also use the material which has been skipped in order to get to Einstein's equation quickly. Since we will discuss now physical applications which make use of all the mathematics which we have skipped, a continuation of this guide would lead to a lot of skipping forward and backward. On the other hand, the reader should have by now enough physical motivation in order to read the mathematical sections which we have skipped without getting bogged down. Still, the reader is advised to read on and to skip back only when needed. On these occasions skipped material should probably be read section-wise.

It appears that all fundamental dynamical equations in physics admit a *Lagrangian formulation*. According to this formulation, a physical system is described by a *Lagrange function* $\mathcal{L}\colon E \to \mathbb{R}$ where $E$ is an appropriate generalization of a vector bundle over spacetime $M$ which contains the possible physical states of the system.

Such a setup is motivated by classical mechanics. One can calculate the movement of a point-particle $\gamma\colon [a,b] \to \mathbb{A}^3$ with mass $m$ which is subject to a conservative[9] force field through the variation of an associated Lagrange function. Let $\mathcal{L}\colon \mathbb{A}^3 \times \mathbb{R}^3 \to \mathbb{R}$ be given by $\mathcal{L}(x,v) = \frac{m}{2}\langle v,v\rangle_{\mathbb{R}^3} - V(x)$. Then a curve $\gamma$ satisfies the equation

$$m\ddot{\gamma}(t) = -\mathrm{grad}V_{|\gamma(t)}$$

(cf. Equation (1.2.7)) if and only if

$$\left(\frac{\mathrm{d}}{\mathrm{d}\tau}\right)_{|\tau=0}\int_{[a,b]} \mathcal{L}(\gamma(t)+\tau h(t), \dot{\gamma}(t)+\tau\dot{h}(t))\mathrm{d}t = 0 \tag{5.3.13}$$

for all smooth maps $h\colon [a,b] \mapsto \mathbb{R}^3$ with $h(a) = h(b) = 0$. In fact, we have

$$\begin{aligned}
&\left(\frac{\mathrm{d}}{\mathrm{d}\tau}\right)_{|\tau=0}\int_{[a,b]} \mathcal{L}(\gamma(t)+\tau h(t), \dot{\gamma}(t)+\tau\dot{h}(t))\mathrm{d}t \\
&\quad= \int_{[a,b]} \left(m\left\langle\dot{\gamma}(t), \dot{h}(t)\right\rangle_{\mathbb{R}^3} - \mathrm{d}V(h(t))\right)\mathrm{d}t \\
&\quad= \int_{[a,b]} \left(m\left(\frac{\mathrm{d}}{\mathrm{d}t}\right)\langle\dot{\gamma}(t), h(t)\rangle_{\mathbb{R}^3} - m\langle\ddot{\gamma}(t), h(t)\rangle_{\mathbb{R}^3}\right. \\
&\qquad \left. - \mathrm{d}V(h(t))\right)\mathrm{d}t \\
&\quad= m\langle\dot{\gamma}(t), h(t)\rangle_{\mathbb{R}^3}\Big|_a^b - \int_{[a,b]} \left\langle m\ddot{\gamma}(t) + \mathrm{grad}V_{|\gamma(t)}, h(t)\right\rangle_{\mathbb{R}^3}\mathrm{d}t.
\end{aligned}$$

the first summand vanishes for all choices of $h$ with $h(a) = h(b) = 0$. Assume, there is a $t_0 \in (a,b)$ with $m\ddot{\gamma}(t_0) + \mathrm{grad}V_{|\gamma(t_0)} \neq 0$. Since the scalar product is non-degenerate there is an $h_0 \in \mathbb{R}^3$ with

$$\left\langle m\ddot{\gamma}(t_0) + \mathrm{grad}V_{|\gamma(t_0)}, h_0\right\rangle_{\mathbb{R}^3} \neq 0$$

(say $> 0$). Let $\tilde{h}\colon [a,b] \to \mathbb{R}^3$ be any smooth function with $\tilde{h}(t_0) = h_0$. By continuity there is a neihbourhood $(t_-, t_+) \subset (a,b)$ of $t_0$ such that $\left\langle m\ddot{\gamma}(t) + \mathrm{grad}V_{|\gamma(t)}, \tilde{h}(t)\right\rangle_{\mathbb{R}^3} > 0$ for all $t \in (t_-, t_+)$. Finally, let

[9] Here "conservative" simply means that there is a function $V\colon \mathbb{A}^3 \to \mathbb{R}$ with $\vec{F} = -\mathrm{grad}V$.

$\psi\colon [a,b] \to \mathbb{R}^+ \cup \{0\}$ be a smooth positive function with non-empty support in $(t_-, t+)$. Setting $h = \psi\tilde{h}$ the integrand $\langle m\ddot{\gamma}(t) + \mathrm{grad}V_{|\gamma(t)}, h(t)\rangle_{\mathbb{R}^3}$ is non-negative and strictly positive in an open set. Hence the integral must be positive in contradiction to Equation (5.3.13). This proves $m\ddot{\gamma}(t) + \mathrm{grad}V_{|\gamma(t)} = 0$ for all $t$.

The existence of a Lagranigian formulation is widely seen as fundamental for general (classical) physical systems. The physical state of an elementary particle[10] is described by the section $\psi\colon M \to E$ of an appropriate vector bundle. Its governing equation should again be determined by the variation of an integral equation whose integrand is built from $\psi$, its derivative, and physical fields which interact with $\psi$. To make this program work one first has to *define* the derivative of a section in an arbitrary vector bundle. It turns out that one can generalize our treatment of derivates of vector fields (cf. Sect. 2.6) and define connections of general vector bundles (Kobayashi and Nomizu 1963). It is also possible to define a notion of curvature for these generalized connections and — analogously to the case of general relativity — one can interpret this curvature $F$ in terms of physical fields which interact with the given elementary particle. To obtain the complete system of equations one writes down a Lagrange function which depends on $\psi$, its derivative, the curvature $F$, and perhaps other physical expressions. We denote all these physical inputs collectively by $\phi\colon M \to E$ and by $\tau \mapsto \phi_\tau = \phi + \tau\hat{\phi}$ a one-parameter family of sections such that $\hat{\phi}$ has compact support. The equations which have to be satisfied by the physical system are then given by

$$\left(\frac{\mathrm{d}}{\mathrm{d}\tau}\right)_{|\tau=0} \int_M (\mathcal{L}\mu_M) \circ \phi_\tau = 0$$

for all variations $\phi_\tau$ of $\phi$. This recipe is referred to as the *variation of the Lagrange function* $\mathcal{L}$. The mathematical properties of equations of this type are similar to the properties one encounters in the theory of minimal submanifolds (cf. Lemma 4.4.8 and the discussion following this lemma).

We will now discuss an example of this formulation, the physical system consisting of electrodynamics and gravity. To keep things simple we will assume that there are no other physical inputs. In particular, there are no electromagnetic sources, i.e., there no charges. Our Lagarange function will consist of two summands,

$$\mathcal{L} = \mathcal{L}_{\mathrm{grav}} + \mathcal{L}_{\mathrm{el}}.$$

where $\mathcal{L}_{\mathrm{grav}}$ stands for the contribution from gravity and $\mathcal{L}_{\mathrm{el}}$ for the contribution from electrodynamics.

[10] It is far beyond the scope of this book to explain what this actually is. Our only purpose is to vaguely set the following discussion into context. Readers who want to know more are referred to books on gauge theory.

*Remark 5.3.3.* If we had included electromagnetic sources we would also have to add at least two more terms:

- A term $\mathcal{L}_{\text{kin}}$ for the elementary particle which is analoguous to the term $m\langle\dot{\gamma},\dot{\gamma}\rangle_{\mathbb{R}^3}$ in the mechanical example above, and
- a term $\mathcal{L}_{\text{int}}$ which describes the interaction of the elementary particle with the electromagnetic field.

The reader may already have suspected that the electro-magnetic field $F$ is in fact the curvature of a generalized connection $\mathfrak{A}$ (Bleecker 1981) The connection $\mathfrak{A}$ can be identified with a 1-form $A$ which, however, is not invariantly defined. (This corresponds to regarding the Christoffel symbols as tensors.). With this 1-form $A$ we have $F = \mathrm{d}A$. The first of Maxwell's equations, $\mathrm{d}F = 0$, is then a trivial consequence. According to the program above we have to vary $\mathcal{L}$ with respect to the 1-form $A$. This will give the second of Maxwell's equations, $\mathrm{div}F = 0$. Gravity depends on two geometric quantities, the torsion-free connection $\nabla$ and the metric $g$. We will use the *Palatini formalism*, i.e., we will independently vary $\mathcal{L}$ *independently* with respect to $\nabla$ and $g$. Observe that this independent variation makes sense only if we do not assume a priori that $\nabla$ is the Levi-Civitā connection. It will turn out that the variation with respect to the connection will fix the Levi-Civitā connection and that the variation with respect to the metric will give Einstein's equation.[11]

The simplest non-trivial, invariant function $\mathcal{L}_{el}$ which can be defined is (modulo constant factors and modulo the addition of a constant term) given by

$$\mathcal{L}_{\text{el}}(A,\nabla,g) = -\frac{1}{16\pi}\langle F,F\rangle = -\frac{1}{16\pi}\langle \mathrm{d}A,\mathrm{d}A\rangle\,,$$

where $\langle F,F\rangle = g^{ac}g^{bd}F_{ab}F_{cd}$.

**Lemma 5.3.1.** *Let $B \in \mathcal{T}_1^0(M)$ be a tensor field with compact support and consider the* variation $\phi_\tau = (A_\tau,\nabla,g) = (A+\tau B,\nabla,g)$. *Then*

$$\left(\frac{\mathrm{d}}{\mathrm{d}\tau}\right)_{|\tau=0}\int_M (\mathcal{L}_{\text{el}}\mu_M)\circ\phi_\tau = \frac{1}{4\pi}\int_M (\widetilde{\mathrm{div}}F)(B^\sharp)\mu_M$$

*holds, where $\widetilde{\nabla}$ is the Levi-Civitā connection of $g$ and $\widetilde{\mathrm{div}}$ is the divergence operator associated with $\widetilde{\nabla}$.*

*Proof.* From $\mathrm{d}B_{bd} = \widetilde{\nabla}_b B_d - \widetilde{\nabla}_d B_b$ we get

[11] Alternatively, one could assume that the connection is the Levi-Civitā connection and only vary the metric. We have chosen the Palatini formalism because this is more akin to the treatment of other gauge theories.

$$\begin{aligned}
-\frac{1}{16\pi}\left(\frac{\mathrm{d}}{\mathrm{d}\tau}\right)_{|\tau=0}&\int_M \langle \mathrm{d}A_\tau, \mathrm{d}A_\tau\rangle \,\mu_M \\
&= -\frac{1}{8\pi}\int_M \langle F, \mathrm{d}B\rangle \,\mu_M \\
&= -\frac{1}{8\pi}\int_M g^{ab}g^{cd}(F_{ac})(\widetilde{\nabla}_b B_d - \widetilde{\nabla}_d B_b)\mu_M \\
&= -\frac{1}{4\pi}\int_M g^{ab}g^{cd}(F_{ac})\widetilde{\nabla}_b B_d \mu_M \\
&= -\frac{1}{4\pi}\int_M (\widetilde{\nabla}_b(g^{ab}g^{cd}(F_{ac})B_d) - g^{ab}g^{cd}\widetilde{\nabla}_b F_{ac}B_d)\mu_M \\
&= -\frac{1}{4\pi}\int_M (\widetilde{\mathrm{div}}(F(\cdot, B^\sharp)^\sharp) - (\widetilde{\mathrm{div}} F^\sharp)(B))\mu_M .
\end{aligned}$$

The first summand vanishes because of the theorem of Gauß and the fact that $B$ has compact support. ∎

Since the $\int_M \mathcal{L}_{\mathrm{el}}(A, \nabla, g)\mu_M$ does not depend on $\nabla$ the variation with respect to $\nabla$ vanishes.

**Lemma 5.3.2.** *Let $h \in \mathrm{sym}(\mathcal{T}_2^0(M))$ be a tensor field with compact support and consider the variation $\phi_\tau = (A, \nabla, g_\tau) = (A, \nabla, g + \tau h)$. Then*

$$\left(\frac{\mathrm{d}}{\mathrm{d}\tau}\right)_{|\tau=0}\int_M (\mathcal{L}\mu_M)\circ\phi_\tau = \frac{1}{2}\int_M (T_{el})_{ab}h^{ab}\mu_M$$

*holds, where $T_{el}$ is given by Equation* (5.2.8)

*Proof.* The equation $0 = \frac{\mathrm{d}}{\mathrm{d}\tau}\left((g_\tau)_{ab}(g_\tau)^{bc}\right)_{|\tau=0} = h_{ab}g^{bc} + g_{ab}\frac{\mathrm{d}}{\mathrm{d}\tau}(g_\tau)^{bc}{}_{|\tau=0}$ implies

$$h^{ac} = -\frac{\mathrm{d}}{\mathrm{d}\tau}(g_\tau)^{ac}.$$

Recall from the proof of Lemma 4.6.20 that the derivative of $\det(\varphi_\tau)$, where $\varphi_\tau$ is a 1-parameter family of matrices, is given by $(\det(\varphi_\tau))^{\cdot} = \mathrm{tr}(\dot\varphi\varphi^{-1})\det\varphi$. This implies

$$(\det(g_\tau))^{\cdot} = \mathrm{tr}(h)\det(g_\tau),$$

where in this formula tr is the metric trace of a covariant $\binom{0}{2}$ tensor. Using these two formulas we calculate

$$
\begin{aligned}
&\frac{\mathrm{d}}{\mathrm{d}\tau}\Big(\int_M (\mathcal{L}_{\mathrm{el}} \circ \phi_\tau)\,(\mu_M)_\tau\Big)_{|\tau=0} \\
&= \frac{-1}{16\pi}\int_M \frac{\mathrm{d}}{\mathrm{d}t}\Big(F_{ac}F_{bd}(g_\tau)^{ab}(g_\tau)^{cd}\sqrt{-\det(g_\tau)}\Big)_{|\tau=0} \mathrm{d}x^1 \wedge \cdots \wedge \mathrm{d}x^n \\
&= \frac{-1}{16\pi}\int_M \Big(-2F_{ac}F_{bd}g^{cd}h^{ab}\sqrt{-\det(g)} \\
&\quad + \frac{-1}{2}\langle F, F\rangle \sqrt{-\det(g)}\mathrm{tr}(h)\Big)\mathrm{d}x^1 \wedge \cdots \wedge \mathrm{d}x^n \\
&= \frac{1}{8\pi}\int_M \Big(F_{ac}F_{bd}g^{cd} - \frac{1}{4}\langle F, F\rangle g_{ab}\Big)h^{ab}\mu_M .
\end{aligned}
$$

∎

*Remark 5.3.4.* We have thus obtained the form of the electro-magnetic energy momentum tensor by variation of the simple Lagrange function

$$-(16\pi)^{-1}\langle F, F\rangle$$

with respect to the electro-magnetic potential $A$. For other matter fields analogous results hold. In this sense it can be said that variational techniques aid in finding the correct energy momentum tensor.

For the graviational term we set

$$\mathcal{L}_{\mathrm{grav}}(A, \nabla, g) = \frac{1}{16\pi}(\mathrm{Ric}_{ab}g^{ab} - 2\Lambda),$$

where Ric is the Ricci tensor with respect to the connection $\nabla$. This is again the (modulo constant factors and summands) simplest invariant function which can be build from the metric $g$ and the connection $\nabla$. Recall that for any two torsion-free connections $\nabla$, $\tilde{\nabla}$ the difference is a $\binom{1}{2}$-tensor field $K$ which is symmetric in its covariant indices. In index-notation, $K$ is given by $\nabla_V W^a = \tilde{\nabla}_V W^a + K^a_{cd}V^c W^d$ for all vector fields $V, W$. We simply write $\nabla = \tilde{\nabla} + K$.

**Lemma 5.3.3.** *Let $C \in \mathcal{T}^1_2(M)$ be a tensor field which is symmetric in its covariant entries and which has compact support. Setting $(\nabla_\tau) = \nabla + \tau C$ and $\phi_\tau = (A, (\nabla_\tau), g)$ we have*

$$
\begin{aligned}
&\frac{\mathrm{d}}{\mathrm{d}\tau}\Big(\int_M (\mathcal{L}_{\mathrm{grav}}\mu_M) \circ \phi_\tau\Big)_{|\tau=0} \\
&\qquad = -\frac{1}{16\pi}\int_M \Big(K^{bd}{}_d\delta^a_c + K^d{}_{dc}g^{ab} - 2K^a{}_c{}^b\Big)C^c{}_{ab}\mu_M ,
\end{aligned}
$$

*where $K$ is defined by $\nabla = \tilde{\nabla} + K$ and $\tilde{\nabla}$ is the Levi-Cività connection of $g$.*

*Proof.* Since the Ricci tensor is the only quantity which involves $(\nabla_\tau)$ we obtain

$$\frac{\mathrm{d}}{\mathrm{d}\tau}\Big(\int_M \mathcal{L}_{\text{grav}} \circ (A, (\nabla_\tau), g)\,(\mu_M)_\tau\Big)_{|\tau=0} = \frac{1}{16\pi}\int_M \frac{\mathrm{d}}{\mathrm{d}\tau}(\mathrm{Ric}_\tau)_{ab|\tau=0}\, g^{ab}\mu_M .$$

Let $x \in M$ and consider a normal coordinate system centered at $x$. From

$$\begin{aligned}(\mathrm{Ric}_\tau)_{ab} &= \partial_c\left(\Gamma^c_{ab} + \tau C^c_{ab}\right) - \partial_b\left(\Gamma^c_{ac} + \tau C^c_{ac}\right)\\ &\quad + \left(\Gamma^d_{ab} + \tau C^d_{ab}\right)\left(\Gamma^c_{dc} + \tau C^c_{dc}\right) - \left(\Gamma^d_{ac} + \tau C^d_{ac}\right)\left(\Gamma^c_{db} + \tau C^c_{db}\right)\end{aligned}$$

and the fact that the Christoffel symbols vanish at $x$ we see without calculation that at $x$ and for $\tau = 0$ the derivative of $\mathrm{Ric}_\tau$ is given by

$$\frac{\mathrm{d}}{\mathrm{d}\tau}(\mathrm{Ric}_\tau)_{ab|\tau=0,x} = \nabla_c C^c_{ab|\tau=0,x} - \nabla_b C^c_{ac|\tau=0,x} .$$

Since this is a tensor equation which is independent of coordinates it must hold at all points of $M$.

We will now re-express $\nabla_c C^c_{ab} - \nabla_b C^c_{ac}$ with respect to the Levi-Cività connection $\widetilde{\nabla}$ and the tensor field $K$. From the definition of $K$ we get

$$\nabla_d C^c_{ab} = \widetilde{\nabla}_d C^c_{ab} + K^c_{de} C^e_{ab} - K^e_{db} C^c_{ae} - K^e_{da} C^c_{be}$$

and therefore

$$\begin{aligned} g^{ab}\left(\nabla_c C^c_{ab} - \nabla_b C^c_{ac}\right) &= g^{ab}\Big(\widetilde{\nabla}_c C^c_{ab} + K^c_{ce} C^e_{ab} - K^e_{cb} C^c_{ae} - K^e_{ca} C^c_{be}\\ &\qquad - \widetilde{\nabla}_b C^c_{ac} - K^c_{be} C^e_{ac} + K^e_{bc} C^c_{ae} + K^e_{ba} C^c_{ce}\Big)\\ &= g^{ab}\left(\widetilde{\nabla}_c C^c_{ab} - \widetilde{\nabla}_b C^c_{ac}\right)\\ &\quad + g^{ab}\left(K^c_{ce} C^e_{ab} - K^e_{ca} C^c_{be} - K^c_{be} C^e_{ac} + K^e_{ba} C^c_{ce}\right)\\ &= \widetilde{\mathrm{div}}(\mathrm{tr}_{2,3} C - \mathrm{tr}_{1,2} C)\\ &\quad + \left(K^c_{ce} g^{ab} - 2 g^{db} K^a_{ed} + K^b_{fh} g^{fh} \delta^a_e\right) C^e_{ab}, \end{aligned}$$

where $\mathrm{tr}_{i,j}$ denotes the (metric) trace over the $i$th and $j$th entry. Since the first summand is a divergence with respect to the Levi-Cività connection its integral vanishes by the theorem of Gauß. ∎

**Lemma 5.3.4.** *Let $h \in \mathrm{sym}(\mathcal{T}^0_2(M))$ be a tensor field with compact support and $g_\tau = g + \tau h$. For $\phi_\tau = (A, \nabla, g_\tau)$ we have*

$$\begin{aligned}\frac{\mathrm{d}}{\mathrm{d}\tau}\Big(\int_M (\mathcal{L}_{\text{grav}}\mu_M) \circ \phi_\tau\Big)_{|\tau=0}\\ = -\frac{1}{16\pi}\int_M \Big(\mathrm{Ric}_{ab} - \frac{1}{2}\mathrm{Ric}_{cd}\, g^{cd} g_{ab} + \Lambda g_{ab}\Big) h^{ab}\mu_M .\end{aligned}$$

*Proof.* We can split the integral into two parts which will be considered separately,

$$\begin{aligned}
&\frac{\mathrm{d}}{\mathrm{d}\tau}\Big(\int_M (\mathcal{L}_{\mathrm{grav}}\mu_M)\circ\phi_\tau\Big)_{|\tau=0}\\
&\quad=\frac{1}{16\pi}\int_M \frac{\mathrm{d}}{\mathrm{d}\tau}\Big((\mathrm{Ric}_{ab}(g_\tau)^{ab}-2\Lambda)\sqrt{-\det(g_\tau)}\Big)_{|\tau=0}\mathrm{d}x^1\wedge\cdots\wedge\mathrm{d}x^n\\
&\quad=\frac{1}{16\pi}\int_M \Big((\mathrm{Ric}_\tau)_{ab}\frac{\mathrm{d}}{\mathrm{d}\tau}\Big((g_\tau)^{ab}\sqrt{-\det(g_\tau)}\Big)_{|\tau=0}\\
&\qquad-2\Lambda\frac{\mathrm{d}}{\mathrm{d}\tau}\sqrt{-\det(g_\tau)}_{|\tau=0}\Big)\mathrm{d}x^1\wedge\cdots\wedge\mathrm{d}x^n.
\end{aligned}$$

Exactly as in the proof of Lemma 5.3.2 we see that the second summand in the integral equals $-\Lambda g^{ab}h_{ab}$. For the first summand we calculate

$$\begin{aligned}
&\frac{\mathrm{d}}{\mathrm{d}\tau}\left((g_\tau)^{ab}\sqrt{-\det(g_\tau)}\right)_{|\tau=0}\\
&\qquad=\frac{\mathrm{d}}{\mathrm{d}\tau}\left((g_\tau)^{ab}\right)_{|\tau=0}\sqrt{-\det(g_\tau)}+(g_\tau)^{ab}\frac{\mathrm{d}}{\mathrm{d}\tau}\sqrt{-\det(g_\tau)}_{|\tau=0}\\
&\qquad=-g^{ac}g^{bd}h_{bd}\sqrt{-\det(g)}+\frac{1}{2}g^{ab}g^{cd}h_{cd}\sqrt{-\det(g)}
\end{aligned}$$

and therefore

$$\begin{aligned}
&\mathrm{Ric}_{ab}\frac{\mathrm{d}}{\mathrm{d}t}\left((g_\tau)^{ab}\sqrt{-\det(g_\tau)}\right)_{|\tau=0}\\
&\qquad\qquad=\Big(-\mathrm{Ric}^{ab}+\frac{1}{2}\mathrm{Ric}_{cd}g^{cd}g^{ab}\Big)h_{ab}\sqrt{-\det(g)}.
\end{aligned}$$

∎

The following corollary is the main result of this subsection.

**Corollary 5.3.1.** *Let $A$ be a 1-form, $\nabla$ be a torsion-free connection, $g$ be a Lorentzian metric and set $F=\mathrm{d}A$.*

*Einstein's equation and Maxwell's equations for a source-free electromagnetical field,*

$$\mathrm{Ric}-\frac{1}{2}\mathrm{Scal}\,g+\Lambda g=\frac{1}{2}\left(g^{cd}F_{ac}F_{bd}-\frac{1}{4}\langle F,F\rangle g_{ab}\right),\quad \mathrm{d}F=0,\quad \mathrm{div}F=0$$

*are equivalent to*

$$\left(\frac{\mathrm{d}}{\mathrm{d}\tau}\right)_{|\tau=0}\int_M(\mathcal{L}_{\mathrm{el}}+\mathcal{L}_{\mathrm{grav}})\circ\phi_\tau(\mu_M)_\tau=0$$

*for all variations $\phi_\tau=(A_\tau,(\nabla_\tau),g_\tau)=(A+\tau B,\nabla+\tau C,g+\tau h)$ where $B,C,h$ are tensor fields with compact support.*

*Proof.* We can consider the variations with respect to $A, \nabla, g$ separately. Since $C$ is arbitrary Lemma 5.3.3 implies at each point $x$

$$\left(K^{bd}{}_d\delta^a_c + K^d{}_{dc}g^{ab} - 2K^a{}_c{}^b\right)C^c{}_{ab} = 0$$

for all tensors $C^c_{ab}$ which are symmetric in $a$ and $b$. This is equivalent to

$$K^{bd}{}_d\delta^a_c + K^{ad}{}_d\delta^b_c + 2K^d{}_{dc}g^{ab} - 2K^a{}_c{}^b - 2K^b{}_c{}^a = 0. \tag{5.3.14}$$

Taking the trace with respect to $b$ and $c$ we get

$$0 = K^{ad}{}_d + nK^{ad}{}_d + 2K^d{}_d{}^a - 2K^a{}_d{}^d - 2K^d{}_d{}^a = (n-1)K^{ad}{}_d.$$

Taking now the trace with respect to $a$ and $b$ we get

$$0 = 2nK^d{}_{dc} - 4K^d{}_{cd} = 2(n-2)K^d{}_{dc}.$$

These equations together with the symmetry of $K^a{}_{bc}$ in $b$ and $c$ imply (for $n > 2$) that all traces of $K$ vanish. Hence Equation (5.3.14) simplifies to $K^a{}_c{}^b = -K^b{}_c{}^a$ and $K_{abc}$ is a tensor with the properties

$$K_{abc} = -K_{bac}, \quad K_{abc} = K_{acb}.$$

We will now show that this tensor vanishes. Since it has the property that it is symmetric in two indices and anti-symmetric in two other indices, the expressions $\mathrm{sym}(K^\flat)$ and $\mathrm{alt}(K^\flat)$ both vanish. This is equivalent to

$$0 = K_{abc} + K_{bca} + K_{cab} + K_{acb} + K_{bac} + K_{cba}$$

and

$$0 = K_{abc} + K_{bca} + K_{cab} - (K_{acb} + K_{bac} + K_{cba}).$$

These equations imply $K_{abc} + K_{bca} + K_{cab} = 0$ and therefore, using the symmetries of $K^\flat$, $0 = K_{abc} + K_{bac} + K_{cab} = K_{cab}$. Hence $K$ vanishes and we have $\nabla = \tilde{\nabla}$.

The equation $\mathrm{d}F = 0$ follows trivially from the definition $F = \mathrm{d}A$. The second part of Maxwell's equations, $\mathrm{div}F = 0$ follows immediately from Lemma 5.3.1 since $B$ is arbitrary.

Since $h$ is arbitrary and we know that $\nabla$ is in fact the Levi-Cività connection the validity of Einstein's equation follows immediately from Lemma 5.3.2 and Lemma 5.3.4. ∎

The process leading to Einstein's equation via Corollary 5.1.1 is referred to as *varying the total Lagrangian* $\mathcal{L}_{\mathrm{el}} + \mathcal{L}_{\mathrm{grav}}$ *with respect to the metric* $g$.

We have chosen electromagnetism for our matter model in order to have a concrete example. To my knowledge all *fundamental*[12] matter models admit a Lagrangian formulation such that

$$\left(\frac{\mathrm{d}}{\mathrm{d}\tau}\right)_{|\tau=0} \int_M \big((\mathcal{L}_{\mathrm{matter}} + \mathcal{L}_{\mathrm{grav}})\mu_M\big) \circ \phi_\tau = 0$$

for all variations of the metric is equivalent to Einstein's equation for the particular matter model.

That the Lagrangian ansatz described in this section works is by no means trivial and I have no explanation for it.

## 5.4 The Einstein equation as a system of partial differential equations

Physicists are accustomed to the fact that (classical) physical systems depend on initial conditions and then evolve in a determined manner which is governed by second order hyperbolic differential equations. Since the energy momentum tensor $T$ contains the metric, the Einstein Equation (5.3.11) cannot be simply solved for a given $T$. Instead, one has to convert the system of Equations (5.3.11) into a system of partial differential equations for $g$ and some matter quantities.

The analogue in relativity would therefore be to fix an $(n-1)$-dimensional Riemannian manifold $(\Sigma, {}^{\Sigma}g)$ which represents an initial instant of time. This manifold will be isometric to a spacelike hypersurface in the solution. Since Einstein's equations are a second order system we would need to prescribe a symmetric $\binom{0}{2}$ tensor field $k$ which specifies the normal derivative of the induced metric ${}^{\Sigma}g$ or, equivalently, the second fundamental form of our hypersurface. We also need to fix functions or tensor fields which represent the initial matter distribution at $\Sigma$, possibly also their normal derivatives.

The character of this system of partial differential equations will crucially depend on the form of matter assumed. In particular, one can choose unphysical matter models which lead to spacetimes in which it is possible for information to travel *faster* than light (cf. Corollary 7.4.1). It is also possible to choose unphysical matter models which do not lead to a hyperbolic system of differential equations.

Another problem lies in the fact that we have always the freedom to change coordinates. Hence the choice of coordinate system may also have an effect on the kind of system of partial differential equations we will end up with.

[12] A perfect fluid is a macroscopic concept and the Lagrangian formulation does not work well in this case. See (De Felice and Clarke 1990, chapter 6.5) for a discussion.

Nevertheless, in most situations of interest, it is possible to obtain a well-posed system of equations. We will show this for the special case that $T = 0$ and $\Lambda = 0$. In order to avoid subtleties arising from the theory of partial differential equations we will assume that our initial data ${}^{\Sigma}g, k$ are analytic and that $\Sigma$ is an analytic manifold. (This restriction allows us to appeal to the relativily elemementary theorem of Cauchy-Kowalewskaya.) We will also fix coordinates in which the equations are especially simple.

In Chap. 6 we will study the more general case of a perfect fluid. However, we will impose strong symmetry assumptions in order to simplify the problem drastically (cf. Sect. 6.2) — the system of equations will be reduced to a system of ordinary differential equations.

Chapter 7 contains an intermediate treatment. We will again consider a perfect fluid but use weaker symmetry assumptions which lead to a system of partial differential with two independent variables. This system of equations is substantially simpler than the general system depending on 4 variables. We will therefore be able to give a smooth (rather than an analytic) existence theorem (cf. Theorem 7.4.1).

*Since the analogous but considerably simpler discussion in Chap. 6 already exhibits some of the key concepts of the initial value problem for Einstein's equation, the reader may wish to skip the rest of this section on first reading.*

Let $(M, g)$ be a Lorentz manifold $\Sigma \subset M$ be a smooth, spacelike hypersurface with normal[13] $\mathfrak{n}$. For each $x \in \Sigma$ consider the geodesic $\gamma_x$ with $\dot{\gamma}_x(0) = \mathfrak{n}_x$. There is a neighbourhood of $\Sigma$ which is foliated by these geodesics. If this neighbourhood is chosen small enough it is also foliated by spacelike hypersurfaces of the form $\Sigma_t = \{\gamma_x(t) | \, x \in \Sigma\}$.

If one views $M$ as being foliated by spacelike hypersurfaces $\Sigma_t$ ($\Sigma_0 = \Sigma$) with induced metric ${}^{\Sigma_t}g$ then one can view the associated second fundamental form $k_t$ as the $t$-derivative of $\frac{1}{2}{}^{\Sigma_t}g$:

**Lemma 5.4.1.** *Let $\Sigma$ be a spacelike hypersurfaces of $(M, g)$ and $\{\Sigma_t\}$ be a foliation of a neighbourhood as constructed above. Let $x^1, \ldots, x^{n-1}$ be a coordinate system of $\Sigma$ centred at $x \in \Sigma$.*

*Then there is a neighbourhood $\mathcal{U}$ of $x \in M$ such that $g = -\mathrm{d}t^2 + \sum_{i=1}^{n-1} g_{ij}(t, x^1, \ldots, x^{n-1})\mathrm{d}x^i \mathrm{d}x^j$.*

*Moreover, the second fundamental form $k_t$ of $\Sigma_t$ is given by $k_t = \frac{1}{2}\pounds_{\partial_t}{}^{\Sigma_t}g$.*

*Proof.* We can find a neighbourhood $\mathcal{U}$ of $x$ such that for every point $y \in \mathcal{U}$ there is exactly one point $\hat{x} \in \Sigma \cap \mathcal{U}$ and one geodesic $\gamma_{\hat{x}}$ through $\hat{x}$ which satisfies $\dot{\gamma}_{\hat{x}}(0) = \mathfrak{n}_{\hat{x}}$ and intersects $\Sigma$ exactly once without leaving $\mathcal{U}$. This gives a chart $(\mathcal{U}, \varphi)$ defined by $\varphi(y) = (t, x^1(\hat{x}), \ldots, x^{n-1}(\hat{x}))$ where $y = \gamma_{\hat{x}}(t)$.

[13] Here we mean: $g(\mathfrak{n}, \mathfrak{n}) = -1, g(\mathfrak{n}, v) = 0$ for all $v \in T\Sigma$.

It follows from our construction that the induced metric on $\Sigma_t$ is given by ${}^{\Sigma_t}g = \sum_{i,j=1}^{n-1} g_{ij}(t, x^1, \ldots, x^{n-1})\mathrm{d}x^i\mathrm{d}x^j$, where $g_{ij}$ are suitable functions. At $t = 0$ we have for each $\hat{x} \in \Sigma$

$$g_{\hat{x}} = -\mathrm{d}t^2 + \sum_{i,j=1}^{n-1} g_{ij}(0, x^1(\hat{x}), \ldots, x^{n-1}(\hat{x}))\mathrm{d}x^i\mathrm{d}x^j$$

since $\dot{\gamma}_{\hat{x}}(0) = \mathfrak{n}_x \perp \Sigma = \Sigma_0$. From

$$\begin{aligned}\dot{\gamma}_{\hat{x}} \bullet \langle \dot{\gamma}_{\hat{x}}, \partial_{x^i} \rangle &= \Big\langle \overbrace{\nabla_{\dot{\gamma}_x}\dot{\gamma}_x}^{=0}, \partial_{x^i} \Big\rangle + \Big\langle \dot{\gamma}_x, \nabla_{\dot{\gamma}_x}\partial_{x^i} \Big\rangle = \Big\langle \partial_t, \nabla_{\partial_t}\partial_{x^i} \Big\rangle \\ &= \Big\langle \partial_t, \nabla_{\partial_{x^i}}\partial_t \Big\rangle = \frac{1}{2}\nabla_{\partial_{x^i}} \langle \dot{\gamma}_{\hat{x}}, \dot{\gamma}_{\hat{x}} \rangle = 0\end{aligned}$$

we get $\dot{\gamma}_{\hat{x}}(t) \perp \Sigma_t$ for all $t$. This implies the claim for the metric components.

From Lemma 4.4.6 we get

$$\begin{aligned}k_t(\partial_{x^i}, \partial_{x^j}) &= \Big\langle \nabla_{\partial_{x^i}}\partial_t, \partial_{x^j} \Big\rangle = \Big\langle \nabla_{\partial_t}\partial_{x^i}, \partial_{x^j} \Big\rangle \\ &= \partial_t \bullet g_{ij} - \Big\langle \partial_{x^i}, \nabla_{\partial_t}\partial_{x^j} \Big\rangle = \partial_t \bullet g_{ij} - k_t(\partial_{x^j}, \partial_{x^i}).\end{aligned}$$

The assertion $k_t = \frac{1}{2}\pounds_{\partial_t}{}^{\Sigma_t}g$ follows now from the symmetry of $k_t$ and from

$$\begin{aligned}(\pounds_{\partial_t}{}^{\Sigma_t}g)_{ij} &= (\pounds_{\partial_t}{}^{\Sigma_t}g)(\partial_{x^i}, \partial_{x^j}) \\ &= \partial_t{}^{\Sigma_t}g_{ij} - {}^{\Sigma_t}g(\overbrace{\pounds_{\partial_t}\partial_{x^i}}^{=0}, \partial_{x^j}) - {}^{\Sigma_t}g(\partial_{x^i}, \overbrace{\pounds_{\partial_t}\partial_{x^j}}^{=0}) = \partial_t g_{ij}.\end{aligned}$$

∎

We denote the Levi-Cività connection induced on $\Sigma_t$ by ${}^{\Sigma_t}\boldsymbol{\nabla}$ and the Ricci tensor of $(\Sigma_t, {}^{\Sigma_t}g)$ by ${}^{\Sigma_t}\mathrm{Ric}$.

**Lemma 5.4.2.** *Einstein's equation with vanishing cosmological constant for vacuum is equivalent to the following system of equations.*

$$\begin{aligned}\partial_t\partial_t{}^{\Sigma_t}g_{ij} &= -2{}^{\Sigma_t}\mathrm{Ric}_{ij} - \Big(\frac{1}{2}\partial_t{}^{\Sigma_t}g_{ij}\partial_t{}^{\Sigma_t}g_{kl} - \partial_t{}^{\Sigma_t}g_{ik}\partial_t{}^{\Sigma_t}g_{jl}{}^{\Sigma_t}g^{jl}\Big){}^{\Sigma_t}g^{kl}, \\ 0 &= {}^{\Sigma_t}\mathrm{Scal} + \frac{1}{4}\Big(\partial_t{}^{\Sigma_t}g_{ij}\partial_t{}^{\Sigma_t}g_{kl} - \partial_t{}^{\Sigma_t}g_{ik}\partial_t{}^{\Sigma_t}g_{jl}\Big){}^{\Sigma_t}g^{ij}{}^{\Sigma_t}g^{kl}, \\ 0 &= -\partial_{x^i}({}^{\Sigma_t}g^{jk}\partial_t{}^{\Sigma_t}g_{jk}) + {}^{\Sigma_t}g^{jk}\partial_{x^j}{}^{\Sigma_t}g_{ik}.\end{aligned}$$

*Proof.* Einstein's equation is given by $\mathrm{Ric} - \frac{\mathrm{Scal}}{2}g = 0$ which is equivalent to $\mathrm{Ric} = 0$. The Gauß equation (Proposition 4.4.1) and $\mathrm{Ric} = 0$ imply

$$\begin{aligned}
{}^{\Sigma_t}\mathrm{Ric}(U,W) &= \langle R(\partial_t,U)W,\partial_t\rangle - \mathrm{tr}(k_t)k_t(U,W) \\
&\quad + {}^{\Sigma_t}g(k_t(U,\cdot)^\sharp, k_t(W,\cdot)^\sharp) \\
&= \langle R(U,\partial_t)\partial_t, W\rangle \\
&\quad - \frac{1}{4}\Big(\partial_t{}^{\Sigma_t}g_{ij}\partial_t{}^{\Sigma_t}g_{kl} - \partial_t{}^{\Sigma_t}g_{ik}\partial_t{}^{\Sigma_t}g_{jl}{}^{\Sigma_t}g^{jl}\Big){}^{\Sigma_t}g^{kl}U^iW^j.
\end{aligned}$$

In order to simplify the term $\langle R(\partial_t,U)W,\partial_t\rangle$ we may assume that $U,W$ can be extended to vector fields $U,W$ which are everywhere tangent to $\Sigma_t$ and commute with $\partial_t$. Using Lemma 4.4.4 we obain

$$\begin{aligned}
\langle R(U,\partial_t)\partial_t, W\rangle &= \Big\langle \nabla_U \overbrace{\nabla_{\partial_t}\partial_t}^{=0} - \nabla_{\partial_t}\nabla_U\partial_t, W\Big\rangle \\
&= -\Big\langle \nabla_{\partial_t}\nabla_U\partial_t, W\Big\rangle = -\Big\langle \nabla_{\partial_t}\nabla_{\partial_t}U, W\Big\rangle \\
&= -\Big\langle \nabla_{\partial_t}k_t(U,\cdot)^\sharp, W\Big\rangle \\
&= -\frac{1}{2}\partial_t \bullet (\mathcal{L}_{\partial_t}{}^{\Sigma_t}g(U,W)) + \frac{1}{2}\mathcal{L}_{\partial_t}{}^{\Sigma_t}g(U,\nabla_{\partial_t}W) \\
&= -\frac{1}{2}\mathcal{L}_{\partial_t}\mathcal{L}_{\partial_t}{}^{\Sigma_t}g(U,W) + \frac{1}{4}\mathcal{L}_{\partial_t}{}^{\Sigma_t}g(U,\mathcal{L}_{\partial_t}{}^{\Sigma_t}g(W,\cdot)^\sharp). \\
&= \Big(-\frac{1}{2}\partial_t\partial_t{}^{\Sigma_t}g_{ij} + \frac{1}{4}\partial_t{}^{\Sigma_t}g_{ik}\partial_t{}^{\Sigma_t}g_{jl}g^{kl}\Big)U^iW^j.
\end{aligned}$$

Hence the spatial components $\mathrm{Ric}_{ij} = 0$ of Einstein's equation are equivalent to the first system of equations in Lemma 5.4.2.

From

$$\begin{aligned}
R_{ijkl}{}^{\Sigma_t}g^{ik}{}^{\Sigma_t}g^{jl} &= R_{ijkl}(g^\sharp + \partial_t\otimes\partial_t)^{ik}(g^\sharp + \partial_t\otimes\partial_t)^{jl} \\
&= \mathrm{Scal} + 2\mathrm{Ric}(\partial_t,\partial_t) = 2(\mathrm{Ric} - \frac{\mathrm{Scal}}{2}g)(\partial_t,\partial_t)
\end{aligned}$$

and the Gauß equation we get

$$\mathrm{Ric}(\partial_t,\partial_t) = -\frac{1}{2}(\mathrm{Scal} + {}^{\Sigma_t}\mathrm{Scal} + \mathrm{tr}(k_t)^2 - \|k_t\|^2).$$

Hence $\mathrm{Ric} = 0$ implies ${}^{\Sigma_t}\mathrm{Scal} + \frac{1}{4}\big(\mathrm{tr}(\mathcal{L}_{\partial_t}{}^{\Sigma_t}g)^2 - \|\mathcal{L}_{\partial_t}{}^{\Sigma_t}g\|^2\big) = 0$.

Let $\{E_1,\ldots,E_{n-1}\}$ be an orthonormal frame of $\Sigma_t$ and $U$ be a vector field which is tangent to $\Sigma_t$. Then the Codazzi equation (Proposition 4.4.2) implies

$$
\begin{aligned}
\mathrm{Ric}(\partial_t, U) &= \sum_{i=1}^{n-1} \langle R(E_i, \partial_t)U, E_i\rangle \\
&= \sum_{i=1}^{n-1} \Big( \langle \nabla_U (k_t \otimes \partial_t)(E_i, E_i), \partial_t\rangle \\
&\quad - \Big\langle \nabla_{E_i} (k_t \otimes \partial_t)(U, E_i), \partial_t \Big\rangle \Big) \\
&= \sum_{i=1}^{n-1} \Big( \langle (\nabla_U k_t)(E_i, E_i)\partial_t, \partial_t\rangle - \Big\langle (\nabla_{E_i} \bullet k_t)(U, E_i)\partial_t, \partial_t \Big\rangle \Big) \\
&= \sum_{i=1}^{n-1} \Big( \langle ({}^{\Sigma_t}\nabla_U k_t)(E_i, E_i)\partial_t, \partial_t\rangle \\
&\quad - \Big\langle ({}^{\Sigma_t}\nabla_{E_i} \bullet k_t)(U, E_i)\partial_t, \partial_t \Big\rangle \Big) \\
&= -U \bullet \mathrm{tr}(k_t) + {}^{\Sigma_t}\mathrm{div}(k_t)(U). \\
&= \frac{1}{2}\Big( - \partial_{x^i}({}^{\Sigma_t}g^{jk} \partial_t {}^{\Sigma_t}g_{jk}) + {}^{\Sigma_t}g^{jk} \partial_{x^j} {}^{\Sigma_t}g_{ik} \Big) U^i.
\end{aligned}
$$

Hence $\mathrm{Ric}_{ti} = 0$ $(i \in \{1, \ldots, n-1\})$ is equivalent to the last system of equation in the statement of the lemma. ∎

The first system of differential equations in Lemma 5.4.2 consists of $\frac{1}{2}n(n-1)$ coupled differential equations for the $\frac{1}{2}n(n-1)$ unknown functions $g_{ij} = g_{ji}$ $(i, j \in \{1, \ldots, n-1\})$. One would expect that these equations would uniquely determine $g$ and that therefore Einstein's equation would be over-determined. Since over-determined systems of differential equations have only very few solutions (if any at all!) and are in general incompatible with initial value problems, Einstein's equation seems (at first sight) to be very different from other equations in physics. However, the following lemma shows that the over-determinancy of this system is of a very special nature and in fact compatible with a (slightly restricted) initial value problem.

Lemma 5.4.3 and Theorem 5.4.2 below hold for initial data which are not necessarily analytic. However, the proof is then much more difficult since we cannot anymore appeal to the relatively elementary theorem of Cauchy-Kowalewskaya.

**Theorem 5.4.1 (Cauchy-Kowalewskaya).**
*Let $F\colon \mathbb{R}^{2m-1} \to \mathbb{R}^k$ and $f_0\colon \mathbb{R}^{m-1} \to \mathbb{R}^k$ be analytic maps. Then there is a neighbourhood $\mathcal{U} \subset \mathbb{R}^m$ of $x^m = 0$ and a unique analytic map $f\colon \mathcal{U} \to \mathbb{R}^k$ which satisfies the system of partial differential equations*

$$\partial_{x^m} f = F(x^1, \ldots, x^m, \partial_{x^1} f, \ldots, \partial_{x^{m-1}} f).$$

*and the initial conditions $f(0, x^1, \ldots, x^{m-1}) = f_0(x^1, \ldots, x^{m-1})$.*

*Proof (sketch).* The idea of proof is to determine the Taylor series of $f$ at $x \in \{x^m = 0\}$ by successive differentiation of the system of partial differential equations and then to show that this series converges. A formal proof can be found in (Dieudonné 1971). ∎

The theorem of Cauchy-Kowalewskaya rests on the fact that an analytic function is determined by its Taylor series and it does not hold when the word "analytic" is replaced by "smooth". In the non-analytic case the structure of the system of partial differential equations matters for both, existence and uniqueness of solutions. This fact indicates that by restricting to the analytic case one may (in general) obtain results which are misleading because they do not generalize to the smooth case.

**Lemma 5.4.3.** *Let $(M, g)$ be a real-analytic spacetime, $\{\Sigma_t\}_{t\in\mathbb{R}}$ be a foliation as in Lemma 5.4.1, and assume that the spatial metric components $g_{ij}$ $(i, j \in \{1, \ldots, n-1\})$ satisfy*

$$\partial_t \partial_t{}^{\Sigma_t} g_{ij} = -2^{\Sigma_t}\mathrm{Ric}_{ij} - \left(\frac{1}{2}\partial_t g_{ij}\partial_t g_{kl} - \partial_t g_{ik}\partial_t g_{jl} g^{jl}\right) g^{kl}.$$

*If at $\Sigma = \Sigma_0$ the "constraint equations"*

$$0 = {}^{\Sigma}\mathrm{Scal} + (\mathrm{tr}(k))^2 - ||k||^2, \quad 0 = -\mathrm{dtr}(k) + {}^{\Sigma}\mathrm{div}(k)$$

*hold then $(M, g)$ satisfies Einstein's vacuum equations with vanishing cosmological constant,* $\mathrm{Ric} = 0$.

*Proof.* Let Ric be the Ricci tensor associated with $g = -\mathrm{d}t^2 + g_{ij}\mathrm{d}x^i\mathrm{d}x^j$. By assumption, this tensor satisfies $\mathrm{Ric}_{ij} = 0$ for all spatial components. The identy $\mathrm{div}(\mathrm{Ric} - \frac{1}{2}\mathrm{Scal}\, g) = 0$ implies therefore $g^{ab}(\partial_a \mathrm{Ric}_{bc} - 2\Gamma^d_{ac}\mathrm{Ric}_{db}) - \partial_c \mathrm{Scal} = 0$ which is equivalent to

$$0 = -\partial_t \mathrm{Ric}_{tc} - \sum_{i=1}^{n-1} \partial_i \mathrm{Ric}_{ic} - \partial_c \mathrm{Ric}_{tt} - 2\Gamma^d_{ac}\mathrm{Ric}_{db} g^{ab}.$$

Since $\mathrm{Ric}_{ij} = 0$ this is a linear system of $n$ partial differential equations for the unknown functions $\mathrm{Ric}_{tt}$, $\mathrm{Ric}_{ti}$. The constraint equations are equivalent to $\mathrm{Ric}_{tt} = \mathrm{Ric}_{ti} = 0$ at $\Sigma$. Hence we have $\mathrm{Ric}_{tt} = \mathrm{Ric}_{ti} = 0$ everywhere by the uniqueness-part of the theorem of Cauchy-Kowalewskaya. ∎

**Theorem 5.4.2.** *Let $(\Sigma, {}^{\Sigma}g)$ be an $(n-1)$-dimensional real-analytic Riemannian manifold and $k \in \mathrm{sym}(\mathcal{T}^0_2(\Sigma))$ be a real-analytic tensor field which satisfies*

$$0 = {}^{\Sigma}\mathrm{Scal} + (\mathrm{tr}(k))^2 - ||k||^2, \quad 0 = -\mathrm{dtr}(k) + {}^{\Sigma}\mathrm{div}(k).$$

*Then there is an $n$-dimensional real-analytic Lorentzian manifold $(M, g)$ and an immersion $\iota\colon \Sigma \to M$ such that $\iota^* g = {}^{\Sigma}g$ and $\iota^* k_0 = k$, where $k_0$ is the second fundamental form of $\iota(\Sigma)$.*

*Moreover, if $(\tilde{M}, \tilde{g})$ is a second Lorentz manifold with these properties then $\iota(\Sigma) \subset M$ and $\tilde{\iota}(\Sigma) \subset \tilde{M}$ have neighbourhoods which are isometric.*

*Proof.* Fix a coordinate system for $\Sigma$ and consider the system of partial differential equations

$$\begin{aligned}
\partial_t (k_t)_{ij} &= -[{}^{\Sigma_t}\mathrm{Ric}_{ij}]({}^{\Sigma_t}g, ((h_t)[k])_{k=1,\dots,n-1}, ((h_t)[kl])_{k,l=1,\dots,n-1}) \\
&\quad - \left((k_t)_{ij}(k_t)_{kl} - 2(k_t)_{ik}(k_t)_{jl}\,{}^{\Sigma_t}g^{jl}\right){}^{\Sigma_t}g^{kl}, \\
\partial_t {}^{\Sigma_t}g_{ij} &= 2(k_t)_{ij}, \\
\partial_t (h_t)[k]_{ij} &= 2\partial_{x^k}(k_t)_{ij}, \\
\partial_t (h_t)[kl]_{ij} &= 2\partial_{x^k}\partial_{x^l}(k_t)_{ij},
\end{aligned}$$

where $[{}^{\Sigma_t}\mathrm{Ric}_{ij}]({}^{\Sigma_t}g, ((h_t)[k])_{k=1,\dots,n-1}, ((h_t)[kl])_{k,l=1,\dots,n-1})$ is the algebraic expression defined by

$$[{}^{\Sigma_t}\mathrm{Ric}_{ij}]({}^{\Sigma_t}g, (\partial_{x^k}{}^{\Sigma_t}g)_{k=1,\dots,n-1}, (\partial_{x^k}\partial_{x^l}{}^{\Sigma_t}g)_{k,l=1,\dots,n-1}) = {}^{\Sigma_t}\mathrm{Ric}_{ij}.$$

The theorem of Cauchy-Kowalewskaya implies that for any real-analytic set of initial values $\{{}^{\Sigma_0}g_{ij}, k_0, ((h_0)[k])_{k=1,\dots,n-1}, ((h_0)[kl])_{k,l=1,\dots,n-1}\}$, there is a neighbourhood $\mathcal{U}$ of $t = 0$ and a unique solution of the system of partial differential equations which is defined on $\mathcal{U}$ and has these initial values. Since $\partial_t {}^{\Sigma_t}g_{ij} = 2(k_t)_{ij}$ the equation $\partial_t (h_t)[k]_{ij} = 2\partial_{x^k}(k_t)_{ij}$ implies

$$0 = \partial_t (h_t)[k]_{ij} - \partial_{x^k}\partial_t {}^{\Sigma_t}g_{ij} = \partial_t((h_t)[k]_{ij} - \partial_{x^k}{}^{\Sigma_t}g_{ij}).$$

From an integration of this equation we see that $(h_t)[k]_{ij} = \partial_{x^k}{}^{\Sigma_t}g_{ij}$ if $(h_0)[k]_{ij} = \partial_{x^k}{}^{\Sigma_0}g_{ij}$. In the same way we see that $(h_t)[kl]_{ij} = \partial_{x^k}\partial_{x^l}{}^{\Sigma_t}g_{ij}$ if $(h_0)[kl]_{ij} = \partial_{x^k}\partial_{x^l}{}^{\Sigma_0}g_{ij}$. It follows that the solutions ${}^{\Sigma_t}g_{ij}$ of this system of equations also solves

$$\partial_t\partial_t {}^{\Sigma_t}g_{ij} = -2\,{}^{\Sigma_t}\mathrm{Ric}_{ij} - \left(\frac{1}{2}\partial_t {}^{\Sigma_t}g_{ij}\partial_t {}^{\Sigma_t}g_{kl} - \partial_t {}^{\Sigma_t}g_{ik}\partial_t {}^{\Sigma_t}g_{jl}\,{}^{\Sigma_t}g^{jl}\right){}^{\Sigma_t}g^{kl}.$$

if and only if the initial conditions

$$\partial_t {}^{\Sigma_0}g_{ij} = 2(k_0)_{ij}, \quad (h_0)[k]_{ij} = \partial_{x^k}{}^{\Sigma_0}g_{ij}, \quad (h_0)[kl]_{ij} = \partial_{x^k}\partial_{x^l}{}^{\Sigma_0}g_{ij}$$

hold. Now the assertion follows directly from Lemma 5.4.3. ∎

Theorem 9.4.1 is local in character. Also note that the coordinates chosen tend to develop singularities due to focussing effects (cf. Proposition 4.6.1) and observe that the geodesics $t \mapsto (t, x^1, \dots, x^n)$ are length maximising.

A discussion of the smooth case can be found in (Hawking and Ellis 1973, chapter 7). An improved but mathematically more sophisticated theorem is presented in (Hughes, Kato, and Marsden 1977)

# 6. Robertson-Walker cosmology

## 6.1 Homogeneity and isotropy

It is very difficult, and one cannot make with any certainty assertions about the universe as a whole. This is so because we only know a very small portion of the universe. Hence any cosmological model reflects our own prejudice. Nevertheless, there are certain assumptions which seem to have a high degree of plausibility. After having built a cosmological model one can compare it with the few data we do have. Although imperfect, this approach seems to have given us much deeper understanding of the development of the universe than would seem possible at first sight.

The first cosmologists placed the earth at the centre of the universe. Copernicus' revolutionary model gave us a much more humble place in the solar system — the earth was reduced to being just one of its planets. This model of the universe had such a success that nowadays we not only take it for granted but don't even sincerely doubt that there may be other (more advanced) forms of life in the universe. At Copernicus' times, such a thought would have been considered blasphemous. The monk Giordano Bruno (1548–1600) was burned because he asserted the truth of such ideas.[1] Our new modesty leads us to think that our place in spacetime is in no way exceptional, and that there are no exceptional places anywhere in spacetime. We will use this fundamental idea to build a cosmology.

Let $x \in M$ be our event in spacetime $(M, g)$ and $U_x \in T_x M$ be the velocity vector of our world line. If there is not any point (or direction) in spacetime which is special then the universe should be isotropic, i.e., it should not be possible to distinguish any direction in $U_x^\perp$ by physical measurements. Although a glance at the nocturnal sky indicates that this is at odds with experience, on a sufficiently big scale this assumption coincides very well with observation. The galaxies seem to be randomly distributed, and since at night we mainly see a part of a single galaxy (the Milky Way), our first impression is not very representative.

---

[1] He is sometimes styled and an important forerunner of enlightenement. I must confess that I find his book (Bruno 1584) quite unscientific.

The mathematical interpretation of the isotropy assumption is that it is impossible to construct geometrical objects, using $U_x$ and $U_x^\perp$ which are breaking this symmetry. Let $E$ be (any) 3-dimensional subspace of the $(n-1)$-dimensional space $U_x^\perp$ and $u, v, w$ be vectors in $E$. Since $R(v, U)U$ is a vector, isotropy about $U_x$ and the fact that $\langle R(v, U)U, U\rangle = 0$ imply that $R(v, U)U = \mu(x)v$ for some scalar $\mu(x)$. Consider any 2-plane $P = \text{span}\{u, v\}$ in $E$. The sectional curvature $K(P)$ should be independent of $P$ since otherwise there would be a plane $P_0$ of maximal sectional curvature which in turn defines a distinguished direction ${P_0}^\perp$ in $E$. Since by isotropy there should not be any distinguished direction in $E$ we conclude that at $x$ the equation $R(u, v)w = \kappa(x)\left(\langle v, w\rangle u - \langle u, w\rangle v\right) + c(u, v, w)U$ holds. We show now that $c = 0$. The vector

$$R(P) = (\langle u, v\rangle^2 - ||u|| \cdot ||v||)^{-1/2} R(u, v)U_x$$

lies in $U_x^\perp$ and depends only on $P$ (rather than on the representatives $u, v$). Denote by $\pi_P$ the orthogonal projection $U^\perp \to P$. For each vector $\mathfrak{n} \in E$ with $g(\mathfrak{n}, \mathfrak{n}) = 1$ let $P_\mathfrak{n}$ be the plane in $E$ orthogonal to $\mathfrak{n}$. Since $\pi_{P_\mathfrak{n}}(R(P_\mathfrak{n}))$ lies in $P_\mathfrak{n}$ and is therefore orthogonal to $\mathfrak{n}$ we obtain a vector field $\mathfrak{V}\colon \mathfrak{n} \mapsto \pi_{P_\mathfrak{n}}(R(P_\mathfrak{n}))$ on the 2-sphere $\{\mathfrak{n} \in E \,|\, g(\mathfrak{n}, \mathfrak{n}) = 1\}$. Since the 2-sphere is compact the vector field $\mathfrak{V}$ has constant length since otherwise there would be a distinguished direction of maximal length in violation of isotropy. By Theorem 2.5.11 this is impossible unless the vector field vanishes identically.

By isotropy, it should have no component in $P$ since otherwise there would be a distinguished direction in $P$. It follows that $R(P)$ is orthogonal to $P$. Further, its length cannot depend on $P$ since otherwise there would be a distinguished plane (and therefore a distinguished vector) in $E$.

Let $u, v, w \in E$. From $\pi_{\text{span}\{u_1, u_2\}}(R\text{span}\{u_1, u_2\}) = 0$ for all vectors $u_1, u_2 \in E$ we obtain

$$\begin{aligned}
\langle R(u, v)U_x, w\rangle &= \langle R(u, v + w)U_x, w)\rangle - \overbrace{\langle R(u, w)U_x, w)\rangle}^{=0} \\
&= \overbrace{\langle R(u, v + w)U_x, v + w)\rangle}^{=0} - \langle R(u, v + w)U_x, v)\rangle \\
&= \overbrace{-\langle R(u, v)U_x, v)\rangle}^{=0} - \langle R(u, w)U_x, v)\rangle\,.
\end{aligned}$$

This implies that the tensor field $(u, v, w) \mapsto \langle R(u, v)U_x, w\rangle$ is antisymmetric. Since $\langle R(u, v)U_x, w\rangle = -\langle R(u, v)w, U_x\rangle = c(u, v, w)$, $c$ is a 3-form. The first Bianchi identity (Lemma 2.8.2) yields $\langle R(u, v)w, U_x\rangle + \langle R(v, w)u, U_x\rangle + \langle R(w, u)u, U_x\rangle = 3c(u, v, w) = 0$. This motivates the following definition.

**Definition 6.1.1.** *Let $(M,g)$ be a Lorentzian manifold and $U_x \in T_xM$, $\langle U_x, U_x\rangle = -1$. The spacetime $(M,g)$ is called* infinitesimally isotropic about $U_x$ *if at $x$ the curvature tensor satisfies*

$$R(u,v)w = \kappa(x)\left(\langle v,w\rangle u - \langle u,w\rangle v\right)$$
$$R(v,U_x)U_x = \mu(x)v,$$

*for all $u,v,w \in U_x^\perp$, where $\mu(x), \kappa(x) \in \mathbb{R}$ are independent of $u,v,w$. The spacetime $(M,g)$ is called* infinitesimally isotropic *if there exits a normalized, timelike vector field $U$ such that $(M,g)$ is isotropic about $U_x$ for all $x \in M$. If $(M,g)$ is infinitesimally isotropic, $U$ is called a* cosmological observer field.

Given an infinitesimally spacetime, there my not be a unique cosmological observer field. For instance, in Minkowski spacetime all normalized timelike vector fields are cosmological observer fields.

In the following we will assume that spacetime is infinitesimally isotropic. While it can be argued that infinitesimal isotropy about our own velocity vector is backed experimentally fairly well, it is a very questionable extrapolation to assume that spacetime is infinitesimally isotropic. On the other hand, this extrapolation seems to be exactly the lesson learned from Copernicus. Hence to demand that $(M,g)$ is infinitesimally isotropic appears very plausible to us. (Our acceptance of such a postulate is in striking opposition to the response a medieval scholar would have given).

**Lemma 6.1.1.** *Let $(M,g)$ be infinitesimally isotropic about $U_x$. Then*

$$R(u,v)U_x = 0 \text{ and } R(U_x,u)v = -\mu(x)\langle u,v\rangle U_x$$

*for all $u,v \in U_x^\perp$ and the energy momentum tensor is given by*

$$T_x = (\epsilon(x)+p(x))U_x^\flat \otimes U_x^\flat + p(x)\,g_x,$$

*where*

$$8\pi\epsilon(x) = \frac{1}{2}(n-2)(n-1)\kappa(x) - \Lambda,$$
$$8\pi p(x) = (n-2)\left(-\frac{1}{2}(n-3)\kappa(x) + \mu(x)\right) + \Lambda.$$

*Proof.* Let $u,v,w \in U_x^\perp$. The first assertion follows from $\langle R(u,v)U_x, w\rangle = -\langle R(u,v)w, U_x\rangle = 0$ and $\langle R(u,v)U_x, U_x\rangle = 0$. The second assertion is a consequence of $\langle R(U_x,u)v, w\rangle = \langle R(v,w)U_x, u\rangle = 0$ and $\langle R(U_x,u)v, U_x\rangle = \langle R(u,U_x)U_x, v\rangle = \mu(x)\langle u,v\rangle$. Taking the trace of $R$ we obtain

$$\mathrm{Ric}(U_x, U_x) = (n-1)\mu(x), \quad \mathrm{Ric}(U_x, v) = 0,$$
$$\mathrm{Ric}(u, v) = ((n-2)\kappa(x) - \mu(x))\,\langle u, v\rangle\,,$$

and therefore

$$\mathrm{Ric} = ((n-2)\mu + (n-2)\kappa)\, U^\flat \otimes U^\flat + ((n-2)\kappa - \mu)\, g.$$

Taking again the trace we have $\mathrm{Scal} = -(n-2)(\mu+\kappa)+n\,((n-2)\kappa-\mu) = (-2n+2)\mu+(n-1)(n-2)\kappa$. The energy momentum tensor is now given by

$$\begin{aligned} 8\pi T &= \mathrm{Ric} - \frac{1}{2}\mathrm{Scal}\, g + \Lambda g \\ &= (n-2)(\mu+\kappa) U^\flat \otimes U^\flat \\ &\quad + \left((n-2)\kappa - \mu - \frac{1}{2}((-2n+2)\mu + (n-1)(n-2)\kappa) + \Lambda\right) g \\ &= (n-2)(\mu+\kappa)U^\flat \otimes U^\flat + \left((n-2)\frac{3-n}{2}\kappa + (n-2)\mu + \Lambda\right) g. \end{aligned}$$

It follows that the energy density $\epsilon$ and the pressure $p$ are given by $8\pi(\epsilon + p) = (n-2)(\mu+\kappa)$ and $8\pi p = \frac{1}{2}(n-2)(3-n)\kappa + (n-2)\mu + \Lambda$. ∎

**Lemma 6.1.2.** *Let $(M, g)$ be infinitesimally isotropic and $U$ be the cosmological observer field. Further assume that $n > 3$ and that $\epsilon + p \neq 0$. Then $U^\perp$ is an integrable distribution. The hypersurfaces perpendicular to $U$ are totally umbilic (cf. Definition 4.4.7) and $\kappa$ is constant on these hypersurfaces.*

*Proof.* Let $X, Y, Z$ be vector fields which lie in $U^\perp$ at $x$ and satisfy the equation

$$\nabla_U X - \langle X, \nabla_U U\rangle\, U = 0$$

(analogously for $Y, Z$). This can easily be arranged by considering vector fields along a hypersurface $\Sigma$ which lie in $U^\perp$. They may then be uniquely extended into a neighbourhood of $\Sigma$ using the above differential equation. From $\nabla_U \langle X, U\rangle = \langle \nabla_U X - \langle X, \nabla_U U\rangle\, U, U\rangle = 0$ it follows that $X, Y, Z$ are everywhere perpendicular to $U$. Moreover, the derivative $\nabla_U X$ is parallel to $U$ since $\nabla_U X = -\langle \nabla_U X, U\rangle\, U$ (analogously for $Y, Z$). We will now exploit the second Bianchi identify (cf. Lemma 2.8.1),

$$(\nabla_U R)\,(X, Y) + (\nabla_X R)\,(Y, U) + (\nabla_Y R)\,(U, X) = 0.$$

Using the formulas in Definition 6.1.1 and and the property

$$\nabla_U X, \nabla_U Y, \nabla_U Z \parallel U$$

we calculate

$$\begin{aligned}
(\nabla_U R)(X,Y)Z &= \nabla_U (R(X,Y)Z) - R(X,Y)\nabla_U Z \\
&\quad - R(\nabla_U X, Y)Z - R(X, \nabla_U Y)Z \\
&= \mathrm{d}\kappa(U)\left(\langle Y, Z\rangle X - \langle X, Z\rangle Y\right) \\
&\quad + \kappa\left(\langle Y, Z\rangle \nabla_U X - \langle X, Z\rangle \nabla_U Y\right) \\
&\quad - 0 + \mu \langle Y, Z\rangle \nabla_U X - \mu \langle X, Z\rangle \nabla_U Y, \\
(\nabla_X R)(Y,U)Z &= \nabla_X (R(Y,U)Z) - R(\pi_{U^\perp}(\nabla_X Y), U)Z \\
&\quad - R(Y, \nabla_X U)Z - R(Y,U)\nabla_X Z \\
&= \nabla_X (\mu \langle Y, Z\rangle U) - \mu \left\langle Z, \nabla_X Y\right\rangle U) \\
&\quad - \kappa\left(\left\langle \nabla_X U, Z\right\rangle Y - \langle Y, Z\rangle \nabla_X U\right) \\
&\quad - \mu(-\left\langle U, \nabla_X Z\right\rangle)Y - \mu \left\langle Y, \nabla_X Z\right\rangle U \\
&= \mathrm{d}\mu(X)\langle Y, Z\rangle U \\
&\quad + (\kappa + \mu)\left(\langle Y, Z\rangle \nabla_X U - \left\langle Z, \nabla_X U\right\rangle Y\right),
\end{aligned}$$

and

$$\begin{aligned}
(\nabla_Y R)(U,X)Z &= -(\nabla_Y R)(X,U)Z \\
&= -\mathrm{d}\mu(Y)\langle X, Z\rangle U \\
&\quad - (\kappa + \mu)\left(\langle X, Z\rangle \nabla_Y U - \left\langle Z, \nabla_Y U\right\rangle X\right).
\end{aligned}$$

Inserting these equations into the second Bianchi identity we obtain

$$\begin{aligned}
0 &= \mathrm{d}\mu(X)\langle Y, Z\rangle U - \mathrm{d}\mu(Y)\langle X, Z\rangle U \\
&\quad + (\mu + \kappa)\left(\langle Y, Z\rangle \nabla_U X - \langle X, Z\rangle \nabla_U Y\right), \qquad (6.1.1) \\
0 &= \mathrm{d}\kappa(U)\left(\langle Y, Z\rangle X - \langle X, Z\rangle Y\right) - (\mu + \kappa)\big(\langle Y, Z\rangle \nabla_X U \\
&\quad - \langle X, Z\rangle \nabla_Y U + \left\langle \nabla_Y U, Z\right\rangle X - \left\langle \nabla_X U, Z\right\rangle Y\big). \qquad (6.1.2)
\end{aligned}$$

Since $n > 3$ there are pointwise linearly independent vector fields $X, Y$ with $X \perp Z$, $Y \perp Z$. For these vector fields Equation 6.1.2 implies

$$0 = (\mu + \kappa)\left(\left\langle \nabla_Y U, Z\right\rangle X - \left\langle \nabla_X U, Z\right\rangle Y\right).$$

It follows immediately that for orthogonal vector $u, v$ the expression $\left\langle u, \nabla_v U\right\rangle$ vanishes. Hence, restricted to $U^\perp$, the bilinear form $\nabla U^\flat$ is a multiple of $g$ restricted to $U^\perp$. In particular, $\nabla U^\flat$ restricted to $U^\perp$ is symmetric. We will now show that the tensor field $\nabla\big((\kappa + \mu)U\big)$ is symmetric in all of $T_x M$. Since $U$ is symmetric on $U^\perp$ and $\nabla\big((\kappa+\mu)U\big) =$

$(\mathrm{d}\kappa + \mathrm{d}\mu) \otimes U + (\kappa+\mu)\nabla U$ we only have to show $\langle \nabla_X((\kappa+\mu)U), U\rangle = \langle \nabla_U((\kappa+\mu)U), X\rangle$. Setting $Y = Z$ and choosing $X \perp Z$ at $x$ we get from Equation 6.1.1

$$\mathrm{d}\mu(X) = (\mu+\kappa)\left\langle \nabla_U X, U\right\rangle = -(\mu+\kappa)\left\langle X, \nabla_U U\right\rangle .$$

Lemma 5.2.1 implies

$$\mathrm{d}p(X) = -(\epsilon + p)\left\langle X, \nabla_U U\right\rangle ,$$

and therefore $-\frac{1}{2}(n-3)\mathrm{d}\kappa(X) + \mathrm{d}\mu(X) = -(\mu+\kappa)\left\langle X, \nabla_U U\right\rangle$, where we have used the formulas provided by Lemma 6.1.1. Combining this equation with $\mathrm{d}\mu(x) = -(\mu+\kappa)\left\langle X, \nabla_U U\right\rangle$ we obtain $\mathrm{d}\kappa(X) = 0$. Now we can calculate.

$$\begin{aligned}
&\left\langle \nabla_X((\kappa+\mu)U), U\right\rangle = -\mathrm{d}(\kappa+\mu)(x) = -\mathrm{d}\mu(X) = (\mu+\kappa)\left\langle X, \nabla_U U\right\rangle ,\\
&\left\langle \nabla_U((\kappa+\mu)U), X\right\rangle = (\mu+\kappa)\left\langle \nabla_U U, X\right\rangle .
\end{aligned}$$

It follows that $\nabla\big((\kappa+\mu)U\big)$ is symmetric in all of $T_xM$ and that therefore $\mathrm{d}\big((\kappa+\mu)U^\flat\big) = 0$. By the lemma of Poincaré there is a function $t$ with $\mathrm{d}t = (\kappa+\mu)U^\flat$. The hypersurface $\Sigma_{\tilde{t}} = \{x \in M \,|\, t(x) = \tilde{t}\}$ is orthogonal to $U$ for each $\tilde{t}$. Its second fundamental form, $k(u,v) = \left\langle u, \nabla_v U\right\rangle$, is a multiple of $g$ restricted to $\Sigma_{\tilde{t}}$ which implies that $\Sigma_{\tilde{t}}$ is totally umbilic. ∎

**Theorem 6.1.1.** *Let $(M,g)$ be infinitesimally isotropic and $U$ be the cosmological observer field. Assume that $\epsilon + p \neq 0$. Then there is an $(n-1)$-dimensional Riemannian manifold $(\hat{M}, \hat{g}_\varepsilon)$ of constant curvature $\varepsilon \in \{1, 0, -1\}$ such that (locally) $M = \mathbb{R} \times \hat{M}$, $g = -\mathrm{d}t^2 + a^2(t)\hat{g}_\varepsilon$, where $\varepsilon/a^2 = \kappa - \frac{1}{4}(\mathrm{d}\kappa(U)/(\kappa+\mu))^2$.*

*Proof.* It is clear that $g = (\kappa+\mu)^{-2}\mathrm{d}t^2 + \hat{g}_t$ for some $t$-dependent Riemannian metric $\hat{g}_t$. Let $u, v \in U^\perp$ be vectors with $u \perp v$, $\langle u,u\rangle = \langle v,v\rangle = 1$. Then Equation (6.1.2) gives with $v = X, u = Y = Z$ $\mathrm{d}\kappa(U) = -(\kappa+\mu)(\left\langle \nabla_v U, v\right\rangle + \left\langle \nabla_u U, u\right\rangle)$. Hence the second fundamental form of the hypersurfaces perpendicular to $U$ is given by

$$k = \left\langle u, \nabla_v U\right\rangle = -\frac{1}{2}\frac{\mathrm{d}\kappa(U)}{\kappa+\mu}\, g = \frac{1}{2}\partial_t \kappa\, g.$$

The Gauß equation gives

$$\begin{aligned}
R_{\Sigma_t}(u,v)w &= R(u,v)w + k(u,w)k(v,\cdot)^\sharp - k(v,w)k(u,\cdot)^\sharp \\
&= \left(\kappa - \frac{1}{4}\left(\frac{\mathrm{d}\kappa(U)}{\kappa+\mu}\right)^2\right)\left(\langle v,w\rangle u - \langle u,w\rangle v\right).
\end{aligned}$$

By the Lemma of Schur (Proposition 4.3.4) the factor $\kappa - \frac{1}{4}\left(\frac{\mathrm{d}\kappa(U)}{\kappa+\mu}\right)^2$ is constant in the hypersurfaces. In other words, the hypersurfaces $\Sigma_t$ have constant curvature. We can therefore write $g = (\kappa+\mu)^{-2}\mathrm{d}t^2 + a^2(t)\hat{g}_\varepsilon$, where $a\colon \mathbb{R} \to \mathbb{R}^+$ is a function and $\hat{g}_\varepsilon$ the $(n-1)$-dimensioanl metric of constant curvature $\varepsilon \in \{-1, 0, 1\}$.

We will now show that the factor $1/(\kappa+\mu)$ can be absorbed into $\mathrm{d}t$. To this end we have to show that $\mu$ only depends on $t$ (recall from the proof of Lemma 6.1.2 that $\mathrm{d}\kappa(X) = 0$ for each $X \in T_x\Sigma_t$ and that therefore $\kappa$ depends only on $t$). Since $\kappa$ is constant in the hypersurfaces $t = \text{const}$, so is $\partial_t\kappa$. This implies that

$$k = -\frac{\partial_t\kappa}{2}g = -\frac{\partial_t\kappa}{2}a^2\,\hat{g}_\varepsilon$$

depends ony on $t$. On the other hand, a direct calculation gives for $v, w \in U^\perp$

$$\begin{aligned} k(v,w) &= \langle v, \nabla_w U\rangle = (\kappa+\mu)\,\langle v, \nabla_w \partial_t\rangle = (\kappa+\mu) g_{ab} v^a \Gamma^b_{ct} w^c \\ &= (\kappa+\mu)\frac{1}{2}\left(\partial_c g_{at} + \partial_t g_{ca} - \partial_a g_{ct}\right) v^a w^b = \frac{1}{2}(\kappa+\mu)\partial_t g_{ca} v^a w^c \\ &= (\mu+\kappa) a\partial_t a \hat{g}_\varepsilon . \end{aligned}$$

Comparing the two formulas for $k$ we obtain that

$$\mu = -\left(\frac{a\partial_t\kappa}{2\partial_t a} + \kappa\right)$$

depends only on $t$. Hence $(\kappa(t)+\mu(t))^{-1}\mathrm{d}t$ is the differential of a function which we take as our new time coordinate. ∎

**Corollary 6.1.1.** *If $(M,g)$ is infinitesimally isotropic then it is also spatially homogeneous, i.e. for hypersurfaces $\Sigma_t$ orthogonal to $U$ and all $x, y \in \Sigma_t$ there exists for any pair of orthogonal frames of $T_x\Sigma_t$, $T_y\Sigma_t$ an isometry $M \to M$ which maps one of the frames into the other.*

*Proof.* This follows from Corollary 4.5.1 and Lemma 4.5.5 since $(\Sigma_t a^2(t)\hat{g}_\varepsilon$ is a Riemannian manifold of constant curvature. ∎

**Corollary 6.1.2.** *Let $(M,g)$ be infinitesimally isotropic and $U$ be the cosmological observer field. Then there is an interval $I = (t_-, t_+)$ and a spaceform $(\Sigma, \hat{g}_\varepsilon)$ of constant curvature $\varepsilon \in \{-1, 0, 1\}$ such that the universal cover of $(M,g)$ is isometric to $(I \times \Sigma, -\mathrm{d}t^2 + \hat{g}_\varepsilon)$ and $U = \partial_t$.*

*If $n = 4$ then there are local coordinates $(t, r, \theta, \varphi)$ such that*

$$g = -\mathrm{d}t^2 + a^2(t)\left(\frac{1}{1-\varepsilon r^2}\mathrm{d}r^2 + r^2\left(\mathrm{d}\theta^2 + \sin^2(\theta)\mathrm{d}\varphi^2\right)\right). \tag{6.1.3}$$

*Proof.* The first part of the corollary is obvious since $I \times \Sigma$ is simply connected for any spaceform $\Sigma$. The second part follows immediately from the classification of 3-dimensional Riemannian manifolds with constant curvature (cf. Lemma 4.5.5). ∎

A spacetime $(M, g)$ which is locally isometric to a 4-dimensional infinitesimally isotropic Lorentzian manifold is called a *Robertson-Walker spacetime* or *Robertson-Walker cosmology*. The metric given by Equation (6.1.3) is called the *Robertson-Walker metric.*

## 6.2 The initial value problem for infinitesimally isotropic spacetimes

In this section we solve Einstein's equations $\mathrm{Ric} - \frac{1}{2}\mathrm{Scal}\, g + \Lambda g = 8\pi T$ for Robertson-Walker cosmologies. While, in general, Einstein's equations give rise to a system of partial differential equations, in the case at hand we have already shown that the unknown functions depend on only one variable. We will therefore obtain an ordinary instead of a partial system of differential equation. This simplifies the problem greatly. However, even this simple case exhibits typical aspects of Einstein's equation.

We denote the derivative with respect to $t$ with $(\cdot)'$.

**Lemma 6.2.1.** *Let $(M, g)$ be a Robertson-Walker spacetime and $u, v, w$ be tangent to the hypersurfaces $\Sigma_t$ which are orthogonal to $U$. The curvature expressions are given by*

$$R(u,v)w = \left(\frac{(a')^2}{a^2} + \frac{\varepsilon}{a^2}\right)(\langle v,w\rangle u - \langle u,w\rangle v), \quad R(v,w)U = 0,$$

$$R(v,U)U = -\frac{a''}{a}v, \quad R(v,U)w = -\frac{a''}{a}\langle v,w\rangle U,$$

$$\mathrm{Ric}(U,U) = -(n-1)\frac{a''}{a}, \quad \mathrm{Ric}(U,v) = 0,$$

$$\mathrm{Ric}(v,w) = \left(\frac{a''}{a} + (n-2)\left(\frac{(a')^2}{a^2} + \frac{\varepsilon}{a^2}\right)\right)\langle v,w\rangle$$

$$\mathrm{Scal} = (n-1)\left(2\frac{a''}{a} + (n-2)\left(\frac{(a')^2}{a^2} + \frac{\varepsilon}{a^2}\right)\right)$$

*Proof.* The constant curvature metric

$$g_\Sigma = \frac{1}{1-\varepsilon r^2}\mathrm{d}r^2 + r^2\left(\mathrm{d}\theta^2 + \sin^2(\theta)\mathrm{d}\varphi^2\right)$$

has the curvature tensor $R_\Sigma(u,v)w = \varepsilon(\langle v,w\rangle u - \langle u,w\rangle v)$ (Proposition 4.3.3). Hence the formula for $R$ follows from Lemma 4.4.14. The other formulas are direct consequences from Lemma 4.4.15. ∎

**Corollary 6.2.1.** *If the energy momentum tensor is given by*

$$T = (\epsilon + p)U^\flat \otimes U^\flat + pg,$$

*then Einstein's equation is equivalent to*

$$8\pi\epsilon + \Lambda = \frac{1}{2}(n-1)(n-2)\left(\frac{(a')^2}{a^2} + \frac{\varepsilon}{a^2}\right) \tag{6.2.4}$$

*and*

$$8\pi p - \Lambda = -(n-2)\left(\frac{a''}{a} + \frac{n-3}{2}\left(\frac{(a')^2}{a^2} + \frac{\varepsilon}{a^2}\right)\right). \tag{6.2.5}$$

*Proof.* Recall from Lemma 6.2.1 that the Ricci tensor Ric restricted to the spatial subspace ${U_x}^\perp$ is a multiple of the metric $g$ restricted to this subspace. It follows that Einstein's equation $\mathrm{Ric} - \frac{1}{2}\mathrm{Scal}\, g + \Lambda g = 8\pi T$ restricted to this subspace,

$$\Big(\frac{a''}{a} + (n-2)\left(\frac{(a')^2}{a^2} + \frac{\varepsilon}{a^2}\right) \\ - \frac{n-1}{2}\left(2\frac{a''}{a} + (n-2)\left(\frac{(a')^2}{a^2} + \frac{\varepsilon}{a^2}\right)\right) + \Lambda\Big) g_{|\Sigma} = 8\pi p g_{|\Sigma}.$$

is equivalent to Equation (6.2.5). From Lemma 6.2.1 we get $\mathrm{Ric}(v, U) = 0$ for all vectors $v \perp U$ which implies that the only other non-trivial component of Einstein's equation is given by evaluating it on the pair of vectors $(U_x, U_x)$. We obtain

$$\left(-(n-1)\frac{a''}{a} + \frac{n-1}{2}\left(2\frac{a''}{a} + (n-2)\left(\frac{(a')^2}{a^2} + \frac{\varepsilon}{a^2}\right)\right) - \Lambda\right) g_{|\Sigma} = 8\pi\epsilon g_{|\Sigma}$$

which is equivalent to Equation (6.2.4). ∎

Observe that Einstein's equation is *not* a well posed system of differential equations. Instead, we have only two equations for three unknowns, $a, \epsilon, p$. Moreover, only derivatives of the function $a$ appear in our system of equations. The first problem has a direct physical resolution. Just specifying a perfect fluid is not enough to specify a matter-model completely. Rather, perfect fluids give a framework which is fitting for many different matter models. In particular, vacuum is a (very degenerate) perfect fluid, and so is dust. In order to arrive at a determined system of equations we therefore have to specify an *additional* relation between the energy density $\epsilon$, the pressure $p$, and the metric described by $a$. In the following we make a rather simplistic assumption, namely that there is a given *equation of state*: $p = f(\epsilon)$ for some smooth function $f : \mathbb{R} \to \mathbb{R}$.

Having made this assumption, we still have the problem that we have two differential equations for $a$ rather than a system of two differential equation for $a$ and $\epsilon$. We can resolve this problem by replacing one of our equations with an equation of motion (Lemma 5.2.1).

**Corollary 6.2.2.** *Assume that there is a smooth function $f\colon \mathbb{R} \to \mathbb{R}$ with $f(\epsilon) > -\epsilon$ for all $\epsilon \in \mathbb{R}$. Let $a_0 \in \mathbb{R}^+ \setminus \{0\}$ and $\epsilon_0 \in \mathbb{R}$ and assume that $\frac{2(a_0)^2}{(n-1)(n-2)}(8\pi\epsilon_0 + \Lambda) - \varepsilon \geq 0$. Then there exists a unique solution $(a, \epsilon)$ of Einstein's equations such $a(0) = a_0$ and $\epsilon(0) = \epsilon_0$. The functions $a$, $\epsilon$ satisfy*

*(i)* $(n-2)\frac{a''}{a} = -8\pi\left(f(\epsilon) + \frac{n-3}{n-1}\epsilon\right) + \frac{2}{n-1}\Lambda$,

*(ii)* $(n-1)\frac{a'}{a} = -\frac{\epsilon'}{\epsilon+p}$.

*and*

$$a'(0) = \sqrt{\frac{2(a_0)^2}{(n-1)(n-2)}(8\pi\epsilon_0 + \Lambda) - \varepsilon}$$

*Proof.* Assume first that $a$, $\epsilon$ are a solution of Einstein's equation. Equation (i) is a linear combination of Equations (6.2.4) and (6.2.5). Equation (ii) follows from Lemma 5.2.1. Finally, the equation for $a'(0)$ follows immediately from the equation for $\epsilon$ in Corollary 6.2.1.

For the converse notice first that for our initial conditions there is a unique solution $a$, $\epsilon$ which satisfies the system of equations (i), (ii) and $a'(0) = \sqrt{\frac{2(a_0)^2}{(n-1)(n-2)}(8\pi\epsilon_0 + \Lambda) - \varepsilon}$. We have to show that this solution is *also* a solution to the system of equations given in Corollary 6.2.1. It is clear that this system of equations is satisfied at $t = 0$. We will show that the first equation is satisfied for all $t$. Since the second equation is a linear combination of the first equation and equation (i), it must then also be satisfied for all $t$. Defining

$$\phi := 8\pi\epsilon + \Lambda - \frac{1}{2}(n-1)(n-2)\left(\frac{(a')^2}{a^2} + \frac{\varepsilon}{a^2}\right)$$

we have to show that $\phi$ vanishes for all $t$. Taking the derivative of $\phi$ and using equations (i), (ii) gives

$$\begin{aligned}\phi' &= 8\pi\epsilon' - \frac{1}{2}(n-1)(n-2)\frac{2a'a''a^2 - 2aa'\left((a')^2 + \varepsilon\right)}{a^4} \\ &\overset{(i)}{=} -8\pi(n-1)\frac{a'}{a}(\epsilon + p) - (N-1)(N-2)\frac{a'}{a}\left(\frac{a''}{a} - \frac{(a')^2 + \varepsilon}{a^2}\right)\end{aligned}$$

$$\stackrel{(ii)}{=} \frac{a'}{a}\Big( - 8\pi(n-1)\epsilon + (n-1)(n-2)\frac{a''}{a} + 8\pi(n-3)\epsilon + 2\Lambda$$
$$- (N-1)(N-2)\frac{a'}{a}\left(\frac{a''}{a} - \frac{(a')^2+\varepsilon}{a^2}\right)\Big)$$
$$= 2\frac{a'}{a}\phi.$$

Since $\phi(0) = 0$ and 0 is a solution of the differential equation $\phi' = 2\frac{a'}{a}\phi$, the fundamental theorem for ordinary differential equations implies that $\phi$ must vanish vor all $t$. ∎

This solution of Einstein's equation is typical in two aspects. Firstly, it is often advantageous to exchange part of the original set of equations for the equations of motion, $\operatorname{div}(T) = 0$. Secondly, Einstein's equations are not a free system of differential equations but are constrained. Recall that we were not free to choose $a'(0)$ even though we had a second order equation for $a$. In other words, only a restricted set of initial values had the chance to lead to solutions of Einstein's equation. The system of differential equations which was solved was derived from Einstein's equation but not identical to it. We had therefore to show that the solutions to this system are also solutions to Einstein's equation. We did so by deriving an additional linear differential equation and used our constraint (i.e., the choice of $a'(0)$) to show that the solution of this equation implies that the original set of equations is satisfied. This phenomenon has a direct counterpart in more general settings where we have to deal with systems of differential equations.

## 6.3 Geodesics and redshift

In 1929, Hubble discovered that distant galaxies are moving away from us (and each other) at a rate proportional to their distance. This astronomical fact shattered the long cherished idea that our universe was an eternal arena in which the physical processes take place.[2] Hubble used an indirect method for his discovery. Each star has a spectrum of light which contains characteristic gaps due to absorption of light of certain frequencies in the atmosphere of the star. Since we have physical explanations for these absorptions, we can calibrate these patterns and thereby obtain information about the chemical composition of the star's atmosphere. Hubble discovered that for stars in galaxies which are not

[2] Einstein introduced his cosmological constant a decade earlier because he wanted to have static solutions in accordance with the prejudice of his time. Had he not done so, there would have been another striking prediction by general relativity.

too close[3] these gaps are shifted towards smaller frequencies. Morover, this shift is proportional to the distance of the galaxy. He concluded from his observations that all galaxies are moving away from each other. (Everyone is familiar with an analogous effect: If a fast car is approaching one has the impression that the noise of the engine is higher pitched than when it is moving away: In other words, if the source and the detector of a sound move away from each other, the frequency of the sound appears to be smaller).

In this section we will show that Hubble's discovery can be understood within the framework of Robertson-Walker cosmology (cf. Corollary 6.3.2 below). This is one of the great successes of general relativity and the isotropy assumption. Recall from Sect. 1.4.3 that we can describe the world lines of photons by null geodesics. The energy of a photon $\gamma$ measured by an observer $u$ is given by $E = h\nu = -\langle u, \dot{\gamma}\rangle$. Here $\nu$ denotes the frequency of the photon (as measured by $u$) and $h$ denotes Planck's constant. In Robertson-Walker spacetime we have a natural unit vector field $U$ which is approximately tangent to the world lines of the galaxies. We will therefore define the energy of a photon using this distinguished observer. In this section we will always refer to this energy.

Let $\gamma$ be a photon which moves from $x \in M$ to $y \in M$. In general, it is possible that its energy is not constant along the world line of the photon. This is traditionally expressed using the fractional increase $z$ of the associated wavelength $\lambda = 1/\nu = h/E$:

**Definition 6.3.1.** *The* redshift factor $z$ *of a photon originating at* $x$ *and being detected at* $y$ *is given by*

$$z(x, y) = \frac{\lambda(y) - \lambda(x)}{\lambda(x)}.$$

If $(t, \vec{x}), (t, \vec{y}) \in I \times \Sigma$ are the events occupied by two galaxies, then the distance of these events at time $t$ is given by $d((t, \vec{x}), (t, \vec{y})) = a(t) d_\Sigma(\vec{x}, \vec{y})$, where $d_\Sigma(\vec{x}, \vec{y})$ is the distance of $\vec{x}$ and $\vec{y}$ in $(\Sigma, \hat{g}_\varepsilon)$. We will show that there is a constant $H$ such that we have approximately $z(x, y) = H d((t, \vec{x}), (t, \vec{y}))$ for galaxies which are distant enough for Hubble's discovery to hold but still so close that it is sensible to linearize $z$. To this end we must first calculate the null geodesics in Robertson-Walker spacetimes.

**Lemma 6.3.1.** *Let* $(M, g) = ((t_-, t_+) \times \Sigma, -\mathrm{d}t^2 + a^2(t)\hat{g}_\varepsilon)$ *be an infinitesimally isotropic Lorentzian manifold. The curve*

$$s \mapsto \gamma(s) = (t(s), \vec{\gamma}(s))$$

[3] For very nearby galaxies the (local) movement of the galaxy relative to us overshadows this effect.

*is a geodesic if and only if*

$$\frac{\mathrm{d}^2 t}{\mathrm{d}s^2} + \left\langle \dot{\vec{\gamma}}, \dot{\vec{\gamma}} \right\rangle a(t)a' = 0, \quad \hat{\nabla}_{\dot{\vec{\gamma}}} \dot{\vec{\gamma}} + 2\frac{a'}{a}\frac{\mathrm{d}t}{\mathrm{d}s}\dot{\vec{\gamma}} = 0$$

*hold, where* $\hat{\nabla}$ *denotes the induced covariant derivative on* $\Sigma$. *If* $\gamma(s) = (t(s), \vec{\gamma}(s))$ *is a null geodesic then the conservation equation*

$$a(t(s))\frac{\mathrm{d}t}{\mathrm{d}s} = const$$

*holds.*

*Proof.* The first part follows immediately from the corresponding formulas for general warped products (Corollary 4.4.1). Assume now that $\gamma$ is a null geodesic. From the first equation and $a^2 \left\langle \dot{\vec{\gamma}}, \dot{\vec{\gamma}} \right\rangle = (\mathrm{d}t/\mathrm{d}s)^2$ we obtain

$$\frac{\mathrm{d}}{\mathrm{d}s}\left(a(t(s))\frac{\mathrm{d}t}{\mathrm{d}s}\right) = a'(t)\left(\frac{\mathrm{d}t}{\mathrm{d}s}\right)^2 + a(t)\frac{\mathrm{d}^2 t}{\mathrm{d}s^2} = 0.$$

∎

**Corollary 6.3.1.** *Let* $(M, g) = ((t_-, t_+) \times \Sigma, -\mathrm{d}t^2 + a^2(t) g_\varepsilon)$ *be an infinitesimally isotropic Lorentzian manifold. The curve* $s \mapsto \gamma(s) = (t(s), \vec{\gamma}(s))$ *is a null geodesic if*

(i) $\tau \mapsto \vec{\bar{\gamma}}(\tau)$ *is a unit speed geodesic in* $(\Sigma, \hat{g}_\varepsilon)$
(ii) *and there is a constant* $c$ *such that*

$$s = c\int_{t_0}^{t(s)} a(\hat{t})\,\mathrm{d}\hat{t}, \quad \vec{\gamma}(s) = \vec{\bar{\gamma}}\left(\int_{t_0}^{t(s)} \frac{\mathrm{d}\hat{t}}{a(\hat{t})}\right).$$

*Proof.* Assume that $\vec{\bar{\gamma}}$ is a unit speed geodesic and that the integral equations (ii) hold. The curve $\gamma(s) = (t(s), \vec{\gamma}(s))$ is a null curve because of

$$\begin{aligned} \langle \dot{\gamma}((s)), \dot{\gamma}(s) \rangle &= -(\mathrm{d}t/\mathrm{d}s)^2 + a^2 \hat{g}_\varepsilon \left(\frac{\mathrm{d}}{\mathrm{d}\tau}\vec{\bar{\gamma}}, \frac{\mathrm{d}}{\mathrm{d}\tau}\vec{\bar{\gamma}}\right)\frac{1}{a^2}(\mathrm{d}t/\mathrm{d}s)^2 \\ &= (\mathrm{d}t/\mathrm{d}s)^2\left(-1 + \hat{g}_\varepsilon\left(\frac{\mathrm{d}}{\mathrm{d}\tau}\vec{\bar{\gamma}}, \frac{\mathrm{d}}{\mathrm{d}\tau}\vec{\bar{\gamma}}\right)\right) = 0. \end{aligned}$$

It follws from the first integral equation in (ii) that $a\mathrm{d}t/\mathrm{d}s$ is constant. Hence the first equation in Lemma 6.3.1 follows from

$$0 = \frac{\mathrm{d}}{\mathrm{d}s}\left(a(t(s))\frac{\mathrm{d}t}{\mathrm{d}s}\right) = a'(t)\left(\frac{\mathrm{d}t}{\mathrm{d}s}\right)^2 + a(t)\frac{\mathrm{d}^2 t}{\mathrm{d}s^2}$$

and the fact that $\gamma$ is a null curve. The second equation follows also by direct calculation:

$$\hat{\nabla}_{\dot{\vec{\gamma}}}\dot{\vec{\gamma}} = \frac{1}{c^2a^2}\hat{\nabla}_{\frac{\mathrm{d}}{\mathrm{d}\tau}\vec{\gamma}}\left(\frac{1}{a^2}\frac{\mathrm{d}}{\mathrm{d}\tau}\vec{\gamma}\right) = \frac{-2}{c^2a^5}\frac{\mathrm{d}a}{\mathrm{d}\tau}\frac{\mathrm{d}}{\mathrm{d}\tau}\vec{\gamma} = \frac{-2}{a}\frac{\mathrm{d}a}{\mathrm{d}s}\dot{\vec{\gamma}}$$

where we have used $\frac{\mathrm{d}\tau}{\mathrm{d}s} = \frac{1}{a}\frac{\mathrm{d}t}{\mathrm{d}s} = \frac{1}{a^2c}$.

To see that all null geodesics can be described this way it is sufficient to notice that (up to a multiple) any null vector can be realized as $\partial_t + e$ where $e$ is a unit vector tangent to $\Sigma_t$. ∎

**Proposition 6.3.1.** *Let $(M,g)$ be infinitesimally isotropic and $\gamma$ be a photon which moves from $x \in M$ to $y \in M$. Then*

$$z = \frac{a(t(y))}{a(t(x))} - 1.$$

*Proof.* From Lemma 6.3.1 we get that $a(t)\frac{\mathrm{d}t}{\mathrm{d}s}$ is constant. Hence $E = -\langle U, \dot{\gamma}\rangle = \frac{\mathrm{d}t}{\mathrm{d}s}$ implies that $\lambda/a = h/(a(t)\frac{\mathrm{d}t}{\mathrm{d}s}) =: k$ is constant. Inserting $\lambda = ka$ in the definition of $z$ proves the claim. ∎

**Corollary 6.3.2.** *Let $x_0 = (t_0, \vec{x_0}) \in M$. Then the frequencies emitted by* nearby *galaxies (situated at $y = (t, \vec{y}) \in M$) appear to be red-shifted at $x_0$ by*

$$z \approx H_{t_0}\, d((t_0, \vec{x_0}), (t_0, \vec{y_0})),$$

*where $H_{t_0} = a'(t_0)/a(t_0)$ is the* Hubble "constant" *at $x_0$.*

*Proof.* We assume that $d_\Sigma(\vec{x_0}, \vec{y_0}) \ll 1$ which in turn implies $|t - t_0| \ll 1$. Hence we obtain $a(t) \approx a(t_0) + (t - t_0)a'(t_0)$ and therefore

$$\begin{aligned} z &= \frac{a(t_0)}{a(t)} - 1 \approx \frac{a(t_0)}{a(t_0) + (t - t_0)a'(t_0)} - 1 \\ &= \frac{1}{1 + (t - t_0)H_{t_0}} - 1 \approx -(t - t_0)H_{t_0}. \end{aligned}$$

Since the speed of light is 1 and $t < t_0$ we have $t - t_0 \approx -d((t_0, \vec{x_0}), (t_0, \vec{y_0})$ which implies the assertion ∎

## 6.4 The age of the universe and the big bang

At our time $t_0$ the Hubble constant $H_{t_0} = a'(t_0)/a(t_0)$ is positive.

In the introduction of the preceding section we have given a heuristic argument which indicates that universe is expanding because of the red-shift of light emitted from nearby galaxies. In this section we we will show that the observation of Hubble implies that there has been a big bang (if infinitesimal isotropy holds). In order to do so we need an assumption which eliminates the unphysical cases that

- $a$ is simply not defined on its maximal domain or that
- $a$ is not everywhere differentiable where it is defined.

We will therefore assume that $\epsilon, p \colon I \to \mathbb{R}$ can be continuously extended, unless they become unbounded or $a \to 0$. If this assumption would not hold then there would exist an extension $(\hat{I} \times \Sigma, -\mathrm{d}t^2 + \hat{g}_\varepsilon)$ of spacetime such that at $\partial(I \times \Sigma) \subset \hat{I} \times \Sigma$ matter would miraculously disapear or spring into existence.

**Theorem 6.4.1.** *Let $(M, g) = ((t_-, t_+) \times \Sigma, -\mathrm{d}t^2 + a^2(t)\hat{g}_\varepsilon)$ be an infinitesimally isotropic, $C^{2-}$-maximally extended Lorentz manifold of dimension $n \geq 3$. If*

*(i) there is a $t_0 \in (t_-, t_+)$ with $H_{t_0} > 0$*
*(ii) $\epsilon$, $p$ are continuous on $(t_-, t_+)$,*
*(iii) $\epsilon + \Lambda/(8\pi) \geq 0$,*
*(iv) there exist constants $c_\pm$ such that $\epsilon$ and $p$ satisfy*

$$-\frac{n-3}{n-1} < c_- \leq \frac{p - \Lambda/(8\pi)}{\epsilon + \Lambda/(8\pi)} \leq c_+,$$

*then $t_- > -\infty$. In addition, we have $\lim_{t\to t_-} a(t) = 0$ and $\lim_{t\to t_-} a'(t) = \lim_{t\to t_-} a^2(t)(\epsilon(t) + \Lambda/(8\pi)) = \infty$. For $t_+$ there are the following possibilities.*

| $\varepsilon$ | $t_+$, | $\lim_{t\to t_+} a(t)$, | $\lim_{t\to t_+} a'(t)$ | $\lim_{t\to t_+} a^2(t)(\epsilon(t) + \Lambda/(8\pi))$ |
|---|---|---|---|---|
| $-1$ | $\infty$ | $\infty$ | 1 | 0 |
| 0 | $\infty$ | $\infty$ | 0 | 0 |
| 1 | *finite* | 0 | $-\infty$ | $\infty$ |

*Proof.* If $\epsilon$ can be extended beyond $t_-$ or $t_+$ as a bounded function so can $p$. For given $\epsilon, p$, Corollary 6.2.2 (i) can be viewed as a linear differential equation of second order for $a$. Consequently, if $t_\pm$ is finite, $a$ could be extended as a $C^{2-}$ function if it is not infinite (Dieudonné 1960, Remark 10.4.6). This implies $\liminf_{t\to t_\pm} a(t) = 0$ or $\limsup_{t\to t_\pm} a(t) = \infty$ since $(M, g)$ is $C^{2-}$-maximally extended by assumption. Conditions (iii) and (iv) imply that $8\pi\left(p + \frac{n-3}{n-1}\epsilon\right) - 2\Lambda/(n-1)$ is positive Hence $a''(t) \leq 0$ for all $t$ by Corollary 6.2.2 (i). The function $a'$ is therefore monotone and we can replace lim inf (and lim sup) by lim for $a$ and $a'$.

First we investigate what happens near $t_-$. The inequality $H_{t_0} > 0$ implies the inequality $a'(t_0) > 0$. Since $a''(t) \leq 0$ for all $t$, the graph of $a$ lies below the graph of the map $t \mapsto a(t_0) + a'(t_0)(t - t_0)$. This linear graph intersects the $(a = 0)$-axis at $t_0 - \frac{a(t_0)}{a'(t_0)} < t_0$ which implies that $t_-$ is finite: $t_- \in [t_0 - a(t_0)/a'(t_0), t_0)$. Since $p - \Lambda/(8\pi) \geq c_-(\epsilon + \Lambda/(8\pi))$, there is a $\delta > 0$ with

$$\begin{aligned}\epsilon + p &\geq (1 + c_-)(\epsilon + \Lambda/(8\pi)) > (1 - \frac{n-3-\delta}{n-1})(\epsilon + \Lambda/(8\pi)) \\ &= \frac{2+\delta}{n-1}(\epsilon + \Lambda/(8\pi)).\end{aligned}$$

Hence Corollary 6.2.2 (ii) implies $(\epsilon + \Lambda/(8\pi))' = -(n-1)(\epsilon + p)a'/a < -(2+\delta)(\epsilon+\Lambda/(8\pi))a'/a$ and therefore $((\epsilon+\Lambda/(8\pi))a^{2+\delta})' < 0$. The equation $\lim_{t\to t_\pm} a(t) = 0$ implies now immediately that $\lim_{t\to t_-} a^2(\epsilon(t) + \Lambda/(8\pi)) = \infty$. From the equation for the energy density $\epsilon$ in Corollary 6.2.1 we infer that $(a')^2$ diverges also.

For $t \to t_+$ there are several possibilities. If $a$ has no maximum then $\lim_{t\to t_+} a(t) = \infty$. $a'' \leq 0$ implies that $t_+ = \infty$. Since $((\epsilon + \Lambda/(8\pi))a^{2+\delta})' < 0$, the function $(\epsilon + \Lambda/(8\pi))a^{2+\delta}$ is decreasing which implies $\lim_{t\to t_+}(\epsilon(t) + +\Lambda/(8\pi))a^2(t) = 0$. The equation for the energy density $\epsilon$ in Corollary 6.2.1 implies now that $\varepsilon \leq 0$. The assertions about $\lim_{t\to t_+} a'(t)$ follow from the same equation. Note that $a$ must have a maximum if $\varepsilon = 1$.

If $a$ has a maximum at some $t_1 \in (t_-, t_+)$, $8\pi\epsilon(t_1) + \Lambda = \frac{1}{2}(n-1)(n-2)\varepsilon/a^2(t_1)$ whence $\varepsilon = 1$. Since $a''(t_1) < 0$ we have $a'(t_2) < 0$ for some $t_2 \in (t_1, t_+)$ and we can — by time reversal — apply the same argument as for $t \to t_-$. This proves $\lim_{t\to t_+} a(t) = 0$, $\lim_{t\to t_+} a'(t) = -\infty$, and $\lim_{t\to t_+} a^2(t)(\epsilon(t) + \Lambda/(8\pi)) = \infty$. ∎

**Corollary 6.4.1.** *If our universe is described by a Robertson-Walker model without cosmological constant, then it is younger than* $1/H_{t_0} = (18 \pm 2) \cdot 10^9$ *years.*

Part of Theorem 6.4.1 can be generalised to locally spatially homogeneous universes which are not necessarily isotropic (Rendall 1994).

While Einstein's equation itself does not indicate any special value for $\Lambda$, it is known that $\Lambda \neq 0$ is incompatible with the Newtonian limit for weak gravitation. This implies that $\Lambda$ must be very small if non-zero (cf. (Hawking and Ellis 1973, p. 362), (Sandage 1968)).

*Example 6.4.1.* The inequality $-\frac{n-3}{n-1} < c_- \leq \frac{p-\Lambda/(8\pi)}{\epsilon+\Lambda/(8\pi)}$ is sharp. This can be seen from 3-dimensional dust spacetimes. We solve the system of differential equations (i), (ii) of Corollary 6.2.2. Both equations decouple since $p = 0$ and $n = 3$. Equation (ii) implies that there is a constant

$k \in \mathbb{R}$ with $\epsilon = ka^2$. Equation (i) reduces to $a'' = -\Lambda a$, whence $a(t) = k_+ e^{\sqrt{-\Lambda}t} + k_- e^{-\sqrt{-\Lambda}t}$. From our initial data $a_0, \epsilon_0$ we obtain $a'(0) = \sqrt{(8\pi\epsilon_0 + \Lambda)(a_0)^2 - \varepsilon}$ and therefore

$$a(t) = a_0 \cos(\sqrt{\Lambda}t) + \frac{\sqrt{(8\,\pi\,\epsilon_0 + \Lambda)\,(a_0)^2 - \varepsilon}\,\sin(\sqrt{\Lambda}t)}{\sqrt{\Lambda}}$$

for $\Lambda > 0$,

$$a(t) = \frac{\left(a_0\,\Lambda - \sqrt{-\Lambda}\sqrt{(8\,\pi\,\epsilon_0 + \Lambda)\,(a_0)^2 - \varepsilon}\right)\,e^{\sqrt{-\Lambda}t}}{2\,\Lambda} + \frac{\left(a_0\,\Lambda + \sqrt{-\Lambda}\sqrt{(8\,\pi\,\epsilon_0 + \Lambda)\,(a_0)^2 - \varepsilon}\right)\,e^{-\sqrt{-\Lambda}t}}{2\,\Lambda}$$

for $\Lambda < 0$, and

$$a(t) = a_0 + \sqrt{8\,(a_0)^2\pi\,\epsilon_0 - \varepsilon}\;t$$

for $\Lambda = 0$. If $\varepsilon = 1$, $\Lambda < 0$, and the initial data $a_0$, $\epsilon_0$ are chosen such that $(\epsilon_0 + \Lambda)\,(a_0)^2 = 1$ then we obtain solutions without singularities.

*Remark 6.4.1.* Observe that the metric of Robertson-Walker spacetimes differs from the *non-singular* Lorentzian manifold

$$\left((t_-, t_+) \times \Sigma, -\mathrm{d}t^2 + \frac{1}{1 - \varepsilon r^2}\mathrm{d}r^2 + r^2\left(\mathrm{d}\theta^2 + \sin^2(\theta)\mathrm{d}\varphi^2\right)\right)$$

only by an overall-factor $a^2$. (We say that these pairs of spacetimes are *conformally equivalent*). This implies that there are past light cones which do not intersect. In particular, there are regions in the universe filled with particles which may never have had a chance to interact. This raises a serious problem. In physical theories, homogeneous states are usually obtained by the statistical description of microscopic states. In particular, homogeneity is always a result of prolonged interaction. If there are regions in the universe which cannot have interacted with each other in the past, then we need a new explanation why they are so similar that we can describe the universe as homogeneous. Physicists are currently trying to find an answer to this question by arguing that during the early phase of the universe the expansion was much faster than would be plausible if one considers ordinary matter. Under this assumption, these regions would have had a chance to interact after all. However, the physicists favouring this *inflational universe* are still[4] very far from a satisfactory physical explanation for the occurence of this "inflationary" phase in the history of the universe.

[4] I write this in 1998.

## 6.5 A simple model for the universe we live in

In this subsection we will discuss two especially simple cases in which Einstein's equation can be explicitly solved. They are of special importance since they are in accordance with qualitative expectations of the development of the universe. At present, there seem to be two principal types of matter which fill the universe. To simplify matters we may assume that today most of the matter consists of galaxies[5]. Galaxies do not move much relative to each other and are too far apart from each other to interact. It is therefore a good description to model them by pressure-less dust. The universe is also filled with radiation which was the dominating form of matter during the early stages of the Universe. This '*cosmic microwave background radiation* has been discovered by Penzias and Wilson (1965) [6]. The most important feature of this microwave background radiation is that it is (almost) completely isotropic and therefore cannot be explained by a confined source which is located somewhere in the universe. The radiation is now very weak (having a temperature of about 2.7 Kelvin) and its energy is completely dominated by the energy contribution from the galaxies. The microwave background radiation can be described by a photon gas. The energy momentum tensor of a photon gas is traceless by Lemma 5.2.3 which, toghether with infinitesimal isotropy, implies $\epsilon = (n-1)p$.

We will not consider the composed system which consists of dust and radiation but only the much simpler cases where we have pure dust or pure radiation. At the end of the section we will give a heuristic justification.

**Lemma 6.5.1.** *Let* $(M,g) = ((t_-,t_+) \times \Sigma, -\mathrm{d}t^2 + a^2(t)g_\varepsilon)$ *be an infinitesimally isotropic Lorentzian manifold of dimension* $n \geq 3$ *and assume that the fuction* $a$ *is non-constant. Then the following statements are equivalent.*

*(i)* $p = 0$
*(ii)* $\epsilon a^{n-1} = m$ *is constant,*
*(iii)* $(n-1)(n-2)\left((a')^2 + \varepsilon\right) = 16\pi m a^{3-n} + 2\Lambda$.

*Proof.* The equation $p = 0$ is equivalent to $(n-1)a'/a + \epsilon'/e = 0$ because of Corollary 6.2.2 and $a' \neq 0$. An integration shows that this equation is equivalent to $\epsilon a^{n-1} = m$ for some constant $m$. The equivalence of (ii) and (iii) is clear from the formula for $\epsilon$ in Lemma 6.2.1. ∎

---

[5] However, it is expected that most of the matter is "dark" which is an euphemism for "we cannot directly observe it and do not know anything about it".

[6] The discoverers where concerned with the development of a new satellite communication system and found that their new high precision antenna seemed to have an unexplainable background noise.

In the physically interesting case $n = 4$, $\Lambda = 0$, the differential equations in Lemma 6.5.1 (iii) can be solved explicitly and their solutions are given by

| $\varepsilon$ | $a$ | $t$ |
|---|---|---|
| $-1$ | $\frac{1}{2}c(1 - \cosh(\vartheta))$ | $-\frac{1}{2}c(\vartheta - \sinh(\vartheta))$ |
| $0$ | $(9c/4)^{1/3}t^{2/3}$ | |
| $1$ | $\frac{1}{2}c(1 - \cos(\vartheta))$ | $\frac{1}{2}c(\vartheta - \sin(\vartheta))$ |

where $c$ is a constant of integration.

**Lemma 6.5.2.** *Let $(M, g) = ((t_-, t_+) \times \Sigma, -dt^2 + a^2(t)g_\varepsilon)$ be an infinitesimally isotropic Lorentzian manifold of dimension $n \geq 3$ and assume that the function $a$ is non-constant. Then the following statements are equivalent.*

*(i)* $\epsilon = (n-1)p$
*(ii)* $\epsilon a^n = m$ *is constant,*
*(iii)* $(n-1)(n-2)\left((a')^2 + \varepsilon\right) = 16\pi m a^{2-n} + 2\Lambda.$

*Proof.* The equation $\epsilon = (n-1)$ is equivalent to $(n-1)\frac{a'}{a} + \frac{n-1}{n}\frac{\epsilon'}{e} = 0$ because of Corollary 6.2.2 and$a' \neq 0$. Integrating this equation we obtain $\epsilon a^n = m$ for some constant $m$. The equivalence of (ii) and (iii) is clear from the formula for $\epsilon$ in Lemma 6.2.1. ∎

In the case $n = 4$, $\Lambda = 0$ the resulting differential equations can be solved explicitly and the solutions are given by

| $\varepsilon$ | $t \mapsto a(t)$ |
|---|---|
| $-1$ | $c\sqrt{-1 + (1 + t/c)^2}$ |
| $0$ | $(4c^2)^{1/4}\sqrt{t}$ |
| $1$ | $c\sqrt{1 - (1 - t/c)^2}$ |

where $c$ is a constant of integration.

We will make the assumption that the interaction between both types of matter is neglectible. Observe that due to the formulas (ii) in Lemmas 6.5.1 and 6.5.2, radiation dominates at early times ($a \ll 1$) and dust dominates at late times (($a \gg 1$). Hence it seems to be a good approximation to use the radiation model for the early universe and the dust model for the present universe.

# 7. Spherical symmetry

This chapter serves two purposes. Firstly, a large isometry group simplifies the problem of solving Einstein's equation considerably. Virtually all explicitly known solutions of Einstein's equations for physically plausible matter fields have a high degree of symmetry. Secondly, a spherically symmetric spacetimes are very good descriptions of non-rotating, isolated stars and therefore of astrophysical interest. (If the star rotates the rotation axis breaks the symmetry). In Sect. 7.2 we will see that there is a unique 1-parameter family of spherically symmetric solutions to Einsteins's equation for vacuum with vanishing cosmological constant. The parameter can be interpreted as the mass of the isolated star. If the mass of the sun is chosen, one obtains an excellent model of the gravitational field in our solar system. Some aspects of this model have been verified experimentally. These solutions also form the basis for much of our intuition of black holes.

In this chapter we will also discuss the initial value problem for the case that the energy momentum tensor represents a perfect fluid together with a non-interacting electric field (cf. Sect. 7.4). While in this more general case we will not obtain explicit solutions we will nevertheless arrive at a non-trivial existence theorem for the considered class of spacetimes. This section will hardly be of primary interest to a geometrically oriented reader. Since the discussion uses elements of the theory of systems of hyperbolic partial differential equations even physically oriented readers may wish to skip the proofs on first reading.

The validity of the physical conclusions from this (and also the following) chapter depends very much on the question of whether the corresponding properties of our explicit solutions are stable under perturbations. We know only very little about the stability of Einstein's equations. Since they are highly non-linear it is well possible that these properties have little to do with our actual universe which has only "approximate isometries".

## 7.1 Pseudo-Riemannian manifolds with spherical symmetry

A property in $\mathbb{R}^3$ is spherically symmetric if it is invariant under rotations about the origin. The rotational isometries defined below form a group which is (locally) isometric to the rotation group SO(3). We take this as our main justification of the following definition.

**Definition 7.1.1.** *A pseudo-Riemannian manifold* $(M, g)$ *is called* spherically symmetric *if it has a dense open subset* $M^\circ$ *such that* $(M^\circ, g)$ *can locally be written as a warped product* $(\Sigma \times S^2, g_\Sigma + r^2 \mathrm{d}\Omega^2)$, *where* $\mathrm{d}\Omega^2$ *is the metric of the 2-dimensional unit sphere,* $(\Sigma, g_\Sigma)$ *an* $(n-2)$-*dimensional pseudo-Riemannian manifold, and* $r \colon \Sigma \to \mathbb{R}$ *a positive function.*

*The sets* $\{x\} \times S^2$ $(x \in \Sigma)$ *are called* spheres of symmetry *and those isometries which map all spheres of symmetry into themselves are called* rotational isometries. *The set* $C = M \setminus M^\circ$ *is the* centre of symmetry.

**Lemma 7.1.1.** *Let* $(M, g)$ *be a spherically symmetric spacetime and* $(\Sigma, g_\Sigma)$ *be the 2-dimensional Lorentzian manifold orthogonal to the spheres of symmetry. For each frame* $\{U, Q\}$ *of* $(\Sigma, g_\Sigma)$ *with*

$$\langle U, U \rangle = -1, \quad \langle U, Q \rangle = 0, \quad \langle Q, Q \rangle = 1,$$

*there are adapted coordinates with respect to which*

$$g = -e^{2\nu(t,q)} \mathrm{d}t^2 + e^{2\lambda(t,q)} \mathrm{d}q^2 + r^2(t,q)(\mathrm{d}\theta^2 + \sin^2\theta \, \mathrm{d}\varphi^2)$$

*and* $U = e^{-\nu}\partial_t$, $Q = e^{-\lambda}\partial_q$.

*In these coordinates the energy momentum tensor* $T$ *is given by*

$$\begin{aligned} 8\pi T(U,U) &= \frac{1}{r^2}\left(1 + (U \bullet r)^2 - (Q \bullet r)^2\right) \\ &\quad + \frac{2}{r}\left(-Q \bullet Q \bullet r + (U \bullet \lambda)(U \bullet r)\right) - \Lambda, \\ 8\pi T(U,Q) &= -\frac{2}{r}\left(U \bullet Qr - (Q \bullet \nu)(U \bullet r)\right), \\ 8\pi T(Q,Q) &= -\frac{1}{r^2}\left(1 + (U \bullet r)^2 - (Q \bullet r)^2\right) \\ &\quad + \frac{2}{r}\left(-U \bullet U \bullet r + (Q \bullet \nu)(Q \bullet r)\right) + \Lambda, \\ 8\pi T(V,X) &= 0 \quad \text{for all } X \in T\Sigma \text{ and } V \in T\Sigma^\perp, \end{aligned}$$

$$\begin{aligned}
8\pi T(\tfrac{\partial_\theta}{r}, \tfrac{\partial_\theta}{r}) &= 8\pi T(\tfrac{\partial_\varphi}{r\sin(\theta)}, \tfrac{\partial_\varphi}{r\sin(\theta)}) \\
&= Q \bullet Q \bullet \nu + (Q \bullet \nu)^2 - U \bullet U \bullet \lambda - (U \bullet \lambda)^2 \\
&\quad + \frac{1}{r}( - U \bullet U \bullet r + Q \bullet Q \bullet r - (Q \bullet \nu)(Q \bullet r) \\
&\quad - (U \bullet \lambda)(U \bullet r)) + \Lambda, \\
8\pi T(\tfrac{\partial_\theta}{r}, \tfrac{\partial_\varphi}{r\sin(\theta)}) &= 0.
\end{aligned}$$

*Proof.* The existence of the adapted coordinates $(t, q, \theta, \varphi)$ is clear from Corollary 2.4.2. Hence we only need to calculate $8\pi T = \mathrm{Ric} - \frac{1}{2}\mathrm{Scal}\, g + \Lambda g$. By Lemma 4.4.15 we have for $X, Y$ tangent to $\Sigma$ and $V, W$ orthogonal to $\Sigma$

$$\begin{aligned}
\mathrm{Ric}(X, Y) &= \frac{1}{2}\mathrm{Scal}_\Sigma \langle X, Y\rangle - \frac{2}{r}\nabla\nabla r(X, Y), \\
\mathrm{Ric}(X, V) &= 0 \\
\mathrm{Ric}(V, W) &= \frac{1}{r^2}\langle V, W\rangle - \left(\frac{\Delta r}{r} + \frac{1}{r^2}\langle \mathrm{grad}(r), \mathrm{grad}(r)\rangle\right)\langle V, W\rangle, \\
\mathrm{Scal} &= \mathrm{Scal}_\Sigma + \frac{2}{r^2} - \frac{4}{r}\Delta r - \frac{2}{r^2}\langle \mathrm{grad}(r), \mathrm{grad}(r)\rangle .
\end{aligned}$$

Since $\nabla\nabla r(X, Y) = X \bullet Y \bullet r - \nabla_X Y \bullet r$ we obtain

$$\begin{aligned}
\nabla\nabla r(U, U) &= U \bullet U \bullet r - (Q \bullet \nu)(Q \bullet r), \\
\nabla\nabla r(U, Q) &= U \bullet Q \bullet r - (Q \bullet \nu)(U \bullet r) = Q \bullet U \bullet r - (U \bullet \lambda)(Q \bullet r), \\
\nabla\nabla r(Q, Q) &= Q \bullet Q \bullet r - (U \bullet \lambda)(U \bullet r)
\end{aligned}$$

which in turn implies $\Delta r = -U \bullet U \bullet r + Q \bullet Q \bullet r + (Q \bullet \nu)(Q \bullet r) - (U \bullet \lambda)(U \bullet r)$. Hence we get

$$\begin{aligned}
8\pi T(U, U) &= -\frac{1}{2}\mathrm{Scal}_\Sigma - \frac{2}{r}(U \bullet U \bullet r - (Q \bullet \nu)(Q \bullet r)) + \frac{1}{2}(\mathrm{Scal}_\Sigma \\
&\quad + \frac{2}{r^2} - \frac{4}{r}(-U \bullet U \bullet r + Q \bullet Q \bullet r + (Q \bullet \nu)(Q \bullet r) - \\
&\quad (U \bullet \lambda)(U \bullet r)) - \frac{2}{r^2}(-(U \bullet r)^2 + (Q \bullet r)^2)) - \Lambda \\
&= \frac{1}{r^2}(1 + (U \bullet r)^2 - (Q \bullet r)^2) \\
&\quad - \frac{2}{r}(Q \bullet Q \bullet r - (U \bullet \lambda)(U \bullet r)) - \Lambda, \\
8\pi T(U, Q) &= -\frac{2}{r}(U \bullet Q \bullet r - (Q \bullet \nu)(U \bullet r)),
\end{aligned}$$

$$
\begin{aligned}
8\pi T(Q,Q) &= \frac{1}{2}\mathrm{Scal}_\Sigma - \frac{2}{r}(Q \bullet Q \bullet r - (U \bullet \lambda)(U \bullet r)) - \frac{1}{2}(\mathrm{Scal}_\Sigma \\
&\quad + \frac{2}{r^2} - \frac{4}{r}(-U \bullet U \bullet r + Q \bullet Q \bullet r + (Q \bullet \nu)(Q \bullet r) - \\
&\quad (U \bullet \lambda)(U \bullet r)) - \frac{2}{r^2}(-(U \bullet r)^2 + (Q \bullet r)^2)) + \Lambda \\
&= -\frac{1}{r^2}(1 + (U \bullet r)^2 - (Q \bullet r)^2) \\
&\quad - \frac{2}{r}(U \bullet U \bullet r - (Q \bullet \nu)(Q \bullet r)) + \Lambda.
\end{aligned}
$$

Let $X \in T_x\Sigma$ and $V \in (T_x\Sigma)^\perp$. Since $\mathrm{Ric}(V,X) = 0$ and $\langle X, V\rangle = 0$ it is clear that $8\pi T(X,V) = 0$. Since there is an isometry which maps $\frac{\partial_\theta}{r}$ into $\frac{\partial_\varphi}{r\sin(\theta)}$ we have $8\pi T(\frac{\partial_\theta}{r}, \frac{\partial_\theta}{r}) = 8\pi T(\frac{\partial_\varphi}{r\sin(\theta)}, \frac{\partial_\varphi}{r\sin(\theta)})$. The Ricci tensor restricted to $(T_x\Sigma)^\perp$ is a multiple of the metric which implies $T(\frac{\partial_\theta}{r}, \frac{\partial_\varphi}{r\sin(\theta)}) = 0$. Finally, we the last component of $T$ which needs to be calculated is given by

$$
\begin{aligned}
8\pi T(\tfrac{\partial_\theta}{r}, \tfrac{\partial_\theta}{r}) &= \frac{1}{r^2} - \Big(\frac{1}{r}\big(-U \bullet U \bullet r + Q \bullet Q \bullet r + (Q \bullet \nu)(Q \bullet r) \\
&\quad - (U \bullet \lambda)(U \bullet r)\big) + \frac{1}{r^2}\big(-(U \bullet r)^2 + (Q \bullet r)^2\big)\Big) \\
&\quad - \frac{1}{2}(\mathrm{Scal}_\Sigma + \frac{2}{r^2} - \frac{4}{r}(-U \bullet U \bullet r + Q \bullet Q \bullet r \\
&\quad + (Q \bullet \nu)(Q \bullet r) - (U \bullet \lambda)(U \bullet r)) \\
&\quad - \frac{2}{r^2}(-(U \bullet r)^2 + (Q \bullet r)^2)) + \Lambda \\
&= \frac{1}{r}\big(-U \bullet U \bullet r + Q \bullet Q \bullet r \\
&\quad + (Q \bullet \nu)(Q \bullet r) - (U \bullet \lambda)(U \bullet r)\big) \\
&\quad - \big(U \bullet U \bullet \lambda + (U \bullet \lambda)^2 - Q \bullet Q \bullet \nu - (Q \bullet \nu)^2\big) + \Lambda,
\end{aligned}
$$

where in the last equation we have used

$$
\mathrm{Scal}_\Sigma = 2(U \bullet U \bullet \lambda + (U \bullet \lambda)^2 - Q \bullet Q \bullet \nu - (Q \bullet \nu)^2)
$$

(cf. Proposition 4.3.5). ∎

We will now re-arrange these equations in a form which is more practical if one wants to solve them. We will not do this in complete generality but rather assume the following genericity assumption on the matter model. Recall that for any normalized timelike vector $v$ the number $T(v,v)$ represents the energy density measured by $u$. This number should be positve. For null vectors $N$ we obtain then $T(N,N) \geq 0$ by continuity. Given a non-extreme matter distribution it is therefore plausible to expect $T(N,N) > 0$ for all null vectors $N$. In this case it is always possible to diagonalize $g$ and $T$ simultaneously (Greub 1981, Chapter IX §3).

**Lemma 7.1.2.** *Let $(M, g)$ be a spherically symmetric Lorentzian manifold and $T$ be a symmetric $\binom{0}{2}$ tensor field which is also spherically symmetric. Assume that $T$ and $g$ can be simultaneously diagonalized. Then there exist coordinates $(t, q, \theta, \varphi)$ such that*

$$g = -e^{2\nu(t,q)}dt^2 + e^{2\lambda(t,q)}dq^2 + r^2(t,q)(d\theta^2 + \sin^2\theta\, d\varphi^2), \tag{7.1.1}$$

$$T = \epsilon U^\flat \otimes U^\flat + p_{\mathrm{rad}} Q^\flat \otimes Q^\flat + p_{\mathrm{sph}} r^2(t,q)(d\theta^2 + \sin^2\theta\, d\varphi^2), \tag{7.1.2}$$

*where $U := e^{-\nu}\partial_t$ and $Q := e^{-\lambda}\partial_q$ are invariantly defined if $\epsilon \neq -p_{\mathrm{rad}}$.*

*Moreover, Einstein's equation,* $\mathrm{Ric} - \mathrm{Scal}/2\, g + \Lambda g, = 8\pi T$ *is equivalent to the system of differential equations*

$$Q \bullet Q \bullet r = (U \bullet r)\,(U \bullet \lambda) + \frac{1 + (U \bullet r)^2 - (Q \bullet r)^2}{2r} - 4\pi r(\epsilon + \frac{\Lambda}{8\pi}), \tag{7.1.3}$$

$$Q \bullet U \bullet r = (Q \bullet r)\,(U \bullet \lambda), \tag{7.1.4}$$

$$U \bullet U \bullet r = (Q \bullet r)\,(Q \bullet \nu) - \frac{1 + (U \bullet r)^2 - (Q \bullet r)^2}{2r} - 4\pi r(p_{\mathrm{rad}} - \frac{\Lambda}{8\pi}), \tag{7.1.5}$$

$$U \bullet U \bullet \lambda = -(U \bullet \lambda)^2 + Q \bullet Q \bullet \nu + (Q \bullet \nu)^2 + \frac{1 + (U \bullet r)^2 - (Q \bullet r)^2}{r^2} - 4\pi(\epsilon - p_{\mathrm{rad}} + 2p_{\mathrm{sph}}), \tag{7.1.6}$$

*and the equation of motion,* $\mathrm{div}(T) = 0$, *is equivalent to*

$$U \bullet \epsilon = -(\epsilon + p_{\mathrm{rad}})U \bullet \lambda - 2(\epsilon + p_{\mathrm{sph}})\frac{U \bullet r}{r}, \tag{7.1.7}$$

$$Q \bullet p_{\mathrm{rad}} = -(\epsilon + p_{\mathrm{rad}})Q \bullet \nu - 2(p_{\mathrm{rad}} - p_{\mathrm{sph}})\frac{Q \bullet r}{r}. \tag{7.1.8}$$

*Proof.* Lemma 7.1.1 implies that $T$ satisfies

- $T(X, V) = 0$ for all $X \in T_x\Sigma$, $V \in (T_x M)^\perp$
- and $T(V, V) = T(W, W)$ for all unit vectors $V, W \in ({}_x\Sigma)^\perp$.

Since by assumption there is a frame which diagonalizes the energy momentum tensor $T$ and the metric $g$ simultaneously there must exist vector fields $Q, U$ tangent to $\Sigma$ and functions $r, \epsilon, p_{\mathrm{rad}}, p_{\mathrm{sph}} : \Sigma \to \mathbb{R}$ such that

$$g = -U^\flat \otimes U^\flat + Q^\flat \otimes Q\flat + r^2 \mathrm{d}\Omega^2,$$
$$T = \epsilon U^\flat \otimes U^\flat + p_{\mathrm{rad}} Q^\flat \otimes Q^\flat + p_{\mathrm{sph}} r^2 \mathrm{d}\Omega^2.$$

The existence of adapted coordinates follows now from Corollary 2.4.2. That Einstein's equations are equivalent to Equations (7.1.3)–(7.1.6) follows immediately from the definitions of $\epsilon, p_{\mathrm{rad}}, p_{\mathrm{sph}}$ and Lemma 7.1.1. Observe that for any vector fields $X, Y$ tangent to $\Sigma$ the decomposition $\nabla_X Y = {}^{\Sigma}\boldsymbol{\nabla}_X Y$ and that for any vector fields $V, W$ tangent to the spheres of symmetry the decomposition $\nabla_V W = -\frac{1}{r}\langle V, W\rangle \operatorname{grad}(r) + {}^{S^2}\boldsymbol{\nabla}_V W$ holds. Using Proposition 4.3.5 we obtain

$$
\begin{aligned}
\operatorname{div}(T) &= (\mathrm{d}\epsilon(U) + \epsilon\operatorname{div}(U))U^\flat + \epsilon(\nabla_U U)^\flat + (\mathrm{d}p_{\mathrm{rad}}(Q) + p_{\mathrm{rad}}\operatorname{div}(Q))Q^\flat \\
&\quad + p_{\mathrm{rad}}(\nabla_Q Q)^\flat + (p_{\mathrm{sph}}\operatorname{div}(\tfrac{1}{r}\partial_\theta))\, r\,\mathrm{d}\theta \\
&\quad + p_{\mathrm{sph}}\left(\nabla_{\frac{1}{r}\partial_\theta}\frac{1}{r}\partial_\theta\right)^\flat + p_{\mathrm{sph}}\operatorname{div}(\frac{1}{r\sin(\theta)}\partial_\varphi) r\sin(\theta)\mathrm{d}\varphi \\
&\quad + p_{\mathrm{sph}}\left(\nabla_{\frac{1}{r\sin(\theta)}\partial_\varphi}\frac{1}{r\sin(\theta)}\partial_\varphi\right)^\flat \\
&= (\mathrm{d}\epsilon(U) + \epsilon\big((U\bullet\lambda) + \frac{2}{r}(U\bullet r)\big) + p_{\mathrm{rad}}(U\bullet\lambda))U^\flat \\
&\quad + (\mathrm{d}p_{\mathrm{rad}}(Q) + p_{\mathrm{rad}}\big((Q\bullet\nu) + \frac{2}{r}(Q\bullet r)\big) + \epsilon(Q\bullet\nu))Q^\flat \\
&\quad + p_{\mathrm{sph}}\Big(\frac{\cos(\theta)}{\sin(\theta)}\mathrm{d}\theta - \frac{1}{r}\mathrm{d}r + 0 + 0 - \frac{1}{r}\mathrm{d}r - \frac{\cos(\theta)}{\sin(\theta)}\mathrm{d}\theta\Big) \\
&= (\mathrm{d}\epsilon(U) + \epsilon\big((U\bullet\lambda) + \frac{2}{r}(U\bullet r)\big) + p_{\mathrm{rad}}(U\bullet\lambda) + 2p_{\mathrm{sph}}\frac{U\bullet r}{r})U^\flat \\
&\quad + \Big(\mathrm{d}p_{\mathrm{rad}}(Q) + p_{\mathrm{rad}}\big((Q\bullet\nu) + \frac{2}{r}(Q\bullet r)\big) + \epsilon(Q\bullet\nu) \\
&\quad - 2p_{\mathrm{sph}}\frac{Q\bullet r}{r}\Big)Q^\flat,
\end{aligned}
$$

where we have used Proposition 4.3.5 to calculate

$$
\begin{aligned}
{}^{S^2}\operatorname{div}(\partial_\theta) &= \mathrm{d}\Omega^2({}^{S^2}\boldsymbol{\nabla}_{\partial_\theta}\partial_\theta, \partial_\theta) + \mathrm{d}\Omega^2({}^{S^2}\boldsymbol{\nabla}_{\frac{\partial_\varphi}{\sin(\theta)}}\partial_\theta, \frac{\partial_\varphi}{\sin(\theta)}) \\
&== \frac{\cos(\theta)}{\sin(\theta)}, \\
{}^{S^2}\operatorname{div}(\frac{\partial_\varphi}{\sin(\theta)}) &= \mathrm{d}\Omega^2({}^{S^2}\boldsymbol{\nabla}_{\partial_\theta}\frac{\partial_\varphi}{\sin(\theta)}, \partial_\theta) = 0.
\end{aligned}
$$

∎

The functions $U\bullet\lambda$ and $Q\bullet\nu$ are well defined invariants since the commutator of $U$ and $Q$ is given by $[U, Q] = (Q\bullet\nu)\,U - (U\bullet\lambda)\,Q$.

For spherically symmetric spacetimes we can define an invariant function which can be interpreted as a mass.

**Definition 7.1.2.** *Let* $(M, g)$ *be a spherically symmetric spacetime. Then its* mass function $m$ *is defined by* $m := \frac{r}{2}\left(1 - \langle \operatorname{grad}(r), \operatorname{grad}(r)\rangle\right)$

The term "mass function" can be motivated under the additional assumptions that there exists a centre of symmetry, that $Q \bullet r \geq 0$, and that it is possible to diagonalize $g$ and $T$ simultaneously. Recall from special relativity that mass and energy are equivalent concepts and that the energy of a point particle measured by an observer depends on this observer.

In order to determine the mass of a material object consisting of particles in spacetime we would first fix a spacelike hypersurface representing an instant of time. The mass of each particle which intersects this hypersurface would be measured by the infinitesimal observer represented by the normal of the hypersurface (cf. Sect. 1.4.3). The sum over all these numbers is then the mass of the material object with respect to the chosen hypersurface.

Since we assume that we can simultaneously diagonalize $g$ and $T$ there must be a timelike eigenvector of the linear map $T^{ab}g_{bc}$. Observe that it is unique if $\epsilon \neq -p_{\mathrm{rad}}$. In this case it is orthogonal to the spacelike hypersurfaces $t = \text{const}$. In other words, the hypersurfaces $t = \text{const}$ are invariantly defined. This indicates that we should use this family of hypersurfaces in order to define mass.

Multiplying Equation (7.1.3) with $Q{\bullet}r$ and inserting Equation (7.1.4) we obtain $8\pi(\epsilon + \frac{\Lambda}{8\pi})r^2 Q \bullet r = Q \bullet (r(1 + (U \bullet r)^2 - (Q \bullet r)^2)) = 2Q \bullet m$ and therefore

$$m(t_0, q_0) = 4\pi \int_0^{r(t_0,q_0)} \left(\epsilon + \frac{\Lambda}{8\pi}\right) r^2 \,\mathrm{d}r \qquad \text{where } t = t_0 \text{ is fixed.} \tag{7.1.9}$$

This integral can also be written as a volume integral,

$$\begin{aligned} m(t_0, q_0) &= \int_B \left(\epsilon + \frac{\Lambda}{8\pi}\right) r^2 \sin(\theta)\,\mathrm{d}r \wedge \mathrm{d}\theta \wedge \mathrm{d}\varphi \\ &= \int_B \left(\epsilon + \frac{\Lambda}{8\pi}\right) r^2 \sin(\theta)(Q \bullet r)\mathrm{e}^{-\lambda}\,\mathrm{d}q \wedge \mathrm{d}\theta \wedge \mathrm{d}\varphi \\ &= \int_B \left(\epsilon + \frac{\Lambda}{8\pi}\right) (Q \bullet r)(U \lrcorner \mu_B), \end{aligned}$$

where $B$ is the ball $\{x \in M \,|\, t(x) = t_0, q(x) < q_0\}$. Equations $Q \bullet r = 1$ and $\Lambda = 0$ would imply that we had just an integral over the energy density $\epsilon$ which — in view of $E = mc^2$ — can also be interpreted as a mass density. If, in addition, $p_{\mathrm{rad}} = p_{\mathrm{sph}} = 0$ then $T$ represents a smooth 3-parameter family of freely falling particles and we would obtain the smooth analogue to the motivation using individual particles above.

In general, however, $Q \bullet r \neq 1$. This reflects that one also has to take into account the energy contribution of the gravitational field.

**Lemma 7.1.3.** *Let $m$ be the mass function of $(M, g)$. Then*

$$U \bullet m = -4\pi r^2 (U \bullet r) \left( p_{\text{rad}} - \frac{\Lambda}{8\pi} \right)$$

$$Q \bullet m = 4\pi r^2 (Q \bullet r) \left( \epsilon + \frac{\Lambda}{8\pi} \right)$$

*Proof.* We have $U \bullet Q \bullet r = Q \bullet U \bullet r + [U, Q] \bullet r = Q \bullet U \bullet r + (Q \bullet \nu)(U \bullet r) - (U \bullet \lambda)(Q \bullet r)$. By Lemma 7.1.2 we can calculate

$$\begin{aligned}
Q \bullet m &= Q \bullet (\frac{r}{2} (1 + (U \bullet r)^2 - (Q \bullet r)^2)) = \frac{Q \bullet r}{r} m \\
&\quad + r (yQ \bullet U \bullet r - (Q \bullet r) Q \bullet Q \bullet r) \\
&= \frac{Q \bullet r}{r} m + r((U \bullet r)(Q \bullet r) U \bullet \lambda - (Q \bullet r)(U \bullet r) U \bullet \lambda \\
&\quad - (Q \bullet r) m + 4\pi r (Q \bullet r) \left( \epsilon + \frac{\Lambda}{8\pi} \right)) \\
&= 4\pi (Q \bullet r) r^2 \left( \epsilon + \frac{\Lambda}{8\pi} \right)
\end{aligned}$$

and

$$\begin{aligned}
U \bullet m &= \frac{U \bullet r}{r} m + r ((U \bullet r) U \bullet U \bullet r - (Q \bullet r) U \bullet Q \bullet r) \\
&= \frac{U \bullet r}{r} m + r((U \bullet r)(Q \bullet r) Q \bullet \nu - (U \bullet r) m \\
&\quad - 4\pi r (U \bullet r) \left( p_{\text{rad}} - \frac{\Lambda}{8\pi} \right) - (Q \bullet r) Q \bullet U \bullet r \\
&\quad - (Q \bullet r)(U \bullet r) Q \bullet \nu + (Q \bullet r)^2 (U \bullet \lambda)) \\
&= \frac{U \bullet r}{r} m + r \left( -m(U \bullet r) - 4\pi r (U \bullet r) \left( p_{\text{rad}} - \frac{\Lambda}{8\pi} \right) \right) \\
&= -4\pi r^2 (U \bullet r) \left( p_{\text{rad}} - \frac{\Lambda}{8\pi} \right).
\end{aligned}$$

∎

While orthogonal coordinates are often very useful *double null coordinates* introduced below are much better adapted to the geometry of spherically symmetric spacetimes.

**Lemma 7.1.4 (double null coordinates).** *Let $(M, g)$ be a 4-dimensional spherically symmetric Lorentz manifold. Then there exist local coordinates $(u, v, \theta, \varphi)$ and functions $G \colon (u, v) \mapsto F(u, v) \in \mathbb{R}$, $r \colon (u, v) \mapsto r(u, v) \in \mathbb{R}$ such that*

$$g = G(u,v)\mathrm{d}u\mathrm{d}v + r^2(u,v)\left(\mathrm{d}\theta^2 + \sin^2(\theta)\mathrm{d}\varphi^2\right).$$

*The function $G$ is unique up to transformations of the form $u \mapsto \hat{u}(u)$ $v \mapsto \hat{v}(v)$ and interchanging of the coordinates.*

*In these coordinates, the Christoffel symbols are given by*

$$\Gamma^u_{uu} = \partial_u(\ln r), \quad \Gamma^\theta_{u\theta} = \Gamma^\varphi_{u\varphi} = \partial_u(\ln G), \quad \Gamma^u_{\theta\theta} = \frac{1}{\sin^2(\theta)}\Gamma^u_{\varphi\varphi} = -\frac{2r\partial_v r}{G},$$

$$\Gamma^v_{vv} = \partial_v(\ln r), \quad \Gamma^\theta_{v\theta} = \Gamma^\varphi_{v\varphi} = \partial_v(\ln G), \quad \Gamma^v_{\theta\theta} = \frac{1}{\sin^2(\theta)}\Gamma^v_{\varphi\varphi} = -\frac{2r\partial_u r}{G},$$

$$\Gamma^\varphi_{\theta\varphi} = \frac{\cos(\theta)}{\sin(\theta)}, \quad \Gamma^\theta_{\varphi\varphi} = -\sin(\theta)\cos(\theta)$$

*Proof.* Since in a two-dimensional Lorentz manifold there exist for each point exactly two linearly independent, lightlike directions, the existence assertion follows immediately from Definition 7.1.1 and Corollary 2.4.2.

Let $\hat{u}$, $\hat{v}$ be coordinates and $\hat{G}(\hat{u},\hat{v})$, $\hat{r}(\hat{u},\hat{v})$ be coordinates with $g = G(\hat{u},\hat{v})\mathrm{d}\hat{u}\mathrm{d}\hat{v} + \hat{r}^2(\hat{u},\hat{v})\left(\mathrm{d}\theta^2 + \sin^2(\theta)\mathrm{d}\varphi^2\right)$. Since the warped product is invariantly defined we have $G(u,v)\mathrm{d}u\mathrm{d}v = G(\hat{u},\hat{v})\mathrm{d}\hat{u}\mathrm{d}\hat{v}$. At each point of a 2-dimensional Lorentzian manifold there are exactly two null directions whence we can assumme (without loss of generality[1]) that there exist functions $f_u$, $f_v$ with $\partial_{\hat{u}} = f_u\partial_u$ and $\partial_{\hat{v}} = f_v\partial_v$. Since the commutator of Gaußian vector fields vanishes we obtain

$$0 = [f_u\partial_u, f_v\partial_v] = f_u(\partial_u f_v)\partial_v - f_v(\partial_v f_u)\partial_u$$

and therefore $\partial_u f_v = \partial_v f_u = 0$. This implies $\hat{u} = \int f_u(u)\mathrm{d}u$ and $\hat{v} = \int f_v(v)\mathrm{d}u$.

It is straightforward to calculate the Christoffel symbols using the formula $\Gamma^a_{bc} = \frac{1}{2}g^{ad}(\partial_b g_{dc} + \partial_c g_{bd} - \partial_d g_{bc})$. ∎

*Remark 7.1.1.* The function $r\colon M \to \mathbb{R}$ gives the area of the orbits $S_x$ via the equation $\mathrm{Area}(S_x) = 4\pi r^2$ and is therefore invariantly defined. That $G$ is almost an invariant is one of the two main reasons why *double null coordinates* $(u,v)$ are a very practical choice. The other reason is that in these coordinates the causal structure of $(M,g)$ is explicitly described.

## 7.2 The Schwarzschild solution

In this section we will solve Einstein's equation for the case of a spherically symmetric vacuum spacetime. These solutions describe the gravitational field caused by a single non-rotating star which is situated in empty space. As the sun rotates rather slowly and space is almost empty, these solutions describe gravitation in the solar system very well.

[1] Otherwise we exchance $\hat{u}$ and $\hat{v}$.

**Theorem 7.2.1 (Birkhoff).** *Let $(M, g)$ be a spherically symmetric vacuum spacetime. Then either $r = \frac{1}{\Lambda}$ is constant and the mass function is given by $m = 1/(2\sqrt{\Lambda})$ or there is a constant $m_0$ and a dense, open subset $M^\circ \subset M$ such that each $x \in M^\circ$ admits a local coordinate neighbourhood with local coordinates $(t, r, \varphi, \theta)$ which satisfy*

$$g = -\left(1 - \frac{2m_0}{r} - \frac{r^2\Lambda}{3}\right) \mathrm{d}t^2 + \frac{\mathrm{d}r^2}{\left(1 - \frac{2m_0}{r} - \frac{r^2\Lambda}{3}\right)} + r^2 \left(\mathrm{d}\theta^2 + \sin^2(\theta)\mathrm{d}\varphi^2\right).$$

*Proof.* Lemma 7.1.3 implies

$$U \bullet m = U \bullet \left(\frac{\Lambda r^3}{6}\right) \text{ and } Q \bullet m = Q \bullet \left(\frac{\Lambda r^3}{6}\right).$$

Hence there is a aconstant $m_0$ such that $m = m_0 + \frac{\Lambda r^3}{6}$. If $(Q \bullet r)^2 = (U \bullet r)^2$ in an open set then from the definition of $m$ we get $\frac{r}{2} = m_0 + \frac{\Lambda r^3}{6}$ which in turn implies that $r = \text{const}$. Since $T = 0$ we can choose coordinates which simultaneously diagonalize $g$ and $T$. Equation (7.1.3) implies then $r^2 = 1/\Lambda$ and we obtain $m = r/2 = 1/(2\sqrt{\Lambda})$.

Let us now assume that $g^\sharp(\mathrm{d}r, \mathrm{d}r) = -(U \bullet r)^2 + (Q \bullet r)^2 \neq 0$. There are orthogonal coordinates $(\tilde{q}, \tilde{t})$ such that $r = \tilde{q}$. Since $T = 0$ this choice of coordinates trivially diagonalizes $g$ and $T$ simultaneously and we can assume without loss of generality that in Lemma 7.1.2 we have $t = \tilde{t}$, $q = \tilde{q}$. We immediately obtain $U \bullet r = \mathrm{e}^{-\nu}\partial_t q = 0$. Hence Equation (7.1.4) yields $U \bullet \lambda = 0$ and Equation (7.1.3) implies

$$Q \bullet Q \bullet r = \frac{m}{r^2} - \frac{\Lambda r}{2} = \frac{m_0}{r^2} - \frac{\Lambda r}{3}.$$

Since $Q = \mathrm{e}^{-\lambda}\partial_r$ this equation is equivalent to

$$\mathrm{e}^{-2\lambda}\partial_r\lambda = -\frac{m_0}{r^2} + \frac{\Lambda r}{3}$$

and can be integrated to give

$$\mathrm{e}^{-2\lambda} = A(t) - \frac{2m_0}{r} - \frac{r^2\Lambda}{3} = A(t) - \frac{2m}{r},$$

where $A(t)$ is an integration constant. Equation (7.1.5) implies $\mathrm{e}^{-2\lambda}\partial_r\nu = \frac{m_0}{r^2} - \frac{\Lambda r}{3}$ and therefore $\mathrm{e}^{-2\lambda}\partial_r\nu = -\mathrm{e}^{-2\lambda}\partial_r\lambda$ which in turn yields $\nu = B(t) - \lambda$. After a re-parametrization of $t$ we can choose $B = 0$. $m = \frac{r}{2}(1 - (Q \bullet r)^2) = \frac{r}{2}(1 - \mathrm{e}^{-2\lambda}) = \frac{r}{2}(1 - A + 2m/r)$ implies $A(t) = 1$ and the assertion is proved. ∎

Observe that any spherically symmetric vaccum spacetime is automatically static in the region $g^\sharp(\mathrm{d}r, \mathrm{d}r) > 0$. In the region $g^\sharp(\mathrm{d}r, \mathrm{d}r) < 0$ it is not static but has a fourth spacelike Killing vector field.

This spacetime has (for $\Lambda = 0$) first been obtained by Schwarzschild (1916) who solved the static, spherically symmetric vacuum equation. Birkhoff then showed that staticity was not needed as an assumption.

**Definition 7.2.1.** *A spherically symmetric vacuum spacetime with vanishing cosmological constant $(M, g)$ is called a* Schwarzschild spacetime[2]. *The coordinates $(t, r, \varphi, \theta)$ are called* Schwarzschild coordinates.

In the rest of this section we will assume $\Lambda = 0$.

The regions $r < 2m$ and $r > 2m$ cannot be matched naively using these coordinates (cf. Fig. 7.2.1 which represents the causal structure), and for some time it has been believed that there is a physical singularity at $r = 2m$. Below we will geometrically determine more useful double null coordinates of the solution. This will show that $r = 2m$ is a spurious singularity and that there exists a unique, inextensible solution of the spherically symmetric vacuum equation.

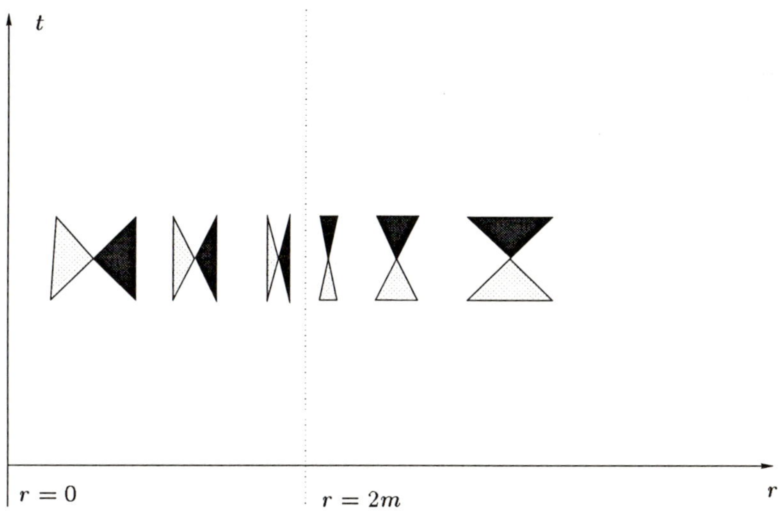


**Fig. 7.2.1.** Schwarzschild spacetime in Schwarzschild coordinates

**Proposition 7.2.1.** *With*

$$f\colon \mathbb{R}^+ \setminus \{0\} \to (-\infty, 1), \qquad r \mapsto f(r) = -\frac{r}{2m}\left(1 - \frac{2m}{r}\right) e^{r/(2m)},$$

$$F\colon \mathbb{R}^+ \setminus \{0\} \to \mathbb{R}^+ \setminus \{0\} \qquad r \mapsto F(r) = \frac{32m^3 e^{-r/(2m)}}{r},$$

*let* $B_{\mathrm{schw}} = \{(X,Y) \in \mathbb{R}^2 | \, XY < 1\}$ *and* $g_{B_{\mathrm{schw}}} = F \circ f^{-1}(XY)\, \mathrm{d}X\mathrm{d}Y$.

[2] In the literature this name is usually reserved for a subset of the maximally extended Schwarzschild spacetime, the shaded region in Fig. 7.2.2.

*Then the Lorentzian manifold*

$$\left(\mathbb{R}^2 \times S^2, g_{B_{\text{schw}}} + (f^{-1}(XY))^2\left(\mathrm{d}\theta^2 + \sin^2(\theta)\mathrm{d}\varphi^2\right)\right)$$

*satisfies* Ric $= 0$. *The coordinates* $(t, r)$ *of Theorem 7.2.1 and* $(X, Y)$ *are related through*

$$\ln\left(\frac{-X}{Y}\right) = \frac{t}{2m}, \qquad XY = f(r).$$

*Proof.* We can restrict to the base manifold $B_{\text{schw}}$. For any null vector field $N$ which is orthogonal to the the spheres of symmetry the equations $-(1-2m/r)(N^t)^2+(1-2m/r)^{-1}(N^r)^2 = 0$ holds, whence we have $N^t = \pm(1-2m/r)^{-1}N^r$. Double null coordinates can now be obtained through an integration of these two vector fields. Since $\int_0^r (1-2m/r)^{-1}\mathrm{d}r = r + 2m\ln\left(\frac{r}{2m} - 1\right)$ we define our coordinates by

$$\hat{X} = t + \left(r + 2m\ln\left(\frac{r}{2m} - 1\right)\right), \qquad \hat{Y} = t - \left(r + 2m\ln\left(\frac{r}{2m} - 1\right)\right).$$

This gives $\mathrm{d}\hat{X}\mathrm{d}\hat{Y} = (\mathrm{d}t + (1-2m/r)^{-1}\mathrm{d}r)(\mathrm{d}t - (1-2m/r)^{-1}\mathrm{d}r) = \mathrm{d}t^2 - (1-\frac{2m}{r})^{-2}\mathrm{d}r^2$ and therfore $g_{B_{\text{schw}}} = (1-\frac{2m}{r})\mathrm{d}\hat{X}\mathrm{d}\hat{Y}$. From $\hat{X} - \hat{Y} = 2(r + 2m\ln(\frac{r}{2m} - 1))$ we obtain

$$\mathrm{e}^{(\hat{X}-\hat{Y})/(4m)} = \mathrm{e}^{r/(2m)}\left(\frac{r}{2m} - 1\right) = \frac{r}{2m}\mathrm{e}^{r/(2m)}\left(1 - \frac{2m}{r}\right) = -f(r)$$

which implies

$$g_{B_{\text{schw}}} = \left(1 - \frac{2m}{r}\right)\mathrm{d}\hat{X}\mathrm{d}\hat{Y} = \frac{2m}{r}\mathrm{e}^{-r/(2m)}\mathrm{e}^{(\hat{X}-\hat{Y})/(4m)}\mathrm{d}\hat{X}\mathrm{d}\hat{Y}.$$

We set $X = \mathrm{e}^{-\hat{X}/(4m)}$, $Y = -\mathrm{e}^{\hat{Y}/(4m)}$ and finally obtain

$$g_{B_{\text{schw}}} = \frac{32m^3\mathrm{e}^{-r/(2m)}}{r}\mathrm{d}X\mathrm{d}Y, \qquad f(r) = XY.$$

Furthermore, $\partial_X$ and $\partial_Y$ are both future oriented (this has been the reason for choosing the minus sign in the coordinate transformation for $Y$). It remains to show that the inverse of $f$ exists for all $r > 0$. But this follows immediately from $f'(r) = -\frac{r}{4m^2}\mathrm{e}^{\frac{r}{2m}} < 0$. ∎

The coordinates provided by Proposition 7.2.1 are called *Kruskal-Szekeres-coordinates* and the corresponding spacetime is often called *Kruskal-Szekeres-spacetime*. This spacetime is locally isometric to the metric given in Theorem 7.2.1 but the global structure is different from the global structure obtained by using Schwarzschild coordinates (cf. Fig. 7.2.2). Nevertheless, in this book we will refer to the inextensible spacetime given in Proposition 7.2.1 as *Schwarzschild spacetime*.

*Remark 7.2.1.* The motivation for the Schwarzschild spacetime is to describe the exterior of a non-rotating star. During the lifetime of the star

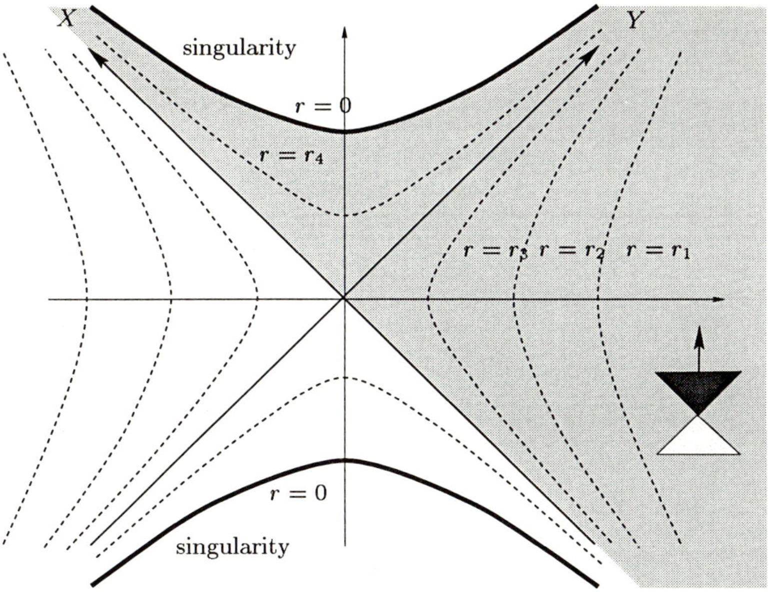


**Fig. 7.2.2.** Schwarzschild spacetime. Radial null geodesics are the straight lines $X$ = const and $Y$ = const. The region covered by Schwarzschild coordinates is shaded

the radius may change (typically, it may shrink and perhaps even reach 0). If we denote the $r$-component of the star at $t$ by $r_{\mathrm{star}}(t)$ then we will need for each $t$ only the part $r > r_{\mathrm{star}}(t)$ of the shaded region in Fig. 7.2.2. The white region can be completely discarded for physical purposes.

We will now investigate this solution in more detail and get a first glimpse at what is known as a *black hole*: In Proposition 7.2.3 below we will show that this spacetime is singular but inextensible. But first we need to calculate its geodesics.

**Lemma 7.2.1.** *Let $(M, g)$ be a pseudo-Riemannian manifold and $(x^1, \dots, x^n)$ be coordinates such that $g_{ab}$ is diagonal,*

$$g_{ab}\mathrm{d}x^a\mathrm{d}x^b = \sum_{a=1}^{n} g_a \mathrm{d}x^a\mathrm{d}x^a.$$

*Then the geodesics $s \mapsto \gamma(s)$ of $(M, g)$ are given by*

$$\frac{\mathrm{d}}{\mathrm{d}s}\left(g_a \circ \gamma(s)\, \dot\gamma^a(s)\right) = \frac{1}{2}\sum_{b=1}^{n} \partial_a g_b \circ \gamma(s) \left(\dot\gamma^b(s)\right)^2 \text{ (no summation over } a\text{).}$$

*Proof.* We suspend the summation convention if the repeated index is $a$. Then we have $g_{ab} = g_a\delta^a_b$ and obtain

$$\left\langle \nabla_{\dot{\gamma}}\dot{\gamma}, \partial_a \right\rangle = \nabla_{\dot{\gamma}} \langle \dot{\gamma}, \partial_a \rangle - \left\langle \dot{\gamma}, \nabla_{\dot{\gamma}}\partial_a \right\rangle = \frac{\mathrm{d}}{\mathrm{d}s}(g_a \dot{\gamma}^a) - \dot{\gamma}^b \dot{\gamma}^c \left\langle \partial_b, \nabla_{\partial_c}\partial_a \right\rangle$$
$$= \frac{\mathrm{d}}{\mathrm{d}s}(g_a \dot{\gamma}^a) - \dot{\gamma}^b \dot{\gamma}^c \frac{1}{2}\left(\partial_c g_a \delta^a_b + \partial_a g_{bc} - \partial_b g_a \delta^a_c\right)$$
$$= \frac{\mathrm{d}}{\mathrm{d}s}(g_a \dot{\gamma}^a) - \frac{1}{2}\dot{\gamma}^b \dot{\gamma}^c \partial_a g_{bc} = \frac{\mathrm{d}}{\mathrm{d}s}(g_a \dot{\gamma}^a) - \frac{1}{2}\left(\dot{\gamma}^b\right)^2 \partial_a g_b .$$

∎

**Proposition 7.2.2.** *Let $s \mapsto \tilde{\gamma}(s)$ be a geodesic in Schwarzschild space-time*

$$\left(\mathbb{R}^2 \times S^2, g_{B_{\mathrm{schw}}} + f^{-2}(XY)\left(\mathrm{d}\theta^2 + \sin^2(\theta)\mathrm{d}\varphi^2\right)\right)$$

*with $\langle \dot{\tilde{\gamma}}, \dot{\tilde{\gamma}} \rangle = \eta \in \{-1, 0, 1\}$ and assume that $\tilde{\gamma}(0) \notin \{x | \, r(x) = 2m\}$. Then there exists a rotational isometry $\phi$ such that $\gamma = \phi \circ \tilde{\gamma}$ is given by*

$$\left(1 - \frac{2m}{r}\right)\frac{\mathrm{d}t}{\mathrm{d}s} = E, \qquad r^2 \frac{\mathrm{d}\varphi}{\mathrm{d}s} = L, \qquad \theta = \frac{\pi}{2}$$
$$E^2 = \left(\frac{\mathrm{d}r}{\mathrm{d}s}\right)^2 + \left(1 - \frac{2m}{r}\right)\left(-\eta + L^2/r^2\right),$$

*where $E, L$ are constants.*

*Proof.* We will use the coordinates provided by Theorem 7.2.1. Since the metric is diagonal we can apply Lemma 7.2.1 and obtain for the $(t, \theta, \varphi)$-components of $\tilde{\gamma}$

$$\frac{\mathrm{d}}{\mathrm{d}s}\left(\left(1 - \frac{2m}{r}\right)\frac{\mathrm{d}t}{\mathrm{d}s}\right) = 0,$$
$$\frac{\mathrm{d}}{\mathrm{d}s}\left(r^2 \sin^2(\theta)\frac{\mathrm{d}\varphi}{\mathrm{d}s}\right) = 0,$$
$$\frac{\mathrm{d}}{\mathrm{d}s}\left(r^2 \frac{\mathrm{d}\theta}{\mathrm{d}s}\right) = r^2 \sin(\theta)\cos(\theta)\frac{\mathrm{d}\varphi}{\mathrm{d}s}.$$

There is an rotational isometry $\phi\colon x \mapsto \tilde{x}$ such that $\theta \circ \phi \circ \tilde{\gamma}(0) = \pi/2$ and $\mathrm{d}\theta(\phi_* \dot{\tilde{\gamma}}(0)) = 0$. Then $\theta \circ \gamma = \pi/2$ is the unique solution of the last equation. The first two equations can immediately be integrated. To derive the fourth equation in the assertion of the proposition it is more convenient to use the conservation property $\langle \dot{\gamma}, \dot{\gamma} \rangle = \eta$ than to use the $r$-component of the geodesic equation in Lemma 7.2.1. In fact, it follows directly from

$$\eta = -\left(1 - \frac{2m}{r}\right)\left(\frac{\mathrm{d}t}{\mathrm{d}s}\right)^2 + \left(1 - \frac{2m}{r}\right)^{-1}\left(\frac{\mathrm{d}r}{\mathrm{d}s}\right)^2 + r^2\left(\frac{\mathrm{d}\varphi}{\mathrm{d}s}\right)^2$$

after inserting the equations for $\mathrm{d}t/\mathrm{d}s$ and $\mathrm{d}\varphi/\mathrm{d}s$. ∎

**Lemma 7.2.2.** *Let $(M, g)$ be a spacetime which is locally extensible. Then there is a null geodesic in $(M, g)$ which is incomplete and extensible.*

*Proof.* Let $(\bar{M}, \bar{g})$ be a local extension of $(M, g)$, $x \in \bar{M} \setminus M$, and $y \in M$. Then there is a (not necessarily future or time oriented) broken null geodesic from $x$ to $y$. This broken geodesic $\gamma$ must intersect $\partial M \subset \bar{M}$ in a point $z = \gamma(0)$. Without loss of generality we can assume that $\gamma(t) \in M$ for $t < 0$, $|t|$ sufficiently small. Since $z \notin M$ this geodesic is incomplete and extensible. ∎

**Proposition 7.2.3.** *Schwarzschild spacetime is inextensible and geodesically incomplete. A future directed null geodesic is incomplete if and only if it enters the region $\{1 - 2m/r < 0\}$. It then approaches $r = 0$ and the Kretschmann scalar given by $R^{abcd}R_{abcd} = 48m^2/r^6$ diverges along this curve.*

*Proof.* Lemma 4.4.14 implies $R_{r\theta r\theta} = \frac{1}{\sin^2(\theta)} R_{r\varphi r\varphi} = \frac{m}{2m-r}$, $R_{rtrt} = -2\frac{m}{r^3}$, $R_{\theta\varphi\theta\varphi} = 2rm\sin^2(\theta)$, $R_{\theta t\theta t} = \frac{1}{\sin^2(\theta)} R_{\varphi t\varphi t} = -\frac{m}{r^2}(2m - r)$, and $R_{abcd} = 0$ for all other components which are not related to these components by the general symmetries of the Riemann tensor. It follows that $R^{abcd}R_{abcd} = 48m^2/r^6$ and therefore that any curve $\gamma(s)$ with $r \circ \gamma(s) \to 0$ is inextensible. A curve $\gamma = (\gamma_{B_{\text{schw}}}, \gamma_{S^2})$ in $B_{\text{schw}} \times S^2$ is extensible if and only if $\gamma_{B_{\text{schw}}}$ is extensible in $B_{\text{schw}}$ and $\gamma_{S^2}$ is extensible in $S^2$. By Lemma 7.2.2 we only have to study null geodesics in order to prove that $(M, g)$ is inextensible. If $\gamma_{B_{\text{schw}}}$ is extensible then $\mathrm{d}r/\mathrm{d}s$ is bounded by Proposition 7.2.2 and $r \not\to 0$. By Corollary 4.4.1 $\gamma_{S^2}$ is a pregeodesic with bounded acceleration in a compact manifold and therefore also extensible. Hence we can restrict to $(B_{\text{schw}}, g_{B_{\text{schw}}})$ and study null geodesics in this 2-dimensionl spacetime. In Kruskal coordinates $(X, Y)$ these geodesics are given by $X = \text{const}$ or $Y = \text{const}$. Because of the reflection iometries $(X, Y) \mapsto (Y, X)$ and $(X, Y) \mapsto (-X, -Y)$ we only need to consider future directed geodesics of the form $X = \text{const}$, $Y > 0$. The region $Y > 0$ is the disjoint union of three different subsets,

$$\text{(i)}\quad r/2m > 1, \qquad \text{(ii)}\quad r/2m = 1, \qquad \text{(iii)}\quad r/2m < 1,$$

each of them being invariant under future directed null geodesics $X = \text{const}$, $Y > 0$.

We have to estimate the affine parameter of our null geodesics. If $\gamma$ is a null geodesic given by $X = \text{const}$, $Y > 0$ then there is a function $Y \mapsto h(Y)$ with $\dot{\gamma} = h(Y)\partial_Y$ and $\nabla_{h(Y)\partial_Y}(h(Y)\partial_Y) = h(Y)(h'(Y)\partial_Y + h(Y)\nabla_{\partial_Y}\partial_Y) = h(Y)(h'(Y) + h(Y)\Gamma^Y_{YY})\partial_Y$. From $\Gamma^Y_{YY} = \partial_Y \ln(g_{XY})$, we obtain therefore $h(Y) = c\,(g_{XY})^{-1}$, where $c$ is a constant.

In region (ii) $g_{XY}$ is constant which implies that $\gamma$ satisfies $\frac{\mathrm{d}}{\mathrm{d}s} Y \circ \gamma(s) = h(Y) = \text{const}$ and is therefore future complete.

Now consider regions (i), (iii). Since

$$\mathrm{d}r/\mathrm{d}s = \sqrt{E^2 - L^2/r^2\,(1-2m/r)} \to |E| > 0$$

for $r \to \infty$ the parameter $s$ diverges if and only if $r$ diverges.

In case (i) we have $(1-2m/r) < 0$ and the square root is well defined for all $s$. The equation

$$\mathrm{d}Y/\mathrm{d}r = h(y)\,\mathrm{d}s/\mathrm{d}r = -\frac{r}{32m^3}\mathrm{e}^{r/(2m)}\left(E^2 - L^2/r^2\,(1-2m/r)\right)^{-1/2}$$

implies that $r$ diverges if $Y$ diverges[3]. Hence $s$ diverges for $Y \to \infty$ and the geodesic $\gamma$ must be future complete.

In case (iii) it is clear from $XY = f(r) = -\frac{r}{2m}\left(1-\frac{2m}{r}\right)\mathrm{e}^{r/(2m)}$ that our future directed null geodesics $X = \text{const}$, $Y > 0$ are approaching $r = 0$ and are therefore inextensible and incomplete. ∎

The region $r < 2m$, $X > 0$ is the simplest model of a *black hole*. A black hole is loosely characterised by the fact that a light ray which enters it cannot leave it any more but instead reaches the edge of the universe before the affine parameter of the corresponding null geodesic has reached the value $\infty$.[4] Since a black hole does not emit a single light ray one is tempted to say that it is black, whence the name coined by J. A. Wheeler. However, this name is slightly misleading, since the black hole is not in the past of any observer who is situated outside this region. Rather than appearing black it is simply invisible.

An observer who enters the region does not have a very low life expectency. The longest timelike curve within the black hole region is given by $X = Y$, $X \in [0,1]$. In Schwarzschild coordinates this corresponds to the path $t = 0$, $r \in (0, 2m)$. Hence the observer's life is bounded by

$$\Delta s = \int_{2m}^{0} \sqrt{-g_{B_{\mathrm{schw}}}(\partial_r, \partial_r)}\,\mathrm{d}r = \int_{2m}^{0} \sqrt{2m/r - 1}^{\,-1}\,\mathrm{d}r = \pi m.$$

### 7.2.1 Experimental tests for the Schwarzschild solution

In this section we will investigate the region $2m/r < 1$ which may be considered as the exterior of a non-rotating, spherically symmetric star

---

[3] This property could also have been seen geometrically: The lines $X = \text{const} < 0$, $Y > 0$ intersect all the hyperbolas $r = \text{const} > 2m$.

[4] A widely accepted general definition of black holes does not exist. The definition we have just given has the disadvantage that any Robertson-Walker solution which satisfies the assumptions of Theorem 6.4.1 and $\varepsilon = 1$ is a giant black hole. In this special case one would have to replace the condition that the null geodesics in the black hole don't reach the affine parameter $\infty$ by the condition that they don't end in the cosmological future singularity given by $t = t_+$.

of mass $m$. The discussion applies in particuar to the gravitational field produced by the sun which was Schwarzschild's motivation for solving Einstein's equation in this special case.

Since the exterior region contains the timelike Killing field $\partial_t$ orthogonal to the spheres of symmetry it admits a natural infinitesimal split $\mathbb{R}\partial_t \oplus (\partial_t)^\perp$ of spacetime into space and time. Moreover, the distribution $(\partial_t)^\perp$ is integrable whence we obtain geometrically defined hypersurfaces of constant time. These hypersurfaces are given by

$$\Sigma_{t_0} = \{(t, r, \theta, \varphi) \,|\, t = t_0, r > 2m\}.$$

Since $\partial_t$ is a Killing vector field the pullback of the metric to $\Sigma_t$ does not depend on $t$ and we can identify all $\Sigma_t$ through projection along the $t$ coordinate. A timelike curve in spacetime corresponds to a curve in the Riemannian manifold $(\Sigma_0, (1-2m/r)^{-1}\mathrm{d}r^2 + r^2(\mathrm{d}\theta^2 + \sin^2(\theta)\mathrm{d}\varphi^2))$ which represents space.

In our case, we may imagine the non-rotating star to be the sun with radius $r_{\mathsf{sun}} > 2m$. It is located in the centre $r = 0$ of the coordinate system but the Schwarzschild solution is of course only valid for $r > r_{\mathsf{sun}} > 2m$.[5] It follows that the region $r < 2m$ can be excluded from our discussion and we can utilize the spacetime split introduced above. It is natural to identify this spacetime split with the infinitesimal splits defined by our own world lines. "Space" has then its intuitive meaning. While in general timelike geodesics represent freely falling particles, in our case they should be interpreted as planets (or perhaps asteroids and satellites).

Because of Proposition 7.2.2 we can assume that the movement of a single planet or light ray is contained in the plane $\theta = \pi/2$.

**Lemma 7.2.3.** *Let $\gamma(s) = (t(s), r(s), \theta(s), \varphi(s))$ be a geodesic. Then we have*

$$\left(\frac{1}{r^2}\frac{\mathrm{d}r}{\mathrm{d}\varphi}\right)^2 + \left(\frac{-\eta}{L^2} + \frac{1}{r^2}\right)\left(1 - \frac{2m}{r}\right) = \frac{E^2}{L^2}.$$

*Proof.* This follows by dividing the equations for $\mathrm{d}r/\mathrm{d}s$ and $\mathrm{d}\varphi/\mathrm{d}s$ in Proposition 7.2.2. ∎

**Corollary 7.2.1.** *Let $\gamma(s) = (t(s), r(s), \theta(s), \varphi(s))$ be a geodesic. Then $\varrho(s) = 1/r(s)$ satisfies*

$$\frac{\mathrm{d}^2\varrho}{\mathrm{d}\varphi^2} + \varrho = \frac{-\eta m}{L^2} + 3m\varrho^2.$$

---

[5] Hence it does not matter that the Schwarzschild metric is not defined at the centre $r = 0$ where the sun is located.

*Proof.* Substituting $\varrho(s) = 1/r(s)$ in Lemma 7.2.3 gives $(\mathrm{d}\varrho/\mathrm{d}\varphi)^2 + (-\eta/L^2 + \varrho^2)(1 - 2m\varrho) = E^2/L^2$. Differentiating this equation implies the assertion. ∎

**Bending of light rays.** Since (null) geodesics are influenced by curvature, according to general relativity, light rays should appear bent near regions where gravity is large. In particular a light ray passing the sun at a short distance should appear to be slightly bent. The experimental verification of this effect was one of the first tests of the theory.

To describe this effect we need to determine the angle $\alpha$ under which a central object appears to an observer in Schwarzschild spacetime. This angle can then be compared with the corresponding angle determined by the background metric $\mathrm{d}r^2 + r^2\mathrm{d}\Omega^2$ of space (cf. Fig. 7.2.3)

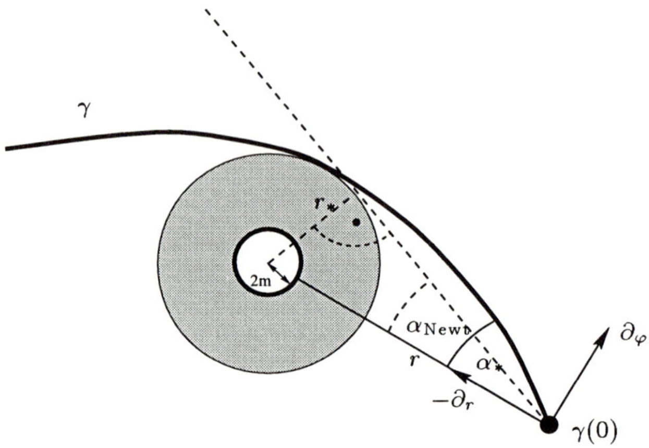


**Fig. 7.2.3.** The size of a central star in Schwarzschild spacetime

**Lemma 7.2.4.** *Let $\gamma$ be a lightlike geodesic and $\gamma(0) = (t_0, r_0, \theta_0, \varphi_0)$. Then $\alpha = \angle(-\partial_r, \dot{\gamma}(0))$ satisfies $r_0 \sin(\alpha) = (|L|/E)\sqrt{1 - 2m/r}$.*

*Proof.* Since $\dot{\gamma}(s) = \mathrm{d}t/ds\partial_t + \dot{\vec{\gamma}}$ the null condition $-(1-2m/r)|\mathrm{d}t/\mathrm{d}s|^2 + |\dot{\vec{\gamma}}|^2 = 0$ implies $|\dot{\vec{\gamma}}| = |\sqrt{1-2m/r}\mathrm{d}t/\mathrm{d}s| = (1 - 2m/r)^{-1/2}E$. Since $\partial_\varphi \perp \partial_r$ the angle $\alpha$ is given by $\sin(\alpha) = |\mathrm{d}\varphi \mathrm{d}s \partial_\varphi|/|\dot{\vec{\gamma}}(s)| = |L/r|/((1 - 2m/r)^{-1/2}E)$. ∎

**Corollary 7.2.2.** *Let $\gamma$ be a lightlike geodesic with past endpoint $\gamma(0) = (t_0, r_0, \theta_0, \varphi_0)$. If $\gamma$) passes the boundary of a centred star of radius $r_* > 3m$, then the angle $\alpha_*$ defined in Fig. 7.2.3 satisfies*

$$\sin(\alpha_*) = \frac{r_*}{r_0}\sqrt{\frac{1 - 2m/r_0}{1 - 2m/r_*}}.$$

*Proof.* Assume that $\gamma$ passes the boundary of the star at $s = s_*$. Then $(\mathrm{d}r/\mathrm{d}s)_{s_*} = 0$ since $r$ has a minimum there. Proposition 7.2.2 implies

$$r_* E/|L| = \sqrt{1 - 2m/r_*}$$

and the assertion follows by inserting this equation into Lemma 7.2.4. ∎

Observe that the angle $\alpha_*$ of the star is larger than one would expect in non-relativistic physics where $\sin(\alpha_*)$ would just be given by $r_*/r_0$. This effect has been verified by a British team lead by *Arthur Eddington* (1882–1944) which measured the bending of light rays close to the sun during the total eclipse in 1919. They used the limiting behaviour given below.

**Proposition 7.2.4.** *Let $\gamma$ be an inextensible null geodesic which does not enter the region $2m/r > 1$. Then there is a minimal radius $r_0 = \min\{r \circ \gamma(s) \big| \, s \in \mathbb{R}\}$ along $\gamma$. Furthermore, there are two lines with respect to the flat metric $g_{\text{flat}} = \mathrm{d}r^2 + r^2 \mathrm{d}\Omega^2$ which are asymptotes of $\gamma$ and intersect at an angle $\Delta = 4m/r_0 + o(\frac{m}{r_0})$*

*Proof.* The minimal value $r_0 = r \circ \gamma(s_0)$ exists since $\gamma$ is inextensible and $r \circ \gamma > 2m$ by assumption. We may choose our spherical coordinates $\theta, \varphi$ so that the the light ray lies in the plane $\theta = \pi/2$ and the equation $\varphi \circ \gamma(s_0) = 0$ holds. Proposition 7.2.2 implies that for $s \to \pm\infty$ the coordinate $\varphi$ converges to limits $\varphi_\pm$. The lines (with respect to $g_{\text{flat}}$) which pass through the origin under these angles $\varphi_\pm$ are therefore parallel to asymptotes of $\gamma$. The differential equation provided by Corollary 7.2.1 can be solved exactly and has the solution

$$\varphi(\varrho) = \int_{1/r_0}^{\varrho} - \frac{1}{\sqrt{-\hat{\varrho}^2 + 2\, m_0 \hat{\varrho}^3 + (r_0)^{-2} - 2\, m_0 (r_0)^{-3}}} d\hat{\varrho}$$

We are interested in situations where the ratio $m/r_0$ is small. Since the angle $\Delta = 2 \lim_{\varrho \to 0} \varphi(\varrho)$ vanishes when $m = 0$ we will linearize $\varphi(\varrho)$ with respect to the parameter $x = m/r_0$ and then take the limit $\varrho \to 0$. Differentiating the function $m/r_0 \mapsto \varphi$ gives

$$\frac{\partial \varphi}{\partial (m_0/r_0)} = r_0 \int_{1/r_0}^{\varrho} \frac{\hat{\varrho}^3 - {r_0}^3}{(-\hat{\varrho}^2 + 2\, m_0 \hat{\varrho}^3 + (r_0)^{-2} - 2\, m_0 (r_0)^{-3})^{3/2}} d\hat{\varrho}$$

and therefore

$$\left( \frac{\partial \varphi}{\partial (m_0/r_0)} \right)_{\varrho = 0, m_0/r_0 = 0} = r_0 \int_{1/r_0}^{0} \frac{\hat{\varrho}^3 - {r_0}^{-3}}{((r_0)^{-2} - \hat{\varrho}^2 +)^{3/2}} d\hat{\varrho} = 2$$

Hence we have $\Delta = 4m_0/r_0 + o(m_0/r_0)$. ∎

**The perihelion precession of Mercury.** Mercury moves around the sun decribing an orbit which is similar to an ellipse but not closed.[6] Already in the 19th century one has measured the angle between consecutive local minima of the distance between Mercury and the sun and has tried to explain this angle within the Newtonian theory of gravity.[7] While such a "precession" occurs if one takes into account the the gravititional fields caused by the other planets, this does not give a quantitative explanation of the measured value. The first outstanding success of general relativity was Einstein's demonstration that his theory could explain this discrepancy.[8]

In order to calculate the "missing angle" we have to compare the Newtonian solution of the two-body problem (the sun, Mercury) with timelike geodesics (Mercury) in the Schwarzschild solution which describes the sun.

In Newtonian gravitation, a particle in the gravitational field of a spherically symmetric star moves according to the ordinary differential eqation

$$\frac{\mathrm{d}^2\vec{\gamma}(s)}{\mathrm{d}s^2} = -\frac{m}{\|\vec{\gamma}(s)\|^3}\vec{\gamma}(s). \tag{7.2.10}$$

**Lemma 7.2.5.** *Let $m > 0$ and $\vec{\gamma}\colon \mathbb{R} \to \mathbb{R}^3$ be a solution of the differential equation (7.2.10). Then the curve $\vec{\gamma}$ is contained in a plane and Equation (7.2.10) is equivalent to*

$$\frac{\mathrm{d}\varphi \circ \vec{\gamma}(s)}{\mathrm{d}s} = L\varrho^2, \qquad \frac{\mathrm{d}^2\varrho_N}{\mathrm{d}\varphi^2} + \varrho_N = m/L^2. \tag{7.2.11}$$

*where $(1/\varrho_N, \varphi)$ are polar coordinates of this plane and $L$ is a constant.*

*Proof.* Equation (7.2.10) implies that $\vec{\gamma}(s) \times \frac{\mathrm{d}}{\mathrm{d}s}\vec{\gamma}(s)$ ("$\times$" being the vector cross product in $\mathbb{R}^3$) is constant with respect to $s$. Hence $\vec{\gamma}$ is contained in the plane spanned by $\frac{\mathrm{d}}{\mathrm{d}s}\vec{\gamma}(0)$ and $\vec{\gamma}(0)$. If $(r, \varphi)$ are polar coordinates of this plane ($x^1 = r\cos\varphi, x^2 = r\sin\varphi$), Equation (7.2.10) is equivalent to

$$\frac{\mathrm{d}^2 r(s)}{\mathrm{d}s^2} - r(s)\left(\frac{\mathrm{d}\varphi(s)}{\mathrm{d}s}\right)^2 = -\frac{m}{r^2(s)}, \qquad r(s)\frac{\mathrm{d}^2\varphi(s)}{\mathrm{d}s^2} + 2\frac{\mathrm{d}r(s)}{\mathrm{d}s}\frac{\mathrm{d}\varphi(s)}{\mathrm{d}s} = 0.$$

---

[6] This is actually true for all planets, but in the case of Mercury the effect is especially pronounced.

[7] In order to do so, Astronomers have assumed the existence of a further planet. However, this planet has never been seen.

[8] He did this using the equation $\mathrm{Ric} = 8\pi T$ before he arrived at his final theory with $\mathrm{Ric} - \mathrm{Scal}/2\, g = 8\pi T$. This was possible since for these calculations only the vacuum equation is needed.

The second equation implies that $r^2 \frac{d\varphi(s)}{ds} = L$ is constant. Setting $\varrho_N = 1/r$, the first equation is therefore equivalent to $d^2\varrho_N/d\varphi^2 + \varrho_N = m/L^2$. ∎

Equation (7.2.11) is an inhomogenous linear differential equation with constant coefficients. It is easy to see that (in the generic case $L \neq 0$) there exist constants $c_1, c_2$ such that $\varrho_N(\varphi) = \frac{m}{L^2} + c_1 \sin(\varphi) + c_2 \cos(\varphi)$. Our polar coordinates are only fixed up to a rotation in the plane. Hence we can assume without loss of generality that there is a constant $e > 0$ such that

$$\varrho_N(\varphi) = \frac{m}{L^2}(1 + e\cos(\varphi)).$$

This solution is periodic. We could now attempt solve the corresponding equation in the Schwarzschild solution, and to calculate the difference of angle $\varphi$ (modulo $2\pi$) between two consecutive minima of the coordinate radius as a Taylor polynomial in $m_0/r_0$. However, it is difficult to use this strategy in practice because it would involve integrals which are quite complicated.

We will therefore employ a different method and obtain approximate solutions from approximate differential equations. Observe that Equation (7.2.11) is the Newtonian analogue to Corollary 7.2.1 and that both equations differ only by the quadratic term $m\varrho^2 = m/r^2$ which is very small. The idea is now to view the Newtonian solution as an approximation to the relativistic equation. Inserting the Newtonian solution into the quadratic term gives a third equation

$$\frac{d^2\varrho_{\text{approx}}}{d\varphi^2} + \varrho_{\text{approx}} = \frac{m}{L^2} + 3m\left(\frac{m}{L^2}(1 + e\cos(\varphi))\right)^2$$

which is also a linear inhomogeneous differential equation with constant coefficients. It appears to be a better approximation than the first differential equation since the term 0 has been replaced by the term $3m(\varrho_N)^2$ which should be a better approximation for $3m\varrho^2$. While this argument is only heuristic a real justification appears to be too complicated to be worthwile in our context.

This third equation gives

$$\varrho_{\text{approx}} = \frac{m}{L^2}(1 + e\cos(\varphi)) + \frac{3m^3}{L^4}\left(1 + \frac{e^2}{2} - \frac{e^2}{6}\cos(2\varphi) + e\varphi\sin(\varphi)\right).$$

To calculate the angle at the perihelion we have to calculate the minima of the function $\varrho(\varphi)$. The equation

$$\frac{d\varrho_{\text{approx}}}{d\varphi} = -\frac{me}{L^2}\sin(\varphi) + \frac{3m^3 e}{L^4}\left(\frac{e}{3}\sin(2\varphi) + \sin(\varphi) + \varphi\cos(\varphi)\right).$$

gives that $\varrho_{\text{approx}}$ has a perihelion at $\varphi_0 = 0$ — as was to be expected. A comparison with $\varrho_N$ indicates that the next perihelion should be at

$\varphi_1 = 2\pi + \delta$ where $\delta$ is small. Hence we can neglect $\frac{e}{3}\sin(2\delta) + \sin(\delta)$ with respect to $(2\pi + \delta)\cos(2\pi + \delta)$ and obtain

$$0 \approx -\frac{me}{L^2}\sin(\delta) + \frac{3m^3 e}{L^2}(2\pi + \delta)\cos(\delta).$$

With $\tan(\delta) \approx \delta$ and neglecting $\delta$ with respect to $2\pi$, this equation implies

$$\delta \approx \frac{6\pi m^2}{L^2},$$

which gives a correction to the Newtonian value in very good agreement with observation.

## 7.3 Quasi-linear hyperbolic systems of equations in two independent variables

*In this section we prove a theorem about hyperbolic systems of partial differential equation in two independent variables which will be applied in Sect. 7.4.*

*The material is very technical and of a different mathematical topic than the rest of this book. The reader may wish to skip this section on first reading.*

For the following theorem we need some notation. If $f\colon \mathbb{R}^l \to \mathbb{R}^k$ we call $j^0(f)\colon \mathbb{R}^l \mapsto \mathbb{R}^l \times \mathbb{R}^k$, $x \mapsto (x, f(x))$ the *0-jet of $f$*. The canonical projection $\mathbb{R}^2 \to \mathbb{R}$, $(t, q) \mapsto q$ is denoted by $\mathrm{pr}_2$.

**Definition 7.3.1.** *Let $h \in C^1(\mathbb{R}^2 \times \mathbb{R}^k, \mathbb{R}^k)$, $A \in C^1(\mathbb{R}^2 \times \mathbb{R}^k, \mathrm{Lin}(\mathbb{R}^k, \mathbb{R}^k))$. The system of differential equations*

$$\partial_t f + A \circ j^0(f)\partial_q f = h \circ j^0(f)$$

*is a* quasi-linear system of hyperbolic equations in two variables *if for every 0-jet $(t, q, F) \in \mathbb{R}^2 \times \mathbb{R}^k$ the linear map $A(t, q, F)$ has $k$ linearly independent left eigenvectors. The directions $\mathbb{R}(\partial_t + \lambda_i \partial_q)$ where $\lambda_i$ are the left eigenvalues of $A$ are called* characteristic directions. *The (unparametrized) integral curves of the characteristic directions*[9] *are called the* characteristics *of the system of differential equations (and the given solution).*

The aim of this section is to prove the following fundamental existence and uniqueness theorem for quasi-linear systems of hyperbolic equations in two variables.

---

[9] Here we mean integral curves of vector fields which are tangent to the characteristic directions

**Theorem 7.3.1.** *Let* $h \in C^\infty(\mathbb{R}^2 \times \mathbb{R}^k, \mathbb{R}^k)$, $A \in C^\infty(\mathbb{R}^2 \times \mathbb{R}^k, \mathrm{Lin}(\mathbb{R}^k, \mathbb{R}^k))$ *such that*

$$\partial_t f + A \circ j^0(f) \partial_q f = h \circ j^0(f)$$

*is a quasi-linear system of hyperbolic equations in two variables. For any function* $f_0 \in C^\infty([a,b], \mathbb{R}^k)$ *there is an open neigbourhood* $\mathcal{U}$ *of* $\{0\} \times (a,b) \subset \mathbb{R}^2$ *and a unique smooth solution* $f : \mathcal{U} \to \mathbb{R}^k$ *of the system of differential equations such that* $f(0,q) = f_0(q)$ *for all* $q \in (a,b)$.

The main part of the proof of Theorem 7.3.1 is contained in the following lemma.

**Lemma 7.3.1.** *Let* $h, \lambda \in C^\infty(\mathbb{R}^2 \times \mathbb{R}^k, \mathbb{R}^k)$ *and* $a < b \in \mathbb{R}$. *Assume that at any point,* $(t,q) \in \mathbb{R}^2$ *at least two of the numbers* $\lambda^i(t,q)$ *are different. For any function* $f_0 \in C^\infty([a,b], \mathbb{R}^k)$ *there is an open neigbourhood* $\mathcal{U}$ *of* $\{0\} \times (a,b) \subset \mathbb{R}^2$ *and a smooth map* $f : \mathcal{U} \to \mathbb{R}^k$ *such that*

*(i)* $f(0,q) = f_0(q)$ *for all* $q \in (a,b)$,
*(ii)* $\partial_t f^i + \lambda^i \circ j^0(f))\partial_q f^i = h^i \circ j^0(f)$. $(i \in \{1, \dots, k\})$.

*Moreover, the solution is unique.*

*Proof.* We will first transform the system of differential equations into a system of integral equations and then employ an iteration technique in order to solve the system of integral equations.

Assume that $f$ is a solution to our system of partial differential equations. For $(s,p) \in \mathbb{R}^2$ and $i \in \{1, \dots, k\}$ we denote by $t \mapsto \gamma^i_{(s,p)}(t)$ the integral curve of the vector field $\partial_t + \lambda^i \circ j^0(f) \partial_q$ with $\gamma^i_{(s,p)}(0) = (s,p)$. From the definition of $\gamma^i_{(s,p)}$ and

$$\begin{aligned}\frac{\mathrm{d}}{\mathrm{d}t}\left(f^i \circ \gamma^i_{(s,p)}(t)\right) &= \mathrm{d}f^i(\partial_t + \lambda^i \circ j^0(f)\partial_q) = \partial_t f^i + \lambda^i \circ j^0(f) \partial_q f^i \\ &= h^i \circ j^0(f)\end{aligned}$$

we obtain the system of integral equations

$$f^i(s,p) = f^i(0, \gamma^i_{(s,p)}(0)) + \int_0^s h^i \circ j^0(f) \circ \gamma^i_{(s,p)}(\tau) \mathrm{d}\tau, \tag{7.3.12}$$

$$\gamma^i_{(s,p)}(t) = (s,p) + (t, \int_s^t \lambda^i \circ j^0(f) \circ \gamma^i_{(s,p)}(\tau) \mathrm{d}\tau). \tag{7.3.13}$$

Conversely, if there are continuous maps $f^i, \gamma^i_{(s,t)}$ $(i \in \{1, \dots, k\})$ which satisfy this system of integral equations then they also satisfy the system of differential equations (ii). This follows since differentiation of Equation (7.3.12) implies the differential equation (ii).

In order to solve the system of integral equations (7.3.12), (7.3.12) we will employ an iteration procedure. Let

$$F_0^i(s,p) = f_0^i(p), \quad \Gamma^i_{(s,p),0}(t) = (s+t,p),$$

and

$$F^i_{m+1}(s,p) = f_0^i(\Gamma^i_{(s,p),m}(0)) + \int_0^s h^i \circ j^0(F_m) \circ \Gamma^i_{(s,p),m}(\tau)\mathrm{d}\tau,$$

$$\Gamma^i_{(s,p),m+1}(t) = (s,p) + (t, \int_s^t \lambda^i \circ j^0(F_m) \circ \Gamma^i_{(s,p),m}(\tau)\mathrm{d}\tau).$$

We will show that these sequences of functions have well defined limits. These limiting functions will then solve our system of differential equations. We will prove the existence of unique limits by showing that the squences

$$F^i_m(s,p) = F^i_0(s,p) + \sum_{j=1}^m (F^i_j(s,p) - F^i_{j-1}(s,p)),$$

$$\mathrm{pr}_2(\Gamma^i_{(s,p),m}(t)) = \mathrm{pr}_2(\Gamma^i_{(s,p),0}(t)) + \sum_{j=1}^m (\mathrm{pr}_2(\Gamma^i_{(s,p),j}(t)) - \mathrm{pr}_2(\Gamma^i_{(s,p),j-1}(t))$$

can be majorised by absolutely converging series which in turn implies that they converge absolutely.

To achieve this it is important to obtain first bounds on $F^i_j$ and $\Gamma^i_{(s,p),j}$. For any $\tilde{\alpha} > \alpha > 0$ let

$$\begin{aligned}
C_1 &= \sup\left\{||f_0(q)|| + ||\mathrm{D}f_0(q)|| \;|q \in [a,b]\right\},\\
C_2 &= \sup\left\{||h(j^0(\tilde{f})(t,q))|| + ||\lambda(j^0(\tilde{f})(t,q))|| + ||\mathrm{D}h(j^0(\tilde{f})(t,q))||\right.\\
&\quad + ||\mathrm{D}\lambda(j^0(\tilde{f})(t,q))|| \;\big|\sup_i |\tilde{f}^i(t,q)| < 2C_1,\\
&\quad \left.(t,q) \in [-\tilde{\alpha},\tilde{\alpha}] \times [a,b]\right\},\\
C &= C_1 + C_2,
\end{aligned}$$

and

$$\begin{aligned}
\mathcal{U}_{\tilde{\alpha},\alpha} = \Big\{&(t,q) \in (-\alpha,\alpha) \times (a,b) | a + \frac{t}{C} < q < b - \frac{t}{C},\\
&a - \frac{t}{C} < q < b + \frac{t}{C}\Big\}.
\end{aligned}$$

We will solve the initial value problem in the region $\mathcal{U}_{\tilde{\alpha},\alpha}$, if $\alpha$ is chosen small enough.[10] Our bounds imply

$$\left|\mathrm{pr}_2(\frac{\mathrm{d}}{\mathrm{d}t}\Gamma^i_{(s,p),m}(t))\right| = \left|\lambda^i \circ j^0(F_{m-1}) \circ \Gamma^i_{(s,p),m-1}(t)\right| \leq C \tag{7.3.14}$$

[10] The choice of $\tilde{\alpha}$ is less significant. The only purpose of its introduction is to guarantee the existence of the constant $C_2$.

for $\Gamma^i_{(s,p),m-1}(t) \in U_\alpha$. Let $(s,p) \in \mathcal{U}_{\tilde{\alpha},\alpha}$. Inequality 7.3.14 and the definition of $\mathcal{U}_{\tilde{\alpha},\alpha}$ imply $\Gamma^i_{(s,p),m}(t) \subset \mathcal{U}_{\tilde{\alpha},\alpha}$. Hence during the iteration process we do not leave the region $\mathcal{U}_{\tilde{\alpha},\alpha}$ and our bounds $C_1, C_2$ are valued at all $\Gamma^i_{(s,p),m}(t)$. Let

$$\delta^F_m = \sup\left\{\left|\frac{\partial}{\partial p}F^i_m(s,p)\right| \,\middle|\, t \in [-\alpha - s, \alpha - s], i \in \{1,\ldots,k\}\right\},$$

$$\delta^\Gamma_m = \sup\left\{\left|\frac{\partial}{\partial p}\mathrm{pr}_2(\Gamma^i_{(s,p),m}(t))\right| \,\middle|\, t \in [-\alpha - s, \alpha - s], i \in \{1,\ldots,k\}\right\}.$$

Assume that $\sup\{|F^i_j(s,p)| \mid (s,p) \in \mathcal{U}_{\tilde{\alpha},\alpha}, j \in \{0,\ldots,m\}\} < 2C_1$. From

$$\begin{aligned}\frac{\partial}{\partial p}F^i_{m+1}(s,p) &= \frac{\partial}{\partial p}\left(f^i_0 \circ \Gamma^i_{(s,p),m}(0)\right) \\ &\quad + \int_0^s \frac{\partial}{\partial p}\left(h \circ j^0(F_m) \circ \Gamma^i_{(s,p),m}\right)(\tau)\mathrm{d}\tau \\ &= \frac{\partial f^i_0}{\partial q}\frac{\partial}{\partial p}\mathrm{pr}_2(\Gamma^i_{(s,p),m}(0)) + \int_0^s \Big(\mathrm{D}_2 h\,\frac{\partial}{\partial p}\mathrm{pr}_2(\Gamma^i_{(s,p),m}(\tau)) \\ &\quad + \mathrm{D}_3 h\,\mathrm{D}_2 F_m\,\frac{\partial}{\partial p}\mathrm{pr}_2(\Gamma^i_{(s,p),m}(\tau))\Big)\mathrm{d}\tau\end{aligned}$$

and

$$\begin{aligned}\frac{\partial}{\partial p}\mathrm{pr}_2(\Gamma^i_{(s,p),m+1}(t)) = 1 + \int_s^t &\mathrm{D}_2\lambda^i\,\frac{\partial}{\partial p}\mathrm{pr}_2(\Gamma^i_{(s,p),m}(\tau)) \\ &+ \mathrm{D}_3\lambda^i\,\mathrm{D}_2 F_m\,\frac{\partial}{\partial p}\mathrm{pr}_2(\Gamma^i_{(s,p),m}(\tau))\mathrm{d}\tau\end{aligned}$$

we get

$$\delta^F_{m+1} \le C\delta^\Gamma_m + \alpha(C\delta^\Gamma_m + C\delta^F_m\delta^\Gamma_m), \quad \delta^\Gamma_{m+1} \le 1 + \alpha(C\delta^\Gamma_m + C\delta^F_m\delta^\Gamma_m).$$

Let $\alpha < \frac{1}{1+2C}$. Then these inequalities imply $\delta^F_{m+1} \le 3C$, $\delta^\Gamma_{m+1} = 2$ if $\delta^F_m \le 3C$ and $\delta^\Gamma_m = 2$. Since $\delta^F_0 \le C$ and $\delta^\Gamma_0 = 1$ these bounds are valid for all $m$. We estimate for $\sup\{|F^i_j(s,p)| \mid (s,p) \in \mathcal{U}_{\tilde{\alpha},\alpha}, j \in \{0,\ldots,m\}\} < 2C_1$

$$\begin{aligned}&|F^i{}_{m+1}(s,p) - F^i_m(s,p)| \\ &\quad \le |f^i_0(\Gamma^i_{(s,p),m}(0)) - f^i_0(\Gamma^i_{(s,p),m-1}(0))| \\ &\qquad + \int_0^s |h \circ j^0(F_m) \circ \Gamma^i_{(s,p),m}(\tau) - h \circ j^0(F_{m-1}) \circ \Gamma^i_{(s,p),m-1}(\tau)|\mathrm{d}\tau\end{aligned}$$

$$
\begin{aligned}
&\leq C|\Gamma^i_{(s,p),m}(0)) - \Gamma^i_{(s,p),m-1}(0))| \\
&\quad + \int_0^s \sup\Big\{ \max\{\|(\mathrm{D}h)(j^0(F_m)\circ\Gamma^i_{(s,p),m}(t))\| \\
&\quad\ , \|(\mathrm{D}h)(j^0(F_{m-1})\circ\Gamma^i_{(s,p),m-1}(t))\|\}, \\
&\quad\ t\in[0,s]\Big\}\|j^0(F_m)\circ\Gamma^i_{(s,p),m}(\tau) - j^0(F_{m-1})\circ\Gamma^i_{(s,p),m-1}(\tau)\|\mathrm{d}\tau \\
&\leq C|\Gamma^i_{(s,p),m}(0)) - \Gamma^i_{(s,p),m-1}(0))| \\
&\quad + C\int_0^s \Big(|\mathrm{pr}_2(\Gamma^i_{(s,p),m}(\tau)) - \mathrm{pr}_2(\Gamma^i_{(s,p),m-1}(\tau))| \\
&\quad + \sum_{j=1}^k |F^j_m(s,p)\circ\Gamma^i_{(s,p),m} - F^j_{m-1}(s,p)\circ\Gamma^i_{(s,p),m-1}|\Big)\mathrm{d}\tau \\
&\leq C|\Gamma^i_{(s,p),m}(0)) - \Gamma^i_{(s,p),m-1}(0))| \\
&\quad + C\int_0^s \Big(|\mathrm{pr}_2(\Gamma^j_{(s,p),m}(\tau)) - \mathrm{pr}_2(\Gamma^j_{(s,p),m-1}(\tau))| \\
&\quad + \sum_{j=1}^k \big(|F^j_m(s,p)\circ\Gamma^i_{(s,p),m} - F^j_{m-1}(s,p)\circ\Gamma^i_{(s,p),m}| \\
&\quad + |F^j_{m-1}(s,p)\circ\Gamma^i_{(s,p),m} - F^j_{m-1}(s,p)\circ\Gamma^i_{(s,p),m-1}|\big)\Big)\mathrm{d}\tau \\
&\leq C|\Gamma^i_{(s,p),m}(0)) - \Gamma^i_{(s,p),m-1}(0))| \\
&\quad + C\int_0^s \Big((1+3kC)|\mathrm{pr}_2(\Gamma^j_{(s,p),m}(\tau)) - \mathrm{pr}_2(\Gamma^j_{(s,p),m-1}(\tau))| \\
&\quad + \sum_{j=1}^k |F^j_m(s,p)\circ\Gamma^i_{(s,p),m} - F^j_{m-1}(s,p)\circ\Gamma^i_{(s,p),m}|\Big)\mathrm{d}\tau,
\end{aligned}
$$

where in the last inequality we have used

$$
\begin{aligned}
|F^j_{m-1}(s,p)\circ\Gamma^i_{(s,p),m} &- F^j_{m-1}(s,p)\circ\Gamma^i_{(s,p),m-1}| \\
&\leq \sup\|\mathrm{D}_2 F^j_{m-1}(s,p)\|\,|\Gamma^i_{(s,p),m} - \Gamma^i_{(s,p),m-1}| \\
&\leq 3C|\Gamma^i_{(s,p),m} - \Gamma^i_{(s,p),m-1}|.
\end{aligned}
$$

Analogously we obtain the estimate

$$
\begin{aligned}
&|\Gamma^i{}_{(s,p),m+1} - \Gamma^i_{(s,p),m}| \\
&\quad \leq C\int_0^s \Big((1+3kC)|\mathrm{pr}_2(\Gamma^j_{(s,p),m}(\tau)) - \mathrm{pr}_2(\Gamma^j_{(s,p),m-1}(\tau))| \\
&\quad\quad + \sum_{j=1}^k |F^j_m(s,p)\circ\Gamma^i_{(s,p),m} - F^j_{m-1}(s,p)\circ\Gamma^i_{(s,p),m}|\Big)\mathrm{d}\tau.
\end{aligned}
$$

Let

$$\epsilon_m^F = \sup\left\{|F_m^i(s,p) - F_{m-1}^i(s,p)| \,\big|(s,p) \in \mathcal{U}_{\tilde{\alpha},\alpha}, i \in \{1,\dots,k\}\right\},$$
$$\epsilon_m^\Gamma = \sup\left\{|\mathrm{pr}_2(\Gamma_{(s,p),m}^i(t)) - \mathrm{pr}_2(\Gamma_{(s,p),m-1}^i(t))| \,\big|(s,p) \in \mathcal{U}_{\tilde{\alpha},\alpha},\right.$$
$$\left. t \in [-\alpha - s, \alpha - s], i \in \{1,\dots,k\}\right\}.$$

From our estimates we obtain

$$\epsilon_{m+1}^F \le (1+\alpha(1+3kC))C\epsilon_m^\Gamma + \alpha kC\epsilon_m^F, \quad \epsilon_{m+1}^\Gamma \le \alpha C(1+3kC)\epsilon_m^\Gamma + \alpha Ck\epsilon_m^F,$$

We will now show that these inequalities imply

$$\epsilon_m^F \le (2kC\sqrt{\alpha})^m \text{ and } \epsilon_m^\Gamma \le (2kC\sqrt{\alpha})^m\sqrt{\alpha}$$

if $\alpha$ is chosen small enough and

$$\sup\{|F_j^i(s,p)| \,|\, (s,p) \in \mathcal{U}_{\tilde{\alpha},\alpha}, j \in \{0,\dots,m\}\} < 2C_1$$

holds. In fact, we have $\epsilon_1^F \le \alpha C$ and $\epsilon_1^\Gamma \le \alpha C$, so the inequalities hold for $m = 1$ if $\alpha < 1/(2k)^2 < 1$. If they hold for $\epsilon_m^F, \epsilon_m^\Gamma$ we get

$$\begin{aligned}
\epsilon_{m+1}^F &\le (1+\alpha(1+3kC))C(2kC\sqrt{\alpha})^m\sqrt{\alpha} + \alpha kC(2kC\sqrt{\alpha})^m \\
&= (1+\alpha(1+3kC) + k\sqrt{\alpha})\frac{1}{2k}(2kC\sqrt{\alpha})^{m+1}, \\
\epsilon_{m+1}^\Gamma &= \alpha C(1+3kC)(2kC\sqrt{\alpha})^m\sqrt{\alpha} + \alpha Ck(2kC\sqrt{\alpha})^m \\
&= (\alpha(1+3kC) + k\sqrt{\alpha})\frac{1}{2k}(2kC\sqrt{\alpha})^{m+1}
\end{aligned}$$

Hence it is sufficient to choose $\alpha < \min\{1/(2k)^2, (2k-1)^2/(1+3kC+k)^2\}$.

Since $|F_m^i(s,p) - F_{m-1}^i(s,p)| < \epsilon_m^F$ and

$$|\mathrm{pr}_2(\Gamma_{(s,p),m}^i(t)) - \mathrm{pr}_2(\Gamma_{(s,p),m-1}^i(t))| < \epsilon_m^\Gamma$$

the sequences

$$F_m^i(s,p) = F_0^i(s,p) + \sum_{j=1}^m (F_j^i(s,p) - F_{j-1}^i(s,p)),$$

$$\begin{aligned}
\mathrm{pr}_2(\Gamma_{(s,p),m}^i(t)) = \mathrm{pr}_2(\Gamma_{(s,p),0}^i(t)) &+ \sum_{j=1}^m (\mathrm{pr}_2(\Gamma_{(s,p),j}^i(t)) \\
&- \mathrm{pr}_2(\Gamma_{(s,p),j-1}^i(t)))
\end{aligned}$$

converge if the series $\sum_{j=1}^\infty \epsilon_j^\Gamma$ and $\sum_{j=1}^\infty \epsilon_j^F$ exist. These series are majorised by a convergent geometrical series if $\alpha < 1/(2kC)^2$. Moreover, it is easy to check that

$$\sup\{|F_j^i(s,p)|| (s,p) \in \mathcal{U}_{\tilde{\alpha},\alpha}, j \in \{0,\dots,m\}\} \leq C_1 + \sum_{j=1}^{m} \epsilon_j^F < 2C_1$$

if

$$\alpha < \frac{(C_1)^2}{4k^2C^2(1+C_1)^2}$$

holds for every $m$. This proves uniform convergence of $F_m^i$, $\Gamma^i_{(s,p),m}$ if $\alpha$ is chosen small enough.

This solution is clearly differentiable with respect to $t$ and continuous with respect to $q$. In order to show that it is $C^r$ with respect to $q$ and $t$ we can apply the same to argument to the prolonged system of differential equations which is obtained by differentiation of (ii) as follows. For $l$ ($l \in \{1,\dots,r-1\}$) differentiate equations (ii) $l$ times with respect to $t$ and replace $(\partial_t)^{\tilde{l}} f^i$ ($\tilde{l} \in \{1,\dots,l\}$) by $g_t^{i,\tilde{l}}$ and $(\partial_t)^l f$ by $\partial_t g_t^{i,l-1}$. This gives $k(r-1)$ equations of the form

$$\partial_t g_t^{i,l} + \lambda^i \partial_q g_t^{i,l} = G_t^{l,i}(f, g_t^{1,1},\dots,g_t^{k,l})$$

which we add to our system of differential equatons. Differentiating equations (ii) $l$ times with respect to $q$ we obtain equations of the form $\partial_t(\partial_q)^l f^i + \lambda^i \partial_q (\partial_q)^l f^i = G_q^{l,i}(f, \partial_q f,\dots,(\partial_q)^l f)$. We therefore also add the $kr$ equations

$$\partial_t g_q^{i,l} + \lambda^i \partial_q g_q^{i,l} = G_q^{l,i}(f, g_q^{1,1},\dots,g_q^{k,l})$$

to our system of differential equatons. Choosing the additional initial conditions

$$\begin{aligned}
(g_t^{i,l})_0(q) &= (-\lambda^i \partial_q g_t^{i,l-1} + G_t^{l,i}(f, g_t^{1,1},\dots,g_t^{k,l}))_{0,q},\\
(g_t^{i,1})_0(q) &= -\lambda^i(0,q,f_0(q)\partial_q f_0 + h^i(0,q,f_0(q)),\\
(g_q^{i,l})_0(q) &= (\partial_q)^l f_0(q)
\end{aligned}$$

we obtain a system of differential equations which is of the same form as (i), (ii). By construction, the first components $f^i$ of the solution of this system coincide with the solution of our original system. Moreover, $g_q^{i,l} = (\partial_q)^l f^i$, $g_t^{i,l} = (\partial_t)^l f^i$ are continous for all $l$ which implies $f^i \in C^r(U_{\tilde{\alpha},\alpha}, \mathbb{R})$.

We will now prove uniqueness of the solution. Assume that $(f^i, \gamma^i_{(s,y)})$, $(\hat{f}, \hat{\gamma}^i_{(s,y)})$ both satisfy the system of integral equations (7.3.12), (7.3.13). Then we get

$$\begin{aligned}
&|f^i(\gamma^i_{(s,y)}) - \hat{f}^i(\hat{\gamma}^i_{(s,y)})|\\
&= \int_0^s (h^i(\gamma^i_{(s,y)}(\tau), f^i(\gamma^i_{(s,y)})(\tau)) - h^i(\hat{\gamma}^i_{(s,y)}(\tau), \hat{f}^i(\hat{\gamma}^i_{(s,y)}(\tau))))\mathrm{d}\tau\\
&\leq C\int_0^s (|\gamma^i_{(s,y)}(\tau) - \hat{\gamma}^i_{(s,y)}(\tau)| + |f^i(\gamma^i_{(s,y)})(\tau)) - \hat{f}^i(\hat{\gamma}^i_{(s,y)}(\tau))|)\mathrm{d}\tau
\end{aligned}$$

and, analogously,

$$|\gamma^i_{(s,y)} - \hat{\gamma}^i_{(s,y)}|$$
$$\leq C \int_0^s (|\gamma^i_{(s,y)}(\tau) - \hat{\gamma}^i_{(s,y)}(\tau)| + |f^i(\gamma^i_{(s,y)})(\tau)) - \hat{f}^i(\hat{\gamma}^i_{(s,y)}(\tau))|)\mathrm{d}\tau.$$

Hence writing $d(s,y) = |\gamma^i_{(s,y)} - \hat{\gamma}^i_{(s,y)}| + |f^i(\gamma^i_{(s,y)}) - \hat{f}^i(\hat{\gamma}^i_{(s,y)})|$ we obtain

$$\max_{\tau \in 0,s}\{d(\tau,y)\} \leq \max_{\tau \in 0,s}\{2C \int_0^s d(\tau,y)\mathrm{d}\tau\}$$
$$\leq \max_{\tau \in 0,s}\{2Cs \max_{\tau \in 0,s}\{d(\tau,y)|\}\} = 2Cs \max_{\tau \in 0,s}\{d(\tau,y)|\}.$$

This equation can only hold if $\mathrm{d}(s,y) = 0$ for $s < 1/(2C)$. ∎

*Proof of Theorem 7.3.1.* Let $\{l_1(t,q,F), \ldots, l_k(t,q,F)\}$ be a basis of left eigenvectors of $A(t,q,F)$ and denote their eigenvalues by $\lambda^1(t,q,F), \ldots, \lambda^k(t,q,F)$. Multiplying the the differential equation from the left by $l^i$ we obtain $l^i \partial_t f + \lambda^i l^i \partial_q f = l^i h$ or, equivalently,

$$\partial_t(l^i f) + \lambda^i \partial_q(l^i f) = (l^i h) + \left(\frac{\mathrm{d}}{\mathrm{d}t} l^i\right) f + \left(\frac{\mathrm{d}}{\mathrm{d}q} l^i\right) \lambda_i f$$
$$= (l^i h) + (D_1 l_i + D_3 l_i \partial_t f) f + \lambda_i (D_2 l_i + D_3 l_i \partial_q f) f$$
$$= (l^i h) + (D_1 l_i) f + \lambda_i (D_2 l_i) f + (D_3 l_i h) f.$$

This system of differential equations is of the diagonal form treated in the Lemma 7.3.1 above. Hence the assertion follows directly from this lemma. ∎

**Corollary 7.3.1.** *In addition to the assumptions of Theorem 7.3.1 let* $[\alpha,\beta] \subset [a,b]$ *and* $t_0 > 0$. *The solution* $f$ *of the quasi-linear system of hyperbolic equations at* $(t_0, q_0) \in \mathbb{R}^2$ *depends only on the initial data restricted to* $[\alpha,\beta]$ *if and only if all characteristics intersect* $\{0\} \times \mathbb{R}$ *in the subset* $\{0\} \times [\alpha,\beta]$.

*Proof.* This follows immediately from the integral representation (7.3.12), (7.3.13). ∎

*The following corollary will only be used in Sect. 9.5.1.*

**Corollary 7.3.2.** *In addition to the assumptions of Theorem 7.3.1 assume that* $A$ *is constant and that the map* $h$ *is linear in* $f$.

*The characteristic directions do not depend on the solution. Moreover, let* $\mathfrak{D}([a,b])$ *be the set of all* $(t,q) \in \mathbb{R} \times \mathbb{R}$ *such that all characteristics through* $(t,q)$ *intersect the set* $\{0\} \times [a,b]$. *Then there is a unique solution* $f$ *which is defined on all of* $\mathfrak{D}^+([a,b])$

*Proof.* It is clear that the characteristics do not depend on the solution since $A$ does not depend on it. We denote the characteristic through $(s,p)$ to the $i$th eigenvalue $\lambda^i$ by $\gamma^i_{(s,p)}$. The non-trivial part of the corollary is to prove that the solution which (by Theorem 7.3.1 may only be defined locally) extends to all of $\mathfrak{D}^+([a,b])$. This can be achieved by applying a slightly refined version of Theorem 7.3.1 repeatedly. To this end we will need better estimates in the proof of Theorem 7.3.1.

Assume that the initial data satisfy $|f^i_0(q)| < C_1$ for all $q \in [a,b]$. Since $h$ is linear with respect to $f$ there are functions $h^i_l$ such that $h^i(j^0(f)(t,q)) = \sum_{l=1^k} h^i_l(t,q) f^l(t,q)$. It follows that there is a constant $K \in \mathbb{R}$ such that

$$K > \sup\{\sum_{i,l=1}^{k} (|h^i_l(t,q)| + |\lambda^i| + \|\mathrm{D}h^i_l(t,q)\|) \mid (t,q)\mathfrak{D}([a,b])\}.$$

Observe that $K$ is well defined since $\mathfrak{D}([a,b])$ is compact. Unlike in the proof of Theorem 7.3.1 the inequality defining $K$ is independent of the solution. We set $\mathcal{U}_\alpha = \{(t,q) \in \mathfrak{D}([a,b]) \mid -\alpha < t < \alpha\}$.

Recall that $\Gamma^i_{(s,p)m} = \gamma^i_{(s,p)}$ for all $m$. The estimate for $|F^i_{m+1}(s,p) - F^i_m(s,p)|$ in the proof of Theorem 7.3.1 simplifies to

$$\begin{aligned} |F^i_{m+1}(s,p) - F^i_m(s,p)| &\le \overbrace{|f^i_0(\gamma^i_{(s,p)}(0)) - f^i_0(\gamma^i_{(s,p)}(0))|}^{=0} \\ &\quad + \int_0^s \sum_{l=1}^{k} \left|(h^i_l F^l_m - h^i_l F^l_{m-1}) \circ \gamma^i_{(s,p)}(\tau)\right| \mathrm{d}\tau \\ &\le \alpha k K \epsilon^F_m, \end{aligned}$$

where

$$\epsilon^F_m = \sup\{|F^l_m(s,p) - F^l_{m-1}(s,p)| \mid (s,p) \in \mathfrak{D}([a,b]), l \in \{1,\dots,k\}\}.$$

This estimate implies the recursive inequatlity $\epsilon^F_{m+1} \le \alpha k K \epsilon^F_m$. From $\epsilon^F_1 \le \alpha k K C_1$ we get $\epsilon^F_m \le C_1(\alpha k K)^m$ and therefore $\sum_{j=1}^m \epsilon^F_j \le \frac{\alpha k K C_1}{1-\alpha k K}$. Using $F^i_m(s,p) = F^i_0(s,p) + \sum_{j=1}^m (F^i_j(s,p) - F^i_{j-1}(s,p))$ we obtain the bound

$$|F^i_m(s,p)| \le C_1 + \sum_{j=1}^{m} \epsilon^F_j \le \frac{C_1}{1-\alpha k K}$$

It follows that for $\alpha < 1/(2kK)$ there is a solution defined in all of $\mathcal{U}_\alpha$. Since the number $1/(2kK)$ is independent of the solution we obtain our global solution by successively solving $2kK \max\{s \mid \exists p \text{ with } (s,p) \in \mathfrak{D}([a,b])\}$ initial value problems. The proof of uniqueness and differentiablility is exactly as in the proof of Theorem 7.3.1. ∎

*Remark 7.3.1.* Observe that Corollary 7.3.2 is still correct if we replace the initial curve $\{0\} \times [a, b]$ by an arbitrary curve $\mathcal{C}$ which is intersected by each characteristic at most once and replace $\mathfrak{D}([a, b])$ by set $\mathfrak{D}(\mathcal{C})$ of all $(t, q) \in \mathbb{R} \times \mathbb{R}$ such that all characteristics through $(t, q)$ intersect the curve $\mathcal{C}$.

## 7.4 The initial value problem for spherically symmetric perfect fluid spacetimes with non-interacting electromagnetic fields

*In this section we discuss the initial value problem (cf. Equations (7.1.3)–(7.1.6)) in some generality for a spherically symmetric spacetime which represents a perfect fluid.*

*Since the section is quite technical and requires the results of Sect. 7.3 the reader may wish to skip it on first reading.*

The Schwarzschild solution is a good description for the exterior of an isolated, spherically symmetric, non-rotating star. Here we wish to solve Einstein's equation for the interior of such a star. The complete model of an isolated, spherically symmetric, non-rotating star is then usually obtained by matching the interior and the exterior solutions at the boundary of the star. This will be done in Sect. 7.5 for the special case of static stars.

The system of Equations (7.1.3)–(7.1.6) is highly non-linear and rather complicated. Observe that the assumption of a perfect fluid ($p_{\text{rad}} = p_{\text{sph}}$) allows us to integrate Equations (7.1.7) and (7.1.8) directly substantially simplifies the problem. This simplification is unaffected when we include a non-interacting electromagnetic field.

In the following we will first study electromagnetic fields in spherically symmetric spacetimes and then discuss the initial value problem for a spherically symmetric spacetime which admits a perfect fluid and a non-interacting electromagnetic field.

*Readers who have not read Sect. 5.2.3 on Maxwell's equation may wish to skip the material up to Lemma 7.4.2 and assume* $T_{\text{el}} = 0$, $e = b = 0$ *in the following discussion.*

Recall from Sect. 5.2.3 that the source free Maxwell equations for an electromagnetic field are given by

$$\mathrm{d}F = 0,$$
$$\operatorname{div}(F) = 0,$$

where $F$ is a 2-form. The electromagnetic part of the energy momentum tensor reads then $(T_{\text{el}})_{ab} = \frac{1}{4\pi}(g^{cd}F_{ac}F_{bd} - \frac{1}{4}\langle F, F\rangle g_{ab})$. Given a spherically symmetric spacetime $(M, g)$ and any 2-form $F$ satisfying the source-free Maxwell equations one could define $T_{\text{matter}} = T - T_{\text{el}}$. Of course, the matter represented by $T_{\text{matter}}$ would in general be quite exotic. Moreover, it is possible that neither $T_{\text{matter}}$ nor $T_{\text{el}}$ are spherically symmetric. This discussion indicates that we should impose additional conditions in order to describe *physical* electromagnetic fields. Given that $(M, g)$ is spherically symmetric, the most natural additional assumption on the energy momentum tensor $T_{\text{el}}$ would be to demand that it is invariant under rotational isometries and that $F$ is well defined in sufficiently large open sets containing complete spheres of symmetry.

**Lemma 7.4.1.** *Let $(\Sigma \times S^2, g_\Sigma + r^2 \mathrm{d}\Omega^2)$ be a spherically symmetric, 4-dimensional Lorentz manifold, $\mu_\Sigma$ be the volume form of $(\Sigma, g_\Sigma)$, and $\mu_{S^2}$ the volume form of $(S^2, \mathrm{d}\Omega^2)$. If $F \in \Omega^2(\Sigma \times S^2)$ satisfies*

$$\mathrm{d}F = 0, \quad \mathrm{div}(F) = 0,$$

*and $(T_{\text{el}})_{ab} = \frac{1}{4\pi}(g^{cd} F_{ac} F_{bd} - \frac{1}{4} \langle F, F \rangle g_{ab})$ is spherically symmetric, then there exist constants $e, b$ with*

$$F = \frac{e}{r^2}(\pi_\Sigma)^* \mu_\Sigma + b(\pi_{S^2})^* \mu_{S^2}.$$

*The corresponding energy momentum tensor is given by*

$$T_{\text{el}} = \frac{1}{8\pi}\Big( - \frac{e^2 + b^2}{r^4} \left( -U^\flat \otimes U^\flat + Q^\flat \otimes Q^\flat \right) + \frac{e^2 + b^2}{r^4} r^2 \mathrm{d}\Omega^2 \Big).$$

*Proof.* We consider the orthonormal frame $\{U, Q, E_2, E_3\}$, where

$$E_2 = \frac{1}{r}\partial_\theta \text{ and } E_3 = \frac{1}{r\sin(\theta)}\partial_\varphi.$$

Spherical symmmetry of $T_{\text{el}}$ implies

$$T_{\text{el}}(E_2, E_2) = T_{\text{el}}(E_3, E_3)$$

and

$$T_{\text{el}}(U, E_2) = T_{\text{el}}(U, E_3) = T_{\text{el}}(Q, E_2) = T_{\text{el}}(Q, E_3) = T_{\text{el}}(E_2, E_3) = 0.$$

All other components of $T_{\text{el}}$ are unconstrained (cf. Lemma 7.1.1).

Since $4\pi (T_{\text{el}})_{ab} = g^{cd} F_{ac} F_{bd} - \frac{1}{4} \langle F, F \rangle g_{ab}$, we get

$$\begin{aligned}
0 &= 4\pi T_{\text{el}}(E_2, E_2) - 4\pi T_{\text{el}}(E_3, E_3) \\
&= -F(E_2, U)^2 + F(E_2, Q)^2 + F(E_2, E_3)^2 \\
&\quad - (-F(E_3, U)^2 + F(E_3, Q)^2 + F(E_3, E_2)^2) \\
&= -F(E_2, U)^2 + F(E_2, Q)^2 + F(E_3, U)^2 - F(E_3, Q)^2
\end{aligned}$$

and

$$0 = 4\pi T_{\rm el}(E_2, E_3) = -F(E_2, U)F(E_3, U) + F(E_2, Q)F(E_3, Q).$$

Multiplying the first equation with $F(E_3, U)^2$ and inserting the second equation we obtain

$$\begin{aligned}0 &= -F(E_2, Q)^2F(E_3, Q)^2 + F(E_2, Q)^2F(E_3, U)^2 \\ &\quad + (F(E_3, U)^2 - F(E_3, Q)^2)F(E_3, U)^2 \\ &= (F(E_3, U)^2 - F(E_3, Q)^2)(F(E_3, U)^2 + F(E_2, Q)^2).\end{aligned}$$

If $F(E_3, U)^2 - F(E_3, Q)^2 \neq 0$ we have $F(E_3, U) = F(E_2, Q) = 0$. Inserting this into $T_{\rm el}(E_2, E_2) - T_{\rm el}(E_3, E_3) = 0$ gives then $-F(E_2, U)^2 - F(E_3, Q)^2 = 0$ which in turn implies $F(E_2, U) = F(E_3, Q) = 0$. In particular, we have shown $F(E_3, U) = F(E_3, Q) = 0$ which contradicts the assumption $F(E_3, U)^2 - F(E_3, Q)^2 \neq 0$.

If $F(E_3, U)^2 - F(E_3, Q)^2 = 0$ we get $F(E_3, U) = \eta F(E_3, Q)$, where $\eta \in \{-1, 1\}$. In the first case the equation $T_{\rm el}(E_2, E_3) = 0$ implies that $F(E_2, U) = \eta F(E_2, Q)$ or $F(E_3, U) = F(E_3, Q) = 0$. In the second case we obtain the same conclusion from the equation $0 = 4\pi T_{\rm el}(E_2, E_2) - 4\pi T_{\rm el}(E_3, E_3)$.

The equations $T_{\rm el}(Q, E_2) = T_{\rm el}(Q, E_3) = 0$ imply now

$$\begin{aligned}0 &= -F(Q, U)(\eta F(E_2, Q)) + (-F(E_3, Q))F(E_2, E_3), \\ &= \eta F(U, Q)F(E_2, Q) - F(E_2, E_3)F(E_3, Q) \\ 0 &= -F(Q, U)(\eta F(E_3, Q)) + (-F(E_2, Q))F(E_3, E_2) \\ &= \eta F(U, Q)F(E_3, Q) + F(E_2, E_3)F(E_2, Q)\end{aligned}$$

This is a linear system of equations for $F(E_2, Q), F(E_3, Q)$. Since the determinant of the associated matrix is $F(U, Q)^2 + F(E_2, E_3)^2$ we have either $F(E_2, Q) = F(E_3, Q) = 0$ or $F(U, Q) = F(E_2, E_3) = 0$.[11] We have therefore two possible cases. There are functions $\tilde{A}, \tilde{B} \colon M \to \mathbb{R}$ such that either $F = \tilde{A}r(U^\flat + \eta Q^\flat) \wedge \mathrm{d}\theta + \tilde{B}r\sin(\theta)(U^\flat + \eta Q^\flat) \wedge \mathrm{d}\varphi$ or $F = \tilde{A}U^\flat \wedge Q^\flat + \tilde{B}r^2\sin(\theta)\mathrm{d}\theta \wedge \mathrm{d}\varphi$.

In the first case let $x \in \Sigma \times S^2$. Then

$$F_{(\pi_\Sigma(x), y)}(-U + \eta Q, \cdot) = 2r(x)(\tilde{A}(\pi_\Sigma(x), y)\mathrm{d}\theta + \tilde{B}(\pi_\Sigma(x), y)\sin(\theta)\mathrm{d}\varphi)$$

defines a 1-form on the sphere of symmetry $S_x := \{\pi_\Sigma(x)\} \times S^2$. Since $S^2$ does not admit any non-vanishing vector fields, this 1-form must vanish at some point $y_0$ of $S_x$. This implies $F_{(\pi_\Sigma(x), y_0)} = 0$ and therefore $(T_{\rm el})_{(\pi_\Sigma(x), y_0)} = 0$. Since $T_{\rm el}$ is spherically symmetric we obtain

[11] The equations $T_{\rm el}(U, E_2) = T_{\rm el}(U, E_3) = 0$ do not give any more information.

$(T_{\mathrm{el}})_{(\pi_\Sigma(x),y)} = 0$ for all $(\pi_\Sigma(x), y) \in S_x$. By arbitrariness of $x$ we finally have $T_{\mathrm{el}} = 0$.

In the second case the electromagnetic field is given by $F = \tilde{A}U^\flat \wedge Q\flat + \tilde{B}r^2 \sin(\theta)\mathrm{d}\theta \wedge \mathrm{d}\varphi$. Since $\mathrm{d}(U^\flat \wedge Q^\flat) = 0$ and $\mathrm{d}(\sin(\theta)\mathrm{d}\theta \wedge \mathrm{d}\varphi) = 0$ we obtain

$$\begin{aligned}\mathrm{d}F &= \partial_\theta \tilde{A}\mathrm{d}\theta \wedge U^\flat \wedge Q^\flat + \partial_\varphi \tilde{A}\mathrm{d}\varphi \wedge U^\flat \wedge Q^\flat \\ &\quad + \partial_t(r^2\tilde{B}) \sin\theta \mathrm{d}t \wedge \mathrm{d}\theta \wedge \mathrm{d}\varphi + \partial_q(r^2\tilde{B}) \sin\theta \mathrm{d}q \wedge \mathrm{d}\theta \wedge \mathrm{d}\varphi.\end{aligned}$$

It follows that $\mathrm{d}F = 0$ is satisfied if and only if $\tilde{A}$ depends only on $t$ and $q$ whereas $r^2\tilde{B}$ depends only on $\theta$ and $\varphi$. Since $\mathrm{div}(U) = U \bullet \lambda + 2U \bullet \ln r$, $\mathrm{div}(Q) = Q \bullet \nu + 2Q \bullet \ln r$, and $[U, Q] = (Q \bullet \nu)U - (U \bullet \lambda)Q$ we get

$$\begin{aligned}\mathrm{div}(F^\sharp) &= \mathrm{div}(-\tilde{A}(U \otimes Q - Q \otimes U) + \frac{\tilde{B}}{r^2 \sin^2(\theta)}(\partial_\theta \otimes \partial_\varphi - \partial_\varphi \otimes \partial_\theta) \\ &= -\mathrm{d}\tilde{A}(U)Q + \mathrm{d}\tilde{A}(Q)U - \tilde{A}(\mathrm{div}(U)Q - \mathrm{div}(Q)U + [U, Q]) \\ &\quad + \mathrm{d}\left(\frac{\tilde{B}}{r^2 \sin(\theta)}\right)(\partial_\theta)\partial_\varphi - \mathrm{d}\left(\frac{\tilde{B}}{r^2 \sin(\theta)}\right)(\partial_\varphi)\partial_\theta \\ &\quad + \frac{\tilde{B}}{r^2 \sin(\theta)}(\mathrm{div}(\partial_\theta)\partial_\varphi - \mathrm{div}(\partial_\varphi)\partial_\theta) \\ &= \tilde{A}(Q \bullet \ln \tilde{A} + Q \bullet \nu + 2Q \bullet \ln r - Q \bullet \nu)U \\ &\quad + \tilde{A}( - U \bullet \ln \tilde{A} - U \bullet \lambda - 2U \bullet \ln r + U \bullet \lambda)Q \\ &\quad + (\partial_\theta \left(\frac{\tilde{B}}{r^2 \sin(\theta)}\right) + \frac{\cos(\theta)}{\sin(\theta)} \frac{\tilde{B}}{r^2 \sin(\theta)})\partial_\varphi + \partial_\varphi \left(\frac{\tilde{B}}{r^2 \sin(\theta)}\right) \partial_\theta\end{aligned}$$

Hence the equation $\mathrm{div}(F) = 0$ is equivalent to

$$U \bullet \ln(\tilde{A}r^2) = 0, \quad Q \bullet \ln(\tilde{A}r^2) = 0, \quad \partial_\varphi \tilde{B} = 0, \quad \frac{\partial_\theta B}{\sin(\theta)} = 0. \tag{7.4.15}$$

It follows that there are constants $e, b \in \mathbb{R}$ with $F = \frac{e}{r^2}U^\flat \wedge Q^\flat + b\sin(\theta)\mathrm{d}\theta \wedge \mathrm{d}\varphi$. We calculate $\langle F, F\rangle = -\frac{2e^2}{r^4} + \frac{2b^2}{r^4}$ and get

$$\begin{aligned}4\pi T &= \frac{e^2}{r^4}(U^\flat \otimes U^\flat - Q^\flat \otimes Q^\flat) + \frac{b^2}{r^2}(\mathrm{d}\theta^2 + \sin^2\theta \mathrm{d}\varphi^2) \\ &\quad - \frac{1}{4}\left(-\frac{2e^2}{r^4} + \frac{2b^2}{r^4}\right)(-U^\flat \otimes U^\flat + Q^\flat \otimes Q^\flat + r^2(\mathrm{d}\theta^2 + \sin^2\theta \mathrm{d}\varphi^2) \\ &= -\frac{1}{r^4}\left(\frac{e^2}{2} + \frac{b^2}{2}\right)(-U^\flat \otimes U^\flat + Q^\flat \otimes Q^\flat) \\ &\quad + \frac{1}{r^4}\left(\frac{e^2}{2} + \frac{b^2}{2}\right) r^2(\mathrm{d}\theta^2 + \sin^2\theta \mathrm{d}\varphi^2).\end{aligned}$$

∎

Observe that assuming $\mathrm{d}F = A$ where $A$ is some spherically independent 1-form would have lead to $b = 0$ since there is no non-vanishing vector field tangent to $S^2$.

**Lemma 7.4.2.** *Let* $T = \epsilon U^\flat \otimes U^\flat + p_{\mathrm{rad}} Q^\flat \otimes Q^\flat + p_{\mathrm{sph}} r^2(t,q)(d\theta^2 + \sin^2\theta\, d\varphi^2) + T_{\mathrm{el}}$. *Then Einstein's equations are equivalent to*

$$U \bullet U \bullet r = -(Q \bullet r)\frac{Q \bullet p_{\mathrm{rad}}}{\epsilon + p_{\mathrm{rad}}} - 2\frac{(Q \bullet r)^2}{r}\frac{p_{\mathrm{rad}} - p_{\mathrm{sph}}}{\epsilon + p_{\mathrm{rad}}} - \frac{m}{r^2} - 4\pi r\left(p_{\mathrm{rad}} - \frac{\Lambda}{8\Pi}\right) + \frac{e^2 + b^2}{2\pi r^3}, \tag{7.4.16}$$

$$U \bullet Q \bullet r = -(U \bullet r)\frac{Q \bullet p_{\mathrm{rad}}}{\epsilon + p_{\mathrm{rad}}} - 2\frac{(U \bullet r)(Q \bullet r)}{r}\frac{p_{\mathrm{rad}} - p_{\mathrm{sph}}}{\epsilon + p_{\mathrm{rad}}}, \tag{7.4.17}$$

$$U \bullet \epsilon = -(\epsilon + p_{\mathrm{rad}})\frac{Q \bullet U \bullet r}{Q \bullet r} - 2(\epsilon + p_{\mathrm{sph}})\frac{U \bullet r}{r}, \tag{7.4.18}$$

$$Q \bullet m = 4\pi r^2 (Q \bullet r)(\epsilon + \frac{\Lambda}{8\pi} + \frac{e^2 + b^2}{8\pi r^4}), \tag{7.4.19}$$

*where* $m = \frac{r}{2}(1 + (U \bullet r)^2 - (Q \bullet r)^2)$.

*As a consequence, of Einstein's equation the equations of motion,*

$$U \bullet \epsilon = -(\epsilon + p_{\mathrm{rad}}) U \bullet \lambda - 2(\epsilon + p_{\mathrm{sph}})\frac{U \bullet r}{r},$$

$$Q \bullet p_{\mathrm{rad}} = -(\epsilon + p_{\mathrm{rad}}) Q \bullet \nu - 2(p_{\mathrm{rad}} - p_{\mathrm{sph}})\frac{Q \bullet r}{r}.$$

*hold.*

*Proof.* The energy momentum tensor is given by

$$T = (\epsilon + \frac{e^2 + b^2}{8\pi r^4}) U^\flat \otimes U^\flat + (p_{\mathrm{rad}} - \frac{e^2 + b^2}{8\pi r^4}) Q^\flat \otimes Q^\flat + (p_{\mathrm{sph}} + \frac{e^2 + b^2}{8\pi r^4}) r^2 \left(d\theta^2 + \sin^2\theta\, d\varphi^2\right).$$

Writing

$$\tilde{\epsilon} = \epsilon + \frac{e^2 + b^2}{8\pi r^4}, \quad \tilde{p_{\mathrm{rad}}} = p_{\mathrm{rad}} - \frac{e^2 + b^2}{8\pi r^4} \quad \tilde{p_{\mathrm{sph}}} = p_{\mathrm{sph}} + \frac{e^2 + b^2}{8\pi r^4},$$

We can apply Lemma 7.1.2 with $\epsilon, p_{\mathrm{rad}}, p_{\mathrm{sph}}$ replaced by $\tilde{\epsilon}, \tilde{p_{\mathrm{rad}}}, \tilde{p_{\mathrm{sph}}}$. Observe that the equations of motions (7.1.7), (7.1.8) are a consequence of they sytem of differential equations (7.1.3)–(7.1.6). Equation (7.1.6) is the only equation in the system (7.1.3)–(7.1.6) which involves the function $p_{\tilde{\text{sph}}}$. Since this equation as well as Equation (7.1.7) can be solved for $p_{\tilde{\text{sph}}}$, we can replace Equation (7.1.6) by Equation (7.1.7) in our system. Using Equation (7.1.8) (which is now a consequence of Equations (7.1.3)–(7.1.5), (7.1.7)), the definition of $m$, and the commutator relation $[U, Q] = (Q \bullet \nu)U - (U \bullet \lambda)Q$ imply

$$\begin{aligned}
U \bullet U \bullet r &= -(Q \bullet r)\frac{Q \bullet p_{\tilde{\text{rad}}} + 2(p_{\tilde{\text{rad}}} - p_{\tilde{\text{sph}}})(Q \bullet r)/r}{\tilde{\epsilon} + p_{\tilde{\text{rad}}}} - \frac{m}{r^2} \\
&\quad - 4\pi r(p_{\tilde{\text{rad}}} - \frac{\Lambda}{8\Pi}) \\
&= -(Q \bullet r)\frac{Q \bullet p_{\tilde{\text{rad}}}}{\tilde{\epsilon} + p_{\tilde{\text{rad}}}} - 2\frac{(Q \bullet r)^2}{r}\frac{p_{\tilde{\text{rad}}} - p_{\tilde{\text{sph}}}}{\tilde{\epsilon} + p_{\tilde{\text{rad}}}} - \frac{m}{r^2} \\
&\quad - 4\pi r(p_{\tilde{\text{rad}}} - \frac{\Lambda}{8\Pi}), \\
U \bullet Q \bullet r &= Q \bullet U \bullet r + (Q \bullet \nu)(U \bullet r) - (U \bullet \lambda)(Q \bullet r) \\
&= -(U \bullet r)\frac{Q \bullet p_{\tilde{\text{rad}}} + 2(p_{\tilde{\text{rad}}} - p_{\tilde{\text{sph}}})(Q \bullet r)/r}{\tilde{\epsilon} + p_{\tilde{\text{rad}}}} \\
&= -(U \bullet r)\frac{Q \bullet p_{\tilde{\text{rad}}}}{\tilde{\epsilon} + p_{\tilde{\text{rad}}}} - 2\frac{(U \bullet r)(Q \bullet r)}{r}\frac{p_{\tilde{\text{rad}}} - p_{\tilde{\text{sph}}}}{\tilde{\epsilon} + p_{\tilde{\text{rad}}}}.
\end{aligned}$$

It follows from the calculation in the proof of Lemma 7.1.3 that we can replace Equation (7.1.3) by $Q \bullet m = 4\pi r^2 (Q \bullet r)\,\tilde{\epsilon}$.

Equations (7.4.16)–(7.4.19) follow now from

$$\begin{aligned}
\tilde{\epsilon} + p_{\tilde{\text{rad}}} &= \epsilon + p_{\text{rad}}, \\
Q \bullet p_{\tilde{\text{rad}}} &= Q \bullet p_{\text{rad}} - Q\left(\frac{e^2 + b^2}{8\pi r^4}\right) \\
&= Q \bullet p_{\text{rad}} + 4\frac{e^2 + b^2}{8\pi r^5}(Q \bullet r) \\
p_{\tilde{\text{rad}}} - p_{\tilde{\text{sph}}} &= p_{\text{rad}} - p_{\text{sph}} - 2\frac{e^2 + b^2}{8\pi r^4}, \\
4\pi r p_{\tilde{\text{rad}}} &= 4\pi r p_{\text{rad}} - \frac{e^2 + b^2}{2\pi r^3}.
\end{aligned}$$

Finally observe that as a consequence of these equations the Equations (7.1.7) and (7.1.8) hold unchanged, even if $e^2 + b^2 \neq 0$. ∎

The system of differential Equations (7.4.16)–(7.4.19) is singular at $r = 0$. It can be shown that in the case $e = b = 0$ this singularity is only a coordinate singularity, provided, the initial data can be smoothly extended to $r < 0$ as symmetric functions. A proof, however, would require the solution of a mixed initial-value-boundary-value problem (cf. (Courant and Hilbert 1962; Müller zum Hagen, Yodzis, and Seifert 1974)). If $e^2+b^2 \neq 0$ then there is physical singularity at $r = 0$. This follows immediately from the fact that the invariant function $g[^0_2](T,T)$ blows up as $r \to 0$. In the following theorem we will solve the initial value problem in a region which does not contain the centre of symmetry $r = 0$.

The initial value problem is not yet "well posed". In order to transform the system of differential equations into a quasi-linear hyperbolic system of equations we write $y = U \bullet r$ and $\Gamma = Q \bullet r$. Adding these two equations two our system of equations we obtain

$$U \bullet r = y, \tag{7.4.20}$$

$$\begin{aligned} U \bullet y = & -\Gamma \frac{Q \bullet p_{\text{rad}}}{\epsilon + p_{\text{rad}}} - 2\frac{\Gamma^2}{r}\frac{p_{\text{rad}} - p_{\text{sph}}}{\epsilon + p_{\text{rad}}} - \frac{m}{r^2} \\ & - 4\pi r\left(p_{\text{rad}} - \frac{\Lambda}{8\pi}\right) + \frac{e^2 + b^2}{2\pi r^3}, \end{aligned} \tag{7.4.21}$$

$$U \bullet \Gamma = -y\frac{Q \bullet p_{\text{rad}}}{\epsilon + p_{\text{rad}}} - 2\frac{y\Gamma}{r}\frac{p_{\text{rad}} - p_{\text{sph}}}{\epsilon + p_{\text{rad}}}, \tag{7.4.22}$$

$$U \bullet \epsilon = -(\epsilon + p_{\text{rad}})\frac{Q \bullet y}{\Gamma} - 2(\epsilon + p_{\text{sph}})\frac{y}{r}, \tag{7.4.23}$$

and

$$Q \bullet r = \Gamma, \tag{7.4.24}$$

$$Q \bullet m = 4\pi r^2 \Gamma\left(\epsilon + \frac{\Lambda}{8\pi} + \frac{e^2 + b^2}{8\pi r^4}\right). \tag{7.4.25}$$

The first 4 equations constitute a quasi-linear hyperbolic system of equations for $r, y, \Gamma, \epsilon$. It will turn out below that last 2 equations will hold everywhere if they hold initially. This leaves us with two undetermined functions, $p_{\text{rad}}$, and $p_{\text{sph}}$. In order to arrive at a well posed system of equations we could either augment the system with two more differential equations which relate the pressures $p_{\text{rad}}, p_{\text{sph}}$ to our remaining quantities or we could impose functional relationships. We will opt for the latter possibility and assume $p_{\text{rad}} = p_{\text{sph}}$ since in this case we can solve Equations (7.1.7), (7.1.8) explicitly which greately simplifies the problem. Furthermore, we will assume a functional relationship $p_{\text{rad}} = p(\epsilon)$ which describes the physical properties of our fluid. This equation is referred to as an *equation of state.*

**Theorem 7.4.1.** *Let $M = \mathbb{R} \times \mathbb{R}^+ \setminus \{0\} \times S^2$, coordinized by functions $t, q, \theta, \varphi$, $\Lambda \in \mathbb{R}$, and let $p\colon \mathbb{R} \to \mathbb{R}$ be a given monotonically increasing smooth function.*

*For any $e, b \in \mathbb{R}$ and any smooth functions $\hat{r}\colon \mathbb{R}^+ \to \mathbb{R}$, $\hat{\epsilon}\colon \mathbb{R}^+ \to \mathbb{R}^+$ such that $\hat{\epsilon}(q) + p \circ \hat{\epsilon}(q) > 0$ for all $q \in \mathbb{R}^+$ there is a neighbourhood $\mathcal{U}$ of the hypersurface $\{t_0\} \times \mathbb{R}^+ \times S^2$ and a unique Lorentz metric $g$ on $\mathcal{U}$ such that*

*(i) g satisfies the spherically symmetric Einstein equation with cosmological constant $\Lambda$ for a perfect fluid with equation of state $p(\epsilon)$, a source-free electromagnetic field with parameters $e, b$;*
*(ii)* $\epsilon_{|t=t_0} = \hat{\epsilon}$;
*(iii)* $r_{|t=t_0} = \hat{r}$.

*Proof.* It is sufficient to prove that for all $r_1 > 0, r2 > r_1$ there is a neighbourhood $\mathcal{U}_{r_1,r_2}$ of $\{t_0\} \times (r_1, r_2) \times S^2$ and a unique solution $g$ defined on $\mathcal{U}_{r_1,r_2}$.

In a first step we will show that — up to reparametrization $\lambda \to h(q)\lambda$, $\nu \to H(q)\nu$ — the initial value problem has a unique solution such that $(r, \epsilon)$ coincides with $(\hat{r}, \hat{\epsilon})$ at $t = t_0$. To this end we introduce two new dependent variables $\Gamma = Q \bullet r$ and $y = U \bullet r$ and augment the system of equations by the definition of $y$. This gives the system of Equations (7.4.20)–(7.4.25). The additional initial values $\hat{\Gamma}$ and $\hat{y}$ for $\Gamma, y$ are calculated from the necessary initial, "constraint equations"

$$\frac{\hat{r}}{2}\left(1 + \hat{y}^2 - \hat{\Gamma}^2\right) = 4\pi \int_0^{\hat{r}} \hat{\epsilon}\tilde{r}^2 \, \mathrm{d}\tilde{r} + \frac{\hat{r}^3}{6}\Lambda - \frac{1}{2}\frac{e^2 + b^2}{r},$$
$$\hat{\Gamma} = Q \bullet \hat{r}.$$

In the first step we will show that there is a unique solution to the initial value problem (7.4.20)–(7.4.23). In the second step we will show that this solution also satisfies Equations (7.4.24),(7.4.25).

Equations ordinary differential equations (7.1.7)), (7.1.8) are consequences of Equations (7.4.20)–(7.4.23) and can be solved independently. In order to simplify the formulas, we define the "baryon number density" $n$ by

$$\frac{\mathrm{d}\,\epsilon}{\mathrm{d}\,n} = \frac{\epsilon + p(\epsilon)}{n}$$

and the asymptotic behaviour $\epsilon(n)/n \to 1$ $(\epsilon \to 0)$. It follows that $\int_0^{\epsilon} \frac{\mathrm{d}\tilde{\epsilon}}{\tilde{\epsilon}+p(\tilde{\epsilon})} = \ln(n)$ which in turn implies

$$\frac{U \bullet \epsilon}{\epsilon + p(\epsilon)} = U \bullet \ln(n)$$

and

$$\begin{aligned}
Q \bullet \ln\left(\frac{\epsilon + p(\epsilon)}{n}\right) &= \frac{n}{\epsilon + p(\epsilon)} Q \bullet \left(\frac{\epsilon + p(\epsilon)}{n}\right) \\
&= \frac{n}{\epsilon + p(\epsilon)}\left(\frac{Q \bullet \epsilon + \frac{\mathrm{d}p}{\mathrm{d}\epsilon} Q \bullet \epsilon}{n} - \frac{(\epsilon + p(\epsilon))\frac{\mathrm{d}n}{\mathrm{d}\epsilon} Q \bullet \epsilon}{n^2}\right) \\
&= \frac{Q \bullet \epsilon}{\epsilon + p(\epsilon)}\left(1 + \frac{\mathrm{d}p}{\mathrm{d}\epsilon} - 1\right) = \frac{Q \bullet p(\epsilon)}{\epsilon + p(\epsilon)}.
\end{aligned}$$

Since Equations (7.1.7), (7.1.8) are equivalent to

$$U \bullet \ln(n) = \frac{U \bullet \epsilon}{\epsilon + p} = -U \bullet \lambda - U \bullet \ln(r^2),$$
$$Q \bullet \ln\left(\frac{\epsilon + p}{n}\right) = \frac{Q \bullet p}{\epsilon + p} = -Q \bullet \nu,$$

we get

$$\mathrm{e}^{-\lambda} = h(q) r^2 n \text{ and } \mathrm{e}^{-\nu} = \frac{\epsilon + p}{H(t) n},$$

where $h, H$ are constants of integration. The system of equations (7.4.20)–(7.4.23) is equivalent to

$$\begin{pmatrix} \dot{r} \\ \dot{y} \\ \dot{\Gamma} \\ \dot{\epsilon} \end{pmatrix} + \begin{pmatrix} 0 & 0 & 0 & 0 \\ 0 & 0 & 0 & \Gamma A \\ 0 & 0 & 0 & yA \\ 0 & B & 0 & 0 \end{pmatrix} \begin{pmatrix} r' \\ y' \\ \Gamma' \\ \epsilon' \end{pmatrix}$$
$$= \begin{pmatrix} \frac{Hny}{\epsilon+p} \\ \frac{Hn}{\epsilon+p}\left(-\frac{m}{r^2} + \frac{e^2+b^2}{2\pi r^3} - \frac{r}{2}b^2 - 4\pi r\left(p - \frac{\Lambda}{8\pi}\right)\right) \\ 0 \\ -2Hn\frac{y}{r} \end{pmatrix},$$

where $\dot{a} := \partial_t a$, $a' := \partial_q a$, and

$$A = hH\left(\frac{nr}{\epsilon + p}\right)^2 \frac{\mathrm{d}p}{\mathrm{d}\epsilon}, \qquad B = hH\frac{(nr)^2}{\Gamma}.$$

The matrix

$$\begin{pmatrix} 0 & 0 & 0 & 0 \\ 0 & 0 & 0 & \Gamma A \\ 0 & 0 & 0 & yA \\ 0 & B & 0 & 0 \end{pmatrix}$$

has left eigenvalues $\alpha_i$ and left eigenvectors $l_i$ given by

$$\alpha_{1/2} = 0, \qquad \alpha_{3/4} = \pm\sqrt{AB\Gamma},$$
$$l_1 = (1, 0, 0, 0), \qquad l_2 = (0, y, -\Gamma, 0), \qquad l_{3/4} = \left(0, 1, 0, \pm\sqrt{\frac{A\Gamma}{B}}\right).$$

The left eigenvectors $l_1, l_2, l_3$, and $l_4$ are linearly independent unless $\Gamma = 0$, $\epsilon + p(\epsilon) = 0$ or $\frac{\mathrm{d}p}{\mathrm{d}\epsilon} = 0$. Hence our system of differential equations is hyperbolic and admits a unique, local solution in $\mathcal{U}_{r_1, r_2}$ (cf. Theorem 7.3.1). By the uniqueness of the solution we can choose a collection of such intervals $(r_1, r_2)$ which cover all of $\mathbb{R}^+$ and patch the solutions with respect to these intervals together. This gives a unique solution in a neighbourhood of the entire hypersurface $\{t_0\} \times \mathbb{R}^+ \times S^2$.

We will now show that this solution also satisfies Equations (7.4.24) and (7.4.25) everywhere. We have chosen our initial data $\Gamma$, $y$ so that they hold at $t - t_0$. From

$$\begin{aligned} U \bullet (\Gamma - Q \bullet r) &= U \bullet \Gamma - Q \bullet y - [U, Q] \bullet r \\ &= -y \frac{Q \bullet p}{\epsilon + p} + \frac{\Gamma}{\epsilon + p} U \bullet \epsilon + 2 \frac{\Gamma y}{r} - (Q \bullet \nu) y \\ &\quad + (U \bullet \lambda)(Q \bullet r) + (U \bullet \lambda)\Gamma - (U \bullet \lambda)\Gamma \\ &= -y \left( \frac{Q \bullet p}{\epsilon + p} + Q \bullet \nu \right) + \Gamma \left( \frac{U \bullet \epsilon}{\epsilon + p} + U \bullet \lambda + 2\frac{y}{r} \right) \\ &\quad + U \bullet \lambda (Q \bullet r - \Gamma) = U \bullet \lambda (Q \bullet r - \Gamma) \end{aligned}$$

we obtain a linear differential equation for the function $(\Gamma - Q \bullet r)$. Since $\Gamma - Q \bullet r = 0$ initially this equation therefore holds everywhere. The other constraint equation (7.4.25) is slightly more complicated. Observe first that Equations (7.1.7), (7.1.8) imply $[U, Q] = -\frac{Q \bullet p}{\epsilon + p} U + \left( \frac{U \bullet \epsilon}{\epsilon + p} + \frac{2y}{r} \right) Q$. The proof of Lemma 7.1.3 shows that the equation $U \bullet m = 4\pi r^2 \Gamma (\epsilon + \frac{e^2 + b^2}{8\pi r^2})$ is a consequence of Equations (7.4.20)–(7.4.23). We have

$$\begin{aligned} &U \bullet Q \bullet \left( 4\pi \int_0^r \left( \epsilon + \frac{\Lambda}{8\pi} \right) \tilde{r}^2 \, d\tilde{r} - \frac{1}{2} \frac{e^2 + b^2}{r} - m \right) \\ &\quad = U \bullet \left( 4\pi \left( \epsilon + \frac{\Lambda}{8\pi} \right) r^2 \Gamma + \frac{e^2 + b^2}{2r^2} \Gamma \right) - Q \bullet U \bullet m \\ &\quad\quad + \frac{Q \bullet p}{\epsilon + p} (U \bullet m) - \overbrace{\left( \frac{U \bullet \epsilon}{\epsilon + p} + \frac{2y}{r} \right)}^{1} (Q \bullet m) \\ &\quad = 8\pi r y \Gamma \left( \epsilon + \frac{\Lambda}{8\pi} \right) - \frac{e^2 + b^2}{r^2} y \Gamma \\ &\quad\quad + 4\pi r^2 \left( \epsilon + \frac{\Lambda}{8\pi} + \frac{e^2 + b^2}{8\pi r^4} \right) (U \bullet \Gamma) \\ &\quad\quad + 4\pi r^2 (U \bullet \epsilon) \Gamma + 4\pi Q \bullet \left( r^2 y \left( p - \frac{\Lambda}{8\pi} - \frac{e^2 + b^2}{8\pi r^4} \right) \right) \\ &\quad\quad - \frac{Q \bullet p}{\epsilon + p} \left( 4\pi r^2 y \left( p - \frac{\Lambda}{8\pi} - \frac{e^2 + b^2}{8\pi r^4} \right) \right) + \overbrace{\frac{Q \bullet y}{\Gamma}}^{1} (Q \bullet m) \\ &\quad = \overbrace{8\pi r y \Gamma \left( \epsilon + \frac{\Lambda}{8\pi} \right)}^{2} \overbrace{- \frac{e^2 + b^2}{r^3} y \Gamma}^{5} \end{aligned}$$

$$\overbrace{+4\pi r^2\left(\epsilon+\frac{\Lambda}{8\pi}+\frac{e^2+b^2}{8\pi r^4}\right)\left(-y\frac{Q\bullet p}{\epsilon+p}\right)}^{4}$$

$$+\,4\pi r^2\left(\overbrace{-(\epsilon+p)\frac{Q\bullet y}{\Gamma}}^{3}\overbrace{-2(\epsilon+p)\frac{y}{r}}^{2}\right)\Gamma$$

$$+\,8\pi r\Gamma y\left(\overbrace{p-\frac{\Lambda}{8\pi}}^{2}\overbrace{-\frac{e^2+b^2}{8\pi r^4}}^{5}\right)$$

$$\overbrace{+4\pi r^2(Q\bullet y)\left(p-\frac{\Lambda}{8\pi}-\frac{e^2+b^2}{8\pi r^4}\right)}^{3}$$

$$\overbrace{+4\pi r^2 y(Q\bullet p)}^{4}\overbrace{+2y\Gamma\frac{e^2+b^2}{r^3}}^{5}$$

$$\overbrace{-\frac{Q\bullet p}{\epsilon+p}\left(4\pi r^2 y\left(p-\frac{\Lambda}{8\pi}-\frac{e^2+b^2}{8\pi r^4}\right)\right)}^{4}\overbrace{+\frac{Q\bullet y}{\Gamma}(Q\bullet m)}^{3}$$

$$=\overbrace{8\pi r\Gamma y\left(\epsilon+\frac{\Lambda}{8\pi}-(\epsilon+p)+p-\frac{\Lambda}{8\pi}\right)}^{2}$$

$$\overbrace{+(Q\bullet y)\Big(-4\pi r^2\frac{\epsilon+p}{\Gamma}}^{3}$$

$$\overbrace{+4\pi r^2(Q\bullet y)\left(p-\frac{\Lambda}{8\pi}-\frac{e^2+b^2}{8\pi r^4}\right)+\frac{Q\bullet m}{\Gamma}\Big)}^{3}$$

$$\overbrace{+4\pi r^2 y(Q\bullet p)\Big(-\left(\epsilon+\frac{\Lambda}{8\pi}+\frac{e^2+b^2}{8\pi r^4}\right)\left(\frac{1}{\epsilon+p}\right)}^{4}$$

$$\overbrace{+1-\frac{1}{\epsilon+p}\left(p-\frac{\Lambda}{8\pi}-\frac{e^2+b^2}{8\pi r^4}\right)\Big)}^{4}\overbrace{+\frac{e^2+b^2}{r^3}y\Gamma\,(-1-1+2)}^{5}$$

$$=\frac{Q\bullet y}{\Gamma}\left(-4\pi r^2\left(\epsilon+\frac{\Lambda}{8\pi}+\frac{e^2+b^2}{8\pi r^4}\right)+Q\bullet m\right).$$

We get a linear differential equation for $4\pi r^2\Gamma(\epsilon+\frac{\Lambda}{8\pi}+\frac{e^2+b^2}{8\pi r^4})-Q\bullet m$ and $4\pi r^2\Gamma(\epsilon+\frac{\Lambda}{8\pi}+\frac{e^2+b^2}{8\pi r^4})=Q\bullet m$ holds everywhere since it is initially satisfied. ∎

Observe that the monotonicity condition on $p$ is necessary to obtain a hyperbolic system of differential equations. The function $\sqrt{\frac{\mathrm{d}p}{\mathrm{d}\epsilon}}$ can be physically interpreted as the velocity of sound. This indicates that $\frac{\mathrm{d}p}{\mathrm{d}\epsilon}>0$ is physically well justified assumption.

**Corollary 7.4.1.** *The characteristic directions of the Einstein equation are*

$$U,\ z=U+\sqrt{\frac{dp}{d\epsilon}}Q,\ \underline{z}=U-\sqrt{\frac{dp}{d\epsilon}}L$$

*It follows that for* $\frac{\mathrm{d}p}{\mathrm{d}\epsilon}>1$ *information can travel faster than light.*

*Proof.* $AB\Gamma=h^2H^2(nr)^4/(\epsilon+p)^2\frac{\mathrm{d}p}{\mathrm{d}\epsilon}=\mathrm{e}^{-2\nu}\mathrm{e}^{2\lambda}\frac{\mathrm{d}p}{\mathrm{d}\epsilon}$ implies $\partial_t\pm\sqrt{AB\Gamma}\partial_q=\mathrm{e}^{\nu}\left(U\pm\sqrt{\frac{\mathrm{d}p}{\mathrm{d}\epsilon}}Q\right)$. The second assertion follows from Corollary 7.3.1. ∎

*Remark 7.4.1.* The system of differential equations is especially simple in the case of dust: $p(\epsilon)=0$. Then it reduces to the following system of ordinary differential equations.

$$U\bullet r=y,$$
$$U\bullet y=-\frac{1+y^2-\Gamma^2}{2r}+\frac{e^2}{2r^3}+\frac{r\Lambda}{2},$$

where $\Gamma$ does not depend on $t$. The equation for the energy density $\epsilon$ decouples and can be calculated from $U\bullet\epsilon=-\epsilon\frac{Q\bullet y}{\Gamma}-2\epsilon\frac{y}{r}$.

## 7.5 Static perfect fluid stars

Most stars do not change very much over long time spans. It appears therefore reasonable to assume that their interior can be described by static, spherically symmetric solution to Einstein's equation. Static solutions should be an even better description once these stars have burned all of their fuel. In this section we will show that the assumption of staticity has a striking consequence: there is an absolute upper bound for the mass of a static, spherically symmetric star. Further, this bound is so small that it is exceeded by a multitude of known stars, which indicates that many of these stars will collapse into singularities once their fuel is exhausted.

In this section we will model a non-rotating star by a spherically symmetric, static perfect fluid spacetime. Under the assumption of staticity, Einstein's equation for a perfect fluid reduces to the following ordinary differential equation.

**Theorem 7.5.1.** *Let $(M, g)$ be a spherically symmetric 4-dimensional spacetime which is $C^1$ and piecewise smooth, and assume that there exists a timlike Killing vector field $U$ such that the energy momentum tensor is given by $T = (\epsilon + p)U^\flat \otimes U^\flat + p\, g$, where $\epsilon$, $p$ are given, smooth function.*

*If the solution has a well defined centre of symmetry $r = 0$ then $p$ satisfies the* Tolman-Oppenheimer-Volkoff equation

$$\frac{\mathrm{d}p}{\mathrm{d}r} = -(p+\epsilon)\frac{m(r) + 4\pi r^3 (p - \frac{\Lambda}{8\pi})}{r(r - 2m(r))},$$

*where $m(r) = 4\pi \int_0^r \left(\epsilon(\hat{r}) + \frac{\Lambda}{8\pi}\right) \hat{r}^2 \,\mathrm{d}\hat{r}$.*

*Conversely, let $\epsilon\colon \mathbb{R}^+ \to \mathbb{R}^+$, $p\colon \mathbb{R}^+ \to \mathbb{R}^+$ given continuous functions such that*

*(i) $\epsilon$ and $p$ can be extended to $\mathbb{R}^-$ as smooth, even functions,*
*(ii) $\epsilon$ and $p$ vanish for $r > r_0$,*
*(iii) $\epsilon(r) + p(r) > 0$ for $r < r_0$,*
*(iv) $\epsilon$ and $p$ are smooth for $r < r_0$,*
*(v) $\epsilon$ and $p$ satisfy the Tolman-Oppenheimer-Volkoff equation, where*

$$m(r) = 4\pi \int_0^r \left(\epsilon(\hat{r}) + \frac{\Lambda}{8\pi}\right) \hat{r}^2 \,\mathrm{d}\hat{r}.$$

*Then there exists a unique 4-dimensional, spherically symmetric Lorentzian $C^1$-manifold $(M, g)$ which is piecewise smooth and satisfies*

*(a) $\mathrm{Ric} - \frac{1}{2}\mathrm{Scal}\, g = 8\pi(\epsilon \circ r + p \circ r)U^\flat \otimes U^\flat + (p \circ r)\, g$,*
*(b) $U$ is a timlike Killing vector field.*

*There is an $m_0 > 0$ such that for $r > r_0$ this spacetime is isometric to a spherically symmetric vacuum spacetime with cosmological constant $\Lambda$ and mass $m_0 + \frac{\Lambda r^3}{6}$.*

*Proof.* Equations (7.1.8) and (7.1.5) imply

$$(Q \bullet r)\frac{Q \bullet p}{\epsilon + p} = -\left(\frac{m}{r^2} + 4\pi r(p - \frac{\Lambda}{8\pi})\right).$$

Since by definition of $m$, $Q \bullet r = \sqrt{1 - \frac{2m}{r}}$, the validity of the Tolman-Oppenheimer-Volkoff equation follows from $\frac{\mathrm{d}p}{\mathrm{d}r} = \frac{Q \bullet p}{Q \bullet r}$.

For the converse we have to check that the Tolman-Oppenheimer-Volkoff equation implies that there exist functions $\nu(r), \lambda(r)$ which satisfy Equations (7.1.3)–(7.1.6).

Observe first that Equation (7.1.4) is trivially satisfied. Equation (7.1.3) is equivalent to

$$\frac{1}{r}\frac{\partial m}{\partial r}Q \bullet r = \frac{1}{r}Q \bullet (\frac{r(1-(Q \bullet r)^2}{2}$$

which implies $m(r) = \frac{r}{2}(1-(Q \bullet r)^2) + A(t) = \frac{r}{2}(1-\mathrm{e}^{-2\lambda}) + A(t)$, where $A(t)$ is a constant of integration. From $m(0) = 0$ we get $A(t) = 0$ and therefore

$$\mathrm{e}^{2\lambda(r)} = \left(1 - \frac{2m}{r}\right)^{-1}. \tag{7.5.26}$$

The 4-dimensional solution should be smooth at $r = 0$. Since a rotation by $\pi$ is an isometry it is clear that the metric must be invariant under the transformation $r \mapsto -r$. But this is equivalent to $\lambda$ and $\nu$ being even functions of $r$. We also need that $\lambda$ satisfies $\lambda(0) = 0$ because for any other value we would get a conical singularity. In fact, consider a centred sphere with area $A(r)$ and (geodesic) radius $R(r)$. In the limit $r \to 0$ we obtain

$$\lim_{R\to 0}\frac{A(r)}{R^2(r)} = \lim_{r\to 0}\frac{\frac{4}{3}\pi r^2}{(\int_0^r (Q \bullet \hat{r})^{-1}\,\mathrm{d}\hat{r})^2} = \frac{4\pi}{3}\frac{2}{2(Q \bullet r)^{-2}{}_{|r=0}} = \frac{4\pi}{3}\mathrm{e}^{-2\lambda(0)}$$

which reduces to the Euclidean relation in the tangent space at the centre of symmetry if and only if $\lambda(0) = 0$.[12]

To see that our conditions are sufficient for the smooth extendability of $g$ to $r = 0$ consider the coordinate transformation $x^1 = r\cos\theta\cos\phi$, $x^2 = r\cos\theta\sin\phi$, $x^3 = r\sin\theta$. Then the metric is of the form

$$g = \mathrm{e}^{2\nu}\mathrm{d}t^2 + \frac{\mathrm{e}^{2\lambda}-1}{\sum_{i=1}^3 (x^i)^2}\sum_{i,j=1}^3 x^i x^j \mathrm{d}x^i \mathrm{d}x^j + \sum_{i=1}^3 (\mathrm{d}x^i)^2$$

Observe first that there are smooth functions $\tilde{\nu}, \tilde{\lambda}$ in a neighbourhood of $(x^1, x^2, x^3) = 0$ with

$$\tilde{\nu}(x^1,x^2,x^3) = \nu(\sqrt{\sum_{i=1}^3 (x^i)^2}), \qquad \tilde{\lambda}(x^1,x^2,x^3) = \lambda(\sqrt{\sum_{i=1}^3 (x^i)^2})$$

if and only if $\nu$ and $\lambda$ are even functions. Assume now that $\lambda$ is even. Then the Taylor series of $\mathrm{e}^{2\lambda} - 1$ is a series in the variable $\sum_{i=1}^3 (x^i)^2$. The equation $\lambda(0) = 0$ implies therefore that the quotient

[12] If $\lambda(0) > 0$ we would get the analogue of the tip of a 3-dimenional cone. This can be visualized in the 2-dimensional case with the sphere being replaced by a circle.

$$(x^1, x^2, x^3) \to \frac{e^{2\lambda \sqrt{\sum_{i=1}^{3}(x^i)^2}} - 1}{\sum_{i=1}^{3}(x^i)^2}$$

well defined at $(x^1, x^2, x^3) = 0$ and smooth.

We obtain $\lambda(0) = 0$ from Equation (7.5.26) since $m$ vanishes to third order at $r = 0$. The function $\lambda$ is even because of (7.5.26) and the fact that the integrand $\epsilon + \frac{\Lambda}{8\pi}$ of $m$ is even.

Equations (7.1.5) simplifies to

$$e^{-2\lambda}\partial_r \nu - \frac{m}{r^2} - 4\pi r \left(p - \frac{\Lambda}{8\pi}\right) = 0$$

which, using $e^{-2\lambda} = 1 - 2m/r$, is equivalent to

$$\partial_r \nu = \frac{m + 4\pi r^3 \left(p - \frac{\Lambda}{8\pi}\right)}{r(r - 2m)}.$$

We can therefore determine $Q$ up to a constant of intergration. Observe that $\nu$ is even since the function

$$r \mapsto \frac{m(r) + 4\pi r^3 \left(p - \frac{\Lambda}{8\pi}\right)}{r(r - 2m(r))}$$

to be integrated is uneven. The Tolman-Oppenheimer-Volkov equation implies now the equation of motion (7.1.8). Since this equation is independent from Equations (7.1.3)–(7.1.5) but Equation (7.1.8) is a consequence of Equations (7.1.3)–(7.1.6) we can derive Equation (7.1.6) from the system of Equations (7.1.3)–(7.1.5), (7.1.8). Let

$$m_0 = \lim_{r \to r_0, r < r_0} m(r) - \frac{\Lambda {r_0}^3}{6}.$$

If we extend $\lambda$ beyond $r_0$ using $e^{2\lambda(r)} = (1 - \frac{2m_0}{r} - \frac{\lambda r^2}{3})^{-1}$ for $r \geq r_0$ then $\lambda$ is continuous. It is $C_1$ if and only if

$$\lim r \to r_0, r < r_0 Q \bullet m(r) = \frac{\Lambda r^2}{2}.$$

Since Equations (7.1.3)–(7.1.6) are satified we can appeal to Lemma 7.1.3 and infer that $\lambda$ is $C^1$ if and only if $4\pi {r_0}^2(\epsilon(r_0) + \frac{\Lambda}{8\pi}) = \frac{\Lambda {r_0}^2}{2}$, which is equivalent to $\epsilon(r_0) = 0$. We choose the constant of integration for $\nu$ such that we have $e^{2\nu(r_0)} = 1 - \frac{2m_0}{r_0} - \frac{\Lambda {r_0}^2}{3}$ and extend the function $\nu$ such that $e^{2\nu(r)} = 1 - \frac{2m_0}{r} - \frac{\Lambda r^2}{3}$ for $r > r_0$ It follows that $\nu$ is continuous. The function $\nu$ is $C^1$ if and only if

$$\lim_{r \to r_0, r < r_0} \partial_r (e^{2\nu}) = \frac{2m_0}{(r_0)^2} - \frac{2\Lambda r_0}{3}.$$

We calculate

$$\begin{aligned}\lim_{r\to r_0, r<r_0} \partial_r (\mathrm{e}^{2\nu}) &= 2(\mathrm{e}^{2\nu(r_0)}) \lim_{r\to r_0, r<r_0} \partial_r \nu \\ &= 2(1-\frac{2m(r_0)}{r_0})\frac{m(r_0)+4\pi r_0{}^3\,(p(r_0)-\frac{\Lambda}{8\pi})}{r_0(r_0-2m(r_0))} \\ &= \frac{2}{r_0{}^2}(m(r_0)+4\pi r_0{}^3\left(p(r_0)-\frac{\Lambda}{8\pi}\right)) \\ &= \frac{2}{r_0{}^2}\left(m_0+\frac{\Lambda r_0{}^3}{6}+4\pi r_0{}^3 p(r_0)-\frac{\Lambda r_0{}^3}{2}\right).\end{aligned}$$

Hence $\nu$ is $C^1$ if and only if $p(r_0)=0$.

Uniqueness follows from Theorem 7.4.1. ∎

Observe that we have used all our freedom in order to construct a spherically symmetric $C^1$ spacetime with exterior vacuum solution. In general, it is impossible to achieve higher differentiability. Even if staticity is not assumed, a *smooth matching* of the interior solution to the vacuum spacetime would imply properties of the boundary which are so strong that they could be used in order to calculate the collapse of the star explicity (Kriele 1995).

Independently of the pressure, there is an upper mass limit for spherically symmetric stars.

**Theorem 7.5.2.** *Let $(M,g)$ be a spherically symmetric static, perfect fluid spacetime with positive, radially decreasing energy density ($\epsilon \geq 0$, $\frac{\mathrm{d}\epsilon}{\mathrm{d}r} \leq 0$).*

*(i) Let $r_c > 0$. Then the mass associated with the radius $r_c$ satisfies the inequality $m(r_c) \leq 5r_c/9$. If, in addition, we assume that $p - \frac{\Lambda}{8\pi} \geq 0$ then the more stringent inequality $m(r_c) \leq 4r_c/9$ holds.*

*(ii) Assume that there is an equation of state $p\colon \epsilon \mapsto p(\epsilon)$ and let $\epsilon_c > 0$. Then there is an $m_c \in \mathbb{R}$ which only depends on the equation of state for low energy densities, $p\colon [0,\epsilon_c] \to \mathbb{R}$ such that $m(r) \leq m_c$ for all $r \in \mathbb{R}^+$.*

*Proof.* Equation (7.1.6) is equivalent to $Q \bullet Q \bullet \nu + (Q \bullet \nu)^2 = -2m/r^3 + 4\pi(\epsilon + p)$ which (together with Equation (7.1.8)) implies

$$\begin{aligned}Q \bullet \left(\frac{1}{r}\mathrm{e}^{\nu} Q \bullet \nu\right) &= \mathrm{e}^{\nu}\left(\frac{1}{r}\left(\frac{-2m}{r^3}+4\pi(\epsilon+p)\right)-\frac{1}{r^2}\Gamma Q \bullet \nu\right) \\ &= \mathrm{e}^{\nu}\left(\frac{-3m}{r^4}+\frac{4\pi}{r}\left(\epsilon+\frac{\Lambda}{8\pi}\right)\right),\end{aligned}$$

where for the last equality we have used Equation (7.1.5). Since $\mathrm{d}\epsilon/\mathrm{d}r \leq 0$ we have

$$m(r) = 4\pi \int_0^r \left(\epsilon(\hat{r}) + \frac{\Lambda}{8\pi}\right) \hat{r}^2 \mathrm{d}\hat{r}$$
$$\geq 4\pi \left(\epsilon(r) + \frac{\Lambda}{8\pi}\right) \int_0^r \hat{r}^2 \mathrm{d}\hat{r} = 4\pi \left(\epsilon(r) + \frac{\Lambda}{8\pi}\right) r^3/3.$$

Hence $Q \bullet \left(\frac{1}{r}\mathrm{e}^{\nu} Q \bullet \nu\right) \leq 0$ and an integration yields

$$\frac{1}{r}\mathrm{e}^{\nu(r)}(Q \bullet \nu)_{|r} \geq \frac{1}{r_c}\mathrm{e}^{\nu(r_c)}(Q \bullet \nu)_{|r_c}$$

for $r \leq r_c$. From $Q \bullet \nu = \frac{\partial \nu}{\partial r} Q \bullet r$ we obtain

$$\frac{\partial \mathrm{e}^{\nu}}{\partial r} \geq \frac{r\mathrm{e}^{\nu(r_c)}}{r_c(Q \bullet r)} (Q \bullet \nu)_{r_c}.$$

We re-express $Q \bullet r$ in terms of $m$ using $m = \frac{r}{2}(1 - (Q \bullet r)^2)$ and integrate the resulting equation. This gives

$$\mathrm{e}^{\nu(r_c)} - \mathrm{e}^{\nu(0)} \geq \frac{\mathrm{e}^{\nu(r_c)} (Q \bullet \nu)_{r_c}}{r_c} \int_0^{r_c} \frac{r}{\sqrt{1 - \frac{2m(r)}{r}}} \mathrm{d}r. \tag{7.5.27}$$

In order to estimate this integral we show first that $m(r) \geq m(r_c) \left(\frac{r}{r_c}\right)^3$ for all $r \in (0, r_c)$. Comparing the derivative of both functions we obtain that the function

$$f(r) = \frac{\mathrm{d}}{\mathrm{d}r}\left(m(r) - \frac{m(r_c) r^3}{{r_c}^3}\right) = 4\pi\epsilon(r) r^2 - 3\frac{m(r_c) r^2}{{r_c}^3}$$
$$= r^2 (4\pi\epsilon(r) - 3m(r_c)/{r_c}^3)$$

satisfies $\frac{\mathrm{d}f(r)}{\mathrm{d}r} = \frac{2}{r} f(r) + 4\pi r^2 \frac{\mathrm{d}\epsilon(r)}{\mathrm{d}r} \leq \frac{2}{r} f(r)$. Thus if there is an $r_1 \in (0, r_c)$ with $f(r_1) < 0$ then $f(r) < 0$ for all $r \in (r_1, r_c)$. Because of $m(0) = 0$ the existence of an $r_2 \in (0, r_c)$ with $m(r_2) < m(r_c) \left(\frac{r_2}{r_c}\right)^3$ implies the existence of an $r_1 \in (0, r_2)$ with $f(r_1) < 0$. Since $f(r) \leq 0$ for all $r \in (r_1, r_c)$ we obtain $m(r) < m(r_c) \left(\frac{r}{r_c}\right)^3$ for all $r \in (r_2, r_c)$ in contradiction to $m(r_c) = m(r_c) \left(\frac{r_c}{r_c}\right)^3$. Hence such an $r_2$ cannot exist which implies our estimate $m(r) \geq m(r_c) \left(\frac{r}{r_c}\right)^3$ for all $r \in (0, r_c)$.

Inserting this estimate into Inequality (7.5.27) we obtain

$$\mathrm{e}^{\nu(0)} \leq \mathrm{e}^{\nu(r_c)} \left(1 - \frac{(Q \bullet \nu)_{r_c}}{r_c} \int_0^{r_c} \frac{r}{\sqrt{1 - \frac{2m(r)}{r}}} \mathrm{d}r\right)$$
$$\leq \mathrm{e}^{\nu(r_c)} \left(1 - \frac{(Q \bullet \nu)_{r_c}}{r_c} \frac{{r_c}^3}{2m(r_c)} \left(1 - \sqrt{1 - \frac{2m(r)}{r}}\right)\right),$$

and therefore (using Equation (7.1.5))

$$\begin{aligned} 0 &\le m(r_c)\sqrt{1-\frac{2m(r_c)}{r_c}}\mathrm{e}^{\nu(0)-\nu(r_c)} \\ &\le m(r_c)\sqrt{1-\frac{2m(r_c)}{r_c}} - \frac{m(r_c)+4\pi {r_c}^3\left(p-\frac{\Lambda}{8\pi}\right)}{2}\left(1-\sqrt{1-\frac{2m(r_c)}{r_c}}\right) \\ &= \frac{r_c}{2}\left(1-{\Gamma_c}^2\right)\Gamma_c - \frac{1}{2}\left(\frac{r_c}{2}\left(1-{\Gamma_c}^2\right)+4\pi {r_c}^3\left(p-\frac{\Lambda}{8\pi}\right)\right)(1-\Gamma_c), \end{aligned} \tag{7.5.28}$$

where we have set $\Gamma_c = \sqrt{1-\frac{2m(r_c)}{r_c}}$. The right hand side is a third order polynomial in $\Gamma_c$. The zeros of this polynomial are

$$1, \frac{1}{3}\left(-1 \pm 2\sqrt{1+6\pi(r_c)^2\left(p-\frac{\Lambda}{8\pi}\right)}\right).$$

It follows that the right hand side of Inequality (7.5.28) vanishes for

$$m(r_c) \in \left\{0, \frac{2}{9}r_c\left(1-6\pi(r_c)^2\left(p-\frac{\Lambda}{8\pi}\right) \pm \sqrt{1+6\pi(r_c)^2\left(p-\frac{\Lambda}{8\pi}\right)}\right)\right\}.$$

Since $m(r_c)$ is positive we obtain the mass-bound

$$\begin{aligned} m(r_c) &\le \frac{2}{9}r_c\left(1-6\pi(r_c)^2\left(p-\frac{\Lambda}{8\pi}\right) + \sqrt{1+6\pi(r_c)^2\left(p-\frac{\Lambda}{8\pi}\right)}\right) \\ &= \frac{2}{9}r_c\left(2-x^2+x\right) \end{aligned}$$

where $x = \sqrt{1+6\pi(r_c)^2\left(p-\frac{\Lambda}{8\pi}\right)}$. Since the function $x \mapsto 2-x^2+x$ has its minimum at $x = 1/2$ it follows that $m(r_c) \le 5r_c/9$. If we impose in addition the energy condition $p-\frac{\Lambda}{8\pi} \ge 0$ then we get $m_c \le 4r_c/9$. This proves the first assertion.

For the second assertion observe that $m(r_c) \ge \frac{4}{3}\pi(r_c)^3\left(\epsilon(r_c)+\frac{\Lambda}{8\pi}\right)$ since $\epsilon$ is radially decreasing. This inequality and the upper bound for $m(r_c)$ restrict all possible values $(r_c, m(r_c))$ to a compact subset $\mathcal{C} \subset \mathbb{R}^2$ which depends only on $\epsilon(r_c), p(r_c)$, and $\Lambda$. We can now solve Einstein's equations for $r > r_c$ using the known equation of state. Clearly, $m(r)$ depends continuously on the data $r_c, m(r_c)$. This implies that $m_c := \sup\{m(r) \,|\, (r_c, m(r_c)) \in \mathcal{C}\}$ is finite. ∎

Theorem 7.5.2 gives an important indication for the existence of singularities in our universe. In Sect. 7.2 we have seen that all non-trivial, maximally extended, non-flat spherically symmetric vacuum solutions with non-constant $r$ of Einstein's equation fail to be static in a subset of spacetime and contain a region where curvature diverges. We saw that it is possible to enter this region but impossible to leave it. Moreover, once having entered the region any observer will fall into the singularity (where curvature is infinite) within the finite time span $\pi m$ where $m$ is the Schwarzschild mass. The question arises whether this property of the vacuum solution also occurs for real stars which have non-vanishing energy momentum tensor. Since in Newtonian gravity we also have a central singularity in the vacuum case which can be avoided if the matter of the star is not assumed to be concentrated in a single point, it is tempting to argue that the property of the Schwarzschild solution is an artefact of the vacuum equation.

Theorem 7.5.2 indicates that this is not the case. We have proved that there is an upper limit for the concentration of matter in static, spherically symmetric perfect fluid stars, $m/r \leq 5/9$, if the energy density of the star decreases outward and is positive. These physical assumptions are very weak and seem to be satisfied for all known objects. Moreover, we have seen that for any star which is governed by an equation of state there is an absolute mass limit. What is more, this mass limit can be estimated using only the equation of state for low energies. This means that we get bounds even if we do not know the physical configuration of extremely dense stellar cores.

It has been shown[13] that there are stars which exceed the mass limits given in this section. This indicates that these stars will collapse into black holes once they have exhausted their nuclear fuel. In Chap. 9 we will give a very general argument to the same extent which does not rely on spherical symmetry. It should be noted however, that all these arguments in favour of the existence of black holes have loop holes. In this section, we heavily rely on spherical symmetry and the assumption of a perfect fluid. Moreover, it is conceivable that there are non-singular solutions which fail to be static. There are other loop holes in Chap. 9 which we will address then.

[13] The argument uses input from physics which is beyond the scope of this book, cf. (Hartle 1978)

# 8. Causality

*In this chapter we link the concept of causality to the conformal structure induced by the metric and present some elementary causal properties and their interpretation.*

*In Minkowski spacetime, causality is trivial since lightlike geodesics are straight lines. Lemma 3.1.4 shows that the* local *causal structure of arbitrary Lorentzian manifolds is the same as the causal structure of Minkowski spacetime. All non-trivial aspects of causality are therefore global in character. In this chapter we will also discuss in detail the possibility of "causality violation" due to the global geometry of spacetime.*

*Chapter 8 requires Sect. 3.1 and develops rather specialized mathematical techniques. It contains a number of technical results which are needed in Chap. 9 where the existence of singularities in generic spacetimes is proved. We will restrict to those results which are necessary to prove and interpret these singularity theorems. For a more comprehensive mathematical treatment of causality see (Beem and Ehrlich 1981; Hawking and Ellis 1973; Penrose 1972).*

According to our experience no signal is faster than light (photons). As we assume that photons move along null geodesics, the integrated light cone $C_x = \exp_x(\{v_x \in T_xM \,|\, g(v_x, v_x) = 0, v_x$ future oriented$\}$ should (at least locally) determine which events can in principle be influenced from a given event $x$. It is therefore plausible to *identify* the conformal structure $\mathfrak{C}$ of spacetime with the causal structure of the universe.

**Postulate 8.0.1 (Causality and conformal structure coincide).** *$x \in M$ can causally influence $y \in M$ if and only if $y \in J^+(x)$. Material objects can reach $y$ from $x$ if and only if $y \in I^+(x)$.*

A proper justification of Postulate 8.0.1 would require a corresponding theory of physical particles and fields. This is far beyond the scope of this book. While Postulate 8.0.1 will not be important for our theorems, it is crucial for their physical interpretation.

*Remark 8.0.1.* For arbitrary matter models, Einstein's equation does not respect the link between the light cone structure and causality. For instance, a spherically symmetric fluid with equation of state $\epsilon \mapsto p(\epsilon)$ satisfying $\mathrm{d}p/\mathrm{d}\epsilon > 1$ has spacelike characteristics (cf. Corollary 7.4.1). Consequently, the characteristics of the initial value problem associated with Einstein's equation is spacelike whence perturbations of the initial data propagate *faster than light*. For this reason one usually regards these matter models as unphysical. In fact, to date all classical (i.e., non-quantum) matter models which describe real matter have causal characteristics.

The *local causal structure* of any Lorentzian manifold is trivial, i.e. the same as in Minkowski spacetime. This follows immediately from Lemma 3.1.4 which is fundamental to this chapter. The following technical corollary will also be useful.

**Corollary 8.0.1.** *Let $(M, g)$ be a Lorentzian manifold and $\mathcal{C}$ a convex neighbourhood of $x \in \mathcal{C}$. Let $K \subset \mathcal{C}$ compact and $\gamma$ be a causal curve in $K$. Then $\gamma$ is extensible.*

*Proof.* Let $\gamma\colon [a,b) \to \mathcal{C}$ be a future directed causal curve in $K$. The curve $\gamma$ can be future extended if $\lim_{t\to b}\gamma(t)$ exists. In order to see that this limit exists, let $\{\gamma(t_i)\}_{i\in\mathbb{N}}$, $\{\gamma(s_j)\}_{j\in\mathbb{N}}$ be convergent sequences with $\lim_{i\to\infty} t_i = \lim_{j\to\infty} s_j = b$ and $x, y$ be their limit points. For any $i$ there is a $j > i$ with $\gamma(t_j) \in J^+(\gamma(s_i), \mathcal{C})$ and for any $j$ there is an $i > j$ with $\gamma(s_i) \in J^+(\gamma(t_j), \mathcal{C})$. Hence we obtain $x \in J^+(y, \mathcal{C})$ and $y \in J^+(x, \mathcal{C})$. Hence by Lemma 3.1.4 (i) there are two future directed causal vectors $v, w$ with $x = \exp_y(v)$ and $y = \exp_x(w)$. Traversing the geodesics $t \mapsto \exp_y(tv)$ backwards we see that at $x$ there is also a causal past directed vector $u$ with $\exp_x(u) = y$. Since the exponential map $\exp_x$ is a diffeomeorphsim of an open set $\tilde{\mathcal{C}} \subset T_xM$ to $\mathcal{C}$ we must have $w = u$. But this is only possible if both vectors vanish. ∎

## 8.1 Causality conditions

*It is easy to mathematically construct a spacetime with closed timelike curves. At first glance one is tempted to rule out such spacetimes since it seems possible to perform experiments in them which lead to* logical *contradictions. In this section we will investigate this issue in some detail. We will also define a slightly stronger "causality condition" which will play an important rôle in subsequent sections.*

In a general Lorentzian manifold, it is possible for closed timelike curves to exist.

**Definition 8.1.1.** *Let $x \in M$. We say that* causality *(resp.,* chronology*) is violated at $x$ if and only if there exists a closed, non-trivial causal (resp., timelike) curve from $x$ to $x$. The* chronology violating set *is given by*

$$\{x \in M \,|\, x \in I^+(x)\}$$

*and the causality violating set by*

$$\{x \in M \mid \exists \text{ a non-trivial causal curve } \gamma \text{ from } x \text{ to } x\}.$$

*A Lorentzian manifold* $(M, g)$ *is* causal *(resp.* chronological*) if the causality violating set (resp., chronology violating set) is empty. If* $(M, g)$ *is chronological (resp. causal), we sometimes say that the* chronology condition *(resp.* causality condition*) holds.*

The term 'causality violation' is somewhat misleading: the possibility of closed causal curves is not contradictory itself and there are no mathematical arguments against causality violation.

The idea that there may be closed timelike curves in our universe is not new: The concept of cyclic time was a widespread idea in ancient Greek philosophy (Kanitscheider 1984, p. 45). These Greek philosophers accepted our fundamental experience of local linearity of time but they compactified the time line to a time circle. Its circumference was identified with the time of one revolution of the universe (according to their model of planetary motion or just according to 'arbitrary' laying down[1]). The sheer length of this period is sufficient to explain why nobody of us ever has reentered her/his own past.

We can easily obtain a spacetime whose causal structure is analogous to the causal structure of this ancient greek model. Just take a horizontal strip of 2-dimensional Minkowski space and identify the upper and the lower boundary.

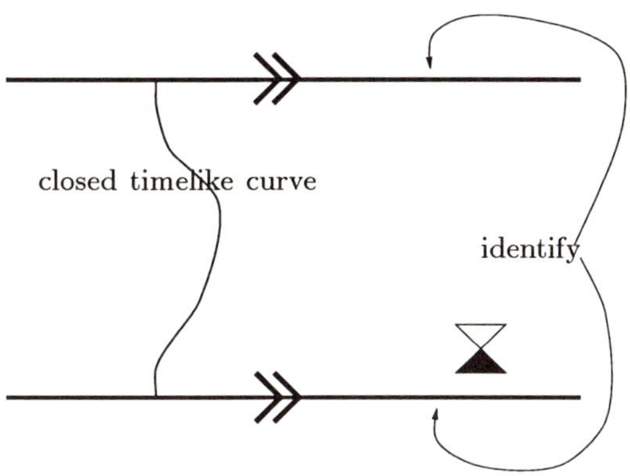


**Fig. 8.1.1.** A strip of two-dimensional Minkowski space where future and past boundaries are identified.

2

[1] Plato, for instance, chose 10,000 years (Kanitscheider 1984, p. 55)

Another very instructive example is the Lorentzian manifold

$$(\mathbb{R} \times S^1, 2\mathrm{d}\varphi \mathrm{d}t + t\mathrm{d}\varphi^2)$$

first given in (Misner 1967) (cf. Fig. 8.1.2). The chronology violating set is given by $\{(t, \varphi) \mid t < 0\}$.

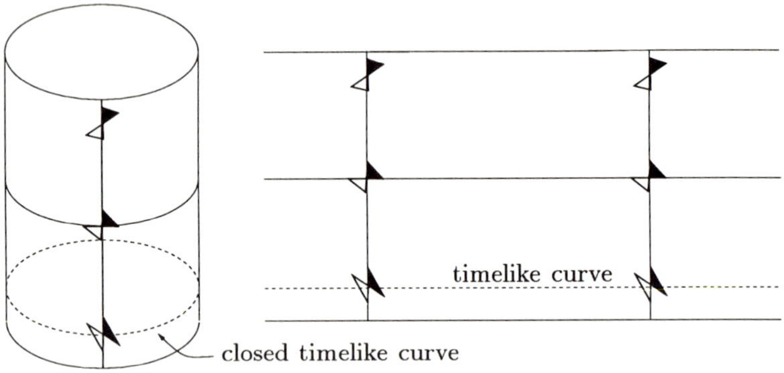


**Fig. 8.1.2.** Misner's spacetime $(S^1 \times \mathbb{R}, 2\mathrm{d}t\mathrm{d}\varphi + t\mathrm{d}\varphi^2)$

The "time compactification" arising in these examples is trivial in the sense that in both examples there is a locally isometric Lorentzian manifold which satisfies the chronology condition. In Lorentzian geometry however, there also exist non-trivial examples where causality violation arises geometrically and not merely topologically. An example which will also be of importance in Chap. 9 is the Gödel solution (Gödel 1949).

*Example 8.1.1.* The Gödel solution describes a solutions of Einstein's equation with dust and postive cosmological constant, $\mathrm{Ric} - \frac{1}{2}g + \Lambda g = 8\pi\epsilon u^\flat \otimes u^\flat$ where $u$ is a timelike unit vector field and $\epsilon = \Lambda/(4\pi)$. The metric is given by

$$g = \frac{2}{\Lambda}\Big( - \mathrm{d}t^2 + \mathrm{d}r^2 + (\sinh^2(r) - \sinh^4(r))\mathrm{d}\varphi^2 + 2\sqrt{2}\sinh^2(r)\mathrm{d}\varphi\mathrm{d}t\Big) + \mathrm{d}z^2,$$

where we have $r > 0$ and identify $\varphi$ with $\varphi + 2\pi$. The vector field $\partial_\varphi$ has closed integral curves and it is spacelike for $r < \operatorname{arsinh}(1)$. For $r = \operatorname{arsinh}(1)$ the integral curves of $\partial_\varphi$ are lightlike (but not null geodesics) and for $r > \operatorname{arsinh}(1)$ they are timelike. Since $\sinh^2(r)$ is an even function of $r$ it follows that at $r = 0$ we have only the usual coordinate singularity associated with polar coordinates. Hence spacetime has the topology $\mathbb{R}^4$ and is in particular simply connected. It follows that chronological violation is an inherent property of the solution.

A physically interesting solution of Einstein's vacuum equation with vanishing cosmological constant is the Kerr solution. For details cf. (O'Neill 1995), (Wald 1984), (Hawking and Ellis 1973).

Despite the existence of these examples most physicists regard causality violation as 'unphysical'. The reason for this rejection of causality violation is the following thought experiment:

> Suppose, you are travelling in spacetime and reach a point in your own past before your departure. Now you decide not to travel after all and instead to stay home. Contradiction.

At a first glance, the possibility of "free will" seems to be at the centre of the issue. However, following (Wheeler and Feynman 1949) Clarke (1977) has re-formulated the thought experiment in terms of a simple machine and has argued that the thought experiment is fallacious: Assume that there is a gun directed at a target in spacetime. This target is connected with a shutter which, if closed, blocks off the path between the gun and the target: If the gun is triggered, the bullet will hit the target which in turn will cause the shutter to fall. A second shot will now be blocked by the shutter and therefore cannot hit the target (c.f. Fig. 8.1.3). Now assume that the configuration is located in a region with causality violation such that the shutter falls along a closed timelike curve so that

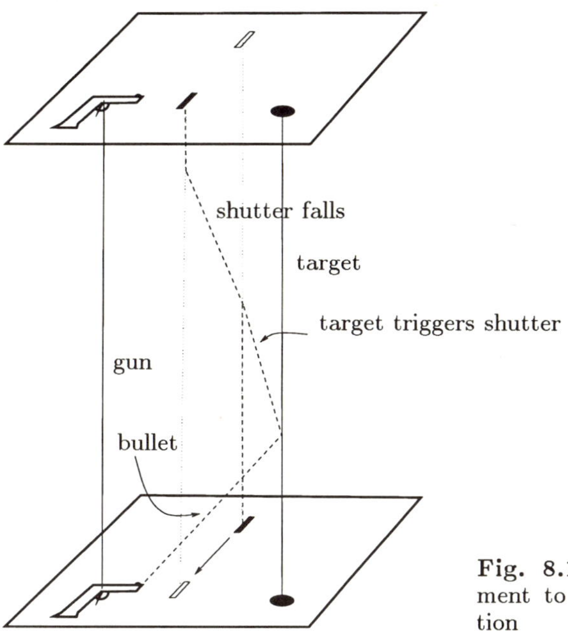


**Fig. 8.1.3.** A gedanken experiment to disprove causality violation

it blocks the bullet *before* the gun had been triggered. Again we seem to arrive at a contradiction: If the shutter is open the bullet can hit the target. But the target closes the shutter which in turn blocks the path of the bullet.

This paradox may be resolved as follows. The angle $\alpha$ under which the bullet is deflected by the shutter depends continuously on the shutter's position $x$ at the time the bullet passes the shutter. For simplicity we assume that the shutter will descend with constant velocity $v$. This velocity is continuously related to the angle $\alpha$. If the length of the closed causal curve is $T$ we obtain the relation $x = Tv(\alpha(x))$. This equation has at least one solution $x_0$ which leads to a contradiction-free situation. In physical terms this can be explained as follows: The original contradiction is due to the fact that the shutter is thought to be either up or down. However, the position of the shutter depends continuously on the parameters of the system. What happens is that while the shutter descends it grazes the bullet and thereby deflects it so that the mechanism works only imperfectly. As a consequence, the shutter is released rather late and not yet in place when the bullet hits it again due to causality violation. Hence it grazes the bullet and we are in a paradox-free time loop.

This scenario appears to be highly non-generic but Clarke argues that exactly this is the effect of causality violation: It picks out those non-generic data which are in accordance with the causal anomaly. The gist of the argument rests on the assumption that physical processes are continuous, an assumption which does not hold for quantum mechanical systems. These systems may be in discrete pure states such as spin up or spin down. However, if one tries to set up a quantum thought experiment one is faced with the fact that all predictions are probabilistic which invalidates the whole thought experiment from the outset.

There are also arguments against Clarke's resolution of the paradox. Instead of releasing the shutter directly when the target is hit we may have a device which automatically releases the shutter a certain time after the impact. This can be achieved with an electronic switch rather than a mechanical connection between target and shutter. It seems now much less probable that this device always releases the shutter such that it grazes the bullet when coming down. For Clarke's argument to work the bullet must comes out of the gun so slowly that it just touches the target but does not really hit it. Otherwise it cannot be explained that the second device is not successful in releasing the shutter at the pre-set time which would lead to a contradiction.

Whether Clarke's argument is correct or not — we are only able to conduct local experiments. But causality violation is a global effect, and so the lack of experience cannot give evidence of its absence. Any objection against causality violation rests on an (unjustified) extrapolatiom of every-day experience.

There is another point which should be addressed. Causality violation seems to constrain free will. While this is not really a physical problem, such an effect would have some bearing on philosophical and moral questions. But an almost trivial observation resolves any possible argument concerning free will at once: If we want to incorporate the notion of free will into a physical description we have to view it at least as a quantum effect (or caused by another yet undiscovered 'mechanism'), but certainly not as something fitting into the framework of classical physics. We only can expect that general relativity is a classical limit of such a theory. It is therefore quite possible that 'free will' is something like a second order effect and that the classical "limit-spacetime" of our world contains closed timelike curves even though we still enjoy free will.

With this discussion in mind we should always be very watchful if in order to obtain physical results the seemingly innocent assumption of chronology has to be made.

**Lemma 8.1.1.** *The chronology (resp., causality) violating set consists of connected components of the form* $I^+(x_i) \cap I^-(x_i)$ *(resp.,* $J^+(x_i) \cap J^-(x_i)$*)* $(i = 1, \ldots)$.

*Proof.* We only show the lemma for chronology violation. The proof for causality violation is completely analogous. Let $C$ be a connected component of the chronology violating set and $x \in C$. Since $C$ is connected there is for each pair of points $\{x, y\} \subset C$ a (not necessarily causal) curve $\gamma \subset C$ which connects $x$ and $y$. Since for all $z \in C$ the set $I^+(z)$ is a neighbourhood of $z$ and the curve $\gamma$ is compact, there are finitely many $z_i \in \gamma$ such that $z_{i+1} \in I^+(z_i)$ and the neighborhoods $I^+(z_i)$ cover $\gamma$. It follows that there is a timelike curve from $x$ to $y$. By the same argument there is a timelike curve form $y$ to $x$. Hence $C \subset I^+(x) \cap I^-(x)$ and the assertion follows since $I^+(x) \cap I^-(x)$ is connected. ∎

The following proposition shows that a compact spacetime cannot be chronological.

**Proposition 8.1.1.** *If* $M$ *is compact then the chronology violating set of* $M$ *is non-empty.*

*Proof.* We can cover $M$ with finitely many sets of the form $I^+(x_i)$ $(i = 1, \ldots, k)$. If $x_1$ is not contained in $I^+(x_1)$ there is an $l \in \{2, \ldots, k\}$ and a permutation $\sigma$ with $x_1 \in I^+(x_{\sigma(l)})$, $x_1 \notin \bigcup_{i=1}^{l-1} I^+(x_{\sigma(i)})$. This implies $x_l \notin \bigcup_{i=1}^{l-1} I^+(x_{\sigma(i)})$ since otherwise we would have

$$x_1 \in I^+(x_{\sigma(l)}) \subset I^+(\bigcup_{i=1}^{l-1} I^+(x_{\sigma(i)})) = \bigcup_{i=1}^{l-1} I^+(x_{\sigma(i)})$$

in contradiction to the definition of $x_l$. If $x_l \notin I^+(x_l)$ we can apply the same argument to $x_l$ instead of $x_1$. Since there are only finitely many

$x_i$, one of the $x_i$ must be in its own future for otherwise we would have that $x_{\sigma(k)}$ is in none of the $I^+(x_l)$. ∎

Proposition 8.1.1 is often taken as a reason for dismissing compact spacetimes as unphysical.

While the chronology condition and the causality condition are very intuitive, from a technical point of view, a slightly stronger condition is advantageous:

**Definition 8.1.2.** *The* strong causality condition *holds at $x \in M$ if for any neighbourhood $\mathcal{V}$ of $x$ there is a neighbourhood $\mathcal{U} \subset \mathcal{V}$ of $x$ such that any causal curve intersects $\mathcal{U}$ at most once.*

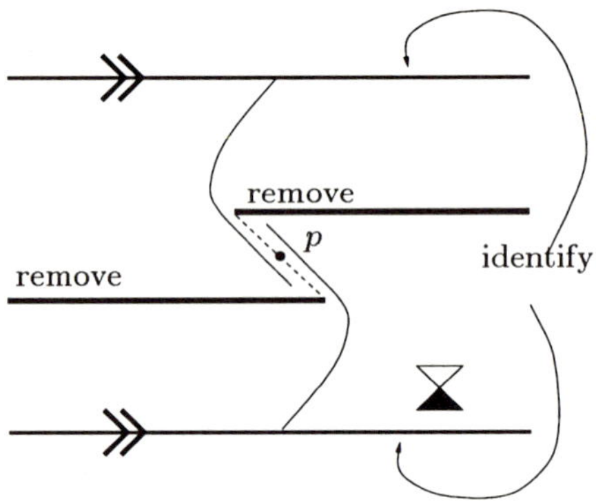


**Fig. 8.1.4.** A spacetime which is causal but fails to be strongly causal

In other words, if the strong causality condition does not hold at $x$, there are timelike curves starting at $x$ which come arbitrarily close to $x$ after leaving a giving convex neighbourhood. Hence the chronology condition is almost violated. In the next section we will see the importance of this causality condition.

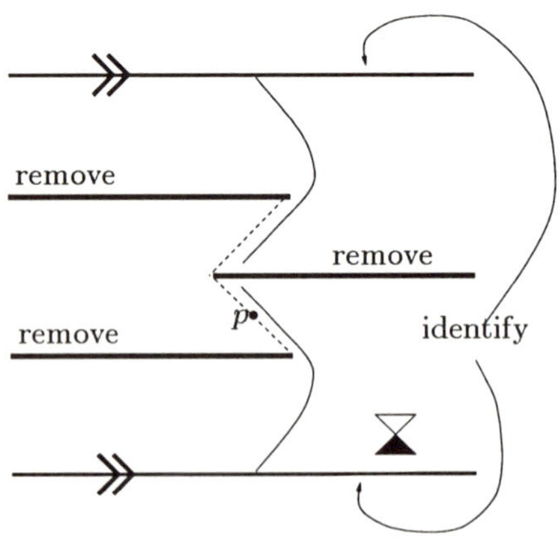


**Fig. 8.1.5.** A spacetime which is strongly causal. An infinitesimally small perturbation of the metric results in a spacetime with chronology violation

Finally, we wish to introduce *global hyperbolicity*[3], the strongest causality condition which is often assumed. Its relevance stems from the fact that in a globally hyperbolic spacetime the set of causal curves connecting two given points is compact in a natural topology. We will use related properties in the next section.

**Definition 8.1.3.** *A subset of $A \subset M$ is said to be* globally hyperbolic *if $A$ is strongly causal and for any two points $x, y \in A$ the set $J^+(x) \cap J^-(y)$ is compact.*

## 8.2 Cluster and limit curves

*In this section we study sequences of causal curves and there limits. The results of this section will be fundamental to what follows. It is based on (Beem and Ehrlich 1981).*

It is practical to generalize the concept of a timelike or causal curve to continuous curves.

**Definition 8.2.1.** *A continuous curve $\gamma$ is called* causal *(resp.,* timelike*) if every point $x$ on $\gamma$ has a convex neighbourhood $\mathcal{C}$ such that any point $y \neq x$ on $\gamma \cap \mathcal{C}$ can be connected by a causal (resp., timelike) $C^1$ curve which is contained in $\mathcal{C}$.*

Clearly, this definition coincides for $C^1$-curves with our previous Definition 3.1.3 (iii).

**Lemma 8.2.1.** *Let $x \in M$. There is a convex coordinate neighbourhood of $\mathcal{C}$ of $x$, a constant $k > 0$, and coordinates $(x^0, \ldots, x^{n-1})$ such that all causal curves $\gamma$ in $\mathcal{C}$ can be parametrized by $t = x^0$ and the coordinate inequality*

$$\sqrt{\sum_{a=0}^{n} \left(\gamma^a(t) - \gamma^a(s)\right)^2} \leq k|t - s|$$

*holds for all $t, s$.*

*Proof.* We choose coordinates $(x^0, \ldots, x^{n-1})$ in a convex neighbourhood $\mathcal{C}$ of $x$ with compact closure such that $\mathrm{d}x^0$ is timelike and all $\mathrm{d}x^i$ $(i \in \{1, \ldots, n-1\})$ are spacelike. Then any causal curve in $\mathcal{C}$ can be parametrized by $x^0$. Let $\mu$ be a causal $C^1$-curve with $\mu(t) = \gamma(t)$ and $\mu(s) = \gamma(s)$ (cf. Fig. 8.2.1). Since the closure of $\mathcal{C}$ is compact there exists a constant $k_0 > 0$ such that all causal vectors are also causal with respect to the flat metric $-k_0 \mathrm{d}t^2 + \sum_{i=1}^{n-1} (\mathrm{d}x^i)^2$. In particular, $\mu$ satisfies $k_0 = k_0(\dot{\mu}^0)^2 \geq \sum_{i=1}^{n-1} (\dot{\mu}^i)^2$. If we denote the standard basis of $\mathbb{R}^n$ by $\{e_0, \ldots, e_{n-1}\}$ and write $\|v^a e_a\|_2 = \sqrt{\sum_{a=0}^{n-1} (v^a)^2}$, we obtain

[3] This name has been coined by Leray (1953) in connection with systems of partial differential equations.

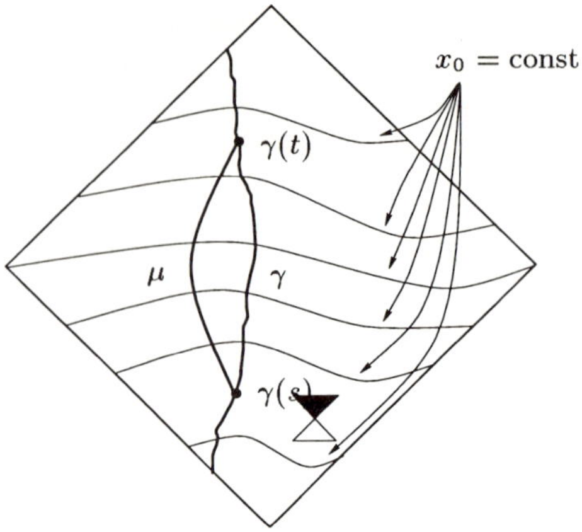


**Fig. 8.2.1.** The proof of Lemma 8.2.1

$$\begin{aligned}||\gamma^a(t)e_a - \gamma^a(s)e_a||_2 = ||\mu^a(t)e_a - \mu^a(s)e_a||_2 &= ||\int_s^t \dot{\mu}^a(\tau)\mathrm{d}\tau e_a||_2 \\ &\le \int_s^t ||\dot{\mu}^a(\tau)e_a||_2 \mathrm{d}\tau \le \sqrt{1+k_0}(t-s).\end{aligned}$$

∎

**Corollary 8.2.1.** *It follows that causal curves are Lipschitz and therefore differentiable almost everywhere.*

For $x, y \in M$ let $C^0_{\mathsf{causal}}(x,y)$ be the space of continuous causal curves from $x$ to $y$ and $C^1_{\mathsf{time}}(x,y)$ be the space of timelike curves from $x$ to $y$ which are $C^1$. We will now specify a natural topology for the space of causal curves.

**Definition 8.2.2.** *Let $\gamma\colon [a,b] \to M$, $\gamma_i\colon [a,b] \to M$ $(i \in \mathbb{N})$ be curves. The sequence $\{\gamma_i\}_{i\in\mathbb{N}}$* converges to $\gamma$ in the $C^0$-topology *if for every neighbourhood $\mathcal{V}$ of $\gamma$ in $M$ there exists an $i_0 \in \mathbb{N}$ such that $\gamma_i \subset V$ for all $i > i_0$. The curve $\gamma$ is called the* limit curve *of the sequence $\{\gamma_i\}_{i\in\mathbb{N}}$.*

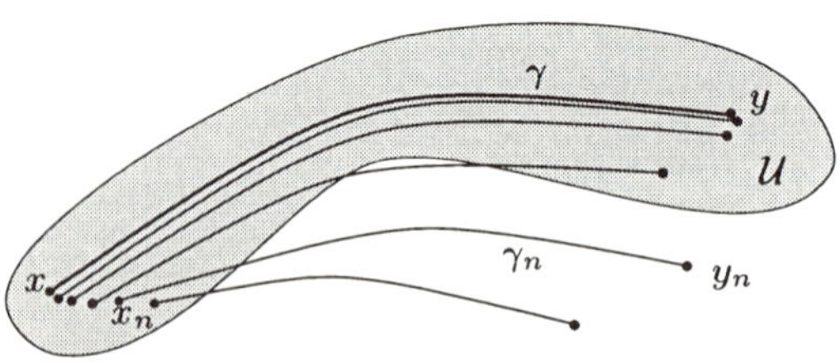


**Fig. 8.2.2.** A limit curve $\gamma$ of a sequence of curves $\gamma_n$

Our terminology is slightly at odds with the traditional definition of "limit curve" in general relativity but closer to generic mathematical terminology. Often, not limits of curves (with respect to a natural topology) but curves which are better thought of as a set of pointwise accumulation points are called "limit curves". We will reserve the term "cluster curve" for such accumulation curves:

**Definition 8.2.3.** *Let $\gamma\colon [a,b] \to M$, $\gamma_i : [a,b] \to M$ $(i \in \mathbb{N})$ be curves. $\gamma$ is said to be a* cluster curve *of the sequence $\{\gamma_i\}_{i\in\mathbb{N}}$ if there exists a subsequence $\{\gamma_{i_j}\}_{j\in\mathbb{N}}$ such that for all $x \in \gamma$ each neighbourhood of $x$ intersects all but finitely many of the curves $\gamma_{i_j}$. Following Beem and Ehrlich (1981) we will say that the sequence $\{\gamma_{i_j}\}_{j\in\mathbb{N}}$* distinguishes *the cluster curve $\gamma$.*

It will turn out that for strongly causal spacetimes cluster and limit curves are essentially the same (cf. Theorem 8.2.2) below. There is no

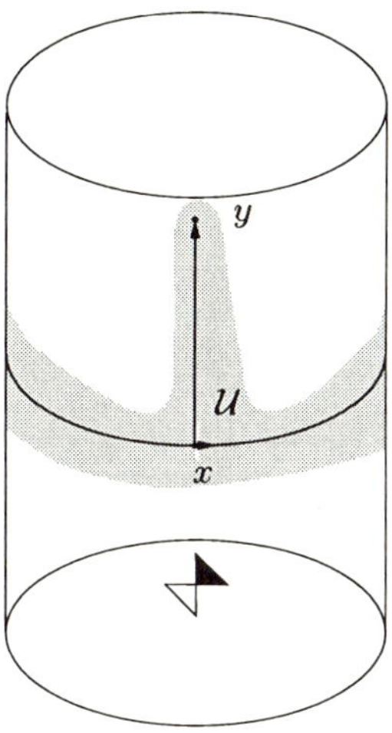


**Fig. 8.2.3.** An example where a limit curve from $x$ to $y$ is not a cluster curve for a sequence of points from $x$ to $y$

strict logical relation between limit curves and cluster curves. Consider in the flat cylinder $(S^1 \times \mathbb{R}, \mathrm{d}\varphi \mathrm{d}t)$ (cf. Fig. 8.2.3) the sequence of identical curves $\gamma_n$ which connect $x$ with $y$ and satisfy $\varphi = \text{const}$. The curve $\gamma$ which first traverses $t = t(x)$ and then connects $x$ with $y$ satisfying $\varphi = \text{const}$ is a limit curve of $\gamma_n$ but it is not a cluster curve. In Fig. 8.2.4 we have a cluster curve $\gamma$ of a sequence of causal curves which is not a limit curve. In general, a cluster curve of a sequence of causal curves may even be spacelike (cf. Fig. 8.2.5

**Proposition 8.2.1.** *If $(M,g)$ is strongly causal and $\gamma$ is a cluster curve of a sequence $\{\gamma_i\}_{i\in\mathbb{N}}$ of causal curves, then $\gamma$ is causal.*

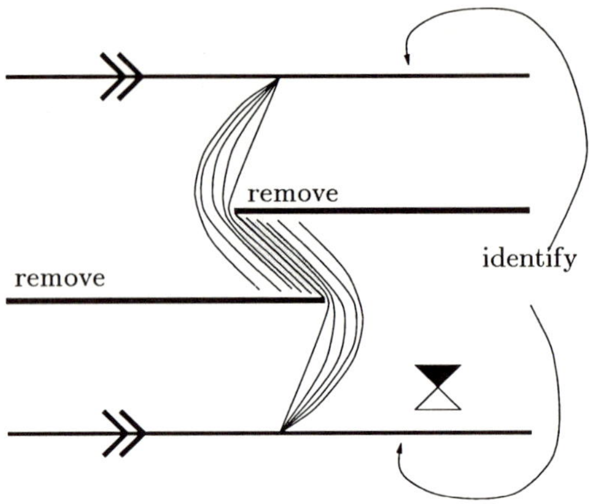


**Fig. 8.2.4.** A spacetime which is causal but fails to be strongly causal

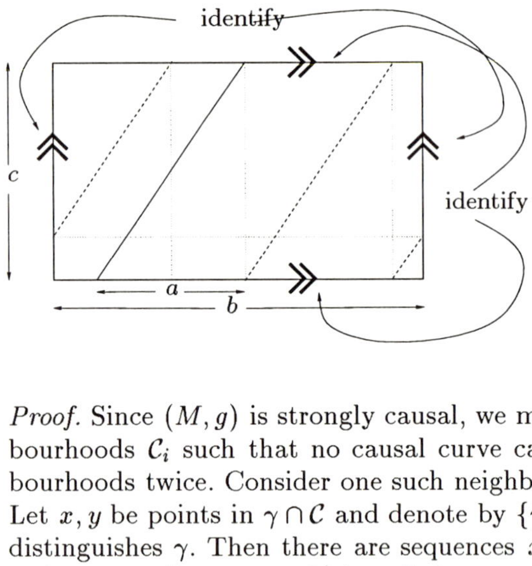


**Fig. 8.2.5.** Asume that $b/c$ is rational and $a/b$ is irrational. Then the projection of the line with slope $c/a$ from $\mathbb{R}^2$ to the torus depicted in the figure is a dense curve $\gamma$. Hence *every* curve is a cluster curve of $\gamma$

*Proof.* Since $(M, g)$ is strongly causal, we may cover $\gamma$ by convex neighbourhoods $\mathcal{C}_i$ such that no causal curve can enter any of these neighbourhoods twice. Consider one such neighbourhood and denote it by $\mathcal{C}$. Let $x, y$ be points in $\gamma \cap \mathcal{C}$ and denote by $\{\gamma_{i_j}\}_{j \in \mathbb{N}}$ a subsequence which distinguishes $\gamma$. Then there are sequences $x_j, y_j \in \gamma_{i_j} \cap \mathcal{C}$ with $x_j \to x$ and $y_j \to y$. Since $y_{i_j} \in J^+(x_{i_j}, \mathcal{C})$ Lemma 3.1.4 implies the existence of a causal vector $v_j$ with $\exp_{x_{i_j}}(v_j) = y_{i_j}$. These vectors have an accumulation point $v$ with $\exp_x(v) = y$. The vector $v$ must be causal since the set of causal vectors is closed. But this implies $y \in J^+(x, \mathcal{C})$. If $x$ and $y$ are arbitrary points on $\gamma$, we can find finitely many neighbourhoods $\mathcal{C}_i$ such that the segment from $x$ to $y$ is covered by $\bigcup_i \mathcal{C}_i$. We can now apply the preceding argument finitely often to conclude that $x$ and $y$ are causally related. ∎

The basic Lemma 8.2.4 below is an application of the theorem of Ascoli which we will present first.

Let $A \subset \mathbb{R}^k$ be a compact set. The space of continuous functions

$$C^0(A, \mathbb{R}^l) = \{f \colon A \to \mathbb{R}^k \,|\, f \text{ is continuous}\}$$

can then be regarded as a normed vector space in a natural way. Just set $\|f\|_\infty = \sup_{x\in A}\{|f(x)|\}$. Moreover, this norm is complete, i.e., every Cauchy sequence $\{f_i\} \subset C^0(A,\mathbb{R}^l)$ converges to a function $f \in C^0(A,\mathbb{R}^l)$. ($f$ can be constructed pointwise using the completeness of $\mathbb{R}^l$.)

**Lemma 8.2.2.** *Let $B \subset C(A,\mathbb{R}^l)$ be a closed set and assume that for every $\epsilon > 0$ there are finitely many balls $B^1_\epsilon(x_1), \ldots, B^{j(\epsilon)}_\epsilon(x_{j(\epsilon)})$ with radius $\epsilon$ and $B \subset \bigcup_{i=1}^{j(\epsilon)} B^i_\epsilon(x_j)$. Then $B$ is compact.*

*Proof.* If the lemma is not true then there are open sets $\mathcal{U}_\iota$ $(\iota \in I)$ covering $B$ such that no finite subset $\mathcal{U}_{\iota_1}, \ldots, \mathcal{U}_{\iota_k}$ covers $B$. Let

$$\{B^1_1(x_1), \ldots, B^{j(1)}_1(x_{j(1)})$$

be a finite set of balls of radius 1 which cover $B$. By our assumption one of these balls cannot be covered by finitely many $\mathcal{U}_\iota$. (Otherwise we would obtain a finite cover of $B$ by finitely many sets which are in turn finitely covered by sets $U_\iota$). We denote this ball by $B_0$. Assume that we have constructed balls $\{B_i\}_{i=0,\ldots,k-1}$ such that

(i) Any two consecutive balls intersect,
(ii) Each ball $B_i$ has radius $2^{-i}$,
(iii) None of these balls can be covered by finitely many $\mathcal{U}_\iota$.

There are balls $B^1_{2^{-k}}(x_1), \ldots, B^{k(2^{-k})}_{2^{-k}}(x_{k(2^{-k})})$ which cover $B$ and therfore also $B_{k-1}$. Since $B_{k-1}$ cannot be covered by finitely many $\mathcal{U}_\iota$ there must exist at least one $B^m_{2^{-k}}(x_m)$ which intersects $B_{k-1}$ and cannot be covered by finitely many $\mathcal{U}_\iota$. Denoting $B^m_{2^{-k}}(x_m)$ by $B_k$ we have inductively defined a sequence $\{B_i\}_{i\in\mathbb{N}\cup 0}$ of balls which satisfy (i)–(iii). Denote the centres of these balls by $y_i$. For any natural numbers $m < n$ we obtain

$$\|y_n - y_m\| \leq \sum_{i=m+1}^{n} \|y_i - y_{i-1}\| \leq \sum_{i=m+1}^{n} (2^{-i} + 2^{-i+1})) \leq 2 \cdot 2^{-m}.$$

Hence $\{y_i\}_{i\in\mathbb{N}\cup 0}$ is a Cauchy sequence. Denoting its limit by $y$ there is an $\mathcal{U}_{\iota_0}$ which contains $y$ and a number $r \in \mathbb{N}$ such that the ball $B_{2^{-r}}(y)$ is contained in $\mathcal{U}_{\iota_0}$. But this implies $B_{2^{-r-1}}(y_{r+1}) \subset \mathcal{U}_\iota$ in contradiction to (iii). ∎

**Theorem 8.2.1 (Ascoli).** *Let $A \subset \mathbb{R}^k$ be a compact subset and $f_i : A \to \mathbb{R}^l$ $(i \in \mathbb{N})$ be an equi-continuous sequence of continuous functions such that for all $a \in A$ the set $\overline{\bigcup_{i\in\mathbb{N}} f_i(a)}$ is compact. Then there is a continuous function $f\colon A \to M$ and a subsequence $\{f_{i_j}\}_{j\in\mathbb{N}}$ of $\{f_i\}_{i\in\mathbb{N}}$ which converges uniformly to $f$.*

*Proof.* We show first that the subset $\overline{\bigcup_{i=1}^{\infty}\{f_i\}}$ is compact in the normed space $(C^0(A,\mathbb{R}^l), \|\cdot\|_\infty)$. By Lemma 8.2.2 we only have to show that for any $\epsilon > 0$ there is a finite number of balls with diameter less than $\epsilon$ which cover $\overline{\bigcup_{i=1}^{\infty}\{f_i\}}$. Let $\epsilon > 0$ and $a \in A$. Since $\{f_i\}_{i\in\mathbb{N}}$ is equi-continuous, there is for each $a \in A$ a neighbourhood $\mathcal{U}_a$ of $a$ such that for all $f_j$ and all $y \in \mathcal{U}_a$ the inequality $\|f_j(a) - f_j(y)\|_\infty < \epsilon/4$ holds. Since $A$ is compact, we can cover $A$ with finitely many such neighbourhoods $\mathcal{U}_{a_l}$ $(l \in \{1,\dots,k\})$. Since the union

$$K = \bigcup_{l=1}^{k}\overline{\bigcup_{i=1}^{\infty}\{f_i(a_l)\}}$$

is compact it can be covered by finitely many open balls of radius $\epsilon/4$. Denote their centres by $x_s \in K$ $(s \in \{1,\dots,r\})$. We will now construct finitely many neighbourhoods of diameter less than $\epsilon$ in $C^0(A,\mathbb{R}^l)$ which cover all of $\overline{\bigcup_{i=1}^{\infty}\{f_i\}}$. These neighbourhood will be defined by the requirement that the functions $f$ in each such neighbourhood have values $x = f(a)$ close to $x_s$ for all $a$ near some $a_i$. More precisely, consider the finite set of all maps $\sigma\colon \{1,\dots,k\} \to \{1,\dots r\}$ and let

$$\mathcal{V}_\sigma = \left\{ h \in C^0(A,\mathbb{R}^l) \cap \overline{\bigcup_{i=1}^{\infty}\{f_i\}} \,\middle|\, \|h(a_l) - x_{\sigma(l)}\| < \epsilon/4 \quad \forall l \in \{1,\dots,k\} \right\},$$

$\hat{f} \in \overline{\bigcup_{i=1}^{\infty}\{f_i\}}$. Since $K$ is covered by finitely many balls $x_1,\dots,x_r$ of radius $\epsilon_4$ there is for each $l$ an $x_{s_l}$ such that $\|\hat{f}(a_l) - x_{s_l}\| < \epsilon/4$. Defining $\hat{\sigma}(l) = s_l$ we see that $\hat{f} \in \mathcal{V}_{\hat{\sigma}}$. Hence the sets $\mathcal{V}_\sigma$ cover all of $\overline{\bigcup_{i=1}^{\infty}\{f_i\}}$. Since the sets $\mathcal{U}_{a_1},\dots,\mathcal{U}_{a_k}$ cover $A$ there is for $h, \hat{h} \in \mathcal{V}_\sigma$ and $a \in A$ an $l \in \{1,\dots,k\}$ such that $\|h(a) - h(a_l)\| < \epsilon/4$ and $\|\hat{h}(a) - \hat{h}(a_l)\| < \epsilon/4$. This implies

$$\begin{aligned}\|h(a) - \hat{h}(a)\| &\le \|h(a) - h(a_l)\| + \|\hat{h}(a) - \hat{h}(a_l)\| \\ &\quad + \|h(a_l) - x_{\sigma(l)}\| + \|x_{\sigma(l)} - \hat{h}(a_l)\| \\ &\le 4\cdot\epsilon/4 = \epsilon.\end{aligned}$$

Hence each set $\mathcal{V}_\sigma$ is contained in a ball of radius $\epsilon$. This implies that $\overline{\bigcup_{i=1}^{\infty}\{f_i\}}$ is covered by finitely many balls of radius $\epsilon$ and therefore compact.

The assertion follows now since in a compact subset of a normed space every sequence has a convergent subsequence. ∎

**Lemma 8.2.3.** *Let $\mathcal{C}$ be a convex coordinate neighbourhood with compact closure and let $\{\gamma_i\}_{i\in\mathbb{N}}$ be a sequence of causal curves in $\overline{\mathcal{C}}$ which are inextensible in $\overline{\mathcal{C}}$. If $x \in M$ is an accumulation point of this sequence, then there is a causal cluster curve $\gamma$ through $x$ which is inextensible in $\mathcal{C}$.*

*Proof.* By considering a subsequence we can assume without loss of generality that for each $i$ there is a $t_i$ such that $\gamma_i(t_i) \to x$. Any cluster curve $\gamma$ of this sequence intersects $x$. We choose the same coordinates as in Lemma 8.2.1 and view the curves $\gamma_i$ as continuous maps from finite intervals $[a_i, b_i]$ to $\mathbb{R}^n$. In order to apply the Theorem of Ascoli 8.2.1 we trivially enlarge the domains $[a_i, b_i]$ to $[\inf_{i\in\mathbb{N}}(a_i), \sup_{i\in\mathbb{N}}(b_i)]$ by setting $\gamma_i(t) = \gamma_i(a_i)$ for $t \in [a, a_i]$ and $\gamma_i(t) = \gamma_i(b_i)$ for $t \in [b_i, b]$. This family of maps is equi-continuous by Lemma 8.2.1. Since $\mathcal{C}$ has compact closure so has the set $\bigcup_{i\in\mathbb{N}} \gamma_i(t)$ for all $t$. The theorem of Ascoli 8.2.1 implies that a subsequence of these curves converges uniformly to a continuous curve $\gamma$. To see that $\gamma$ is causal let $t, s \in [a, b]$, $s < t$. Since $\gamma_i(t) \in J^+(\gamma_i(s), \mathcal{C})$ for all $i$, there is a causal geodesic $\mu_i$ from $\gamma_i(s)$ to $\gamma_i(t)$ for all $i$. By the continuous dependence of the solutions of ordinary differential equations with respect to initial values, there is a limit geodesic $\mu$ starting at $\gamma(s)$. It is clearly causal and has future end point $\gamma(t)$ by the convexity of $\mathcal{C}$. This implies $\gamma(t) \in J^+(\gamma(s), \mathcal{C})$. Since $s$ and $t$ were arbitrary, $\gamma$ must be a causal curve. Since $\gamma(a)$ is an accummulation point of $\gamma_i(a) \in \overline{\mathcal{C}} \setminus \mathcal{C}$ this point must also lie in $\overline{\mathcal{C}} \setminus \mathcal{C}$. Analogously for $\gamma(b)$. This implies that $\gamma$ is inextensible in $\mathcal{C}$. ∎

**Lemma 8.2.4.** *Let $\{\gamma_i\}_{i\in\mathbb{N}}$ be a sequence of future inextensible causal curves. If $x \in M$ is an accumulation point of this sequence, then there is a causal, future inextensible cluster curve $\gamma$ through $x$.*

*Proof.* Let $\mathcal{C}_x$ be a convex neighbourhood as in Lemma 8.2.3. Then there exists a subsequence which distinguishes a cluster curve which is inextensible in $\mathcal{C}_x$. We may now take the intersection $x_1$ of the future boundary of $\mathcal{C}_x$ with this cluster curve. Applying the same argument to a convex neighbourhood $\mathcal{C}_{x_1}$ of $x_1$ and the subsequence, we obtain an extension of the cluster curve to $\mathcal{C}_x \cup \mathcal{C}_{x_1}$. Now we can proceed by induction. ∎

**Proposition 8.2.2.** *Assume that $(M, g)$ is globally hyperbolic and let $\{x_i\}_{i\in\mathbb{N}}$, $\{y_i\}_{i\in\mathbb{N}}$ be sequences of points converging to $x, y \in M$ such that $y_i \in J^+(x_i)$, $y \in J^+(x)$. Let $\gamma_i$ be a causal curve from $x_i$ to $y_i$. Then the sequence $\{\gamma_i\}_{i\in\mathbb{N}}$ has a cluster curve which connects $x$ with $y$.*

*Proof.* Let $\tilde{x} \in I^-(x)$ and $\tilde{y} \in I^+(y)$. The set $J^+(\tilde{x}) \cap J^-(\tilde{y})$ is then a neighbourhood of $x, y$ and we can assume without loss of generality that all $\gamma_i$ lie in $J^+(\tilde{x}) \cap J^-(\tilde{y})$. Since $J^+(\tilde{x}) \cap J^-(\tilde{y})$ is compact, we can extend all $\gamma_i$ such that they are inextensible in $J^+(\tilde{x}) \cap J^-(\tilde{y})$. By compactness and strong causality of $J^+(\tilde{x}) \cap J^-(\tilde{y})$ we can cover this set by finitely many convex neighbourhoods as provided by Lemma 8.2.3 such that no $\gamma_i$ intersects any neighbourhood twice. Now the claim is an immediate consequence of Lemma 8.2.3. ∎

**Theorem 8.2.2.** *Let $(M,g)$ be strongly causal. For given points $x, y \in M$ let $\gamma_i\colon [a,b] \to M$ be a sequence of causal geodesics with*

$$\lim_{i\to\infty} \gamma_i(a) = x \text{ and } \lim_{i\to\infty} \gamma_i(b) = y.$$

*Then $\gamma\colon [a,b] \to M$, $\gamma \in C^0_{\mathsf{causal}}(x,y)$ is a cluster curve of the sequence $\{\gamma_i\}_{i\in\mathbb{N}}$ if and only if there is a subsequence $\{\gamma_{i_j}\}_{j\in\mathbb{N}}$ of $\{\gamma_i\}_{i\in\mathbb{N}}$ which converges to $\gamma$ in the $C^0$ topology of curves.*

*Proof.* Assume first that $\gamma$ is a cluster curve of $\{\gamma_i\}_{i\in\mathbb{N}}$. Since the image of $\gamma$ is compact, there are for each neighbourhood $\mathcal{U}$ of $\gamma$ finitely many convex sets $\{\mathcal{C}_k\}_{k=1,\dots,l}$ such that

(i) $\gamma \subset \bigcup_{k=1}^{l} \mathcal{C}_k \subset \mathcal{U}$,
(ii) no causal curve can enter any $\mathcal{C}_k$ twice.

We have to show that there is a subsequence $\{\gamma_{i_j}\}_{j\in\mathbb{N}}$ such that for sufficiently large $j$ all $\gamma_{i_j}$ are contained in $\mathcal{U}$. We can choose a finite sequence of points $\{\gamma(t_k)\}_{k=1,\dots,k}$ with $\gamma(t_k) \in \mathcal{C}_k \cap \mathcal{C}_{k+1}$. By the definition of a cluster sequence (and the fact that there are only finitely many such points) there is a subsequence $\{\gamma_{i_j}\}_{j\in\mathbb{N}}$ such that each $\gamma_{i_j}$ intersects all of the sets $\mathcal{C}_k \cap \mathcal{C}_{k+1}$. Since all $\mathcal{C}_k$ are convex and no causal curve can re-enter any $\mathcal{C}_k$, $\gamma_{i_j}$ is contained in $\bigcup_{k=1}^{l} \mathcal{C}_k \subset \mathcal{U}$.

Conversely, assume (without loss of generality) that $\{\gamma_i\}_{i\in\mathbb{N}}$ converges to $\gamma$ in the $C^0$ topology of curves. Let $h$ be a fixed Riemannian metric $h$ on $M$ and let for $r \in \mathbb{N}$ $\mathcal{U}_{\frac{1}{r}}$ be a neighbourhood of $\gamma$ which is the finite union of convex sets $\{\mathcal{C}_{k,\frac{1}{r}}\}_{k=1,\dots,l_r}$ such that no causal curve can enter any $\mathcal{C}_{k,\frac{1}{r}}$ twice and all $\mathcal{C}_{k,\frac{1}{r}}$ have diameter less than $\frac{1}{r}$ with respect to $h$. We can extend the causal curves $\gamma_i$ to inextensible curves in $\bigcup_{k=1}^{l} \mathcal{C}_{k,\epsilon}$. Now let $\{\mathcal{U}_{\frac{1}{r}}\}_{r\in\mathbb{N}}$ be a sequence of neighbourhoods of $\gamma$ such that $\bigcap_{r=1}^{\infty} \mathcal{U}_{\frac{1}{r}} = \gamma$. Since $\{\gamma_i\}_{i\in\mathbb{N}}$ converges to $\gamma$ in the $C^0$ topology of curves, we obtain a subsequence of curves $\{\gamma_{i_r}\}_{r\in\mathbb{N}}$ with $\gamma_{i_r} \in \mathcal{U}_{\frac{1}{r}}$ for each $r \in \mathbb{N}$. Lemma 8.2.4 implies that this subsequence has a causal cluster curve $\lambda$. Since $\lambda$ lies in the intersection of all $\mathcal{U}_{\frac{1}{r}}$ and this intersection equals the set traversed by $\gamma$, we obtain $\lambda \subset \gamma$. The other inclusion $\gamma \subset \lambda$ follows since $\lambda$ is inextensible and $\gamma$ does not have self-intersections. ∎

In view of Corollary 8.2.1 we can define the *length* $L(\gamma)$ of a causal curve $\gamma$ simply by setting

$$L(\gamma) = \int \sqrt{-\langle\dot\gamma(t), \dot\gamma(t)\rangle}\, \mathrm{d}t. \tag{8.2.1}$$

Since in any neighbourhood $\mathcal{U}$ of any causal curve $\gamma \subset C^0_{\mathsf{causal}}(x,y)$ there is a broken null geodesic from $x$ to $y$, the length functional cannot be lower semi-continuous with respect to the $C^0$ topology of curves.

If $(M, g)$ is not chronological, then the length functional is not upper semi-continuous either. To see this consider a timelike curve $\gamma$ which has self intersections. Then in any neighbourhood of $\gamma$ there are timelike $C^1$ curves which contain a closed segment which can be repeatedly traversed. In particular, we see that $L = \infty$ for all such curves while $\gamma$ has finite length. On the other hand, if $(M, g)$ is causal, we have the following proposition.

**Proposition 8.2.3.** *Let $x, y \in M$ and assume that $M$ is causal. The length functional $L$ is upper semi-continuous in $C^0_{\text{causal}}(x, y)$ with respect to the $C^0$ topology of causal curves.*

For the proof of Proposition 8.2.3 we will need the following result.

**Lemma 8.2.5.** *Let $(M, g)$ be causal and $\gamma \in C^1_{\text{time}}(x, y)$ with $g(\dot\gamma, \dot\gamma) = -1$. Then there is a neighbourhood $\mathcal{U}$ of $\gamma$ in $M$ and an extension of the curve parameter $t$ of $\gamma$ to a function $t\colon \mathcal{U} \to \mathbb{R}$ such that $\mathrm{d}t$ is timelike and* $\operatorname{grad}(t)_{|\gamma} = -\dot\gamma$.

*Proof.* Let $\mathcal{U}$ be a neighbourhood of $\gamma$ in $M$ which is the union of a finite number of convex sets with compact closure. Let $\tilde\gamma$ be a timelike extension of $\gamma$ which has no endpoint in $\mathcal{U}$ (cf. Corollary 8.0.1). Since $(M, g)$ is causal $\gamma$ cannot have self intersections. Hence we can choose $\mathcal{U}$ to be simply connected. Let $\{E_0(t), \ldots, E_{n-1}(t)\}$ be an orthonormal frame along $\tilde\gamma$ with $E_0 = \dot\gamma$. The map $f\colon \mathcal{V} \subset \mathbb{R}^n \to \mathcal{U}$, $(t, x^1, \ldots, x^{n-1}) \mapsto \exp_{\tilde\gamma(t)}\left(\sum_{i=1}^{n-1} x^i E_i(t)\right)$ is a local diffeomorphism near $\gamma$ since exp is a local diffeomorphism near the origin. Hence we can choose $\mathcal{U}$ and $\mathcal{V}$ so small that $f$ is a diffeomorphism. This and the fact that $\mathcal{U}$ is simply connected (otherwise $t$ may be a cyclic coordinate and only defined modulo an additive constant) implies that we can extend the function $t$ to all of $\mathcal{U}$. Since $\mathrm{d}t$ is timelike at $\tilde\gamma$, by choosing $\mathcal{U}$ small enough, $\mathrm{d}t$ is timelike everywhere in $\mathcal{U}$. Finally observe that $\mathrm{d}t(E_i) = \mathrm{d}t(\frac{\partial}{\partial x^i} \exp_{\gamma(t)}\left(\sum_{i=1}^{n-1} x^i E_i(t)\right) = \mathrm{d}t(\partial_{x^i}) = 0$. Hence $\operatorname{grad}(t)$ is orthogonal to $E_i$ and therefore parallel to $E_0 = \dot\gamma_0$ From $1 = \mathrm{d}t(\dot\gamma_0) = g(\operatorname{grad}(t), \dot\gamma)$ we obtain $\operatorname{grad}(t)_{|\gamma} = -\dot\gamma_0$. ∎

*Proof of Proposition 8.2.3.* Since $C^0_{\text{causal}}(x, y) \subset \overline{C^1_{\text{time}}(x, y)}$, we need only show that the length functional is upper semi-continuous in $C^1_{\text{time}}(x, y)$. Let $\gamma\colon [a, b] \to M$ be a curve in $C^1_{\text{time}}(x, y)$ and $\mathcal{U}$ be the neighbourhood given by Lemma 8.2.5. Let $\mu$ be any timelike curve which connects $x$ and $y$ and is contained in $\mathcal{U}$. Since $\mathrm{d}t(\dot\mu(t)) = 1$ we obtain $\dot\mu(t) = \frac{1}{g(\operatorname{grad}(t), \operatorname{grad}(t))}(\operatorname{grad}(t) + v(t))$ where $v(t)$ is orthogonal to $\operatorname{grad}(t)$. This implies

$$\begin{aligned}
-g_{\mu(t)}(\dot{\mu},\dot{\mu}) &= -\left(\frac{1}{g(\mathrm{grad}(t),\mathrm{grad}(t))}\right)^2 g_{\mu(t)}(\mathrm{grad}(t),\mathrm{grad}(t)) \\
&\quad -\left(\frac{1}{g(\mathrm{grad}(t),\mathrm{grad}(t))}\right)^2 g_{\mu(t)}(v(t),v(t)) \\
&\leq -\left(\frac{1}{g(\mathrm{grad}(t),\mathrm{grad}(t))}\right)^2 g_{\mu(t)}(\mathrm{grad}(t),\mathrm{grad}(t)) \\
&= -\frac{1}{g(\mathrm{grad}(t),\mathrm{grad}(t))}.
\end{aligned}$$

On the other hand, we have $1 = -g_{\gamma(t)}(\dot{\gamma},\dot{\gamma}) = -g_{\gamma(t)}(\mathrm{grad}(t),\mathrm{grad}(t))$. Since $x \mapsto g_x(\mathrm{grad}(t),\mathrm{grad}(t))$ is continuous we may choose for a given $\epsilon > 0$ the set $\mathcal{U}$ so small that $-1-\epsilon < g(\mathrm{grad}(t),\mathrm{grad}(t)) < -1+\epsilon$. This implies $-g_{\mu(t)}(\dot{\mu},\dot{\mu}) \leq \frac{1}{1-\epsilon} = -\frac{1}{1-\epsilon} g_{\gamma(t)}(\partial_t,\partial_t)$ and therefore $L(\mu) \leq \frac{1}{\sqrt{1-\epsilon}} L(\gamma)$. Since $\mu \in C^1_{\mathsf{time}}(x,y)$ was arbitrary the length function $L$ is upper semi-continuous on this set. ∎

*Remark 8.2.1.* Hawking and Ellis (1973) give a slightly different definition for the length of a causal curve. They first define the length of a timelike curve $\gamma \in C^1_{\mathsf{time}}(x,y)$ by Equation 8.2.1. For a causal curve $\mu$ they set

$$\tilde{L}(\mu) = \inf\{\ell(\mathcal{U}) \,|\, \mathcal{U} \text{ is a neighbourhood of } \mu\},$$

where

$$\ell(\mathcal{U}) = \sup\{L(\gamma) \,|\, \gamma \subset \mathcal{U} \text{ is timelike and } C^1\}.$$

It follows that this length functional $\tilde{L}$ is also upper semi-continuous in $C^0_{\mathsf{causal}}(x,y)$ with respect to the $C^0$ topology of causal curves. Hence in causal spacetimes, $L$ and $\tilde{L}$ coincide. However, if $(M,g)$ is not chronological, $\tilde{L}$ does not coincide with the original (and more intuitive) definition since for any timelike curve $\gamma$ with self-intersection one has $\tilde{L}(\gamma) = \infty$.

## 8.3 Achronal submanifolds and Cauchy developments

Given a set $A$ the set $I^+(A)$ consists of those events which can be influenced by $A$. It is clear that its boundary $\partial I^+(A)$ is of special importance. In this section we show that $\partial I^+(A)$ must be a Lipschitz manifold. We will also investigate the "domain of dependence" of $A$, i.e., the set of those events which are completely determined by the physical data of $A$.

For the following, definition, standard examples are spacelike submanifolds and $\partial I^+(B)$ where $B$ is any subset.

**Definition 8.3.1.** *A subset $A \subset M$ is called* achronal *if there is no timelike curve which intersects it twice. A set $F$ is called a* future set *if $I^+(F) \subset F$.* Past sets *are defined analogously.*

**Lemma 8.3.1.** *Let $F$ be a future set. Then $\partial F$ is an achronal Lipschitz hypersurface.*

*Proof.* Let $x \in \partial F$. Since $I^+(x) \subset F$ is open, $\partial F$ must be achronal. It also follows that $I^-(x) \subset M \backslash F$. Consider a convex coordinate neighbourhood $\mathcal{U}$ of $x$ as given by Lemma 8.2.1. The integral curves $\mu_{(x^1,\ldots,x^{n-1})}$ of $\partial_{x^0}$ through $(0, x^1, \ldots, x^{n-1})$ are timelike. Since the integral curve of $\partial_{x^0}$ through $x$ intersects both $I^-(x, \mathcal{C})$ and $I^+(x, \mathcal{C})$ there is a neighbourhood $\mathcal{V} \subset \mathcal{U}$ of $x$ such that all integral curves $\mu_{(x^1,\ldots,x^{n-1})}$ of $\partial_{x^0}$ which intersect $\mathcal{V}$ intersect $I^-(x, \mathcal{C})$ and $I^+(x, \mathcal{C})$. It follows that each of these curves $\mu_{(x^1,\ldots,x^{n-1})}$ intersecta $\partial F$ in a unique point $y(x^1, \ldots, x^{n-1})$. Since $\partial F$ is achronal, there is a constant $k > 0$ such that

$$|x^0 \circ y(x^1, \ldots, x^{n-1}) - x^0 \circ y(\hat{x}^1, \ldots, \hat{x}^{n-1})| \leq k \sqrt{\sum_{i=1}^{n-1} (x^i - \hat{x}^i)^2}$$

for all $(x^1, \ldots, x^{n-1}), (\hat{x}^1, \ldots, \hat{x}^{n-1})$. Hence the function

$$(x^1, \ldots, x^{n-1}) \mapsto y(x^1, \ldots, x^{n-1})$$

is Lipschitz. ∎

**Definition 8.3.2.** *The* null-boundary *of a future set $F$ is the set*

$$\partial^{\text{null}} F = \{x \in \partial F \,|\, \exists \text{ a neighbourhood } \mathcal{U} \text{ of } x \text{ with } I^+(F \setminus \mathcal{U}) = I^+(F)\}.$$

*The* acausal boundary *of $F$ is $\partial^{\text{ac}} F = \partial F \setminus \partial^{\text{null}} F$.*

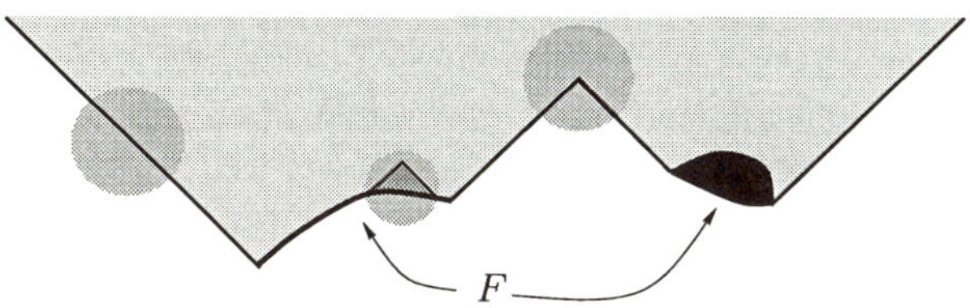


**Fig. 8.3.1.** The definition of null boundary and achronal boundary

This definition is justified by the following lemma.

**Lemma 8.3.2.** *Let $F$ be a future set and $x \in \partial^{\text{null}} F$. Then there is an achronal null geodesic generator through $x$ with future endpoint $x$. This generator is past inextensible or has past endpoint in $\partial^{\text{ac}} F$.*

*Proof.* Let $\mathcal{U}$ be a neighbourhood of $x$ such that $I^+(F) = I^+(F \setminus \mathcal{U})$. There is a sequence of points $\{x_i\}_{i\in\mathbb{N}} \subset I^+(F)$ with $x_i \to x$. Let $\gamma_i$ be a causal curve from some $y_i \in I^+(F) \setminus U$ to $x_i$ and let $\gamma$ be a causal cluster curve with future endpoint $x$. This curve cannot intersect $I^+(F)$ since then we would have $x \in I^+(F)$. It follows that $\gamma$ lies in the achronal set $\partial F$ and is therefore an achronal null geodesic. If $\gamma$ has a past endpoint $z \in \partial^{\text{null}} F$, we can repeat the argument above to obtain a cluster curve $\mu$ which has future endpoint $z$ and which is also an achronal null geodesic. Since the concatenation of $\mu$ and $\gamma$ must be achronal, they are both part of the same null geodesic. It follows that this null geodesic is either past inextensible or has past endpoint in $\partial^{\text{ac}} F$. ∎

**Definition 8.3.3.** *Let $A$ be an achronal set. Then the* edge *of $A$,* edge$(A)$ *is the set of all points $x \in \bar{A}$ such that for any neighbourhood $\mathcal{U}$ of $x$ each pair of points $x^\pm$ in $I^\pm(x,\mathcal{U})$ can be joined by a timelike curve which is contained in $\mathcal{U}$ and does not intersect $A$.*

**Corollary 8.3.1.** *Let $A$ be a set. Then $\partial J^+(A)$ is an achronal Lipschitz hypersurface. The set $\partial J^+(A) \setminus \overline{A}$ is generated by null geodesics without conjugate points. These generators are past inextensible or have past endpoint in* edge$(A)$.

*Proof.* The first property is clear since $J^+(A)$ is a future set. Let $x \in \partial J^+(A) \setminus \overline{A}$ and $\gamma$ be the maximally past extended null geodesic generator with future endpoint $x$. Let $y$ be the past endpoint of $\gamma$. Since $\gamma$ is a generator of $\partial J^+(A) \setminus \overline{A}$ there is a neighbourhood $\mathcal{W}$ of $\gamma \setminus \{y\}$ which does not intersect $\overline{A}$.

Assume that $y \in \overline{A} \setminus \text{edge}(A)$. Then there is a neighbourhood convex $\mathcal{U}$ of $y$ such that for every pair of points $z^\pm$ in $I^\pm(y,\mathcal{U})$ every causal curve $\lambda$ which connects $z^-$ and $z^+$ intersects $A$. Let $z^- \in I^-(x,\mathcal{U})$, $z^0 \in I^+(z^-,\mathcal{U}) \cap I^-(z^+,\mathcal{U}) \cap \gamma \cap \mathcal{W}$, and $\lambda_1$ a timelike curve from $z^-$ to $z^0$. There is a point $z^+ \in I^+(y,\mathcal{U}) \cap \mathcal{W}$ and a timelike curve $\lambda_2 \subset \mathcal{W}$ from$z^0$ to $z^+$. The concatenation $\lambda$ of $\lambda_1$ and $\lambda_2$ interesects $A$ since $y \in \overline{A} \setminus \text{edge}(A)$. By the construction of $\mathcal{W}$ it is clear that $\lambda_2$ cannot intersect $A$. Hence there is a point $z \in \lambda_2 \cap A$. Consequently, we obtain $x \in J^+(z_0) \subset I^+(z) \subset I^+(A)$ in contradiction to $x \in \partial J^+(A)$. This proves $y \notin \overline{A} \setminus \text{edge}(A)$.

Assume now that $y$ is past endpoint of the null geodesic generator $\gamma$ and is not contained in $\overline{A}$. Then $y$ has a neighbourhood $\mathcal{U}$ which does not intersect $\overline{A}$. This implies that for each $z \in I^+(A) \cap \mathcal{U}$ there is a timelike curve from $A$ to $z$ which initially does not lie in $\mathcal{U}$. Hence we have shown $I^+(A) = I^+(I^+(A)) = I^+(I^+(A) \setminus \mathcal{U})$, i.e., $y \in \partial^{\text{null}} J^+(A)$. This is impossible by Lemma 8.3.2. ∎

**Lemma 8.3.3.** *If $A$ is a spacelike hypersurface,* edge($A$) *is a subset of the boundary of $A$.*

*Proof.* Let $x \in A \setminus \partial A$. Since $A$ is spacelike there exists a neighbourhood $\mathcal{U}$ of $x$ such that $(I^+(x,\mathcal{U}) \cup I^-(x,\mathcal{U})) \cap A = \emptyset$. Since $A$ is a hypersurface it divides $\mathcal{U}$ (if chosen small enough) into two disconnected components, one of them containing $I^+(x,\mathcal{U})$ and the other $I^-(x,\mathcal{U})$. But this implies that every causal curve fron $I^-(x,\mathcal{U})$ to $I^+(\mathcal{U})$ must interesect $A$. ∎

Another important set is the *future horismos* of $A \subset M$. It consists of those points which can be reached from $A$ via causal but not via timelike curves.

**Definition 8.3.4.** *The* future horismos *of a subset $A \subset M$ relative to a neighbourhood $\mathcal{U}$ of $A$ is the set $E^+(A,\mathcal{U}) = J^+(A,\mathcal{U}) \setminus I^+(A,\mathcal{U})$.*

**Lemma 8.3.4.** *Let $A \subset M$ be a spacelike submanifold and $x \in M$. Then $x \in E^+(A,\mathcal{U})$ holds if and only if there is a null geodesic from $A$ to $x$ which is completely contained in $\mathcal{U}$ and does not have focal points before $x$.*

*Proof.* This follows immediately from Corollary 8.3.1, Theorem 4.6.1, Lemma 4.6.15, and the fact that $E^+(A,\mathcal{U})$ is a subset of $\partial J^+(A,\mathcal{U})$. ∎

If $A$ is a set we are also interested in those points which are completely determined by the physical data at $A$. If Postulate 8.0.1 holds then this set of these points is described by the following definition.

**Definition 8.3.5.** *Let $A$ be a set. Then the* future Cauchy development *$D^+(A)$ is the set of all points $x \in M$ such that all past inextensible causal curves through $x$ intersect $A$. The* past Cauchy development *is defined analogously and denoted by $D^-(A)$. The union $D(A) = D^+(A) \cup D^-(A)$ is called the* Cauchy development *of $A$.*

**Lemma 8.3.5.** *For any achronal set $A$ we have $I^-(\overline{D^+(A)}) \cap \overline{I^+(A)} \subset D^+(A)$.*

*Proof.* If $x \in I^-(\overline{D^+(A)}) \cap \overline{I^+(A)}$ then there is a $y \in D^+(A) \cap I^+(x)$. Let $\gamma$ be a timelike curve from $x$ to $y$. If $x \notin D^+(A)$ then there is a past inextensible curve $\mu$ with future endpoint $x$ which does not intersect $A$. Since the concatenation of $\mu$ and $\gamma$ is a past inextensible curve with future endpoint $y$ the curve $\gamma$ intersects $A$ at some point $z$. From $z \in I^+(x)$ and $x \in \overline{I^+(A)}$ we obtain $z \in I^+(A)$ which implies that there is a timelike curve with future endpoint $y \in I^+(z)$ which intersects $A$ twice. This gives a contradiction to the achronality of $A$. ∎

The following theorem shows that global hyperbolicity (Definition 8.1.3) and Cauchy developments are strongly linked.

**Theorem 8.3.1.** *Let $A$ be an achronal set. Then* $\operatorname{int}(D(A))$ *is globally hyperbolic or empty.*

First we need to establish the following two lemmas.

**Lemma 8.3.6.** *Any past inextensible causal curve which passes through* $x \in \operatorname{int}(D^+(A))$ *intersects* $I^-(A)$.

*Proof.* Let $\gamma$ be a past inextensible causal curve with future end point $x_0 = x$. The set $I^+(\gamma) \cap A$ is empty unless $\gamma$ intersects $I^-(A)$. We choose a Riemannian metric $h$ on $M$ and denote for $z \in M$ the neighbourhood $\{\tilde{z} \in M \,|\, \operatorname{dist}_h(z, \tilde{z}) < \epsilon\}$ by $B_\epsilon(x)$. Let $\{x_i\} \subset \gamma$, $x_{i+1} \in J^-(x_i)$ be a sequence without past accumulation point and let $y_0 \in I^+(x_0) \cap \operatorname{int}(D^+(A))$. There is a point $y_1 \in I^-(y_0) \cap I^+(x_1, B_1(x_1))$. We inductively define a sequence $\{y_i\}_{i \in \mathbb{N}}$ with $y_i \in I^-(y_{i-1}) \cap I^+(x_i, B_{\frac{1}{i}}(x_i))$ and connect the $y_i$ by timelike curve segments. The concatenation of these curve segements is an inextensible timelike curve $\mu \subset I^+(\gamma)$. Since $y_0 \in \operatorname{int}(D^+(A))$, the timelike curve $\mu$ must intersect $A$ at some point $y$. There is an $y_j \in \mu \cap I^-(y)$ which implies $x_j \in I^-(y)$. This gives a contradiction to $I^+(\gamma) \cap A = \emptyset$. ∎

The spacetime depicted in Fig. 8.1.4 violates strong causality along an achronal null geodesic. The following lemma shows that for chronological spacetimes the set where strong causality is violated is always generated by null geodesics.

**Lemma 8.3.7.** *Assume that $(M, g)$ is chronological. If it fails to be strongly causal at $x \in M$, then there is an achronal, inextensible null geodesic $\mu$ through $x$ along which strong causality is violated.*

*The endpiece of $\mu$ to the past (resp. future) of $x$ is a cluster curve of curves which have intersected (resp. will intersect) arbitrarily small neighbourhoods of $x$ before (resp. afterwards)*

*Proof.* Let $\mathcal{U}_i$ be a basis of convex neighbourhoods of $x$ and $\{\mu_i\}_{i \in \mathbb{N}}$ be a sequence of inextensible timelike curves such that $\mu_i$ intersects $\mathcal{U}_i$ at least twice. Let $x_i, y_i \in \mathcal{U}_i \cap \mu_i$, $y_i \in J^+(x_i)$ be points such that the segment of $\mu_i$ between $x_i$ and $y_i$ leaves $\mathcal{U}_i$. We denote by $\mu_i[\rightarrow y_i]$ (resp. $\mu_i[x_i \rightarrow]$) the past (resp. future)inextensible endpiece of $\mu_i$ with future (resp. past) endpoint $y_i$ (resp. $x_i$).

Let $\mu^+[x \rightarrow]$ be a future inextensible causal cluster curve of $\{\mu_i[x_i \rightarrow]\}_{i \in \mathbb{N}}$ with past endpoint $x$ and $\{\mu_{i_j}[x_{i_j} \rightarrow]\}_{j \in \mathbb{N}}$ be a a distinguishing subsequence. If there is a point $z \in \mu^+[x \rightarrow] \cap I^+(x)$ then there are neighbourhoods $\mathcal{V}_x$, $\mathcal{V}_z$ of $x$ and $z$ such that

$$\mathcal{V}_z \subset I^+(\hat{x}) \text{ for all } \hat{x} \in \mathcal{V}_x \text{ and } \mathcal{V}_x \subset I^-(\hat{z}) \text{ for all } \hat{z} \in \mathcal{V}_z .$$

Let $j_0 \in \mathbb{N}$ be a number such that $y_{i_j} \in \mathcal{V}_x$ for all $j > j_0$. By the definition of a cluster curve there is a $j_1 > j_0$ with $\mu_{i_{j_1}}[x_{i_{j_1}} \rightarrow] \cap \mathcal{V}_z \neq \emptyset$.

Let $\tilde{z} \in \mu_{i_{j_1}}[x_{i_{j_1}} \rightarrow] \cap \mathcal{V}_z$. We obtain a closed timelike curve by first traversing $\mu_{i_{j_1}}[x_{i_{j_1}} \rightarrow]$ from $\tilde{z}$ to $y_{i_{j_1}}$ and then connecting $y_{i_{j_1}} \in \mathcal{V}_x$ with $\tilde{z} \in \mathcal{V}_z$ through a timelike curve. Since this contradicts the chronology of $(M, g)$ we have shown that $\mu^+[x \rightarrow] \cap I^+(x) = \emptyset$ and therefore that the cluster curve $\mu^+[x \rightarrow]$ is achronal.

Let $\mu^-[\rightarrow x]$ be a future inextensible causal cluster curve of $\{\mu_{i_j}[\rightarrow y_{i_j}]\}_{j \in \mathbb{N}}$ with future endpoint $x$. As for $\mu^+[x \rightarrow]$ we see that $\mu^-[\rightarrow x]$ is achronal.

The concatenation $\mu$ of $\mu^-[\rightarrow x]$ and $\mu^+[x \rightarrow]$ if future and past inextensible. If it is not achronal there exist points $z_- \in \mu^-[\rightarrow x]$ and $z_+ \in \mu^+[x \rightarrow]$ with $z_+ \in I^+(z_-)$ and therefore also neighbourhoods $\mathcal{V}_{z_-}$ and $\mathcal{V}_{z_+}$ of $z_-$ and $z_+$ such that all $\hat{z}_- \in calV_{z_-}$ and all $\hat{z}_+ \in calV_{z_+}$ are chronologically related. Let $\{\mu_{i_j}[\rightarrow y_{i_j}]\}_{j \in \mathbb{N}}$ be a subsequence of $\{\mu_i[\rightarrow y_i]\}_{i \in \mathbb{N}}$ which distinguishes $\mu^-[\rightarrow x]$. Hence there is a $j_0 \in \mathbb{N}$ such that $\mu_{i_j}[\rightarrow y_{i_j}]$ interesects $\mathcal{V}_{z_-}$ for all $j > j_0$. Consider now the subsequence $\{\mu_{i_j}[\rightarrow y_{i_j}]\}_{j > j_0}$ and recall that the curves $\mu_{i_j}[\rightarrow y_{i_j}]$ and $\mu_{i_j}[x_{i_j} \rightarrow]$ coincide. We can assume (without loss of generality) that the sequence $\{\mu_{i_j}[x_{i_j} \rightarrow]\}_{j > j_0}$ distinguishes $\mu^+[x \rightarrow]$. Hence there is a $j_1 > j_0$ and a curve $\mu_{i_{j_1}}[x_{i_{j_1}} \rightarrow]$ which interesects $\mathcal{V}_{z^+}$. Since $\mu^+[x \rightarrow]$ is a cluster curve with past endpoint $x = \lim x_{i_j}$ this neighbourhood is interesected before $\mu_{i_{j_1}}[x_{i_{j_1}} \rightarrow] = \mu_{i_{j_1}}[\rightarrow y_{i_{j_1}}]$ intersects $\mathcal{V}_{z^-}$. Again we get a contradiction to the chronology of $(M, g)$. This proves the first assertion of the lemma.

The second assertion follows directly from our construction. ∎

*Proof of Theorem 8.3.1.* We show first that $\text{int}(D(A))$ is strongly causal. By the definition of Cauchy development there cannot be any closed causal curve in $\text{int}(D(A))$. Assume that the strong causality condition is violated at $x \in \text{int}(D(A))$ and let $\mathcal{U}$ be an arbitrarily small convex neighbourhood of $x$ in $\text{int}(D(A))$. Then there exists a sequence of future directed timelike curves $\gamma_i$ which have $x$ as an accumulation point and intersect $\mathcal{U}$ twice. There is an inextensible cluster curve $\gamma$ through $x$. This curve must intersect $A$ since $x \in \text{int}(D(A))$. By Lemma 8.3.6, $\gamma$ intersects both $I^-(A)$ and $I^+(A)$. Since these sets are open they are both intersected by $\gamma_i$ for $i$ large enough. From Lemma 8.3.7 we see that $\gamma_i$ first intersects $I^+(A)$ and then $I^-(A)$. Hence we can prolong these curves such that they intersect $A$ more than once. This gives a contradiction to the achronality of $A$.

We show now that for any $x, y \in \text{int}(D(A))$ the set $J^+(x) \cap J^-(y)$ is compact. Since there are no closed causal curves in $\text{int}(D(A))$ the equation $x = y$ implies $J^+(x) \cap J^-(y) = \{x\}$ which is compact. Hence we can assume that $y \in J^+(x) \setminus \{y\}$. We may also assume that $y \in \text{int}(D^+(A))$ since otherwise $x \in \text{int}(D^-(A))$ and we could apply the time reversed argument. Let $\{z_i\}_{i \in \mathbb{N}}$ be a sequence in $J^+(x) \cap J^-(y)$. We need to show that it contains a convergent subsequence. For each $z_i$ there is a causal curve $\gamma_i$ from $x$ through $z_i$ to $y$. If a subsequence $\{z_{i_j}\}_{j\in\mathbb{N}}$ of $\{z_i\}_{i\in\mathbb{N}}$ lies in $D^+(A) \cup A$, we consider the cluster curve with future end point $y$. Otherwise a subsequence $\{z_{i_j}\}_{j\in\mathbb{N}}$ of $\{z_i\}_{i\in\mathbb{N}}$ lies in $D^-(A)$ and we can consider the cluster curve with past endpoint $x$. Again by time-reversal, we can restrict to the first possibility. Since $y \in D^+(A)$, the cluster curve must intersect $A$. Hence the segment from $A$ to $y$ is compact and has a compact neighbourhood $\mathcal{U}$. Since this segment is a cluster curve, infinitely many of the $z_{i_j}$ lie in $\mathcal{U}$ and must therefore have a convergent subsequence. ∎

**Lemma 8.3.8.** *Let $A$ be a set. Then*

$$\overline{D^+(A)} = \{x \in M \mid \text{ every past inextensible timelike curve with future endpoint } x \text{ intersects } A\}$$

*Proof.* "$\subset$":. Let $x \in \overline{D^+(A)} \setminus A$ and $\gamma$ be a past inextensible timelike curve with future endpoint $x$. Let $\mathcal{U}$ be a convex neighbourhood of $x$ which does not intersect $A$ and $y \in \gamma \cap \mathcal{U}$ be a point different from $x$. Then $I^+(y,\mathcal{U})$ is a neighbourhood of $x$ and must therefore intersect $D^+(A)$. Let $\mu$ be a timelike curve from $y$ to some $z \in D^+(A) \cap \mathcal{U}$ which is contained in $\mathcal{U}$. Since it does not intersect $A$ and the concatenation of $\mu$ and $\gamma$ is past inextensible, $\gamma$ must intersect $A$.

"$\supset$:" let $x \notin A$ be a point such that every past inextensible timelike curve with future endpoint $x$ intersects $A$. Now let $\gamma$ be a timelike curve with future endpoint $x$ and $\{x_i\}_{i\in\mathbb{N}}$ be a sequence in $\gamma$ which converges to $x$. Let $\lambda_i$ be a causal curve which is past inextensible and has future endpoint $x_i$. Since any causal curve which is not an achronal null geodesic can be perturbed slightly such that the resulting curve is timelike, the concatenation of $\lambda_i$ and the future endpiece $\gamma[x_i \rightarrow]$ of $\gamma$ must intersect $A$. Hence for $i$ sufficiently large, $\lambda_i$ must intersect $A$. Since $\lambda_i$ was arbitrary it follows that $x_i \in D^+(A)$ and, consequently, $x \in \overline{D^+(A)}$. ∎

**Definition 8.3.6.** *The future boundary of $D^+(A)$,*

$$H^+(A) = \overline{D^+(A)} \setminus I^-(D^+(A))$$

*is called the* future Cauchy horizon. *The* past Cauchy horizon $H^-(A)$ *is defined analogously.*

The future Cauchy horizon marks the limit of the set which can be predicted from knowing data at $A$. It has the following fundamental properties.

**Lemma 8.3.9.** *Let $x \in H^+(A) \setminus \overline{A}$. Then $x \in I^+(A)$.*

*Proof.* There is a neighbourhood $\mathcal{U}$ of $x$ which does not intersect $A$. Since for any $y \in I^-(x,\mathcal{U})$ the set $I^+(y,\mathcal{U})$ is a neighbourhood of $x$ it must intersect $D^+(A)$. Let $z \in I^+(y,\mathcal{U}) \cap D^+(A)$. It follows that $y \in D^+(A) \subset I^+(A)$ since otherwise we could construct a past inextensible curve through $z$ which which does not intersect $A$. ∎

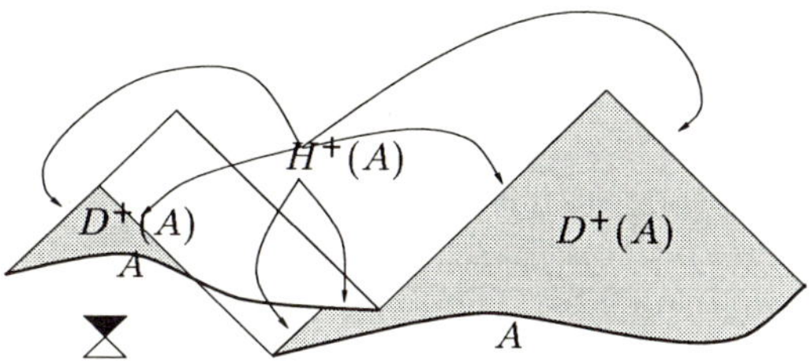


**Fig. 8.3.2.** The Cauchy horizon for a set which fails to be achronal

**Proposition 8.3.1.** *Let $A$ be a closed achronal set. Its future horizon $H^+(A)$ is generated by achronal null geodesics which have either no past endpoint or past endpoint in* edge$(A)$.

*Proof.* The set $P = D^+(A) \cup I^-(A)$ is a past set since $I^-(D^+(A)) \subset D^+(A) \cup I^-(A)$. Let $y \in H^+(A) = \overline{D^+(A)} \setminus I^-(D^+(A))$. If $y \in I^-(A)$ then there exists a timelike curve which intersects $A$ twice in contradiction to the achronality of $A$. Hence $H^+(A)$ is a subset of the boundary of $P$ and therefore a subset of an achronal Lipschitz manifold.

Let $x \in H^+(A)\backslash\text{edge}(A)$ and let $\{x_i\}_{i\in\mathbb{N}}$ be a sequence in $I^+(x)$ which converges to $x$. By the definition of $H^+(A)$ none of the $x_i$ is contained in $D^+(A)$. Hence for each $x_i$ there is a past inextensible causal curve $\gamma_i$ which does not intersect $A$. Let $\gamma$ be a causal cluster curve with future endpoint $x$. We will show that $x$ has a convex neighbourhood $\mathcal{C}$ such that $\gamma \cap \mathcal{C} \subset H^+(A)$.

Since Lemma 8.3.9 implies that $x \in H^+(A) \subset A \cup I^+(A)$ there are two possible cases, $x \in A \setminus \text{edge}(A)$ or $x \in I^+(A)$. If $x \in A \setminus \text{edge}(A)$ we choose $\mathcal{C}$ such that every timelike curve from $I^-(x,\mathcal{C})$ to $I^+(x,\mathcal{C})$ which is contained in $\mathcal{C}$ intersects $A$. If $x \in I^+(A)$ we choose $\mathcal{C} \subset I^+(A)$.

It follows that $\gamma \cap \mathcal{C} \subset I^+(A) \cup A$. This is clear if $x \in I^+(A)$. Assume therefore that $x \in A$ and that there is an $y \in \gamma \cap \mathcal{C} \setminus A$. Since $A$ is closed there is a convex neighbourhood $\mathcal{U} \subset \mathcal{C}$ of $y$ which does not intersect $A$. This neighbourhood intersect $\gamma_i$ (for $i$ large enough). It also intersects $I^-(x)$ since $x \in A$ and $y \in J^-(A)$. Hence there is a timelike curve $\lambda$ in $\mathcal{U}$ from $I^-(x)$ to $\gamma_i$. Since $\gamma_i$ does not intersect $A$ we can concatenate $\lambda$ with a the future endpiece of $\gamma_i$ to obtain a timelike curve from $I^-(x)$ to $x_i \in I^+(x)$ which is contained in $\mathcal{C}$ and does not intersect $A$. This gives a contradiction to $x \in A \setminus \text{edge}(A)$ and we have proved $\gamma \cap \mathcal{C} \subset I^+(A) \cup A$.

Assume that there is an $y \in (\mu \cap \mathcal{C}) \setminus \overline{D^+(A)}$. If the segment $(\gamma \cap \mathcal{C})[y \to x]$ of $\gamma \cap \mathcal{C}$ between $y$ and $x$ would intersect $A$ then we would obtain a contradiction to the achronality of $A$ from $y \in (I^+(A) \cup A) \setminus \overline{D^+(A)} \subset I^+(A)$ and $(\gamma \cap \mathcal{C})[y \to x] \in J^+(y)$. Let $\mathcal{U} \subset \mathcal{C}$ be a neighbourhood of $y$ which does not intersect $\overline{D^+(A)}$ and consider a point $z \in (I^-(y, \mathcal{U}) \cap I^+(A)) \setminus \overline{D^+(A)}$. Then the concatenation $\lambda$ of a timelike curve from $z$ to $y$ in $\mathcal{U}$ and the segement $(\gamma \cap \mathcal{C})[x \to y]$ does not intersect $A$. Since $x \in H^+(A)$ and $A$ is closed a slight deformation of $\lambda$ results in a causal curve from $z$ to some point in $\tilde{x} \in D^+(A)$ which does not intersect $A$. We could now prolong this curve to the past of $z \in M \setminus \overline{D^+(A)}$ thereby obtaining a past inextensible curve which does not intersect $A$. This gives a contradiction to $\tilde{x} \in D^+(A)$, whence we have proved $(\gamma \cap \mathcal{C}) \subset \overline{D^+(A)}$.

Assume that there is an $y \in (\gamma \cap \mathcal{C}) \cap I^-(D^+(A))$ and let $z \in I^+(y) \cap D^+(A)$. Let $\lambda$ be a timelike curve from $y$ to $z$. This curve cannot intersect $A$ to the future of $y$ because of $y \in I^+(A) \cup A$ and the achronality of $A$. If $y \notin A$ then there is a neighbourhood $\mathcal{U}$ of $y$ with $\mathcal{U} \subset I^-(z)$ which does not intersect $A$. Since $\gamma$ is a cluster curve of $\{\gamma_i\}_{i \in \mathbb{N}}$ there is an $i$ and a point $\hat{y} \in \gamma_i \cap \mathcal{U}$. It follows that there is a causal curve $\tilde{\lambda}$ from $\hat{y} \in \gamma_i$ to $z$ which does not intersect $A$. If $y \in A$ then $x \in A \setminus \mathrm{edge}(A)$ by the construction of $\mathcal{C}$. The point $y$ has a convex neighbourhood of $\mathcal{U} \subset \mathcal{C} \cap I^-(z)$ which is intersected by infinitely many $\gamma_i$. Let $i \in \mathbb{N}$ with $\gamma_i \cap \mathcal{U} \neq \emptyset$, $\hat{y} \in \gamma_i \cap \mathcal{U}$, and $\tilde{\lambda}$ be a timelike curve from $\hat{y}$ to $z$. Consider a timelike curve $\mu$ from $I^-(x) \cap \mathcal{C}$ to $\hat{y}$ which is contained in $\mathcal{C}$. The concatenation of $\mu$ and the part of $\gamma_i$ to the future of $\hat{y}$ intersects $A$ because of $x \notin \mathrm{edge}(A)$ and $x_i \in I^+(x)$. The equation $\gamma_i \cap A = \emptyset$ implies that $\mu$ intersects $A$. Hence $\tilde{\lambda}$ cannot intersect $A$ by the achronality of $A$. We have shown that in either case, $y \notin A$ and $y \in A$, the concatenation of the past endpiece of $\gamma_i$ with future endpoint $\hat{y}$ and $\tilde{\lambda}$ is past inextensible and does not intersect $A$. This gives a contradiction to $z \in D^+(A)$. Consequently, we have shown $(\gamma \cap \mathcal{U}) \cap I^-(\overline{D^+(A)}) = \emptyset$

We have $\gamma \cap \mathcal{C} \subset \overline{D^+(A)} \setminus I^-(\overline{D^+(A)}) = H^+(A)$. The past endpoint $\hat{x}$ of $\gamma \cap \mathcal{C}$ lies in $H^+(A)$ since $H^+(A)$ is closed. If $\hat{x} \notin \mathrm{edge}(A)$ we can repeat the construction thereby obtaining a curve $\hat{\gamma} \cap \hat{\mathcal{C}} \subset H^+(A)$ with future endpoint $\hat{x}$. The concatenation of $\gamma \cap \mathcal{C}$ and $\hat{\gamma} \cap \hat{\mathcal{C}}$ gives a curve $\gamma_1 \subset H^+(A)$ with future endpoint $x$. Repeating this construction inductively we obtain a causal curve $\gamma_\infty \subset H^+(A)$ with future endpoint $x$ which has either no past endpoint or has past endpoint in $\mathrm{edge}(A)$. Since $H^+(A)$ is achronal this curve must be an achronal null geodesic. ∎

# 9. Singularity theorems

*In this chapter we prove and investigate "singularity theorems". These theorems are usually interpreted as an indication that black holes exist and that there has been a big bang — or at least that there are regions in spacetime where general relativity breaks down. They are one of the main motivations for attempting to quantize general relativity. While there is a lot of evidence in favour of this interpretation we will see that there are also open problems which have to be addressed in order to justify this interpretation.*

In Chaps. 7 and 6 we have seen that spacetimes describing a single, non-rotating star and the simplest cosmological models of our universe contain regions where the curvature diverges. One may think that these singularities are only an artefact of our high symmetry assumptions, but in this section, we will give an indication that a physically realistic spacetime *must* contain such singularities. More precisely, we will show that there exist causal, inextensible geodesics which are incomplete. Recall that a freely falling particle is represented by a timelike geodesic. If the geodesic cannot be extended to a complete one (i.e. if its future endless continuation or its past endless continuation is of finite length), then either the particle suddenly ceases to exist or the particle suddenly springs into existence[1]. In either case this can only happen if spacetime admits a "singularity" at the end (or beginning) of the history of the particle. This singularity may be a curvature singularity, there may be a topological obstruction, or spacetime may simply cease to be sufficiently smooth. However, the Schwarzschild and Robertson Walker solutions indicate that these singularities are accompanied with diverging curvature. (But cf. Sect. 9.5.1 below where we present a spacetime which indicates that such singularities are very mild). We will prove a singularity theorem which only establishes the existence of incomplete *causal* geodesics rather than incomplete *timelike* geodesics. While the innocent looking extension to null geodesics is necessary for the proof, the name *singularity theorem* is in this case somewhat misleading, because there exist

---

[1] This should not be confused with pair creation or pair annihilation of particles and anti particles, because during these processes nothing really ceases or starts to exist. These quantum mechanical phenomenons are merely changes of state.

perfectly regular spacetimes with incomplete null geodesics contained in compact subsets. On the other hand, it has been argued that such examples are very special and that in stable, physically realistic spacetimes this phenomenon does not occur (cf. (Hawking and Ellis 1973)).

## 9.1 Energy conditions

In general, a maximally extended Lorentzian manifold need not contain incomplete causal geodesics. In order to prove a singularity theorem, we will have to make some physical assumptions.

There are two sorts of fundamental physical experience which come to mind. Firstly, energy density as measured by the energy momentum tensor is positive. Secondly, gravitation is attractive.

Recall that the energy density measured by an observer $\gamma$ (with $g(\dot{\gamma}, \dot{\gamma}) = -1$) is given by $\epsilon = T(\dot{\gamma}, \dot{\gamma})$. We feel that this energy density should be positive. Recall also that in the motivation of the energy momentum tensor (cf. Sect. 5.1) we have obtained the energy density $\epsilon = T(U, U)$ as an average of a *positive* mass distribution. For our purpose this should be enough of a motivation of the following definition

**Definition 9.1.1.** *We say that the* weak energy condition *holds at* $x \in M$ *if*

$$T(u, u) \geq 0 \textit{ for all causal vectors } u \in T_x M.$$

For a physical verification of the weak energy condition one would have to consider all realistic physical matter models. This is beyond the scope of this book but so far the available evidence points to the fact that the weak energy condition does hold.

Gravity is attractive if and only if any two nearby freely falling observers will be forced to approach each other under the influence of the underlying spacetime geometry. This can be formulated infinititesimally in a rigorous manner. A freely falling observer is modelled by a timelike geodesic $\gamma\colon [a, b] \to M$. Let $f\colon (-\delta, \delta) \times [a, b] \to M$ a geodesic variation of $\gamma$ and $J = f_s(0, \cdot)$ be the variation vector field. Observe that $J$ is a Jacobi vector field. From Taylor's theorem we get with respect to any coordinate system $f^i(s, t) = f^i(0, t) + sJ^i(t) + O(s^2)$. This coordinate expression can be interpreted in Newtonian terms as follows. The observers $\gamma$ and $f(s, \cdot)$ have (up to first order) the same rest space and are separated by the space vector $sJ^i$. Hence up to first order it makes sense to speak of the (Newtonian) force $F$ with which $\gamma$ acts on $f(s, \cdot)$. This force is approximately given by $F^i = -ms\ddot{J}^i$ where $m$ is the mass of the observer $f(s, \cdot)$. (The minus sign is inserted because the force vector points from $f(s, t)$ to $\gamma(t)$ and $J$ points into the opposite direction). Clearly there cannot be a direct translation of Newtonian concepts to general relativity. But for small relative velocities (as in this case, $(f_t(s,t))^i \approx \dot{\gamma}^i(t)$ for $s \ll 1$) there is a well defined infinitesimal limit. In fact, the location of an infinitesimally nearby observer is characterized by the Jacobi vector field $J$ orthogonal to $\gamma$ and the force it is acted on is given by $F = -m\nabla_{\dot{\gamma}}\nabla_{\dot{\gamma}}J$. Since $J$ is a Jacobi vector field, this is exactly the force in the following definition.

**Definition 9.1.2.** *Let $\gamma$ be a timelike geodesic with $\langle\dot{\gamma},\dot{\gamma}\rangle = -1$. A* neigbouring freely falling observer $J$ of mass $m$ *is a pair $(J, m)$, where $J$ is a Jacobi field along $\gamma$ with values in $\dot{\gamma}^{\perp}$ and $m$ is a positive number.*

*The* tidal force *which acts between the observer $\gamma$ of mass $m$ and its neighbouring freely falling observer $J$ is given by*

$$F = mR(J,\dot{\gamma})\dot{\gamma}.$$

The component of $F$ pointing towards the observer $\gamma$ can is given by

$$\left\langle F, -\frac{1}{\sqrt{\langle J,J\rangle}}J\right\rangle = -\frac{m}{\sqrt{\langle J,J\rangle}}\left\langle R(J,\dot{\gamma})\dot{\gamma}, J\right\rangle.$$

Hence the assertion that $\gamma$ attracts the infinitesimally neighbouring observer corresponding to the Jacobi field $J$ is equivalent to the assertion that the sectional curvature of the plane spanned by $J$ and $\dot{\gamma}$ is non-positive.[2] This motivates the following definition.

**Definition 9.1.3.** *We say that at $x \in M$* gravity is attractive in all directions *if and only the sectional curvature of all timelike planes in $T_xM$ is non-positive.*

The requirement that gravity is attractive in all directions is very strong. Also note that apart from our experience in weak gravitational fields we have not much evidence that gravity is really attractive in all directions. On the other hand, it is clear that gravity must be attractive on average. The reason is that gravity is much weaker than all the other fundamental physical interactions. Electromagnetism and gravitation are the only long range interactions. On large scales, electromagnetism is not of primary importance because it is attractive or repulsive depending on the configuration. If this would also be true for gravity, it would be of even less importance for astrophysical applications. We know however that this is not the case. There are of course several ways of defining averages of gravitation. We will restrict attention to our single observer $\gamma$ together with the associated space of neighbouring, freely falling observers. The average of the tidal forces in every direction is given by integrating the

[2] Observer that we neglect the contribution of $\gamma$ and the neighbouring observer to the gravitational field. They are thought to be of negligible mass.

tidal force component in every direction over the unit sphere $S^{n-2}$. For $n = 4$ we obtain

$$
\begin{aligned}
&-\frac{3m}{4\pi}\int_{S^2\subset T_xM}\langle R(\cdot,\dot\gamma)\dot\gamma,\cdot\rangle\,\mu_{S^2}\\
&\quad=-\frac{3m}{4\pi}\int_{-\pi/2}^{\pi/2}\int_0^{2\pi}\Big\langle R\big(\cos\theta(\cos\varphi e_1+\sin\varphi e_2)+\sin\theta e_3,\dot\gamma\big)\dot\gamma,\\
&\qquad\cos\theta(\cos\varphi e_1+\sin\varphi e_2)+\sin\theta e_3\Big\rangle\cos\theta\,\mathrm{d}\varphi\,\mathrm{d}\theta\\
&\quad=-\frac{m}{4/3\pi}\int_{-\pi/2}^{\pi/2}\int_0^{2\pi}\Big(\cos^2\varphi\cos^2\theta\,\langle R(e_1,\dot\gamma)\dot\gamma,e_1\rangle\\
&\qquad+2\cos\varphi\sin\varphi\cos^2\theta\,\langle R(e_1,\dot\gamma)\dot\gamma,e_2\rangle\\
&\qquad+2\cos\varphi\cos\theta\sin\theta\,\langle R(e_1,\dot\gamma)\dot\gamma,e_3\rangle+\sin^2\varphi\cos^2\theta\,\langle R(e_2,\dot\gamma)\dot\gamma,e_2\rangle\\
&\qquad+2\sin\varphi\cos\theta\sin\theta\,\langle R(e_2,\dot\gamma)\dot\gamma,e_3\rangle\\
&\qquad+\sin^2\theta\,\langle R(e_3,\dot\gamma)\dot\gamma,e_3\rangle\Big)\cos\theta\,\mathrm{d}\varphi\,\mathrm{d}\theta\\
&\quad=-\frac{m}{4/3\pi}\int_{-\pi/2}^{\pi/2}(\pi\cos^3\theta\,\langle R(e_1,\dot\gamma)\dot\gamma,e_1\rangle+\pi\cos^3\theta\,\langle R(e_2,\dot\gamma)\dot\gamma,e_2\rangle\\
&\qquad+2\pi\cos\theta\sin^2\theta\,\langle R(e_3,\dot\gamma)\dot\gamma,e_3\rangle)\cos\theta\,\mathrm{d}\theta\\
&\quad=-\sum_{i=1}^{3}R(e_i,\dot\gamma)\dot\gamma,e_i=-\mathrm{Ric}(\dot\gamma,\dot\gamma)
\end{aligned}
$$

This motiviates the definition

**Definition 9.1.4.** *The* timelike convergence condition *holds if*

$$\mathrm{Ric}(u,u)\geq 0 \textit{ for all causal vectors } u.$$

Using Einstein's equation this condition can be re-expressed in terms of the energy momentum tensor:

$$T(u,u)-\frac{1}{n-2}\left(\mathrm{tr}(T)-\frac{\Lambda}{4\pi}\right)g(u,u)\geq 0 \text{ for all causal vectors } u.$$

While this inequality is not implied by the weak energy condition (nor does imply it), the timelike convergence condition is often also called the *strong energy condition*. The following lemma gives a partial motivation for this terminology.

**Lemma 9.1.1.** *Assume that $g$ and $T$ can be simultaneously diagonalized so that the energy density $\epsilon$ and the principal pressures $p_i$ of $T$ are defined. Then the weak energy condition is equivalent to*

$$\epsilon\geq 0,\qquad \epsilon+p_i\geq 0 \textit{ for all } i\in\{1,\ldots,n-1\},$$

*and the timelike convergence condition is equivalent to*

$$(n-3)\epsilon + \sum_{i=1}^{n-1} p_i - \frac{\Lambda}{4\pi} \geq 0, \qquad \epsilon + p_i \geq 0 \textit{ for all } i \in \{1, \ldots, n-1\}.$$

*Proof.* Let $\{e_0, \ldots, e_{n-1}\}$ be an orthonormal basis of $T_xM$ which diagonalizes $T$. Any causal vector $u$ can be written as $u = c(e_0 + \sum_{i=1}^{n-1} c^i e_i)$, where $c, c^i$ numbers with $\sum_{i=1}^{n-1}(c^i)^2 \leq 1$. Hence the weak energy condition is equivalent to

$$\begin{aligned} 0 \leq T(e_0 + \sum_{i=1}^{n-1} c^i e_i, e_0 + \sum_{i=1}^{n-1} c^i e_i) &= \epsilon + \sum_{i=1}^{n-1}(c^i)^2 p_i \\ = \left(1 - \sum_{i=1}^{n-1}(c^i)^2\right)\epsilon &+ \sum_{i=1}^{n-1}(c^i)^2\,(\epsilon + p_i)\,. \end{aligned}$$

The condition for the weak energy condition is sufficient since the factors

$$\left(1 - \sum_{i=1}^{n-1}(c^i)^2\right) \text{ and } (c^i)^2$$

are positive. It is necessary since we can choose the numbers $c^i = 0$ or for any given $j$ the numbers $c^i = \delta^i_j$.

The strong energy condition is equivalent to

$$\begin{aligned} 0 &\leq (n-2)\mathrm{Ric}(e_0 + \sum_{i=1}^{n-1} c^i e_i, e_0 + \sum_{i=1}^{n-1} c^i e_i) \\ &= (n-2)T(e_0 + \sum_{i=1}^{n-1} c^i e_i, e_0 + \sum_{i=1}^{n-1} c^i e_i) \\ &\quad - \left(\mathrm{tr}(T) - \frac{\Lambda}{4\pi}\right) g(e_0 + \sum_{i=1}^{n-1} c^i e_i, e_0 + \sum_{i=1}^{n-1} c^i e_i) \\ &= (n-2)\left(\epsilon + \sum_{i=1}^{n-1}(c^i)^2 p_i\right) - \left(-1 + \sum_{i=1}^{n-1}(c^i)^2\right)\left(-\epsilon + \sum_{j=1}^{n-1} p_j - \frac{\Lambda}{4\pi}\right) \\ &= (n-2)\sum_{i=1}^{n-1}(c^i)^2\,(\epsilon + p_i) \\ &\quad + \left(1 - \sum_{i=1}^{n-1}(c^i)^2\right)\left((n-3)\epsilon + \sum_{j=1}^{n-1} p_j - \frac{\Lambda}{4\pi}\right). \end{aligned}$$

Hence the assertion for the timelike convergence condition follows by the same argument. ∎

It seems physically more plausible to demand that the matter distribution of our universe satisfies the inequality $\epsilon \geq 0$ rather than the inequality $\epsilon + \sum_{i=1}^{n-1} p_i - \frac{\Lambda}{4\pi} \geq 0$. In this sense the timelike convergence condition is "physically" a stronger condition on the matter distribution than the weak energy condition. Since the cosmological constant is close to zero and the energy density is much larger than the principal pressures[3] the timelike convergence condition is still a very plausible assumption to make.

A further condition we want to impose is the *genericity condition.* It states that along each causal geodesic $\gamma$ there exists a point at which

$$\dot{\gamma}^c \dot{\gamma}^d \dot{\gamma}_{[a} R_{b]cd[e} \dot{\gamma}_{f]} \neq 0$$

holds. This condition is only of technical nature, because the set of metrics which satisfy the strong energy condition is dense in the set of metrics which satisfy both the strong energy and the genericity condition. Here we impose the (very fine) $C^2$-Whitney topology (for details and proofs see (Lerner 1973)).

The mathematical significance of the strong energy condition in conjunction with the genericity condition is clear from the following corollary to Proposition 4.6.3.

**Corollary 9.1.1.** *Assume that $(M, g)$ is a Lorentzian manifold and let $\gamma\colon (a, b) \to M$ be an inextensible causal geodesic. If* $\mathrm{Ric}(\dot{\gamma}(t), \dot{\gamma}(t)) \geq 0$ *for all $t$ and the genericity condition holds along $\gamma$ then either*

*(i) $\gamma$ is incomplete, or*
*(ii) $\gamma$ contains a pair of conjugate points.*

*Proof.* We have to show that the genericity condition implies the existence of a $t_0$ such that the map

$$R\colon (\dot{\gamma}(t_0))^\perp \to (\dot{\gamma}(t_0))^\perp, \quad v \mapsto Rv := R(v, \dot{\gamma})\dot{\gamma}$$

is not identically zero.

If $\gamma$ is timelike and $\dot{\gamma}^c(t)\dot{\gamma}^d(t)\dot{\gamma}_{[a}(t) R_{b]cd[e}\dot{\gamma}_{f]}(t) \neq 0$ then we have in particular $\dot{\gamma}^c(t)\dot{\gamma}^d(t) R_{bcde} \neq 0$. Since a symmetric bilinear form is determined by its associated quadratic form there is a vector $\xi \in (\dot{\gamma}(t))^\perp$ with $g(R(\xi, \dot{\gamma}(t))\xi, \dot{\gamma}(t)) \neq 0$. Consequently, $R(\xi, \dot{\gamma}(t))\dot{\gamma}(t) \neq 0$.

Suppose that $\gamma$ is null and that $\dot{\gamma}^c(t)\dot{\gamma}^d(t)\dot{\gamma}_{[a}(t) R_{b]cd[e}\dot{\gamma}_{f]}(t) \neq 0$. We choose a basis $\{e_i\}_{i=1,\dots,n}$ of $T_{\gamma(t)}M$ such that $e_n = \dot{\gamma}(t_0)$ and $\langle e_j, e_k \rangle = \delta_{jk}$, $\langle e_j, e_r \rangle = 0$, $\langle e_r, e_s \rangle = \delta_{rs} - 1$ for $j, k \in \{1, \dots, n-2\}$, $r, s \in \{n-1, n-2\}$. In this and the associated dual basis we have $\dot{\gamma}^a = \delta^a_n$ and $\dot{\gamma}_a = -\delta^{n-1}_a$. This implies

[3] Recall that in physical units where the velocity of light $c$ is not normalized to 1 the numerical value of the energy density increases by a factor $c^2$ in comparison to the principal pressures.

$$\begin{aligned}4\dot\gamma^c(t)\dot\gamma^d(t)\dot\gamma_{[a}(t)R_{b]cd[e}\dot\gamma_{f]}(t) &= 4\delta^{n-1}_{[a}R_{b]nn[e}\delta^{n-1}_{f]}\\ &= \delta^{n-1}_a R_{bnne}\delta^{n-1}_f - \delta^{n-1}_b R_{anne}\delta^{n-1}_f\\ &\quad + \delta^{n-1}_b R_{annf}\delta^{n-1}_e - \delta^{n-1}_a R_{bnnf}\delta^{n-1}_e.\end{aligned}$$

If this expression does not vanish then either $a$ or $b$ must be equal to $n-1$ and so must be either $e$ or $f$. For definiteness, assume that $b = f = n-1$. The formula then simplifies to

$$\delta^{n-1}_a R_{(n-1)nne} - R_{anne} + R_{ann(n-1)}\delta^{n-1}_e - \delta^{n-1}_a R_{(n-1)nn(n-1)}\delta^{n-1}_e.$$

This clearly vanishes if $n \in \{a, e\}$. The first two terms and the last two terms cancel pairwise if $a = n-1$. If $e = n-1$ the second summand cancels the third one and the first summand the fourth summand. Hence $\dot\gamma^c(t)\dot\gamma^d(t)\dot\gamma_{[a}(t)R_{b]cd[e}\dot\gamma_{f]}(t) \neq 0$ implies that $a, e \in \{1, \ldots, n-2\}$ and $R_{anne} \neq 0$. But this implies in turn that $R(\cdot, \dot\gamma)\dot\gamma\colon (\dot\gamma(t))^\perp \to (\dot\gamma(t))^\perp$ does not vanish. ∎

## 9.2 Closed trapped surfaces

As a further preparation, consider an isolated, dense object, say a star, in spacetime. If it produces enough gravitation, it will not only attract material objects but even the light rays it sends out. (Recall that they are modelled by null geodesics and therefore perceptible to the curvature of spacetime). Exactly this situation happens in the Schwarzschild spacetime. To be more concrete, let $\mathcal{T}$ be an $(n-2)$-dimensional spacelike submanifold of $M$. We may think of $\mathcal{T}$ as the surface of the star at a fixed time. We can now send out light orthogonal to this surface, either in direction to the centre of the star or into the opposite direction. Everyday experience suggests that the light congruence directed to the centre should converge while the light congruence directed into the opposite direction should diverge. However, this does not take into account the extrinsic curvature of the spacelike hypersurface which represents the instant of time. There are many examples where both congruences converge, for instance the surfaces $r = \text{const} < 2m$, $t = \text{const}$ in the Schwarzschild solution. These surfaces are in the black hole region of the Schwarzschild solution and the general interpretation is that the gravitation of the black hole is so strong that it forces even initially outgoing light rays to converge.

Since the normal bundle of $\mathcal{T}$ in $M$ is a Lorentzian plane at each point, there exist two future directed null vector fields $N_+, N_-$ along $\mathcal{T}$ which are orthogonal to $\mathcal{T}$ and satisfy $\langle N_-, N_+\rangle = -1$. They are unique up to transformations of the form $N_\pm \mapsto \alpha^{\pm 1} N_\pm$, where $\alpha \in C^\infty(\mathcal{T}, \mathbb{R}^+ \setminus \{0\})$.

Let $x \in \mathcal{T}$ and $\{e_A\}_{A=2,\dots,n-1}$ be an orthonormal basis of $T_x\mathcal{T}$. Then the requirement that both light concruences immediately start to converge can be expressed by the inequalities

$$\sum_{A=2}^{n-1} g(\nabla_{e_A} N^+, e_A) < 0, \qquad \sum_{A=2}^{n-1} g(\nabla_{e_A} N^-, e_A) < 0. \tag{9.2.1}$$

This requirement can be formulated in a manifestly invariant way by using the mean curvature vector field $H$ of $\mathcal{T}$ (cf. Definition 4.4.2).

**Definition 9.2.1.** *A* closed (future) trapped $(n-2)$-surface *(respectively* closed (future) strictly trapped $(n-2)$-surface, closed (future) marginally trapped $(n-2)$-surface*) is a closed $(n-2)$-dimensional spacelike submanifold $\mathcal{T}$ such that the mean curvature vector field $H$ of $\mathcal{T}$ is past pointing and causal (respectively timelike, null).*

The mean curvature vector field is defined by $H = \frac{1}{n-2}\sum_{A=2}^{n-1} I\!I(e_A, e_A)$. Any vector $v \in T_xM$ can be decomposed as $v = v^A e_A + v^- N_- + v^+ N_+$, where $v^A = \langle v, e_A\rangle$ and $v_\pm = -\langle v, N_\mp\rangle$. Using this decomposition , we obtain

$$\begin{aligned} H &= \frac{1}{n-2}\sum_{A=2}^{n-1}\left(-\left\langle \nabla_{e_A} e_A, N_-\right\rangle N_+ - \left\langle \nabla_{e_A} e_A, N_+\right\rangle N_-\right) \\ &= \frac{1}{n-2}\left(\mathrm{tr}_{\mathcal{T}}(\chi_-)N_+ + \mathrm{tr}_{\mathcal{T}}(\chi_+)N_-\right), \end{aligned} \tag{9.2.2}$$

where $\chi_\pm = \nabla N_\pm^\flat$ are the *null second fundamental forms.* Like $N_\pm$, these null second fundamental forms are uniquely defined up to transformations of the form $\chi_\pm \mapsto \alpha^{\pm 1}\chi_\pm$, where $\alpha \in C^\infty(\mathcal{T}, \mathbb{R}^+ \setminus \{0\})$. The *null expansions* are $\theta^\pm = g^{AB}\left(\chi_\pm\right)_{AB}$. It is clear that $\mathcal{T}$ is a strictly closed trapped surface if and only if both null expansions are everywhere negative on $\mathcal{T}$ and therefore equivalent to Inequalities (9.2.1).

## 9.3 The singularity theorem of Hawking and Penrose

We are now ready to state the main result of this chapter,

**Theorem 9.3.1.** *A spacetime $(M, g)$ is not causal geodesically complete if*

*(i) the strong energy condition and the genericity condition hold,*
*(ii) The chronolology conditions holds,*
*(iii) There exists at least one of the following:*
  *(a) a strictly closed trapped surface,*

*(b) a compact achronal set without edge,*
*(c) a point $x$ such that along every past (or every future) inextensible null geodesic from $x$ the expansion of the null geodesics starting at $x$ becomes negative.*

Strictly closed trapped surfaces are expected to surround very dense stars. The paradigmatic example is the Schwarzschild solution. Further evidence is provided by some theorems which prove the existence of strictly closed trapped surfaces if the concentration of matter is high (Schoen and Yau 1983; Bizon, Malec, and O'Murchadha 1988). Condition (b) is satisfied for spatially closed universes such as the Robertson Walker spacetimes with positive curvature. Condition (c) seems to be satisfied for our point in the universe (assuming a spacetime which differs only slightly from a Robertson Walker cosmology). This indicates that there was a big bang or that there will be a big chrunch. (For more details cf. (Hawking and Ellis 1973, p. 358)).

Theorem 9.3.1 will follow as a corollary of the following proposition.

**Proposition 9.3.1.** *The following three conditions cannot hold all together.*

*(i) every inextensible causal geodesic contains a pair of conjugate points,*
*(ii) $(M, g)$ is strongly causal*
*(iii) there is an achronal set $A$ such that $E^+(A)$ or $E^-(A)$ is compact.*

*Proof that Theorem 9.3.1 follows from Proposition 9.3.1.* Assume, that $(M, g)$ is causally geodesically complete and satisfies the chronology condition. By Corollary 9.1.1, the strong energy condition and the genericity condition imply that any inextensible causal geodesic has a pair of conjugate points. In particular, there do not exist maximal, inextensible causal geodesics. It follows that $(M, g)$ must be strongly causal, since otherwise it would contain an inextensible achronal null geodesic by Lemma 8.3.7.

If $(M, g)$ contains a strictly closed trapped surface $\mathcal{T}$, then $E^+(\mathcal{T}) \subset \partial J^+(\mathcal{T})$ is generated by null geodesics. These null geodesics are orthogonal to $\mathcal{T}$ and the definition of a strictly closed trapped surface implies that each of them has a focal point (cf. Proposition 4.6.2). Since $\mathcal{T}$ is compact and $E^+(\mathcal{T})$ is generated by null geodesics without focal points it follows that $E^+(\mathcal{T})$ is also be compact. An analogous argument shows that in case (c) the set $E^+(x)$ is compact.

If $(M, g)$ contains a compact achronal set $A$ without edge, then $E^+(A) = A$. This follows since $E^+(A) = J^+(A) \setminus I^+(A)$ and Corollary 8.3.1 imply that through every point $x \in E^+(A) \setminus A$ there is a generator of $E^+(A)$ which intersects edge$(A)$. Hence the set $E^+(A)$ is also compact.

Properties (i) – (iii) of Proposition 9.3.1 would therefore have to hold under the conditions of Theorem 9.3.1 and the additional assumption that all causal geodesics are complete. ∎

The idea for the proof of Proposition 9.3.1 is simple but the proof itself is quite involved. We will therefore first give an outline and then establish a sequence of lemmas which will imply the proposition.

Suppose, (i), (ii), (iii) in Proposition 9.3.1 hold and assume without loss of generality that $E^+(A)$ is compact. We will show that the horizon $H^+(E^+(A))$ is non-compact or empty. Every non-vanishing vector field $U$ must have a future inextensible integral curve $\gamma$ in $D^+(E^+(A))$. Otherwise we could map the compact set $E^+(A)$ along the integral curve of $U$ onto $H^+(E^+(A))$ which in turn would have to be compact (and non-empty), too. We apply a similar construction to the past of $E^+(A) \cap J^-(\gamma)$ and obtain an inextensible causal curve $\mu$ which is wholly contained in $D(E^+(A))$. This curve can then be used to construct an inextensible maximal causal geodesic in contradiction to (i).

In order to carry out this program we will prove the following facts.

1. $H^+(E^+(A)) \subset H^+(\partial J^+(A))$,
2. The Cauchy horizon $H^+(E^+(A))$ is non-compact or empty.
3. There is a future inextensible timelike curve $\gamma \subset D^+(E^+(A))$.
4. Set $\mathcal{F} := E^+(A) \cap \overline{J^-(\gamma)}$. Then there is a past inextensible curve $\lambda \subset D^-(E^-(\mathcal{F}))$.
5. There is an inextensible causal geodesic without conjugate points in $D(E^-(\mathcal{F}))$.

The last property 4 is in contradiction with (i) of Proposition 9.3.1.

**Lemma 9.3.1.** *Let $A$ be a closed achronal set. Then the inclusion*

$$H^+(E^+(A)) \subset H^+(\partial J^+(A))$$

*holds.*

*Proof.* Let $x \in H^+(E^+(A)) \setminus H^+(\partial J^+(A))$. From $E^+(A) \subset \partial J^+(A)$ we obtain $D^+(E^+(A)) \subset D^+(\partial J^+(A))$ and therefore $x \in I^-(D^+(\partial J^+(A)))$. Hence there is a $y \in I^+(x) \cap D^+((\partial J^+(A)))$.

We will first show that $I^+(x) \cap I^-(y)$ does not intersect $\partial J^+(A)$. Assume that there is a point $z \in \partial J^+(A) \cap I^+(x) \cap I^-(y)$. Then the open set $I^-(z)$ is a neighbourhood of $x \in H^+(E^+(A))$ and intersects therefore $D^+(E^+(A))$. Since every past inextensible timelike curve with future endpoint in $D^+(E^+(A))$ intersects $E^+(A) \subset \partial J^+(A)$ we would find a point $\hat{z} \in I^-(z) \cap \partial J^+(A) \subset I^-(\partial J^+(A)) \cap \partial J^+(A)$ in contradiction to the achronality of $\partial J^+(A)$.

Since $I^-(y)$ is a neighbourhood of $x \in H^+(E^+(A))$ and $I^+(x) \cap I^-(y)$ does not intersect $\partial J^+(A)$ there is a past inextensible timelike curve

$\gamma$ which has future endpoint $y$ and does not intersect $E^+(A)$. From $y \in D^+((\partial J^+(A)))$ we see that $\gamma$ does intesect $\partial J^+(A)$ at some point $z$. Let $\mu$ be the generator of $\partial J^+(A)$ with future endpoint $z$. By Corollary 8.3.1 this generator is either past inextensible or intersects edge$(A)$. We will show that both cases lead to a contradiction.

Assume first that there is a point $\hat{z} \in$ edge$(A)$ which is intersected by $\mu$. This point also lies in $A$ since $A$ is closed. Hence $\mu$ is contained in $J^+(A)$ which in turn implies $z \in J^+(A) \cap \partial J^+(A) = E^+(A)$. This is a contradiction to the construction of $\gamma$.

Assume that $\mu$ is past inextensible and does not intersect $A$. Since $\gamma$ is timelike and has future endpoint in $D^+(\partial J^+(A))$ it intersects the set $\text{int}(D^+(\partial J^+(A)))$. This implies that $\mu$ intersects $I^-(\partial J^+(A))$ (cf. Lemma 8.3.6). The inclusion $\mu \subset \partial J^+(A)$ gives a contradiction to the achronality of $\partial J^+(A)$. ∎

**Lemma 9.3.2.** *Let $A$ be a closed achronal set such that $\overline{J^+(A)}$ is strongly causal. Then $H^+(\overline{E^+(A)})$ is non-compact or empty.*

*Proof.* Suppose that $H^+(E^+(A))$ is non-empty but compact. Since $\overline{J^+(A)}$ is strongly causal, $H^+(E^+(A))$ can be covered by a finite number of convex neighbourhoods $\mathcal{U}_i$ with compact closure such that no $\mathcal{U}_i$ is intersected twice by any causal curve. Let $z_1 \in H^+(E^+(A))$ and $\mathcal{U}_{i(1)}$ be one of the convex neighbourhoods $\mathcal{U}_i$ with $z_1 \in \mathcal{U}_{i(1)}$. Because of Lemma 9.3.1 there is a point $x_1 \in J^+(A) \cap (\mathcal{U}_{i(1)} \setminus \overline{D^+(\partial J(A))})$. By Lemma 8.3.8, there is a timelike past inextensible curve $\alpha_1$ through $x_1$ which does not intersect $\overline{D^+(\partial J(A))}$. Hence $\alpha_1$ neither intersects $\partial J^+(A)$ nor $\overline{D^+(E^+(A))}$. Since $\alpha_1$ does not intersect $\partial J^+(A)$ it is contained in $\text{int}(J^+(A)) = I^+(A)$. The curve $\alpha_1$ leaves $\mathcal{U}_{i(1)}$ because of its compactness. There is a point $y_1 \in \alpha_1 \setminus \mathcal{U}_{i(1)} \subset I^+(A)$. Let $\beta_1$ be a past directed timelike curve from $y_1$ to $A$. Since $A \subset E^+(A)$ and $E^+(A)$ is an achronal (topological) hypersurface this curve must intersect $D^+(E^+(A))$ and therefore also $H^+(E^+(A))$. Let $z_2 \in \beta_1 \cap H^+(E^+(A))$ and let $\mathcal{U}_{i(2)}$ be one of the convex neighbourhoods $\mathcal{U}_i$ with $z_2 \in \mathcal{U}_{i(2)}$. The neighbourhoods $\mathcal{U}_{i(1)}$ and $\mathcal{U}_{i(2)}$ are different since by construction we have $z_2 \in J^-(z_1)$ and since no $\mathcal{U}_i$ can be entered by any causal curve twice. By induction we obtain an infinite sequence of pairwise disjunct neighbourhoods $\{\mathcal{U}_{i(k)}\}_{k\in\mathbb{N}}$ in contradiction the finite number of sets $\mathcal{U}_i$. ∎

**Lemma 9.3.3.** *Let $A$ be a closed achronal set such that $\overline{J^+(A)}$ is strongly causal and assume that $E^+(A)$ is compact. Then there exists a future inextensible timelike curve $\gamma$ which is wholly contained in $D^+(E^+(A))$.*

*Proof.* Without loss of generality we can assume that $(M, g)$ is time oriented. Hence there exists a timelike, time oriented vector field $V$ on

$M$. Since $E^+(A)$ is an achronal hypersurface all future directed timelike curves with past endpoint in $E^+(A)$ are initially in $\text{int}(D^+(E^+(A)))$. If every integral curve of $V$ intersected $H^+(E^+(A))$ after having intersectes $E^+(A)$, we would obtain a continuous map $E^+(A) \to H^+(E^+(A))$, $x \mapsto F_{t(x)}(x)$ where $F$ is the flow of $V$ and $t(x) \geq 0$ the unique number with $F_{t(x)}(x) \in H^+(E^+(A))$. This map would be surjective because, by Lemma 8.3.8, every past inextensible timelike curve which intersects the event horizon of a closed set must intersect this set as well. Since $E^+(A)$ is compact, $H^+(E^+(A))$ would also be compact in contradiction to Lemma 9.3.2. Hence there is at least one future inextensible integral curve $\gamma$ of $V$ which is contained in $\text{int}(D^+(E^+(A)))$. ∎

**Lemma 9.3.4.** *Let $(M, g)$ be a causal geodesically complete and strongly causal spacetime in which every inextensible causal geodesic has a pair of conjuate points. Let $A$ be a closed achronal set with compact future horismos $E^+(A)$ and let $\gamma$ be a future inextensible timelike curve in $D^+(E^+(A))$.*

*Then there exists a past inextensible timelike curve $\lambda$ which is contained in $D^-(E^-(\mathcal{F}))$, where $\mathcal{F} = E^+(A) \cap \overline{J^-(\gamma)}$.*

*Proof.* We will first show the inclusion $E^-(\mathcal{F}) \subset \mathcal{F} \cup \partial J^-(\gamma)$.

Let $x \in E^-(\mathcal{F}) \setminus \mathcal{F}$. If there was a point $\tilde{x} \in I^-(x) \cap E^+(A)$ then $I^+(\tilde{x})$ would be a neighbourhood of $x$ and therefore intersect $I^-(E^+(A))$ in contradiction to the achronality of $E^+(A)$. Hence $I^-(x) \cap E^+(A) = \emptyset$. If $x \in I^-(\gamma)$ then there is a $z \in I^+(x) \cap I^-(\gamma)$. Denote by $\mu$ a timelike curve from $x$ through $z$ to $\gamma$. This curve must intersect $E^+(A)$ since $\gamma \subset D^+(E^+(A))$ and $I^-(x) \cap E^+(A) = \emptyset$. Since this intersection point is in $E^+(A) \cap I^-(\gamma) \subset \mathcal{F}$ we obtain $x \in I^-(\mathcal{F})$ in contradiction to the assumption $x \in E^-(\mathcal{F})$. Hence we have $x \in J^-(\mathcal{F}) \setminus I^-(\gamma) \subset \overline{J^-(\gamma)} \setminus I^-(\gamma) = \partial J^-(\gamma)$ and the assertion $E^-(\mathcal{F}) \subset \mathcal{F} \cup \partial J^-(\gamma)$ follows.

The set $\mathcal{F}$ is the intersection of a closed and a compact set and therefore compact. Since $\gamma$ is future inextensible, all generators of $\partial J^-(\gamma)$ must be future inextensible as well. Suppose, there was a sequence $\beta_i$ of generators of $E^-(\mathcal{F})$ with diverging affine lengths. Since $\mathcal{F}$ is compact, there would exist a cluster curve $\beta$ of $\{\beta_i\}_{i \in \mathbb{N}}$ which would be past inextensible. But then its geodesic prolongation would be an inextensible generator of $\partial J^-(\gamma)$. By assumption this generator cannot be achronal which gives a contradiction to the to the achronality of $\partial J^-(\gamma)$. Hence $E^-(\mathcal{F})$ is compact and we can apply the time reverse of Lemma 9.3.3. ∎

**Lemma 9.3.5.** *Let $C$ be a compact subset of $M$. If $D^+(C)$ contains a future inextensible timelike curve $\gamma$ and $D^-(C) \cap \overline{J^-(\gamma)}$ contains a past inextensible timelike curve $\lambda$, then $D(C)$ contains an inextensible causal geodesic without conjugate points*

*Proof.* Let $\{y_i\}_{i\in\mathbb{N}}$ be a sequence of points in $\gamma$ without accumulation point such that $y_{i+1} \in I^+(y_i)$. Choose a sequence $\{x_i\}_{i\in\mathbb{N}}$ in $\lambda$ such that $y_i \in I^+(x_i)$ and $x_i \in I^+(x_{i+1})$ for all $i$. For every $i$ we obtain a causal curve $\tilde{\mu}_i$ which joins $x_i$ via $\mathcal{C}$ to $y_i$. This causal curve is contained in a globally hyperbolic set (Theorem 8.3.1) and can therefore be replaced by a maximal geodesic segment $\mu_i$ (Proposition 8.2.2). Without loss of generality we have $\mu_i(0) \in \mathcal{C}$. Then the oriented half lines $\{(\mathbb{R}^+ \setminus \{0\}) \cdot \dot{\mu}_i(0) \,|\, i \in \mathbb{N}\}$ have an accumulation point $\ell$ in the space of causal directions over $\mathcal{C}$, because this space is compact. Any inextensible geodesic $\mu$ with $\dot{\mu}(0) \in \ell$ is a cluster curve of the sequence $\{\mu_i\}_{i\in\mathbb{N}}$. Since any cluster curve of maximal geodesics is maximal, the curve $\mu$ does not have a pair of conjugate points. ∎

*Proof of Proposition 9.3.1.* We only need to choose $\mathcal{C} = E^-(\mathcal{F})$. ∎

### 9.3.1 Applications of the singularity theorem

(i) Consider the Schwarzschild solution. It satisfies all assumptions of Theorem 9.3.1 with the exception of the genericity condition. However, it seems plausible that any generic perturbation of the Schwarzschild solution using a reasonable matter model should result in a spacetime which satisfies all the assumptions. Here it s important that the existence of a strictly closed trapped surface is an *open* condition, i.e., if a spacetime which contains a strictly closed trapped surface is slightly perturbed then this surface is also a strictly closed trapped surface in the perturbed spacetime. Hence Corollary 9.1.1 indicates that the Schwarzschild singularity is stable under (physically reasonable) perturbations. In particular, it is not an artefact of the high symmetry of the Schwarzschild spacetime. This application of Corollary 9.1.1 is one of the main reasons why the existence of black holes is widely accepted.

(ii) Consider a Robertson Walker solution without cosmological constant and spacelike hypersurfaces of constant, positive sectional curvature. These spacelike hypersurfaces $t = \text{const}$ are compact achronal sets without edge. The Ricci tensor, which is given by

$$\mathrm{Ric} = T - \frac{\mathrm{tr}(T)}{2} g = \frac{1}{2}\left(3(\epsilon + p) + (\epsilon - p))U^\flat \otimes U^\flat + (\epsilon - p)g\right),$$

is positive definite for $\epsilon > p \geq 0$. Hence in this case the timelike convergence and the genericity conditions are both satisfied. It follows that all assumption of Theorem 9.3.1 hold *even if the spacetime is slightly perturbed.* This indicates that (at least for closed universes) the big bang is not an artefact of the symmetry properties of the Robertson Walker cosmologies.

(iii) Consider again a Robertson Walker solution $(M, g)$ of arbitrary constant sectional curvature and assume that $\epsilon > p \geq 0$. Then the genericity and the timelike convergence conditions are satisfied. We will now show that $(M, g)$ contains a strictly closed trapped surface. By Corollary 6.1.2 there are coordinates $(t, r, \theta, \varphi)$ and a positive function $t \mapsto a(t)$ such that

$$g = -\mathrm{d}t^2 + a^2(t)\left(\frac{1}{1-\varepsilon r^2}\mathrm{d}r^2 + r^2\left(\mathrm{d}\theta^2 + \sin^2(\theta)\mathrm{d}\varphi^2\right)\right).$$

Consider the codim-2 surface $\mathcal{T}_{\hat{t},\hat{r}}) = \{x \,|\, t(x) = \hat{t}, r(x) = \hat{r}\}$. This surface is clearly compact and spacelike. The vector fields

$$N_\pm = \frac{1}{2}\left(\partial_t \pm \frac{\sqrt{1-\varepsilon r^2}}{a}\partial_r\right)$$

along $\mathcal{T}_{\hat{t},\hat{r}}$ are normalized null vector fields orthogonal to $\mathcal{T}_{\hat{t},\hat{r}}$ and the corresponding 1-forms are given by

$$(N_\pm)^\flat = \frac{1}{2}\left(-\mathrm{d}t \pm \frac{a}{\sqrt{1-\varepsilon r^2}}\mathrm{d}r\right).$$

Hence

$$\begin{aligned}
\theta_\pm = \mathrm{tr}_{\mathcal{T}_{\hat{t},\hat{r}}}(\nabla N_\pm) &= \frac{1}{a^2r^2}\left(\nabla_{\partial_\theta} N_\pm^\flat(\partial_\theta) + \frac{1}{\sin^2(\theta)}\nabla_{\partial_\varphi} N_\pm^\flat(\partial_\varphi)\right) \\
&= \frac{2}{a^2r^2}\nabla_{\partial_\theta} N_\pm^\flat(\partial_\theta) = \frac{2}{a^2r^2}\Gamma^i_{\theta\theta}(N_\pm)_i,
\end{aligned}$$

where for the third equation we have used spherical symmetry. Since for $i \in \{t, r\}$ we have $\Gamma^i_{\theta\theta} = \frac{1}{2}g^{ij}(2\partial_\theta g_{j\theta} - \partial_i g_{\theta\theta}) = -\frac{1}{2}g^{ii}\partial_i(a^2r^2)$ we obtain $\Gamma^t_{\theta\theta} = -\partial_t a/a$ and $\Gamma^r_{\theta\theta} = -1/r$. This implies

$$\theta_\pm = \frac{2}{a^2r^2}\left(\Gamma^t_{\theta\theta}(N_\pm)_t + \Gamma^r_{\theta\theta}(N_\pm)_r\right) = \frac{2}{a^2r^2}\left(-\frac{\partial_t a}{2a} \mp \frac{a}{r\sqrt{1-\varepsilon r^2}}\right).$$

Hence for $a \ll 1$, $\partial_t a > 0$ both expansions, $\theta_\pm$ are negative. Since these inequalities are satisfied near the big bang we can apply the time reverse of Theorem 9.3.1 to infer the existence of a singularity. In fact, in our example this singularity is just the big bang.[4] If we perturb our spacetime slightly all assumptions of Theorem 9.3.1 are still satisfied. Hence we can conclude that the perturbed spacetime also contains an incomplete, inextensible causal geodesic. It is therefore natural to expect that the big bang is stable under perturbations of the metric.

[4] Note, however, that Theorem 9.3.1 does not make any assertion about the location of the singularity. In particular, it does not assert whether it is to the future or to the past of the strictly closed trapped surface.

(iv) Using condition (c) of Theorem 9.3.1 it is also possible to give arguments in favour of the existence of a singularity without assuming that our universe is well described by a Robertson Walker spacetime. This argument requires assumptions on the spectrum of the microwave background radiation and is therefore beyond the scope of this book (cf. (Hawking and Ellis 1973, pp. 354 – 359)).

### 9.3.2 General problems with Theorem 9.3.1

These physical applications of Theorem 9.3.1 suffer from two defects which are often considered to be negligible. Firstly, instead of the formation of singularities spacetime could form closed timelike curves. Secondly, even if singularities occur, Theorem 9.3.1 does not predict their strength. For a long time, it has been thought that these problems are only technical and that the theorem could be sharpened accordingly. Unfortunately, this is not the case. In Sect. 9.4 we will present an example (due to Newman) that the chronology condition is necessary for Theorem 9.3.1. In Sect. 9.5 we will show that Theorem 9.3.1 may only predict the existence of singularities which are too weak to be taken seriously by most physicists. At the time of writing it is not clear whether it is possible to improve on Theorem 9.3.1 if additional physically realistic assumptions are made. It should be remarked that other singularity theorems suffer similar defects.

## 9.4 Singularities and causality violations

The chronology assumption in Theorem 9.3.1 is a global assumption and therefore can not be verified by physical measurements. While it is often considered to be self-evident we have seen in Chap. 8 that this view is to be debated. It is therefore an important question whether Theorem 9.3.1 continues to hold even if the chronology condition is dropped. In this section we will give an example due to R. P. A. C. Newman (1989) which proves that the chronology assumption is essential. We will then quote a generalization of Theorem 9.3.1 which sheds some light on what is going on.

### 9.4.1 The Gödel solution

*The proofs of Propositions 9.4.1 and 9.4.2 consist of straightforward but long calculations. We will not spell them out in all details. However, we will provide enough information such that a careful reader equipped with pen & paper (better: with access to a symbolic computing program such as REDUCE, MAPLE or MATHEMATICA) should be able to fill in the missing details.*

In 1949, the famous mathematician Kurt Gödel published a new solution to Einstein's equation for a dust matter model with a cosmological constant. His solution is completely homogenous and has the property that there is a closed timelike curve through each point. Newman's counter example is a modification of the Gödel solution.

**Definition 9.4.1.** *Let* $(t, x, y, z)$ *be standard coordinates of* $\mathbb{R}^4$ *and* $\omega \in \mathbb{R}^+\backslash\{0\}$. *The spacetime* $(\mathbb{R}^4, -\mathrm{d}t^2+dx^2+\frac{1}{2}\mathrm{e}^{2\sqrt{2}\omega x}\mathrm{d}y^2+\mathrm{d}z^2-2\mathrm{e}^{\sqrt{2}\omega x}\mathrm{d}t\mathrm{d}y)$ *is called the* Gödel solution.

**Proposition 9.4.1.** *The Gödel solution is a Lorentzian spacetime and satisfies*

$$\mathrm{Ric} - \frac{1}{2}\mathrm{Scal}g - \omega^2 g = 8\pi\left(\frac{\omega^2}{4\pi}\right)(\partial_t)^\flat \otimes (\partial_t)^\flat .$$

*It corresponds therefore to a dust solution with negative cosmological constant.*

*Proof.* For the first claim observerve that $(\partial_t)^\flat = -(\mathrm{d}t + \mathrm{e}^{\sqrt{2}\omega x}\mathrm{d}y)$ implies

$$g = -(\partial_t)^\flat \otimes (\partial_t)^\flat + dx^2 + \frac{1}{2}\mathrm{e}^{2\sqrt{2}\omega x}\mathrm{d}y^2 + \mathrm{d}z^2.$$

For the second claim we will have to calculate the Ricci tensor. Since the metric depends on only one variable this is a simple task and left to the reader. Here we only note that the values of the Christoffel symbols are

$$\Gamma^t_{tx} = \Gamma^t_{xt} = \sqrt{2}\omega, \Gamma^y_{tx} = \Gamma^y_{xt} = -\sqrt{2}\omega\mathrm{e}^{-\sqrt{2}\omega x}, \Gamma^x_{ty} = \Gamma^x_{yt} = \frac{1}{\sqrt{2}}\omega\mathrm{e}^{\sqrt{2}\omega x},$$
$$\Gamma^t_{xy} = \Gamma^t_{yx} = \frac{1}{\sqrt{2}}\omega\mathrm{e}^{\sqrt{2}\omega x}, \Gamma^x_{yy} = \frac{1}{\sqrt{2}}\omega\mathrm{e}^{2\sqrt{2}\omega x},$$

where it is understood that all other Christoffel symbols vanish. The Ricci tensor reads then

$$\mathrm{Ric} = 2\omega^2\left(\mathrm{d}t^2 + \mathrm{e}^{2\sqrt{2}\omega x}\mathrm{d}y^2 + 2\mathrm{e}^{\sqrt{2}\omega x}\mathrm{d}t\mathrm{d}y\right) = 2\omega^2 (\partial_t)^\flat \otimes (\partial_t)^\flat$$

and we obtain

$$\mathrm{Ric} - \frac{\mathrm{Scal}}{2}g = \omega^2\left(2(\partial_t)^\flat \otimes (\partial_t)^\flat + g\right).$$

∎

The following lemma implies that the Gödel solution is homogenous in space and time.

**Lemma 9.4.1.** *For any two points* $p, q \in M$ *there is an isometry* $\psi \colon M \to M$ *which satisfies* $\psi(p) = q$.

*Proof.* The spacetime $M$ admits the following four 1-parameter groups of isometries[5].

$$\begin{aligned}
\psi_1(\alpha)&: (t,x,y,z) \mapsto (t+\alpha, x, y, z)\\
\psi_2(\alpha)&: (t,x,y,z) \mapsto (t, x+\alpha, y\mathrm{e}^{-\sqrt{2}\omega\alpha}, z)\\
\psi_3(\alpha)&: (t,x,y,z) \mapsto (t, x, y+\alpha, z)\\
\psi_4(\alpha)&: (t,x,y,z) \mapsto (t, x, y, z+\alpha).
\end{aligned}$$

It follows that for any $p, q \in M$ there are numbers $\alpha_1, \ldots, \alpha_4$ such that $p = \psi_4(\alpha_4) \circ \psi_3(\alpha_3) \circ \psi_2(\alpha_2) \circ \psi_1(\alpha_1)(q)$. ∎

The following lemma implies that the Gödel spacetime does not contain singularities.

**Lemma 9.4.2.** *Each inextensible causal geodesic in* $(M, g)$ *is complete.*

*Proof.* We will first partially solve the system of equations for geodesics. To this end it is practical to work with slightly different coordinates. Consider the global coordinate transformation $\psi\colon M \to \mathbb{R}\times\mathbb{R}^+\backslash\{0\}\times\mathbb{R}^2$, $(t,x,y,z) \mapsto (\tilde{t}, \tilde{x}, \tilde{y}, \tilde{z})$ where $t = \tilde{t}$, $\sqrt{2}\omega x = -\ln(\sqrt{2}\omega\tilde{x})$, $y = \sqrt{2}\tilde{y}$, and $z = \tilde{z}$. In these coordinates the metric reads

$$g = -(\mathrm{d}\tilde{t} + \frac{1}{\omega\tilde{x}}\mathrm{d}\tilde{y})^2 + \frac{1}{\omega^2\tilde{x}^2}(\mathrm{d}\tilde{x}^2 + \mathrm{d}\tilde{y}^2) + \mathrm{d}\tilde{z}^2.$$

We have the 1-parameter families of isometris

$$\begin{aligned}
\psi_1(\alpha)&: (\tilde{t},\tilde{x},\tilde{y},\tilde{z}) \mapsto (\tilde{t}+\alpha, \tilde{x}, \tilde{y}, \tilde{z})\\
\psi_2(\alpha)&: (\tilde{t},\tilde{x},\tilde{y},\tilde{z}) \mapsto (\tilde{t}, (1+\alpha)\tilde{x}, (1+\alpha)\tilde{y}, \tilde{z})\\
\psi_3(\alpha)&: (\tilde{t},\tilde{x},\tilde{y},\tilde{z}) \mapsto (\tilde{t}, \tilde{x}, \tilde{y}+\alpha, \tilde{z})\\
\psi_4(\alpha)&: (\tilde{t},\tilde{x},\tilde{y},\tilde{z}) \mapsto (\tilde{t}, \tilde{x}, \tilde{y}, \tilde{z}+\alpha).
\end{aligned}$$

The corresponding Killing vector fields are

$$\xi_1 = \partial_{\tilde{t}}, \quad \xi_2 = \tilde{x}\partial_{\tilde{x}} + \tilde{y}\partial_{\tilde{y}}, \quad \xi_3 = \partial_{\tilde{y}}, \quad \xi_4 = \partial_{\tilde{z}}.$$

These Killing vector fields give four constants of motion along the geodesic, $\langle \xi_i, \dot{\gamma} \rangle = c_i$ is constant. Writing $\gamma(\tau) = (\tilde{t}(\tau), \tilde{x}(\tau), \tilde{y}(\tau), \tilde{z}(\tau))$ we obtain

$$c_1 = -\dot{\tilde{t}} - \frac{\dot{\tilde{y}}}{\omega\tilde{x}}, \quad c_2 = -\frac{\tilde{x}}{2\omega^2\tilde{x}} \quad c_3 = \frac{\tilde{y}\dot{\tilde{x}} - 2\omega\tilde{x}\dot{\tilde{t}} - \dot{\tilde{y}}}{\omega^2 x^2} \quad c_3 = \dot{\tilde{z}}.$$

This is a linear system of equations for $(\dot{\tilde{t}}, \dot{\tilde{x}}, \dot{\tilde{y}}, \dot{\tilde{z}})$, and its solution is given by

[5] The isometry group is five-dimensional, but the additional 1-parameter group of isometries is not important for the argument.

$$\dot{\tilde{t}} = c_1 - 2\omega\tilde{x}c_3, \quad \dot{\tilde{x}} = 2\omega^2\tilde{x}(c_2 - \tilde{y}c_3), \quad \dot{\tilde{y}} = 2\omega^2\tilde{x}(-\frac{c_1}{\omega} + \tilde{x}c_3), \quad \dot{\tilde{z}} = c_4.$$

If we rename $(c_1, c_2, c_3, c_4) = (\frac{\tilde{x}'}{\sqrt{2}C}, \frac{\tilde{y}'}{\sqrt{2}\omega C}, \frac{1}{\sqrt{2}\omega C}, -\frac{d}{C})$ we obtain

$$\dot{\tilde{t}} = \frac{\tilde{x}' - 2\tilde{x}}{\sqrt{2}C}, \quad \dot{\tilde{x}} = \frac{\sqrt{2}\omega\tilde{x}(\tilde{y} - \tilde{y}')}{C},$$
$$\dot{\tilde{y}} = -\frac{\sqrt{2}\omega\tilde{x}(\tilde{x} - \tilde{x}')}{C}, \quad \dot{\tilde{z}} = -\frac{d}{C}. \tag{9.4.3}$$

We can assume without loss of generality that $\langle\dot{\gamma}, \dot{\gamma}\rangle = \eta \in \{-1, 0, 1\}$. Inserting our values for $\dot{\tilde{t}}, \dot{\tilde{x}}, \dot{\tilde{y}}, \dot{\tilde{z}}$ into this equation we obtain

$$(\tilde{y} - \tilde{y}')^2 + (\tilde{x} - \tilde{x}')^2 = \frac{(\tilde{x}')^2}{2} - d^2 + \eta C^2.$$

Hence the projection of $\gamma$ to the $(x, y)$-plane traverses an arc of a circle with centre $(\tilde{x}', \tilde{y}')$ and radius $\sqrt{\frac{(\tilde{x}')^2}{2} - d^2 + \eta C^2}$. Observe that for causal geodesics we have the inequality

$$(\tilde{x} - \tilde{x}')^2 \le \frac{(\tilde{x}')^2}{2}$$

which implies that the circle is wholly contained in a compact subset of $\mathbb{R}^+ \setminus \{0\} \times \mathbb{R}$. In particular, the coordinates $\tilde{x}, \tilde{y}$ in remain bounded along $\gamma$. Equations 9.4.3 imply that $\dot{\gamma}$ is bounded in our coordinate system. Since these coordinates are global, it follows that the affine parameter must range from $-\infty$ to $\infty$. ∎

The Gödel solution is axially symmetric with respect to any point. To see this, we will introduce different coordinates.

**Proposition 9.4.2.** *There is a dense open set $N \subset M$ and coordinates $(s, r, \varphi, \tilde{z}) \in \mathbb{R} \times \mathbb{R}^+ \setminus \{0\} \times S^1 \times \mathbb{R}$ such that $g_{|N}$ is given by*

$$g = 2\omega^{-2}\Big( - \mathrm{d}s^2 + \mathrm{d}r^2 + \sinh^2 r(1 - \sinh^2 r)\mathrm{d}\varphi^2 + \mathrm{d}\tilde{z}^2 - 2\sqrt{2}\sinh^2 r\mathrm{d}\varphi\mathrm{d}s\Big).$$

*Proof.* Observe that the metric is a direct product, $(M, g) = (\mathbb{R}^3 \times \mathbb{R}, h + \mathrm{d}z^2)$ with $h = -\mathrm{d}t^2 + dx^2 + \frac{1}{2}\mathrm{e}^{2\sqrt{2}\omega x}\mathrm{d}y^2 - 2\mathrm{e}^{\sqrt{2}\omega x}\mathrm{d}t\mathrm{d}y$. Hence it is sufficient to show that there is a dense open set $\tilde{N} \subset \mathbb{R}^3$ and coordinates $(s, r, \varphi)$ such that

$$h_{|\tilde{N}} = 2\omega^2\left(-\mathrm{d}s^2 + \mathrm{d}r^2 + \sinh^2 r(1 - \sinh^2 r)\mathrm{d}\varphi^2 - 2\sqrt{2}\sinh^2 r\mathrm{d}\varphi\mathrm{d}s\right).$$

The assertion follows then with $\tilde{z} = \frac{1}{\sqrt{2}\omega} z$. We define the coordinates $(s, r, \varphi)$ via the equations

$$\mathrm{e}^{\sqrt{2}\omega x} = \cosh(2r) + \cos(\varphi)\sinh(2r) \tag{9.4.4}$$

$$\omega y \mathrm{e}^{\sqrt{2}\omega x} = \sin(\varphi)\sinh(2r) \tag{9.4.5}$$

$$\tan\left(\frac{1}{2}(\varphi + \omega t - \sqrt{2}s)\right) = \mathrm{e}^{-2r}\tan\left(\frac{1}{2}\varphi\right). \tag{9.4.6}$$

To show that $h$ has the desired form in these coordinates is "straightforward" but very cumbersome.[6]

We first differentiate Equations (9.4.4)–(9.4.6) to obtain

$$\begin{aligned}
0 &= \sqrt{2}\omega \mathrm{e}^{\sqrt{2}\omega x}\mathrm{d}x - 2(\sinh(2r) + \cos(\varphi)\cosh(2r))\mathrm{d}r \\
&\quad + \sin(\varphi)\sinh(r)\mathrm{d}\varphi \\
0 &= \sqrt{2}\omega^2 \mathrm{e}^{\sqrt{2}\omega x} y\mathrm{d}x + \mathrm{e}^{\sqrt{2}\omega x}\omega \mathrm{d}y - 2\sin(\varphi)\cosh(2r)\mathrm{d}r \\
&\quad - \cos(\varphi)\sinh(2r)\mathrm{d}\varphi \\
0 &= \frac{1}{2}\left(1 + \left(\tan(\varphi/2 + 1/2\omega t - 1/2\sqrt{2}s)\right)^2\right)\omega \mathrm{d}t \\
&\quad - \frac{1}{2}\left(1 + \left(\tan(\varphi/2 + \omega t/2 - s/\sqrt{2})\right)^2\right)\sqrt{2}\mathrm{d}s + 2\mathrm{e}^{-2r}\tan(\varphi/2)\mathrm{d}r \\
&\quad + \frac{1}{2}\Bigg( - \mathrm{e}^{-2r}\left(1 + (\tan(\varphi/2))^2\right) + 1 \\
&\quad + \left(\tan(\varphi/2 + \omega t/2 - s/\sqrt{2})\right)^2\Bigg)\mathrm{d}\varphi.
\end{aligned}$$

In this system of equations we can eliminate $y$ and $\mathrm{e}^{\sqrt{2}\omega x}$ using Equations (9.4.4) and (9.4.5). The system can then be considered as a linear system for $\mathrm{d}t, \mathrm{d}x, \mathrm{d}y$ which only depends on $s, r, \varphi, \mathrm{d}s, \mathrm{d}r, \mathrm{d}\varphi$. Solving this linear system gives (after some simplifications)

$$\begin{aligned}
\mathrm{d}t &= \frac{\sqrt{2}\mathrm{d}s}{\omega} - 4\frac{\sin(\varphi/2)\cos(\varphi/2)\mathrm{e}^{2r}\mathrm{d}r}{\omega\left(\mathrm{e}^{4r}\cos^2(\varphi/2) + \sin^2(\varphi/2)\right)} \\
&\quad + \frac{\left(-\sin^2(\varphi/2) + \mathrm{e}^{2r} - \mathrm{e}^{4r}\cos^2(\varphi/2)\right)\mathrm{d}\varphi}{\omega\left(\mathrm{e}^{4r}\cos^2(\varphi/2) + (\sin(\varphi/2))^2\right)},
\end{aligned}$$

[6] In fact, in his original paper, Gödel chose to derive this form independently of the geometric assumptions which led him to the metric.

$$\begin{aligned}
\mathrm{d}x &= 1/2\frac{\sqrt{2}\left(2\mathrm{e}^{4r}\cos^2(\varphi/2) - 2\sin^2(\varphi/2)\right)\mathrm{d}r}{\omega\left(\mathrm{e}^{4r}\cos^2(\varphi/2) + \sin^2(\varphi/2)\right)} \\
&\quad + 1/2\frac{\sqrt{2}\sin(1/2\varphi)\cos(\varphi/2)\left(1 - \mathrm{e}^{4r}\right)\mathrm{d}\varphi}{\omega\left(\mathrm{e}^{4r}\cos^2(\varphi/2) + \sin^2(\varphi/2)\right)}, \\
\mathrm{d}y &= 4\frac{\sin(\varphi/2)\cos(\varphi/2)\mathrm{e}^{4r}\mathrm{d}r}{\omega\left(\mathrm{e}^{4r}\cos^2(\varphi/2) + \sin^2(\varphi/2)\right)^2} \\
&\quad + 1/2\frac{\left(\sin^2(\varphi/2) + \mathrm{e}^{8r}\cos^2(\varphi/2) - \mathrm{e}^{4r}\right)\mathrm{d}\varphi}{\omega\left(\mathrm{e}^{4r}\cos^2(\varphi/2) + \sin^2(\varphi/2)\right)^2}.
\end{aligned}$$

We can now simply calculate $h = -\mathrm{d}t^2 + dx^2 + \frac{1}{2}\mathrm{e}^{2\sqrt{2}\omega x}\mathrm{d}y^2 - 2\mathrm{e}^{\sqrt{2}\omega x}\mathrm{d}t\mathrm{d}y$ in the coordinates $(s, r, \varphi)$ using our expressions for $\mathrm{d}t$, $\mathrm{d}x$, $\mathrm{d}y$ and Equation (9.4.4) (which is equivalent to

$$\mathrm{e}^{\sqrt{2}\omega x} = \mathrm{e}^{-2r}\left(\mathrm{e}^{4r}\cos^2(\varphi/2) + \left(\sin(\varphi/2)\right)^2\right)).$$

This gives with

$$\begin{aligned}
A(r,\varphi) &= \sin^2(\varphi/2) + \mathrm{e}^{8\,r}\cos^2(\varphi/2) - \mathrm{e}^{4\,r} \\
B(r,\varphi) &= -\sin^2(\varphi/2) + \mathrm{e}^{2\,r} - \mathrm{e}^{4\,r}\cos^2(\varphi/2)
\end{aligned}$$

and using trigonometric identities

$$\begin{aligned}
g &= -2\,\frac{\mathrm{d}s^2}{\omega^2} + \left(-\frac{\sqrt{2}\mathrm{e}^{-2\,r}A(r,\varphi) + 2B(r,\varphi)\sqrt{2}}{\omega^2\left(\mathrm{e}^{4\,r}\cos^2(\varphi/2) + \sin^2(\varphi/2)\right)}\right)\mathrm{d}\varphi\,\mathrm{d}s \\
&\quad + \left(\frac{8\,\mathrm{e}^{-4\,r}\sin^2(\varphi/2)\cos^2(\varphi/2)\left(\mathrm{e}^{4\,r}\right)^2}{\omega^2\left(\mathrm{e}^{4\,r}\cos^2(\varphi/2) + \sin^2(\varphi/2)\right)^2}\right. \\
&\qquad \left. + 2\frac{\left(\sin^2(\varphi/2) - \mathrm{e}^{4\,r}\cos^2(\varphi/2)\right)^2}{\omega^2\left(\mathrm{e}^{4\,r}\cos^2(\varphi/2) + \sin^2(\varphi/2)\right)^2}\right)\mathrm{d}r^2 \\
&\quad + \left(-\frac{\left(B(r,\varphi)\right)^2 + B(r,\varphi)\mathrm{e}^{-2\,r}A(r,\varphi)}{\omega^2\left(\mathrm{e}^{4\,r}\cos^2(\varphi/2) + \left(\sin(\varphi/2)\right)^2\right)^2}\right. \\
&\qquad \left. - \frac{\frac{1}{8}\,\mathrm{e}^{-4\,r}\left(A(r,\varphi)\right)^2 - \frac{1}{2}\cos^2(\varphi/2)\sin^2(\varphi/2)\left(1 - \mathrm{e}^{4\,r}\right)^2}{\omega^2\left(\mathrm{e}^{4\,r}\cos^2(\varphi/2) + \sin^2(\varphi/2)\right)^2}\right)\mathrm{d}\varphi^2 \\
&= 2\omega^{-2}\big(-\mathrm{d}s^2 + \mathrm{d}r^2 + \sinh^2 r(1 - \sinh^2 r)\mathrm{d}\varphi^2 + \mathrm{d}\tilde{z}^2 \\
&\qquad - 2\sqrt{2}\sinh^2 r\mathrm{d}\varphi\mathrm{d}s\big).
\end{aligned}$$

∎

We will now show that there are spacelike submanifolds given by $r =$ const, $s =$ const which have vanishing expansions. They are our candidates for strictly closed trapped surfaces in a suitably perturbed metric.

**Proposition 9.4.3.** *Let $\hat{s} \in \mathbb{R}$, $\hat{r} \in \mathbb{R}^+ \setminus \{0\}$, and $\mathcal{T}_{\hat{s},\hat{r}}$ be the codim-2 surface given by $s = \hat{s}, r = \hat{r}$. Then $\mathcal{T}_{\hat{s},\hat{r}}$ is spacelike if and only if $\hat{r} < \ln(1+\sqrt{2})$. In this case its mean curvature vector field reads*

$$H = \frac{1 - 2\sinh^2(r)}{2\sqrt{2}\sinh(r)\cosh(r)} (N_+ - N_-) ,$$

*where*

$$N_\pm = \frac{\sqrt{1-\sinh^2(r)}}{\sqrt{2}\cosh(r)}\partial_s \pm \frac{1}{\sqrt{2}}\partial_r + \frac{1}{\cosh(r)\sqrt{1-\sinh^2(r)}}\partial_\varphi$$

*are a pair of normalized null vector fields orthogonal to $\mathcal{T}_{\hat{s},\hat{r}}$.*

*Proof.* Since $g = h + \mathrm{d}z^2$ is a direct product it is sufficient to prove the analogus assertion for $h$. For any $x \in \mathcal{T}_{\hat{s},\hat{r}}$ the tangent space of $\mathcal{T}_{\hat{s},\hat{r}}$ is spanned by $\partial_\varphi$. It is clear that $h(\partial_\varphi, \partial_\varphi) > 0$ if and only if $r < \sinh^{-1}(1) = \ln(1+\sqrt{2})$. We first calculate $N_\pm$. From

$$0 = h(N_\pm, \partial_\varphi) = \sinh^2(r)(1-\sinh^2(r))N_\pm^\varphi - \sqrt{2}\sinh^2(r)N_\pm^t$$

we get $N_\pm^\varphi = \frac{\sqrt{2}}{1-\sinh^2(r)}N_\pm^t$. The equation

$$\begin{aligned}
0 &= g(N_\pm, N_\pm) \\
&= -(N_\pm^s)^2 + (N_\pm^r)^2 + \sinh^2(r)\left((1-\sinh^2(r))\right)N_\pm^\varphi)^2 \\
&\quad - 2\sqrt{2}\sinh^2(r)N_\pm^\varphi N_\pm^s \\
&= (N_\pm^r)^2 - \frac{1+\sinh^2(r)}{1-\sinh^2(r)}(N_\pm^t)^2
\end{aligned}$$

implies $N_\pm^r = \pm\frac{\cosh(r)}{\sqrt{(1-\sinh^2(r)}}N_\pm^t$ and therefore

$$N_\pm = \left(\partial_s \pm \frac{\cosh(r)}{\sqrt{1-\sinh^2(r)}}\partial_r + \frac{\sqrt{2}}{1-\sinh^2(r)}\partial_\varphi\right) N_\pm^s .$$

We normalize $N_+, N_-$ by demanding $\langle N_+, N_-\rangle = -1$. This is equivalent to

$$\begin{aligned}
-1 &= \Big( -1 - \frac{\cosh^2(r)}{1-\sinh^2(r)} + \sinh^2(r)\left(1-\sinh^2(r)\right)\frac{2}{1-\sinh^2(r)} \\
&\quad - 2\sqrt{2}\sinh^2(r)\frac{\sqrt{2}}{1-\sinh^2(r)}\Big)(N_\pm^t)^2 \\
&= -\frac{2\cosh^2(r)}{1-\sinh^2(r)}(N_\pm^t)^2,
\end{aligned}$$

whence

$$\begin{aligned}
N^\flat_\pm &= \frac{2}{\omega^2}\Big(\Big(-N^s_\pm - \sqrt{2}\sinh^2(r)N^\varphi_\pm\Big)\,\mathrm{d}s + \big(N^r_\pm\big)\,\mathrm{d}r \\
&\quad + \Big(-\sqrt{2}\sinh^2(r)N^s_\pm + \sinh^2(r)(1-\sinh^2(r))N^\varphi_\pm\Big)\,\mathrm{d}\varphi\Big) \\
&= \frac{2}{\omega^2}\left(\left(-\frac{\sqrt{1-\sinh^2(r)}}{\sqrt{2}\cosh(r)} - \frac{\sqrt{2}\sinh^2(r)}{\cosh(r)\sqrt{1-\sinh^2(r)}}\right)\mathrm{d}s \pm \frac{1}{\sqrt{2}}\mathrm{d}r\right. \\
&\quad \left. + \left(-\sqrt{2}\sinh^2(r)\frac{\sqrt{1-\sinh^2(r)}}{\sqrt{2}\cosh(r)} + \frac{\sinh^2(r)(1-\sinh^2(r))}{\cosh(r)\sqrt{1-\sinh^2(r)}}\right)\mathrm{d}\varphi\right) \\
&= \frac{2}{\omega^2}\left(-\frac{\cosh(r)}{\sqrt{2}\sqrt{1-\sinh^2(r)}}\mathrm{d}s \pm \frac{1}{\sqrt{2}}\mathrm{d}r\right).
\end{aligned}$$

We can now calculate the covariant derivatives $\nabla N_\pm$. Since we are only interested in $\mathrm{tr}_{\mathcal{T}_{\hat{s},\hat{r}}}(\chi_\pm) = \mathrm{tr}_{\mathcal{T}_{\hat{s},\hat{r}}}(\nabla N^\flat_\pm)$ and $T\mathcal{T}_{\hat{s},\hat{r}}$ is spanned by $\partial_\varphi$ we need only calculate the covariant derivative in direction $\partial_\varphi$,

$$\mathrm{tr}_{\mathcal{T}_{\hat{s},\hat{r}}}(\nabla N^\flat_\pm) = h^{\varphi\varphi}\nabla_\varphi N^\flat_\pm(\partial_\varphi) = -h^{\varphi\varphi}\left((N^\flat_\pm)_s\Gamma^s_{\varphi\varphi} + (N^\flat_\pm)_r\Gamma^r_{\varphi\varphi}\right).$$

In order to calculate the Christoffel symbols $\Gamma^s_{\varphi\varphi}$, $\Gamma^r_{\varphi\varphi}$ note first that the inverse of $h$ is given by

$$h^\sharp = -\frac{1-\sinh^2(r)}{\cosh^2(r)}(\partial_s)^2 + (\partial_r)^2 + \frac{1}{\sinh^2(r)\cosh^2(r)}(\partial_\varphi)^2 - \frac{2\sqrt{2}}{\cosh^2(r)}\partial_\varphi\partial_s.$$

It is now a simple exercise to compute

$$\begin{aligned}
\Gamma^r_{\varphi\varphi} &= \frac{1}{2}h^{rr}(\partial_\varphi h_{r\varphi} + \partial_\varphi h_{\varphi r} - \partial_r h_{\varphi\varphi}) \\
&\quad -\frac{1}{2}h^{rr}\partial_r h_{\varphi\varphi} = \frac{1}{2}\partial_r(\sinh^2(r)(1-\sinh^2(r))) \\
&= -\sinh(r)\cosh(r)(1-2\sinh^2(r)), \\
\Gamma^s_{\varphi\varphi} &= \frac{1}{2}h^{ss}(\partial_\varphi h_{\varphi s} + \partial_\varphi h_{s\varphi} - \partial_s h_{\varphi\varphi}) + \frac{1}{2}h^{s\varphi}(\partial_\varphi h_{\varphi\varphi} + \partial_\varphi h_{\varphi\varphi} - \partial_\varphi h_{\varphi\varphi}) \\
&= 0.
\end{aligned}$$

Hence we get

$$\begin{aligned}
\mathrm{tr}_{\mathcal{T}_{\hat{s},\hat{r}}}(\nabla N^\flat_\pm) &= -\frac{1}{\sinh^2(r)\cosh^2(r)}\left(\pm\frac{-\sinh(r)\cosh(r)(1-2\sinh^2(r))}{\sqrt{2}}\right) \\
&= \pm\frac{1-2\sinh^2(r)}{\sqrt{2}\sinh(r)\cosh(r)}.
\end{aligned}$$

and the assertion follows now directly from Equation (9.2.2). ∎

Proposition 9.4.3 implies that the surfaces $\mathcal{T}_{\hat{s},\ln((1+\sqrt{3})/\sqrt{2})}$ are spacelike and have vanishing mean curvature vector field $H$.

### 9.4.2 Newman's example

Consider the following partial compactification of the Gödel solution. $(\tilde{M},\tilde{g}) = (\mathbb{R}^3 \times S^1, h + \mathrm{d}z^2)$ where $z$ is the natural coordinate of $S^1$. (Here we view $S^1$ as the subset $[0,2\pi] \subset \mathbb{R}$, where the points 0 and $2\pi$ are identified). Clearly, $(\tilde{M},\tilde{g})$ is locally isometric to $(M,g)$. Oberve that the corresponding sets $\mathcal{T}_{\hat{s},\hat{r}}$ are compact (and diffeomorphic to tori). In particular, the surfaces $\mathcal{T}_{\hat{s},\ln((1+\sqrt{3})/\sqrt{2})}$ are closed (but not strictly closed) trapped surfaces. It is plausible that a suitable deformation of $(\tilde{M},\tilde{g})$ will result in compact surfaces with past pointing, timelike mean curvature vector field. This would give an example which is causally geodesically complete and satisfies all assumptions of Theorem 9.3.1 — with the exception of chronology. In order to preserve the causal structure of the Gödel spacetime we will deform $\tilde{g}$ by multiplying it with a conformal factor $\Omega^2$, $\Omega\colon \tilde{M} \to \mathbb{R}$. We need to calculate the change of the mean curvature vector field when $\tilde{g}$ is replaced by $\hat{g} = \Omega^2\tilde{g}$.

**Lemma 9.4.3.** *Let $(M,g)$ a pseudo-Riemannian manifold and $\Omega\colon M \to \mathbb{R}$. Let $\hat{g} = \Omega^2 g$ and $\nabla$ (resp., $\hat{\nabla}$) the Levi-Cività connection of $g$ (resp., $\hat{g}$). Then for every vector field $U$ and every 1-form $\lambda$ on $M$ we have*

$$\hat{\nabla}_U\lambda = \nabla_U\lambda - \Omega^{-1}\left(\mathrm{d}\Omega(U)\lambda + \lambda(U)\mathrm{d}\Omega - \lambda(\mathrm{grad}\Omega)g(U,\cdot)\right).$$

*Proof.* Denote the difference tensor of $\hat{\nabla}$ and $\nabla$ by $C$, $C(U,V) = \hat{\nabla}_U V - \nabla_U V$. Let $(x^1,\ldots,x^n)$ be a normal coordinate system with respect to $g$ centred at $x \in M$. Then we have at $\Gamma^i_{jk} = 0$ at $x$ and therfore

$$\begin{aligned}(C(\partial_{x^i},\partial_{x^j}))^k &= (\hat{\nabla}_{\partial_{x^i}}\partial_{x^j})^k - (\nabla_{\partial_{x^i}}\partial_x^j)^k = (\hat{\nabla}_{\partial_{x^i}}\partial_x^j)^k = \hat{\Gamma}^k_{ij}\\ &= \frac{1}{2}\hat{g}^{kl}\left(\partial_{x^i}\hat{g}_{lj} + \partial_{x^j}\hat{g}_{il} - \partial_{x^l}\hat{g}_{ij}\right)\\ &= \frac{1}{2}\hat{g}^{kl}\left(\nabla_{\partial_{x^i}}\hat{g}_{lj} + \nabla_{\partial_{x^j}}\hat{g}_{il} - \nabla_{\partial_{x^l}}\hat{g}_{ij}\right)\end{aligned}$$

at $x$. Since this is a tensor equation we can infer

$$C^k_{ij} = \frac{1}{2}\hat{g}^{kl}\left(\nabla_i\hat{g}_{lj} + \nabla_j\hat{g}_{il} - \nabla_l\hat{g}_{ij}\right)$$

everywhere. From $\hat{g}_{ij} = \Omega^2 g_{ij}$ and $\hat{g}^{ij} = \Omega^{-2}g^{ij}$ we get

$$\nabla_i\hat{g}_{jk} = 2\Omega(\nabla_i\Omega)g_{jk}$$

and therefore

$$C^k_{ij} = \Omega^{-1} g^{kl} \left( g_{lj} \nabla_i \Omega + g_{il} \nabla_j \Omega - g_{ij} \nabla_l \Omega \right).$$

The assertion follows now directly from $\hat{\nabla}\lambda = \nabla\lambda - \lambda(C(\cdot,\cdot))$. ∎

**Lemma 9.4.4.** *Let $\tilde{\mathcal{T}}_{\hat{s},\hat{r}}$ be the projection of $\mathcal{T}_{\hat{s},\hat{r}}$ to $\tilde{M}$ and $\mathcal{U},\mathcal{V}$ be open neighbourhoods of $\tilde{\mathcal{T}}_{\hat{s},\hat{r}}$ which have compact closure of $\tilde{\mathcal{T}}_{\hat{s},\hat{r}}$ and satisfy $\overline{\mathcal{U}} \subset \mathcal{V}$. Then there is a function Let $f\colon \tilde{M} \to \mathbb{R}^+$ such that*

*(i)* $f(x) \in (-\frac{1}{2}, \frac{1}{2})$,
*(ii)* $\operatorname{supp} f \subset \mathcal{V}$,
*(iii)* $f_{|\mathcal{U}}$ *only depends on* $t$, *and*
*(iv)* $\partial_s f(x) < 0$ *for all* $x \in \tilde{\mathcal{T}}_{\hat{s},\hat{r}}$.

*Proof.* This is a simple application of Lemma 2.1.7. ∎

**Proposition 9.4.4.** *Let $\hat{r} = \ln((1+\sqrt{3})/\sqrt{2})$, $\tau_0 \in (0,1)$, $f$ be the function provided by Lemma 9.4.4 and set $\Omega_\tau(x) = 1 + \tau f(x)$ for all $x$, $\tau \in [0,\tau_0)$. Then the the family $\hat{g}_\tau := (\Omega_\tau)^2 \tilde{g}$ is a deformation of $\tilde{g}$ which depends smoothly on $\tau$ and satisfies $\hat{g}_0 = \tilde{g}$. Furthermore, there is a $\tau_1 \in (0,\tau_0)$ such that each $\hat{g}_\tau$ ($\tau \in (0,\tau_1)$) contains a strictly closed trapped surface.*

*Proof.* The first assertion is trivial and we only have to show the existence of a strictly closed trapped surface for $\tau > 0$ sufficiently small. Consider the surface $\tilde{\mathcal{T}}_{\hat{s},\hat{r}}$. We must show that $\operatorname{tr}_{\tilde{\mathcal{T}}_{\hat{s},\hat{r}}}(\nabla \hat{N}^\flat_\pm)$ are both negative. Here $\hat{N}_\pm = {\Omega_\tau}^{-1} N_\pm$ denotes a pair of normalized null vector fields orthogonal to $\tilde{\mathcal{T}}_{\hat{s},\hat{r}}$. We can restrict attention to $\mathcal{U}$ and obtain

$$\begin{aligned}
\theta_\pm &= \hat{g}^{\varphi\varphi} \hat{\nabla}_{\partial_\varphi} (\hat{N}_\pm)^\flat (\partial_\varphi) + \hat{g}^{zz} \hat{\nabla}_{\partial_z} (\hat{N}_\pm)^\flat (\partial_z) \\
&= \Omega^{-1} \left( g^{\varphi\varphi} \hat{\nabla}_{\partial_\varphi} (N_\pm)^\flat (\partial_\varphi) + g^{zz} \hat{\nabla}_{\partial_z} (N_\pm)^\flat (\partial_z) \right) \\
&= \Omega^{-1} \Bigg( g^{\varphi\varphi} \left( \nabla_{\partial_\varphi} (N_\pm)^\flat (\partial_\varphi) - \Omega^{-1} (-\mathrm{d}\Omega(N_\pm)) g_{\varphi\varphi} \right) \\
&\qquad + g^{zz} \Bigg( \overbrace{\nabla_{\partial_z} (N_\pm)^\flat (\partial_z)}^{=0} - \Omega^{-1} (-\mathrm{d}\Omega(N_\pm)) g_{\partial_z, p_z} \Bigg) \Bigg)
\end{aligned}$$

$$= \Omega^{-1}\Big( \pm \frac{1 - 2\sinh^2(r)}{\sqrt{2}\sinh(r)\cosh(r)} + \Omega^{-1}\frac{1 - \sinh^2(r)}{\cosh^2(r)}\mathrm{d}\Omega(N_\pm)$$
$$+ \Omega^{-1}\mathrm{d}\Omega(N_\pm)\Big)$$
$$= \Omega^{-1}\Big( \pm \frac{1 - 2\sinh^2(r)}{\sqrt{2}\sinh(r)\cosh(r)} + \Omega^{-1}\frac{2}{\cosh^2(r)}(\partial_s\Omega)\frac{\sqrt{1 - \sinh^2(r)}}{\sqrt{2}\cosh(r)}\Big)$$

Inserting $r = \hat{r}$ this gives

$$\theta_\pm = \Omega^{-2}\frac{2}{\cosh^2(r)}(\partial_s\Omega)\frac{\sqrt{1 - \sinh^2(r)}}{\sqrt{2}\cosh(r)} < 0.$$

∎

It is clear that for $\tau$ small enough the genericity and timelike convergence conditions are still satisfied. Hence for $\tau$ small enough the spacetime $(\tilde{M}, \hat{g}_\tau)$ is complete but satisfies all assumption of Theorem 9.3.1 with the exception of the chronology condition.

It is possible estimate which kinds of causality violation can invalidate Theorem 9.3.1. In the proof of Theorem 9.3.1 the future horismos $E^+(\mathcal{T})$ of the strictly closed trapped surface $\mathcal{T}$ played a significant rôle. In a spacetime with chronology violations (and in particular in our example) this set is in general empty. However, there is a generalization of the non-global features of a horismos.

**Definition 9.4.2.** *Let $(M, g)$ be a time oriented spacetime and $\mathcal{D}$ be a compact set which is achronal in some neighbourhood of $\mathcal{U}$ of $\mathcal{D}$.*

*(i) Let $\gamma\colon [0, b)$ be a future directed future inextensible null geodesic starting in $\mathcal{D}$. A point $x = \gamma(t)$ is called a* focal point, *if*

*(a) for all $\gamma(t_+)$ $(t_+ > t)$ there is a timelike curve from $\mathcal{D}$ to $\gamma(t_+$ arbitrarily close to $\gamma$,*

*(b) There is not any $t_- < t$ sucht that $\mathcal{D}$ and $\gamma(t_-)$ can be connected by timelike curves arbitrarily close to $\gamma$.*

*(ii) Denote by $\gamma_x : [0, b(x))$ the maximal geodesic prolongation of the generator of $E^+(\mathcal{D}, \mathcal{U})$ with starting point $x = \gamma_x(0) \in \mathcal{D}$ which does not have a focal point. The* generalized future horismos of $\mathcal{D}$ *is the closure $e^+(\mathcal{D})$ of the set $\{\gamma_x(t) \in M \,|\, x \in \mathcal{D}\}$.*

*(iii) The* generalized future focal set of $\mathcal{D}$ *is defined by*

$$f^+(\mathcal{D}) = \{y \in e^+(\mathcal{D}) \,|\, y \text{ is future endpoint of some generator } \gamma_x \text{ of } e^+(\mathcal{D})\}.$$

*$e^-(\mathcal{D})$ and $f^-(\mathcal{D})$ are defined analogously.*

It is clear that for every spacetime and every compact set $\mathcal{D}$ we have $E^+(\mathcal{D}) \subset e^+(\mathcal{D})$ and $\mathcal{D} \subset e^+(\mathcal{D})$. The future horismos is always a Lipschitz hypersurface with induced degenerate metric. A similar property is also true for $e^+(\mathcal{D})$.

**Lemma 9.4.5.** *Let $\mu \subset e^+(\mathcal{D})$ be a causal curve. Then $\mu$ is either*

- *a subset of a null geodesic generator $\gamma_x$ of $e^+(\mathcal{D})$ or*
- *a causal curve contained in $f^+(\mathcal{D})$ or*
- *a concatenation a subset of a null geodesic generator and of a causal curve contained in $f^+(\mathcal{D})$ .*

*Proof.* If $\mu$ is a non-trivial curve in $e^+(\mathcal{D})$ which is not a part of a generator of $e^+(\mathcal{D})$ and does not lie in $f^+(\mathcal{D})$, then it is intersected transversely by some generator $\gamma_x$ of $e^+(\mathcal{D})$. Hence there exists a broken, causal curve $\lambda$ in $e^+(\mathcal{D}) \setminus f^+(\mathcal{D})$ with past endpoint in $\mathcal{D}$. Let $y \in \lambda$ be a point s after this break and $\gamma_z$ the generator of $e^+(\mathcal{D})$ with future endpoint $y$. Using our broken causal curve we see that there is a timelike curve from $\mathcal{D}$ to $y$ arbitrarily close to $\gamma_z$. This gives a contradiction to the definition of $e^+(\mathcal{D})$. ∎

Hence any closed causal curve arbitrarily close to the generalized future horismos must lie in $f^+(\mathcal{D})$. This is exactly what happens in the example of Newman. The null geodesic generators of $e^+(\mathcal{T}_{\hat{s},\hat{r}})$ end in a caustic $f^+(\mathcal{T}_{\hat{s},\hat{r}})$ which is ruled by closed null curves of the form $s = \textit{const}$, $r = \text{const}$, $z = \text{const}$. Observe that these curves are not null geodesics. The existence of these curves is basically the reason why Theorem 9.3.1 fails in the presence of causality violation. Notice that we can slighly generalize our example such that $f^+(\mathcal{T}_{\hat{s},\hat{r}})$ is not ruled by closed causal curve but only by almost closed causal curves. We simply replace the identification $(s, r, \varphi + 2\pi, z) = (s, r, \varphi, z)$ by an identification $(s, r, \varphi + 2\pi, z) = (s, r, \varphi, z + a)$ such that the quotient $a/2\pi$ is irrational. It is clear that the curves $\gamma$ locally defined by $s = \text{const}$, $r = \text{const}$, $z = \text{const}$ are not closed but satisfy instead:

> For each $\gamma(t)$ and each (small enough) neighbourhood $\mathcal{V}$ of $\gamma(t)$ there is a $t_+ > t$ such that the segment of $\gamma$ between $\gamma(t)$ and $\gamma(t_+)$ leaves $\mathcal{V}$ and then re-enters this set.

All other properties of our example are unchanged since the new spacetime is locally isometric to the old one. In order to state a theorem which justifies the claim that the only impediment to a version of Theorem 9.3.1 in the presence of causality violation is the possible existence of almost closed causal curves in $f^+(\mathcal{T})$ we need the following technical definition.

**Definition 9.4.3.** *Let $\gamma$ be a curve and choose any Riemannian metric $h$ on $M$. Let $\mu$: $(a, b)$ be a reparametrization of $\gamma$ which satisfies*

*$h(\dot{\mu}, \dot{\mu}) = 1$. We call $\gamma$* almost closed *if there exists a vector $u \in \{\dot{\mu}(t) \,|\, t \in (a,b)\}$ such that for every neighbourhood $\mathfrak{U}$ of $u$ in $TM$ there exists a deformation $\lambda$ of $\mu$ in $\pi_{TM}(\mathfrak{U})$ which yields a closed curve and satisfies $\lambda(t) \in \pi_{TM}(\mathfrak{U}) \Rightarrow \dot{\lambda}(t) \in \mathfrak{U}$.*

Observe that this definition is independent of the choice of $h$.

**Theorem 9.4.1.** *A spacetime $(M, g)$ is not causal geodesically complete if*

*(i) the timelike convergence condition and the genericity condition hold,*
*(ii) there exists at least one of the following:*
  *(a) a (locally spacelike) strictly closed trapped surface $\mathcal{T}$,*
  *(b) a compact achronal set $\mathcal{T}$ without edge,*
  *(c) a point $x$ such that along every past (or every future) inextensible null geodesic from $x$ the expansion of the null geodesics starting at $x$ becomes negative,*
*(iii) neither $f^+(\mathcal{T})$ (respectively, $f^+(\{x\})$ nor any $f^-(\mathcal{D})$, where $\mathcal{D}$ is a compact topological submanifold (possibly with boundary) with $\mathcal{D} \cap \mathcal{T} \neq \emptyset$ (respectively, $x \in \mathcal{D}$) contains any almost closed causal curve that is a cluster curve of a sequence of closed timelike curves.*

This is a proper generalization of Theorem 9.3.1. The technical condition (iii) just states the situation which we have already anticipated by analyzing Newman's example. The proof of Theorem 9.4.1 is far too technical to be reproduced here. It basically consists of a cutting and pasting procedure (Kriele 1990).

The closed trapped surface in Newman's counter example has the topology of a torus. In a physically realistical collapse scenario of a star one would rather expect that there exists a closed trapped surface of topology $S^2$ surrounding the collapsing star. This motivates the following conjecture:

*Conjecture 9.4.1.* A 4-dimensional spacetime $(M, g)$ is not causal geodesically complete if

(i) the timelike convergence condition and the genericity condition hold,
(ii) there exists a strictly closed trapped surface of topology $S^2$.

In Newman's example the generalized future focal set of $\mathcal{T}_{\hat{s}, \ln((1+\sqrt{3})/2)}$ is generated by closed null curves. This is impossible if $\mathcal{T}$ has topology $S^2$.

In spite of this small piece of evidence and the importance of Conjecture 9.4.1 for our interpretation of singularity theorems it is completely open whether this conjecture is true or not.

## 9.5 Strength of singularities and cosmic censorship

*In this section we will investigate the character of the singularities predicted by Theorem 9.3.1. We will also give an example (cf. Sect. 9.5.1) which shows that the theorem of Hawking and Penrose may only imply the existence of "singularities" which are so weak that the energy density exists in a distributional sense. Our example is not very physical — for a start, it is 3-dimensional rather than 4-dimensional. On the other hand, it is a good test case for the mechanism behind the singularity theorems.*

The existence of incomplete causal geodesic does not imply that there is a singularity. This is the reason why "singularity theorems" are often referred to as "incompleteness theorems". The standard counter example in general relativity is the Taub-NUT spacetime (cf. (Hawking and Ellis 1973, chapter 5.8). The following two-dimensional example is especially simple.

*Example 9.5.1 (Clifford-Pohl torus).* Let $(M, g) = (\mathbb{R}^2 \backslash \{0\}, \frac{2}{u^2+v^2} dudv)$. Then the curve $\gamma(t) := (\frac{1}{1-t}, 0)$ is an incomplete geodesic and the map $\psi\colon (u, v) \mapsto (2u, 2v)$ is an isometry. Defining

$$x \sim y :\Leftrightarrow \exists k \in \mathbb{Z} \text{ with } \psi^k(x) = y, \qquad \pi\colon x \mapsto [x].$$

we obtain a compact Lorentzian torus $(\pi(M), \pi_* g)$. The curve $\pi(\gamma)$ is an incomplete lightlike geodesic in this compact and therefore *non-singular* spacetime.

While Example 9.5.1 shows that there exist spacetimes which are non-singular and geodesically incomplete, so far there is no example of such a spacetime which satisfies the assumptions of Theorem 9.3.1. In fact, it is believed that the incomplete geodesics predicted by the singularity theorems are of a different nature. The justification for this expectation is the fact that a non-compact or empty "Cauchy horizon" plays an important rôle in the proof of the theorem of Hawking and Penrose. Since the Cauchy horizon is necessarily closed, one would expect that a Cauchy horizon due to compactly imprisoned curves (such as in Example 9.5.1 above) is compact and non-empty. This motivates the following conjecture.

*Conjecture 9.5.1.* If the assumptions of Theorem 9.3.1 are satisfied, then there exists a causal inextensible incomplete geodesic which leaves every compact subset in both future and past direction.

Using an approximation argument one can show that the singularity predicted by the theorem of Hawking & Penrose is not just due to $g$ merely being $C^{2-}$ instead of $C^2$. There also exist upper bounds on the divergence of the curvature along any incomplete geodesic $\gamma$ which are large enough to allow for the possibility of strong curvature singularities,i.e, curvature scalars which not only diverge but whose integral over an appropriate spacetime region diverges as well. Unfortunately, we have no curvature estimates for anything in between these two extremes.

If, in general, the singularity predicted by Theorem 9.3.1 were only weak, this theorem would only predict the existence of "shockwaves" and we would lack evidence for the existence of serious singularities such as black holes. The following conjecture (formulated by Hawking & Ellis (Hawking and Ellis 1973) with respect to a slightly different singularity theorem) is therefore central to our interpretation.

*Conjecture 9.5.2.* Assume that $(M, g)$ is chronological, satisfies the timelike convergence and the genericity condition and contains a closed trapped surface. Then there exists an incomplete, future inextensible geodesic $\gamma$ and a neighbourhood $\mathcal{U}$ of $\gamma$ such that $\mathrm{Vol}(\mathcal{U}) < \infty$ and $\int_{\mathcal{U}} f(R)\mu = \infty$ for some curvature invariant $f(R)$ which is (positively) homogeneous in the Riemann tensor, $f(\lambda R) = |\lambda| f(R)$ for all $\lambda > 0$.

The homogeneity condition is important because otherwise we could take an appropiate power of a weakly diverging curvature invariant in order to obtain a diverging integral: While $\int_0^1 \frac{1}{\sqrt{x}}\mathrm{d}x$ is finite, the integral $\int_0^1 \frac{1}{\sqrt{x}^2}\mathrm{d}x$ is not.

Hawking & Ellis state that while they are convinced of the validity of such a conjecture[7] they are unable to prove it. In Sect. 9.5.1 below we will give a 3-dimensional example which indicates that Conjecture 9.5.2 may not be true in the present form.

If there is a singularity in our universe, we would like to interpret it as a black hole, i.e., we would hope that it is invisible — just as the singularity in the Schwarzschild spacetime. Otherwise we would not have a chance to globally solve Einstein's equation as a Cauchy problem since the singularity (whose data are unknown) would influence the geometry of spacetime to its future. There are also important theorems for our interpretation of black holes which need a assumption similar to cosmic censorship. The prime example is the "area theorem" due to Hawking which states that the area of black holes can only increase[8]) (Wald 1984, theorem 12.2.6).

Since it is easy to find examples of inextensible Lorentzian manifolds which contain visible (or "naked") singularities, additional assumptions on spacetime must be made in any conjecture which "censores" naked singularities. The following conjecture is due to Penrose.

---

[7] They state their conjecture with respect to a different singularity theorem.

[8] This is only true in classical general relativity without taking quantum effects into account

*Conjecture 9.5.3 ((strong) cosmic censorship).* If $(M, g)$ is qualitatively stable and its matter model $T$ is physically reasonable then no future incomplete, future inextensible causal geodesic $\gamma$ lies in the past of any $x \in M$.

It is not sufficent just to demand a "physically reasonable" matter model because of the Reissner-Nordström solution,

$$g = -\left(1 - \frac{2m}{r} + \frac{e^2}{r^2}\right) \mathrm{d}t^2 + \left(1 - \frac{2m}{r} + \frac{e^2}{r^2}\right)^{-1} \mathrm{d}r^2 + r^2 \left(\mathrm{d}\theta^2 + \sin^2 \theta \mathrm{d}\varphi^2\right).$$

The energy momentum tensor $T = \frac{1}{8\pi}(\mathrm{Ric} - (\mathrm{Scal}/2)g)$ is given by

$$T = \frac{e^2}{8\pi r^4} \left(U^\flat \otimes U^\flat - Q^\flat \otimes Q^\flat + r^2 \left(\mathrm{d}\theta^2 + \sin^2 \theta \mathrm{d}\varphi^2\right)\right),$$

where we have set

$$U = \left(1 - \frac{2m}{r} + \frac{e^2}{r^2}\right)^{\frac{-1}{2}} \partial_t, \quad Q = \left(1 - \frac{2m}{r} + \frac{e^2}{r^2}\right)^{\frac{1}{2}} \partial_r .$$

This spherically symmetric spacetime satisfies Einstein's equations for an electromagnetic field (cf. Lemma 7.4.1) which is certainly "physically reasonable". On the other hand, unlike in the case of the Schwarzschild solution ($e = 0$) where the hypersurfaces $r = \text{const} \ll 1$ are spacelike, the hypersurfaces $r = \text{const} \ll 1$ in the Reissner-Nordström solution are timelike. It is easy to see that there exist points $x, y \in M$ such that $I^+(x) \cap I^-(y)$ contains timelike future inextensible curves which approach $r = 0$. (For a more thorough discussion of the Reissner-Nordström spacetime including its global properties cf. (Hawking and Ellis 1973, chapter 5.5)). This Reissner-Nordström spacetime therefore violates Conjecture 9.5.3 if it is qualitatively stable. Calculations by Simpson and Penrose (1973) and McNamara (1978) indicate that this is not the case for an intuitive notion of stability. It is generally believed that a generic, physically acceptable perturbation of the Reissner-Nordström spacetime results in a spacetime which is qualitatively more similar to the Schwarzschild spacetime, *even though the Reissner-Nordström spacetime itself can be thought of as a perturbation of the Schwarzschild spacetime.*

### 9.5.1 A simple, 3-dimensional example

Let $(M, g)$ be a 3-dimensional spacetime and assume that the energy momentum tensor is given by $T = \epsilon U^\flat \otimes U^\flat$ where $U$ be the spacetime velocity of the dust particles and $\epsilon$ their energy density. We are seeking solutions of Einstein's equation

$$\mathrm{Ric} - \frac{1}{2}\mathrm{Scal}g = 8\pi\epsilon U^\flat \otimes U^\flat,$$

where $\epsilon\colon M \to \mathbb{R}$ is a function. In general, this is still too difficult even though we assume $n = 3$. Assuming that there is a foliation of spacelike hypersurfaces orthogonal to $U$ simplifies the problem dramatically.

**Lemma 9.5.1.** *The vector field $U$ is irrotational (i.e., $\mathrm{d}U^\flat = 0$) if and only if the Pfaffian system $\{U^\flat\}$ is integrable.*

*Proof.* By Lemma 2.5.8 the integrability of $\{U^\flat\}$ is equivalent to the equation $\mathrm{d}U^\flat \wedge U^\flat = 0$. Recall from Lemma 5.2.1 that $U$ satisfies the geodesic equation $\nabla_U U = 0$. Let $X$ be any vector field. Then we have $\mathrm{d}U^\flat(X,U) = \nabla_X U^\flat(U) - \nabla_U U^\flat(X) = g(\nabla_X U, U) - g(\nabla_U U, X) = 0 - 0$ since $g(U,U) = -1$. It follows that $\mathrm{d}U^\flat$ is completely determined by evaluating it on vectors orthogonal to $U$. Let $v, w$ be two vectors orthogonal to $U$. Then we have $\mathrm{d}U^\flat \wedge U^\flat(v,w,U) = -\mathrm{d}U^\flat(v,w)$ and the equivalence follows. ∎

**Lemma 9.5.2.** *Let $(M,g)$ be an irrotational, 3-dimensional dust space-time and let $\Sigma$ be a hypersurface which is orthogonal to $U$. If at $p \in \Sigma$ the second fundamental form of $\Sigma$ is not a multiple of the metric, then $p$ has a neighbourhood with coordinates $(t,x,y)$ such that $g = -\mathrm{d}t^2 + V^2\mathrm{d}x^2 + W^2\mathrm{d}y^2$, where $V, W$ are functions of $t, x, y$ and $T = \epsilon(t,x,y)\mathrm{d}t \otimes \mathrm{d}t$.*

*These coordinates are unique up to transformations of the type $x \mapsto X(x)$, $y \mapsto Y(y)$, $t \mapsto t +$ const. and interchanging of $x$ and $y$.*

*Proof.* Since $\{U^\flat\}$ satisfies $\mathrm{d}U^\flat = 0$ there is a function $t$ with $\mathrm{d}t = U^\flat$. We can write $g = -\mathrm{d}t^2 + \sum_{i,j=1}^{2} {}^{(2)}g_{ij}(t,x^1,x^2)\mathrm{d}x^i\mathrm{d}x^j$, where for each $t$ the bilinear form ${}^{(2)}g(t,\cdot,\cdot)$ is a Riemannian 2-metric and $\Sigma$ is given by $t = t_0$. Since $\partial_t{}^{(2)}g$ is not umbilic at $p$, there exists a frame $\{e_1, e_2\}$ of $\Sigma$ in a neighbourhood of $p$ such that ${}^{(2)}g$ and $\partial_t{}^{(2)}g$ are both diagonal with respect to this frame. Let $\{\omega^1, \omega^2\}$ be the dual frame. It follows that there exist coordinates $(x,y)$ such that at $t = t_0$ both ${}^{(2)}g$ and $\partial_t{}^{(2)}g$ are diagonal with respect to $\partial_x, \partial_y$. In fact, we only need to show that there exist multiples $\alpha_1 e_1$, $\alpha_2 e_2$ of $e_1, e_2$ such that $[\alpha_1 e_1, \alpha_2 e_2] = 0$. This is equivalent to $\mathrm{d}\alpha_i(e_{i+1 \bmod 2}) + (-1)^i \alpha\omega^i([e_1,e_2]) = 0$ (no summation over $i$) which is a system of ordinary differential equations and can be solved by Theorem 2.4.1. With respect to the coordinates $(t,x,y)$ the equations $T_{ij} = 0$ $(i,j \in \{x,y\})$ imply

$$\begin{aligned}\partial_t\partial_t{}^{(2)}g_{ij} - \mathrm{tr}(\partial_t\partial_t{}^{(2)}g){}^{(2)}g_{ij} &= -\frac{1}{2}\mathrm{tr}(\partial_t{}^{(2)}g)\partial_t{}^{(2)}g_{ij} + {}^{(2)}g^{kl}\partial_t{}^{(2)}g_{ik}\partial_t{}^{(2)}g_{jl} \\ &\quad + \frac{1}{4}\left(\left(\mathrm{tr}(\partial_t{}^{(2)}g)\right)^2 - 3\left|\partial_t{}^{(2)}g\right|^2\right){}^{(2)}g_{ij}.\end{aligned}$$

Since at $t = t_0$ the bilinear forms ${}^{(2)}g_{ij}$ and $\partial_t {}^{(2)}g_{ij}$ are diagonal, it follows that the right hand side is also diagonal at $t = t_0$. Since the system has a unique solution and there exists a solution when ${}^{(2)}g, \partial_t {}^{(2)}g$ are simultaneously diagonal, ${}^{(2)}g, \partial_t {}^{(2)}g$ must be diagonal for all $t$. Hence we have existence. For uniqueness observe that the frame $\{e_1, e_2\}$ is unique up to multiples and permutation, and that the coordinate $t$ is already chosen so that it is unique up to an additive constant. Thus any other coordinates $X, Y$ with the same properties must satisfy either $\partial_x \parallel \partial_X$ and $\partial_y \parallel \partial_Y$ or $\partial_x \parallel \partial_Y$ and $\partial_y \parallel \partial_X$. This proves the lemma. ∎

**Corollary 9.5.1.** *Let $(\Sigma, {}^{(2)}g)$ be a 2-dimensional Riemannian manifold and $k$ be a symmetric $\binom{0}{2}$ tensor field which is not proportional to ${}^{(2)}g$. Then the initial value problem for irrotational, 3-dimensional dust spacetimes with initial data $(\Sigma, {}^{(2)}g, k)$ reduces to a constrained system of ordinary differential equations.*

In Theorem 9.5.1 below we will summarize properties of generic, irrotational, 3-dimensional dust spacetimes using standard differential geometric terminology. Consider a 2-dimensional Riemannian manifold $(\Sigma, g_\Sigma)$ and denote the set of unoriented lines in $T\Sigma$ by $\mathbb{P}\Sigma$. Then there exists a natural map $\cdot^\perp \colon \mathbb{P}\Sigma \to \mathbb{P}\Sigma$ which maps an unoriented line $l \in \mathbb{P}\Sigma$ to the line orthogonal to it. We call a section $l$ of $\mathbb{P}\Sigma$ *nowhere geodesic* if for any local, non-vanishing vector field $L$ with $L(p) \in l(p) \quad \forall p$ we have $g_\Sigma(L^\perp, \nabla_L L) \neq 0 \quad \forall p$. This condition does not depend on the chosen representative $L$. It is a local but not necessarily a global genericity condition on $l$. Locally, this condition is slightly stronger than demanding that $l$ does not have any local integral curve which is a geodesic. Let $(\Sigma, g_\Sigma)$ be a 2-dimensional, spacelike submanifold of a Lorentzian 3-manifold $(M, g)$ with future directed normal $\mathfrak{n}$ and second fundamental form $k(X, Y) = -g(\nabla_X Y, \mathfrak{n})$. We denote the bilinear form associated with the square of the corresponding matrix by $k^2$, i.e., $(k^2)_{ij} = (g_\Sigma)^{lm} k_{il} k_{jm}$. We call the eigenvalues $k_1$, $k_2$ of $k$ with respect to $g_\Sigma$ the *principal curvatures of $\Sigma$* and the (unoriented) lines spanned by the eigenvectors the *principal directions of $\Sigma$*.

**Theorem 9.5.1.** *Let $(\mathbb{R}^2, {}^{(2)}g)$ be a Riemannian 2-manifold and $l \colon \mathbb{R}^2 \to$ ¶$\mathbb{R}^2$ which maps each point $p \in \mathbb{R}^2$ into an unoriented line $l(p) \subset T_p\mathbb{R}^2$ and assume that $l$ is nowhere geodesic. Let $\mathcal{C} \subset \mathbb{R}^2$ be a smooth curve which divides $\mathbb{R}^2$ into two disconnected regions such that $l$ and $l^\perp$ intersect $T_p\mathcal{C}$ transversely at each $p \in \mathcal{C}$. Finally, let $K_1, K_2 \colon \mathcal{C} \to \mathbb{R}$ be smooth functions.*

*(i) Let $\mathfrak{D}(\mathcal{C}, l, l^\perp)$ the set of points $p \in \mathbb{R}^2$ such that the integral curves of $l$ and $l^\perp$ through $p$ intersect $\mathcal{C}$. There exists an irrotational, 3-dimensional dust spacetime $(M, g)$ and an isometric embedding*

*$\iota\colon (\mathfrak{D}(\mathcal{C}, l, l^\perp), {}^{(2)}g) \to \Sigma \subset M$ such that the second fundamental form $k$ of $\Sigma$ in $M$ satisfies*

*(a) the principal directions of $\Sigma$ are given by $\iota_* l$, $\iota_* l^\perp$,*

*(b) along $\mathcal{C}$ the submanifold $\Sigma$ has principal curvatures $k_1 = K_1$, $k_2 = K_2$.*

*$\mathcal{C}$ can be chosen such that $\mathfrak{D}(\mathcal{C}, l, l^\perp) = \mathbb{R}^2$. Then $(M, g)$ is inextensible if $(\mathbb{R}^2, {}^{(2)}g)$ is so.*

*(ii) For any $p \in \Sigma$ let*

$$k_-(p) = \min(k_1(p), k_2(p)) \text{ and } k_+(p) = \max(k_1(p), k_2(p)).$$

*The world line of the dust particle through $p$ ends in a curvature singularity in finite proper times $\frac{-1}{k_-(p)}$, $\frac{-1}{k_+(p)}$ if $k_+(p) > 0 > k_-(p)$, at finite proper time $\frac{-1}{k_-(p)}$ if $k_-(p) > 0$, at finite proper time $\frac{-1}{k_+(p)}$ if $k_+(p) < 0$. There are no other singularities.*

*(iii) All singularities are weak in the sense that for all open sets $\mathcal{U}$ with bounded volume, $vol(\mathcal{U}) = \int_{\mathcal{U}} \sqrt{\det(g_{ab})} \mathrm{d}t \mathrm{d}x \mathrm{d}y < \infty$, the spacetime average of the energy density,*

$$\frac{1}{vol(\mathcal{U})} \int_{\mathcal{U}} \epsilon(t, x, y, z) \sqrt{\det(g_{ab})} \mathrm{d}t \mathrm{d}x \mathrm{d}y,$$

*is also bounded.*

*(iv) The spacetime is non-singular if and only if $K_1 = K_2 = 0$.*

*(v) For generic initial data strong cosmic censorship is violated, provided one regards the solution as "qualitatively stable" and "physically reasonable".[9]*

*(vi) Generically, the data ${}^{(2)}g$, $l$, $K_1$, $K_2$ parameterize the set of local, irrotational, 3-dimensional dust spacetimes.*

Properties (iii) and (ii) hold true in a more general context (Kriele and Lim 1995).

*Proof of Theorem 9.5.1.* (i): We can choose coordinates $(x, y)$ such that at each point $p \in \mathbb{R}^2$ the Gaussian vectors $\partial_x$, $\partial_y$ span $l$ and $l^\perp$ (see the proof of Lemma 9.5.2). The metric ${}^{(2)}g$ is diagonal in these coordinates. Let $V_0^2 = {}^{(2)}g_{xx}$ and $W_0^2 = {}^{(2)}g_{yy}$. In view of Lemma 9.5.2 we can set $\Sigma = \{(t, x, y) \,|\, t = 0\}$ and assume that the dust-metric is given by $g = -\mathrm{d}t^2 + V^2(t, x, y)\mathrm{d}x^2 + W^2(t, x, y)\mathrm{d}y^2$, $V(0, x, y) = V_0(x, y)$, $W(0, x, y) = W_0(x, y)$. Since the constraint $T_{ty} = 0$ is equivalent to

$$0 = \frac{\partial^2 V}{\partial t \partial y} W - \frac{\partial W}{\partial t} \frac{\partial V}{\partial y}$$

we obtain either $\frac{\partial V}{\partial y} = 0$ or $W = w \frac{\partial V}{\partial y}$, where $w$ is some non-vanishing function of $(x, y)$. In the coordinates $(x, y)$ we can write $L(x, y) = f(x, y)\partial_x$ and $(L(x, y))^\perp = h(x, y)\partial_y$, where $f$, $h$ are non-vanishing functions. The condition that $l$ is nowhere geodesic reduces then to

[9] The relation of these solutions and cosmic censorship are further discussed in (Kriele 1997).

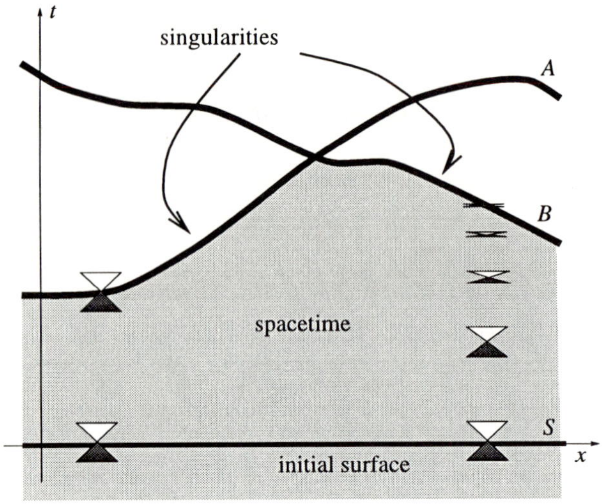


**Fig. 9.5.1.** The singularity structure of spacetime. The case $k_1, k_2 < 0$. The $y$-component of spacetime is suppressed. The singularity $A$ is given by $1 + t\,(\partial V_0/\partial y)^{-1}\,(\partial q/\partial y) = 0$ and the singularity $B$ is given by $V_0 + tq = 0$. Observe that at the singularity $A$ the light cone degenerates in the $y$-direction and that at $B$ degenerates in the $x$-direction. Hence there exist future directed timelike curves emanating from the singularity and cosmic censorship is violated

$f^2 h V_0 \frac{\partial V_0}{\partial y} \neq 0$. Thus we have $\frac{\partial V_0}{\partial y} \neq 0$. Since $0 = T_{yy} = \frac{w^2}{V}\left(\frac{\partial V}{\partial y}\right)^2 \frac{\partial^2 V}{\partial t^2}$ we can write $V(t,x,y) = V_0(x,y) + tq(x,y)$. Now the equations $T_{tx} = 0$ gives

$$0 = \frac{\partial w}{\partial x} V_0 \frac{\partial q}{\partial y} - \frac{\partial w}{\partial x}\frac{\partial V_0}{\partial y} q - w \frac{\partial^2 V_0}{\partial x \partial y} q + w \frac{\partial^2 q}{\partial x \partial y} V_0. \tag{9.5.7}$$

Re-expressing $w(x,y)$ by $W_0 = w\frac{\partial V_0}{\partial y}$ we obtain the linear, hyperbolic partial differential equation

$$\frac{\partial^2 q}{\partial x \partial y} + \left( W_0^{-1} \frac{\partial W_0}{\partial x} - \left(\frac{\partial V_0}{\partial y}\right)^{-1} \frac{\partial^2 V_0}{\partial x \partial y} \right) \frac{\partial q}{\partial y} - \left( V_0^{-1} \frac{\partial V_0}{\partial y} W_0^{-1} \frac{\partial W_0}{\partial x} \right) q = 0 \tag{9.5.8}$$

for the function $q(x,y)$. The lines $l$, $l^\perp$ are transverse to $\mathcal{C}$. Moreover, it follows that each integral curve of $l$ and $l^\perp$ intersects $\mathcal{C}$ at most once. In fact, if a coordinate line $x = x_0$ would intersect $\mathcal{C}$ twice then there would exist a point $(x_m, y_m)$ of $\mathcal{C}$ in between these intersection points which has locally maximal distance to $(x_0, y_m)$ with respect to the flat metric $dx^2 + dy^2$. At this point $\partial_y$ would be tangent to $\mathcal{C}$ in contradiction to our transversality assumption. We will now show that for given initial values this differential equation has a unique solution in $\mathfrak{D}(\mathcal{C}, l, l^{\perp})$ by reducing it to an appropriate system of hyperbolic differential equations.[10] Let $\tau^1 = x + y$ and $\tau^2 = x - y$. With respect to these variables Equation (9.5.8) reduces to an equation of the form

$$\frac{\partial^2 \tilde{q}}{(\partial \tau_1)^2} - \frac{\partial^2 \tilde{q}}{(\partial \tau_2)^2} = h\left(\tau_1, \tau_2, \tilde{q}, \frac{\partial \tilde{q}}{\partial \tau_1}, \frac{\partial \tilde{q}}{\partial \tau_1}\right),$$

where $\tilde{h}\colon \mathbb{R}^5 \to \mathbb{R}$ is a suitable function. Setting $\tilde{r} = \frac{\partial \tilde{q}}{\partial \tau_1}$ and $\tilde{s} = \frac{\partial \tilde{q}}{\partial \tau_2}$ we see that Equation (9.5.8) is equivalent to the hyperbolic system c equations

$$\frac{\partial \tilde{q}}{\partial \tau_1} = \tilde{r}, \quad \frac{\partial \tilde{r}}{\partial \tau_1} = \frac{\partial \tilde{s}}{\partial \tau_2} + h(\tau_1, \tau_2, \tilde{q}, \tilde{r}, \tilde{s}), \quad \frac{\partial \tilde{s}}{\partial \tau_1} = \frac{\partial \tilde{r}}{\partial \tau_2}.$$

The characteristic directions are given by the left eigenvalues of the matrix

$$A = \begin{pmatrix} 0 & 0 & 0 \\ 0 & 0 & 1 \\ 0 & 1 & 0 \end{pmatrix},$$

$l_1 = (1, 0, 0)$, $l_2 = (0, 1, 1)$, $l_3 = (0, 1, -1)$. These vectors are linearly independent and we can apply Corollary 7.3.2 together with Remark 7.3.1 in order to obtain a solution for given initial values. The inital values for $\tilde{q}$ can be calculated from the initial values for $q$. If we parameterize $\mathcal{C}$ by a curve $s \mapsto \lambda(s)$ we get

$$\frac{\partial q \circ \lambda}{\mathrm{d}s} = \frac{\partial q}{\partial \tilde{q}} (\mathrm{d}\tau_1(\dot{\lambda})\tilde{r} + \mathrm{d}\tau_2(\dot{\lambda})\tilde{s})$$

and another linear combination for the normal derivative of $q$ which depends on the coordinate expression for ${}^{(2)}g$. Hence we can calculate the inital values for $(\tilde{q}, \tilde{r}, \tilde{s})$ if we know $q_{|\mathcal{C}}$ and its normal derivative. Since this normal derivative can be calculated from $q_{|\mathcal{C}}$ and $\left(\frac{\partial q}{\partial y}\right)_{|\mathcal{C}}$ we have as our initial data $q_{|\mathcal{C}} = K_1 V_{0|\mathcal{C}}$ and $\left(\frac{\partial q}{\partial y}\right)_{|\mathcal{C}} = K_2 \left(\frac{\partial V_0}{\partial y}\right)_{|\mathcal{C}}$. Since $T_{xy} = 0$ and $T_{xx} = 0$ hold automatically this leads to a solution of Einstein's equation which is given by

$$g = -\mathrm{d}t^2 + (V_0 + tq)^2 \mathrm{d}x^2 + W_0^2 \left(1 + t\left(\frac{\partial V_0}{\partial y}\right)^{-1} \frac{\partial q}{\partial y}\right)^2 \mathrm{d}y^2. \tag{9.5.9}$$

[10] Alternatively, we could directly appeal to standard theorems. Since the symbol of the hyperbolic equation (9.5.8) is constant it has a unique, global solution in $\mathfrak{D}(\mathcal{C}, l, l^{\perp})$ for the initial value problem with respect to the initial curve $\mathcal{C} \subset \Sigma$ (Garabedian 1986, section 4.2).

It follows immediately that the principal curvatures are given by

$$k_1(x,y) = q(x,y)/V_0(x,y) \text{ and } k_2(x,y) = \left(\frac{\partial V_0}{\partial y}\right)^{-1} \frac{\partial q}{\partial y}.$$

If one chooses $\mathcal{C} = \{(x,y) | x = y\}$ then $\mathfrak{D}(\mathcal{C}, l, l^{\perp}) = \mathbb{R}^2$. Since the solution is inextensible in $t$-direction $(M, g)$ is inextensible provided $\mathfrak{D}(\mathcal{C}, l, l^{\perp}) = \mathbb{R}^2$ and $(\mathbb{R}^2, {}^{(2)}g)$ is inextensible.

(ii): The energy density is given by

$$\epsilon(t,x,y) = \frac{E(x,y)}{(V_0 + tq)\left(\frac{\partial V_0}{\partial y} + t\frac{\partial q}{\partial y}\right)}, \tag{9.5.10}$$

where

$$\begin{aligned} E(x,y) = {} & \frac{\partial q}{\partial y} q + \left(\frac{\partial V_0}{\partial y}\right)^2 \frac{\partial W_0}{\partial y} {W_0}^{-3} - \frac{\partial V_0}{\partial y} \frac{\partial^2 V_0}{\partial y^2} {W_0}^{-2} \\ & - \frac{\partial V_0}{\partial y} \frac{\partial^2 W_0}{\partial x^2} {V_0}^{-1} {W_0}^{-1} + \frac{\partial V_0}{\partial y} \frac{\partial W_0}{\partial x} \frac{\partial V_0}{\partial x} {V_0}^{-2} {W_0}^{-1}. \end{aligned}$$

Since $g$ is given by Equation (9.5.9) we have ${}^{(2)}g = {V_0}^2 \mathrm{d}x^2 + W_0^2 \mathrm{d}y^2$ and $k = qV_0 \mathrm{d}x^2 + W_0^2 \left(\frac{\partial V_0}{\partial y}\right)^{-1} \frac{\partial q}{\partial y} \mathrm{d}y^2$. It follows from Equation (9.5.10) that $\epsilon$ may become infinite at $V_0 + tq = 0$, $\frac{\partial V_0}{\partial y} + t\frac{\partial q}{\partial y} = 0$ and that generically it will be infinite at these points. The first part of (ii) follows immediately. To see that there is no other singularity observe that for 3-dimensional spacetimes the Riemann tensor is completely determined by the energy momentum tensor and that therefore all Riemann tensor components are bounded where $\epsilon$ is bounded.

(iii): This follows since

$$\begin{aligned} \epsilon\sqrt{-\det(g_{ab})} &= \epsilon W_0 \left(\frac{\partial V_0}{\partial y}\right)^{-1} (V_0 + tq)\left(\frac{\partial V_0}{\partial y} + t\frac{\partial q}{\partial y}\right) \\ &= E W_0 \left(\frac{\partial V_0}{\partial y}\right)^{-1} \end{aligned}$$

is finite.

(iv): Since the initial data for Equation (9.5.8) are given by $q_{|\mathcal{C}} = K_1 V_0$ and $\left(\frac{\partial q}{\partial y}\right)_{|\mathcal{C}} = K_2 \frac{\partial V_0}{\partial y}$ the claim follows from the uniqueness theorem for PDEs of the type (9.5.8).

(v): Since $g_{tt} = -1$ and $g_{xx}$ or $g_{yy}$ converges to zero it is easy to see that at these singularities strong cosmic censorship is violated unless the singularity is given by $t = \text{const.}$ (cf. Fig. 9.5.1).

(vi): Implicitly we have assumed that the second fundamental form and the metric can be diagonalized with respect to a smooth frame. This may not be possible at umbilic points but points at which this problem occurs are isolated and therefore not important for local genericity. Since $\frac{\partial V}{\partial y} \neq 0$ is also a local genericity condition, locally almost every irrotational dust spacetime can be obtained in this way. Given $\mathcal{C}$, our initial data are invariants. Hence (vi) follows. ∎

**Corollary 9.5.2.** *Almost all 3-dimensional, irrotational dust spacetimes can be obtained by quadratures.*

*Proof.* In Theorem 9.5.1 we have solved Einstein's equation as an initial value problem where $V_0$, $W_0$, $q_{|\mathcal{C}}$, $\left(\frac{\partial q}{\partial y}\right)_{|\mathcal{C}}$ are given. Now we consider $V_0$, $q$, $W_{0|x=x_0}$ as given initial data. Then we can solve Equation (9.5.8) and obtain the explicit solution

$$W_0 = W_{0|x=x_0}(y) e^{\int_{x_0}^{x} \frac{V_0}{\partial V_0/\partial y} \frac{V_0 \partial q/\partial y - q \partial V_0/\partial y}{\partial^2 V_0/\partial x \partial y \partial q/\partial y - \partial^2 q/\partial x \partial y \partial V_0/\partial y} \mathrm{d}\xi} . \tag{9.5.11}$$

∎

*Remark 9.5.1.* In Theorem 9.5.1 we have only considered the case where $\frac{\partial V_0}{\partial y} \neq 0$. There are additional restrictions at points with $\frac{\partial V_0}{\partial y} = 0$. It follows from Equation (9.5.8) that at these points either $\frac{\partial q}{\partial y} = 0$ or all $y$-derivatives and mixed derivatives of $V$ vanish. For completeness we will discuss both cases in more detail.

If $\frac{\partial V_0}{\partial y}$ does not vanish identically then we can use Equation 9.5.11 to solve Einstein's equation. Since $W_0$ is smooth and does not vanish, the integrand in Equation 9.5.11 must also be smooth. It follows that in a neighbourhood of $\{(x,y) | \frac{\partial V_0}{\partial y} = 0 \text{ at } (x,y)\}$ the $y$-derivative of $q$ must satisfy $q_y = \frac{\partial V_0/\partial y}{\partial^2 V_0/\partial x \partial y}\left(\beta \frac{\partial V_0}{\partial y} + \frac{\partial^2 q}{\partial x \partial y}\right)$, where $\beta$ is any function of $(x,y)$. If this is satisfied then we obtain a local, non-singular solution. Otherwise we simply have specified singular initial data.

The non-generic case $\frac{\partial V_0}{\partial y} \equiv 0$ can be easily solved. If $\frac{\partial V}{\partial y} = 0$ then $T_{xx} = \frac{V^2}{W}\frac{\partial^2 W}{\partial t^2}$ and $T_{yy} = \frac{W^2}{V}\frac{\partial^2 V}{\partial t^2}$. Thus we can write $V(t,x) = V_0(x) + tq(x)$ and $W(t,x,y) = W_0(x,y) + ts(x,y)$. Now $T_{tx} = 0$ implies

$$\frac{\partial s}{\partial x} V_0 - \frac{\partial W_0}{\partial x} q = 0. \tag{9.5.12}$$

Since the initial metric induced on $\Sigma$ is given by

$$g_\Sigma = V_0{}^2(x)\mathrm{d}x^2 + W_0{}^2(x,y)\mathrm{d}y^2$$

we have $V_0(x) \neq 0 \quad \forall x$. Thus we can use our coordinate freedom to normalize $V_0(x) = 1$. Equation (9.5.12) can be immediately integrated for any given $q, W_0$. We have automatically $T_{xy} = 0$ and hence for any functions $q(x), W_0(x,y), w(y)$ we obtain a solution

$$g = -\mathrm{d}t^2 + (1 + tq(x))^2\mathrm{d}x^2 + (W_0(x,y) + ts(x,y))^2\mathrm{d}y^2,$$

were $s(x,y) = \int_{x_0}^{x} q(\xi)\frac{\partial W_0(\xi,y)}{\partial \xi}\mathrm{d}\xi + w(y)$. The space of solutions is parameterized by the functions $q(x), W_0(x,y), w(y)$ modulo coordinate transformations $(x,y) \mapsto (x + x_0, Y(y))$. The energy density is given by

$$\epsilon(t,x,y) = -\frac{-\frac{\partial^2 W_0}{\partial x^2} + qs}{(1+tq)(W_0 + ts)}. \tag{9.5.13}$$

Hence we obtain the same type of singularities as in Theorem 9.5.1.

We can express the metric provided by Theorem 9.5.1 (or the preceding remark) in a more geometrical form.

**Corollary 9.5.3.** *Let $(M,g)$ be a 3-dimensional spacetime. If there exists a timelike unit length vector field $U$ and a function $\epsilon$ such that $T = \epsilon U^\flat \otimes U^\flat$ and $\mathrm{d}U^\flat \wedge U^\flat = 0$, then there exist coordinates $(t,x,y)$ with $U^\flat = \mathrm{d}t$ and*

$$\begin{aligned} g = -\mathrm{d}t^2 &+ e^{2a(x,y)}(1 + tk_1(x,y))^2\mathrm{d}x^2 \\ &+ e^{2b(x,y)}(1 + tk_2(x,y))^2\mathrm{d}y^2 \end{aligned}$$

*where $a, b$ are free functions and $k_1, k_2$ satisfy*

$$\begin{aligned} \partial_y k_1 &= (k_2 - k_1)\partial_y a, \\ \partial_x\partial_y k_1 &= (k_2 - k_1)(\partial_x\partial_y a - \partial_y a \partial_x b) - \partial_x k_1 \partial_y a. \end{aligned}$$

*The energy density is given by*

$$\epsilon = \frac{k_1 k_2 + e^b e_1 \bullet e_1 \bullet e^{-b} + e^a e_2 \bullet e_2 \bullet e^{-a}}{(1 + tk_1)(1 + tk_2)},$$

*where $e_1 = e^{-a}\partial_x, \quad e_2 = e^{-b}\partial_y$.*

We will now show that our examples do not satisfy the genericity condition. It turns out that otherwise they would provide counterexamples to Conjecture 9.5.2 (cf. corollary 9.5.5 below).

**Lemma 9.5.3.** *Let $(M,g)$ be a 3-dimensional pseudo-Riemannian manifold. Then the Riemann tensor is completely determined by the Ricci tensor and given by*

$$R_{ijkl} = 2(g_{i[k}R_{l]j} - g_{j[k}R_{l]i}) - \mathrm{Scal}\, g_{i[k}g_{l]j}\,.$$

*Proof.* Proposition 4.3.2 implies that for every pair of tensors $G_x, S_x \in \text{sym}(T_2^0(T_xM))$ there is a metric $g$ such that $g_x = G_x$ and $S_x = \text{Ric}_x$. In fact, we can choose coordinates $(x^1, x^2, x^3)$ such that $(G_x)_{ab} = \eta_a \delta_{ab}$ (no summation) where $\eta_a \in \{-1, 1\}$. Then we simply set

$$g_{ab}(x^1, x^2, x^3) = (G_x)_{ab} - \frac{1}{3} \sum_{c,d=1}^{3} \frac{1}{3} (S_x)_{ab} \eta_c \delta_{cd} x^c x^d.$$

From the first, the third and the fourth symmetry in Proposition 4.3.1 we obtain that at a given point $x$ the Riemann tensor of a 3-dimensional pseudo-Riemannian manifold is already specified by the 6 components

$$R_{1212}, R_{1213}, R_{1223}, R_{1313}, R_{1323}, R_{2323}.$$

Since $\text{sym}(T_2^0(T_xM))$ is a 6-dimenional vector space and every tensor $S_x \in \text{sym}(T_2^0(T_xM))$ can be realised as the Ricci tensor of a metric, the map $\text{tr}\colon R_x \mapsto \text{tr}(R_x) = \text{Ric}_x$ is linear isomorphism. The tensor $r(\text{Ric})_{ijkl} = 2(g_{i[k}R_{l]j} - g_{j[k}R_{l]i}) - \text{tr}(\text{Ric}) g_{i[k}g_{l]j}$ satisfies the equations given in Proposition 4.3.1 and $\text{tr}(r(\text{Ric})) = \text{Ric}$. Hence it is the Riemann tensor corresponding to Ric. ∎

**Corollary 9.5.4.** *If $(M, g)$ is a 3-dimensional, irrotational dust spacetime then the genericity condition does not hold.*

*Proof.* The Ricci tensor is given by $\text{Ric} = \epsilon(e_1{}^\flat \otimes e_1{}^\flat + e_2{}^\flat \otimes e_2{}^\flat)$. It follows directly from Lemma 9.5.3 that the components $R_{tjkt}$ $(j, k \in \{t, x, y\})$ of the Riemann tensor vanish. Hence for any fixed numbers $(x_0, y_0)$ the genericity condition is violated along the timelike geodesic $t \mapsto (t, x_0, y_0)$. ∎

**Corollary 9.5.5.** *There is a 3-dimensional spacetime $(M, g)$ which*

*(i) is chronological,*
*(ii) is geodesically inextensible,*
*(iii) satisfies the timelike convergence condition*
*(iv) contains a closed trapped surface,*
*(v) and contains an incomple, future inextensible geodesic $\gamma$ and a neighbourhood $\mathcal{U}$ of $\gamma$ such that $\text{Vol}(\mathcal{U}) < \infty$ and $\int_{\mathcal{U}} f\mu < \infty$ for any polynomial curvature invariant $f$ which is linear in the Riemann tensor.*

*Proof.* Consider a spacetime $(\mathbb{R}^2, g)$ as given by Corollary 9.5.3. In order to obtain a closed trapped surface we let $M = \mathbb{R} \times S^1 \times \mathbb{R} = \{(t, x \bmod 1, y)\}$ and $b(x, y) = 1, a(x, y) = a(y)$. Then for each function $k_1(y)$, the metric

$$g = -\mathrm{d}t^2 + \mathrm{e}^{2a}(1 + tk_1)^2\mathrm{d}x^2 + (1 + t(k_1 + (k_1)'/a'))^2\mathrm{d}y^2$$

is a solution with

$$\epsilon = \frac{k_1((k_1 + \partial_y k_1/\partial_y a) + \mathrm{e}^a e_2 \bullet e_2 \bullet \mathrm{e}^{-a}}{(1 + tk_1)(1 + t(k_1 + (k_1)'/a'))}.$$

Let $\mathcal{T} := \{t, x, y | t = 0, y = 0\}$ and $S = S_{11}\omega^1 \otimes \omega^1$ be the second fundamental form of $\mathcal{T} \subset \{t = 0\}$. Then up to a positive factor, the expansions $\theta^{\pm}$ are given by $\theta^{\pm} = k_1 \mp S_{11}$. It follows that $\mathcal{T}$ is a closed trapped surface if $k_1(0) < -|S_{11}|$. This can always be arranged since $k_1$ can be freely specified. The spacetime $(M, g)$ has a singularity at $y = 0, t = -1/k_1(0)$. However, this singularity is so weak that it satisfies $\mathrm{Vol}(\mathcal{U}) < \infty \Rightarrow \int_{\mathcal{U}} |\epsilon| \mu_M < \infty$. ∎

Corollary 9.5.5 implies that Conjecture 9.5.2 without the genericity conditions does not hold. In order to estimate the physical relevance of our example we have to examine its special features.

(i) The genericity condition is not satified along the geodesic $t \mapsto (t, x_0, y_0)$. Its usage in the proof of Theorem 9.3.1 is to ensure the existence of a singularity along it. Since there actually develops a singularity at $1 + tk_1(0) = 0$, the failure of $(M, g)$ to satisfy the genericity condition does not appear to be grave. Moreover, it seems very likely that a perturbation of $(M, g)$ through dust spacetimes which are not strictly irrational will not suffer from this defect. On the other hand, we don't know much about the global properties of these perturbed spacetimes. In particular, at this point of time[11] we cannot exclude the possibility that they form stronger singularities then the special spacetimes we have examined.

(ii) Our example is 3-dimensional rather than 4-dimensional. Here it is important to note that Theorem 9.3.1 does hold for 3-dimensional spacetimes as well as for 4-dimensional ones. Moreover, there do exist 4-dimensional, spherically symmetric dust spacetimes which have similar singularities (Müller zum Hagen, Yodzis, and Seifert 1974). However, these 4-dimensional examples also contain much stronger singularities in the centre of symmetry. One may speculate whether these strong (central) singularities are a typical feature for 4-dimensional spacetimes. In the absence of independent evidence disqualifying 3-dimensional models it seems fair to state that our example indicates otherwise.

(iii) The closed trapped surface we have constructed has not much to do with the existence of singularities. In fact, the general solution shows that the singularities depend solely on the principal pressures

[11] I am writing this in 1998

$k_1$, $k_2$. This indicates that the example is more appropriate to illustrate condition (iii)(b) of Theorem 9.3.1 rather than condition (iii) (a). However, the hyperbolic nature of Equation (9.5.8) greatly restricts the existence of global solutions with compact hypersurface $\Sigma$. Still, as an immediate consequence of Corollary 9.5.3 we have the existence of a 3-dimensional dust spacetime for any given 2-dimensional Riemannian manifold $(\Sigma, g_\Sigma)$ by choosing $k = \lambda g_\Sigma$, where $\lambda$ is a constant. The energy density $\epsilon$ is positive if $|\lambda|$ is sufficiently large.

Our construction of closed trapped surface requires (at least) a periodicity with respect to $x$. Observe that perturbation of our initial data are very restricted since the differential Equation (9.5.8) does not need to respect this artificial periodicity. It is possible, however, to construct closed trapped surfaces with are stable with respect to arbitrary perturbations of the initial data. Choose a 2-dimensional Riemannian manifold $(\Sigma, g_\Sigma)$ and a closed curve $\mathcal{T} \subset \Sigma$. If $\lambda << 1$ is small enough and $k = \lambda g_\Sigma$ then $\mathcal{T}$ is a closed trapped surface. This construction does not rely on periodicity and is therefore stable with respect to perturbations of initial conditions.

The trapped surface has topology $S^1$ which is qualitatively different from $S^2$ and more akin to the torus $S^1 \times S^1$. This is inevitable if one works with 3-dimensional spacetimes. Unlike in the case of Conjecture 9.4.1 there there does not seem to exist evidence that the topology of the closed trapped surface matters in our context.

(iv) One can argue that dust, arising as an idealization from the energy momentum tensor for collisionless gas is not a very realistic matter model. Moreover, even in the corresponding Newtonian theory, congruences of dust tend to form weak singularities. (Rein, Rendall, and Schaeffer 1995) has shown for spherically symmetric 4-dimensional spacetimes representing a collisionless gas that one does not obtain weak singularities *before* a central singularity has formed. This is in striking contrast to the analogous situation in the case of dust (Müller zum Hagen, Yodzis, and Seifert 1974). One may therefore be tempted to disregard our example as typical for a notoriously ill behaved matter model. However, since in our class of examples we only obtain weak singularities, which are forced on us by the singularity Theorem 9.3.1[12], we are still led to conclude that singularities (mathematically) due to the singularity theorems may be very weak and completely different from what one may expect at first sight.

Hence whether or not our example is physically realistic, it indicates that existing singularity theorems are *not sufficient* to conclude the existence of black holes or the big bang.

[12] Here we assume that in our case the violation of the genericity condition is irrelevant

One additional, physically motivated assumption could be that the principal pressures diverge comparably to the energy density. It would therefore be of interest to study a similar example with $T = (1+\beta)\epsilon\omega \otimes \omega + \beta\epsilon g$ ($\beta \in \mathbb{R}$). For these 3-dimensional spacetimes there exist coordinates $(t, x, y)$ as above:

$$g = -\mathrm{d}t^2 + V^2\mathrm{d}x^2 + W^2\mathrm{d}y^2,$$
$$T = \epsilon\mathrm{d}t^2 + \beta\epsilon(\mathrm{d}x^2 + \mathrm{d}y^2).$$

We have some control over the location of the singularities (they must occur before proper time $2/(k_1 + k_2)$, where $k_1, k_2$ denote the principal curvatures of the initial hypersurface). This class of solution may still be manageable and can give us an important clue as to whether the singularity theorems really give evidence for the existence of physical singularities.

# References


Abraham, R. and J. E. Marsden (1967). *Foundations of Mechanics* (first ed.). New York: W. A. Benjamin Inc.

Abraham, R. and J. E. Marsden (1978). *Foundations of Mechanics* (second ed.). New York: Addison-Wesley Publishing.

Alexandrov, A. D. (1950). On Lorentz transformations. *Uspehi Mat. Nauk 5*, 187. (Russian).

Alexandrov, A. D. (1975). Mappings of spaces with families of cones and space-time transformations. *Annali di Matematica 103*, 229–257.

Beem, J. K. and P. Ehrlich (1981). *Global Lorentzian Geometry.* New York and Basel: Marcel Dekker.

Benz, W. (1992). *Geometrische Transformationen unter besonderer Berücksichtigung der Lorentztransformationen.* Bibliographisches Institut Wissenschaftsverlag.

Berger, M. (1987). *Geometry I.* Berlin, Heidelberg, New-York: Springer-Verlag.

Bizon, P., E. Malec, and N. O'Murchadha (1988). Trapped surfaces in spherical stars. *Phys. Rev. Lett. 61*, 1147–1150.

Bleecker, D. (1981). *Gauge Theory and Variational Principles.* Reading, Massachusetts: Addison-Wesly.

Bott, R. and L. W. Tu (1982). *Differential Forms in Algebraic Geometry.* New York: Springer-Verlag.

Bruno, G. (1584). *De l'Infinito Universo et Mondi.* Venice (stated).

Bryant, R. L., S. S. Chern, R. B. Gardner, H. L. Goldschmidt, and P. A. Griffiths (1991). *Exterior Differential Systems.* Berlin: Springer-Verlag.

Clarke, C. J. S. (1977). Time in general relativity. In *Minnesota Studies in the Philosophy of Science VIII*, pp. 94–108.

Copernicus, N. (1543). *De Revolutionibus Orbium Coelestium Libri VI.* Nürnberg.

Courant, R. and D. Hilbert (1962). *Methods of Mathematical Physics. Volume II.* New York: John Wiley and Sons.

De Felice, F. and C. J. S. Clarke (1990). *Relativity on Curved Manifolds.* Cambridge: Cambridge University Press.

Descartes, R. (1637). *Discours de la Méthode.* Appeard anonymously. The book *La géométrie* is part of this work.

Dieudonné, J. (1960). *Foundations of Modern Analysis.* New York: Academic Press.

Dieudonné, J. (1971). *Elements d'Analyse Tome IV*, Volume 4. Paris: Gauthier-Villars. There is also an English translation.

Ehlers, J. (1973). Survey of general relativity. In W. Israel (Ed.), *Relativity, Astrophysics, and Cosmology*, pp. 1–125. Dordrecht: Reidel.

Ehlers, J., F. Pirani, and A. Schild (1972). The geometry of free fall and light propagation. In L. O'Raifeartaigh (Ed.), *General relativity*, pp. 63–84. Oxford: Clarendon. In honour of J. Synge.

Einstein, A. (1905). Zur Elektrodynamik bewegter Körper. *Annalen d. Physik 17*, 891–921.

Eötvös (1896). Untersuchungen über Gravitation und Erdmagnetismus. *Annalen d. Physik 59*, 354–400.

Galilei, G. (1610). *Sidereus Nuncius.* Venice.

Galilei, G. (1632). *Dialogo.* Florence.

Garabedian, P. R. (1986). *Partial Differential Equations* (2 ed.). New York: Chelsea Publishing Company.

Gödel, K. (1949). An example of a new type of cosmological solution of Einstein's field equations of gravitation. *Rev. Mod. Phys. 21*, 447–450.

Greub, W. (1981). *Linear Algebra* (4 ed.). New York, Heidelberg, Berlin: Springer Verlag.

Guillemin, V. and A. Pollack (1974). *Differential Topology.* Englewood Cliffs, New Jersey: Prentice-Hall.

Hartle, J. B. (1978). Bounds on the mass and moment of inertia of non-rotating neutron stars. *Phys. Rep. 46*(6), 201–247.

Hasse, W. (1995). Private communication.

Hawking, S. W. and G. F. R. Ellis (1973). *The Large Scale Structure of Space-Time.* Cambridge: Cambridge University Press.

Hirsch, M. W. (1976). *Differential Topology.* New York, Heidelberg, and Berlin: Springer-Verlag.

Hughes, T. J. R., T. Kato, and J. E. Marsden (1977). Well posed quasi-linear 2nd. order hyperbolical systems with applications to non-linear elastodynamics and general relativity. *Arch. Rat. Mech. Anal. 63*, 273–294.

Huygens, C. (1690). *Tractatus de Lumine.*

Kanitscheider, B. (1984). *Kosmologie.* Stuttgart: Phillip Reclam jun.

Kant, I. (1781). *Kritik der reinen Vernunft.*

Karcher, H. (1994). Geometrische Eigenschaften relativistischer Modelle. Lecture notes taken by students, individual copies.

Kobayashi, S. and K. Nomizu (1963). *Foundations of Differential Geometry*, Volume I. New York: John Wiley & Sons.

Kriele, M. (1990). A generalization of the singularity theorem of Hawking and Penrose to spacetimes with causality violations. *Proc. Roy. Soc. Lond. A 431*, 451–464.

Kriele, M. (1995). On the collapse of a spherically symmetric star. *J. Math. Phys. 36*, 3676–3693.

Kriele, M. (1997). A stable class of spacetimes with naked singularities. In P. Chruschiel and A. Krolak (Eds.), *Mathematics of Gravitation, Lorentzian Geometry and Einstein Equations*, Volume 47 (I), pp. 169–178. Warsaw: Banach Center Publications.

Kriele, M. and G. Lim (1995). Physical properties of geometrical singularities. *Class. Quantum Grav. 12*, 3019–3035.

Leray, J. (1953). *Hyperbolic Differential Equations.* Lecture Notes. Princeton: Institute for Advanced Study.

Lerner, D. E. (1973). The space of Lorentz metrics. *Commun. Math. Phys. 32*, 19–38.

Lovelock, D. (1972). The four-dimensionality of space and the Einstein tensor. *J. Math. Phys. 13*, 874–876.

McNamara, J. M. (1978). Instability of black hole inner horizons. *Proc. Roy. Soc. Lond. A 358*, 499–517.

Michelson, A. A. (1881). *American J. of Science 22*, 120.
Michelson, A. A. and E. W. Morley (1887). *American J. of Science 34*, 333.
Minkowski, H. (1909). Raum und Zeit. *Physikalische Zeitschrift 10*, 104–111. Lecture held at the 80th Naturforscher Versammlung in Köln (1908).
Misner, C. W. (1967). Taub-NUT space as a counterexample to almost anything. In J. Ehlers (Ed.), *Relativity Theory and Astrophysics I: Relativity and Cosmology*, Volume 8 of *Lectures in Applied Mathematics*, pp. 160–169. American Mathematical Society.
Müller zum Hagen, H., P. Yodzis, and H.-J. Seifert (1974). On the occurrence of naked singularities in general relativity. II. *Commun. Math. Phys. 37*, 29–40.
Newman, R. P. A. C. (1989). Black holes without singularities. *Gen. Rel. Grav. 18*(11), 981–995.
Nitsche, J. C. C. (1975). *Vorlesungen über Minimalflächen*. Berlin: Springer-Verlag.
O'Neill, B. (1983). *Semi-Riemannian Geometry. With applications to general relativity*. New York: Academic Press.
O'Neill, B. (1995). *The Geometry of Kerr Black Holes*. Wellesley: A. K. Peters.
Osserman, R. (1995). *Poetry of the Universe*. New York: Anchor Books.
Penrose, R. (1972). *Techniques of Differential Topology in Relativity*. Philadelphia: SIAM.
Penzias, A. A. and R. W. Wilson (1965). A measurement of excess antenna temperature at 4080mc/s. *Astrophysical Journal 142*, 419–421.
Rein, G., A. D. Rendall, and J. Schaeffer (1995). A regularity theorem for solutions of the spherically symmetric vlasov-Einstein system. *Commun. Math. Phys. 168*, 467–478.
Rendall, A. D. (1994). Global properties of locally spatially homogeneous cosmologi al models with matter. 18 pages, Albert-Einstein-Institute Potsdam, Germany Report Nr. MPA-AR-94-4. See also: xxx Physics e-print archive Nr. gr-qc/9409009.
Sachs, R. K. and H. Wu (1979). *General Relativity for Mathematicians*. New York, Berlin, Heidelberg: Springer.
Sandage, A. (1968). Observational cosmology. *Observatory 88*, 91–106.
Schoen, R. and S.-T. Yau (1983). The existence of a black hole due to condensation of matter. *Commun. Math. Phys. 90*, 575–579.
Schwarzschild, K. (1916). Über das Gravitationsfeld eines Massenpunktes nach der Einsteinschen Theorie. *Sitzungsberichte Deut. Akad. Wiss. Berlin, Kl. Math.-Phys.-Tech.*, 189–196.
Simpson, M. and R. Penrose (1973). Internal instability in a Reissner-Nordström black hole. *Int. J. theor. Phys. 7*, 183–197.
Wald, R. (1984). *General Relativity*. Chicago: Chicago University Press.
Weinberg, S. (1972). *Gravitation and Cosmology*. New York: John Wiley.
Weyl, H. (1923). *Raum Zeit Materie* (5 ed.). Berlin: Springer.
Wheeler, J. A. and R. P. Feynman (1949). Classical electrodynamics in terms of direct interparticle action. *Reviews of Modern Physics 21*, 425–433.
Wolf, J. A. (1977). *Spaces of Constant Curvature*. Berkely: Publish or Perish.

# Index